FLUID MECHANICS

IN SI UNITS

R. C. Hibbeler

SI Conversion by
Kai Beng Yap

PEARSON

Boston Columbus Indianapolis New York San Francisco Hoboken
Amsterdam Cape Town Dubai London Madrid Milan Munich Paris Montréal Toronto
Delhi Mexico City São Paulo Sydney Hong Kong Seoul Singapore Taipei Tokyo

Vice President and Editorial Director, ECS: *Marcia Horton*
Acquisitions Editors: *Norrin Dias and Tacy Quinn*
Acquisitions Editor, Global Editions: *Murchana Borthakur*
Editorial Assistant: *Sandra L. Rodriguez*
Executive Marketing Manager: *Tim Galligan*
Marketing Assistant: *Jon Bryant*
Senior Managing Editor: *Scott Disanno*
Project Manager: *Rose Kernan*
Project Editor, Global Editions: *Donald Villamero*
Operations Specialist: *Linda Sager*
Senior Production Manufacturing Controller, Global Editions: *Trudy Kimber*
Media Production Manager, Global Editions: *Vikram Kumar*
Art Director: *Marta Samsel*
Photo Researcher: *Marta Samsel*
Cover Image: *Modern urban wastewater treatment plant.*
© *Hxdyl/Shutterstock*

Pearson Education Limited
Edinburgh Gate
Harlow
Essex CM20 2JE
England

and Associated Companies throughout the world

Visit us on the World Wide Web at:
www.pearsonglobaleditions.com

© 2017 by R. C. Hibbeler

British Library Cataloguing-in-Publication Data

A catalogue record for this book is available from the British Library.

10 9 8 7 6 5 4 3 2 1

ISBN 10: 1-292-08935-0
ISBN 13: 978-1-292-08935-5

Printed in Malaysia (CTP-VVP)

To the Student

With the hope that this work will stimulate
an interest in Fluid Mechanics
and provide an acceptable guide to its understanding.

This book has been written and revised several times over a period of nine years, in order to further improve its contents and account for the many suggestions and comments from my students, university colleagues, and reviewers. It is hoped that this effort will provide those who use this work with a clear and thorough presentation of both the theory and application of fluid mechanics. To achieve this objective, I have incorporated many of the pedagogic features that I have used in my other books. These include the following:

Organization and Approach. Each chapter is organized into well-defined sections that contain an explanation of specific topics, illustrative example problems, and at the end of the chapter, a set of relevant homework problems. The topics within each section are placed into subgroups defined by boldface titles. The purpose of this organization is to present a structured method for introducing each new definition or concept, and to make the book a convenient resource for later reference and review.

Procedures for Analysis. This unique feature provides the student with a logical and orderly method to follow when applying the theory that has been discussed in a particular section. The example problems are then solved using this outlined method in order to clarify its numerical application. Realize, however, that once the relevant principles have been mastered, and enough confidence and judgment has been obtained, the student can then develop his or her own procedures for solving problems.

Important Points. This feature provides a review or summary of the most important concepts in a section, and highlights the most significant points that should be remembered when applying the theory to solve problems. A further review of the material is given at the end of the chapter.

Photos. The relevance of knowing the subject matter is reflected by the realistic applications depicted in the many photos placed throughout the book. These photos are often used to show how the principles of fluid mechanics apply to real-world situations.

Fundamental Problems. These problem sets are selectively located just after the example problems. They offer students simple applications of the concepts and therefore provide them with the chance to develop their problem-solving skills before attempting to solve any of the standard problems that follow. Students may consider these problems as extended examples, since they all have complete solutions and answers given in the back of the book. Additionally, the fundamental problems offer students an excellent means of preparing for exams, and they can be used at a later time to prepare for the Fundamentals in Engineering Exam.

Homework Problems.   The majority of problems in the book depict realistic situations encountered in engineering practice. It is hoped that this realism will both stimulate interest in the subject, and provide a means for developing the skills to reduce any problem from its physical description to a model or symbolic representation to which the principles of fluid mechanics may then be applied.

An attempt has been made to arrange the problems in order of increasing difficulty. Except for every fourth problem, indicated by an asterisk (*), the answers to all the other problems are given in the back of the book.

Accuracy.   Apart from my work, the accuracy of the text and problem solutions have all been thoroughly checked by other parties. Most importantly, Kai Beng Yap, Kurt Norlin along with Bittner Development Group, as well as James Liburdy, Jason Wexler, Maha Haji, and Brad Saund.

Contents

The book is divided into 14 chapters. Chapter 1 begins with an introduction to fluid mechanics, followed by a discussion of units and some important fluid properties. The concepts of fluid statics, including constant accelerated translation of a liquid and its constant rotation, are covered in Chapter 2. In Chapter 3, the basic principles of fluid kinematics are covered. This is followed by the continuity equation in Chapter 4, the Bernoulli and energy equations in Chapter 5, and fluid momentum in Chapter 6. In Chapter 7, differential fluid flow of an ideal fluid is discussed. Chapter 8 covers dimensional analysis and similitude. Then the viscous flow between parallel plates and within pipes is treated in Chapter 9. The analysis is extended to Chapter 10 where the design of pipe systems is discussed. Boundary layer theory, including topics related to pressure drag and lift, is covered in Chapter 11. Chapter 12 discusses open channel flow, and Chapter 13 covers a variety of topics in compressible flow. Finally, turbomachines, such as axial and radial flow pumps and turbines are treated in Chapter 14.

Alternative Coverage.   After covering the basic principles of Chapters 1 through 6, at the discretion of the instructor, the remaining chapters may be presented in *any sequence*, without the loss of continuity. If time permits, sections involving more advanced topics, may be included in the course. Most of these topics are placed in the later chapters of the book. In addition, this material also provides a suitable reference for basic principles when it is discussed in more advanced courses.

Acknowledgments

I have endeavored to write this book so that it will appeal to both the student and instructor. Through the years many people have helped in its development, and I will always be grateful for their valued suggestions and comments. During the past years, I have had the privilege to teach my students during the summer at several German universities, and in particular I would like to thank Prof. H. Zimmermann at the University of Hanover, Prof F. Zunic of the Technical University in Munich, and Prof. M. Raffel at the Institute of Fluid Mechanics in Goettingen, for their assistance. In addition, I. Vogelsang and Prof. M. Geyh of the University of Mecklenburg have provided me with logistic support in these endeavors. I would also like to thank Prof. K.Cassel at Illinois Institute of Technology, Prof. A. Yarin at the University of Illinois-Chicago, and Dr. J. Gotelieb for their comments and suggestions. In addition, the following individuals have contributed important reviewer comments relative to preparing this work:

S. Kumpaty, *Milwaukee School of Engineering*
N. Kaye, *Clemson University*
J. Crockett, *Brigham Young University*
B. Wadzuk, *Villanova University*
K. Sarkar, *University of Delaware*
E. Petersen, *Texas A&M University*
J. Liburdy, *Oregon State University*
B. Abedian, *Tufts University*
S. Venayagamoorthy, *Colorado State University*
D. Knight, *Rutgers University*
B. Hodge, *Mississippi State University*
L. Grega, *The College of New Jersey*
R. Chen, *University of Central Florida*
R. Mullisen, *Cal Poly Institute*
C. Pascual, *Cal Poly Institute*

There are a few people that I feel deserve particular recognition. A long-time friend and associate, Kai Beng Yap, was of great help in checking the entire manuscript, and helping to further check all the problems. And a special note of thanks also goes to Kurt Norlin for his diligence and support in this regard. During the production process I am also thankful for the support of my long time Production Editor, Rose Kernan, and my Managing Editor, Scott Disanno. My wife, Conny, and daughter, Mary Ann, have been a big help with the proofreading and typing needed to prepare the manuscript for publication.

Lastly, many thanks are extended to all my students who have given me their suggestions and comments. Since this list is too long to mention, it is hoped that those who have helped in this manner will accept this anonymous recognition.

I value your judgment as well, and would greatly appreciate hearing from you if at any time you have any comments or suggestions that may help to improve the contents of this book.

Russell Charles Hibbeler
hibbeler@bellsouth.net

Global Edition

The publishers would like to thank the following for their contribution to the Global Edition:

Contributor

Kai Beng Yap

Kai is currently a registered Professional Engineer who works in Malaysia. He has BS and MS degrees in Civil Engineering from the University of Louisiana, Lafayette, Louisiana; and he has done further graduate work at Virginia Polytechnic Institute in Blacksberg, Virginia. His professional experience has involved teaching at the University of Louisiana, and doing engineering consulting work related to structural analysis and design and its associated infrastructure.

Reviewers

Jitendra Singh Rathore, Department of Mechanical Engineering, *Birla Institute of Technology and Science*
M. Haluk Aksel, Department of Mechanical Engineering, *Middle East Technical University*
Suresh Babu, Centre for Nano Sciences and Technology, *Pondicherry University*

For pressure of 47.2 kPa, $h_1 = 6$ km,

$$V_A = V_p$$

At 6 km, $\rho_a = 0.59$ kg/m³

Applying the Bernoulli equation at A and B,

$$\frac{P_A}{\rho} + \frac{V_A^2}{2} + g\, z_A = \frac{P_B}{\rho} + \frac{V_B^2}{2} + g\, z_B$$

$$\frac{47.2(10^3) \, N/m^2}{0.59 \, kg/m^3} + \frac{V_p^2}{2} + 0 = \frac{49.6(10^3) \, N/m^2}{0.59 \, kg/m^3} + 0 + 0$$

$$V_p = 90.2 \text{ m/s}$$

your answer specific feedback

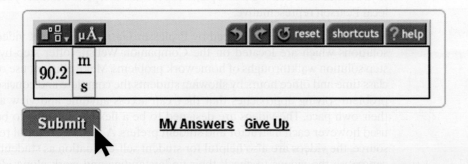

Incorrect; Try Again

It appears you have used the wrong value of density from the table. Check the density corresponding to the altitude.

www.MasteringEngineering.com

Resources for Instructors

- **MasteringEngineering.** This online Tutorial Homework program allows you to integrate dynamic homework with automatic grading and adaptive tutoring. MasteringEngineering allows you to easily track the performance of your entire class on an assignment-by-assignment basis, or the detailed work of an individual student.

- **Instructor's Solutions Manual.** An instructor's solutions manual was prepared by the author. The manual includes homework assignment lists and was also checked as part of the accuracy checking program. The Instructor Solutions Manual is available at www.pearsonglobaleditions.com.

- **Presentation Resource.** All art from the text is available in PowerPoint slide and JPEG format. These files are available for download from the Instructor Resource Center at www.pearsonglobaleditions.com. If you are in need of a login and password for this site, please contact your local Pearson representative.

- **Video Solutions.** Developed by Professor Garret Nicodemus, video solutions which are located on the Companion Website offer step-by-step solution walkthroughs of homework problems. Make efficient use of class time and office hours by showing students the complete and concise problem solving approaches that they can access anytime and view at their own pace. The videos are designed to be a flexible resource to be used however each instructor and student prefers. A valuable tutorial resource, the videos are also helpful for student self-evaluation as students can pause the videos to check their understanding and work alongside the video. Access the videos at www.pearsonglobaleditions.com and follow the links for the *Fluid Mechanics* text.

Resources for Students

- **MasteringEngineering.** Tutorial homework problems emulate the instructor's office-hour environment.

- **Companion Website.** The Companion Website, located at www.pearsonglobaleditions.com/hibbeler includes opportunities for practice and review including:

- **Video Solutions.** Complete, step-by-step solution walkthroughs of representative homework problems. Videos offer: students need it with over 20 hours helpful review.

- **Animations.** Help students visualize the relation between mathematical models and fluid mechanics concepts, breaking down complicated concepts with the aid of numerical calculations and diagrams.

An access code for the *Fluid Mechanics* companion website was included with this text. To redeem the code and gain access to the site, go to www.pearsonglobaleditions.com/hibbeler and follow the directions on the access code card.

CONTENTS

1
Fundamental Concepts 3

Chapter Objectives 3

1.1 Introduction 3

1.2 Characteristics of Matter 5

1.3 The International System of Units 6

1.4 Calculations 8

1.5 Problem Solving 10

1.6 Basic Fluid Properties 12

1.7 Viscosity 17

1.8 Viscosity Measurement 22

1.9 Vapor Pressure 26

1.10 Surface Tension and Capillarity 27

2
Fluid Statics 45

Chapter Objectives 45

2.1 Pressure 45

2.2 Absolute and Gage Pressure 48

2.3 Static Pressure Variation 50

2.4 Pressure Variation for Incompressible Fluids 51

2.5 Pressure Variation for Compressible Fluids 53

2.6 Measurement of Static Pressure 56

2.7   Hydrostatic Force on a Plane Surface—Formula Method   64

2.8   Hydrostatic Force on a Plane Surface—Geometrical Method   70

2.9   Hydrostatic Force on a Plane Surface—Integration Method   75

2.10   Hydrostatic Force on an Inclined Plane or Curved Surface Determined by Projection   78

2.11   Buoyancy   85

2.12   Stability   88

2.13   Constant Translational Acceleration of a Liquid   91

2.14   Steady Rotation of a Liquid   96

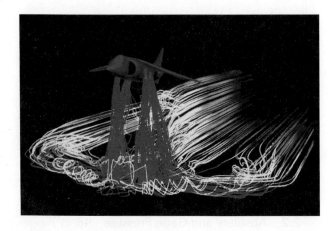

3
Kinematics of Fluid Motion   129

Chapter Objectives   129

3.1   Fluid Flow Descriptions   129

3.2   Types of Fluid Flow   131

3.3   Graphical Descriptions of Fluid Flow   134

3.4   Fluid Acceleration   142

3.5   Streamline Coordinates   149

4
Conservation of Mass   165

Chapter Objectives   165

4.1 Finite Control Volumes   165

4.2 The Reynolds Transport Theorem   168

4.3 Volumetric Flow, Mass Flow, and Average Velocity   174

4.4 Conservation of Mass   178

5
Work and Energy of Moving Fluids   207

Chapter Objectives   207

5.1 Euler's Equations of Motion   207

5.2 The Bernoulli Equation   211

5.3 Applications of the Bernoulli Equation   214

5.4 Energy and Hydraulic Grade Lines   226

5.5 The Energy Equation   234

6
Fluid Momentum 269

Chapter Objectives   269

6.1   The Linear Momentum Equation   269

6.2   Applications to Bodies at Rest   271

6.3   Applications to Bodies Having Constant Velocity   281

6.4   The Angular Momentum Equation   286

6.5   Propellers and Wind Turbines   294

6.6   Applications for Control Volumes Having Accelerated Motion   299

6.7   Turbojets and Turbofans   300

6.8   Rockets   301

7
Differential Fluid Flow 323

Chapter Objectives   323

7.1   Differential Analysis   323

7.2   Kinematics of Differential Fluid Elements   324

7.3   Circulation and Vorticity   328

7.4   Conservation of Mass   332

7.5   Equations of Motion for a Fluid Particle   334

7.6   The Euler and Bernoulli Equations   336

7.7   The Stream Function   340

7.8   The Potential Function   345

7.9   Basic Two-Dimensional Flows   349

7.10 Superposition of Flows   360

7.11 The Navier–Stokes Equations   370

7.12 Computational Fluid Dynamics   374

8
Dimensional Analysis and Similitude   393

Chapter Objectives   393

8.1 Dimensional Analysis   393

8.2 Important Dimensionless Numbers   396

8.3 The Buckingham Pi Theorem   399

8.4 Some General Considerations Related to Dimensional Analysis   408

8.5 Similitude   409

9
Viscous Flow within Enclosed Surfaces   433

Chapter Objectives   433

9.1 Steady Laminar Flow between Parallel Plates   433

9.2 Navier–Stokes Solution for Steady Laminar Flow between Parallel Plates   439

9.3 Steady Laminar Flow within a Smooth Pipe   444

9.4 Navier–Stokes Solution for Steady Laminar Flow within a Smooth Pipe   448

9.5 The Reynolds Number   450

9.6 Fully Developed Flow from an Entrance   455

9.7 Laminar and Turbulent Shear Stress within a Smooth Pipe   457

9.8 Turbulent Flow within a Smooth Pipe   460

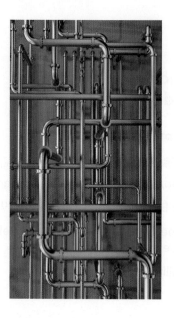

10
Analysis and Design for Pipe Flow 479

Chapter Objectives 479

10.1 Resistance to Flow in Rough Pipes 479

10.2 Losses Occurring from Pipe Fittings and Transitions 490

10.3 Single-Pipeline Flow 496

10.4 Pipe Systems 502

10.5 Flow Measurement 508

11
Viscous Flow over External Surfaces 525

Chapter Objectives 525

11.1 The Concept of the Boundary Layer 525

11.2 Laminar Boundary Layers 531

11.3 The Momentum Integral Equation 540

11.4 Turbulent Boundary Layers 544

11.5 Laminar and Turbulent Boundary Layers 546

11.6 Drag and Lift 552

11.7 Pressure Gradient Effects 554

11.8 The Drag Coefficient 558

11.9 Drag Coefficients for Bodies Having Various Shapes 562

11.10 Methods for Reducing Drag 569

11.11 Lift and Drag on an Airfoil 572

12
Open-Channel Flow 601

Chapter Objectives 601

12.1 Types of Flow in Open Channels 601

12.2 Open-Channel Flow Classifications 603

12.3 Specific Energy 604

12.4 Open-Channel Flow over a Rise or Bump 612

12.5 Open-Channel Flow under a Sluice Gate 616

12.6 Steady Uniform Channel Flow 620

12.7 Gradual Flow with Varying Depth 627

12.8 The Hydraulic Jump 634

12.9 Weirs 639

13
Compressible Flow 657

Chapter Objectives 657

13.1 Thermodynamic Concepts 657

13.2 Wave Propagation through a Compressible Fluid 666

13.3 Types of Compressible Flow 669

13.4 Stagnation Properties 673

13.5 Isentropic Flow through a Variable Area 680

13.6 Isentropic Flow through Converging and Diverging Nozzles 685

13.7 The Effect of Friction on Compressible Flow 694

13.8 The Effect of Heat Transfer on Compressible Flow   704

13.9 Normal Shock Waves   710

13.10 Shock Waves in Nozzles   713

13.11 Oblique Shock Waves   718

13.12 Compression and Expansion Waves   723

13.13 Compressible Flow Measurement   728

14
Turbomachines   747

Chapter Objectives   747

14.1 Types of Turbomachines   747

14.2 Axial-Flow Pumps   748

14.3 Radial-Flow Pumps   754

14.4 Ideal Performance for Pumps   756

14.5 Turbines   761

14.6 Pump Performance   767

14.7 Cavitation and the Net Positive Suction Head   770

14.8 Pump Selection Related to the Flow System   772

14.9 Turbomachine Similitude   774

Appendix

A. Physical Properties of Fluids   790

B. Compressible Properties of a Gas ($k = 1.4$)   793

Fundamental Solutions   803

Answers to Selected Problems   818

Index   831

FLUID
MECHANICS

IN SI UNITS

Chapter 1

(©AZybr/Shutterstock)

Fluid mechanics plays an important role in the design and analysis of pressure vessels, pipe systems, and pumps used in chemical processing plants.

Fundamental Concepts

CHAPTER OBJECTIVES

■ To provide a description of fluid mechanics and indicate its various branches.

■ To explain how matter is classified as a solid, liquid, or gas.

■ To discuss the system of units for measuring fluid quantities, and establish proper calculation techniques.

■ To define some important fluid properties, such as density, specific weight, bulk modulus, and viscosity.

■ To describe the concepts of vapor pressure, surface tension, and capillarity.

1.1 Introduction

Fluid mechanics is a study of the behavior of fluids that are either at rest or in motion. It is one of the primary engineering sciences that has important applications in many engineering disciplines. For example, aeronautical and aerospace engineers use fluid mechanics principles to study flight, and to design propulsion systems. Civil engineers use this subject to design drainage channels, water networks, sewer systems, and water-resisting structures such as dams and levees. Fluid mechanics is used by mechanical engineers to design pumps, compressors, turbines, process control systems, heating and air conditioning equipment, and to design wind turbines and solar heating devices. Chemical and petroleum engineers apply this subject to design equipment used for filtering, pumping, and mixing fluids. And finally, engineers in the electronics and computer industry use fluid mechanics principles to design switches, screen displays, and data storage equipment. Apart from the engineering profession, the principles of fluid mechanics are also used in biomechanics, where it plays a vital role in the understanding of the circulatory, digestive, and respiratory systems, and in meteorology to study the motion and effects of tornadoes and hurricanes.

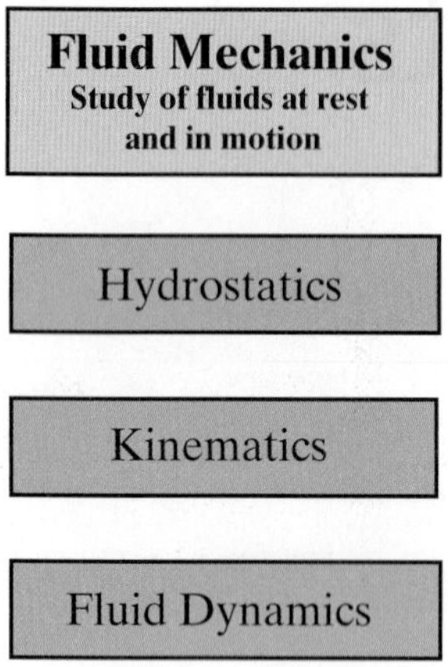

Fluid Mechanics
Study of fluids at rest and in motion

Hydrostatics

Kinematics

Fluid Dynamics

Fig. 1–1

Branches of Fluid Mechanics.   The principles of fluid mechanics are based on Newton's laws of motion, the conservation of mass, the first and second laws of thermodynamics, and laws related to the physical properties of a fluid. The subject is divided into three main categories, as shown in Fig. 1–1.

- *Hydrostatics* considers the forces acting on a fluid at rest.
- *Fluid kinematics* is the study of the geometry of fluid motion.
- *Fluid dynamics* considers the forces that cause acceleration of a fluid.

Historical Development.   A fundamental knowledge of the principles of fluid mechanics has been of considerable importance throughout the development of human civilization. Historical records show that through the process of trial and error, early societies, such as the Roman Empire, used fluid mechanics in the construction of their irrigation and water supply systems. In the middle of the 3rd century B.C., Archimedes discovered the principle of buoyancy, and then much later, in the 15th century, Leonardo Da Vinci developed principles for the design of canal locks and other devices used for water transport. However, the greatest discoveries of basic fluid mechanics principles occurred during the 16th and 17th centuries. It was during this period that Evangelista Torricelli designed the barometer, Blaise Pascal formulated the law of static pressure, and Isaac Newton developed his law of viscosity to describe the nature of fluid resistance to flow.

In the 1700s, Leonhard Euler and Daniel Bernoulli pioneered the field of *hydrodynamics*, a branch of mathematics dealing with the motion of an idealized fluid, that is, one having a constant density and providing no internal frictional resistance. Unfortunately, hydrodynamic principles could not be used by engineers to study some types of fluid motion, since the physical properties of the fluid were not fully taken into account. The need for a more realistic approach led to the development of *hydraulics*. This field uses empirical equations found from fitting curves to data determined from experiments, primarily for applications involving water. Contributors included Gustave Coriolis, who developed water turbines, and Gotthilf Hagen and Jean Poiseuille, who studied the resistance to water flowing through pipes. In the early 20th century, hydrodynamics and hydraulics were essentially *combined* through the work of Ludwig Prandtl, who introduced the concept of the boundary layer while studying aerodynamics. Through the years, many others have also made important contributions to this subject, and we will discuss many of these throughout the text.*

---

*References [1] and [2] provide a more complete description of the historical development of this subject.

1.2 Characteristics of Matter

In general, matter can be classified by the state it is in—as a solid, a liquid, or a gas.

Solid. A *solid* maintains a definite shape and volume, Fig. 1–2a. It maintains its shape because the molecules or atoms of a solid are densely packed and are held tightly together, generally in the form of a lattice or geometric structure. The spacing of atoms within this structure is due in part to large cohesive forces that exist between molecules. These forces prevent any relative movement, except for any slight vibration of the molecules themselves. As a result, when a solid is subjected to a load it will not easily deform, but once in its deformed state, it will continue to support the load.

Liquid. A *liquid* is composed of molecules that are more spread out than those in a solid. Their intermolecular forces are weaker, so liquids do not hold their shape. Instead, they *flow* and take the shape of their container, Fig. 1–2b. Although liquids can easily deform, their molecular spacing allows them to resist compressive forces when they are confined.

Gas. A *gas* is a substance that fills the entire volume of its container, Fig. 1–2c. Gases are composed of molecules that are much farther apart than those of a liquid. As a result, the molecules of a gas are free to travel away from one another until a force of repulsion pushes them away from other gas molecules or from the molecules on the surface of a solid or liquid boundary.

Definition of a Fluid. Liquids and gases are classified as **fluids** *because they are substances that continuously deform or flow when subjected to a shear or tangential force.* This behavior is shown on small fluid elements in Fig. 1–3, where a plate moves over the top surface of the fluid. The deformation of the fluid will continue as long as the shear force is applied, and once it is removed, the fluid will keep its new shape rather than returning to its original one. In this text we will only concentrate on those substances that exhibit *fluid behavior*, meaning any substance that will flow because it cannot support a shear loading, *regardless* of how *small* the shear force is, or how *slowly* the "fluid" deforms.

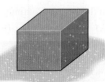

Solids maintain a
constant shape

(a)

Liquids take the shape of
their container

(b)

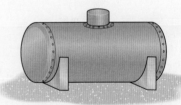

Gases fill the entire volume of
their container

(c)

Fig. 1–2

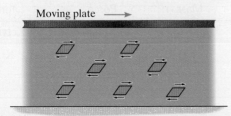

Moving plate ⟶

All fluid elements *deform* when subjected to shear

Fig. 1–3

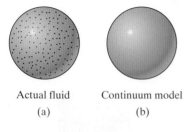

Actual fluid   Continuum model
   (a)   (b)

Fig. 1–4

Continuum.   Studying the behavior of a fluid by analyzing the motion of all its many molecules would be an impossible task, Fig. 1–4a. Fortunately, however, almost all engineering applications involve a volume of fluid that is much greater than the distance between adjacent molecules of the fluid, and so it is reasonable to assume the fluid is uniformly dispersed and continuous throughout this volume. Under these circumstances, we can then consider the fluid to be a ***continuum***, that is, a continuous distribution of matter leaving no empty space, Fig. 1–4b. This assumption allows us to use *average properties* of the fluid at any point within the volume the fluid occupies. For those situations where the molecular distance does become important, which is on the order of a billionth of a meter, the continuum model does not apply, and it is necessary to employ statistical techniques to study the fluid flow, a topic that will not be considered here. See Ref. [3].

1.3   The International System of Units

There are five basic quantities primarily used in fluid mechanics: length, time, mass, force, and temperature. Of these, length, time, mass, and force are all *related* by Newton's second law of motion, $F = ma$. As a result, the *units* used to define the size of these quantities cannot *all* be selected arbitrarily. The equality $F = ma$ is maintained when *three* of these units are arbitrarily defined, and the fourth unit is then *derived* from the equation.

The International System of units, abbreviated as SI after the French term *Système International d'Unités*, is a modern version of the metric system that has received worldwide recognition. As shown in Table 1–1, the SI system specifies length in meters (m), time in seconds (s), and mass in kilograms (kg). The unit of force, called a newton (N), is *derived* from $F = ma$, where 1 newton is equal to the force required to give 1 kilogram of mass an acceleration of 1 m/s^2 $\left(\text{N} = \text{kg} \cdot \text{m/s}^2 \right)$, Fig. 1–5a.

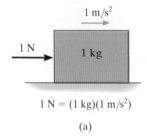

$1 \text{ N} = (1 \text{ kg})(1 \text{ m/s}^2)$

(a)

Fig. 1–5

TABLE 1–1 International System of Units						
Quantity	Length	Time	Mass	Force	Temperature	
SI Units	meter	second	kilogram	Newton*	Kelvin	Celsius
	m	s	kg	N $\left(\dfrac{\text{kg} \cdot \text{m}}{\text{s}^2} \right)$	K	°C

*Derived unit.

Weight. To determine the weight of a fluid in newtons at the "standard location," where the acceleration due to gravity is $g = 9.81$ m/s², and the mass of the fluid is m (kg), we have

$$W \text{ (N)} = [m \text{ (kg)}] (9.81 \text{ m/s}^2) \qquad (1\text{–}1)$$

And so a fluid having a mass of 1 kg has a weight of 9.81 N, 2 kg of fluid has a weight of 19.62 N, and so on.

Temperature. The absolute temperature is the temperature measured from a point where the molecules of a substance have so called "zero energy".* The unit for *absolute temperature* in the SI system is the kelvin (K). This unit is expressed *without* reference to degrees, so 7 K is stated as "seven kelvins." Although not officially an SI unit, an equivalent size unit measured in degrees Celsius (°C) is often used. This measurement is referenced from the freezing and boiling points of water, where the freezing point is at 0°C (273 K) and the boiling point is at 100°C (373 K), Fig. 1–5b. For conversion,

$$T_K = T_C + 273 \qquad (1\text{–}2)$$

Equations 1–1 and 1–2 will be used in this text since they are suitable for most engineering applications. However, use the exact value of 273.15 K in Eq. 1–2 for more accurate work. Also, at the "standard location," the more exact value $g = 9.807$ m/s² or the *local* acceleration due to gravity should be used in Eq. 1–1.

Prefixes. When a numerical quantity is either very large or very small, the units used to define its size should be modified by using a prefix. The range of prefixes used for problems in this text is shown in Table 1–2. Each

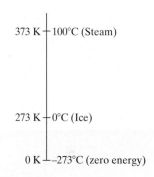

373 K — 100°C (Steam)

273 K — 0°C (Ice)

0 K — –273°C (zero energy)

The Kelvin and Celsius scales

(b)

Fig. 1–5 (cont.)

TABLE 1–2 Prefixes			
	Exponential Form	Prefix	SI Symbol
Submultiple			
0.001	10^{-3}	milli	m
0.000 001	10^{-6}	micro	μ
0.000 000 001	10^{-9}	nano	n
Multiple			
1 000 000 000	10^{9}	Giga	G
1 000 000	10^{6}	Mega	M
1 000	10^{3}	kilo	k

*This is actually an unreachable point according to the law of quantum mechanics.

represents a multiple or submultiple of a unit that moves the decimal point of a numerical quantity either forward or backward by three, six, or nine places. For example, 5 000 000 g = 5000 kg (kilogram) = 5 Mg (Megagram), and 0.000 006 s = 0.006 ms (millisecond) = 6 μs (microsecond).

As a general rule, quantities defined by several units that are multiples of one another are separated by a dot to avoid confusion with prefix notation. Thus, m · s is a meter-second, whereas ms is a millisecond. And finally, the exponential power applied to a unit having a prefix refers to *both the unit and its prefix*. For example, $ms^2 = (ms)^2 = (ms)(ms) = (10^{-3}s)(10^{-3}s) = 10^{-6}s^2$.

1.4   Calculations

Application of fluid mechanics principles often requires algebraic manipulations of a formula followed by numerical calculations. For this reason it is important to keep the following concepts in mind.

Dimensional Homogeneity.   The terms of an equation used to describe a physical process must be ***dimensionally homogeneous***, that is, each term must be expressed in the *same units*. Provided this is the case, then all the terms of the equation can be *combined* when numerical values are substituted for the variables. For example, consider the Bernoulli equation, which is a specialized application of the principle of work and energy. We will study this equation in Chapter 5, but it can be expressed as

$$\frac{p}{\gamma} + \frac{V^2}{2g} + z = \text{constant}$$

Here, the pressure p is expressed in N/m^2, the specific weight γ is in N/m^3, the velocity V is in m/s, the acceleration due to gravity g is in m/s^2, and the elevation z is in meters, m. *Regardless of how this equation is algebraically arranged, it must maintain its dimensional homogeneity.* In the form stated, each of the three terms is in meters, as noted by a cancellation of units in each fraction.

$$\frac{N/m^2}{N/m^3} + \frac{(m/s)^2}{m/s^2} + m$$

Because almost all problems in fluid mechanics involve the solution of dimensionally homogeneous equations, a *partial check* of the algebraic manipulation of any equation can therefore be made by checking to see if all the terms have the *same units*.

Calculation Procedure.

When performing numerical calculations, *first* represent all the quantities in terms of their base or derived units by converting any prefixes to powers of 10. Then do the calculation, and finally express the result using a *single prefix*. For example, $3 \text{ MN}(2 \text{ mm}) = \left[3(10^6)\text{ N}\right]\left[2(10^{-3})\text{ m}\right] = 6(10^3)\text{ N}\cdot\text{m} = 6\text{ kN}\cdot\text{m}$.

In the case of fractional units, with the exception of the kilogram, the prefix should always be in the numerator, as in MN/s or mm/kg. Also, after the calculation, it is best to keep numerical values between 0.1 and 1000; otherwise, a suitable prefix should be chosen.

Accuracy.

Numerical work in fluid mechanics is almost always performed using pocket calculators and computers. It is important, however, that the answers to any problem be reported with justifiable accuracy using an appropriate number of significant figures. As a general rule, always retain more digits in your calculations than are given in the problem data. Then round off your final answer to *three significant figures*, since data for fluid properties and many experimental measurements are often reported with this accuracy. We will follow this procedure in this text, where the intermediate calculations for the example problems will often be worked out to four or five significant figures, and then the answers will generally be reported to *three* significant figures.

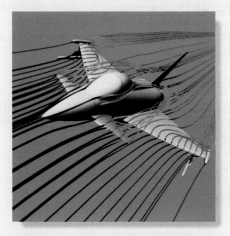

Complex flows are often studied using a computer analysis; however, it is important to have a good grasp of the principles of fluid mechanics to be sure reasonable predictions have been made. (© CHRIS SATTLBERGER/Science Source)

1.5 Problem Solving

At first glance, the study of fluid mechanics can be rather daunting, because there are many aspects of this field that must be understood. Success at solving problems, however, will depend on your attitude and your willingness to both focus on class lectures and to carefully read the material in the text. Aristotle once said, "What we have to learn to do, *we learn by doing*," and indeed *your ability to solve problems* in fluid mechanics depends upon a thoughtful preparation and neat presentation.

In any engineering subject, it is very important that you follow a logical and orderly procedure when solving problems. In the case of fluid mechanics this should include the sequence of steps outlined below:

General Procedure for Analysis

Fluid Description.

Fluids can behave in many different ways, and so at the outset it is important to *identify the type of fluid flow* and specify the fluid's *physical properties*. Knowing this provides a means for the proper selection of equations used for an analysis.

Analysis.

This generally involves the following steps:

- Tabulate the problem data and draw, to a reasonably large scale, any necessary diagrams.

- Apply the relevant principles, generally in mathematical form. When substituting numerical data into any equations, be sure to include their units, and check to be sure the terms are dimensionally homogeneous.

- Solve the equations, and report any numerical answers to three significant figures.

- Study the answer with technical judgment and common sense to determine whether or not it seems reasonable.

When applying this procedure, do the work as neatly as possible. Being neat generally stimulates clear and orderly thinking, and vice versa.

Important Points

- Solids have a definite shape and volume, liquids take the shape of their container, and gases fill the entire volume of their container.

- Liquids and gases are fluids because they continuously deform or flow when subjected to a shear force, no matter how small this force is.

- For most engineering applications, we can consider a fluid to be a continuum, and therefore use its average properties to model its behavior.

- Weight is measured in newtons and is determined from W (N) $=$ $[m$ (kg)] (9.81 m/s^2).

- Certain rules must be followed when performing calculations and using prefixes. First convert all numerical quantities with prefixes to their base units, then perform the calculations, and finally choose an appropriate prefix for the result.

- The derived equations of fluid mechanics are all dimensionally homogeneous, and thus each term in an equation has the same units. Careful attention should therefore be paid to the units when entering data and then solving an equation.

- As a general rule, perform calculations with sufficient numerical accuracy, and then round off the final answer to three significant figures.

EXAMPLE 1.1

Evaluate $(80 \text{ MN/s})(5 \text{ mm})^2$, and express the result with SI units having an appropriate prefix.

SOLUTION

We first convert all the quantities with prefixes to powers of 10, perform the calculation, and then choose an appropriate prefix for the result.

$$(80 \text{ MN/s})(5 \text{ mm})^2 = [80(10^6) \text{ N/s}][5(10^{-3}) \text{ m}]^2$$

$$= [80(10^6) \text{ N/s}][25(10^{-6}) \text{ m}^2]$$

$$= 2(10^3) \text{ N} \cdot \text{m}^2/\text{s} = 2 \text{ kN} \cdot \text{m}^2/\text{s} \qquad Ans.$$

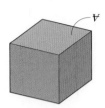

Density is
mass/volume

Fig. 1–6

1.6   Basic Fluid Properties

Assuming the fluid to be a continuum, we will now define some important physical properties that are used to describe it.

Density.   The *density* ρ (rho) refers to the mass of the fluid that is contained in a unit of volume, Fig. 1–6. It is measured in kg/m³ and is determined from

$$\rho = \frac{m}{V} \tag{1–3}$$

Here m is the mass of the fluid, and V is its volume.

Liquid.   Through experiment it has been found that a liquid is practically incompressible, that is, the density of a liquid varies little with pressure. It does, however, have a slight but greater variation with temperature. For example, water at 4°C has a density of $\rho_w = 1000$ kg/m³, whereas at 100°C, $\rho_w = 958.1$ kg/m³. For most practical applications, provided the temperature range is small, we can therefore consider *the density of a liquid to be essentially constant.*

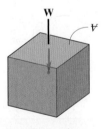

Specific weight is
weight/volume

Fig. 1–7

Gas.   Unlike a liquid, temperature and pressure can markedly affect the density of a gas, since it has a higher degree of compressibility. For example, air has a density of $\rho = 1.23$ kg/m³ when the temperature is 15°C and the atmospheric pressure is 101.3 kPa [1 Pa (pascal) = 1 N/m²]. But at this same temperature, and at twice the pressure, the density of air *doubles* and becomes $\rho = 2.46$ kg/m³.

Appendix A lists typical values for the densities of common liquids and gases. Included are tables of specific values for water at different temperatures, and air at different temperatures and elevations.

Specific Weight.   The *specific weight* γ (gamma) of a fluid is its weight per unit volume, Fig. 1–7. It is measured in N/m³. Thus,

$$\gamma = \frac{W}{V} \tag{1–4}$$

Here, W is the weight of the fluid, and V is its volume.

Since weight is related to mass by $W = mg$, then substituting this into Eq. 1–4, and comparing this result with Eq. 1–3, the specific weight is related to the density by

$$\gamma = \rho g \qquad (1-5)$$

Typical values of specific weights for common liquids and gases are also listed in Appendix A.

Specific Gravity.
The **specific gravity** S of a substance is a dimensionless quantity that is defined as the ratio of its density or specific weight to that of some other substance that is taken as a "standard." It is most often used for liquids, and water at an atmospheric pressure of 101.3 kPa and a temperature of 4°C is taken as the standard. Thus,

$$S = \frac{\rho}{\rho_w} = \frac{\gamma}{\gamma_w} \qquad (1-6)$$

The density of water for this case is $\rho_w = 1000 \text{ kg/m}^3$, and its specific weight is 9.81 kN/m³. So, for example, if an oil has a density of $\rho_o = 880 \text{ kg/m}^3$, then its specific gravity will be $S_o = 0.880$.

Ideal Gas Law.
In this text we will consider every gas to behave as an **ideal gas**.* Such a gas is assumed to have enough separation between its molecules so that the molecules have no attraction to one another. Also, the gas must not be near the point of condensation into either a liquid or a solid state.

From experiments, mostly performed with air, it has been shown that ideal gases behave according to the **ideal gas law**. It can be expressed as

The volume, pressure, and temperature of the gas in this tank are related by the ideal gas law.

$$p = \rho R T \qquad (1-7)$$

Here, p is the **absolute pressure**, or force per unit area, referenced from a perfect vacuum, ρ is the density of the gas, R is the gas constant, and T is the **absolute temperature**. Typical values of R for various gases are given in Appendix A. For example, for air, $R = 286.9 \text{ J/(kg} \cdot \text{K)}$, where 1 J (joule) $= 1$ N $\cdot$ m.

*Nonideal gases and vapors are studied in thermodynamics.

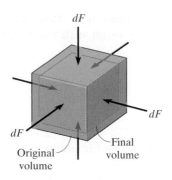

Bulk modulus

Fig. 1–8

Bulk Modulus.

The *bulk modulus of elasticity*, or simply the **bulk modulus**, is a measure of the amount by which a fluid offers a resistance to compression. To define this property, consider the cube of fluid in Fig. 1–8, where each face has an area A and is subjected to an incremental force dF. The intensity of this force per unit area is the *pressure*, $dp = dF/A$. As a result of this pressure, the original volume V of the cube will *decrease* by dV. This incremental pressure, divided by this decrease in volume per unit volume, dV/V, defines the bulk modulus, namely,

$$E_V = -\frac{dp}{dV/V} \tag{1–8}$$

The minus sign is included to show that the *increase* in pressure (positive) causes a decrease in volume (negative).

The units for E_V are the same as for pressure—that is, force per area—since the volume ratio is dimensionless. Typical unit is N/m^2 or Pa.

Liquid. Because the density of a liquid changes very little with pressure, its bulk modulus is very high. For example, sea water at atmospheric pressure and room temperature has a bulk modulus of about $E_V = 2.20$ GPa.* If we use this value and consider the deepest region of the Pacific Ocean, where the water pressure is 110 MPa, then Eq. 1–8 shows that the fractional compression of water is only $\Delta V/V = \left[110(10^6)\,Pa \right]/\left[2.20(10^9)\,Pa \right] = 5.0\%$. For this reason, we can assume that for most practical applications, *liquids can be considered incompressible*, and, as stated previously, their density remains constant.**

---

*Of course, solids can have much higher bulk moduli. For example, the bulk modulus for steel is 160 GPa.

**The *compressibility* of a flowing liquid must, however, be considered for some types of fluid analysis. For example, "water hammer" is created when a valve on a pipe is suddenly closed. This causes an abrupt local change in density of the water near the valve, which generates a pressure wave that travels down the pipe and produces a hammering sound when the wave encounters a bend or other obstruction in the pipe. See Ref. [7].

Gas.   A gas, because of its low density, is thousands of times more compressible than a liquid, and so its bulk modulus will be much smaller. For a gas, however, the relation between the applied pressure and the volume change depends upon the process used to compress the gas. Later, in Chapter 13, we will study this effect as it relates to compressible flow, where changes in pressure become significant. However, if the gas flows at *low velocities*, that is, less than about 30% the speed of sound in the gas, then only *small changes* in the gas pressure occur, and so, even with its low bulk modulus, at constant temperature a gas, like a liquid, can in this case also be considered incompressible.

Important Points

- The mass of a fluid is often characterized by its *density* $\rho = m/V$, and its weight is characterized by its *specific weight* $\gamma = W/V$, where $\gamma = \rho g$.

- The *specific gravity* is a ratio of the density or specific weight of a liquid to that of water, defined by $S = \rho/\rho_w = \gamma/\gamma_w$. Here $\rho_w = 1000 \text{ kg/m}^3$ and $\gamma_\omega = 9.81 \text{ kN/m}^3$.

- For many engineering applications, we can consider a gas to be *ideal*, and can therefore relate its *absolute pressure* to its *absolute temperature* and density using the ideal gas law, $p = \rho RT$.

- The *bulk modulus* of a fluid is a measure of its resistance to compression. Since this property is very high for liquids, we can generally consider liquids as incompressible fluids. Provided a gas has a low velocity of flow—less than 30% of the speed of sound—and has a constant temperature, then the pressure variation within the gas will be low, and we can, under these circumstances, also consider it to be incompressible.

1

EXAMPLE | 1.2

Air contained in the tank, Fig. 1–9, is under an absolute pressure of 60 kPa and has a temperature of 60°C. Determine the mass of the air in the tank.

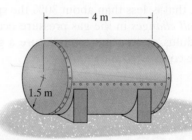

Fig. 1–9

SOLUTION

We will first find the density of the air in the tank using the ideal gas law, Eq. 1–7, $p = \rho RT$. Then, knowing the volume of the tank, we can determine the mass of the air. The *absolute temperature* of the air is

$$T_K = T_C + 273 \text{ K} = 60°C + 273 \text{ K} = 333 \text{ K}$$

From Appendix A, the gas constant for air is $R = 286.9 \text{ J}/(\text{kg} \cdot \text{K})$. Then,

$$p = \rho RT$$
$$60(10^3) \text{ N/m}^2 = \rho(286.9 \text{ J/kg} \cdot \text{K})(333 \text{ K})$$
$$\rho = 0.6280 \text{ kg/m}^3$$

The mass of air within the tank is therefore

$$\rho = \frac{m}{V}$$

$$0.6280 \text{ kg/m}^3 = \frac{m}{\left[\pi\left(1.5 \text{ m}\right)^2(4 \text{ m}) \right]}$$

$$m = 17.8 \text{ kg} \qquad \qquad Ans.$$

Many people are often surprised by how large the mass of a gas contained within a volume can be. For example, if we repeat the calculations for the mass of air in a typical classroom that measures 4 m by 6 m by 3 m, at a standard room temperature of 20°C and pressure of 101.3 kPa, the result is 86.8 kg. The weight of this air is 851 N. It is no wonder that the flow of air can cause the lift of an airplane and structural damage to buildings.

EXAMPLE | 1.3

An amount of glycerin has a volume of 1 m³ when the pressure is 120 kPa. If the pressure is increased to 400 kPa, determine the change in volume of this cubic meter. The bulk modulus for glycerin is $E_V = 4.52$ GPa.

SOLUTION

We must use the definition of the bulk modulus for the calculation. First, the pressure increase applied to the cubic meter of glycerin is

$$\Delta p = 400 \text{ kPa} - 120 \text{ kPa} = 280 \text{ kPa}$$

Thus, the change in volume is

$$E_V = -\frac{\Delta p}{\Delta V / V}$$

$$4.52(10^9)\text{N/m}^2 = -\frac{280(10^3)\text{N/m}^2}{\Delta V / 1 \text{ m}^3}$$

$$\Delta V = -61.9(10^{-6})\text{m}^3 \qquad Ans.$$

This is indeed a very small change. Since ΔV is directly proportional to the change in pressure, doubling the pressure change will then double the change in volume. Although E_V for water is about half that of glycerin, even for water the volume change will still remain very small!

1.7   Viscosity

Viscosity is a property of a fluid that measures the *resistance to movement* of a very thin layer of fluid over an adjacent one. This resistance occurs only when a tangential or shear force is applied to the fluid, Fig. 1–10a. The resulting deformation occurs at different rates for different types of fluids. For example, water or gasoline will shear or flow faster (low viscosity) than tar or syrup (high viscosity).

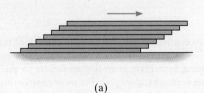

(a)

Fig. 1–10

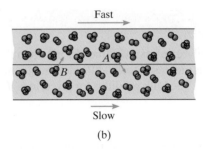

Fast

Slow

(b)

Fig. 1–10 (cont.)

Physical Cause of Viscosity.

The resistance that gives rise to viscosity in a fluid can be understood by considering the two layers of fluid in Fig. 1–10*b* sliding past one another. Since the molecules composing the fluid are always in continuous motion, then when molecule *A* in the *faster* top layer happens to travel down to the *slower* bottom layer, it will have a component of motion to the right. Collisions that occur with any slower-moving molecule of the bottom layer will cause it to be *pushed along* due to the momentum exchange with *A*. The reverse effect occurs when molecule *B* in the bottom layer migrates upward. Here this slower-moving molecule will *retard* a faster-moving molecule through their momentum exchange. On a grand scale, both of these effects cause resistance or viscosity.

Newton's Law of Viscosity.

To show on a small scale how fluids behave when subjected to a shear force, let us now consider a thin layer of fluid that is confined between a fixed surface and a very wide horizontal plate, Fig. 1–11*a*. When a very *small* horizontal force **F** is applied to the plate, it will cause elements of the fluid to distort as shown. After a brief acceleration, the viscous resistance of the fluid will bring the plate into equilibrium, such that the plate will begin to move with a *constant velocity* **U**. During this motion, the molecular adhesive force between the fluid particles in contact with *both* the fixed surface and the plate creates a "*no-slip condition*," such that the fluid particles at the *fixed surface* remain *at rest*, while those on the plate's bottom surface move with the same velocity as the plate.* In between these two surfaces, very thin layers of fluid are dragged along, so that the velocity profile *u* across the thickness of the fluid will be parallel to the plate, and can vary, as shown in Fig. 1–11*b*.

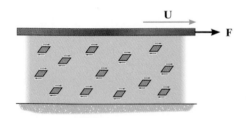

U

F

Distortion of fluid elements due to shear

(a)

Fig. 1–11

*Recent findings have confirmed that this "no-slip condition" is not always true. A fast-moving fluid flowing over *an extremely smooth surface* develops no adhesion. Also, surface adhesion can be reduced by adding *soap-like molecules* to the fluid, which coats the surface, thereby making it extremely smooth. For *most* engineering applications, however, the layer of fluid molecules adjacent to a solid boundary *will adhere to the surface*, and so these special cases with slipping at the boundary will *not* be considered in this text. See Ref. [11].

Shear Stress. The motion just described is a consequence of the shearing effect within the fluid caused by the plate. This effect subjects each element of fluid to a *shear stress* τ (tau), Fig. 1–11c, that is defined as a tangential force ΔF that acts on an area ΔA of the element. It can be expressed as

$$\tau = \lim_{\Delta A \to 0} \frac{\Delta F}{\Delta A} = \frac{dF}{dA} \qquad (1\text{–}9)$$

Shear Strain. Since a fluid will flow, this shear stress will cause each element to deform into the shape of a parallelogram, Fig. 1–11c, and during the short time Δt, the resulting deformation is defined by its *shear strain*, specified by the small angle $\Delta \alpha$ (alpha), where

$$\Delta \alpha \approx \tan \Delta \alpha = \frac{\delta x}{\Delta y}$$

A solid would hold this angle under load, but a fluid element will *continue to deform*, and so in fluid mechanics, the *time rate of change in this shear strain (angle)* becomes important. Since the top of the element moves at a rate of Δu relative to its bottom, Fig. 1–11b, then $\delta x = \Delta u \, \Delta t$. Substituting this into the above equation, the time rate of change of the shear strain becomes

$$\frac{\Delta \alpha}{\Delta t} = \frac{\Delta u}{\Delta y}$$

And in the limit, as $\Delta t \to 0$,

$$\frac{d\alpha}{dt} = \frac{du}{dy}$$

The term on the right is called the *velocity gradient* because it is an expression of the change in velocity u with respect to y.

In the late 17th century, Isaac Newton proposed that the shear stress in the fluid is directly proportional to this shear strain rate or velocity gradient. This is often referred to as *Newton's law of viscosity*, and it can be written as

$$\tau = \mu \frac{du}{dy} \qquad (1\text{–}10)$$

The constant of proportionality μ (mu) is a *physical property of the fluid* that measures the *resistance* to fluid movement. Although it is sometimes called the *absolute or dynamic viscosity*, we will refer to it simply as the *viscosity*. From the equation, μ has units of $N \cdot s/m^2$.

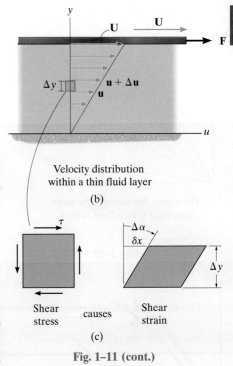

Velocity distribution
within a thin fluid layer

(b)

Shear
stress   causes   Shear
strain

(c)

Fig. 1–11 (cont.)

1

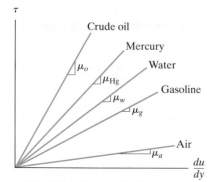

The higher the viscosity, the more difficult it is for a fluid to flow.

Fig. 1–12

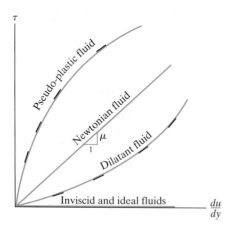

Fig. 1–13

Velocity profile for a real fluid

(a)

Velocity profile for an inviscid or ideal fluid

(b)

Fig. 1–14

Newtonian Fluids. Experiments have shown that many common fluids obey Newton's law of viscosity, and any fluid that does so is referred to as a **Newtonian fluid**. A plot showing how the shear stress and shear-strain rate (velocity gradient) behave for some common Newtonian fluids is shown in Fig. 1–12. Notice how the slope (viscosity) increases, from air, which has a very low viscosity, to water, and then to crude oil, which has a much higher viscosity. In other words, *the higher the viscosity, the more resistant the fluid is to flow*.

Non-Newtonian Fluids. Fluids whose very thin layers exhibit a *nonlinear* behavior between the applied shear stress and the shear-strain rate are classified as **non-Newtonian fluids**. There are basically two types, and they behave as shown in Fig. 1–13. For each of these fluids, the *slope of the curve* for any specific shear-strain rate defines the **apparent viscosity** for that fluid. Those fluids that have an increase in apparent viscosity (slope) with an increase in shear stress are referred to as shear-thickening or **dilatant fluids**. Examples include water with high concentrations of sugar, and quicksand. Many more fluids, however, exhibit the opposite behavior and are called shear-thinning or **pseudo-plastic fluids**. Examples include blood, gelatin, and milk. As noted, these substances flow slowly at low applications of shear stress (large slope), but rapidly under a higher shear stress (smaller slope).

Finally, there exist other classes of substances that have *both* solid and fluid properties. For example, paste and wet cement hold their shape (solid) for small shear stress, but can easily flow (fluid) under larger shear loadings. These substances, as well as other unusual solid–fluid substances, are studied in the field of *rheology*, not in fluid mechanics. See Ref. [8].

Inviscid and Ideal Fluids. Many applications in engineering involve fluids that have *very low viscosities*, such as water and air, [$1.00(10^{-3})$ N · s/m^2 and $18.1(10^{-6})$ N · s/m^2, at 20°C] and so we can sometimes *approximate* them as inviscid fluids. By definition, an **inviscid fluid** has *zero viscosity, $\mu = 0$,* and as a result it offers *no resistance to shear stress*, Fig. 1–13. In other words, it is *frictionless*. Hence, if the fluid in Fig. 1–11 is inviscid, then when the force **F** is applied to the plate, it will cause the plate to continue to *accelerate*, since no shear stress can be developed within an inviscid fluid to offer a restraining frictional resistance to the bottom of the plate. If in addition to being inviscid, the fluid is also assumed to be incompressible, then it is called an **ideal fluid**. By comparison, if any *real fluid* flows slowly through a pipe, it will have a velocity profile that looks something like that in Fig. 1–14a, whereas an inviscid or ideal fluid will have a uniform velocity profile, Fig. 1–14b.

Pressure and Temperature Effects. Through experiment it has been found that the viscosity of a fluid is actually *increased* with pressure, although this effect is quite small and so it is generally neglected for most engineering applications. Temperature, however, affects the viscosity of fluids to a much greater extent. In the case of a *liquid*, an *increase in temperature* will *decrease its viscosity*, as shown in Fig. 1–15 for water and mercury; Ref. [9]. This occurs because a temperature increase will cause the molecules of the liquid to have more vibration or mobility, thus breaking their molecular bonds and allowing the layers of the liquid to "loosen up" and slip more easily. If the fluid is a gas, an *increase in temperature* has the opposite effect, that is, the *viscosity will increase* as noted for air and carbon dioxide in Fig. 1–15; Ref. [10]. Since gases are composed of molecules that are much farther apart than for a liquid, their intermolecular attraction to one another is *smaller*. When the temperature increases the molecular motion of the gas will increase, and this will increase the momentum *exchange* between successive layers. It is this additional resistance, developed by molecular collisions, that causes the viscosity to increase.

Attempts have been made to use empirical equations to fit the experimental curves of viscosity versus temperature for various liquids and gases, such as those shown in Fig. 1–15. For liquids, the curves can be represented using *Andrade's equation*.

$$\mu = Be^{C/T} \text{ (liquid)}$$

And for gases, the *Sutherland equation* works well.

$$\mu = \frac{BT^{3/2}}{(T + C)} \text{ (gas)}$$

In each of these cases T is the *absolute temperature*, and the constants B and C can be determined if specific values of μ are known for two different temperatures.*

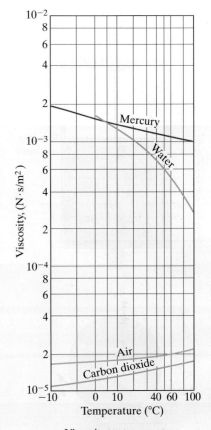

Viscosity vs. temperature

Fig. 1–15

Kinematic Viscosity.

Another way to express the viscosity of a fluid is to represent it by its **kinematic viscosity**, v(nu), which is the ratio of its dynamic viscosity to its density.

$$v = \frac{\mu}{\rho} \tag{1–11}$$

The units are m^2/s.** The word "kinematic" is used to describe this property because force is not involved in the dimensions. Typical values of the dynamic and kinematic viscosities are given in Appendix A for some common liquids and gases, and more extensive listings are also given for water and air.

---

*See Probs. 1–30 and 1–33.

**In the standard metric system (not SI), grams and centimeters (100 cm = 1 m) are used. In this case the dynamic viscosity μ is expressed using a unit called a *poise*, where poise = 1 g/(cm·s), and the kinematic viscosity v is measured in *stokes*, where 1 stoke = 1 cm^2/s.

1.8   Viscosity Measurement

The viscosity of a Newtonian liquid can be measured in several ways. One common method is to use a **rotational viscometer**, sometimes called a *Brookfield viscometer*. This device, shown in the photo on the next page, consists of a solid cylinder that is suspended within a cylindrical container as shown in Fig. 1–16a. The liquid to be tested fills the small space between these two cylinders, and as the container is forced to rotate with a very slow constant angular velocity ω, it causes the contained cylinder to twist the suspension wire a small amount before it attains equilibrium. By measuring the angle of twist of the wire, the torque M in the wire can be calculated using the theory of mechanics of materials. This torque resists the moment caused by the shear stress exerted by the liquid on the surface of the suspended cylinder. Once this torque is known, we can then find the viscosity of the fluid using Newton's law of viscosity.

To demonstrate how this is done, consider only the effect of shear stress developed on the vertical surface of the cylinder.* We require M, the torque in the wire, to balance the moment of the resultant shear force the liquid exerts on the cylinder's surface about the axis of the cylinder, Fig. 1–16b. This gives $F_s = M/r_i$. Since the area of the surface is $(2\pi r_i)h$, the shear stress acting on the surface is

$$\tau = \frac{F_s}{A} = \frac{M/r_i}{2\pi r_i h} = \frac{M}{2\pi r_i^2 h}$$

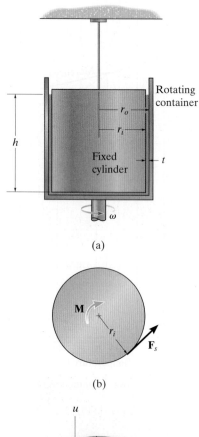

Rotating container

Fixed cylinder

(a)

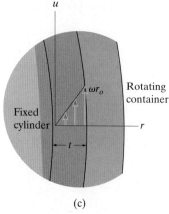

(b)

(c)

Fig. 1–16

The angular rotation of the container causes the liquid in contact with its wall to have a speed of $U = \omega r_o$, Fig. 1–16c. Since the suspended cylinder is held stationary by the wire once the wire is fully twisted, and because the gap t is *very small*, the velocity gradient across the thickness t of the liquid can be assumed to be constant. If this is the case, it can then be expressed as

$$\frac{du}{dr} = \frac{\omega r_o}{t}$$

Using Newton's law of viscosity,

$$\tau = \mu \frac{du}{dr}; \qquad\qquad \frac{M}{2\pi r_i^2 h} = \mu \frac{\omega r_o}{t}$$

Solving for μ in terms of the measured properties, the viscosity is then

$$\mu = \frac{Mt}{2\pi \omega r_i^2 r_o h}$$

---

*An extended analysis includes the frictional resistance of the liquid on the cylinder's bottom surface. See Probs. 1–50 and 1–51.

The viscosity of a liquid can also be obtained by using other methods. For example, W. Ostwald invented the *Ostwald viscometer* shown in the photo at the bottom of the page. Here the viscosity is determined by measuring the time for a liquid to flow through the short, small-diameter tube, and then correlating this time with the time for another liquid of known viscosity to flow through this same tube. The unknown viscosity is then determined by direct proportion. Another approach is to measure the speed of a small sphere as it falls through the liquid that is to be tested. It will be shown in Sec. 11.8 that this speed can be related to the viscosity of the liquid. Such an approach works well for transparent liquids, such as honey, which have a very high viscosity. In addition, many other devices have been developed to measure viscosity, and the details on how they work can be found in books related to this subject. For example, see Ref. [14].

Brookfield viscometer

Important Points

- A *Newtonian fluid*, such as water, oil, or air, develops shear stress within successive thin layers of the fluid that is directly proportional to the velocity gradient that occurs between the fluid layers, $\tau = \mu \, (du/dy)$.

- The shear resistance of a Newtonian fluid is measured by the proportionality constant μ, called the viscosity. The higher the viscosity, the greater the resistance to flow caused by shear.

- A non-Newtonian fluid has an apparent viscosity. If the apparent viscosity increases with an increase in shear stress, then the fluid is a dilatant fluid. If the apparent viscosity decreases with an increase in shear stress, then it is a pseudo-plastic fluid.

- An inviscid fluid has no viscosity, and an ideal fluid is both inviscid and incompressible; that is, $\mu = 0$ and $\rho = $ constant.

- The viscosity varies only slightly with pressure; however, for increasing temperature, μ will decrease for liquids, but it will increase for gases.

- The kinematic viscosity ν is the ratio of the two fluid properties ρ and μ, where $\nu = \mu/\rho$.

Ostwald viscometer

- It is possible to obtain the viscosity of a liquid in an indirect manner by using a rotational viscometer, an Ostwald viscometer, or by several other methods.

EXAMPLE 1.4

The plate in Fig. 1–17 rests on top of the thin film of water, which is at a temperature of 25°C. When a small force **F** is applied to the plate, the velocity profile across the thickness of the fluid can be described as $u = (40y - 800y^2)$ m/s, where y is in meters. Determine the shear stress acting on the fixed surface and on the bottom of the plate.

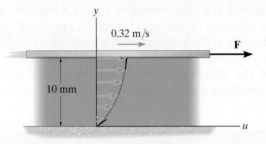

Fig. 1–17

SOLUTION

Fluid Description. Water is a Newtonian fluid, and so Newton's law of viscosity applies. The viscosity of water at 25°C is found from Appendix A to be $\mu = 0.897(10^{-3})$ N·s/m².

Analysis. Before applying Newton's law of viscosity, we must first obtain the velocity gradient.

$$\frac{du}{dy} = \frac{d}{dy}(40y - 800y^2) \text{ m/s} = (40 - 1600y) \text{ s}^{-1}$$

Therefore, at the fixed surface, $y = 0$,

$$\tau = \mu\frac{du}{dy}\bigg|_{y=0} = (0.897(10^{-3}) \text{ N·s/m}^2)(40 - 0) \text{ s}^{-1}$$

$$\tau = 35.88(10^{-3}) \text{ N/m}^2 = 35.9 \text{ mPa} \qquad\qquad Ans.$$

And, at the bottom of the moving plate, $y = 0.01$ m,

$$\tau = \mu\frac{du}{dy}\bigg|_{y=0.01 \text{ m}} = [0.897(10^{-3}) \text{ N·s/m}^2](40 - 1600(0.01)) \text{ s}^{-1}$$

$$\tau = 21.5 \text{ mPa} \qquad\qquad Ans.$$

By comparison, the *larger shear stress* develops on the fixed surface rather than on the bottom of the plate since the *velocity gradient* or slope du/dy is *large* at the fixed surface. Both of these slopes are indicated by the short dark lines in Fig. 1–17. Also, notice that the equation for the velocity profile must satisfy the boundary condition of no slipping, i.e., at the fixed surface $y = 0, u = 0$, and with the movement of the plate at $y = 10$ mm, $u = U = 0.32$ m/s.

EXAMPLE 1.5

The 100-kg plate in Fig. 1–18a is resting on a very thin film of SAE 10W-30 oil, which has a viscosity of $\mu = 0.0652$ N·s/m². Determine the force **P** that must be applied to the center of the plate to slide it over the oil with a constant velocity of 0.2 m/s. Assume the oil thickness is 0.1 mm, and the velocity profile across this thickness is linear. The bottom of the plate has a contact area of 0.75 m² with the oil.

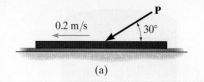

(a)

SOLUTION

Fluid Description. The oil is a Newtonian fluid, and so Newton's law of viscosity can be applied.

Analysis. First we draw the free-body diagram of the plate in order to relate the shear force **F** caused by the oil on the bottom of the plate to the applied force **P**, Fig. 1–18b. Because the plate moves with constant velocity, the force equation of equilibrium in the horizontal direction applies.

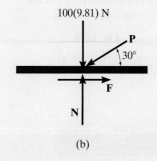

$$\overset{+}{\rightarrow}\Sigma F_x = 0; \qquad F - P\cos 30° = 0$$

$$F = 0.8660P$$

The effect of this force *on the oil* is in the opposite direction, and so the *shear stress* on the top of the oil acts to the left. It is

(b)

$$\tau = \frac{F}{A} = \frac{0.8660P}{0.75 \text{ m}^2} = (1.155P) \text{ m}^{-2}$$

Since the velocity profile is assumed to be linear, Fig. 1–18c, the velocity gradient is constant, $du/dy = U/t$, and so

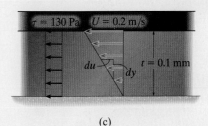

$$\tau = \mu \frac{du}{dy} = \mu \frac{U}{t}$$

$$(1.155P) \text{ m}^{-2} = \left(0.0652 \text{ N·s/m}^2\right)\left[\frac{0.2 \text{ m/s}}{0.1(10^{-3}) \text{ m}}\right]$$

$$P = 113 \text{ N} \qquad\qquad Ans.$$

(c)

Fig. 1–18

Notice that the constant velocity gradient will produce a constant shear-stress distribution across the thickness of the oil, which is $\tau = \mu\,(U/t) = 130$ Pa, Fig. 1–18c.

1

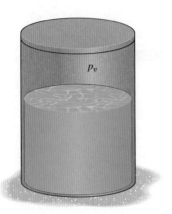

The vapor pressure p_v will form within the top space of the closed tank that was originally a vacuum

Fig. 1–19

1.9 Vapor Pressure

Consider a liquid contained in a closed tank, as in Fig. 1–19. Because the temperature of the liquid will cause continuous thermal agitation of the liquid's molecules, some of these molecules near the surface will acquire enough kinetic energy to break their molecular bonds with adjacent molecules, and will move upward or evaporate into the empty space of the tank. When a state of equilibrium is reached, the number of molecules that evaporate from the liquid will equal the number of molecules that condense back to it. The empty space is then said to be **saturated**. By bouncing off the walls of the tank and the liquid surface, the evaporated molecules create a pressure within the tank. This pressure is called the **vapor pressure**, p_v. Any increase in liquid temperature will increase the rate of evaporation, and also the kinetic energy of the liquid's molecules, so higher temperatures will cause higher vapor pressures.

The liquid will begin to boil when the absolute pressure at its surface is at or lower than its *vapor pressure*. For example, if water at sea level is brought to a temperature of 100°C, then at this temperature its vapor pressure will equal the atmospheric pressure, which is 101.3 kPa, and so the water will boil. In a similar manner, if the atmospheric pressure at the water surface is *reduced*, such as at the top of a mountain, then boiling occurs at this lower pressure, when the temperature is less than 100°C. Specific values of the vapor pressure for water at various temperatures are given in Appendix A. Notice that as the temperature increases, so does the vapor pressure due to the increase in the thermal agitation of its molecules.

Cavitation. When engineers design pumps, turbines, or piping systems, it is important that they do not allow the liquid at any point within the flow to be subjected to a pressure *equal to or less than* its vapor pressure. If this occurs as stated above, rapid evaporation or boiling will occur within the liquid. The resulting bubbles will migrate to regions of higher pressure, and then suddenly collapse, creating a phenomenon known as **cavitation**. The repeated pounding caused by this effect against the surface of a propeller blade or pump casing can eventually wear down its surface, and so it is important to avoid its occurrence. Later, in Chapter 14, we will study the significance of cavitation in greater detail.

1.10 Surface Tension and Capillarity

A liquid maintains its form because its molecules are attracted to one another by **cohesion**. It is this force that enables liquids to resist tensile stress and thereby creates *surface tension* in the liquid. On the other hand, if liquid molecules are attracted to those of a different substance, the force of attraction is known as **adhesion**, and this force, along with that of cohesion, gives rise to *capillarity*.

Surface Tension. The phenomenon of surface tension can be explained by visualizing the cohesive forces acting on two molecules (or particles) in a liquid, shown in Fig. 1–20a. The molecule located deep within the liquid has the same cohesive forces acting on it by all the surrounding molecules. Consequently, there is no resultant force acting on it. However, the molecule located on the surface of the liquid has cohesive forces that come only from molecules that are next to it on the surface and from those below it. This will produce a net resultant downward force, and the effect of all such forces will produce a *contraction* of the surface. In other words, the resultant cohesive force attempts to pull downward on the surface.

To separate the molecules at the surface requires a tensile force. We call this tensile force per unit length in any direction along the surface the **surface tension**, σ (sigma), Fig. 1–20b. It has units of N/m, and for any liquid, its value depends primarily upon the temperature. The higher the temperature, the more thermal agitation occurs, and so the surface tension becomes smaller. For example, water at 10°C has $\sigma = 74.2$ mN/m, whereas at the higher temperature of 50°C, $\sigma = 67.9$ mN/m. Values of σ, such as these, are sensitive to impurities, so care should be taken when using published values.

Because cohesion resists any increase in the surface area of a liquid, it actually tries to *minimize* the size of the surface. Separating the molecules and thus breaking the surface tension requires work, and the energy produced by this work is called **free-surface energy**. For example, suppose a small element of the surface is subjected to the surface tension force $F = \sigma \Delta y$ along one of its sides, as shown in Fig. 1–20c. If the surface stretches δx, then the increase in the area is $\Delta y \, \delta x$. The force **F** does work of $F \, \delta x$, and so the work done per area increase is therefore

$$\frac{F \, \delta x}{\Delta y \, \delta x} = \frac{\sigma \Delta y \, \delta x}{\Delta y \, \delta x} = \sigma$$

In other words, the surface tension can also be thought of as the amount of free-surface energy required to increase a unit surface area of a liquid.

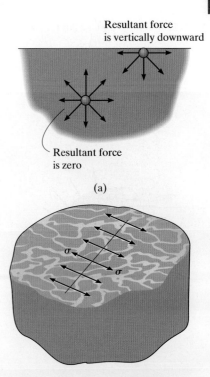

Resultant force is vertically downward

Resultant force is zero

(a)

Surface tension is the force per unit length needed to separate the molecules on the surface

(b)

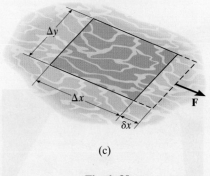

(c)

Fig. 1–20

Like rain drops, the water ejected from this fountain forms spherical droplets due to the cohesive force of surface tension.

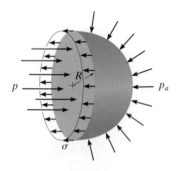

Fig. 1–21

Liquid Drops. Cohesion is responsible for the formation of liquid droplets that naturally form when a liquid is sprayed into the atmosphere. The cohesion minimizes the shape of any water droplet, and so it forms a sphere. We can determine the pressure that cohesion causes within a droplet provided we know the surface tension σ for the liquid. To do this, consider the free-body diagram of half the drop, Fig. 1–21. If we neglect gravity and the effects of atmospheric drag as the drop falls, then the only forces acting are those due to atmospheric pressure, p_a, on its *outside* surface; surface tension, σ, around the *surface* of the drop where it is sectioned; and the internal pressure, p, on the sectioned area. As will be explained in the next chapter, the resultant horizontal forces due to p_a and p are determined by multiplying each pressure by the *projected area* of the drop, that is, πR^2, and the resultant force of the surface tension is determined by multiplying σ by the circumferential distance around the drop, $2\pi R$. For horizontal equilibrium, we therefore have

$$\xrightarrow{+}\Sigma F_x = 0; \qquad p\left(\pi R^2\right) - p_a\left(\pi R^2\right) - \sigma(2\pi R) = 0$$

$$p = \frac{2\sigma}{R} + p_a$$

Here the internal pressure is composed of two parts, one due to surface tension and the other due to atmospheric pressure. For example, mercury at a temperature of 20°C has a surface tension of $\sigma = 486$ mN/m. If the mercury forms into a 2-mm-diameter drop, its surface tension will create an internal pressure of $p_{st} = 2(0.486 \text{ N/m})/(0.001 \text{ m}) = 972$ Pa within the drop, in addition to the pressure caused by the atmosphere.

Capillarity.   The capillarity of a liquid depends upon the comparison between the forces of adhesion and cohesion. If the force of a liquid's adhesion to the molecules of the surface of its container is *greater* than the force of cohesion between the liquid's molecules, then the liquid is referred to as a **wetting liquid**. In this case, the **meniscus** or surface of the liquid, such as water in a narrow glass container, will be concave, Fig. 1–22*a*. If the adhesive force is *less* than the cohesive force, as in the case of mercury, then the liquid is called a **nonwetting liquid**. The meniscus forms a convex surface, Fig. 1–22*b*.

Mercury is a nonwetting liquid as noted by the way its edge curls inward.

Wetting liquids will rise up along a narrow tube, Fig. 1–23a, and we can determine this height h by considering a free-body diagram of the portion of the liquid suspended in the tube, Fig. 1–23b. Here the free surface or meniscus makes a contact angle θ between the sides of the tube and the liquid surface. This angle defines the direction of the force of adhesion, which is the effect of the surface tension σ of the liquid as it holds the liquid surface up against the wall of the tube. The resultant of this force, which acts around the inner circumference of the tube, is therefore $\sigma(2\pi r)\cos\theta$. The other force is the weight of the suspended liquid, $W = \rho g V$, where $V = \pi r^2 h$. For vertical equilibrium, we require

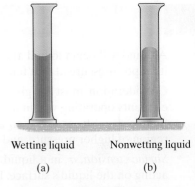

Wetting liquid   Nonwetting liquid

(a)   (b)

Fig. 1–22

$$+\uparrow \Sigma F_y = 0; \qquad \sigma(2\pi r)\cos\theta - \rho g(\pi r^2 h) = 0$$

$$h = \frac{2\sigma\cos\theta}{\rho g r}$$

Experiments have shown that the contact angle between water and glass is $\theta \approx 0°$, and so for water the meniscus surface, shown in Fig. 1–23a, actually becomes somewhat hemispherical. By carefully measuring h, the above equation can then be used with $\theta = 0°$ to determine the surface tension σ for water at various temperatures.

In the next chapter we will show how to determine pressure by measuring the height of a liquid in a glass tube. When it is used for this purpose, however, errors due to the additional height caused by capillarity within the tube can occur. To minimize this effect, notice that h in the above result is inversely proportional to the density of the liquid and the radius of the tube. The smaller they are, the higher h becomes. For example, for a 3-mm-diameter tube containing water at 20°C, where $\sigma = 72.7$ mN/m and $\rho = 998.3$ kg/m³, we have

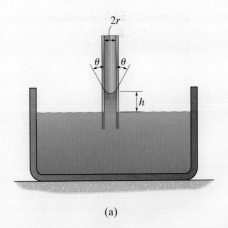

(a)

$$h = \frac{2(0.0727 \text{ N/m})\cos 0°}{(998.3 \text{ kg/m}^3)(9.81 \text{ m/s}^2)(0.0015 \text{ m})} = 9.90 \text{ mm}$$

This is somewhat significant, and so for experimental work it is generally preferable to use tubes having a diameter of 10 mm or greater, since at 10 mm, $h \approx 3$ mm, and the effect of capillarity is minimized.

Throughout our study of fluid mechanics, we will find that, for the most part, the forces of cohesion and adhesion will be small compared to the effects of gravity, pressure, and viscosity. Surface tension generally becomes important, however, when we want to study phenomena related to bubble formation and growth, examine movement of liquids through porous media such as soil, or consider the effects of liquid films on surfaces.

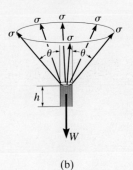

(b)

Fig. 1–23

Important Points

- A liquid will begin to boil at a specific temperature when the pressure within it, or at its surface, is equal to its vapor pressure at that temperature.

- Consideration must be given to the possibility of *cavitation* when designing mechanical or structural elements operating within a fluid environment. This phenomenon is caused when the pressure within the fluid is equal to or less than the vapor pressure, causing boiling, migration of the resulting bubbles to a region of higher pressure, and then their sudden collapse.

- *Surface tension*, σ, in a liquid is caused by molecular cohesion. It is measured as a force per unit length acting on the liquid's surface. It becomes smaller as the temperature rises.

- Capillarity of a *wetting liquid*, such as water in a narrow glass tube, creates a concave surface since the force of adhesion to the walls of the tube will be greater than the force caused by the cohesion of the liquid. For a *nonwetting liquid*, such as mercury, the surface is convex since the force of cohesion will be greater than that of adhesion.

References

1. G. A. Tokaty, *A History and Philosophy of Fluid Mechanics*, Dover Publications, New York, NY, 1994.
2. R. Rouse and S. Ince, *History of Hydraulics*, Iowa Institute of Hydraulic Research, Iowa City, IA, 1957.
3. A. S. Monin and A. M. Yaglom, *Statistical Fluid Mechanics, Mechanics of Turbulence*, Vol. 1, Dover Publications, New York, NY, 2007.
4. *Handbook of Chemistry and Physics*, 62nd ed., Chemical Rubber Publishing Co., Cleveland, OH, 1988.
5. *The U.S. Standard Atmosphere (1976)*, U.S. Government Printing Office, Washington, DC, 1976.
6. *Handbook of Tables for Applied Engineering Science*, Chemical Rubber Publishing Co., Cleveland, OH, 1970.
7. B. S. Massey and J. Ward-Smith, *Mechanics of Fluids*, 9th ed., Spon Press, London and New York, 2012.
8. C. W. Macosko, *Rheology: Principles, Measurements, and Applications*, VCH Publishers New York, NY, 1994.
9. P. M. Kampmeyer, "The Temperature Dependence of Viscosity for Water and Mercury," *Journal of Applied Physics*, 23, 99, 1952.
10. R. D. Trengove and W. A. Wrakeham, "The Viscosity of Carbon Dioxide, Methane, and Sulfur Hexafluoride in the Limit of Zero Density," *Journal of Physics and Chemistry*, Vol. 16, No 2, 1987.
11. S. Granick and E. Zhu, *Physical Review Letters*, August 27, 2001.
12. D. Blevins, *Applied Fluid Dynamics Handbook*, Van Nostrand Reinhold, New York, NY, 1984.
13. R. L. Mott, *Applied Fluid Mechanics*, Prentice Hall, Upper Saddle River, NJ, 2006.
14. R. Goldstein, *Fluid Mechanics Measurements*, 2nd ed., Taylor and Francis, Washington, DC, 1996.
15. P. R. Lide, W. M. Haynes, eds. *Handbook of Chemistry and Physics*, 90th ed., Boca Raton, FL, CRC Press.

PROBLEMS

Sec. 1.1–1.6

1–1. Evaluate each of the following to three significant figures, and express each answer in SI units using an appropriate prefix: (a) $(425 \text{ mN})^2$, (b) $(67\,300 \text{ ms})^2$, (c) $[723(10^6)]^{1/2} \text{ mm}$.

1–2. Evaluate each of the following to three significant figures, and express each answer in SI units using an appropriate prefix: (a) $749 \text{ μm}/63 \text{ ms}$, (b) (34 mm) $(0.0763 \text{ Ms})/263 \text{ mg}$, (c) $(4.78 \text{ mm})(263 \text{ Mg})$.

1–3. Represent each of the following quantities with combinations of units in the correct SI form, using an appropriate prefix: (a) $GN \cdot μm$, (b) $kg/μm$, (c) N/ks^2, (d) $kN/μs$.

*1–4. Convert the following temperatures: (a) 20°C to Kelvin, (b) 500 K to degrees Celsius.

1–5. Mercury has a specific weight of $133 \text{ kN}/m^3$ when the temperature is 20°C. Determine its density and specific gravity at this temperature.

1–6. If air within the tank is at an absolute pressure of 680 kPa and a temperature of 70°C, determine the weight of the air inside the tank. The tank has an interior volume of 1.35 m^3.

1–7. The bottle tank has a volume of 0.12 m^3 and contains oxygen at an absolute pressure of 12 MPa and a temperature of 30°C. Determine the mass of oxygen in the tank.

*1–8. The bottle tank has a volume of 0.12 m^3 and contains oxygen at an absolute pressure of 8 MPa and temperature of 20°C. Plot the variation of the pressure in the tank (vertical axis) versus the temperature for $20°C \leq T \leq 80°C$. Report values in increments of $\Delta T = 10°C$.

Probs. 1–7/8

1–9. Determine the specific weight of carbon dioxide when the temperature is 100°C and the absolute pressure is 400 kPa.

1–10. Dry air at 25°C has a density of $1.23 \text{ kg}/m^3$. But if it has 100% humidity at the same pressure, its density is 0.65% less. At what temperature would dry air produce this same smaller density?

1–11. The tank contains air at a temperature of 15°C and an absolute pressure of 210 kPa. If the volume of the tank is 5 m^3 and the temperature rises to 30°C, determine the mass of air that must be removed from the tank to maintain the same pressure.

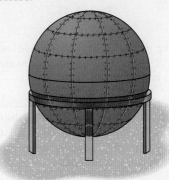

Prob. 1–11

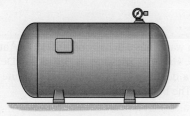

Prob. 1–6

*1–12. Water in the swimming pool has a measured depth of 3.03 m when the temperature is 5°C. Determine its approximate depth when the temperature becomes 35°C. Neglect losses due to evaporation.

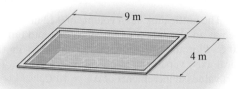

Prob. 1–12

1–13. The tank initially contains carbon dioxide at an absolute pressure of 200 kPa and a temperature of 50°C. As more carbon dioxide is added, the pressure increases at 25 kPa/min. Plot the variation of the pressure in the tank (vertical axis) versus the temperature for the first 10 minutes. Report the values in increments of two minutes. Assume the tank expands to keep the density constant.

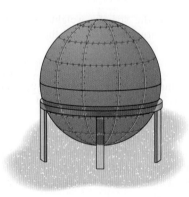

Prob. 1–13

1–14. A 2-kg mass of oxygen is held at a constant temperature of 50° and an absolute pressure of 220 kPa. Determine its bulk modulus.

1–15. The tank contains 2 kg of air at an absolute pressure of 400 kPa and a temperature of 20°C. If 0.6 kg of air is added to the tank and the temperature rises to 32°C, determine the pressure in the tank.

*1–16. The 8-m-diameter spherical balloon is filled with helium that is at a temperature of 28°C and an absolute pressure of 106 kPa. Determine the weight of the helium contained in the balloon. The volume of a sphere is $V = \frac{4}{3}\pi r^3$.

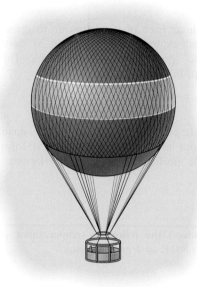

Prob. 1–16

1–17. What is the increase in the density of helium when the absolute pressure changes from 230 kPa to 450 kPa, while the temperature *remains constant* at 20°C? This is called an *isothermal process*.

1–18. Sea water has a density of 1030 kg/m³ at its surface, where the absolute pressure is 101 kPa. Determine its density at a depth of 7 km, where the absolute pressure is 70.4 MPa. The bulk modulus is 2.33 GPa.

1–19. The tank is fabricated from steel that is 20 mm thick. If it contains carbon dioxide at an absolute pressure of 1.35 MPa and a temperature of 20°C, determine the total weight of the tank. The density of steel is 7.85 Mg/m³, and the inner diameter of the tank is 3 m. *Hint*: The volume of a sphere is $V = \frac{4}{3}\pi r^3$.

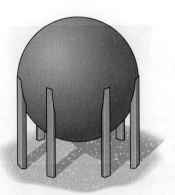

Prob. 1–19

*1–20. Water at 20°C is subjected to a pressure increase of 44 MPa. Determine the percent increase in its density. Take $E_V = 2.20 \, \text{GPa}$.

1–21. The rain cloud has an approximate volume of 1150 km³ and an average height, top to bottom, of 105 m. If a cylindrical container 2 m in diameter collects 50 mm of water after the rain falls out of the cloud, estimate the total mass of rain that fell from the cloud.

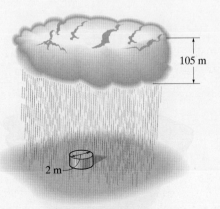

105 m

2 m

Prob. 1–21

1–22. The container is filled with water at a temperature of 25°C and a depth of 2.5 m. If the container has a mass of 30 kg, determine the combined weight of the container and the water.

1 m

2.5 m

Prob. 1–22

1–23. If 4 m³ of helium at 100 kPa of absolute pressure and 20°C is subjected to an absolute pressure of 600 kPa while the temperature remains constant, determine the new density and volume of the helium.

*1–24. A solid has a density of 4500 kg/m³. When a change in pressure of 5.50 MPa is applied, the density increases to 4750 kg/m³. Determine the approximate bulk modulus.

1–25. If the bulk modulus for water at 20°C is 2.20 GPa, determine the change in pressure required to reduce its volume by 0.3%.

1–26. The density of sea water at its surface is 1020 kg/m³, where the absolute pressure is 101.3 kPa. If at a point deep under the water the density is 1060 kg/m³, determine the absolute pressure in MPa at this point. Take $E_V = 2.33 \, \text{GPa}$.

Sec. 1.7–1.8

1–27. Two measurements of shear stress on a surface and the rate of change in shear strain at the surface for a fluid have been determined by experiment to be $\tau_1 = 0.14$ N/m^2, $(du/dy)_1 = 13.63$ s^{-1} and $\tau_2 = 0.48$ N/m^2, $(du/dy)_2 = 153$ s^{-1}. Classify the fluid as Newtonian or non-Newtonian.

*1–28. When the force **P** is applied to the plate, the velocity profile for a Newtonian fluid that is confined under the plate is approximated by $u = (12y^{1/4})$ mm/s, where y is in mm. Determine the shear stress within the fluid at $y = 8$ mm. Take $\mu = 0.5(10^{-3})$ N·s/m^2.

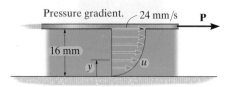

Pressure gradient.   24 mm/s   **P**

16 mm

y u

Prob. 1–28

1–29. At a particular temperature the viscosity of an oil is $\mu = 0.354$ N·s/m^2. Determine its kinematic viscosity. The specific gravity is $S_o = 0.868$.

1–30. Determine the constants B and C in Andrade's equation for water, if it has been experimentally determined that $\mu = 1.00(10^{-3})$ N·s/m^2 at a temperature of 20°C and that $\mu = 0.554(10^{-3})$ N·s/m^2 at 50°C.

1–31. The viscosity of water can be determined using the empirical Andrade's equation with the constants $B = 1.732(10^{-6})$ N·s/m^2 and $C = 1863$ K. With these constants, compare the results of using this equation with those tabulated in Appendix A for temperatures of $T = 10$°C and $T = 80$°C.

*1–32. The constants $B = 1.357(10^{-6})$ N·s/(m^2·K$^{1/2}$) and $C = 78.84$ K have been used in the empirical Sutherland equation to determine the viscosity of air at standard atmospheric pressure. With these constants, compare the results of using this equation with those tabulated in Appendix A for temperatures of $T = 10$°C and $T = 80$°C.

1–33. Determine the constants B and C in the Sutherland equation for air if it has been experimentally determined that at standard atmospheric pressure and a temperature of 20°C, $\mu = 18.3(10^{-6})$ N·s/m^2, and at 50°C, $\mu = 19.6(10^{-6})$ N·s/m^2.

1–34. An experimental test using human blood at $T = 30$°C indicates that it exerts a shear stress of $\tau = 0.15$ N/m^2 on surface A, where the measured velocity gradient at the surface is 16.8 s^{-1}. Since blood is a non-Newtonian fluid, determine its *apparent viscosity* at the surface.

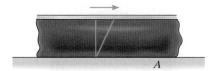

A

Prob. 1–34

1–35. When the force of 3 mN is applied to the plate, the line AB in the liquid remains straight and has an angular rate of rotation of 0.2 rad/s. If the surface area of the plate in contact with the liquid is 0.6 m^2, determine the approximate viscosity of the liquid.

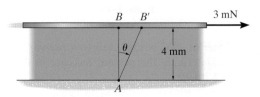

B B' 3 mN

θ 4 mm

A

Prob. 1–35

*1–36. The conical bearing is placed in a lubricating Newtonian fluid having a viscosity μ. Determine the torque **T** required to rotate the bearing with a constant angular velocity of ω. Assume the velocity profile along the thickness t of the fluid is linear.

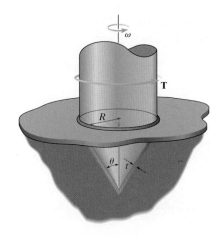

ω

T

R

θ t

Prob. 1–36

1–37. When the force **P** is applied to the plate, the velocity profile for a Newtonian fluid that is confined under the plate is approximated by $u = (12y^{1/4})$ mm/s, where y is in mm. Determine the minimum shear stress within the fluid. Take $\mu = 0.5(10^{-3})$ N·s/m².

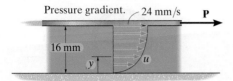

Pressure gradient.　24 mm/s　**P**

16 mm

y　　u

Prob. 1–37

1–38. The velocity profile for a thin film of a Newtonian fluid that is confined between the plate and a fixed surface is defined by $u = (10y - 0.25y^2)$ mm/s, where y is in mm. Determine the shear stress that the fluid exerts on the plate and on the fixed surface. Take $\mu = 0.532$ N·s/m².

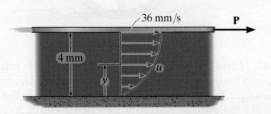

36 mm/s　**P**

4 mm

u

y

Prob. 1–38

1–39. If a force of $P = 2$ N causes the 30-mm-diameter shaft to slide along the lubricated bearing with a constant speed of 0.5 m/s, determine the viscosity of the lubricant and the constant speed of the shaft when $P = 8$ N. Assume the lubricant is a Newtonian fluid and the velocity profile between the shaft and the bearing is linear. The gap between the bearing and the shaft is 1 mm.

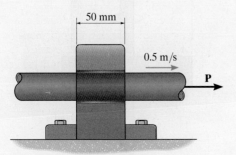

50 mm

0.5 m/s

P

Prob. 1–39

*1–40. The velocity profile of a Newtonian fluid flowing over a fixed surface is approximated by $u = U \sin\left(\dfrac{\pi}{2h}y\right)$. Determine the shear stress in the fluid at $y = h$ and at $y = h/2$. The viscosity of the fluid is μ.

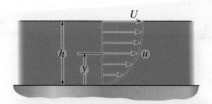

U

h

y　u

Prob. 1–40

1–41. The tank containing gasoline has a long crack on its side that has an average opening of 10 μm. The velocity through the crack is approximated by the equation $u = 10(10^9)\left[10(10^{-6})y - y^2\right]$ m/s, where y is in meters, measured upward from the bottom of the crack. Find the shear stress at the bottom, at $y = 0$, and the location y within the crack where the shear stress in the gasoline is zero. Take $\mu_g = 0.317(10^{-3})$ N·s/m².

1–42. The tank containing gasoline has a long crack on its side that has an average opening of 10 μm. If the velocity profile through the crack is approximated by the equation $u = 10(10^9)\left[10(10^{-6})y - y^2\right]$ m/s, where y is in meters, plot both the velocity profile and the shear stress distribution for the gasoline as it flows through the crack. Take $\mu_g = 0.317(10^{-3})$ N·s/m².

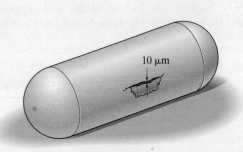

10 μm

Probs. 1–41/42

1–43. The velocity profile for a thin film of a Newtonian fluid that is confined between the plate and a fixed surface is defined by $u = (10y - 0.25y^2)$ mm/s, where y is in mm. Determine the force **P** that must be applied to the plate to cause this motion. The plate has a surface area of 5000 mm² in contact with the fluid. Take $\mu = 0.532$ N·s/m².

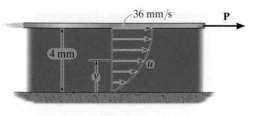

Prob. 1–43

*****1–44.** The 0.15-m-wide plate passes between two layers, A and B, of oil that has a viscosity of $\mu = 0.04$ N·s/m². Determine the force **P** required to move the plate at a constant speed of 6 mm/s. Neglect any friction at the end supports, and assume the velocity profile through each layer is linear.

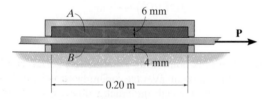

Prob. 1–44

1–45. The 0.15-m-wide plate passes between two layers A and B of different oils, having viscosities of $\mu_A = 0.03$ N·s/m² and $\mu_B = 0.01$ N·s/m². Determine the force **P** required to move the plate at a constant speed of 6 mm/s. Neglect any friction at the end supports, and assume the velocity profile through each layer is linear.

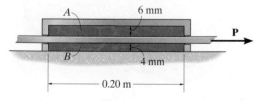

Prob. 1–45

1–46. The tape is 10 mm wide and is drawn through an applicator, which applies a liquid coating (Newtonian fluid) that has a viscosity of $\mu = 0.830$ N·s/m² to each side of the tape. If the gap between each side of the tape and the applicator's surface is 0.8 mm, determine the torque **T** at the instant $r = 150$ mm that is needed to rotate the wheel at 0.5 rad/s. Assume the velocity profile within the liquid is linear.

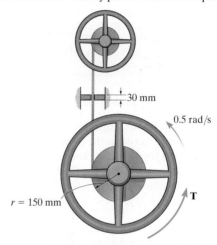

Prob. 1–46

1–47. Disks A and B rotate at a constant rate of $\omega_A = 50$ rad/s and $\omega_B = 20$ rad/s, respectively. Determine the torque **T** required to sustain the motion of disk B. The gap, $t = 0.1$ mm, contains SAE 10 oil for which $\mu = 0.02$ N·s/m². Assume the velocity profile is linear.

*****1–48.** If disk A is stationary, $\omega_A = 0$, and disk B rotates at $\omega_B = 20$ rad/s, determine the torque **T** required to sustain the motion. Plot your results of torque (vertical axis) versus the gap thickness for $0 \le t \le 0.1$ mm. The gap contains SAE 10 oil for which $\mu = 0.02$ N·s/m². Assume the velocity profile is linear.

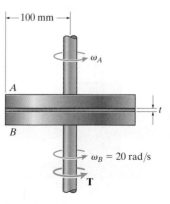

Probs. 1–47/48

1–49. The very thin tube A of mean radius r and length L is placed within the fixed circular cavity as shown. If the cavity has a small gap of thickness t on each side of the tube, and is filled with a Newtonian liquid having a viscosity μ, determine the torque **T** required to overcome the fluid resistance and rotate the tube with a constant angular velocity of ω. Assume the velocity profile within the liquid is linear.

*1–52. Water at A has a temperature of 15°C and flows along the top surface of the plate C. The velocity profile is approximated as $u_A = 10 \sin(2.5\pi y)$ m/s, where y is in meters. Below the plate the water at B has a temperature of 60°C and a velocity profile of $u_B = 4(10^3)(0.1y - y^2)$, where y is in meters. Determine the resultant force per unit length of plate C the flow exerts on the plate due to viscous friction. The plate is 3 m wide.

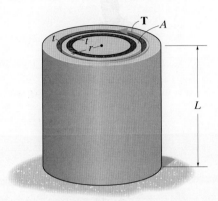

Prob. 1–49

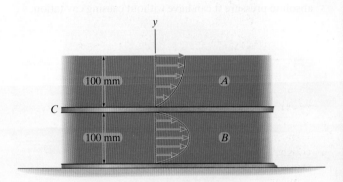

Prob. 1–52

1–50. The shaft rests on a 2-mm-thin film of oil having a viscosity of $\mu = 0.0657$ N·s/m². If the shaft is rotating at a constant angular velocity of $\omega = 2$ rad/s, determine the shear stress in the oil at $r = 50$ mm and $r = 100$ mm. Assume the velocity profile within the oil is linear.

1–51. The shaft rests on a 2-mm-thin film of oil having a viscosity of $\mu = 0.0657$ N·s/m². If the shaft is rotating at a constant angular velocity of $\omega = 2$ rad/s, determine the torque **T** that must be applied to the shaft to maintain the motion. Assume the velocity profile within the oil is linear.

1–53. The read–write head for a hand-held music player has a surface area of 0.04 mm². The head is held 0.04 μm above the disk, which is rotating at a constant rate of 1800 rpm. Determine the torque **T** that must be applied to the disk to overcome the frictional shear resistance of the air between the head and the disk. The surrounding air is at standard atmospheric pressure and a temperature of 20°C. Assume the velocity profile is linear.

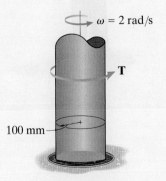

Probs. 1–50/51

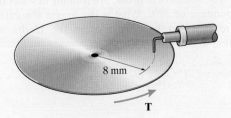

Prob. 1–53

Sec. 1.9–1.10

1–54. The city of Denver, Colorado, is at an elevation of 1610 m above sea level. Determine how hot one can prepare boiling water to make a cup of tea.

1–55. As water at 40°C flows through the transition, its pressure will begin to decrease. Determine the lowest absolute pressure it can have without causing cavitation.

Prob. 1–55

*1–56. Steel particles are ejected from a grinder and fall gently into a tank of water. Determine the largest average diameter of a particle that will float on the water. Take $\rho = 7850 \ \text{kg/m}^3$ and $\sigma = 0.0726 \ \text{N/m}$. Assume that each particle has the shape of a sphere, where $V = \frac{4}{3}\pi r^3$.

1–57. The blades of a turbine are rotating in water that has a temperature of 30°C. What is the lowest absolute water pressure that can be developed at the blades so that cavitation will not occur?

1–58. Water at 15°C is flowing through a garden hose. If the hose is bent, a hissing noise can be heard. Here cavitation has occurred in the hose because the velocity of the flow has increased at the bend, and the pressure has dropped. What would be the highest absolute pressure in the hose at this location in the hose?

Prob. 1–58

1–59. Water at 25°C is flowing through a garden hose. If the hose is bent, a hissing noise can be heard. Here cavitation has occurred in the hose because the velocity of the flow has increased at the bend, and the pressure has dropped. What would be the highest absolute pressure in the hose at this location in the hose?

Prob. 1–59

*1–60.   Determine the distance h that the column of mercury in the tube will be depressed as a function tube's diameter D when the tube is inserted into the mercury at a room temperature of 20°C. Plot the relationship of h (vertical axis) versus D for 1 mm $\leq D \leq$ 4 mm. Give values for increments of ΔD = 0.025 in. Discuss this result.

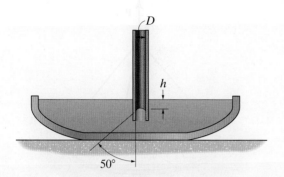

Prob. 1–60

1–61.   Water in the glass tube is at a temperature of 40°C. Plot the height h of the water as a function of the tube's inner diameter D for 0.5 mm $\leq D \leq$ 3 mm. Use increments of 0.5 mm. Take σ = 69.6 mN/m.

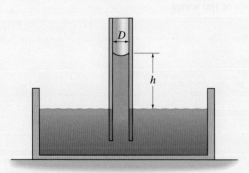

Prob. 1–61

1–62.   When a can of soda water is opened, small gas bubbles are produced within it. Determine the difference in pressure between the inside and outside of a bubble having a diameter of 0.5 mm. The surrounding temperature is 20°C. Take σ = 0.0735 N/m.

1–63.   Determine the greatest distance h that the column of mercury in the tube will be depressed when the tube is inserted into the mercury at a room temperature of 20°C. Set D = 3 mm

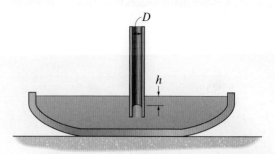

Prob. 1–63

*1–64.   The tube has an inner diameter d and is immersed in water at an angle θ from the horizontal. Determine the average length L to which water will rise along the tube due to capillary action. The surface tension of the water is σ and its density is ρ.

1–65.   The tube has an inner diameter of d = 2 mm and is immersed in water. Determine the average length L to which the water will rise along the tube due to capillary action as a function of the angle of tilt, θ. Plot this relationship of L (vertical axis) versus θ for 10° $\leq \theta \leq$ 30°. Give values for increments of $\Delta\theta$ = 5°. The surface tension of the water is σ = 75.4 mN/m, and its density is ρ = 1000 kg/m³.

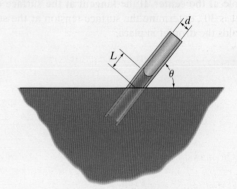

Probs. 1–64/65

1–66. The marine water strider, *Halobates*, has a mass of 0.36 g. If it has six slender legs, determine the minimum contact length of all of its legs combined to support itself in water having a temperature of $T = 20°C$. Take $\sigma = 72.7\,\text{mN/m}$, and assume the legs are thin cylinders that are water repellent.

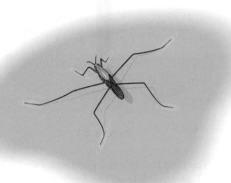

Prob. 1–66

1–67. Many camera phones now use liquid lenses as a means of providing a quick auto-focus. These lenses work by electrically controlling the internal pressure within a liquid droplet, thereby affecting the angle of the meniscus of the droplet, and so creating a variable focal length. To analyze this effect, consider, for example, a segment of a spherical droplet that has a base diameter of 3 mm. The pressure in the droplet is 105 Pa and is controlled through a tiny hole at the center. If the tangent at the surface of the droplet is 30°, determine the surface tension at the surface that holds the droplet in place.

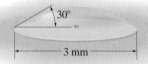

Prob. 1–67

*1–68.** The ring has a weight of 0.2 N and is suspended on the surface of the water, for which $\sigma = 73.6\,\text{mN/m}$. Determine the vertical force **P** needed to pull the ring free from the surface. *Note*: This method is often used to measure surface tension.

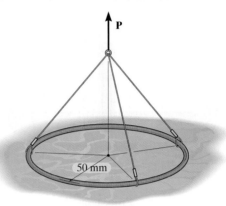

Prob. 1–68

1–69. The ring has a weight of 0.2 N and is suspended on the surface of the water. If it takes a force of $P = 0.245\,\text{N}$ to lift the ring free from the surface, determine the surface tension of the water.

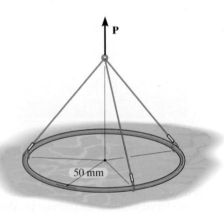

Prob. 1–69

CONCEPTUAL PROBLEMS

P1–1. The air pressure for the bicycle tire is 220 kPa. Assuming the volume of air in the tire remains constant, find the pressure difference that occurs between a typical summer and winter day. Discuss the forces that are responsible for supporting the tire and the bicycle when a rider is on the bicycle.

P1–1

P1–2. Water poured from this pitcher tends to cling to its bottom side. Explain why this happens and suggest what can be done to prevent it.

P1–2

P1–3. If a drop of oil is placed on a water surface, then the oil will tend to spread out over the surface as shown. Explain why this happens.

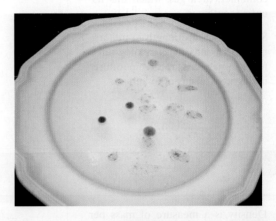

P1–3

P1–4. The shape of this municipal water tank has the same shape as a drop of water resting on a nonabsorbent surface, such as wax paper. Explain why engineers design the tank in this manner.

P1–4

CHAPTER REVIEW

Matter can be classified as a solid, which maintains its shape; a liquid, which takes the shape of its container; or a gas, which fills its entire container.

SI units measure length in meters, time in seconds, mass in kilograms, force in newtons, and temperature in degrees Celsius or in kelvins.

Density is a measure of mass per unit volume.

$$\rho = \frac{m}{V}$$

Specific weight is a measure of weight per unit volume.

$$\gamma = \frac{W}{V}$$

Specific gravity is the ratio of either a liquid's density or its specific weight to that of water, where $\rho_w = 1000 \text{ kg/m}^3$.

$$S = \frac{\rho}{\rho_w} = \frac{\gamma}{\gamma_w}$$

The ideal gas law relates the *absolute pressure* of the gas to its density and *absolute temperature*.

$$p = \rho R T$$

The bulk modulus is a measure of a fluid's resistance to compression.

$$E_V = -\frac{dp}{dV/V}$$

The viscosity of a fluid is a measure of its resistance to shear. The higher the viscosity, the higher the resistance.

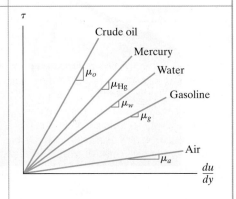

For Newtonian fluids, the shear stress in a fluid is directly proportional to the time rate of change in its shear strain, which is defined by the velocity gradient, du/dy. The constant of proportionality, μ, is called the dynamic viscosity or simply the viscosity.	$$\tau = \mu\frac{du}{dy}$$	
The kinematic viscosity is a ratio of the fluid's viscosity to its density.	$$\nu = \frac{\mu}{\rho}$$	
The viscosity is measured indirectly by using a rotational viscometer, an Ostwald viscometer, or other device.		
A liquid will boil if the pressure above or within it is equal to or below its vapor pressure. This can lead to cavitation, which produces bubbles that can migrate to regions of higher pressure and then collapse.		
Surface tension on a liquid surface develops because of the cohesive (attractive) forces between its molecules. It is measured as a force per unit length.		
The capillarity of a liquid depends on the comparative forces of adhesion and cohesion. Wetting liquids have a greater force of adhesion to their contacting surface than the liquid's cohesive force. The opposite effect occurs for nonwetting liquids, where the cohesive force is greater than the adhesive force.		

Chapter 2

(© Jim Lipschutz/Shutterstock)

These heavily reinforced gates have been designed to resist the hydrostatic loadings within a canal. Along with another set, they form a "lock," controlling the level of water within the confined region, thus allowing ships to pass from one elevation to another.

Fluid Statics

CHAPTER OBJECTIVES

- To discuss pressure and show how it varies within a static fluid.
- To present the various ways of measuring pressure in a static fluid using barometers, manometers, and pressure gages.
- To show how to calculate a resultant hydrostatic force and find its location on a submerged surface.
- To present the topics of buoyancy and stability.
- To show how to calculate the pressure within a liquid subjected to a constant acceleration, and to a steady rotation about a fixed axis.

2.1 Pressure

In general, fluids can exert both normal and shear forces on their surfaces of contact. However, if the fluid is *at rest* relative to the surface, then the viscosity of the fluid will have no shearing effect on the surface. Instead, the only force the fluid exerts is a normal force, and the effect of this force is called *pressure*. From a physical point of view, the pressure of a fluid on the surface is the result of the impulses exerted by vibrating fluid molecules as they contact and bounce off the surface.

Pressure is defined as the force acting normal to an area divided by this area. If we assume the fluid to be a continuum, then *at a point* within the fluid the area can approach zero, Fig. 2–1a, and so the pressure becomes

$$p = \lim_{\Delta A \to 0} \frac{\Delta F}{\Delta A} = \frac{dF}{dA} \qquad (2\text{–}1)$$

If the surface has a finite area and the pressure is *uniformly distributed* over this area, Fig. 2–1b, then the *average pressure* is

$$p_{\text{avg}} = \frac{F}{A} \qquad (2\text{–}2)$$

Pressure can have units of pascals Pa (N/m^2).

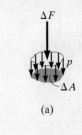

(a)

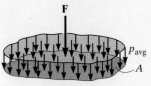

Average pressure
(b)

Fig. 2–1

45

Pascal's Law.

In the 17th century the French mathematician Blaise Pascal was able to show that *the intensity of the pressure acting at a point in a fluid is the same in all directions.* This statement is commonly known as **Pascal's law**, even though Giovanni Benedotti and Simon Stevin had deduced it previously in the late 16th century.

Pascal's law seems intuitive, since if the pressure at the point was larger in one direction than in the opposite direction, the inbalance would cause movement or agitation of the fluid, something that is not observed. We can formally prove Pascal's law by considering the equilibrium of a small triangular element located within a fluid, Fig. 2–2a. Provided the fluid is at rest (or moving with constant velocity), the only forces acting on its free-body diagram are due to pressure and gravity, Fig. 2–2b. The force of gravity is the result of the specific weight γ of the fluid multiplied by the volume of the element. In accordance with Eq. 2–1, the *force* created by the pressure is determined by multiplying the pressure by the area of the element face upon which it acts. In the y-z plane there are three pressure forces. Assuming the inclined face has a length Δs, the dimensions of the other faces are $\Delta y = \Delta s \cos \theta$ and $\Delta z = \Delta s \sin \theta$, Fig. 2–2a. Therefore, applying the force equations of equilibrium in the y and z directions, we have

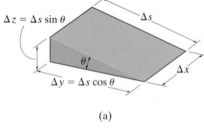

(a)

$$\Sigma F_y = 0; \quad p_y(\Delta x)(\Delta s \, \sin \theta) - \left[p(\Delta x \Delta s)\right] \sin \theta = 0$$

$$\Sigma F_z = 0; \quad p_z(\Delta x)(\Delta s \, \cos \theta) - \left[p(\Delta x \Delta s)\right] \cos \theta$$
$$- \gamma\left[\frac{1}{2}\Delta x(\Delta s \, \cos \theta)(\Delta s \, \sin \theta)\right] = 0$$

Dividing by $\Delta x \Delta s$ and letting $\Delta s \to 0$, so the element reduces in size, we obtain

$$p_y = p$$

$$p_z = p$$

By a similar argument, the element can be rotated 90° about the z axis, and $\Sigma F_x = 0$ can be applied to show $p_x = p$. Since the angle θ of the inclined face is *arbitrary*, this indeed shows that the pressure at a point is the *same in all directions* for any fluid that has no relative motion between its adjacent layers.*

Since the pressure at a point is transmitted throughout the fluid by action, equal but opposite force reaction, to each of its neighboring points, then it follows from Pascal's law that any *pressure increase* Δp at one point in the fluid will cause the *same increase* at all the other points within the fluid. This principle has widespread application for the design of hydraulic machinery, as noted in the following example.

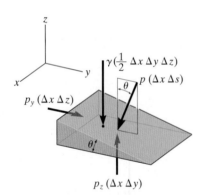

Free-body diagram

(b)

Fig. 2–2

---

*It is also possible to show that Pascal's law applies even if the fluid is accelerating. See Prob. 2–1.

EXAMPLE | 2.1

The mechanics of a pneumatic jack used in a service station is shown in Fig. 2–3. If the car and lift have a weight of 25 kN, determine the force that must be developed by the air compressor at B to raise the lift at a constant velocity. Air is in the line from B to A. The air line at B has an inner diameter of 15 mm, and the post at A has a diameter of 280 mm.

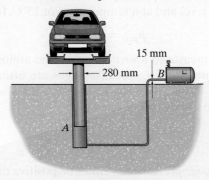

Fig. 2–3

SOLUTION

Fluid Description.   The weight of the air can be neglected.

Analysis.   Due to equilibrium, the force created by air pressure at A is equal and opposite to the weight of the car and lift. The average pressure at A is therefore

$$p_A = \frac{F_A}{A_A}; \qquad \frac{25(10^3)\text{ N}}{\pi(0.140 \text{ m})^2} = 406.00(10^3) \text{ Pa}$$

Since the weight of the air is neglected, the pressure at each point is the same in all directions (Pascal's law), and this same pressure is transmitted to B. Therefore, the force at B is

$$p_B = \frac{F_B}{A_B}; \qquad 406.00(10^3) \text{ N/m}^2 = \frac{F_B}{\pi(0.0075 \text{ m})^2}$$

$$F_B = 71.7 \text{ N} \qquad\qquad Ans.$$

This 71.7-N *force* will lift the 25-kN load, even though the pressure at A and B is the same.

The principles on which this example is based also extend to many hydraulic systems, where the working fluid is oil. Typical applications include jacks, construction equipment, presses, and elevators. The pressures in these systems often range from 8 MPa when used with small vehicles, all the way to 60 MPa for hydraulic jacks. Any compressor or pump used for these applications has to be designed so that its washers and seals maintain these high pressures over an extended period of time.

2.2   Absolute and Gage Pressure

If a fluid such as air were removed from its container, a vacuum would exist and the pressure within the container would be zero. This is commonly referred to as **zero absolute pressure**. Any pressure that is measured above this value is referred to as the **absolute pressure**, p_{abs}. For example, **standard atmospheric pressure** is the absolute pressure that is measured at sea level and at a temperature of 15°C. Its value is

$$p_{atm} = 101.3 \text{ kPa}$$

Any pressure measured above or below the atmospheric pressure is called the **gage pressure**, p_g, because gages are often used to measure pressure relative to the atmospheric pressure. The absolute pressure and the gage pressure are therefore related by

$$\boxed{p_{abs} = p_{atm} + p_g} \tag{2–3}$$

Realize that the *gage pressure* can either be positive or negative, Fig. 2–4. For example, if the absolute pressure is $p_{abs} = 301.3$ kPa, then the gage pressure becomes $p_g = 301.3$ kPa $- 101.3$ kPa $= 200$ kPa. Likewise, if the absolute pressure is $p_{abs} = 51.3$ kPa, then the gage pressure is $p_g = 51.3$ kPa $- 101.3$ kPa $= -50$ kPa, a negative value producing a suction, since it is *below* atmospheric pressure.

In this text we will always measure the gage pressure relative to standard atmospheric pressure; however, for greater accuracy the *local* atmospheric pressure should be used, and from that the local gage pressure can be determined. Also, unless otherwise stated, all pressures reported in the text and in the problems will be considered as gage pressures. If absolute pressure is intended, it will be specifically stated or denoted as, for example, 5 Pa (abs.).

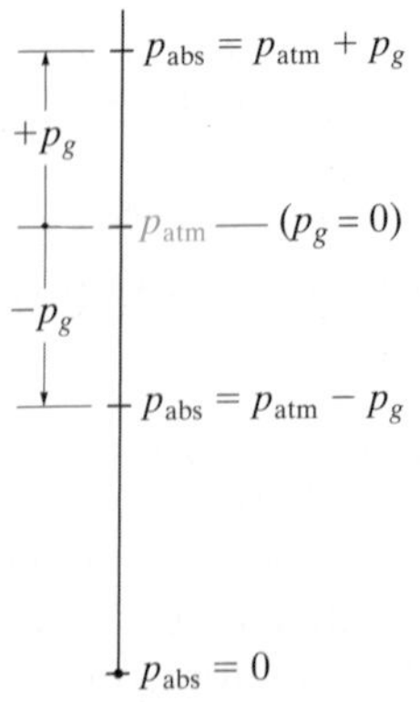

Pressure Scale

Fig. 2–4

EXAMPLE | 2.2

The air pressure within the bicycle tire is determined from a gage to be 70 kPa, Fig. 2–5. If the local atmospheric pressure is 104 kPa, determine the absolute pressure in the tire.

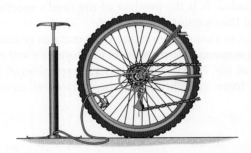

Fig. 2–5

SOLUTION

Fluid Description. The air remains static under constant pressure.

Analysis. Before the tire was filled with air, the pressure within it was atmospheric, 104 kPa. Therefore, after the tire is filled, the absolute pressure in the tire is

$$p_{abs} = p_{atm} + p_g$$

$$p_{abs} = 104 \text{ kPa} + 70 \text{ kPa}$$

$$= 174 \text{ kPa} \qquad \qquad \textit{Ans.}$$

A point to remember is that a newton is about the weight of an apple, and so when this weight is distributed over a square meter, a pascal is actually a very small pressure $\left(\text{Pa} = \text{N/m}^2\right)$. For this reason, for engineering work, pressures measured in pascals are almost always accompanied by a prefix.

2.3   Static Pressure Variation

In this section we will determine how the pressure varies within a static fluid due to the *weight* of the fluid. To do this, we will consider small, slender horizontal and vertical fluid elements having cross-sectional areas ΔA, and lengths of Δy and Δz, respectively. The free-body diagrams, showing only the forces acting in the y and z directions on each element, are shown in Fig. 2–6. For the element extending in the z direction, the weight is included. It is the product of the fluid's specific weight γ and the volume of the element, $\Delta V = \Delta A \, \Delta z$.

The gradient or change in pressure from one side of each element to its opposite side is assumed to *increase* in the *positive y and z directions*, and is expressed as $(\partial p / \partial y)\Delta y$ and $(\partial p/\partial z)\Delta z$, respectively.* If we apply the equation of force equilibrium to the horizontal element, Fig. 2–6a, we obtain

$$\Sigma F_y = 0; \qquad p(\Delta A) - \left(p + \frac{\partial p}{\partial y}\Delta y\right)\Delta A = 0$$

$$\partial p = 0$$

This same result will also occur in the x direction, and since the change in pressure is zero, it indicates that the *pressure remains constant in the horizontal plane*. In other words, the pressure will *only* be a function of z, $p = p(z)$, and so we can now express its change as a total derivative. From Fig. 2–6b,

$$\Sigma F_z = 0; \qquad p(\Delta A) - \left(p + \frac{\partial p}{dz}\Delta z\right)\Delta A - \gamma(\Delta A \, \Delta z) = 0$$

$$dp = -\gamma dz \qquad (2\text{–}4)$$

The negative sign indicates that the pressure will *decrease as one moves upwards in the fluid*, positive z direction.

The above two results apply to both incompressible and compressible fluids, and in the next two sections we will treat each of these types of fluids separately.

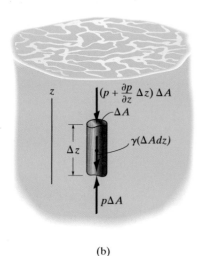

(a)

(b)

Fig. 2–6

---

*This is the result of a Taylor series expansion about a point, for which we have omitted the higher-order terms, $\frac{1}{2}\left(\frac{\partial^2 p}{\partial y^2}\right)\Delta y^2 + \cdots$ and $\frac{1}{2}\left(\frac{\partial^2 p}{\partial z^2}\right)\Delta z^2 + \cdots$, because they will drop out as $\Delta y \to 0$ and $\Delta z \to 0$. Also, the partial derivative is used here because the pressure is *assumed* to be changing in every coordinate direction, i.e., the pressure is assumed to be different at each point, and so $p = p(x,y,z)$.

2.4 Pressure Variation for Incompressible Fluids

If the fluid is assumed to be *incompressible*, as in the case of a liquid, then its specific weight γ is constant since its volume does not change. Consequently, Eq. 2–4, $dp = -\gamma dz$, can be integrated vertically from a reference level $z = z_0$, where $p = p_0$, to a higher level z, where the pressure is p, Fig. 2–7a. Hence,

$$\int_{p_0}^{p} dp = -\gamma \int_{z_0}^{z} dz$$

$$p = p_0 + \gamma(z_0 - z)$$

For convenience, the reference level is usually established at the *free surface of the liquid, $z_0 = 0$,* and the coordinate z is directed *positive downward*, Fig. 2–7b. If this is the case, then the pressure at a distance h *below* the surface becomes

$$p = p_0 + \gamma h \qquad (2\text{–}5)$$

If the pressure at the surface is the atmospheric pressure, $p_0 = p_{\text{atm}}$, then the term γh represents the *gage pressure* in the liquid. Therefore,

$$\boxed{p = \gamma h} \qquad (2\text{–}6)$$

Incompressible fluid

Like diving into a swimming pool, this result indicates that the weight of the water will cause the gage pressure to *increase linearly* as one descends deeper into the water.

Water tanks are used in many municipalities in order to provide a constant pressure within the water distribution system. This is especially important when demand is high in the early morning and early evening.

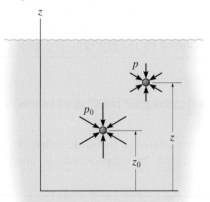

(a)

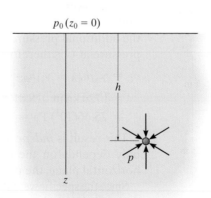

Pressure increases with depth
$p = \gamma h$

(b)

Fig. 2–7

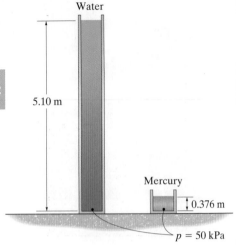

Water

5.10 m

Mercury

0.376 m

$p = 50$ kPa

Pressure heads

Fig. 2–8

Pressure Head. If we solve Eq. 2–6 for h, we obtain

$$h = \frac{p}{\gamma} \tag{2–7}$$

Here h is referred to as the **pressure head**, since it indicates the height of a column of liquid that produces the (gage) pressure p. For example, if the gage pressure is 50 kPa, then the pressure heads for water $(\gamma_w = 9.81\,\text{kN/m}^3)$ and mercury $(\gamma_{Hg} = 133\,\text{kN/m}^3)$ are

$$h_w = \frac{p}{\gamma_w} = \frac{50(10^3)\ \text{N/m}^2}{9.81(10^3)\ \text{N/m}^3} = 5.10\ \text{m}$$

$$h_{Hg} = \frac{p}{\gamma_{Hg}} = \frac{50(10^3)\ \text{N/m}^2}{133(10^3)\ \text{N/m}^3} = 0.376\ \text{m}$$

As shown in Fig. 2–8, there is a significant difference in these pressure heads since the densities (or specific weights) of these liquids are so different.

EXAMPLE | 2.3

2 m

A

Gasoline

1 m

B

Glycerin

1.5 m

C

Fig. 2–9

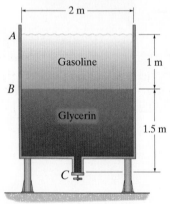

The tank and drainpipe in Fig. 2–9 are filled with gasoline and glycerin to the depths shown. Determine the pressure on the drain plug at C. Report the answer as a pressure head in meter of water. Take $\rho_{ga} = 726\ \text{kg/m}^3$ and $\rho_{gl} = 1260\ \text{kg/m}^3$.

SOLUTION

Fluid Description. Each of the liquids is assumed to be incompressible.

Analysis. Notice that the gasoline will "float" on the glycerin since it has a lower specific weight. To obtain the pressure at C, we need to find the pressure at depth B caused by the gasoline, and then add to it the additional pressure from B to C caused by the glycerin. The *gage* pressure at C is therefore

$$p_C = \gamma_{ga}h_{AB} + \gamma_{gl}h_{BC}$$
$$= (726\ \text{kg/m}^3)(9.81\ \text{m/s}^2)(1\ \text{m}) + (1260\ \text{kg/m}^3)(9.81\ \text{m/s}^2)(1.5\ \text{m})$$
$$= 25.66(10^3)\ \text{Pa} = 25.7\,\text{kPa}$$

This result is *independent* of the *shape* or size of the tank; rather, it only depends on the *depth* of each liquid. In other words, in any horizontal plane, the pressure is constant.

Since the specific weight of water is $\gamma_w = \rho_{wg} = (1000\ \text{kg/m}^3)(9.81\ \text{m/s}^2) = 9.81(10^3)\ \text{N/m}^3$, the pressure head, in meter of water at C, is

$$h = \frac{p_C}{\gamma_w} = \frac{25.66(10^3)\ \text{N/m}^2}{9.81(10^3)\ \text{N/m}^3} = 2.62\ \text{m} \qquad \text{Ans.}$$

In other words, the tank would have to be filled with water to this depth to create the *same pressure* at C caused by both the gasoline and glycerin.

2.5 Pressure Variation for Compressible Fluids

When the fluid is *compressible*, as in the case of a gas, then its specific weight γ will not be constant throughout the fluid. Therefore, to obtain the pressure, we must integrate Eq. 2–4, $dp = -\gamma\,dz$. This requires that we express γ as a function of p. Using the ideal gas law, Eq. 1–11, $p = \rho RT$, where $\gamma = \rho g$, we have $\gamma = pg/RT$. Then

$$dp = -\gamma\,dz = -\frac{pg}{RT}\,dz$$

or

$$\frac{dp}{p} = -\frac{g}{RT}\,dz$$

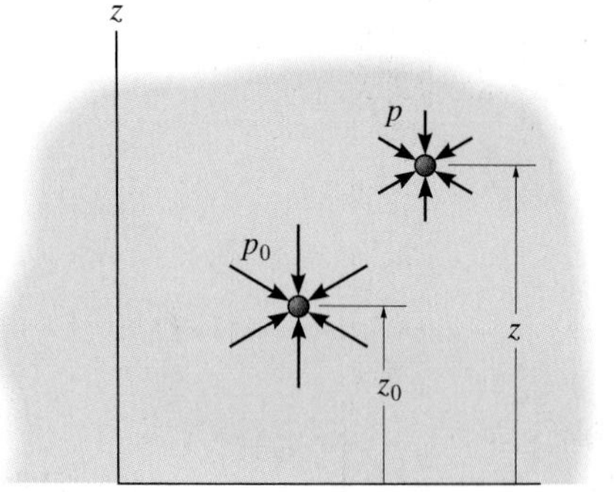

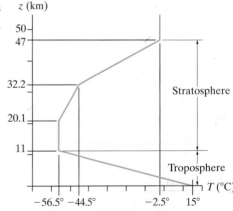

Fig. 2–10

Remember that here p and T must represent the *absolute pressure* and *absolute temperature*. Integration can now be carried out, provided we can express T as a function of z.

Constant Temperature. If the temperature throughout the gas remains constant (isothermal) at $T = T_0$, then assuming the pressure at a reference location $z = z_0$ is $p = p_0$, Fig. 2–10, we have

$$\int_{p_0}^{p} \frac{dp}{p} = -\int_{z_0}^{z} \frac{g}{RT_0}\,dz$$

$$\ln\frac{p}{p_0} = -\frac{g}{RT_0}(z - z_0)$$

or

$$p = p_0 e^{-\left(\frac{g}{RT_0}\right)(z - z_0)} \qquad (2\text{–}8)$$

Approximate temperature distribution in the U.S. standard atmosphere

Fig. 2–11

This equation is often used to calculate the pressure within the lowest region of the stratosphere. As shown on the graph of the U.S. standard atmosphere, Fig. 2–11, this region starts at an elevation of about 11.0 km and reaches an elevation of about 20.1 km. Here the temperature is practically *constant*, at $-56.5°C$ (216.5 K).

EXAMPLE 2.4

The natural gas in the storage tank is contained within a flexible membrane and held under *constant pressure* using a weighted top that is allowed to move up or down as the gas enters or leaves the tank, Fig. 2–12a. Determine the required weight of the top if the (gage) pressure at the outlet A is to be 600 kPa. The gas has a constant temperature of 20°C.

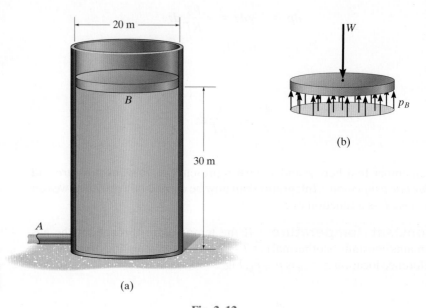

Fig. 2–12

SOLUTION

Fluid Description. We will compare the results of considering the gas to be both incompressible and compressible.

Analysis. Here the pressure at A is a gage pressure, and so there are two forces that act on the free-body diagram of the top, Fig. 2–12b. They are the pressure p_B of gas in the tank and the weight of the top W. We require

$$+\uparrow \Sigma F_y = 0; \qquad p_B A_B - W = 0$$

$$p_B \left[\pi (10 \text{ m})^2 \right] - W = 0$$

$$W = \left[314.16 \, p_B \right] \text{N} \qquad (1)$$

Incompressible gas.   If the gas is considered *incompressible*, the pressure at the outlet A can be related to the pressure at B using Eq. 2–5. From Appendix A for natural gas, $\rho_g = 0.665 \text{ kg/m}^3$, and since $\gamma_g = \rho_g g$, we have

$$p_A = p_B + \gamma_g h$$

$$600(10^3) \text{ N/m}^2 = p_B + (0.665 \text{ kg/m}^3)(9.81 \text{ m/s}^2)(30 \text{ m})$$

$$p_B = 599\,804 \text{ Pa}$$

Substituting into Eq. 1 yields

$$W = \left[314.16(599\,804)\right] \text{N} = 188.4 \text{ MN} \qquad \textit{Ans.}$$

Compressible gas.   If the gas is assumed to be compressible, then since its temperature is constant, Eq. 2–8 applies. From Appendix A, for natural gas, $R = 518.3 \text{ J/(kg·K)}$, and the absolute temperature is $T_0 = 20 + 273 = 293 \text{ K}$. Thus

$$p_B = p_A e^{-\left(\frac{g}{RT_0}\right)(z_B - z_A)}$$

$$= 600(10^3)e^{-\left(\frac{9.81}{[518.3(293)]}\right)(30-0)}$$

$$= 598\,838 \text{ Pa}$$

From Eq. 1,

$$W = \left[314.16(598\,838)\right] \text{N} = 188.1 \text{ MN} \qquad \textit{Ans.}$$

By comparison, there is a difference of less than 0.2% between these two results. Furthermore, notice that the pressure difference between the top B and bottom A of the tank is actually *very small*. For incompressible gas, $(600 \text{ kPa} - 599.8 \text{ kPa}) = 0.2 \text{ kPa}$, and for compressible gas, $(600 \text{ kPa} - 598.8 \text{ kPa}) = 1.2 \text{ kPa}$. For this reason, it is generally satisfactory to *neglect* the change in pressure due to the weight of the gas, and consider the pressure within any gas to be *essentially constant* throughout its volume. If we do this, then $p_B = p_A = 600 \text{ kPa}$, and from Eq. 1, $W = 188.5 \text{ MN}$.

2.6   Measurement of Static Pressure

There are several ways in which engineers measure the absolute and gage pressures at points within a static fluid. Here we will discuss some of the more important ones.

Barometer.   Atmospheric pressure can be measured using a simple device called a ***barometer***. It was invented in the mid-17th century by Evangelista Torricelli, using mercury as a preferred fluid, since it has a high density and a very small vapor pressure. In principle, the barometer consists of a closed-end glass tube that is first entirely filled with mercury. The tube is then submerged in a dish of mercury and then turned upside down, Fig. 2–13. Doing this causes a slight amount of the mercury to empty from the closed end, thereby creating a small volume of mercury vapor in this region. For moderate seasonal temperatures, however, the vapor pressure created is practically zero, so that at the mercury level, $p_A = 0$.*

As the atmospheric pressure, p_{atm}, pushes down on the surface of the mercury in the dish, it causes the pressure at points B and C to be the same, since they are at the same horizontal level. If the height h of the mercury column in the tube is measured, the atmospheric pressure can be determined by applying Eq. 2–5 between A and B.

$$p_B = p_A + \gamma_{Hg} h$$

$$p_{atm} = 0 + \gamma_{Hg} h = \gamma_{Hg} h$$

Normally the height h is stated in either millimeters or inches of mercury. For example, standard atmospheric pressure, 101.3 kPa, will cause the mercury column $\left(\gamma_{Hg} = 133\,290 \text{ N/m}^3\right)$ to rise $h \approx 760$ mm in the tube.

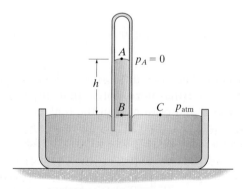

Simple barometer

Fig. 2–13

*To increase the accuracy, the vapor pressure of the mercury that exists within this space should be determined at the temperature recorded when the measurement is made.

Manometer.   A *manometer* consists of a transparent tube that is used to determine the gage pressure in a liquid. The simplest type of manometer is called a *piezometer*. The tube is open at one end to the atmosphere, while the other end is inserted into a vessel, where the pressure of a liquid is to be measured, Fig. 2–14. Any pressure within the vessel will push the liquid up the tube. If the liquid has a specific weight γ, and the pressure head h is measured, then the pressure at point A is $p_A = \gamma h$. Piezometers do not work well for measuring large gage pressures, since h would be large. Also, they are not effective at measuring high negative (suction) gage pressures, since air may leak into the vessel through the point of insertion.

When negative gage pressures or moderately high pressures are to be found, a simple *U-tube manometer*, as shown in Fig. 2–15, can be used. Here one end of the tube is connected to the vessel containing a fluid of specific weight γ, and the other end is open to the atmosphere. To measure relatively high pressures, a liquid with a high specific weight γ', like mercury, is placed in the U-tube. The pressure at point A in the vessel is the same as at point B in the tube, since both points are on the same level. The pressure at C is therefore $p_C = p_A + \gamma h_{BC}$. This is the same pressure as at D, again, because C and D are on the same level. Finally, since $p_C = p_D = \gamma' h_{DE}$, then

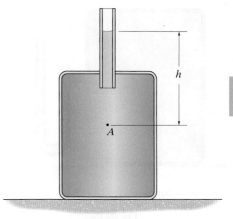

Piezometer

Fig. 2–14

$$\gamma' h_{DE} = p_A + \gamma h_{BC}$$

or

$$p_A = \gamma' h_{DE} - \gamma h_{BC}$$

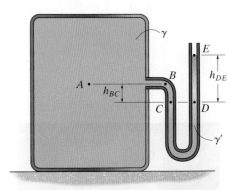

Simple manometer

Fig. 2–15

Note that if the fluid in the vessel is a gas, then its specific weight will be very small compared to that of the manometer liquid, so $\gamma \approx 0$ and the above equation becomes $p_A = \gamma' h_{DE}$.

To *reduce errors* when using a manometer, it is best to use a manometer fluid that has a low specific gravity, such as water, when the anticipated pressure is low. This way the fluid is pushed higher in the manometer so the reading of the pressure head is more sensitive to pressure differences. Also, as noted in Sec. 1.10, errors can be reduced in reading the meniscus, that is, the curved surface caused by capillary attraction, if the tube generally has a diameter of 10 mm or greater. Finally, for very sensitive work, a more exact specific weight of the fluid can be specified if its temperature is known.

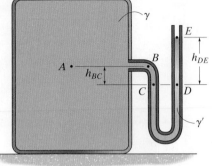

Simple manometer

Fig. 2–15

Manometer Rule. The previous result can also be determined in a more direct manner by using the ***manometer rule***, something that works for all types of manometers. It can be stated as follows:

Start at a point in the fluid where the pressure is to be determined, and proceed to add to it the pressures algebraically from one vertical fluid interface to the next, until you reach the liquid surface at the other end of the manometer.

As with any fluid system, a pressure term will be *positive* if it is *below* a point since it will cause an *increase*, and it will be *negative* if it is *above* a point since it will cause a *decrease*. Thus, for the manometer in Fig. 2–15, we start with p_A at point A, then add γh_{BC}, and finally subtract $\gamma' h_{DE}$. This algebraic sum is equal to the pressure at E, which is zero; that is, $p_A + \gamma h_{BC} - \gamma' h_{DE} = 0$. Therefore, $p_A = \gamma' h_{DE} - \gamma h_{BC}$, as obtained previously.

As another example, consider the manometer in Fig. 2–16. Starting at A, and realizing that the pressure at C is zero, we have

$$p_A - \gamma h_{AB} - \gamma' h_{BC} = 0$$

so that

$$p_A = \gamma h_{AB} + \gamma' h_{BC}$$

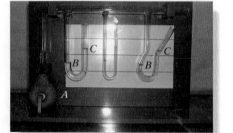

Increasing the pressure by squeezing the bulb at A will cause the elevation difference BC to be the same in each tube, regardless of the shape of the tube.

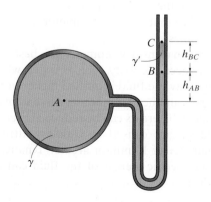

Fig. 2–16

Differential Manometer. A *differential manometer* is used to determine the difference in pressure between two points in a *closed* fluid system. For example, the differential manometer in Fig. 2–17 measures the difference in static pressure between points A and D in the pipe assembly containing a liquid that is flowing through it. Following the path through the manometer from A to B to C to D, and summing the pressures as outlined by the manometer rule, we have

$$p_A + \gamma h_{AB} - \gamma' h_{BC} - \gamma h_{CD} = p_D$$

$$\Delta p = p_D - p_A = \gamma h_{AB} - \gamma' h_{BC} - \gamma h_{CD}$$

Since $h_{BC} = h_{AB} - h_{CD}$, then

$$\Delta p = -(\gamma - \gamma') h_{BC}$$

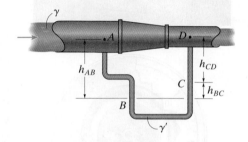

Fig. 2–17

This result represents the difference in either the absolute or the gage pressures between points A and D. Notice that if γ' is chosen close to γ, then the difference $(\gamma - \gamma')$ will be rather small, and a large value of h_{BC} will occur, making it more accurate to detect Δp.

Small differences in pressure can also be detected by using an inverted U-tube manometer as in Fig. 2–18. Here the manometer fluid has a smaller specific weight γ' than that of the contained fluid γ. An example would be oil versus water. Starting from point A and going to point D, we have

$$p_A - \gamma h_{AB} + \gamma' h_{BC} + \gamma h_{CD} = p_D$$

$$\Delta p = p_D - p_A = -\gamma h_{AB} + \gamma' h_{BC} + \gamma h_{CD}$$

Since $h_{BC} = h_{AB} - h_{CD}$, then

$$\Delta p = -(\gamma - \gamma') h_{BC}$$

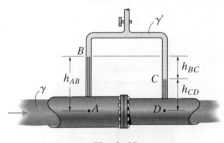

Fig. 2–18

If air is used as the lighter fluid, then it can be pumped into the top portion of the tube and the valve closed to make a suitable liquid level. In this case, $\gamma' \approx 0$ and $\Delta p = -\gamma h_{BC}$.

Of course, none of the above results should be memorized. Rather, the principle of writing the pressures at the different levels in the manometer circuit should be understood.

There have been many other modifications to the U-tube manometer that have improved its accuracy for measuring pressures or their differences. For example, one common modification is to make one of the tubes inclined, the principles of which are discussed in Example 2.7. Another is to use a micro-manometer as discussed in Prob. 2–55.

2

Bourdon Gage. If the gage pressures are *very high*, then a manometer may not be effective, and so the measurement can be made using a ***Bourdon gage***, Fig. 2–19. Essentially, this gage consists of a coiled metal tube that is connected at one end to the vessel where the pressure is to be measured. The other end of the tube is closed so that when the pressure in the vessel is increased, the tube begins to uncoil and respond elastically. Using the mechanical linkage attached to the end of the tube, the dial on the face of the gage gives a direct reading of the pressure, which can be calibrated in various units, such as kPa.

Bourdon gage

Fig. 2–19

Pressure Transducers. An electromechanical device called a ***pressure transducer*** can be used to measure pressure as a digital readout. It has the advantage of producing a quick response to changes in pressure, and providing a continuous readout over time. Figure 2–20 illustrates the way it works. When end A is connected to a pressure vessel, the fluid pressure will deform the thin diaphragm. The resulting strain in the diaphragm is then measured using the attached electrical strain gage. Essentially, the changing length of the thin wires composing the strain gage will change their resistance, producing a change in electric current. Since this change in current is directly proportional to the strain caused by the pressure, the current can be converted into a direct reading of the pressure.

Pressure transducers can also be used to give a direct reading of absolute pressure, if the volume B behind the diaphragm is *sealed* so it is in a vacuum. If this volume is open to the atmosphere, then the gage pressure is recorded. Finally, if regions A and B are connected to two different fluid pressures, then a differential pressure can be recorded.

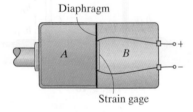

Diaphragm

Strain gage

Pressure transducer

Fig. 2–20

Other Pressure Gages. Apart from the gages discussed above, there are several other methods for measuring pressure. One of the more accurate gages available is a ***fused quartz force-balance Bourdon tube***. Within it, pressure causes elastic deformation of a coiled tube that is detected optically. The tube is then restored to an original position by a magnetic field that is measured and correlated to the pressure that caused the deformation. In a similar manner, ***piezoelectric gages***, such as a quartz crystal, can change their electric potential when subjected to small pressure changes, and so the pressure can be correlated to this potential change and presented as a digital readout. The same type of gage can also be constructed using thin silicon wafers. Their deformation causes a measured change in capacitance or vibrational frequency that gives immediate response to sudden pressure changes. Further details on these gages and others like them, along with their specific applications, can be found in the literature or from Ref. [5]–[11].

Important Points

- The pressure at a point in a fluid is *the same in all directions*, provided the fluid has no relative motion. This is Pascal's law. As a consequence, any *increase in pressure* Δp at one point in the fluid will cause *the same increase in pressure* Δp at some other point.

- Absolute pressure is the pressure above that of a vacuum. Standard atmospheric pressure, which is measured at sea level and a temperature of 15°C, is 101.3 kPa.

- Gage pressure is measured as the pressure that is above (positive) or below (negative) atmospheric pressure.

- When the weight of a static fluid is considered, *the pressure in the horizontal direction is constant*; however, in the *vertical direction*, it increases with depth.

- If a fluid is essentially *incompressible*, as in the case of a liquid, then its specific weight is constant, and the pressure can be determined using $p = \gamma h$.

- If a fluid is considered *compressible*, as in the case of a gas, then the variation of the fluid's specific weight (or density) with pressure must be taken into account to obtain an accurate measurement of pressure.

- For small changes in elevation, the static *gas pressure* in a tank, vessel, manometer, pipe, and the like can be considered *constant* throughout its volume, since the specific weight of a gas is very small.

- The pressure p at a point can be represented by its *pressure head*, which is the height h of a column of fluid needed to produce the pressure, $h = p/\gamma$.

- Atmospheric pressure can be measured using a *barometer*.

- Manometers can be used to measure small pressures in pipes or tanks, or differential pressures between points in two pipes. Pressures at any two points in the manometer can be related using the manometer rule.

- High pressures are generally measured using a Bourdon gage or pressure transducer. Besides these, many other types of pressure gages are available for specific applications.

EXAMPLE | 2.5

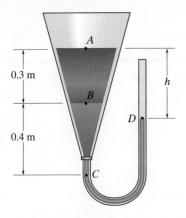

0.3 m

0.4 m

Fig. 2–21

The funnel in Fig. 2–21 is filled with oil and water to the levels shown, while portion CD of the tube contains mercury. Determine the distance h the mercury level is from the top of the oil surface for equilibrium. Take $\rho_o = 880 \text{ kg/m}^3$, $\rho_w = 1000 \text{ kg/m}^3$, $\rho_{Hg} = 13\,550 \text{ kg/m}^3$.

SOLUTION

Fluid Description. The fluids are liquids, so we will consider them to be incompressible.

Analysis. We can treat the system of fluids as a "manometer" and write an equation using the manometer rule from A to D, noting that the (gage) pressures at A and D are both zero. We have

$$0 + \rho_o g h_{AB} + \rho_w g h_{BC} - \rho_{Hg} g h_{CD} = 0$$

$$0 + (880 \text{kg/m}^3)(9.81 \text{ m/s}^2)(0.3 \text{ m}) + (1000 \text{ kg/m}^3)(9.81 \text{ m/s}^2)(0.4 \text{ m})$$
$$- (13\,550 \text{ kg/m}^3)(9.81 \text{ m/s}^2)(0.3 \text{ m} + 0.4 \text{ m} - h) = 0$$

Thus,

$$h = 0.651 \text{ m}$$ *Ans.*

EXAMPLE | 2.6

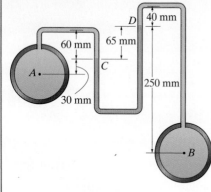

D   40 mm

60 mm   65 mm

C

A

250 mm

30 mm

B

Fig. 2–22

Determine the difference in pressure between the centerline points A and B in the two pipelines in Fig. 2–22 if the manometer liquid CD is in the position shown. The density of the liquid in AC and DB is $\rho = 800 \text{ kg/m}^3$, and in CD, $\rho_{CD} = 1100 \text{ kg/m}^3$.

SOLUTION

Fluid Description. The liquids are assumed incompressible.

Analysis. Starting at point B and moving through the manometer to point A, using the manometer rule, we have

$$p_B - \rho g h_{BD} + \rho_{CD} g h_{DC} + \rho g h_{CA} = p_A$$

$$p_B - (800 \text{ kg/m}^3)(9.81 \text{ m/s}^2)(0.250 \text{ m}) + (1100 \text{ kg/m}^3)(9.81 \text{ m/s}^2)(0.065 \text{ m})$$
$$+ (800 \text{ kg/m}^3)(9.81 \text{ m/s}^2)(0.03 \text{ m}) = p_A$$

Thus,

$$\Delta p = p_A - p_B = -1.03 \text{ kPa}$$ *Ans.*

Since the result is negative, the pressure at A is less than that at B.

EXAMPLE 2.7

The inclined-tube manometer shown in Fig. 2–23 is used to measure small pressure changes. Determine the difference in pressure between points A and E if the manometer liquid, mercury, is in the position shown. The pipe at A contains water, and the one at E contains natural gas. For mercury, $\rho_{Hg} = 13\,550$ kg/m³.

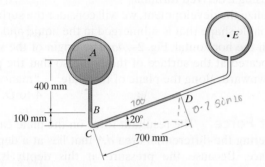

Fig. 2–23

SOLUTION

Fluid Description.   We will consider the fluids to be incompressible and neglect the specific weight of the natural gas. As a result, the pressure at E is essentially the same as that at D.

Analysis.   Applying the manometer rule between points A and D, we have

$$p_A + \gamma_w h_{AB} + \gamma_{Hg} h_{BC} - \gamma_{Hg} h_{CD} = p_E$$

$$p_A + (1000 \text{ kg/m}^3)(9.81 \text{ m/s}^2)(0.4 \text{ m}) + (13\,550 \text{ kg/m}^3)(9.81 \text{ m/s}^2)(0.1 \text{ m})$$

$$- (13\,550 \text{ kg/m}^3)(9.81 \text{ m/s}^2)(0.7 \sin 20° \text{ m}) = p_E$$

$$p_A - p_E = 14.61\,(10^3) \text{ Pa} = 14.6 \text{ kPa} \qquad\qquad Ans.$$

Notice that any slight pressure change will cause the distance Δ_{CD} to be significantly altered, since the elevation change Δh_{CD} depends on the factor ($\sin 20°$). In other words, $\Delta_{CD} = \Delta h_{CD}/\sin 20° = 2.92\Delta h_{CD}$. In practice, angles less than about 5° are not practical. This is because at such small angles, the exact location of the meniscus is hard to detect, and the effect of surface tension will be magnified if there are any surface impurities within the tube.

2.7   Hydrostatic Force on a Plane Surface—Formula Method

When designing gates, vessels, dams, or other bodies that are submerged in a liquid, it is important to be able to obtain the resultant force caused by the pressure loading of the liquid, and to specify the location of this force on the body. In this section we will show how this is done on a *plane surface* by using a derived formula.

To generalize the development, we will consider the surface to be a flat plate of variable shape that is submerged in the liquid and oriented at an angle θ with the horizontal, Fig. 2–24a. The origin of the x, y coordinate system is located at the surface of the liquid so that the positive y axis extends downward, along the plane of the plate.

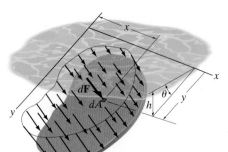

(a)

Resultant Force.   The resultant force on the plate can be found by first considering the differential area dA that lies at a depth h from the liquid surface. Because the pressure at this depth is $p = \gamma h$, the differential force acting on the area is

$$dF = p\,dA = (\gamma h)\,dA = \gamma\,(y\sin\theta)\,dA$$

The resultant force acting on the plate is equivalent to the sum of all these forces. Therefore, integrating over the entire area A, we have

$$F_R = \Sigma F; \qquad F_R = \int_A \gamma y \sin\theta\, dA = \gamma \sin\theta \int_A y\, dA = \gamma \sin\theta\,(\overline{y}A)$$

The integral $\int_A y\, dA$ represents the "moment of the area" about the x axis. Here it has been replaced by $\overline{y}A$, where $\overline{y}$ is the distance from the x axis to the *centroid* C, or geometric center of the area, Fig. 2–24b*. Since the *depth* of the centroid is $\overline{h} = \overline{y}\sin\theta$, we can write the above equation as

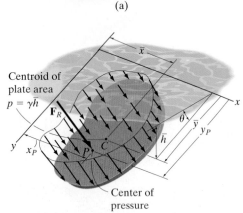

Centroid of plate area

$p = \gamma\overline{h}$

Center of pressure

$F_R = \gamma\overline{h}\,A$

$\mathbf{F}_R$ acts through the center of pressure

(b)

Fig. 2–24

$$\boxed{F_R = \gamma\overline{h}A} \qquad (2\text{–}9)$$

This result indicates that *the magnitude of the resultant force on the plate is the product of the pressure acting at the plate's centroid, $\gamma\,\overline{h}$, and the area A of the plate.* This force has a *direction* that is *perpendicular* to the plate since the entire pressure distribution acts in this direction.

*See *Engineering Mechanics: Statics*, 13th ed., R. C. Hibbeler, Pearson Education.

Location of the Resultant Force.

The resultant force of the pressure distribution acts through a point on the plate called the **center of pressure**, P, shown in Fig. 2–24b. The location of this point, (x_P, y_P), is determined by a balance of moments that requires the moment of the entire pressure loading about the y axis and about the x axis, Fig. 2–24a, to be equal to the moment of the resultant force about each of these axes, Fig. 2–24b.

These lock gates have been designed to resist the hydrostatic pressure of the water in the canal.

The y_P Coordinate.

We require

$$(M_R)_x = \Sigma M_x; \qquad y_P F_R = \int_A y \, dF$$

Since $F_R = \gamma \sin \theta \, (\bar{y} A)$ and $dF = \gamma \, (y \sin \theta) \, dA$, then

$$y_P \left[\gamma \sin \theta (\bar{y} A)\right] = \int_A y \left[\gamma (y \sin \theta) \, dA\right]$$

Canceling $\gamma \sin \theta$, we get

$$y_P \, \bar{y} A = \int_A y^2 \, dA$$

Here the integral represents the *area moment of inertia I_x* for the area about the x axis.* Thus,

$$y_P = \frac{I_x}{\bar{y} A}$$

Values for the area moment of inertia are normally referenced from an axis passing through the *centroid* of the area, referred to as $\bar{I}_x$. Examples for some common shapes, along with the location of their centroids, are given on the inside back cover. With these values we can then use the *parallel-axis theorem** to obtain I_x, that is, $I_x = \bar{I}_x + A\bar{y}^2$, and write the above equation as

$$\boxed{y_P = \bar{y} + \frac{\bar{I}_x}{\bar{y} A}} \qquad (2\text{–}10)$$

The circular access door of this industrial tank is subject to the pressure of the fluid within the tank. The resultant force and its location can be determined using Eqs. 2–9 and 2–10.

Notice that the term $\bar{I}_x / \bar{y} A$ will always be positive, and so the distance y_P to the center of pressure will always be *below* $\bar{y}$, the distance to the centroid of the plate, Fig. 2–24b.

*Ibid.

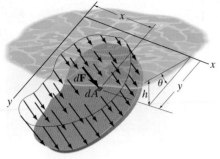

(a)

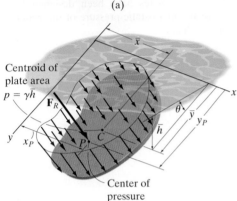

Centroid of
plate area
$p = \gamma \bar{h}$

Center of
pressure

$F_R = \gamma \bar{h} A$

$\mathbf{F}_R$ acts through the center of pressure

(b)

Fig. 2–24

The x_P Coordinate. The lateral position of the center of pressure, x_P, can be determined by a balance of moments about the y axis, Figs. 2–24a and 2–24b. We require

$$(M_R)_y = \Sigma M_y; \qquad\qquad x_P F_R = \int_A x \, dF$$

Again, using $F_R = \gamma \sin\theta(\bar{y}A)$ and $dF = \gamma(y \sin\theta) \, dA$, we have

$$x_P[\gamma \sin\theta(\bar{y}A)] = \int_A x \, [\gamma(y \sin\theta) \, dA]$$

Canceling $\gamma \sin\theta$ gives

$$x_P \bar{y} A = \int_A xy \, dA$$

This integral is referred to as the *product of inertia* I_{xy} for the area.* Thus,

$$x_P = \frac{I_{xy}}{\bar{y}A}$$

If we apply the *parallel-plane theorem*,* $I_{xy} = \bar{I}_{xy} + A\,\bar{x}\bar{y}$, where $\bar{x}$ and $\bar{y}$ locate the area's centroid, then this result can also be expressed as

$$\boxed{x_P = \bar{x} + \frac{\bar{I}_{xy}}{\bar{y}A}} \qquad (2\text{–}11)$$

For most engineering applications, the submerged area will be *symmetrical* about either an x or a y axis passing through its centroid. If this occurs, as in the case of the rectangular plate in Fig. 2–25, then $\bar{I}_{xy} = 0$ and $\bar{x} = 0$. The above result then becomes $x_P = 0$, which simply indicates that the center of pressure P will lie on the $\bar{y}$ centroidal axis, as shown in the figure.

*Ibid.

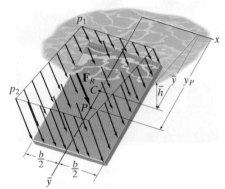

$F_R = \gamma \bar{h} A$

$\mathbf{F}_R$ acts through the center of pressure

Fig. 2–25

Important Points

- A liquid creates a pressure loading that acts perpendicular to a submerged surface. Since the liquid is assumed to be incompressible, then the pressure intensity increases linearly with depth, $p = \gamma h$.

- The resultant force of the pressure on a *plane surface* having an area A and submerged in a liquid can be determined from $F_R = \gamma \bar{h} A$, where $\bar{h}$ is the depth of the area's centroid C, measured from the liquid's surface.

- The resultant force acts through the center of pressure P, determined from $x_P = \bar{x} + \bar{I}_{xy}/(\bar{y}A)$ and $y_P = \bar{y} + \bar{I}_x/(\bar{y}A)$. If the submerged area has an axis of symmetry along the y axis, then $\bar{I}_{xy} = 0$, and so $x_P = 0$. In this case P is located on the $\bar{y}$ centroidal axis of the area.

EXAMPLE 2.8

Determine the force that water pressure exerts on the inclined side plate *ABDE* of the storage tank, and find its location measured from *AB*, Fig. 2–26*a*.

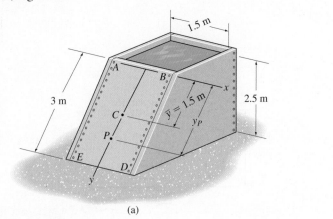

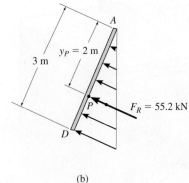

(a)   (b)

Fig. 2–26

SOLUTION

Fluid Description.   We consider water to be incompressible, where $\rho_w = 1000 \ \text{kg/m}^3$.

Analysis.   The centroid of the plate's area is located at its *midpoint*, $\bar{y} = 1.5$ m from its top *AB*, Fig. 2–26*a*. The *depth* of the water at this point is therefore the mean depth, $\bar{h} = 1.25$ m. Thus,

$$F_R = \gamma_w \bar{h} A = (1000 \ \text{kg/m}^3)(9.81 \ \text{m/s}^2)(1.25 \ \text{m})[(1.5 \ \text{m})(3 \ \text{m})]$$
$$= 55.18 \ (10^3) \ \text{N} = 55.2 \ \text{kN} \qquad \qquad Ans.$$

From the inside back cover, for a rectangular area, $\bar{I}_x = \dfrac{1}{12} b a^3$. Here $b = 1.5$ m and $a = 3$ m. Thus,

$$y_P = \bar{y} + \frac{\bar{I}_x}{\bar{y}A} = 1.5 \ \text{m} + \frac{\dfrac{1}{12}(1.5 \ \text{m})(3 \ \text{m})^3}{(1.5 \ \text{m})[(1.5 \ \text{m})(3 \ \text{m})]} = 2 \ \text{m} \quad Ans.$$

Since the rectangle is symmetrical about its centroidal y axis, Fig. 2–26*a*, then $\bar{I}_{xy} = 0$ and Eq. 2–11 gives

$$x_P = \bar{x} + \frac{\bar{I}_{xy}}{\bar{y}A} = 0 + 0 = 0 \qquad \qquad Ans.$$

The results are shown from a side view of the plate in Fig. 2–26*b*.

EXAMPLE | **2.9**

The bin in Fig. 2–27a contains water. Determine the resultant force that the water pressure exerts on the circular plate, and find its location.

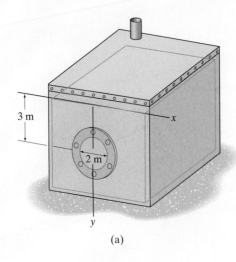

(a)

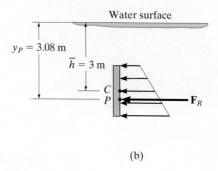

(b)

Fig. 2–27

SOLUTION

Fluid Description.   We assume the water to be incompressible. For water, $\rho_w = 1000$ kg/m³.

Analysis.   From Fig. 2–27b, the resultant force is

$$F_R = \gamma_w \bar{h} A = (1000 \text{ kg/m}^3)(9.81 \text{ m/s}^2)(3 \text{ m})[\pi(1 \text{ m})^2] = 92.46 \text{ kN}$$

Using the table on the inside back cover for a circle, the location of the resultant force is determined from

$$y_P = \bar{y} + \frac{\bar{I}_x}{\bar{y}A} = 3 \text{ m} + \frac{\frac{\pi}{4}(1 \text{ m})^4}{(3 \text{ m})[\pi(1 \text{ m})^2]} = 3.08 \text{ m} \qquad Ans.$$

Since $I_{xy} = 0$, due to symmetry,

$$x_P = \bar{x} + \frac{\bar{I}_{xy}}{\bar{y}A} = 0 + 0 = 0 \qquad Ans.$$

EXAMPLE 2.10

Determine the magnitude and location of the resultant force acting on the triangular end plate of the settling tank in Fig. 2–28a. The tank contains kerosene.

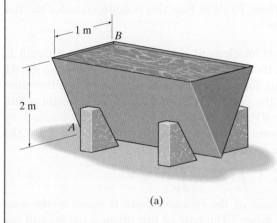

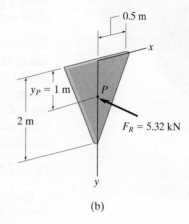

(a)

(b)

Fig. 2–28

SOLUTION

Fluid Description. The kerosene is considered an incompressible fluid for which $\gamma_k = (814 \text{ kg/m}^3)(9.81 \text{ m/s}^2) = 7985 \text{ N/m}^3$ (Appendix A).

Analysis. From the inside back cover, for a triangle,

$$\bar{y} = \bar{h} = \frac{1}{3}(2 \text{ m}) = 0.6667 \text{ m}$$

$$\bar{I}_x = \frac{1}{36}ba^3 = \frac{1}{36}(1 \text{ m})(2 \text{ m})^3 = 0.2222 \text{ m}^4$$

Therefore,

$$F_R = \gamma_k \bar{h} A = \left(7985 \text{ N/m}^3\right)(0.6667)\left[\frac{1}{2}(1 \text{ m})(2 \text{ m})\right]$$

$$= 5323.56 \text{ N} = 5.32 \text{ kN} \qquad Ans.$$

$$y_P = \bar{y} + \frac{\bar{I}_x}{\bar{y}A} = 0.6667 \text{ m} + \frac{0.2222 \text{ m}^4}{(0.6667 \text{ m})\left[\frac{1}{2}(1 \text{ m})(2 \text{ m})\right]} = 1.00 \text{ m} \quad Ans.$$

The triangle is symmetrical about the y axis, so $I_{xy} = 0$. Thus,

$$x_P = \bar{x} + \frac{\bar{I}_x}{\bar{y}A} = 0 + 0 = 0 \qquad Ans.$$

These results are shown in Fig. 2–28b.

2.8 Hydrostatic Force on a Plane Surface—Geometrical Method

Rather than using the equations of the previous section, the resultant force and its location on a flat submerged plate can also be determined using a geometrical method. To show how this is done, consider the flat plate shown in Fig. 2–29a.

Resultant Force. If an element dA of the plate is at a depth h, where the pressure is p, then the force on this element is $dF = p\,dA$. As shown in the figure, this force geometrically represents a differential volume element $d\mathcal{V}$ of the pressure distribution. It has a height p and base dA, and so $dF = d\mathcal{V}$. The resultant force can be obtained by integrating these elements over the entire volume enclosed by the pressure distribution, we have

$$F_R = \Sigma F; \qquad F_R = \int_A p\,dA = \int_{\mathcal{V}} d\mathcal{V} = \mathcal{V} \qquad (2\text{–}12)$$

Therefore, *the magnitude of the resultant force is equal to the total volume of the "pressure prism."* The base of this prism is the area of the plate, and the height varies linearly from $p_1 = \gamma h_1$ to $p_2 = \gamma h_2$, Fig. 2–29a.

Location. To locate the resultant force on the plate, we require the moment of the resultant force about the y axis and about the x axis, Fig. 2–29b to equal the moment created by the entire pressure distribution about these axes, Fig. 2–29a; that is,

$$(M_R)_y = \Sigma M_y; \qquad x_P F_R = \int x\,dF$$

$$(M_R)_x = \Sigma M_x; \qquad y_P F_R = \int y\,dF$$

Since $F_R = \mathcal{V}$ and $dF = d\mathcal{V}$, we have

$$x_P = \frac{\displaystyle\int_A x\,p\,dA}{\displaystyle\int_A p\,dA} = \frac{\displaystyle\int_{\mathcal{V}} x\,d\mathcal{V}}{\mathcal{V}}$$

$$y_P = \frac{\displaystyle\int_A y\,p\,dA}{\displaystyle\int_A p\,dA} = \frac{\displaystyle\int_{\mathcal{V}} y\,d\mathcal{V}}{\mathcal{V}} \qquad (2\text{–}13)$$

These equations locate the x and y coordinates of the *centroid $C_{\mathcal{V}}$* of the pressure prism volume. In other words, *the line of action of the resultant force will pass through both the centroid $C_{\mathcal{V}}$ of the volume of the pressure prism and the center of pressure P on the plate*, Fig. 2–29b.

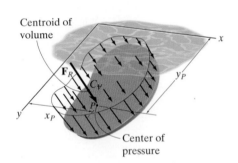

(a)

Centroid of volume

Center of pressure

F_R equals the volume of the pressure diagram, and it passes through the centroid $C_{\mathcal{V}}$ of this volume

(b)

Fig. 2–29

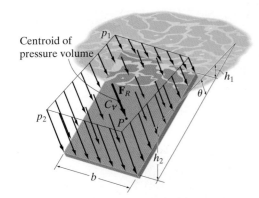

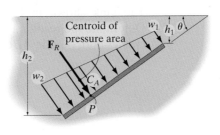

F_R equals the volume of
the pressure diagram, and it passes
through the centroid of this volume

(a)

F_R equals the area of
the w diagram, and it passes
through the centroid of this area

(b)

Fig. 2–30

Plate Having Constant Width.   As a special case, if the plate has a
constant width b, as in the case of a *rectangle*, Fig. 2–30a, then the
pressure loading along the width at depth h_1 and at depth h_2 is constant.
As a result, the loading may be viewed along the side of the plate, in two
dimensions, Fig. 2–30b. The intensity w of this distributed load is
measured as a force/length, and varies linearly from $w_1 = p_1 b = (\gamma h_1)b$
to $w_2 = p_2 b = (\gamma h_2)b$. The magnitude of $\mathbf{F}_R$ is then equivalent to the
trapezoidal area defining the distributed loading, and $\mathbf{F}_R$ has a *line of
action* that passes through both the *centroid* C_A of this area and the
center of pressure P on the plate. Of course, these results are equivalent
to finding the trapezoidal volume of the pressure prism, F_R, and its
centroidal location C_V, as shown in Fig. 2–30a.

Important Points

- The *resultant force* on a *plane surface* can be determined
  graphically by finding the *volume* V of the pressure prism, $F_R = V$.
  The line of action of the resultant force passes through the
  centroid of this volume. It intersects the surface at the center of
  pressure P.

- If the submerged surface has a constant width, then the pressure
  prism can be viewed from the side and represented as a planar
  distributed loading w. The resultant force equals the *area* of this
  loading diagram, and it acts through the centroid of this area.

EXAMPLE | **2.11**

The tank shown in Fig. 2–31a contains water to a depth of 3 m. Determine the resultant force, and its location, that the water pressure creates both on side ABCD of the tank and on its bottom.

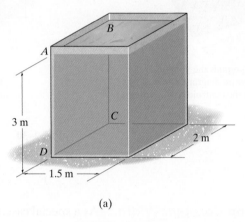

(a)

Fig. 2–31

SOLUTION

Fluid Description. The water is considered to be incompressible, with $\rho_w = 1000 \text{ kg/m}^3$.

Analysis I

Loading. The pressure at the bottom of the tank is

$$p = \rho_w g h = (1000 \text{ kg/m}^3)(9.81 \text{ m/s}^2)(3 \text{ m}) = 29.43 \text{ kPa}$$

Using this value, the pressure distribution along the side and bottom of the tank is shown in Fig. 2–31b.

Resultant Forces. The magnitudes of the resultant forces are equal to the *volumes* of the pressure prisms.

$$(F_R)_s = \frac{1}{2}(3 \text{ m})(29.43 \text{ kN/m}^2)(2 \text{ m}) = 88.3 \text{ kN} \qquad \textit{Ans.}$$

$$(F_R)_b = (29.43 \text{ kN/m}^2)(2 \text{ m})(1.5 \text{ m}) = 88.3 \text{ kN} \qquad \textit{Ans.}$$

These resultants act through the centroids of their respective volumes, and define the location of the center of pressure P for each plate, Fig. 2–31.

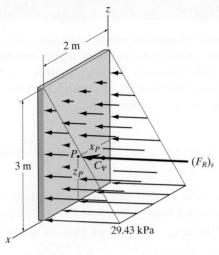

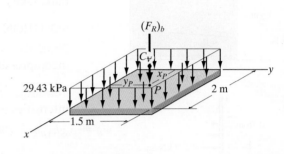

(b)

Location. Using the inside back cover, for the side plate, z_P in Fig. 2–31b is determined for a triangle to be $\frac{1}{3}a$, so that

$$x_P = 1 \text{ m} \qquad\qquad Ans.$$

$$z_P = \frac{1}{3}(3 \text{ m}) = 1 \text{ m} \qquad\qquad Ans.$$

For the bottom plate,

$$x_P = 1 \text{ m} \qquad\qquad Ans.$$

$$y_P = 0.75 \text{ m} \qquad\qquad Ans.$$

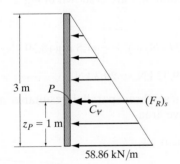

Analysis II

Loading. Since the side and bottom plates in Fig. 2–31a both have a constant width of $b = 2$ m, the pressure loading can also be viewed in two dimensions. The intensity of the loading at the bottom of the tank is

$$w = (\rho_w g h)b$$

$$= (1000 \text{ kg/m}^3)(9.81 \text{ m/s}^2)(3 \text{ m})(2 \text{ m}) = 58.86 \text{ kN/m}$$

The distributions are shown in Fig. 2–31c.

Resultant Forces. Here the resultant forces are equal to the *areas* of the loading diagrams.

$$(F_R)_s = \frac{1}{2}(3 \text{ m})(58.86 \text{ kN/m}) = 88.3 \text{ kN} \qquad\qquad Ans.$$

$$(F_R)_b = (1.5 \text{ m})(58.86 \text{ kN/m}) = 88.3 \text{ kN} \qquad\qquad Ans.$$

Location. These results act through the centroids of their respective areas as shown in Fig. 2–31c.

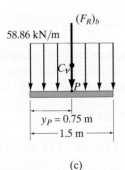

(c)

Fig. 2–31 (cont.)

EXAMPLE 2.12

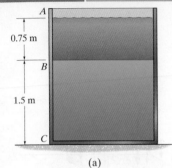

(a)

The storage tank contains oil and water at the depths shown in Fig. 2–32a. Determine the resultant force that both of these liquids together exert on the side ABC of the tank if the side has a width of $b = 1.25$ m. Also, determine the location of this resultant, measured from the top of the tank. Take $\rho_o = 900$ kg/m³, $\rho_w = 1000$ kg/m³.

SOLUTION

Fluid Description. Both the water and the oil are assumed to be incompressible.

Loading. Since the side of the tank has a constant width, the intensities of the distributed loading at B and C, Fig. 2–32b, are

$$w_B = \rho_o g h_{AB} b = \left(900 \text{ kg/m}^3\right)\left(9.81 \text{ m/s}^2\right)(0.75 \text{ m})(1.25 \text{ m}) = 8.277 \text{ kN/m}$$

$$w_C = w_B + \rho_w g h_{BC} b = 8.277 \text{ kN/m} + \left(1000 \text{ kg/m}^3\right)\left(9.81 \text{ m/s}^2\right)(1.5 \text{ m})(1.25 \text{ m})$$

$$= 26.67 \text{ kN/m}$$

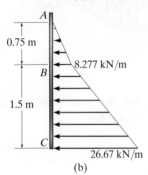

(b)

Resultant Force. The resultant force can be determined by adding the three shaded triangular and rectangular areas shown in Fig. 2–32c.

$$F_R = F_1 + F_2 + F_3$$

$$= \frac{1}{2}(0.75 \text{ m})(8.277 \text{ kN/m}) + (1.5 \text{ m})(8.277 \text{ kN/m}) + \frac{1}{2}(1.5 \text{ m})(18.39 \text{ kN/m})$$

$$= 3.104 \text{ kN} + 12.42 \text{ kN} + 13.80 \text{ kN} = 29.32 \text{ kN} = 29.3 \text{ kN} \qquad Ans.$$

Location. As shown, each of these three parallel resultants acts through the centroid of its respective area.

$$y_1 = \frac{2}{3}(0.75 \text{ m}) = 0.5 \text{ m}$$

$$y_2 = 0.75 \text{ m} + \frac{1}{2}(1.5 \text{ m}) = 1.5 \text{ m}$$

$$y_3 = 0.75 \text{ m} + \frac{2}{3}(1.5 \text{ m}) = 1.75 \text{ m}$$

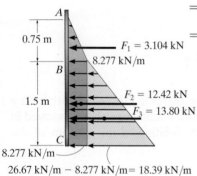

(c)

The location of the resultant force is determined by equating the moment of the resultant about A, Fig. 2–32d, to the sum of the moments of all the component forces about A, Fig. 2–32c. We have

$$y_P F_R = \Sigma \tilde{y} F; \quad y_P (29.32 \text{ kN}) = (0.5 \text{ m})(3.104 \text{ kN})$$

$$+ (1.5 \text{ m})(12.42 \text{ kN}) + (1.75 \text{ m})(13.80 \text{ kN})$$

$$y_P = 1.51 \text{ m} \qquad Ans.$$

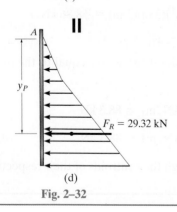

(d)

Fig. 2–32

2.9 Hydrostatic Force on a Plane Surface—Integration Method

If the boundary of the flat plate in Fig. 2–33a can be defined in terms of its x and y coordinates as $y = f(x)$, then the resultant force F_R and its location P on the plate can be determined by direct integration.

Resultant Force. If we consider a differential area strip dA of the plate that is located at a depth h, where the pressure is p, then the force acting on this strip is $dF = p\,dA$, Fig. 2–33a. The resultant force on the entire area is therefore

$$F_R = \Sigma F; \qquad \boxed{F_R = \int_A p\,dA} \qquad (2\text{–}14)$$

The hydrostatic force acting on the elliptical back plate of this water truck can be determined by integration.

Location. Here we require the moment of $\mathbf{F}_R$ about the y and x axes to equal the moment of the pressure distribution about these axes. Provided dF passes through the center (centroid) of dA, having coordinates $(\tilde{x}, \tilde{y})$, then from Fig. 2–33a and Fig. 2–33b,

$$(M_R)_y = \Sigma M_y; \qquad x_P F_R = \int_A \tilde{x}\,dF$$

$$(M_R)_x = \Sigma M_x; \qquad y_P F_R = \int_A \tilde{y}\,dF$$

Or, written in terms of p and dA, we have

$$\boxed{x_P = \frac{\displaystyle\int_A \tilde{x}\,p\,dA}{\displaystyle\int_A p\,dA} \qquad\qquad y_P = \frac{\displaystyle\int_A \tilde{y}\,p\,dA}{\displaystyle\int_A p\,dA}} \qquad (2\text{–}15)$$

Application of these equations is given in the following examples.

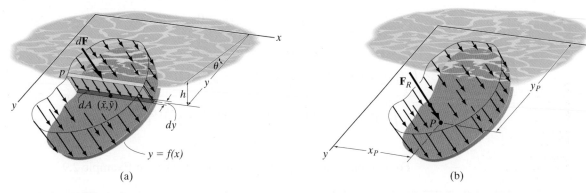

(a)

(b)

Fig. 2–33

EXAMPLE | 2.13

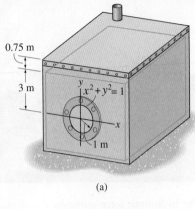

(a)

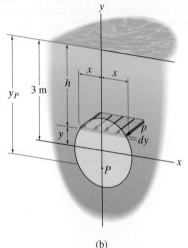

(b)

Fig. 2–34

The bin in Fig. 2–34a contains water. Determine the resultant force, and its location, that the water pressure exerts on the circular plate.

SOLUTION

Fluid Description. We assume the water to be incompressible. For water, $\rho_w = 1000 \text{ kg/m}^3$.

Resultant Force. We can determine the resultant force on the plate by using integration, since the circular boundary can be defined from the center of the plate in terms of the x, y coordinates shown in Fig. 2–34a. The equation is $x^2 + y^2 = 1$. The rectangular horizontal strip shown in Fig. 2–34b has an area $dA = 2x\,dy = 2(1 - y^2)^{1/2}dy$. It is at a depth $h = 3 - y$, where the pressure is $p = \gamma_w h = \gamma_w(3 - y)$. Applying Eq. 2–14, to obtain the resultant force, we have

$$F = \int_A p\,dA = \int_{-1}^{1}\left[(1000 \text{ kg})\left(9.81 \text{ m/s}^2\right)(3 - y)\right](2)\left(1 - y^2\right)^{1/2}dy$$

$$= 19\,620 \int_{-1}^{1}\left[3(1 - y^2)^{1/2} - y(1 - y^2)^{1/2}\right]dy$$

$$= 92.46 \text{ kN} \qquad\qquad\qquad Ans.$$

Location. The location of the center of pressure P, Fig. 2–34b, can be determined by applying Eq. 2–15. Here $dF = p\,dA$ is located at $\tilde{x} = 0$ and $\tilde{y} = 3 - y$ and so

$$x_P = 0 \qquad\qquad Ans.$$

$$y_P = \frac{\displaystyle\int_A (3 - y)p\,dA}{\displaystyle\int_A p\,dA} = \frac{\displaystyle\int_{-1}^{1}(3 - y)(1000 \text{ kg})\left(9.81 \text{ m/s}^2\right)(3 - y)(2)\left(1 - y^2\right)^{1/2}dy}{\displaystyle\int_{-1}^{1}(1000 \text{ kg})\left(9.81 \text{ m/s}^2\right)(3 - y)(2)\left(1 - y^2\right)^{1/2}dy} = 3.08 \text{ m} \quad Ans.$$

For comparison, this problem was also worked in Example 2.9.

EXAMPLE 2.14

Determine the magnitude and location of the resultant force acting on the triangular end plate of the settling tank in Fig. 2–35a. The tank contains kerosene.

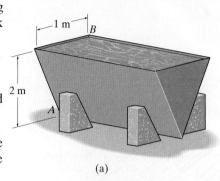

(a)

SOLUTION

Fluid Description. The kerosene is considered an incompressible fluid for which $\gamma_k = (814 \text{ kg/m}^3)(9.81 \text{ m/s}^2) = 7985 \text{ N/m}^3$ (Appendix A).

Resultant Force. The pressure distribution acting on the end plate is shown in Fig. 2–32b. Using the x and y coordinates, and choosing the differential area strip shown in the figure, we have

$$dF = p \, dA = (\gamma_k y)(2x \, dy) = 15.97(10^3) yx \, dy$$

The equation of line AB can be determined using similar triangles:

$$\frac{x}{2 \text{ m} - y} = \frac{0.5 \text{ m}}{2 \text{ m}}$$

$$x = 0.25(2 - y)$$

Hence, applying Eq. 2–14 and integrating with respect to y, from $y = 0$ to $y = 2$ m, yields

$$F = \int_A p \, dA = \int_0^{2m} 15.97(10^3) y[0.25(2 - y)] \, dy$$

$$= 5.323(10^3) \text{ N} = 5.32 \text{ kN} \qquad Ans.$$

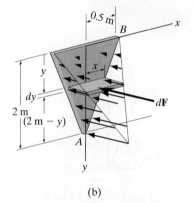

(b)

Location. Because of symmetry along the y axis, $\tilde{x} = 0$ and so

$$x_P = 0 \qquad Ans.$$

Applying Eq. 2–15, with $\tilde{y} = y$, we have

$$y_P = \frac{\int_A \tilde{y}p \, dA}{\int_A p \, dA} = \frac{\int_0^{2 \text{ m}} y[15.97(10^3)y][0.25(2-y)]dy}{5.323(10^3)} = 1.00 \text{ m} \qquad Ans.$$

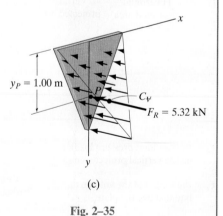

(c)

Fig. 2–35

The results are shown in Fig. 2–35c. They were also obtained in Example 2.10 using the formula method.

2

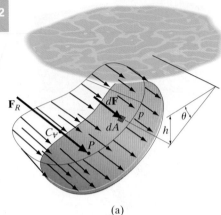

(a)

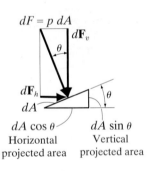

$dF = p\, dA$

$dA \cos \theta$
Horizontal
projected area

$dA \sin \theta$
Vertical
projected area

(b)

2.10 Hydrostatic Force on an Inclined Plane or Curved Surface Determined by Projection

If a submerged surface is curved, then the pressure acting on the surface will change not only its magnitude but also its direction, since it must *always* act normal to the surface. For this case, it is generally best to determine the horizontal and vertical *components* of the resultant force caused by the pressure, and then use vector addition to find the resultant. The method for doing this will now be described with reference to the submerged curved plate in Fig. 2–36a. Realize that this method will also work for an inclined flat plate, as in Fig. 2–30.

Horizontal Component.   The force shown acting on the differential element dA in Fig. 2–36a is $dF = p\, dA$, and so its horizontal component is $dF_h = (p\, dA) \sin \theta$, Fig. 2–36b. If we integrate this result over the entire *area* of the plate, we will obtain the resultant's horizontal component

$$F_h = \int_A p \sin \theta \, dA$$

Since $dA \sin \theta$ is the *projected differential area onto the vertical plane*, Fig. 2–36b, and the pressure p at a point must be the same in all directions, then the above integration over the entire area of the plate can be interpreted as follows: *The resultant horizontal force component acting on the plate is equal to the resultant force of the pressure loading acting on the area of the vertical projection of the plate*, Fig. 2–36c. Since this vertical area is flat or "planar," any of the methods of the previous three sections can be used to determine F_h and its location on this projected area.

F_h = the resultant pressure loading
on the vertical projected area

F_v = the weight of the volume of
liquid above the plate

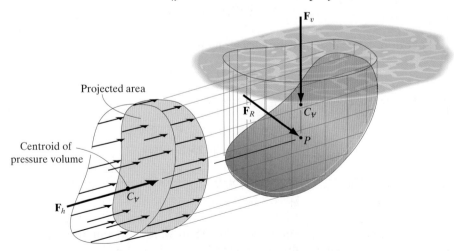

(c)

Fig. 2–36

Vertical Component.

The vertical component of the resultant force acting on the element dA in Fig. 2–36b is $dF_v = (p\,dA)\cos\theta$. This same result can also be obtained by noting that the horizontal projection of dA is $dA\cos\theta$, and so

$$dF_v = p(dA\cos\theta) = \gamma h(dA\cos\theta)$$

Since a vertical column of liquid above dA has a volume of $d\forall = h(dA\cos\theta)$, then $dF_v = \gamma d\forall$. Therefore, the resultant's vertical component is

$$F_v = \int_\forall \gamma\,d\forall = \gamma\forall$$

In other words, *the resultant vertical force acting on the plate is equivalent to the weight of the volume of the liquid acting above the plate*, Fig. 2–36c. This force acts through the centroid $C_\forall$ of the volume, which has the same location as the center of gravity for the weight of the liquid, since the specific weight of the liquid is constant.

Once the horizontal and vertical components of force are known, the magnitude of the resultant force, its direction, and its line of action can be established. As shown in Fig. 2–36c, this force will act through the center of pressure P on the plate's surface.

This same type of analysis can also be applied in cases where the liquid is *below* the plate, rather than above it. For example, consider the curved plate AD of constant width shown in Fig. 2–37. The horizontal component of $\mathbf{F}_R$ is determined by finding the force $\mathbf{F}_h$ acting on the projected area, DE. The vertical component of $\mathbf{F}_R$, however, will act *upward*. To see why, imagine that liquid is also present within the volume $ABCD$. If this were the case, then the *net vertical force* caused by pressure on the top and bottom of AD would be zero. In other words, the vertical pressure components on the top and bottom surfaces of the plate will be equal but opposite and will have the same lines of action. Therefore, if we determine the weight of imaginary liquid contained within the volume $ABCD$, and *reverse the direction of this weight*, we can then establish $\mathbf{F}_v$ acting upward on AD.

Gas.

If the fluid is a *gas*, then *its weight can generally be neglected*, and so the pressure within the gas is constant. The horizontal and vertical components of the resultant force are then determined by projecting the curved surface area onto vertical and horizontal planes and determining the components as shown in Fig. 2–38.

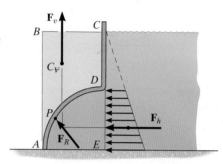

$\mathbf{F}_h$ = the resultant pressure loading on the vertical projected area DE

$\mathbf{F}_v$ = the weight of the volume of imaginary liquid $ADCBA$ above the plate

Fig. 2–37

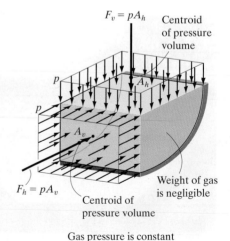

$F_v = pA_h$ Centroid of pressure volume

$F_h = pA_v$ Centroid of pressure volume

Weight of gas is negligible

Gas pressure is constant

Fig. 2–38

2

Important Points

- The *horizontal component* of the resultant force acting on a submerged flat inclined or curved surface is equivalent to the force acting on the *projection* of the area of the surface onto a *vertical plane*. The magnitude of this component and the location of its point of application can be determined using the methods outlined in Sec. 2.7 through 2.9.

- The *vertical component* of the resultant force acting on a submerged flat inclined or curved surface is equivalent to the weight of the volume of liquid acting above the surface. This component passes through the centroid of this volume. If the liquid is confined *below* the flat inclined or curved surface, then the vertical component is *equal but opposite* to the weight of imaginary liquid located within the volume extending *above* the surface to the liquid level.

- Pressure due to a gas is *uniform in all directions* since the weight of the gas is generally neglected. As a result, the horizontal and vertical components of the resultant force of pressure acting on a flat inclined or curved surface can be determined by multiplying the pressure by its associated vertical and horizontal projected areas, respectively. These components act through the centroids of these projected areas.

The formula, geometrical, or integration method can be used to determine the resultant pressure force acting on a surface, such as the endplates of this water trough or oil tank.

EXAMPLE 2.15

The sea wall in Fig. 2–39a is in the form of a semiparabola. Determine the resultant force acting on 1 m of its length. Where does this force act on the wall? Take $\rho_w = 1050 \text{ kg/m}^3$.

SOLUTION

Fluid Description.   We treat the water as an incompressible fluid.

Horizontal Force Component.   The vertical projection of the wall is AB, Fig. 2–39b. The intensity of the distributed load caused by water pressure at point A is

$$w_A = (\rho_w g h)(1 \text{ m}) = (1050 \text{ kg/m}^3)(9.81 \text{ m/s}^2)(8 \text{ m})(1 \text{ m}) = 82.40 \text{ kN/m}$$

Thus,

$$F_x = \frac{1}{2}(8 \text{ m})(82.40 \text{ kN/m}) = 329.62 \text{ kN}$$

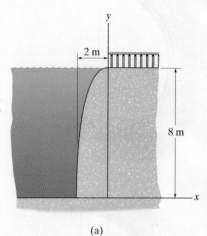

(a)

Using the table on the inside back cover for a triangle, from the surface of the water, this component acts at

$$\bar{y} = \frac{2}{3}(8 \text{ m}) = 5.33 \text{ m} \qquad\qquad Ans.$$

Vertical Force Component.   The vertical force is equivalent to the weight of the water contained within the volume of the exparabolic segment ABC, Fig. 2–39b. From the inside back cover, the area of this segment is $A_{ABC} = \frac{1}{3}ba$. Thus,

$$F_y = (\rho_w g)A_{ABC}(1 \text{ m})$$

$$= \left[1050 \text{ kg/m}^3(9.81 \text{ m/s}^2)\right]\left[\frac{1}{3}(2 \text{ m})(8 \text{ m})\right](1 \text{ m}) = 54.94 \text{ kN}$$

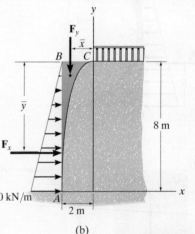

(b)

Fig. 2–39

This force acts through the centroid of the volume (area); that is, from the inside back cover,

$$\bar{x} = \frac{3}{4}(2 \text{ m}) = 1.5 \text{ m} \qquad\qquad Ans.$$

Resultant Force.   The resultant force is therefore

$$F_R = \sqrt{(329.62 \text{ kN})^2 + (54.94 \text{ kN})^2} = 334 \text{ kN} \qquad\qquad Ans.$$

EXAMPLE 2.16

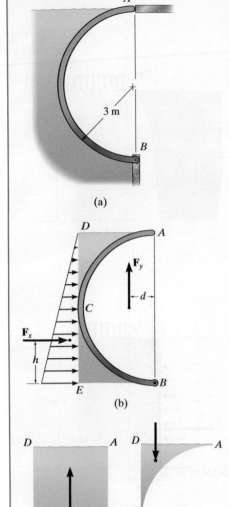

(a)

(b)

Force acting on
segment CB

(c)

Force acting on
segment CA
(d)

Fig. 2–40

The semicircular plate in Fig. 2–40a is 4 m long and acts as a gate in a channel. Determine the resultant force the water pressure exerts on the plate, and then find the components of reaction at the hinge (pin) B and at the smooth support A. Neglect the weight of the plate.

SOLUTION

Fluid Description.   Water is assumed to be an incompressible fluid for which $\gamma_w = (1000 \text{ kg/m}^3)(9.81 \text{ m/s}^2) = 9.81(10^3) \text{ N/m}^3$.

Analysis I

We will first determine the horizontal and vertical components of the resultant force acting on the plate.

Horizontal Force Component.   The vertical projected area of AB is shown in Fig. 2–40b. The intensity of the distributed loading at B (or E) is

$$w_B = \gamma_w h_B b = [9.81(10^3) \text{ N/m}^3](6 \text{ m})(4 \text{ m}) = 235.44(10^3) \text{ N/m}$$

Therefore, the horizontal force component is

$$F_x = \frac{1}{2}[235.44(10^3) \text{ N/m}](6 \text{ m}) = 706.32(10^3) \text{ N} = 706.32 \text{ kN}$$

This force acts at $h = \frac{1}{3}(6 \text{ m}) = 2 \text{ m}$.

Vertical Force Component.   From Fig. 2–40b, note that the force *pushing up* on segment BC is due to the water pressure *under* this segment. It is equal to the imaginary weight of water contained within BCDAB, Fig. 2–40c. And the vertical force *pushing down* on segment AC in Fig. 2–40b is due to the weight of water contained within CDAC, Fig. 2–40d. The *net* vertical force acting on the entire plate is therefore the *difference* in these two weights, namely an upward force equivalent to the volume of water contained within the semicircular region BCAB in Fig. 2–40b. Thus,

$$F_y = \gamma_w \Psi_{BCAB} = [9.81(10^3) \text{ N/m}^3]\left\{ \frac{1}{2}[\pi(3 \text{ m})^2] \right\}(4 \text{ m})$$
$$= 176.58\pi(10^3) \text{ N} = 176.58\pi \text{ kN}$$

The centroid of this semicircular volume of water can be found on the inside back cover.

$$d = \frac{4r}{3\pi} = \frac{4(3 \text{ m})}{3\pi} = \frac{4}{\pi} \text{ m}$$

Resultant Force.   The magnitude of the resultant force is therefore

$$F_R = \sqrt{F_x^2 + F_y^2} = \sqrt{(706.32 \text{ kN})^2 + (176.58\pi \text{ kN})^2} = 898 \text{ kN} \quad Ans.$$

Reactions.   The free-body diagram of the plate is shown in Fig. 2–40e. Applying the equations of equilibrium, we have

$$+\uparrow \Sigma F_y = 0; \quad -B_y + 176.58\pi \text{ kN} = 0$$

$$B_y = 176.58\pi \text{ kN} = 555 \text{ kN} \qquad Ans.$$

$$\zeta + \Sigma M_B = 0; F_A(6 \text{ m}) - (706.32 \text{ kN})(2 \text{ m}) - (176.58\pi \text{ kN})\left(\frac{4}{\pi}\text{ m}\right) = 0$$

$$F_A = 353.16 \text{ kN} = 353 \text{ kN} \qquad Ans.$$

$$\xrightarrow{+} \Sigma F_x = 0 \qquad 706.32 \text{ kN} - 353.16 \text{ kN} - B_x = 0$$

$$B_x = 353.16 \text{ kN} = 353 \text{ kN} \qquad Ans.$$

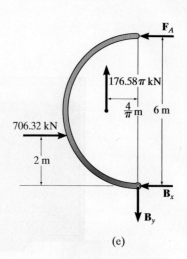

(e)

Analysis II

We can also determine the resultant force components directly using integration. In Fig. 2–40f, notice how the pressure varies over the cross section. To simplify the analysis, we will use polar coordinates because of the circular shape. The elemental strip of width b has an area of $dA = b \, ds = (4 \text{ m})(3 \, d\theta \text{ m}) = 12 \, d\theta \text{ m}^2$. Therefore, the pressure acting on it is

$$p = \gamma_w h = [9.81(10^3) \text{ N/m}^3](3 - 3 \cos \theta) \text{ m}$$

$$= 29.43(10^3)(1 - \cos \theta) \text{ N/m}^2$$

For the horizontal component, $dF_x = p \, dA \sin \theta$, and so

$$F_x = \int_A p \sin \theta \, dA = 29.43(10^3) \int_O^\pi (1 - \cos \theta)(\sin \theta)(12 \, d\theta) = 706.32 \text{ kN}$$

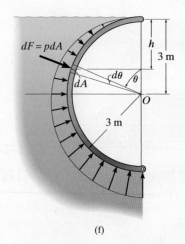

In a similar manner, the y component can be found from $dF_y = p \, dA \cos \theta$. You may wish to evaluate this to verify our previous result for F_y.*

(f)

Fig. 2–40 (cont.)

*Be aware that this method can only be used to determine the *components* of the resultant force. The resultant force *cannot* be found from $F_R = \int_A p \, dA$ because it *does not* account for the *changing direction* of the force.

EXAMPLE | 2.17

The plug in Fig. 2–41a is 50 mm long and has a trapezoidal cross section. If the tank is filled with crude oil, determine the resultant vertical force acting on the plug due to the oil pressure.

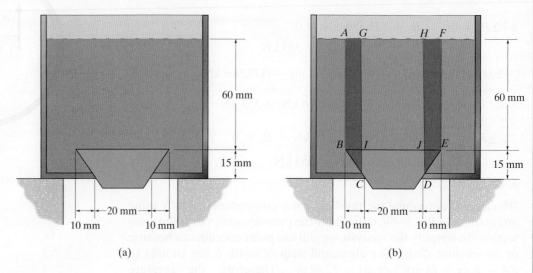

Fig. 2–41

SOLUTION

Fluid Description.   We take the oil to be incompressible, and from Appendix A, $\rho_o = 880 \text{ kg/m}^3$.

Analysis.   With reference to Fig. 2–41b, the force *pushing down* on the plug is due to the weight of oil contained within region *ABEFA*. The force *pushing up* occurs due to pressure along sides *BC* and *ED* and is equivalent to the weight of oil within the dark brown strips, *ABCGA* and *FEDHF*. We have

$$+\downarrow F_R = \rho_o g \left[\Psi_{ABEFA} - 2\Psi_{ABCGA} \right]$$

$$= 880 \text{ kg/m}^3 (9.81 \text{ m/s}^2) \Big[(0.06 \text{ m})(0.04 \text{ m})(0.05 \text{ m})$$

$$- 2 \Big[(0.06 \text{ m})(0.01 \text{ m}) + \frac{1}{2}(0.01 \text{ m})(0.015 \text{ m}) \Big] (0.05 \text{ m}) \Big]$$

$$= 0.453 \text{ N} \qquad\qquad\qquad\qquad\qquad\qquad\qquad\qquad\qquad Ans.$$

Since the result is positive, this force acts *downward* on the plug.

2.11 Buoyancy

The Greek scientist Archimedes (287–212 B.C.) discovered the **principle of buoyancy**, which states that *when a body is placed in a static fluid, it is buoyed up by a force that is equal to the weight of the fluid that is displaced by the body*. To show why this is so, consider the submerged body in Fig. 2–42a. Due to fluid pressure, the vertical resultant force *acting upward* on the bottom surface of the body, *ABC*, is equivalent to the weight of fluid contained above this surface, that is, within the volume *ABCEFA*. Likewise, the resultant force due to pressure acting *downward* on the top surface of the body, *ADC*, is equivalent to the weight of fluid contained within the volume *ADCEFA*. The *difference* in these forces acts upward, and is the **buoyant force**. It is equivalent to the weight of an imaginary amount of fluid contained within the volume of the body, *ABCDA*. This force $\mathbf{F}_b$ acts through the **center of buoyancy**, C_b, which is located at the centroid of the volume of liquid displaced by the body. If the density of the fluid is constant, then this force will remain constant, *regardless of how deep* the body is placed within the fluid.

These same arguments can also be applied to a floating body, as in Fig. 2–42b. Here the displaced amount of fluid is within the region *ABC*, the buoyant force is equal to the weight of fluid within this displaced volume, and the center of buoyancy C_b is at the centroid of this volume.

If a hydrostatic problem that involves buoyancy is to be solved, then it may be necessary to investigate the forces acting on the free-body diagram of the body. This requires the buoyant force to be shown acting upward at the center of buoyancy, while the body's weight acts downward, through its center of gravity.

This cargo ship has a uniform weight distribution and it is empty, as noticed by how high it floats level in the water relative to its waterline.

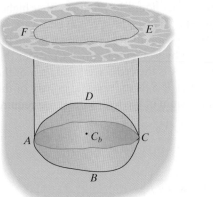

Submerged body

(a)

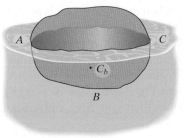

Floating body

(b)

Fig. 2–42

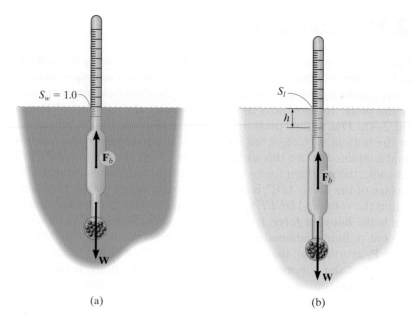

(a)   (b)

Hydrometer

Fig. 2–43

Hydrometer.   The principle of buoyancy can be used in a practical way to measure the specific gravity of a liquid using a device called a **hydrometer**. As shown in Fig. 2–43a, it consists of a hollow glass tube that is weighted at one end. If the hydrometer is placed in a liquid such as pure water, it will float in equilibrium when its weight W equals the weight of displaced water, that is, when $W = \gamma_w V_0$, where V_0 is the volume of water displaced. If the stem is marked at the water level as 1.0, Fig. 2–43a, then this position can indicate water's specific gravity, since for water, $S_w = \gamma_w/\gamma_w = 1.0$, Eq. 1–10.

When the hydrometer is placed in another liquid, it will float either higher or lower, depending upon the liquid's specific weight γ_l relative to water. If the liquid is lighter than water, such as kerosene, then a greater volume of the liquid must be displaced in order for the hydrometer to float. Consider this displaced volume to be $V_0 + Ah$, where A is the cross-sectional area of the stem, Fig. 2–43b. Now $W = \gamma_l(V_0 + Ah)$. If S_l is the specific gravity of the liquid, then $\gamma_l = S_l\gamma_w$, and so for equilibrium of the hydrometer, we require

$$W = \gamma_w V_0 = S_l\gamma_w(V_0 + Ah)$$

Solving for S_l yields

$$S_l = \left(\frac{V_0}{V_0 + Ah}\right)$$

Using this equation, for each depth h, calibration marks can be placed on the stem to indicate the specific gravities, S_l, for various types of liquids. In the past, hydrometers were often used to test the specific weight of acid in automobile batteries. When a battery is fully charged, the hydrometer will float higher in the acid than when the battery is discharged.

EXAMPLE | 2.18

The 500-N flat-bottom container in Fig. 2–44a is 600 mm wide and 900 mm long. Determine the depth the container will float in the water (a) when it carries the 200-N steel block, and (b) when the block is suspended directly beneath the container, Fig. 2–44b. Take $\gamma_{st} = 77.0 \text{ kN/m}^3$.

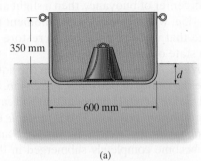

350 mm

600 mm

(a)

Fig. 2–44

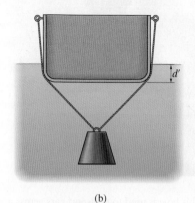

d′

(b)

SOLUTION

Fluid Description. The water is assumed to be incompressible, and has a specific weight of $\gamma_w = (1000 \text{ kg/m}^3)(9.81 \text{ m/s}^2) = 9.81(10^3) \text{ N/m}^3$.

Analysis. In each case, for equilibrium, the weight of the container and block must be equal to the weight of the displaced water, which creates the buoyant force.

Part (a). From the free-body diagram in Fig. 2–44c, we require

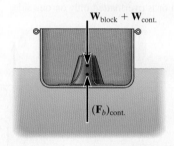

$W_{block} + W_{cont.}$

$(F_b)_{cont.}$

(c)

$$+\uparrow \Sigma F_y = 0; \qquad -(W_{cont.} + W_{block}) + (F_b)_{cont.} = 0$$

$$-(500 \text{ N} + 200 \text{ N}) + [9.81(10^3) \text{ N/m}^3][(0.6 \text{ m})(0.9 \text{ m})d] = 0$$

$$d = 0.1321 \text{ m} = 132 \text{ mm} < 350 \text{ mm} \quad \text{OK} \qquad Ans.$$

Part (b). In this case, Fig. 2–44d, we first have to find the volume of steel used to make the block. Since the specific weight of steel is given, then $V_{st} = W_{st}/\gamma_{st}$. From this the buoyancy force can be determined. Thus,

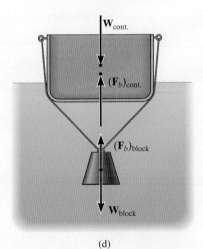

$W_{cont.}$

$(F_b)_{cont.}$

$(F_b)_{block}$

W_{block}

(d)

$$+\uparrow \Sigma F_y = 0; -(W_{cont.} + W_{block}) + (F_b)_{cont.} + (F_b)_{block} = 0$$

$$-(500 \text{ N} + 200 \text{ N}) + [9.81(10^3) \text{ N/m}^3][(0.6 \text{ m})(0.9 \text{ m})d']$$

$$+ [9.81(10^3) \text{ N/m}^3]\left[\frac{200 \text{ N}}{77.0(10^3) \text{ N/m}^3}\right] = 0$$

$$d' = 0.1273 \text{ m} = 127 \text{ mm} \qquad Ans.$$

Notice that here the container floats higher in the water because when the block is supported under the water, its buoyancy force reduces the force needed to support it. Also, note that the answer is independent of the depth the block is suspended in the water.

This boat is used to transport cars across a river. Care must be taken so that it does not become unstable if it makes a sharp turn or is overloaded only on one side.

2.12   Stability

A body can float in a liquid (or gas) in stable, unstable, or neutral equilibrium. To illustrate this, consider a uniform light-weight bar with a weight attached to its end so that its center of gravity is at G, Fig. 2–45.

Stable equilibrium.   If the bar is placed in a liquid so that its center of gravity is below its center of buoyancy, then a slight angular displacement of the bar, Fig. 2–45a, will create a couple moment between the weight and buoyant force that will cause the bar to restore itself to the vertical position. This is a state of *stable equilibrium.*

Unstable equilibrium.   If the bar is placed in the liquid so that its center of gravity is above the center of buoyancy, Fig. 2–45b, then a slight angular displacement will create a couple moment that will cause the bar to rotate *farther* from its equilibrium position. This is *unstable equilibrium.*

Neutral equilibrium.   If the weight is removed and the bar is made heavy enough to become completely submerged in the liquid so that the weight and buoyant force are *balanced*, then its center of gravity and center of buoyancy will coincide, Fig. 2–45c. Any rotation of the bar will cause it to *remain* in the newfound equilibrium position. This is a state of *neutral equilibrium.*

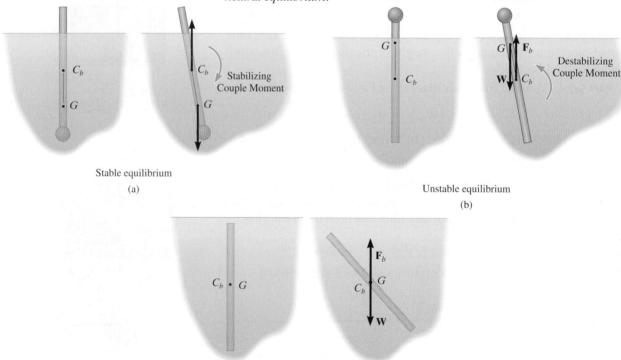

Fig. 2–45

Although the bar in Fig. 2–45b is in unstable equilibrium, some floating bodies can maintain stable equilibrium when their center of gravity is *above* their center of buoyancy. For example, consider the ship in Fig. 2–46a, which has its center of gravity at G and its center of buoyancy at C_b. When the ship undergoes a slight roll, which occurs at the water level about point O, Fig. 2–46b, the new center of buoyancy C_b' will be *to the left* of G. This is because a portion of the displaced water is gained on the left, $OABO$, while an equivalent portion $ODEO$, is lost on the right. This new volume of displaced fluid, $ABFDOA$, is used to locate C_b'. If we establish a vertical line through C_b' (the line of action of $\mathbf{F}_b$), it will intersect the centerline of the ship at point M, which is called the **metacenter**. If M is *above* G, Fig. 2–46b, the clockwise couple moment created by the buoyant force and the ship's weight will tend to restore the ship to its equilibrium position. Therefore, the ship is in *stable equilibrium*.

For a ship with a large, high deck loading, as in Fig. 2–46c, M will be *below* G. In this case, the counterclockwise couple moment created by $\mathbf{F}_b$ and $\mathbf{W}$ will cause the ship to become *unstable* and to roll over, a condition that obviously must be avoided when designing or loading any ship. Realizing this danger, however, maritime engineers design modern day cruise ships so that their centers of gravity are as high as possible above their centers of buoyancy. Doing this will cause the ship to roll back and forth very slowly in the water. When these two points are closer together, the back and forth motion is faster, which can be discomforting to the passengers.

This same rule discussed above also applies to the bar in Fig. 2–45. Because the bar is *thin*, the metacenter M is on the centerline of the bar and *coincides* with C_b. When it is above G it is in stable equilibrium, Fig. 2–45a; when it is below G it is in unstable equilibrium, Fig. 2–45b; and when it is at G it is in neutral equilibrium, Fig. 2–45c.

(a)

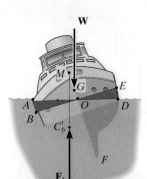

$OM > OG$
Stable equilibrium
(b)

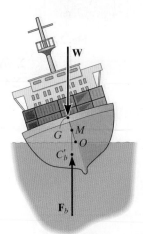

$OG > OM$
Unstable equilibrium
(c)

Fig. 2–46

Important Points

- The buoyant force on a body is equal to the weight of the fluid the body displaces. It acts upward through the center of buoyancy, which is located at the centroid of the displaced volume of fluid.

- A hydrometer uses the principle of buoyancy to measure the specific gravity of a liquid.

- A floating body can be in stable, unstable, or neutral equilibrium. If its metacenter M is above the center of gravity G of the body, then the body will float in stable equilibrium. If M is below G, then the body will be unstable.

EXAMPLE 2.19

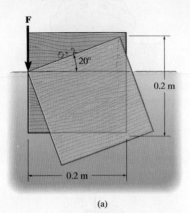

(a)

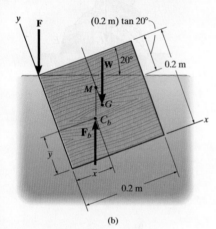

(b)

Fig. 2–47

The wooden block (cube) in Fig. 2–47a is 0.2 m on each side. A vertical force **F** is applied at the center of one of its sides and pushes the edge of the block to the water surface so that it is held at an angle of 20°. Determine the buoyant force on the block, and show that the block will be in stable equilibrium when the force **F** is removed.

SOLUTION

Fluid Description. The water is assumed to be incompressible, and has a density of $\rho_w = 1000 \text{ kg/m}^3$.

Analysis. To find the buoyant force, we must first find the *submerged volume* of the block, shown on the free-body diagram in Fig. 2–47b. It is

$$V_{sub} = (0.2 \text{ m})^3 - \frac{1}{2}(0.2 \text{ m})(0.2 \tan 20°)(0.2 \text{ m}) = 6.544(10^{-3}) \text{ m}^3$$

Then

$$F_b = \rho_w g V_{sub} = 1000(9.81) \text{ N/m}^3 \left[6.5441(10^{-3}) \text{ m}^3 \right]$$
$$= 64.2 \text{ N} \qquad\qquad Ans.$$

This force acts through the centroid of this volume V_{sub} (or area) having coordinates measured from the x, y axes of

$$\bar{x} = \frac{\Sigma \tilde{x} A}{\Sigma A} = \frac{0.1 \text{ m}(0.2 \text{ m})^2 - \frac{2}{3}(0.2 \text{ m})\left(\frac{1}{2}\right)(0.2 \text{ m})(0.2 \text{ m} \tan 20°)}{(0.2 \text{ m})^2 - \left(\frac{1}{2}\right)(0.2 \text{ m})(0.2 \text{ m} \tan 20°)}$$

$$= 0.0926 \text{ m}$$

$$\bar{y} = \frac{\Sigma \tilde{y} A}{\Sigma A} = \frac{0.1 \text{ m}(0.2 \text{ m})^2 - \left[(0.2 \text{ m}) - \left(\frac{1}{3}\right)(0.2 \text{ m} \tan 20°)\right]\left(\frac{1}{2}\right)(0.2 \text{ m})(0.2 \text{ m} \tan 20°)}{(0.2 \text{ m})^2 - \left(\frac{1}{2}\right)(0.2 \text{ m})(0.2 \text{ m} \tan 20°)}$$

$$= 0.0832 \text{ m}$$

This location for F_b will be to the *left* of the block's center of gravity (0.1 m, 0.1 m), and so the clockwise moment of F_b about G will restore the block when the force **F** is removed. Hence the block is in *stable equilibrium*. In other words, the metacenter M will be above G, Fig. 2–47b.

Although it is not part of this problem, the force **F** and the block's weight **W** can be determined by applying the vertical force and moment equilibrium equations to the block.

2.13 Constant Translational Acceleration of a Liquid

In this section we will discuss both horizontal and vertical constant accelerated motion of a container of liquid, and we will study how the pressure varies within the liquid for these two motions.

Constant Horizontal Acceleration.   If the container of liquid in Fig. 2–48a has a *constant velocity*, $\mathbf{v}_c$, then the surface of the liquid will *remain horizontal* since equilibrium occurs. As a result, the pressure exerted on the walls of the container can be determined in the usual manner using $p = \gamma h$. If the container undergoes a *constant acceleration*, $\mathbf{a}_c$, however, then the liquid surface will begin to rotate clockwise about the center of the container and will eventually maintain a fixed tilted position θ, Fig. 2–48b. After this adjustment, all the liquid will behave as though it were a solid. No shear stress will be developed between layers of the liquid since there is no relative motion between the layers. A force analysis using a free-body diagram of vertical and horizontal differential elements of the liquid will now be considered to study the effects of this motion.

Vertical Element.   For this case, the differential element extends downward a distance h from the liquid surface and has a cross-sectional area ΔA, Fig. 2–48c. The two vertical forces acting on it are the weight of the contained liquid, $\Delta W = \gamma \Delta V = \gamma (h \Delta A)$, and the pressure force acting upward on its bottom, $p \Delta A$. Equilibrium exists in the vertical direction since no acceleration occurs in this direction.

$$+\uparrow \Sigma F_y = 0; \qquad\qquad p \Delta A - \gamma (h \Delta A) = 0$$
$$p = \gamma h \qquad\qquad\qquad (2\text{–}16)$$

This result indicates that *the pressure at any depth from the inclined liquid surface is the same as if the liquid were static.*

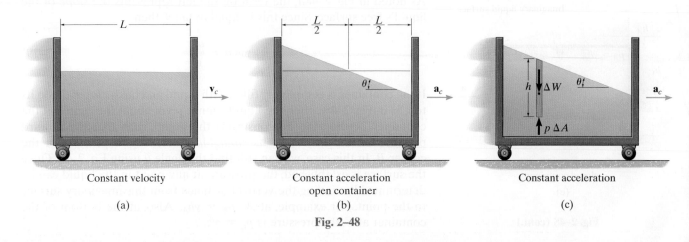

Constant velocity

(a)

Constant acceleration
open container

(b)

Constant acceleration

(c)

Fig. 2–48

2

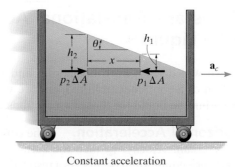

Constant acceleration

(d)

Fig. 2–48 (cont.)

Tank cars carry a variety of liquids. Their endplates must be designed to resist the hydrostatic pressure of the liquid within the car caused by any anticipated acceleration of the car.

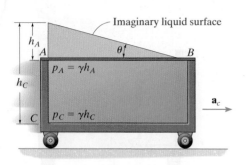

Constant acceleration
closed container

(e)

Fig. 2–48 (cont.)

Horizontal Element.   Here the differential element has a length x and cross-sectional area ΔA, Fig. 2–48d. The only horizontal forces acting on it are caused by the pressure of the adjacent liquid on each of its ends. Since the mass of the element is $\Delta m = \Delta W/g = \gamma(x\Delta A)/g$, the equation of motion becomes

$$\overset{+}{\rightarrow} \Sigma F_x = ma_x; \qquad p_2\Delta A - p_1\Delta A = \frac{\gamma(x\,\Delta A)}{g}a_c$$

$$p_2 - p_1 = \frac{\gamma x}{g}a_c \qquad (2\text{--}17)$$

Using $p_1 = \gamma h_1$ and $p_2 = \gamma h_2$, we can also write this expression as

$$\frac{h_2 - h_1}{x} = \frac{a_c}{g} \qquad (2\text{--}18)$$

As noted in Fig. 2–48d, the term on the left represents the *slope* of the liquid's free surface. Since this is equal to $\tan\theta$, then

$$\boxed{\tan\theta = \frac{a_c}{g}} \qquad (2\text{--}19)$$

If the container is completely filled with liquid and has a *closed lid* on its top, as in Fig. 2–48e, then the liquid cannot pivot about the center of the container. Rather, the lid constrains the liquid, such that the upward pressure on the lid forces its "*imaginary* surface" to pivot from the corner B. In this case, we can still find the angle θ using Eq. 2–19. Once the surface is established, the pressure at any point in the liquid can be determined by finding the vertical distance from this imaginary surface to the point. For example, at A, $p_A = \gamma h_A$. Also, at the bottom of the container at C, the pressure is $p_C = \gamma h_C$.

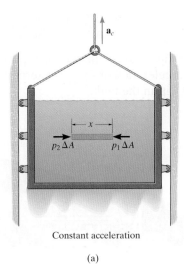

Constant acceleration

(a)

Constant Vertical Acceleration.

When the container is accelerated upward at $\mathbf{a}_c$, the liquid surface maintains its horizontal position; however, the pressure within the liquid will change. To study this effect, we will again select horizontal and vertical differential elements and use their free-body diagrams.

Horizontal Element. Since the horizontal element in Fig. 2–49a is at the same depth in the liquid, the pressure of the adjacent liquid on each of its ends exerts forces as shown. No motion occurs in this direction, so

$$\xrightarrow{+} \Sigma F_x = 0; \qquad p_2 \Delta A - p_1 \Delta A = 0$$

$$p_2 = p_1$$

Hence, as in the static case, for vertical acceleration the pressure is the same at points that lie in the same horizontal plane.

Vertical Element. The forces acting on the vertical element of depth h and cross section ΔA, Fig. 2–49b, consist of the element's weight $\Delta W = \gamma \Delta \mathcal{V} = \gamma(h\,\Delta A)$ and the pressure force on its bottom. Since the mass of the element is $\Delta m = \Delta W/g = \gamma(h\,\Delta A)/g$, application of the equation of motion yields

$$+\uparrow \Sigma F_y = ma_y; \qquad p\Delta A - \gamma(h\,\Delta A) = \frac{\gamma(h\,\Delta A)}{g}a_c$$

$$\boxed{p = \gamma h\left(1 + \frac{a_c}{g}\right)} \qquad (2\text{--}20)$$

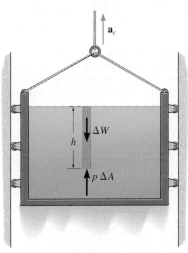

Constant acceleration

(b)

Fig. 2–49

Thus, the pressure within the liquid will *increase* by $\gamma h(a_c/g)$ when the container is accelerated *upward*. If it has a *downward acceleration*, the pressure will *decrease* by this amount. If free-fall occurs, then $a_c = -g$ and the (gage) pressure throughout the liquid will be zero.

EXAMPLE | 2.20

The tank on the truck in Fig. 2–50a is filled to its top with gasoline. If the truck has a constant acceleration of 4 m/s², determine the pressure at points A, B, C, and D within the tank.

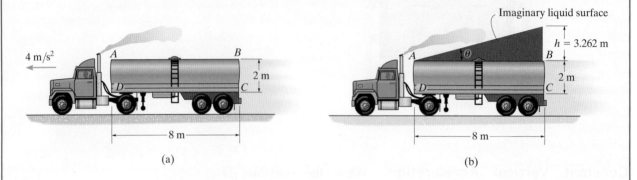

(a)   (b)

Fig. 2–50

SOLUTION

Fluid Description. We assume that gasoline is incompressible, and from Appendix A it has a density of $\rho_g = 729$ kg/m³.

Analysis. When the truck is at rest or moving with a constant velocity, the (gage) pressures at A and B are zero since the gasoline surface remains horizontal. When the truck accelerates, the surface is imagined to pivot at A and tilt back, Fig. 2–50b. We can determine the height h using Eq. 2–18.

$$\frac{h_2 - h_1}{x} = \frac{a_c}{g}$$

$$\frac{h - 0}{8 \text{ m}} = \frac{4 \text{ m/s}^2}{9.81 \text{ m/s}^2}$$

$$h = 3.262 \text{ m}$$

The top of the tank prevents the formation of this sloped surface, and so the imaginary gasoline surface exerts a pressure on the top. This pressure can be obtained using Eq. 2–16, $p = \gamma h$. Thus,

$$p_A = \gamma h_A = (726 \text{ kg/m}^3)(9.81 \text{ m/s}^2)(0) = 0 \qquad \qquad Ans.$$

$$p_B = \gamma h_B = (726 \text{ kg/m}^3)(9.81 \text{ m/s}^2)(3.262 \text{ m}) = 23.2 \text{ kPa} \qquad Ans.$$

$$p_C = \gamma h_C = (726 \text{ kg/m}^3)(9.81 \text{ m/s}^2)(3.262 \text{ m} + 2 \text{ m}) = 37.5 \text{ kPa} \qquad Ans.$$

$$p_D = \gamma h_D = (726 \text{ kg/m}^3)(9.81 \text{ m/s}^2)(2 \text{ m}) = 14.2 \text{ kPa} \qquad Ans.$$

EXAMPLE 2.21

The container in Fig. 2–51a is 1.25 m wide and is filled with crude oil to a height of 2 m. Determine the resultant force the oil exerts on the container's side and on its bottom if a crane begins to hoist it upward with an acceleration of 3 m/s².

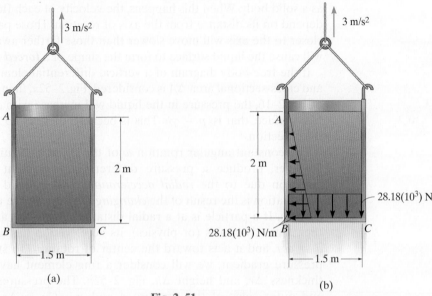

Fig. 2–51

SOLUTION

Fluid Description. The oil is assumed to be incompressible, and from Appendix A its specific weight is $\gamma_o = \rho_o g = (880 \text{ kg/m}^3)(9.81 \text{ m/s}^2) = 8.6328(10^3) \text{ N/m}^3$.

Analysis. The (gage) pressure at A is zero, and hence the pressure at B and C can be determined using Eq. 2–20. Since $a_c = +3 \text{ m/s}^2$, we have

$$p = \gamma_o h\left(1 + \frac{a_c}{g}\right) = [8.6328(10^3) \text{ N/m}^3](2 \text{ m})\left(1 + \frac{3 \text{ m/s}^2}{9.81 \text{ m/s}^2}\right) = 22.55(10^3) \text{ N/m}^2$$

Since the tank has a width of 1.25 m, the intensity of the distributed load at the bottom of the tank, Fig. 2–51b, is

$$w = pb = [22.55(10^3) \text{ N/m}^2](1.25 \text{ m}) = 28.182(10^3) \text{ N/m}$$

Side of Tank. For the triangular distributed load acting on side AB, we have

$$(F_R)_s = \frac{1}{2}[28.182(10^3) \text{ N/m}](2 \text{ m}) = 28.182(10^3) \text{ N} = 28.2 \text{ kN} \quad Ans.$$

Bottom of Tank. The bottom of the tank is subjected to a uniform distributed load. Its resultant force is

$$(F_R)_b = [28.182(10^3) \text{ N/m}](1.5 \text{ m}) = 42.273(10^3) \text{ N} = 42.3 \text{ kN} \quad Ans.$$

2.14   Steady Rotation of a Liquid

If a liquid is placed into a cylindrical container that rotates at a constant angular velocity ω, Fig. 2–52a, the shear stress developed within the liquid will begin to cause the liquid to rotate with the container. Eventually, no relative motion within the liquid will occur, and the system will then rotate as a solid body. When this happens, the velocity of each fluid particle will depend on its distance from the axis of rotation. Those particles that are closer to the axis will move slower than those farther away. This motion will cause the liquid surface to form the shape of a ***forced vortex***.

If the free-body diagram of a vertical differential element of height h and cross-sectional area ΔA is considered, Fig. 2–52a, then, as in the proof of Eq. 2–16, the pressure in the liquid will increase with depth from the free surface, that is, $p = \gamma h$. This is because there is no acceleration in this direction.

The constant angular rotation ω of the cylinder–liquid system does, however, produce a pressure difference or gradient in the radial direction due to the *radial acceleration* of the liquid particles. This acceleration is the result of the *changing direction* of the *velocity* of each particle. If a particle is at a radial distance r from the axis of rotation, then from dynamics (or physics), its acceleration has a magnitude of $a_r = \omega^2 r$, and it acts toward the center of rotation. To study the radial pressure gradient, we will consider a ring element having a radius r, thickness Δr, and height Δh, Fig. 2–52b. The pressures on the inner and outer sides of the ring are p and $p + (\partial p / \partial r)\,\Delta r$, respectively.*

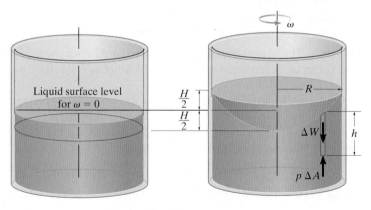

Forced vortex
(a)

Fig. 2–52

*The partial derivative is used here since the pressure is a function of both depth and radius.

Since the mass of the ring is $\Delta m = \Delta W/g = \gamma \, \Delta V/g = \gamma(2\pi r) \, \Delta r \, \Delta h/g$, the equation of motion in the radial direction gives

$$\Sigma F_r = ma_r; \quad -\left[p + \left(\frac{\partial p}{\partial r}\right)\Delta r\right](2\pi r\Delta h) + p(2\pi r\Delta h) = -\frac{\gamma(2\pi r)\Delta r\Delta h}{g}\omega^2 r$$

$$\frac{\partial p}{\partial r} = \left(\frac{\gamma\omega^2}{g}\right)r$$

Integrating, we obtain

$$p = \left(\frac{\gamma\omega^2}{2g}\right)r^2 + C$$

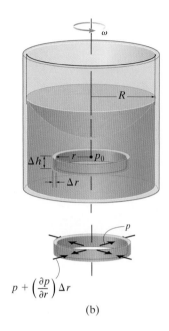

(b)

We can determine the constant of integration provided we know the pressure in the fluid at a specific point. Consider the point on the vertical axis at the free surface, where $r = 0$ and $p_0 = 0$, Fig. 2–52c. Then $C = 0$, and so

$$p = \left(\frac{\gamma\omega^2}{2g}\right)r^2 \qquad (2\text{–}21)$$

The pressure increases with the square of the radius. Since $p = \gamma h$, the equation of the free surface of the liquid, Fig. 2–52c, becomes

$$\boxed{h = \left(\frac{\omega^2}{2g}\right)r^2} \qquad (2\text{–}22)$$

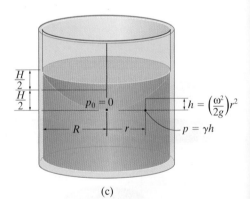

(c)

Fig. 2–52 (cont.)

This is the equation of a parabola. Specifically, the liquid as a whole forms a surface that describes a **_paraboloid of revolution_**. Since the interior radius of the container is R, the height of this paraboloid is $H = \omega^2 R^2/2g$, Fig. 2–52c. The volume of this paraboloid is one-half its base area πR^2 times its height, H. As a result, during the rotation, the high and low points of the liquid surface will be halfway from the liquid surface when the liquid is at rest, Fig. 2–52a.

If the rotating container has a closed lid, then an imaginary free surface of the paraboloid can be established above the lid, and the pressure at any point on the liquid is determined by measuring the depth h from this surface.

2

Important Points

- When an open container of liquid is *uniformly accelerated horizontally*, the surface of the liquid will be inclined at an angle θ determined from $\tan \theta = a_c/g$. The pressure varies linearly with depth from this surface, $p = \gamma h$. If a lid is on the container, then an *imaginary surface* should be established, and the pressure at any point can be determined using $p = \gamma h$, where h is the vertical distance from the imaginary surface to this point.

- When a container of liquid is *uniformly accelerated vertically*, the surface of the liquid remains horizontal. If this acceleration is *upward*, the pressure at a depth h is *increased* by $\gamma h(a_c/g)$; and if the acceleration is *downward*, the pressure is *decreased* by this amount.

- When a cylindrical container of liquid has a *constant rotation about a fixed axis*, the liquid surface forms a *forced vortex* having the shape of a paraboloid. The volume of this paraboloid is one-half the volume of a circumscribed cylinder. Once the surface of the rotating liquid is established, using $h = (\omega^2/2g)r^2$, the pressure varies with depth from this surface, $p = \gamma h$. If a lid is on the container, then an *imaginary liquid surface* should be established, and the pressure at any point can be determined using the vertical distance from this surface.

EXAMPLE | 2.22

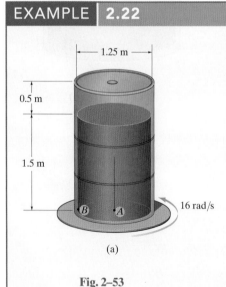

(a)

Fig. 2–53

The closed cylindrical drum in Fig. 2–53a is filled with crude oil to the level shown. If the pressure of the air within the drum remains atmospheric due to the hole in the center of the lid, determine the pressures at points A and B when the drum and oil attain a constant angular velocity of 12 rad/s.

SOLUTION

Fluid Description.   The oil is assumed incompressible, and from Appendix A it has a specific weight of

$$\gamma_o = \rho_o g = \left(880 \text{ kg/m}^3\right)\left(9.81 \text{ m/s}^2\right) = 8.6328(10^3) \text{ N/m}^3$$

Analysis.   Before finding the pressure, we must define the shape of the oil surface. As the drum rotates, the oil takes the shape shown in Fig. 2–53b. Since the volume of open space *within* the drum must remain constant, this volume must be equivalent to the volume of the shaded paraboloid of unknown radius r and height h. Since the volume of a paraboloid is one-half that of a cylinder having the same radius and height, we require

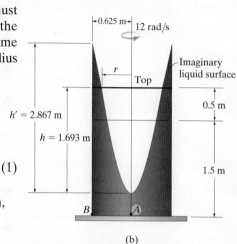

$$V_{\text{cyl}} = V_{\text{parab}}$$

$$\pi(0.625 \text{ m})^2(0.5 \text{ m}) = \frac{1}{2}\pi r^2 h$$

$$r^2 h = 0.3906 \tag{1}$$

Also, from Eq. 2–22, for this contained paraboloid within the drum,

$$h = \left(\frac{\omega^2}{2g}\right)r^2 = \left[\frac{(12 \text{ rad/s})^2}{2(9.81 \text{ m/s}^2)}\right]r^2$$

$$h = 7.3394r^2 \tag{2}$$

Solving Eqs. 1 and 2 simultaneously, we get

$$r = 0.4803 \text{ m}, \quad h = 1.693 \text{ m}$$

Without the lid, the oil would rise to a level h', Fig. 2–53b, which is

$$h' = \left(\frac{\omega^2}{2g}\right)R^2 = \left[\frac{(12 \text{ rad/s})^2}{2(9.81 \text{ m/s}^2)}\right](0.625 \text{ m})^2 = 2.867 \text{ m}$$

Since the free surface of the oil has now been defined, the pressures at A and B are

$$p_A = \gamma h_A = [8.6328(10^3) \text{ N/m}^3](2 \text{ m} - 1.693 \text{ m})$$

$$= 2.648(10^3) \text{ Pa} = 2.65 \text{ kPa} \qquad\qquad Ans.$$

$$p_B = \gamma h_B = [8.6328(10^3) \text{ N/m}^3][2.867 \text{ m} + (2 \text{ m} - 1.693 \text{ m})]$$

$$= 27.40(10^3) \text{ Pa} = 27.4 \text{ kPa} \qquad\qquad Ans.$$

Although it is not part of this problem, realize that if a cap were placed on the hole, and the pressure of the air within the drum were increased to 30 kPa, then this pressure would simply be added to the pressures at A and B.

References

1. I. Khan, *Fluid Mechanics*, Holt, Rinehart and Winston, New York, NY, 1987.
2. A. Parr, *Hydraulics and Pneumatics*, Butterworth-Heinemann, Woburn, MA, 2005.
3. *The U.S. Standard Atmosphere*, U.S Government Printing Office, Washington, DC.
4. K. J. Rawson and E. Tupper, *Basic Ship Theory*, 2nd ed., Longmans, London, UK, 1975.
5. S. Tavoularis, *Measurements in Fluid Mechanics*, Cambridge University Press, New York, NY, 2005.
6. R. C. Baker, *Introductory Guide to Flow Measurement*, John Wiley, New York, NY, 2002.
7. R. W. Miller, *Flow Measurement Engineering Handbook*, 3rd ed., McGraw-Hill, New York, NY, 1996.
8. R. P. Benedict, *Fundamentals of Temperature, Pressure, and Flow Measurement*, 3rd ed., John Wiley, New York, NY, 1984.
9. J. W. Dally et al., *Instrumentation for Engineering Measurements*, 2nd ed., John Wiley, New York, NY, 1993.
10. B. G. Liptak, *Instrument Engineer's Handbook: Process Measurement and Analysis*, 4th ed., CRC Press, Boca Raton, FL, 2003.
11. F. Durst et al., *Principles and Practice of Laser-Doppler Anemometry*, 2nd ed., Academic Press, New York, NY, 1981.

FUNDAMENTAL PROBLEMS

The solutions to all fundamental problems are given in the back of the book.

Sec. 2.1–2.5

F2–1. Water fills the pipe AB such that the absolute pressure at A is 400 kPa. If the atmospheric pressure is 101 kPa, determine the resultant force the water and surrounding air exert on the cap at B. The inner diameter of the pipe is 50 mm.

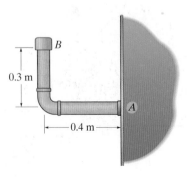

Prob. F2–1

F2–2. The container is partially filled with oil, water, and air. Determine the pressures at A, B, and C. Take $\rho_w = 1000 \text{ kg/m}^3$, $\rho_o = 830 \text{ kg/m}^3$.

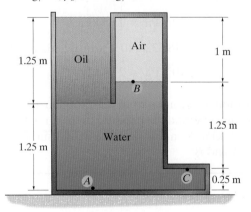

Prob. F2–2

Sec. 2.6

F2–3. The U-tube manometer is filled with mercury, having a density of $\rho_{Hg} = 13\ 550\ kg/m^3$. Determine the differential height h of the mercury when the tank is filled with water.

F2–5. The air pressure in the pipe at A is 300 kPa. Determine the water pressure in the pipe at B.

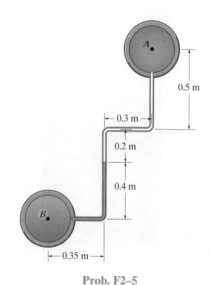

Prob. F2–5

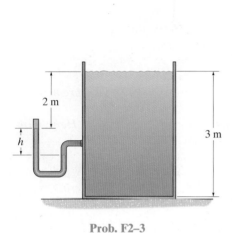

Prob. F2–3

F2–4. The tube is filled with mercury from A to B, and with water from B to C. Determine the height h of the water column for equilibrium.

F2–6. Determine the absolute water pressure in the pipe at B if the tank is filled with crude oil to the depth of 1.5 m. $p_{atm} = 101$ kPa.

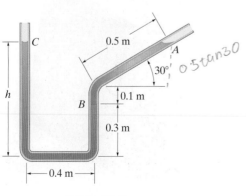

Prob. F2–4

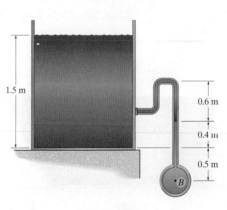

Prob. F2–6

Sec. 2.7–2.9

F2–7. The bin is 1.5 m wide and is filled with water to the level shown. Determine the resultant force on the side AB and on the bottom BC.

F2–9. The 2-m-wide container is filled with water to the depth shown. Determine the resultant force on the side panels A and B. How far does each resultant act from the surface of the water?

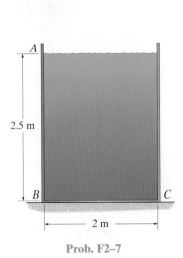

Prob. F2–7

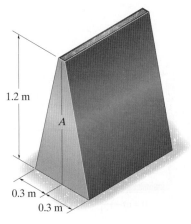

Prob. F2–9

F2–10. Determine the magnitude of the resultant force of the water acting on the triangular end plate A of the trough. Neglect the width of the opening at the top. How far does this force act from the surface of the water?

F2–8. The bin is 2 m wide and is filled with oil to the depth shown. Determine the resultant force acting on the inclined side AB. Take $\rho_o = 900$ kg/m³.

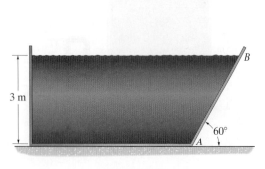

Prob. F2–8

Prob. F2–10

F2–11. Determine the magnitude of the resultant force of the water acting on the circular glass plate that is bolted to the side panel of the tank. Also, determine the location of the center of pressure along this inclined side, measured from the top.

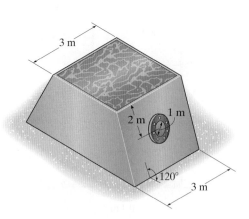

Prob. F2–11

F2–12. The tank is filled with water and kerosene to the depths shown. Determine the total resultant force the liquids exert on side AB of the tank. The tank has a width of 2 m. Take $\rho_w = 1000 \text{ kg/m}^3$, $\rho_k = 814 \text{ kg/m}^3$.

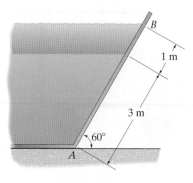

Prob. F2–12

F2–13. The 0.5-m-wide inclined plate holds water in a tank. Determine the horizontal and vertical components of force and the moment that the fixed support at A exerts on the plate.

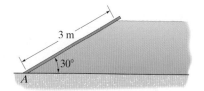

Prob. F2–13

F2–14. Determine the resultant force the oil exerts on the semicircular surface AB. The tank has a width of 3 m. Take $\rho_o = 900 \text{ kg/m}^3$.

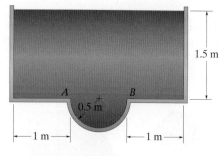

Prob. F2–14

F2–15. Determine the resultant force the water exerts on side AB and on side CD of the inclined wall. The wall is 0.75 m wide.

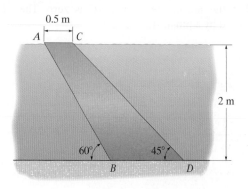

Prob. F2–15

Sec. 2.10

F2–16. The tank has a width of 2 m and is filled with water. Determine the horizontal and vertical components of the resultant force acting on plate *AB*.

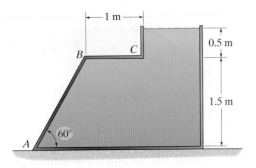

Prob. F2–16

F2–17. Determine the horizontal and vertical components of the resultant force the water exerts on plate *AB* and on plate *BC*. The width of each plate is 1.5 m.

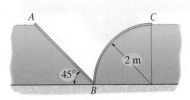

Prob. F2–17

F2–18. The plate *ABC* is 2 m wide. Determine the angle θ so that the normal reaction at *C* is zero. The plate is supported by a pin at *A*.

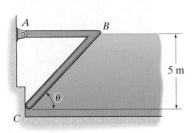

Prob. F2–18

Sec. 2.11–2.12

F2–19. The cylindrical cup *A* of negligible weight contains a 2-kg block *B*. If the water level of the cylindrical tank is $h = 0.5$ m before the cup is placed into the tank, determine h when *A* floats in the water.

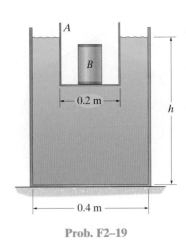

Prob. F2–19

F2–20. The 3-m-wide cart is filled with water to the level of the dashed line. If the cart is given an acceleration of 4 m/s², determine the angle θ of the water surface and the resultant force the water exerts on the wall *AB*.

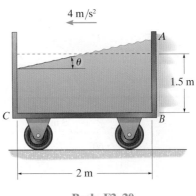

Prob. F2–20

F2–21. The closed tank is filled with oil and given an acceleration of 6 m/s^2. Determine the pressure on the bottom of the tank at points A and B. $\rho_o = 880$ kg/m^3.

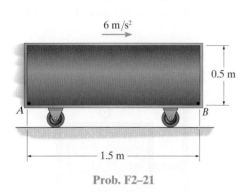

6 m/s^2

0.5 m

A $\qquad$ B

1.5 m

Prob. F2–21

F2–22. The open cylindrical container is filled with water to the level shown. Determine the smallest angular velocity that will cause the water to spill over the sides.

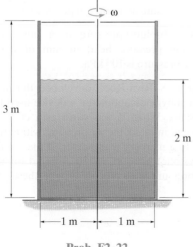

ω

3 m

2 m

1 m $\quad$ 1 m

Prob. F2–22

F2–23. If the open cylindrical container rotates at $\omega = 8$ rad/s, determine the maximum and minimum pressure of the water acting on the bottom of the container.

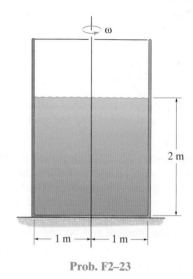

ω

2 m

1 m $\quad$ 1 m

Prob. F2–23

F2–24. The closed drum is filled with crude oil. Determine the pressure on the lid at A when the drum is spinning at a constant rate of 4 rad/s.

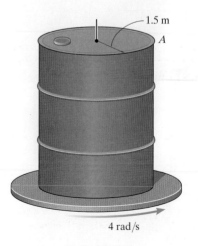

1.5 m

A

4 rad/s

Prob. F2–24

PROBLEMS

The answers to all but every fourth problem are given in the back of the book.

Unless otherwise stated, take the density of water to be $\rho_w = 1000 \text{ kg/m}^3$. Also, assume all pressures are gage pressures.

Sec. 2.1–2.5

2–1. Show that Pascal's law applies within a fluid that is accelerating, provided there is no shearing stresses acting within the fluid.

2–2. The oil derrick has drilled 5 km into the ground before it strikes a crude oil reservoir. When this happens, the pressure at the well head A becomes 25 MPa. Drilling "mud" is to be placed into the entire length of pipe to displace the oil and balance this pressure. What should be the density of the mud so that the pressure at A becomes zero?

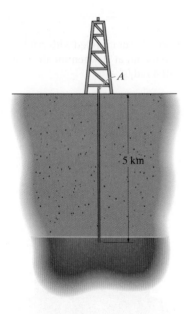

Prob. 2–2

2–3. The pipes connected to the closed tank are completely filled with water. If the absolute pressure at A is 300 kPa, determine the force acting on the inside of the end caps at B and C if the pipe has an inner diameter of 60 mm.

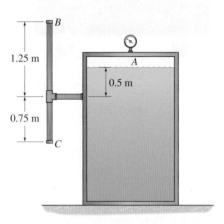

Prob. 2–3

*2–4. The water in a lake has an average temperature of 15°C. If the barometric pressure of the atmosphere is 720 mm of Hg (mercury), determine the gage pressure and the absolute pressure at a water depth of 14 m.

2–5. If the absolute pressure in a tank is 140 kPa, determine the pressure head in mm of mercury. The atmospheric pressure is 100 kPa.

2–6. In 1896, S. Riva-Rocci developed the prototype of the current sphygmomanometer, a device used to measure blood pressure. When it was worn as a cuff around the upper arm and inflated, the air pressure within the cuff was connected to a mercury manometer. If the reading for the high (or systolic) pressure is 120 mm and for the low (or diastolic) pressure is 80 mm, determine these pressures in pascals.

2–7. The structure shown is used for the temporary storage of crude oil at sea for later loading into ships. When it is not filled with oil, the water level in the stem is at B (sea level). Why? As the oil is loaded into the stem, the water is displaced through exit ports at E. If the stem is filled with oil, that is, to the depth of C, determine the height h of the oil level above sea level. Take $\rho_o = 900 \text{ kg/m}^3$, $\rho_w = 1020 \text{ kg/m}^3$.

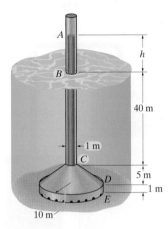

Prob. 2–7

*2–8. The Burj Khalifa in Dubai is currently the world's tallest building. If air at 40°C is at an atmospheric pressure of 105 kPa at the ground floor (sea level), determine the absolute pressure at the top of the building, which has an elevation of 828 m. Assume that the temperature is constant and that air is compressible. Work the problem again assuming that air is incompressible.

2–9. The density ρ of a fluid varies with depth h, although its bulk modulus E_V can be assumed constant. Determine how the pressure varies with h. The density at the surface of the fluid is ρ_0.

2–10. Due to its slight compressibility, the density of water varies with depth, although its bulk modulus $E_V = 2.20$ GPa (absolute) can be considered constant. Accounting for this compressibility, determine the pressure in the water at a depth of 300 m, if the density at the surface of the water is $\rho_0 = 1000 \text{ kg/m}^3$. Compare this result with assuming water to be incompressible.

2–11. In the troposphere, the absolute temperature of the air varies with elevation such that $T = T_0 - Cz$, where C is a constant. If $p = p_0$ at $z = 0$, determine the absolute pressure as a function of elevation.

*2–12. In the troposphere the absolute temperature of the air varies with elevation such that $T = T_0 - Cz$, where C is a constant. Using Fig. 2–11, determine the constants, T_0 and C. If $p_0 = 101$ kPa at $z_0 = 0$, determine the absolute pressure in the air at an elevation of 5 km.

2–13. The density of a nonhomogeneous liquid varies as a function of depth h, such that $\rho = (850 + 0.2h) \text{ kg/m}^3$, where h is in meters. Determine the pressure when $h = 20$ m.

2–14. The underground storage tank used in a service station contains gasoline filled to the level A. Determine the pressure at each of the five identified points. Note that point B is located in the stem, and point C is just below it in the tank. Take $\rho_g = 730 \text{ kg/m}^3$.

2–15. The underground storage tank contains gasoline filled to the level A. If the atmospheric pressure is 101.3 kPa, determine the absolute pressure at each of the five identified points. Note that point B is located in the stem, and point C is just below it in the tank. Take $\rho_g = 730 \text{ kg/m}^3$.

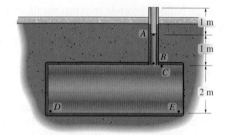

Probs. 2–14/15

*2–16. If the water in the structure in Prob. 2–7 is displaced with crude oil to the level D at the bottom of the cone, then how high h will the oil extend above sea level? Take $\rho_o = 900 \text{ kg/m}^3$, $\rho_w = 1020 \text{ kg/m}^3$.

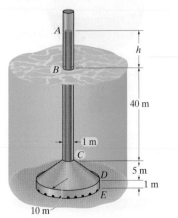

Prob. 2–16

2–17. In the troposphere, which extends from sea level to 11 km, it is found that the temperature decreases with altitude such that $dT/dz = -C$, where C is the constant lapse rate. If the temperature and pressure at $z = 0$ are T_0 and p_0, determine the pressure as a function of altitude.

2–18. At the bottom of the stratosphere the temperature is assumed to remain constant at $T = T_0$. If the pressure is $p = p_0$ where the elevation is $z = z_0$, derive an expression for the pressure as a function of elevation.

2–19. Determine the pressure at an elevation of $z = 20$ km in the stratosphere if the temperature remains constant at $T_0 = -56.5°C$. Assume the stratosphere begins at $z = 11$ km as shown in Fig. 2–11.

*2–20. The density of a nonhomogeneous liquid varies as a function of depth h, such that $\rho = (635 + 60h)$ kg/m³, where h is in meters. Plot the variation of the pressure (vertical axis) versus depth for $0 \le h < 10$ m. Give values for increments of 2 m.

2–21. As the balloon ascends, measurements indicate that the temperature begins to decrease at a constant rate, from $T = 20°C$ at $z = 0$ to $T = 16°C$ at $z = 500$ m. If the absolute pressure and the density of the air at $z = 0$ are $p = 101$ kPa and $\rho = 1.202$ kg/m³, determine these values at $z = 500$ m.

2–22. As the balloon ascends, measurements indicate that the temperature begins to decrease at a constant rate, from $T = 20°C$ at $z = 0$ to $T = 16°C$ at $z = 500$ m. If the absolute pressure of the air at $z = 0$ is $p = 101$ kPa, plot the variation of pressure (vertical axis) versus altitude for $0 \le z \le 3000$ m. Give values for increments of $u\ \Delta z = 500$ m.

Probs. 2–21/22

Sec. 2.6

2–23. The pressure in the tank at the closed valve A is 300 kPa. If the differential elevation in the oil level in $h = 2.5$ m, determine the pressure in the pipe at B. Take $\rho_o = 900$ kg/m³.

*2–24. The pressure in the tank at B is 600 kPa. If the differential elevation of the oil is $h = 2.25$ m, determine the pressure at the closed valve A. Take $\rho_o = 900$ kg/m³.

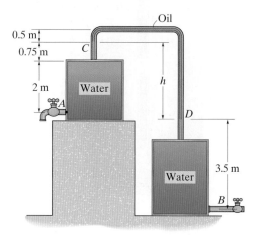

Probs. 2–23/24

2–25. The inverted U-tube manometer is used to measure the difference in pressure between water flowing in the pipes at A and B. If the top segment is filled with air, and the water levels in each segment are as indicated, determine this pressure difference between A and B. $\rho_w = 1000$ kg/m³.

2–26. Solve Prob. 2–25 if the top segment is filled with an oil for which $\rho_o = 800$ kg/m³.

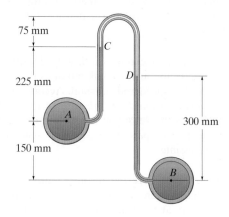

Probs. 2–25/26

2–27. The funnel is filled with oil and water to the levels shown. Determine the depth of oil h' that must be in the funnel so that the water remains at a depth C, and the mercury is at $h = 0.8$ m from the top of the funnel. Take $\rho_o = 900$ kg/m³, $\rho_w = 1000$ kg/m³, $\rho_{Hg} = 13\,550$ kg/m³.

*2–28. The funnel is filled with oil to a depth of $h' = 0.3$ m and water to a depth of 0.4 m. Determine the distance h the mercury level is from the top of the funnel. Take $\rho_o = 900$ kg/m³, $\rho_w = 1000$ kg/m³, $\rho_{Hg} = 13\,550$ kg/m³.

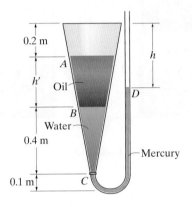

Probs. 2–27/28

2–29. Determine the pressures at points A and B. The containers are filled with water.

2–30. Determine the pressure at point C. The containers are filled with water.

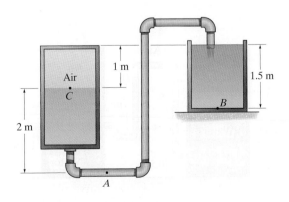

Probs. 2–29/30

2–31. The two pipes contain hexylene glycol, which causes the level of mercury in the manometer to be at $h = 0.3$ m. Determine the differential pressure in the pipes, $p_A - p_B$. Take $\rho_{hgl} = 923$ kg/m³, $\rho_{Hg} = 13\,550$ kg/m³. Neglect the diameter of the pipes.

*2–32. The two pipes contain hexylene glycol, which causes the differential pressure reading of the mercury in the manometer to be at $h = 0.3$ m. If the pressure in pipe A increases by 6 kPa, and the pressure in pipe B decreases by 2 kPa, determine the new differential reading h of the manometer. Take $\rho_{hgl} = 923$ kg/m³, $\rho_{Hg} = 13\,550$ kg/m³. Neglect the diameter of the pipes.

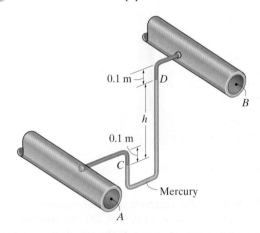

Probs. 2–31/32

2–33. Butyl carbitol, used in the production of plastics, is stored in a tank having the U-tube manometer. If the U-tube is filled with mercury to level E, determine the pressure in the tank at point A. Take $S_{Hg} = 13.55$, and $S_{bc} = 0.957$.

2–34. Butyl carbitol, used in the production of plastics, is stored in a tank having the U-tube manometer. If the U-tube is filled with mercury to level E, determine the pressure in the tank at point B. Take $S_{Hg} = 13.55$, and $S_{bc} = 0.957$.

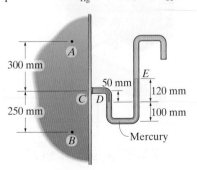

Probs. 2–33/34

2–35. The 150-mm-diameter container is filled to the top with glycerin, and a 50-mm-diameter thin pipe is inserted within it to a depth of 300 mm. If 0.00075 m³ of kerosene is then poured into the pipe, determine the height h to which the kerosene rises from the top of the glycerin.

*2–36. The 150-mm-diameter container is filled to the top with glycerin, and a 50-mm-diameter thin pipe is inserted within it to a depth of 300 mm. Determine the maximum volume of kerosene that can be poured into the pipe so it does not come out from the bottom end. How high h does the kerosene rise above the glycerin?

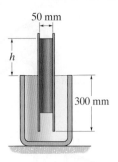

Probs. 2–35/36

2–37. Water in the reservoir is used to control the water pressure in the pipe at A. If $h = 200$ mm, determine this pressure when the mercury is at the elevation shown. Take $\rho_{Hg} = 13\ 550\ \text{kg/m}^3$. Neglect the diameter of the pipe.

2–38. If the water pressure in the pipe at A is to be 25 kPa, determine the required height h of water in the reservoir. Mercury in the pipe has the elevation shown. Take $\rho_{Hg} = 13\ 550\ \text{kg/m}^3$. Neglect the diameter of the pipe.

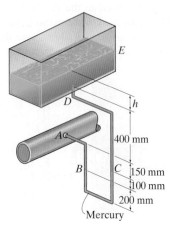

Probs. 2–37/38

2–39. The two tanks A and B are connected using a manometer. If waste oil is poured into tank A to a depth of $h = 0.6$ m, determine the pressure of the entrapped air in tank B. Air is also trapped in line CD as shown. Take $\rho_o = 900\ \text{kg/m}^3$, $\rho_w = 1000\ \text{kg/m}^3$.

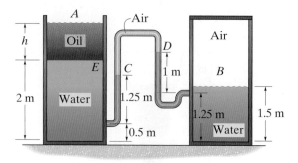

Prob. 2–39

*2–40. The two tanks A and B are connected using a manometer. If waste oil is poured into tank A to a depth of $h = 1.25$ m, determine the pressure of the trapped air in tank B. Air is also trapped in line CD as shown. Take $\rho_o = 900\ \text{kg/m}^3$, $\rho_w = 1000\ \text{kg/m}^3$.

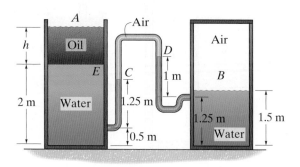

Prob. 2–40

2–41. The *micro-manometer* is used to measure small differences in pressure. The reservoirs R and upper portion of the lower tubes are filled with a liquid having a specific weight of γ_R, whereas the lower portion is filled with a liquid having a specific weight of γ_t, Fig. (*a*). When the liquid flows through the venturi meter, the levels of the liquids with respect to the original levels are shown in Fig. (*b*). If the cross-sectional area of each reservoir is A_R and the cross-sectional area of the U-tube is A_t, determine the pressure difference $p_A - p_B$. The liquid in the venturi meter has a specific weight of γ_L.

2–43. Trichlorethylene, flowing through both pipes, is to be added to jet fuel produced in a refinery. A careful monitoring of pressure is required through the use of the inclined-tube manometer. If the pressure at A is 200 kPa and the pressure at B is 170 kPa, determine the position s that defines the level of mercury in the inclined-tube manometer. Take $S_{Hg} = 13.55$ and $S_t = 1.466$. Neglect the diameter of the pipes.

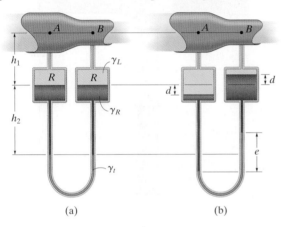

(a)   (b)

Prob. 2–41

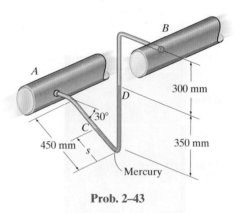

Prob. 2–43

2–42. The Morgan Company manufactures a micro-manometer that works on the principles shown. Here there are two reservoirs filled with kerosene, each having a cross-sectional area of 300 mm². The connecting tube has a cross-sectional area of 15 mm² and contains mercury. Determine h if the pressure difference $p_A - p_B = 40$ Pa. $\rho_{Hg} = 13\,550$ kg/m³, $\rho_{ke} = 814$ kg/m³. *Hint:* Both h_1 and h_2 can be eliminated from the analysis.

2–44. Determine the difference in pressure $p_B - p_A$ between the centers A and B of the pipes, which are filled with water. The mercury in the inclined-tube manometer has the level shown $S_{Hg} = 13.55$.

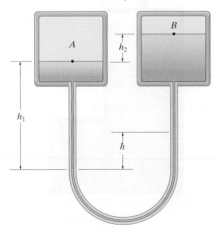

Prob. 2–42

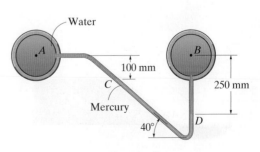

Prob. 2–44

Sec. 2.7–2.9

2–45. The storage tank contains oil and water acting at the depths shown. Determine the resultant force that both of these liquids exert on the side ABC of the tank if the side has a width of $b = 1.25$ m. Also, determine the location of this resultant, measured from the top of the tank. Take $\rho_o = 900$ kg/m^3.

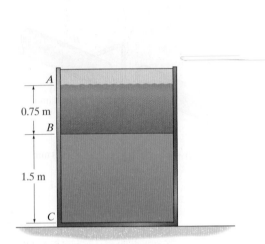

Prob. 2–45

2–46. The uniform swamp gate has a mass of 4 Mg and a width of 1.5 m. Determine the angle θ for equilibrium if the water rises to a depth of $d = 1.5$ m.

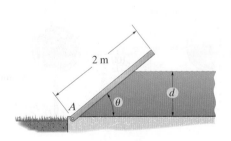

Prob. 2–46

2–47. The vertical pipe segment has an inner diameter of 100 mm and is capped at its end and suspended from the horizontal pipe as shown. If it is filled with water and the pressure at A is 80 kPa, determine the resultant force that must be resisted by the bolts at B in order to hold the flanges together. Neglect the weight of the pipe but not the water within it.

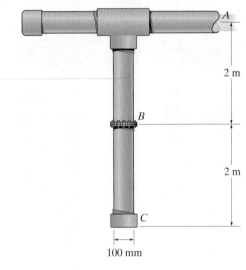

Prob. 2–47

*2–48. The pressure of the air at A within the closed tank is 200 kPa. Determine the resultant force acting on the plates BC and CD caused by the water. The tank has a width of 1.75 m.

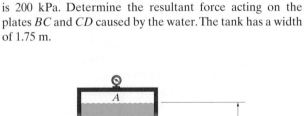

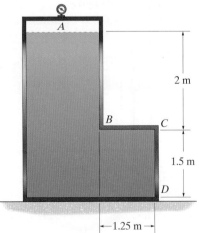

Prob. 2–48

2–49. The tide gate opens automatically when the tide water at *B* subsides, allowing the marsh at *A* to drain. For the water level *h* = 4 m, determine the horizontal reaction at the smooth stop *C*. The gate has a width of 2 m. At what height *h* will the gate be on the verge of opening?

2–50. The tide gate opens automatically when the tide water at *B* subsides, allowing the marsh at *A* to drain. Determine the horizontal reaction at the smooth stop *C* as a function of the depth *h* of the water level. Starting at *h* = 6 m, plot values of *h* for each increment of 0.5 m until the gate begins to open. The gate has a width of 2 m.

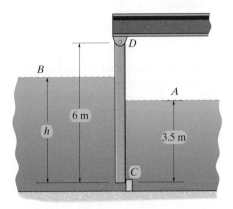

Probs. 2–49/50

2–51. Determine the smallest base length *b* of the concrete gravity dam that will prevent the dam from overturning due to water pressure acting on the face of the dam. The density of concrete is ρ_c = 2.4 Mg/m³. *Hint:* Work the problem using a 1-m width of dam.

*2–52. Determine the smallest base length *b* of the concrete gravity dam that will prevent the dam from overturning due to water pressure acting on the face of the dam. Assume water also seeps under the base of the dam. The density of concrete is ρ_c = 2.4 Mg/m³. *Hint:* Work the problem using a 1-m width of dam.

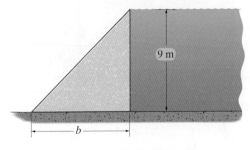

Probs. 2–51/52

2–53. The container in a chemical plant contains carbon tetrachloride, ρ_{ct} = 1593 kg/m³, and benzene, ρ_b = 875 kg/m³. Determine the height *h* of the carbon tetrachloride on the left side so that the separation plate, which is pinned at *A*, will remain vertical.

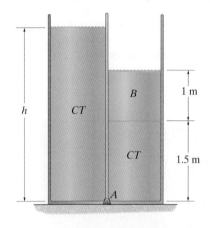

Prob. 2–53

2–54. The tapered settling tank is filled with oil. Determine the resultant force the oil exerts on the trapezoidal clean-out plate located at its end. How far from the top of the tank does this force act on the plate? Use the formula method. Take ρ_o = 900 kg/m³.

2–55. The tapered settling tank is filled with oil. Determine the resultant force the oil exerts on the trapezoidal clean-out plate located at its end. How far from the top of the tank does this force act on the plate? Use the integration method. Take ρ_o = 900 kg/m³.

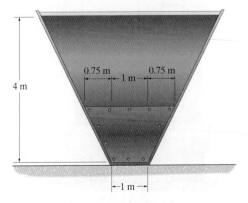

Probs. 2–54/55

*2–56. Access plates on the industrial holding tank are bolted shut when the tank is filled with vegetable oil as shown. Determine the resultant force that this liquid exerts on plate A, and its location measured from the bottom of the tank. Use the formula method. $\rho_{vo} = 932 \text{ kg/m}^3$.

2–57. Access plates on the industrial holding tank are bolted shut when the tank is filled with vegetable oil as shown. Determine the resultant force that this liquid exerts on plate B, and its location measured from the bottom of the tank. Use the formula method. $\rho_{vo} = 932 \text{ kg/m}^3$.

2–58. Solve Prob. 2–57 using the integration method.

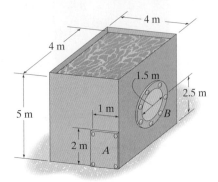

Probs. 2– 56/57/58

2–59. The tank truck is filled to its top with water. Determine the magnitude of the resultant force on the elliptical back plate of the tank, and the location of the center of pressure measured from the top of the tank. Solve the problem using the formula method.

*2–60. Solve Prob. 2–59 using the integration method.

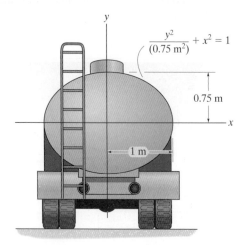

Probs. 2–59/60

2–61. The uniform swamp gate has a mass of 3 Mg and a width of 1.5 m. Determine the depth of the water d if the gate is held in equilibrium at an angle of $\theta = 60°$.

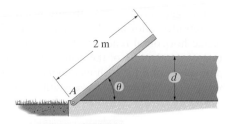

Prob. 2–61

2–62. The tank is filled with water. Determine the resultant force acting on the triangular plate A and the location of the center of pressure, measured from the top of the tank. Solve the problem using the formula method.

2–63. Solve Prob. 2–62 using the integration method.

*2–64. The tank is filled with water. Determine the resultant force acting on the semicircular plate B and the location of the center of pressure, measured from the top of the tank. Solve the problem using the formula method.

2–65. Solve Prob. 2–64 using the integration method.

2–66. The tank is filled with water. Determine the resultant force acting on the trapezoidal plate C and the location of the center of pressure, measured from the top of the tank. Solve the problem using the formula method.

2–67. Solve Prob. 2–66 using the integration method.

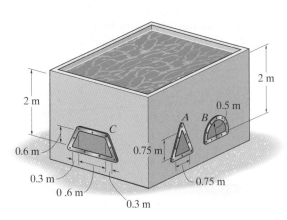

Probs. 2–62/63/64/65/66/67

*2–68. Ethyl alcohol is pumped into the tank, which has the shape of a four-sided pyramid. When the tank is completely full, determine the resultant force acting on each side, and its location measured from the top A along the side. Use the formula method. $\rho_{ea} = 789 \text{ kg/m}^3$.

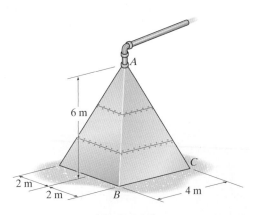

Prob. 2–68

2–69. The tank truck is half filled with water. Determine the magnitude of the resultant force on the elliptical back plate of the tank, and the location of the center of pressure measured from the x axis. Solve the problem using the formula method. *Hint*: The centroid of a semi-ellipse measured from the x axis is $\bar{y} = \frac{4b}{3\pi}$.

2–70. Solve Prob. 2–69 using the integration method.

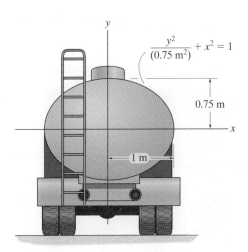

Probs. 2–69/70

Sec. 2.10

2–71. Water is confined in the vertical chamber, which is 2 m wide. Determine the resultant force it exerts on the arched roof AB.

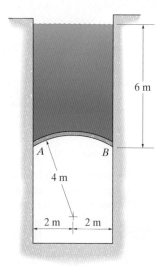

Prob. 2–71

*2–72. The settling tank is 3 m wide and contains turpentine having a density of 860 kg/m^3. If the parabolic shape is defined by $y = (x^2) \text{ m}$, determine the magnitude and direction of the resultant force the turpentine exerts on the side AB of the tank.

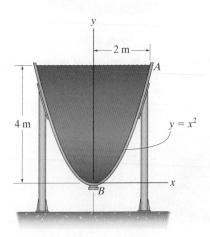

Prob. 2–72

2–73. The bent plate is 1.5 m wide and is pinned at A and rests on a smooth support at B. Determine the horizontal and vertical components of reaction at A and the vertical reaction at the smooth support B for equilibrium. The fluid is water.

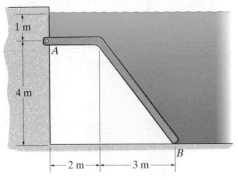

Prob. 2–73

2–74. Determine the horizontal and vertical components of reaction at the hinge A and the normal reaction at B caused by the water pressure. The gate has a width of 3 m.

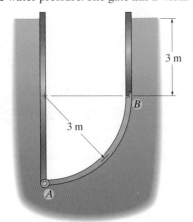

Prob. 2–74

2–75. Determine the magnitude of the resultant force the water exerts on the curved vertical wall. The wall is 2 m wide.

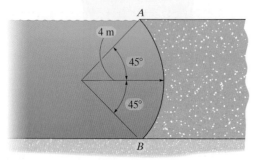

Prob. 2–75

*2–76. The 5-m-wide wall is in the form of a parabola. If the depth of the water is $h = 4$ m, determine the magnitude and direction of the resultant force on the wall.

2–77. The 5-m-wide wall is in the form of a parabola. Determine the magnitude of the resultant force on the wall as a function of depth h of the water. Plot the results of force (vertical axis) versus depth h for $0 \le h \le 4$ m. Give values for increments of $\Delta h = 0.5$ m.

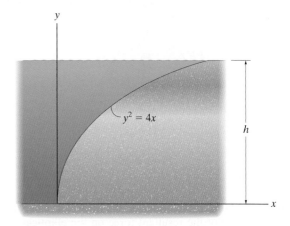

Probs. 2–76/77

2–78. The 5-m-wide overhang is in the form of a parabola, as shown. Determine the magnitude and direction of the resultant force on the overhang.

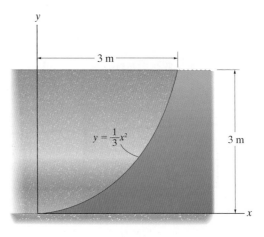

Prob. 2–78

2–79. If the water depth is $h = 2\,\text{m}$, determine the magnitude and direction of the resultant force, due to water pressure acting on the parabolic surface of the dam, which has a width of 5 m.

*2–80. Determine the magnitude of the resultant force due to water pressure acting on the parabolic surface of the dam as a function of the depth h of the water. Plot the results of force (vertical axis) versus depth h for $0 \le h \le 2\,\text{m}$. Give values for increments of $\Delta h = 0.5\,\text{m}$. The dam has a width of 5 m.

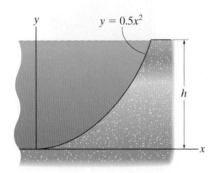

$$y = 0.5x^2$$

Probs. 2–79/80

2–81. Determine the resultant force that water exerts on the overhang sea wall along ABC. The wall is 2 m wide.

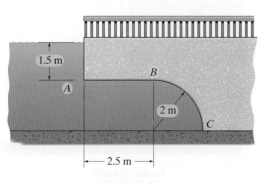

Prob. 2–81

2–82. Determine the magnitude and direction of the resultant hydrostatic force the water exerts on the face AB of the overhang if it is 2 m wide.

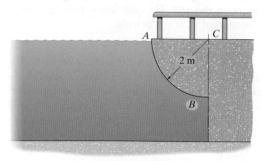

Prob. 2–82

2–83. Determine the resultant force the water exerts on AB, BC, and CD of the enclosure, which is 3 m wide.

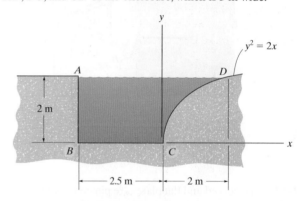

$$y^2 = 2x$$

Prob. 2–83

*2–84. Gate AB has a width of 0.5 m and a radius of 1 m. Determine the horizontal and vertical components of reaction at the pin A and the horizontal reaction at the smooth stop B due to the water pressure.

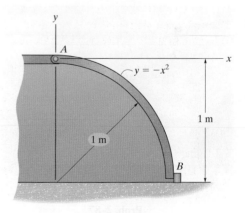

$$y = -x^2$$

Prob. 2–84

2

2–85. The Tainter gate is used to control the flow of water over a spillway. If the water is at its highest level as shown, determine the torque **T** that must be applied at the pin A in order to open the gate. The gate has a mass of 5 Mg and a center of mass at G. It is 3 m wide.

2–86. The Tainter gate is used to control the flow of water over a spillway. If the water is at its highest level as shown, determine the horizontal and vertical components of reaction at pin A and the vertical reaction at the spillway crest B. The gate has a mass of 5 Mg and a center of mass at G. It is 3 m wide. Take $T = 0$.

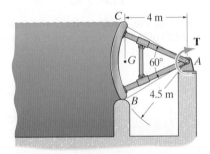

Probs. 2–85/86

2–87. The curved and flat plates are pin connected at A, B, and C. They are submerged in water at the depth shown. Determine the horizontal and vertical components of reaction at pin B. The plates have a width of 4 m.

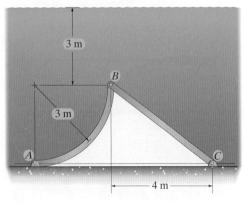

Prob. 2–87

*2–88. The stopper in the shape of a frustum is used to plug the 100-mm-diameter hole in the tank that contains amyl acetate. If the greatest vertical force the stopper can resist is 100 N, determine the depth d before it becomes unplugged. Take $\rho_{aa} = 863 \text{ kg/m}^3$. *Hint:* The volume of a cone is $V = \frac{1}{3}\pi r^2 h$.

2–89. The stopper in the shape of a frustum is used to plug the 100-mm-diameter hole in the tank that contains amyl acetate. Determine the vertical force this liquid exerts on the stopper. Take $d = 0.6 \text{ m}$ and $\rho_{aa} = 863 \text{ kg/m}^3$. *Hint:* The volume of a cone is $V = \frac{1}{3}\pi r^2 h$.

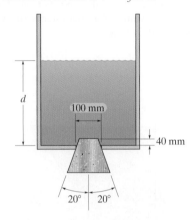

Probs. 2– 88/89

2–90. The Tainter gate for a water channel is 1.5 m wide and in the closed position, as shown. Determine the magnitude of the resultant force of the water acting on the gate. Solve the problem by considering the fluid acting on the horizontal and vertical projections of the gate. Determine the smallest torque **T** that must be applied to open the gate if its weight is 30 kN and its center of gravity is at G.

2–91. Solve the first part of Prob. 2–90 by the integration method using polar coordinates.

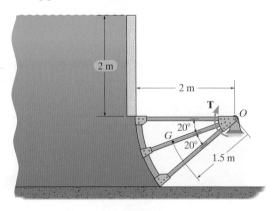

Probs. 2–90/91

Sec. 2.11–2.12

*2–92. Consider an iceberg to be in the form of a cylinder of arbitrary diameter and floating in the ocean as shown. If the cylinder extends 2 m above the ocean's surface, determine the depth of the cylinder below the surface. The density of ocean water is $\rho_w = 1024$ kg/m³, and the density of the ice is $\rho_i = 935$ kg/m³.

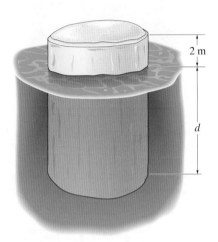

Prob. 2–92

2–93. A flat-bottomed boat has vertical sides and a bottom surface area of 0.75 m². It floats in water such that its draft (depth below the surface) is 0.3 m. Determine the mass of the boat. What is the draft when a 50-kg man stands in the center of the boat?

2–94. The raft consists of a uniform platform having a mass of 2 Mg and four floats, each having a mass of 120 kg and a length of 4 m. Determine the height h at which the platform floats from the water surface. Take $\rho_w = 1$ Mg/m³.

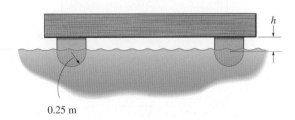

Prob. 2–94

2–95. The cone is made of wood having a density of $\rho_{wood} = 650$ kg/m³. Determine the tension in rope AB if the cone is submerged in the water at the depth shown. Will this force increase, decrease, or remain the same if the cord is shortened? Why? *Hint:* The volume of a cone is $V = \frac{1}{3}\pi r^2 h$.

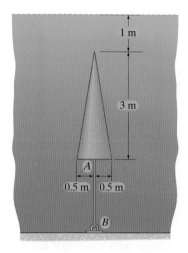

Prob. 2–95

*2–96. The cylinder has a diameter of 75 mm and a mass of 600 g. If it is placed in the tank, which contains oil and water, determine the height h above the surface of the oil at which it will float if maintained in the vertical position. Take $\rho_o = 980$ kg/m³.

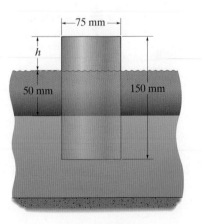

Prob. 2–96

2–97. An open-ended tube having an inner radius r is placed in a wetting liquid A having a density ρ_A. The top of the tube is just below the surface of a surrounding liquid B, which has a density ρ_B, where $\rho_A > \rho_B$. If the surface tension σ causes liquid A to make a wetting angle θ with the tube wall as shown, determine the rise h of liquid A within the tube. Show that the result is independent of the depth d of liquid B.

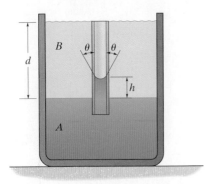

Prob. 2–97

2–98. The container with water in it has a mass of 30 kg. Block B has a density of 8500 kg/m³ and a mass of 15 kg. If springs C and D have an unstretched length of 200 mm and 300 mm, respectively, determine the length of each spring when the block is submerged in the water.

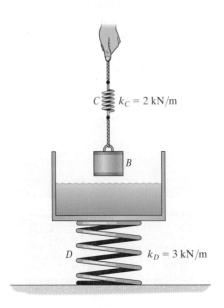

Prob. 2–98

2–99. The cylinder floats in the water and oil to the level shown. Determine the weight of the cylinder. $\rho_o = 910$ kg/m³.

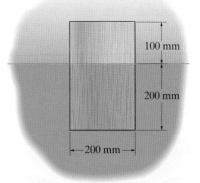

Prob. 2–99

*2–100. A glass having a diameter of 50 mm is filled with water to the level shown. If an ice cube with 25-mm sides is placed into the glass, determine the new height h of the water surface. Take $\rho_w = 1000$ kg/m³ and $\rho_{ice} = 920$ kg/m³. What will the water level h be when the ice cube completely melts?

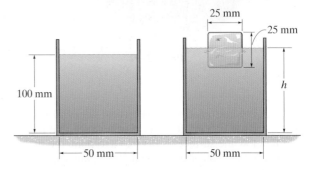

Prob. 2–100

2–101. When loaded with gravel, the barge floats in water at the depth shown. If its center of gravity is located at G determine whether the barge will restore itself when a wave causes it to tip slightly.

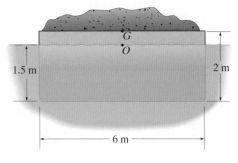

Prob. 2–101

2–102. When loaded with gravel, the barge floats in water at the depth shown. If its center of gravity is located at G, determine whether the barge will restore itself when a wave causes it to roll slightly at 9°.

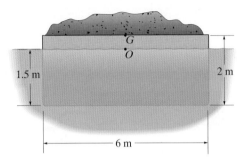

Prob. 2–102

2–103. A boat having a mass of 80 Mg rests on the bottom of the lake and displaces 10.25 m³ of water. Since the lifting capacity of the crane is only 600 kN, two balloons are attached to the sides of the boat and filled with air. Determine the smallest radius r of each spherical balloon that is needed to lift the boat. What is the mass of air in each balloon if the air and water temperature is 12°C? The balloons are at an average depth of 20 m. Neglect the mass of air and of the balloon for the calculation required for the lift. The volume of a sphere is $V = \frac{4}{3}\pi r^3$.

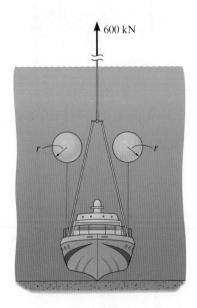

Prob. 2–103

Sec. 2.13–2.14

*2–104. The barrel of oil rests on the surface of the scissors lift. Determine the maximum pressure developed in the oil if the lift is moving upward with (a) a constant velocity of 4 m/s, and (b) an acceleration of 2 m/s². Take $\rho_o = 900$ kg/m³. The top of the barrel is open to the atmosphere.

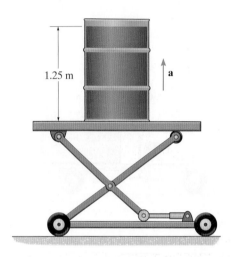

Prob. 2–104

2–105. The closed cylindrical tank is filled with milk, for which $\rho_m = 1030$ kg/m³. If the inner diameter of the tank is 1.5 m, determine the difference in pressure within the tank between corners A and B when the truck accelerates at 0.8 m/s².

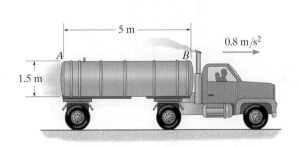

Prob. 2–105

2–106. A large container of benzene is transported on the truck. Determine the level in each of the vent tubes A and B if the truck accelerates at $a = 1.5$ m/s². When the truck is at rest, $h_A = h_B = 0.4$ m.

2–107. A large container of benzene is being transported by the truck. Determine its maximum constant acceleration so that no benzene will spill from the vent tubes A or B. When the truck is rest, $h_A = h_B = 0.4$ m.

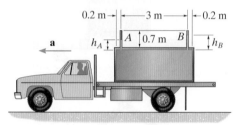

Probs. 2–106/107

*2–108. The truck carries an open container of water as shown. If it has a constant acceleration of 2 m/s², determine the angle of inclination of the surface of the water and the pressure at the bottom corners A and B.

2–109. The truck carries an open container of water as shown. Determine the maximum constant acceleration it can have without causing the water to spill out of the container.

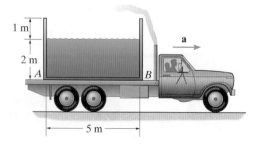

Probs. 2–108/109

2–110. The fuel tank, supply line, and engine for an airplane are shown. If the gas tank is filled to the level shown, determine the largest constant acceleration a that the plane can have without causing the engine to be starved of fuel. The plane is accelerating to the right for this to happen. Suggest a safer location for attaching the fuel line.

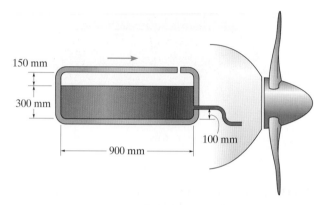

Prob. 2–110

2–111. The cart is given a constant acceleration **a** up the plane, as shown. Show that the lines of constant pressure *within* the liquid have a slope of $\tan\theta = (a\cos\phi)/(a\sin\phi + g)$.

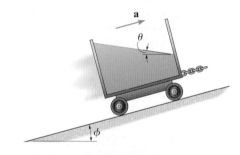

Prob. 2–111

*2–112. The cart is allowed to roll freely down the inclined plane due to its weight. Show that the slope of the surface of the liquid, θ, during the motion is $\theta = \phi$.

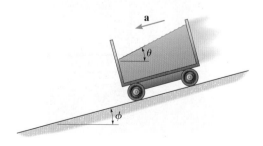

Prob. 2–112

2–113. If the truck has a constant acceleration of 2 m/s², determine the water pressure at the bottom corners B and C of the water tank. There is a small opening at A.

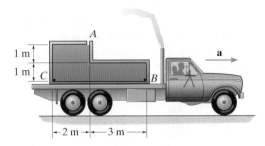

Prob. 2–113

2–114. If the truck has a constant acceleration of 2 m/s², determine the water pressure at the bottom corners A and B of the water tank.

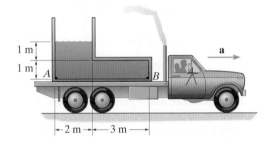

Prob. 2–114

2–115. The drum has a hole in the center of its lid and contains kerosene to a level of 400 mm when $\omega = 0$. If the drum is placed on the platform and it attains an angular velocity of 12 rad/s, determine the resultant force the kerosene exerts on the lid.

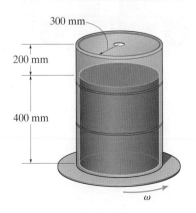

Prob. 2–115

*2–116. The sealed assembly is completely filled with water such that the pressures at C and D are zero. If the assembly is given an angular velocity of $\omega = 15$ rad/s, determine the difference in pressure between points C and D.

2–117. The sealed assembly is completely filled with water such that the pressures at C and D are zero. If the assembly is given an angular velocity of $\omega = 15$ rad/s, determine the difference in pressure between points A and B.

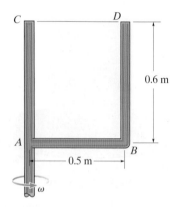

Probs. 2–116/117

2–118. The beaker is filled to a height of $h = 0.1$ m with kerosene and placed on the platform. What is the maximum angular velocity ω it can have so that no kerosene spills out of the beaker?

2–119. The beaker is filled to a height of $h = 0.1$ m with kerosene and placed on the platform. To what height $h = h'$ does the kerosene rise against the wall of the beaker when the platform has an angular velocity of $\omega = 15$ rad/s?

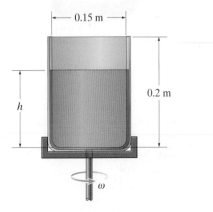

Probs. 2–118/119

*2–120. The drum is filled to the top with oil and placed on the platform. If the platform is given a rotation of $\omega = 12$ rad/s, determine the pressure the oil will exert on the cap at A. Take $\rho_o = 900$ kg/m³.

2–121. The drum is filled to the top with oil and placed on the platform. Determine the maximum rotation of the platform if the maximum pressure the cap at A can sustain before it opens is 40 kPa. Take $\rho_o = 900$ kg/m³.

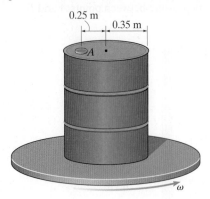

Probs. 2–120/121

2–122. The U-tube is filled with water and A is open while B is closed. If the axis of rotation is at $x = 0.2$ m, determine the constant rate of rotation so that the pressure at B is zero.

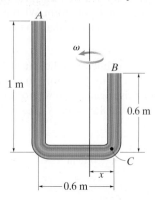

Prob. 2–122

2–123. The U-tube is filled with water and A is open while B is closed. If the axis of rotation is at $x = 0.2$ m and the tube is rotating at a constant rate of $\omega = 10$ rad/s, determine the pressure at points B and C.

*2–124. The U-tube is filled with water and A is open while B is closed. If the axis of rotation is at $x = 0.4$ m and the tube is rotating at a constant rate of $\omega = 10$ rad/s, determine the pressure at points B and C.

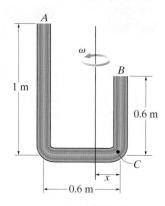

Probs. 2–123/124

2–125. Determine the water pressure at points B and C in the tank if the truck has a constant acceleration $a_c = 2$ m/s². When the truck is at rest, the water level in the vent tube A is at $h_A = 0.3$ m.

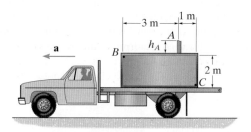

Prob. 2–125

CONCEPTUAL PROBLEMS

P2–1. By moving the handle up and down on this hand pump, one is able to pump water from a reservoir. Do some research to explain how the pump works, and show a calculation that indicates the maximum height to which it can lift a column of water.

P2–1

P2–2. In 1656 Otto Von Guericke placed the two halves of a 300-mm-diameter hollow sphere together and pumped the air out of the inside. He tied one rope at A to a tree and the other to a team of eight horses. Assuming a perfect vacuum was developed within the sphere, do you think the horses could pull the hemispheres apart? Explain. If he used sixteen horses, eight on each side, would this make a difference? Explain.

P2–2

P2–3. Ice floats in the glass when the glass is filled with water. Explain what happens to the water level when the ice melts. Does it go up, go down, or remain the same?

P2–3

P2–4. The beaker of water rests on the scale. Will the scale reading increase, decrease, or remain the same if you put your finger in the water? Explain.

P2–4

CHAPTER REVIEW

Pressure is a normal force acting per unit area. At a point in a fluid, it is the same in all directions. This is known as Pascal's law.		
The absolute pressure is equal to the atmospheric pressure plus the gage pressure.	$p_{abs} = p_{atm} + p_g$	
In a static fluid, the pressure is constant at points that lie in the same horizontal plane. If the fluid is incompressible, then the pressure depends on the specific weight of the fluid, and it increases linearly with depth.	$p = \gamma h$	
If the depth is not great, the pressure within a gas can be assumed constant.		
Atmospheric pressure is measured using a barometer.		
A manometer can be used to measure the gage pressure in a liquid. The pressure is determined by applying the manometer rule. The pressure can also be measured using other devices such as a Bourdon gage or a pressure transducer.		
The resultant hydrostatic force acting on a plane surface area has a magnitude of $F_R = \gamma \bar{h} A$, where $\bar{h}$ is the depth of the *centroid* of the area. The location of $\mathbf{F}_R$ is at the center of pressure $P(x_P, y_P)$.	$x_P = \bar{x} + \dfrac{\bar{I}_{xy}}{\bar{y}A}$ $y_P = \bar{y} + \dfrac{\bar{I}_x}{\bar{y}A}$	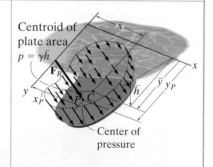 Centroid of plate area $p = \gamma h$ Center of pressure
The resultant hydrostatic force acting on a plane surface area can also be determined by finding the *volume* of its pressure prism. If the surface has a constant width, one can then view the pressure prism perpendicular to its width and find the *area* of the load distribution that is caused by the pressure. The resultant force acts through the centroid of the volume or area.		
Direct integration of the pressure distribution can also be used to determine the resultant force and its location on a plane surface area.		

If the surface is inclined or curved, the resultant hydrostatic force can be determined by first finding its horizontal and vertical components.

The *horizontal component* is found by projecting the surface onto the vertical plane and finding the force acting on this projected area.

The *vertical component* is equal to the weight of the volume of liquid above the inclined or curved surface. If the liquid is *below* this surface, then the weight of imaginary liquid *above* the surface is determined. The vertical component then acts *upward* on the surface because it represents the equivalent pressure force of the liquid below the surface.

The principle of buoyancy states that the buoyant force acting on a body immersed in a fluid is equal to the weight of fluid displaced by the body.

A floating body can be in stable, unstable, or neutral equilibrium. The body will be stable if its metacenter is located above its center of gravity.

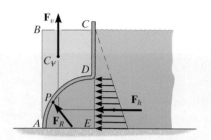

F_h = the resultant pressure loading on the vertical projected area DE

F_v = the weight of the volume of liquid $ADCBA$ above the plate

If an open container of liquid has a constant horizontal acceleration a_c, the surface of the liquid will be inclined at an angle given by $\tan \theta = a_c/g$. If a lid is on the container, then an imaginary liquid surface should be established. In either case, the pressure at any point in the liquid is determined from $p = \gamma h$, where h is the depth, measured from the liquid surface.

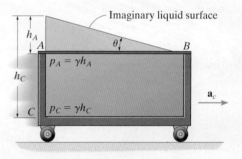

If a container of liquid has a constant vertical upward acceleration a_c then the pressure within the liquid at a depth h will be increased by $\gamma h(a_c/g)$. It is decreased by this amount if the acceleration is downward.

If a container has a constant rotation about a fixed axis, the liquid surface will form a forced vortex having the shape of a paraboloid. The surface is defined by $h = (\omega^2/2g)r^2$. If a lid is on the container, then an imaginary liquid surface can be established. The pressure at any point within the liquid is then determined from $p = \gamma h$, where h is the depth measured from the liquid surface.

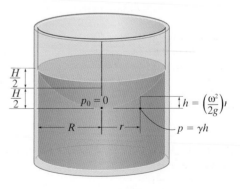

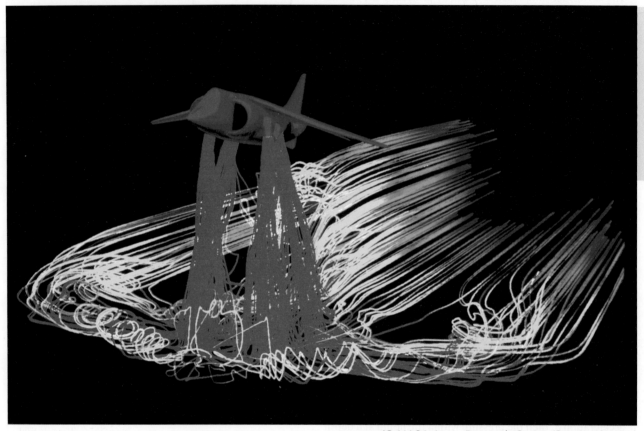

Chapter 3

(© NASA Ames Research Center/Science Source)

Exhaust flow through the engines of this jet is modeled using a computer
program involving computational fluid dynamics.

Kinematics of Fluid Motion

3.1   Fluid Flow Descriptions

In most cases fluids do not remain static but rather they flow. In this chapter we will consider the kinematics of this flow. Specifically, *kinematics* is the study of the geometry of the flow; that is, it provides a description of the position, velocity, and acceleration of a system of fluid particles. A *system* actually consists of a specific quantity of the fluid that is enclosed within a region of space, apart from the fluid outside this region, which is called the *surroundings*, Fig. 3–1a.

It is very important to be able to know the kinematics of a fluid flow because once the flow pattern is established, the pressure or forces that act on a structure or machine submerged in the fluid can then be determined. To completely define the flow pattern, it is necessary to specify the velocity and acceleration of each fluid particle at *each point* within the system, and *at each instant* of time. In fluid mechanics there are two ways of doing this.

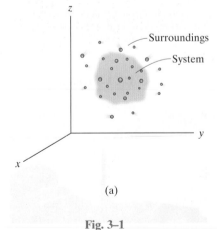

(a)

Fig. 3–1

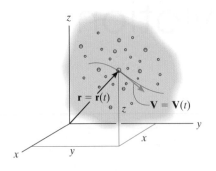

Lagrangian description of motion
follows a *single* fluid particle as it
moves about within the system

(b)

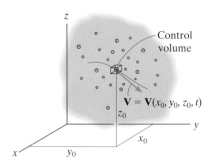

Eulerian description of motion specifies
a point or region within the system, and it
measures the velocity of the particles that
pass through this point or control volume

(c)

Fig. 3–1 (cont.)

Lagrangian description: Particles of smoke
from this stack can be tagged, and the motion
of each is measured from a common origin.

Eulerian description: A control volume is set
up at a specific point, and the motion of
particles passing through it is measured.

Lagrangian Description—System Approach.

The flow within a fluid system can be defined by "tagging" *each fluid particle*, and then specifying its velocity and acceleration as a function of time as the particle moves from one position to the next. This method is typically used in particle dynamics, and it is referred to as a ***Lagrangian description***, named after the Italian mathematician Joseph Lagrange.

If the position of a fluid particle is specified by a position vector **r**, Fig. 3–1b, then **r** will be a function of time, and so its time derivative yields the velocity of the particle, that is,

$$\mathbf{V} = \mathbf{V}(t) = \frac{d\mathbf{r}(t)}{dt} \qquad (3\text{--}1)$$

Here the velocity is *only a function of time*. In other words, the motion is *measured at the particle* and is calculated as the *time rate of change of the particle's position*. The velocity is *not* a function of the particle's position; rather, the position itself is known as a function of time, $\mathbf{r} = \mathbf{r}(t)$, Fig. 3–1b.

Eulerian Description—Control Volume Approach.

The velocity of the fluid particles within a system can also be described by considering a fixed point (x_0, y_0, z_0) surrounded by a differential volume of space. The velocity of all particles that pass through this point or volume can then be measured at this point, Fig. 3–1c. This method is named after the Swiss mathematician Leonard Euler, and is referred to as a ***Eulerian description***.

The volume of space through which the particles flow is called a ***control volume***, and the boundary of this volume is the ***control surface***. To obtain information on the *entire system*, differential-size control volumes must be established at *every point* (x, y, z), and the velocities of particles passing through these control volumes are then measured as time passes. Doing this, we then have a *velocity field* for the system that is defined as a function of *both* space, where each control volume is located, and time. That is,

$$\mathbf{V} = \mathbf{V}(x, y, z, t) \qquad (3\text{--}2)$$

This "vector field" can also be expressed in terms of its Cartesian components.

$$\mathbf{V}(x, y, z, t) = u(x, y, z, t)\mathbf{i} + v(x, y, z, t)\mathbf{j} + w(x, y, z, t)\mathbf{k}$$

where u, v, w are the x, y, z components of the velocity, and $\mathbf{i}, \mathbf{j}, \mathbf{k}$ are the unit vectors that define the positive directions of the x, y, z axes.

In fluid mechanics it is generally easier to use a Eulerian description rather than a Lagrangian description to define the flow. This is because all the particles composing the fluid can have very erratic motion, and the fluid system may not maintain a constant shape. A Eulerian description is localized, in that it specifies a point and measures the motion of the particles passing this point. From a Lagrangian point of view, it is very difficult to account for the position of *all the particles* in the system, from one instant to the next, and then measure the velocities of all these particles as they move about and change the system's shape. A Lagrangian description, however, works well in rigid-body dynamics. Here the body *maintains a fixed shape*, and so the location and motion of the particles composing the body can be readily specified with respect to one another.

3.2   Types of Fluid Flow

Apart from the two ways of describing the motion of fluid particles, there are also various ways to classify the flow of a fluid system. Here we will consider three of them.

Classification of Flow Related to Its Frictional Effects.
When a highly viscid fluid such as oil flows at a very slow rate through a pipe, the paths the particles follow are uniform and undisturbed. In other words, the lamina or thin layers of fluid are "orderly" and so one cylindrical layer slides smoothly relative to an adjacent layer. This behavior is referred to as *laminar flow*, Fig. 3–2a. Increase the velocity or decrease the viscosity, and the fluid particles may then follow erratic paths, which causes a high rate of mixing within the fluid. We refer to this as *turbulent flow*, Fig. 3–2b. Between these two types, we have *transitional flow*, that is, a state in which regions of both laminar and turbulent flow coexist.

In Chapter 9 we will see that one of the most important reasons for classifying the flow in this manner is to determine the amount of energy the fluid loses due to frictional effects. Obtaining this energy loss is necessary when designing pumps and pipe networks for transporting the fluid.

The velocity profile for both laminar and turbulent flow between two surfaces is shown in Fig. 3–3. Notice how laminar flow is shaped entirely by the viscosity of the fluid's sliding layers, whereas turbulent flow "mixes" the fluid both horizontally and vertically, which causes its average velocity profile to flatten out, or become more uniform.

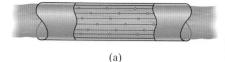

(a)

Laminar flow

Fluid particles follow straight-line paths since fluid flows in thin layers

(b)

Turbulent flow

Fluid particles follow erratic paths which change direction in space and time

Fig. 3–2

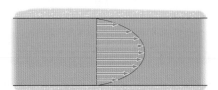

Average velocity profile for laminar flow

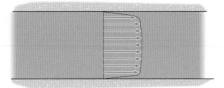

Average velocity profile for turbulent flow

Fig. 3–3

Classifications of Dimensional Flow. A flow can also be classified by how many spatial coordinates are needed to describe it. If all three space coordinates are required, then it is called ***three-dimensional flow***. Examples include the flow of water around a submarine and the airflow around an automobile. Three-dimensional flows are rather complex and therefore difficult to analyze. They are most often studied using a computer, or experimentally using models.

For many problems in engineering, we can simplify the analysis by assuming the flow to be two-, one-, or even nondimensional. For example, the flow through the converging pipe in Fig. 3–4a is ***two-dimensional flow***. The velocity of any particle depends only on its axial and radial coordinates x and r.

A further simplification can be made in the case of unchanging flow through a uniform straight pipe, Fig. 3–4b. This is a case of ***one-dimensional flow***. Its velocity profile changes only in the radial direction.

Finally, if we consider the unchanging flow of an *ideal fluid*, where the viscosity is zero and the fluid is incompressible, then its velocity profile is *constant* throughout, and therefore independent of its coordinate location, Fig. 3–3c. It becomes ***nondimensional flow***.

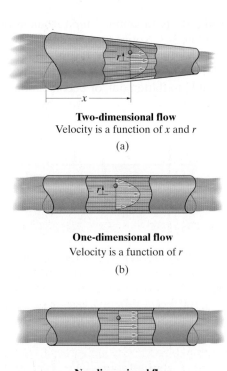

Two-dimensional flow
Velocity is a function of x and r

(a)

One-dimensional flow
Velocity is a function of r

(b)

Nondimensional flow
Velocity is constant

(c)

Fig. 3–4

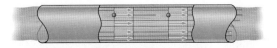

Steady uniform flow
An ideal fluid maintains the same velocity
at all times and at each point

(a)

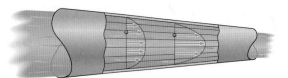

Steady nonuniform flow
The velocity remains constant with time,
but it is different from one location to the next

(c)

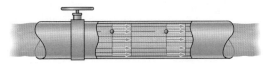

Time *t*

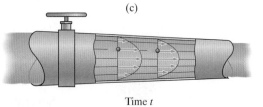

Time *t*

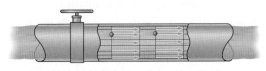

Time *t* + Δ*t*

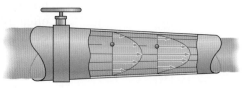

Time *t* + Δ*t*

Unsteady uniform flow
The valve is slowly opened and so at any instant
the velocity of an ideal fluid is the same at all
points, but it changes with time

(b)

Unsteady nonuniform flow
The valve is slowly opened, and because of the
changing cross section of the pipe, the velocity
will be different at each point and at each time

(d)

Fig. 3–5

Classification of Flow Based on Space and Time.   When
the velocity of a fluid at a point *does not change with time*, we refer to the
flow as **steady flow**, and when the velocity *does not change from one
position to the next*, it is referred to as **uniform flow**. In general, there are
four possible combinations of these two flows, and an example of each is
given in Fig. 3–5.

Most engineering applications of fluid mechanics involve steady flow,
and fortunately, it is the easiest to analyze. Furthermore, it is generally
reasonable to assume that any unsteady flow that *varies* for a short time
may, over the long term, be considered a case of steady flow. For example,
flow through the moving parts of a pump is unsteady, but the pump
operation is cyclic or repetitious, so we may consider the flow at the inlet
and outlet of the pump to be "steady in the mean." It may also be possible
to establish steady flow relative to a moving observer. Consider the case
of a car passing through still, smoky air. The airflow will appear steady to
the driver when the car moves through it at constant speed on a straight
road. However, to an observer standing along the roadway, the air would
appear to have unsteady flow during the time the car passes by.

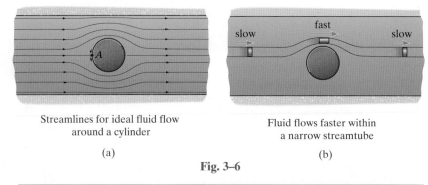

Streamlines for ideal fluid flow
around a cylinder

(a)

Fluid flows faster within
a narrow streamtube

(b)

Fig. 3–6

3.3   Graphical Descriptions of Fluid Flow

Several graphical methods have been devised to visualize the behavior of a flow. For analysis, these include using streamlines or streamtubes, and for experimental work, pathlines and streaklines, and optical methods are often used. We will now give each of these separate treatment.

Streamlines.   A *streamline* is a curve that is drawn through the fluid in such a manner that it indicates the direction of the velocity of particles located on it at a particular *instant of time*. Specifically, the velocity of any particle is always *tangent* to the streamline along which it is traveling at that instant. Consequently, no fluid can flow across a streamline, only *along* the streamline. For example, the streamlines for the steady flow of an *ideal fluid* around a cylinder enclosed within a rectangular duct are shown in Fig. 3–6*a*. In particular, notice the streamline at the center of the flow field. It intersects the cylinder at *A*. This point is called a **stagnation point,** because here the velocity of any particle is *momentarily* reduced to zero when it strikes the cylinder's surface.

It is important to keep in mind that streamlines are used to represent the flow field during each *instant of time*. In the example just given, the direction of the streamlines is *maintained* as time passes and fluid particles move along them; however, sometimes the streamlines can be a function of *both* space and time. For example, this occurs when fluid flows through a rotating pipe. Here the streamlines *change position* from one instant to the next, and so the fluid particles do not move along any streamline that is fixed in position.

Since the particle's velocity is *always tangent* to its streamline, then if the velocity is known, the equation that defines a streamline at a particular instant can always be determined. For example, in the case of two-dimensional flow, the particle's velocity will have two components, u in the x direction and v in the y direction. As noted in Fig. 3–7, we require

$$\frac{dy}{dx} = \frac{v}{u} \qquad (3\text{–}3)$$

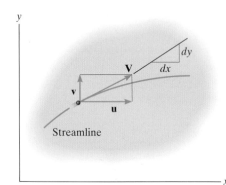

The velocity is always
tangent to the streamline

Fig. 3–7

Integrating this equation will give the *equation of the streamlines*, $y = f(x) + C$. To find the streamline that passes through a particular point (x_0, y_0), we can substitute these coordinates into this equation to evaluate the constant C. This process is demonstrated in Examples 3.1 and 3.2.

Streamtubes. For some types of analysis, it is convenient to consider a *bundle of streamlines* that surround a region of flow, Fig. 3–8. Such a circumferential grouping is called a ***streamtube***. Here the fluid flows through the streamtube as if it were contained within a curved conduit.

In two dimensions, streamtubes can be formed between any two streamlines. For example, consider the streamtube in Fig. 3–6b for flow around the cylinder. Notice that a small element of fluid traveling along this streamtube will move *slower* when the streamlines are *farther apart* (or the streamtube is wider), and move *faster* when they are *closer together* (or the streamtube is narrow). This is a consequence of the conservation of mass, something we will discuss in the next chapter.

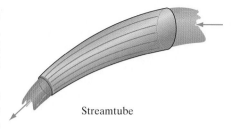

Streamtube

Fig. 3–8

The pathline shows the path of a *single particle* using a time exposure photograph for $0 \leqslant t \leqslant t_1$

(a)

The streakline shows the path of many particles at the *instant* $t = t_1$

(b)

Fig. 3–9

Pathlines. The ***pathline*** for a fluid particle defines the "path" the particle travels over a *period of time*. To obtain the pathline experimentally, a neutrally buoyant *single particle* can be released within the flow stream and a *time-exposed photograph* taken. The line on the photograph then represents the pathline for this particle, Fig. 3–9a.

Streaklines. If smoke is released continuously in a gas, or colored dye is released in a liquid, a "trace" of *all the particles* will be carried along with the flow. This resulting succession of marked particles that have all come from the same point of origin is called a ***streakline***. It can be identified by taking an *instantaneous photograph* of the trace, or "streak," of *all the particles*, Fig. 3–9b.

As long as the flow is *steady*, the streamlines, pathlines, and streaklines will all *coincide*. For example, this occurs for the steady flow of water ejected from the *fixed* nozzle, Fig. 3–10. Here a streamline will maintain its same direction from one instant to the next, and so *every particle* coming from the nozzle will follow this same streamline, thereby producing a coincident pathline and streakline.

This photo shows the streamlines of water particles ejected from a water sprinkler at a given instant.

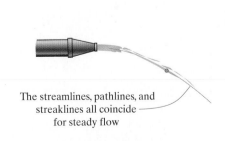

The streamlines, pathlines, and streaklines all coincide for steady flow

Fig. 3–10

Optical Methods.   If the fluid is transparent, such as air and water, then the flow can be visualized indirectly using a shadowgraph. Simply put, a *shadowgraph* is the result of refracted or bent light rays that interact with the fluid and then cast a shadow on a screen in close proximity. Perhaps you have noticed the shadow of a heated air plume rising from a candle, or the shadow of the exhaust from a jet engine cast against a surface. In both of these cases the heat being produced changes the local density of the air, and the light passing through this air is bent. The greater the change in the fluid density, the more the light rays are bent. Shadowgraphs have been used in industry to visualize the air flow around jet aircraft and rockets, as well as to study the flow around devices that produce heat.

One other optical technique that is used to visualize the flow of a transparent fluid is **schlieren photography**. It also is based on detecting the density gradients within a fluid produced by the flow. In this case, a collimated or parallel beam of light shines on the object. The light is focused with a lens, and a knife edge is placed perpendicular to the beam at its focal point. This blocks about half the light and the fluid causes a distortion in the beam due to the density variations within the fluid. The distortion will cast light and dark regions on a nearby background surface. An example is shown in the accompanying photo of a heated source generating a flow of hot air used to suspend a ball. Schlieren photography has been used extensively in aeronautical engineering to visualize the formation of shockwaves and expansion waves formed around jet aircraft and missiles. For further details on these two optical techniques, see Ref. [5].

A schlieren photograph showing a ball that is suspended by an airstream produced by a jet of hot air.

Computational Fluid Dynamics. Although the experimental techniques just discussed have played an important role in studying complicated flow patterns, they have been increasingly replaced with numerical techniques that apply the laws of fluid dynamics using high-speed computers. This is referred to as **computational fluid dynamics** (CFD). There are many types of commercially available CFD programs, and with some of them the engineer can extend an analysis to include heat transfer and multiphase changes within the fluid. Generally the output includes a plot of the streamlines or pathlines, with the velocity field displayed in a color-coded fashion. Digital printouts of steady flow can be made, or if the flow is unsteady, then a video can be produced. We will discuss this important field in greater length in Sec. 7.12.

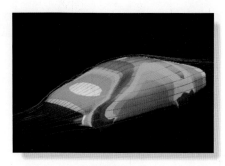

A CFD printout showing the flow around a model of a car. Based on the results, the model can be reshaped to enhance the design. (© Hank Morgan/Science Source)

Important Points

- There are two ways to describe the motion of fluid particles within a flow. A *Lagrangian description*, or system approach, has limited use since it requires tracking the location of *each particle* in the flow and reporting its motion. A *Eulerian description*, or control volume approach, is more practical, since it considers a specific region or point in the flow, and measures the motion of any fluid particles passing through this region or point.

- Fluid flow can be described in various ways. It can be classified as the result of viscous friction, which is, either laminar, transitional, or turbulent. It can be classified as nondimensional, or as one-, two-, or three-dimensional. And finally, for a space–time classification, the flow is *steady* if it does not vary with time, and *uniform* if it does not vary with location.

- A *streamline* contains particles that have velocities that are tangent to the streamline at each instant of time. If the flow is steady, then the particles move along fixed streamlines. However, if unsteady flow causes the direction of the streamlines to change, then the particles will move along streamlines that have a different orientation from one instant to the next.

- Experimentally, the flow of a *single* marked particle can be visualized using a *pathline*, which shows the path taken by the particle using a time-lapse photograph. *Streaklines* are formed when smoke or colored dye is released into the flow from the same point, and an instantaneous photo is taken of the trace of *many* particles.

- Optical methods, such as shadowgraphs and schlieren photographs, are useful for observing flow in transparent fluids caused by heat, or to visualize shock and expansion waves created in high speed flows.

- Computational fluid dynamics uses numerical analysis to apply the laws of fluid dynamics, and thereby produces data to visualize complex flow.

3

EXAMPLE | 3.1

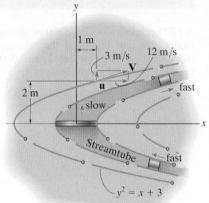

Fig. 3–11

The velocity for the two-dimensional flow shown in Fig. 3–11 is defined by $V = \{6y\mathbf{i} + 3\mathbf{j}\}\,m/s$, where y is in meters. Determine the equation of the streamline that passes through point $(1\,m, 2\,m)$.

SOLUTION

Fluid Description.   This is a Eulerian description since the velocity is reported in terms of its spatial coordinates. In other words, the description reports the velocity of particles that pass through a control volume located at point (x, y) within the flow. Since time is not involved, we have steady flow, where $u = (6y)\,m/s$ and $v = 3\,m/s$.

Analysis.   To find the equations of the streamlines, we must use Eq. 3–3,

$$\frac{dy}{dx} = \frac{v}{u} = \frac{3}{6y}$$

Separating the variables and integrating yields

$$\int 6y\,dy = \int 3\,dx$$

$$3y^2 = 3x + C$$

This is the equation for a parabola. Each selected constant C will produce a unique streamline. For the one passing through the point $(1\,m, 2\,m)$, we require $3(2)^2 = 3(1) + C$, or $C = 9$. Therefore,

$$y^2 = x + 3 \qquad\qquad Ans.$$

A plot of this equation (streamline) is shown in Fig. 3–11. Here particles that pass through a differential size control volume located at the point $(1\,m, 2\,m)$ will have the velocity $V = \{12\mathbf{i} + 3\mathbf{j}\}\,m/s$, as indicated. By selecting other points to evaluate the integration constant C, we can establish the graphical representation of the entire flow field, a portion of which is also shown in Fig. 3–11. Once this is established notice how the fluid element constrained in the selected streamtube travels fast, slows down as it passes the x axis, and then speeds up again.

NOTE: If the velocity were also a function of time, then unsteady flow would occur, and the streamlines may change location from one instant to the next. To obtain them, we would first have to evaluate V at a particular instant of time, and then apply Eq. 3–3 in the same manner as explained here to establish the streamlines or flow field for this instant.

EXAMPLE | 3.2

The velocity components of a particle in the flow field are defined by $u = 3$ m/s and $v = (6t)$ m/s, where t is in seconds. Plot the pathline for the particle if it is released from the origin when $t = 0$. Also draw the streamline for this particle when $t = 2$ s.

SOLUTION

Fluid Description.   Since the velocity is only a function of time, this is a Lagrangian description of the particle's motion. We have nonuniform flow.

Pathline.   The pathline describes the location of the particle at various times. Since the particle is at $(0, 0)$ when $t = 0$, then

$$u = \frac{dx}{dt} = 3 \qquad\qquad v = \frac{dy}{dt} = 6t$$

$$\int_0^x dx = \int_0^t 3\, dt \qquad\qquad \int_0^y dy = \int_0^t 6t\, dt$$

$$x = (3t)\ \text{m} \qquad\qquad y = \left(3t^2\right)\ \text{m} \qquad (1)$$

Eliminating the time between these two parametric equations, we obtain our result

$$y = 3\left(\frac{x}{3}\right)^2 \quad \text{or} \quad y = \frac{1}{3}x^2 \qquad\qquad Ans.$$

The pathline or path taken by the particle is a parabola, shown in Fig. 3–12.

Streamline.   The streamline shows the direction of the velocity of a particle at a particular instant of time. At the instant $t = 2$ s, using Eq. 1, the particle is located at $x = 3(2) = 6$ m, $y = 3(2)^2 = 12$ m. Also it has velocity components of $u = 3$ m/s, $v = 6(2) = 12$ m/s, Fig 3–12. Applying Eq. 3–3 to obtain the equation of the streamline at this instant, we have

$$\frac{dy}{dx} = \frac{v}{u} = \frac{12}{3} = 4$$

$$\int dy = \int 4\, dx$$

$$y = 4x + C$$

Since $x = 6$ m and $y = 12$ m, then $C = -12$ m. Therefore

$$y = 4x - 12$$

This streamline is shown in Fig. 3–12. Note that both the streamline and pathline have the same slope at $(6\text{ m}, 12\text{ m})$. This is to be expected since both must give the same direction for the velocity when $t = 2$ s. At another instant, the particle moves along the pathline, and has another streamline (tangent).

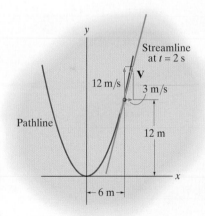

Fig. 3–12

EXAMPLE | 3.3

The velocity of gas particles flowing along the center of the pipe in Fig. 3–13 is defined for $x \geq 1$ m, by the velocity field $V = (t/x)$ m/s, where t is in seconds and x is in meters.* If a particle is at $x = 1$ m when $t = 0$, determine its velocity when it is at $x = 2$ m.

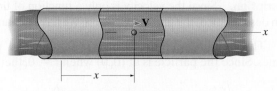

Fig. 3–13

SOLUTION

Fluid Description. Because the velocity is a function of time, the flow is unsteady, and because it is a function of position, it is nonuniform. This flow field $V = V(x, t)$ is a Eulerian description. Here *all particles* passing the location x have a velocity that changes with time, and it is measured as $V = (t/x)$.

Analysis. To find the velocity at $x = 2$ m, we must first find the time for the particle to travel from $x = 1$ m to $x = 2$ m, that is, we must find $x = x(t)$. This is a Lagrangian description since we must follow the motion of a single particle.

The particle's position can be determined from the velocity field, by realizing that for a Lagrangian description, Eq. 3–1 gives

$$V = \frac{dx}{dt} = \frac{t}{x}$$

Separating the variables and integrating yields

$$\int_{1\,\text{m}}^{x} x \, dx = \int_{0}^{t} t \, dt$$

$$\frac{x^2}{2} - \frac{1}{2} = \frac{1}{2}t^2$$

$$x = \left(\sqrt{t^2 + 1}\right) \text{m} \tag{1}$$

---

*Notice that substituting t in seconds and x in meters gives units of s/m; however, here there must be a constant $1\,\text{m}^2/\text{s}^2$ (not shown) that converts these units to velocity, m/s, as stated.

The Lagrangian description of the velocity of the particle can now be determined. Here the velocity must only be a function of time.

$$V = \frac{dx}{dt} = \frac{1}{2}\left(t^2 + 1\right)^{-1/2}(2t)$$

$$= \left(\frac{t}{\sqrt{t^2 + 1}}\right) \text{ m/s} \qquad (2)$$

Using these results, we are now able to follow the particle as it moves along the path during each instant of time. Its position at any instant is determined from Eq. 1, and its velocity at this instant is determined from Eq. 2. Therefore, when the particle is located at $x = 2$ m, the time is

$$2 = \sqrt{t^2 + 1}$$

$$t = 1.732 \text{ s}$$

And at this time, the particle is traveling at

$$V = \left(\frac{1.732}{\sqrt{(1.732)^2 + 1}}\right) \text{ m/s}$$

$$V = 0.866 \text{ m/s} \qquad\qquad Ans.$$

We can check this result, using the Eulerian description. Since we require the particle to be located at $x = 2$ m, when $t = 1.732$ s its velocity is

$$V = \frac{t}{x} = \frac{1.723}{2} = 0.866 \text{ m/s} \qquad\qquad Ans.$$

as expected.

NOTE: In this example the velocity field has only one component, $V = u = t/x, v = 0, w = 0$. In other words, it is one-dimensional flow, and so the streamline does not change direction, rather it remains in the x direction.

3.4   Fluid Acceleration

Once the velocity field $\mathbf{V} = \mathbf{V}(x, y, z, t)$ is established, then it is possible to determine the acceleration field for the flow. One reason for doing this is to apply Newton's second law of motion, $\Sigma\mathbf{F} = m\mathbf{a}$, to relate the acceleration of the fluid particles to the forces produced by the flow. Since the velocity field is a function of *both* space and time (Eulerian description), the *time* rate of change of velocity (acceleration) must account for the changes made in *both* its space and time variables. To show how to obtain these changes, let us consider the unsteady, nonuniform flow of a fluid through the nozzle in Fig. 3–14. For simplicity, we will study the motion of a single particle as it moves through a control volume. Here the velocity is defined by $V = V(x, t)$ along the center streamline. When the particle is at position x, where the first open control surface to the control volume is located, it will have a velocity that will be *smaller* than when it is at position $x + \Delta x$, where it exits the control volume's other open control surface. This is because the nozzle constricts the flow and therefore causes the particle to have a greater speed at $x + \Delta x$. Thus the velocity of the particle will change due to its *change Δx in position* (nonuniform flow). If the valve is opened then the velocity of the particle can *also change within* the control volume due to a *change in time Δt* (unsteady flow). As a result, the total change in $V = V(x, t)$ will be

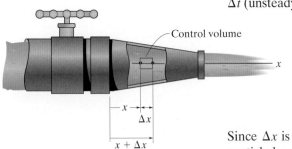

Fig. 3–14

$$\Delta V = \underbrace{\frac{\partial V}{\partial t}\Delta t}_{\substack{\text{Change of } V \\ \text{with time} \\ \text{(unsteady flow)}}} + \underbrace{\frac{\partial V}{\partial x}\Delta x}_{\substack{\text{Change of } V \\ \text{with position} \\ \text{(nonuniform flow)}}}$$

Since Δx is the distance covered in the time Δt, the acceleration of the particle becomes

$$a = \lim_{\Delta t \to 0} \frac{\Delta V}{\Delta t} = \frac{\partial V}{\partial t} + \frac{\partial V}{\partial x}\frac{dx}{dt}$$

Another way to determine this result is to use the *chain rule* of calculus. Since $V = V(x, t)$, we must use partial derivatives to find each change in V, that is, $a = \partial V/\partial t(dt/dt) + \partial V/\partial x(dx/dt)$. Realizing that $dx/dt = V$, in either case, we get

$$a = \frac{DV}{Dt} = \underbrace{\frac{\partial V}{\partial t}}_{\substack{\text{Local} \\ \text{accel.}}} + \underbrace{V\frac{\partial V}{\partial x}}_{\substack{\text{Convective} \\ \text{accel.}}} \tag{3–4}$$

The notation $D(\ )/Dt$ is referred to as a ***material derivative,*** because it gives the time rate of change of a fluid property (in this case the velocity) as the fluid particle (material) passes through the control volume. Let's summarize our result.

Local Acceleration. The first term on the right side, $\partial V/\partial t$, indicates the *time rate of change* of the velocity of the particle, which occurs *within* the control volume. For this reason it is called the **local acceleration**. Opening the valve in Fig. 3–14 causes the flow to increase, producing this local change. For *steady flow* this term will be zero, because the flow will not change with time.

Convective Acceleration. The last term on the right side, $V(\partial V/\partial x)$, is referred to as the **convective acceleration**, since it measures the change in the particle's velocity as the particle moves through the entrance of the control volume, and then through its exit. The conical shape of the nozzle in Fig. 3–14 causes this change. Only when the flow is *uniform*, as in the case of a pipe with a constant cross section, will this term be zero.

Three-dimensional Flow. Now let's generalize these results and consider the case of three-dimensional flow, Fig. 3–15. Here the particle passes through the control volume located at point x, y, z, where it has the velocity

$$\mathbf{V}(x, y, z, t) = u(x, y, z, t)\mathbf{i} + v(x, y, z, t)\mathbf{j} + w(x, y, z, t)\mathbf{k} \quad (3\text{–}5)$$

The particle's velocity can increase (or decrease) due to either a change in its position dx, dy, dz or a change in time dt. To find these changes, we must use the chain rule to obtain a result similar to Eq. 3–4. We have

$$\mathbf{a} = \frac{D\mathbf{V}}{Dt} = \frac{\partial \mathbf{V}}{\partial t} + \frac{\partial \mathbf{V}}{\partial x}\frac{dx}{dt} + \frac{\partial \mathbf{V}}{\partial y}\frac{dy}{dt} + \frac{\partial \mathbf{V}}{\partial z}\frac{dz}{dt}$$

Since $u = dx/dt$, $v = dy/dt$, and $w = dz/dt$, then

$$\mathbf{a} = \frac{D\mathbf{V}}{Dt} = \frac{\partial \mathbf{V}}{\partial t} + \left(u\frac{\partial \mathbf{V}}{\partial x} + v\frac{\partial \mathbf{V}}{\partial y} + w\frac{\partial \mathbf{V}}{\partial z} \right) \quad (3\text{–}6)$$

$$\underbrace{\phantom{\frac{D\mathbf{V}}{Dt}}}_{\substack{\text{Total} \\ \text{accel.}}} \quad \underbrace{\phantom{\frac{\partial \mathbf{V}}{\partial t}}}_{\substack{\text{Local} \\ \text{accel.}}} \quad \underbrace{\phantom{u\frac{\partial \mathbf{V}}{\partial x} + v\frac{\partial \mathbf{V}}{\partial y} + w\frac{\partial \mathbf{V}}{\partial z}}}_{\substack{\text{Convective} \\ \text{accel.}}}$$

If we substitute Eq. 3–5 into this expression, the expansion results in the x, y, z components of acceleration, that is

$$a_x = \frac{\partial u}{\partial t} + u\frac{\partial u}{\partial x} + v\frac{\partial u}{\partial y} + w\frac{\partial u}{\partial z}$$

$$a_y = \frac{\partial v}{\partial t} + u\frac{\partial v}{\partial x} + v\frac{\partial v}{\partial y} + w\frac{\partial v}{\partial z} \quad (3\text{–}7)$$

$$a_z = \frac{\partial w}{\partial t} + u\frac{\partial w}{\partial x} + v\frac{\partial w}{\partial y} + w\frac{\partial w}{\partial z}$$

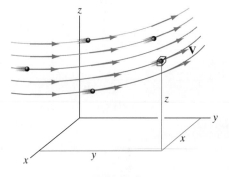

Fig. 3–15

Material Derivative. Besides the velocity field $\mathbf{V} = \mathbf{V}(x, y, z, t)$, other fluid properties can also be described using a Eulerian description. For example, while a liquid is heated in a boiler, there will be an uneven temperature rise of the liquid at each point. This creates a scalar temperature field $T = T(x, y, z, t)$ that changes with *both* position and time. Furthermore, since we have assumed a fluid to be a continuum, then at *each point* the pressure and density of the liquid can *also* be described by scalar pressure and density fields, $p = p(x, y, z, t)$ and $\rho = \rho(x, y, z, t)$. The time rate of change of each of these fields produces local and convective changes related to each control volume. For example, in the case of the temperature field, $T = T(x, y, z, t)$,

$$\frac{DT}{Dt} = \underbrace{\frac{\partial T}{\partial t}}_{\substack{\text{Local} \\ \text{change}}} + \underbrace{u\frac{\partial T}{\partial x} + v\frac{\partial T}{\partial y} + w\frac{\partial T}{\partial z}}_{\substack{\text{Convective} \\ \text{change}}} \tag{3–8}$$

In general, the material derivative can be written in more compact form using vector notation. It is

$$\frac{D(\ )}{Dt} = \frac{\partial (\ )}{\partial t} + (\mathbf{V} \cdot \nabla)(\ ) \tag{3–9}$$

Here the dot product between the velocity vector, $\mathbf{V} = u\mathbf{i} + v\mathbf{j} + w\mathbf{k}$, and the gradient operator del, $\nabla = (\partial (\ )/\partial x)\mathbf{i} + (\partial (\ )/\partial y)\mathbf{j} + (\partial (\ )/\partial z)\mathbf{k}$, yields $\mathbf{V} \cdot \nabla = u(\partial (\ )/\partial x)\mathbf{i} + v(\partial (\ )/\partial y)\mathbf{j} + w(\partial (\ )/\partial z)\mathbf{k}$. You may want to show that the expansion of Eq. 3–9 yields Eqs. 3–7 when $D(\mathbf{V})/Dt$ is determined.

Fluid particles moving upward from this water sprinkler have a decrease in their velocity magnitude. Also the direction of their velocity is changing. Both of these effects produce acceleration.

EXAMPLE | 3.4

As the valve in Fig. 3–16 is being closed, oil particles flowing through the nozzle along the center streamline have a velocity of $V = \left[6\left(1 + 0.4x^2\right)\left(1 - 0.5t\right)\right]$ m/s, where x is in meters and t is in seconds. Determine the acceleration of an oil particle at $x = 0.25$ m when $t = 1$ s.

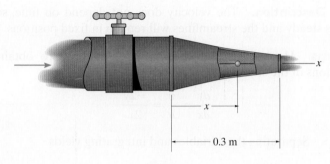

Fig. 3–16

SOLUTION

Fluid Description.   The flow along the streamline is nonuniform and unsteady because its Eulerian description is a function of both x and t.

Analysis.   Here $V = u$. Applying Eq. 3–4 or the first of Eqs. 3–7, we have

$$a = \frac{\partial V}{\partial t} + V\frac{\partial V}{\partial x} = \frac{\partial}{\partial t}\left[6\left(1 + 0.4x^2\right)\left(1 - 0.5t\right)\right]$$

$$+ \left[6\left(1 + 0.4x^2\right)\left(1 - 0.5t\right)\right]\frac{\partial}{\partial x}\left[6\left(1 + 0.4x^2\right)\left(1 - 0.5t\right)\right]$$

$$= \left[6\left(1 + 0.4x^2\right)\left(0 - 0.5\right)\right] + \left[6\left(1 + 0.4x^2\right)\left(1 - 0.5t\right)\right]\left[6\left(0 + 0.4(2x)\right)\left(1 - 0.5t\right)\right]$$

Evaluating this expression at $x = 0.25$ m, $t = 1$ s, we get

$$a = -3.075\,\text{m/s}^2 + 1.845\text{ m/s}^2 = -1.23\text{ m/s}^2 \qquad \textit{Ans.}$$

The local acceleration component $\left(-3.075\,\text{m/s}^2\right)$ is decreasing the velocity of the particle at $x = 0.25$ m since the valve is being closed to decrease the flow. The convective acceleration component $\left(1.845\text{ m/s}^2\right)$ is increasing the velocity of the particle since the nozzle constricts as x increases, and this causes the velocity to increase. The net result, however, causes the particle to decelerate at 1.23 m/s^2.

EXAMPLE 3.5

The velocity for a two-dimensional flow is defined by $\mathbf{V} = \{2x\mathbf{i} - 2y\mathbf{j}\}\,\text{m/s}$, where x and y are in meters. Plot the streamlines for the flow field, and determine the magnitude of the velocity and acceleration of a particle located at the point $x = 1\,\text{m}$, $y = 2\,\text{m}$.

SOLUTION

Flow Description. The velocity does not depend on time, so the flow is steady and the streamlines will remain in fixed positions.

Analysis. Here $u = (2x)\,\text{m/s}$ and $v = (-2y)\,\text{m/s}$. To obtain the equations of the streamlines, we must use Eq. 3–3.

$$\frac{dy}{dx} = \frac{v}{u} = \frac{-2y}{2x}$$

Separating the variables and integrating yields

$$\int \frac{dx}{x} = -\int \frac{dy}{y}$$
$$\ln x = -\ln y + C$$
$$\ln(xy) = C$$
$$xy = C'$$

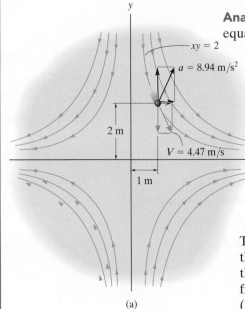

The arbitrary integration constant is C'. Using various values of this constant, the above equation represents a family of streamlines that are hyperbolas, and so the flow field is shown in Fig. 3–17a. To find the streamline passing through point (1 m, 2 m), we require $(1)(2) = C'$ so that $xy = 2$.

Velocity. The velocity of a particle passing through the control volume located at point (1 m, 2 m) has components of

$$u = 2(1) = 2\,\text{m/s}$$
$$v = -2(2) = -4\,\text{m/s}$$

Therefore,

$$V = \sqrt{(2\,\text{m/s})^2 + (-4\,\text{m/s})^2} = 4.47\,\text{m/s} \qquad \textit{Ans.}$$

From the direction of its components, this velocity is shown in Fig. 3–17a. It indicates the direction of the flow along its hyperbolic streamline. Flow along the other hyperbolas can be determined in the same manner, that is, by selecting a point and then showing the vector addition of the velocity components of a particle at this point.

Fig. 3–17

(in figure: $xy = 2$, $a = 8.94\,\text{m/s}^2$, $2\,\text{m}$, $1\,\text{m}$, $V = 4.47\,\text{m/s}$, (a))

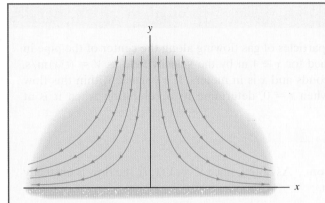

Flow striking a fixed surface

(b)

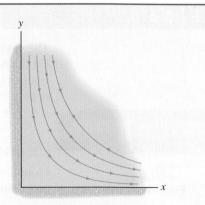

Flow along two
perpendicular surfaces

(c)

It is interesting to note that we can use a portion of this flow pattern to describe, for example, a flow that strikes a fixed surface, as in Fig. 3–17b, a flow within a corner, Fig. 3–17c, or a flow constrained between a corner and a hyperbolic boundary defined by one of the streamlines, Fig. 3–17d. In all of these cases there is a *stagnation point* at the origin $(0, 0)$ since the velocity is zero at this point, that is, $u = 2(0) = 0$ and $v = -2(0) = 0$. As a result, if the fluid contained debris, it would tend to accumulate within the region around the origin.

Acceleration.   The components of acceleration are determined using Eqs. 3–7. Due to the steady flow there is no local acceleration, just convective acceleration.

$$a_x = \frac{\partial u}{\partial t} + u\frac{\partial u}{\partial x} + v\frac{\partial u}{\partial y} = 0 + 2x(2) + (-2y)(0)$$

$$= 4x$$

$$a_y = \frac{\partial v}{\partial t} + u\frac{\partial v}{\partial x} + v\frac{\partial v}{\partial y} = 0 + 2x(0) + (-2y)(-2)$$

$$= 4y$$

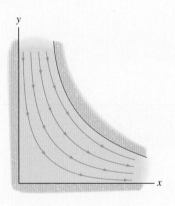

Flow between a corner
and a hyperbolic surface

(d)

At point $(1\,\text{m}, 2\,\text{m})$, a particle will therefore have acceleration components

$$a_x = 4(1) = 4 \ \text{m/s}^2$$
$$a_y = 4(2) = 8 \ \text{m/s}^2$$

The magnitude of the particle's acceleration, shown in Fig. 3–17a, is therefore

$$a = \sqrt{\left(4 \ \text{m/s}^2\right)^2 + \left(8 \ \text{m/s}^2\right)^2} = 8.94 \ \text{m/s}^2 \qquad Ans.$$

EXAMPLE   3.6

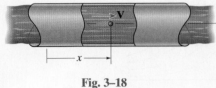

Fig. 3–18

3

The velocity of particles of gas flowing along the center of the pipe in Fig. 3–18 is defined for $x \geq 1$ m by the velocity field as $V = (t/x)\,\text{m/s}$, where t is in seconds and x is in meters. If a particle within this flow is at $x = 1$ m when $t = 0$, determine its acceleration when it is at $x = 2$ m.

SOLUTION

Fluid Description.   As noted, since $V = V(x, t)$, the flow is unsteady and nonuniform.

Analysis.   In Example 3.3 we have obtained the Lagrangian descriptions of the particle's position and velocity, namely

$$x = \left(\sqrt{t^2 + 1} \right) \text{ m}$$

and so

$$V = \frac{t}{x} = \left(\frac{t}{\sqrt{t^2 + 1}} \right) \text{m/s}$$

Also, when the particle arrives at $x = 2$ m, the time was found to be $t = 1.732\,\text{s}$. From the Lagrangian point of view, the acceleration is simply the time derivative of the velocity,

$$a = \frac{dV}{dt} = \frac{(t^2 + 1)^{1/2}(1) - t\left[\dfrac{1}{2}(t^2 + 1)^{-1/2}(2t) \right]}{t^2 + 1} = \left[\frac{1}{(t^2 + 1)^{3/2}} \right] \text{m/s}^2$$

Thus, when $t = 1.732\,\text{s}$,

$$a = \frac{1}{[(1.732)^2 + 1]^{3/2}} = 0.125 \text{ m/s}^2 \qquad\qquad Ans.$$

Now let's check our work by taking the material derivative of the velocity field (Eulerian description). Since this is one-dimensional flow, applying Eq. 3–4, we have

$$a = \frac{DV}{Dt} = \frac{\partial V}{\partial t} + V \frac{\partial V}{\partial x} = \left[\frac{1}{x} + \frac{t}{x}\left(-\frac{t}{x^2} \right) \right] \text{m/s}^2$$

At the control volume, located at $x = 2$ m, the particle, defined by its Lagrangian description, will appear when $t = 1.732\,\text{s}$. Here its acceleration will be

$$a = \frac{1}{2} + \frac{(1.732)^2}{2}\left[-\frac{1}{(2)^2} \right] = 0.125 \text{ m/s}^2 \qquad\qquad Ans.$$

which agrees with our previous result.

3.5 Streamline Coordinates

When the path or streamline for the fluid particles is *known*, such as when the flow is through a fixed conduit, streamline coordinates can be used to describe the motion. To show how these coordinates are established, consider the fluid particle moving along the streamline in Fig. 3–19a. The *origin* of the coordinate axes is placed at the *point* on the streamline where the control volume is located. The *s* axis is *tangent* to the streamline at this point, and is positive in the direction in which the particles are traveling. We will designate this positive direction with the unit vector $\mathbf{u}_s$. Before we establish the *normal axis*, notice that geometrically, *any* streamline curve can be constructed from a series of differential arc segments *ds* as in Fig. 3–19b. Each segment is formed from an arc length *ds* on an associated circle, having a *radius of curvature R* and *center of curvature at O'*. For the case in Fig. 3–19a, the normal *n* axis is perpendicular to the *s* axis at the control volume, with the positive sense of the *n* axis directed *toward* the center of curvature *O'* for the arc *ds*. This positive direction, which is *always on the concave side of the curve*, will be designated by the unit vector $\mathbf{u}_n$. Once the *s* and *n* axes are established in this manner, we can then express the velocity and acceleration of the particle passing through the control volume in terms of these coordinates.

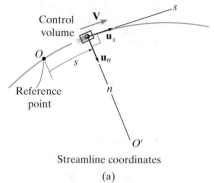

Streamline coordinates

(a)

Velocity. Since the direction of the particle's velocity **V** is *always tangent to the path* that is in the positive *s* direction, Fig. 3–19a, we have

$$\mathbf{V} = V\mathbf{u}_s \tag{3–10}$$

where $V = V(s, t)$.

Acceleration. The acceleration of the particle is the time rate of change of the velocity, and so to determine it using the material derivative, we must account for both the local and the convective changes to the velocity, as the particle moves a distance Δs through the control volume, Fig. 3–19c.

Local Change. If a condition of *unsteady flow* exists, then during the time *dt*, local changes can occur to the particle's velocity *within* the control volume. This local change produces the acceleration components

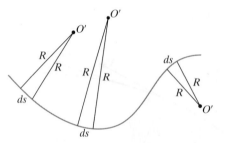

Radius of curvature at various points along the streamline

(b)

$$a_s\Big|_{\text{local}} = \left(\frac{\partial V}{\partial t}\right)_s \quad \text{and} \quad a_n\Big|_{\text{local}} = \left(\frac{\partial V}{\partial t}\right)_n$$

For example, a *local streamline (or tangential)* component of *acceleration* $(\partial V/\partial t)_s$ can occur if the *speed* of the flow in a pipe is increased or decreased by opening or closing a valve. Here the *magnitude* of the particle's velocity within the control volume increases or decreases as a function of time. Also, a *local normal component of acceleration* $(\partial V/\partial t)_n$ can occur in the control volume if the pipe is *rotating*, as in the case of a sprinkler, since this will cause the *direction* of the steamline *and* the velocity of the particle *within* the control volume to change with time.

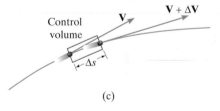

(c)

Fig. 3–19

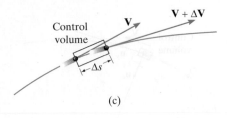

(c)

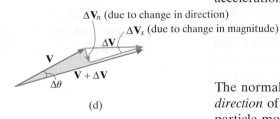

(d)

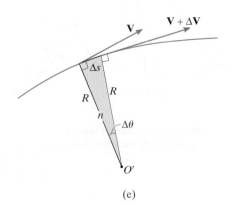

(e)

Fig. 3–19 (cont.)

Convective Change. The velocity of the particle can also change as the particle *moves* Δs from the entrance to the exit control surface, Fig. 3–19c. This convective change, represented here as $\Delta \mathbf{V}$, has components $\Delta \mathbf{V}_s$ and $\Delta \mathbf{V}_n$, Fig. 3–19d. In particular, $\Delta \mathbf{V}_s$ represents the convective *change in the magnitude* of $\mathbf{V}$. It indicates whether the fluid particle speeds up, as when it moves through a converging pipe (or nozzle), or slows down, as when it moves through a diverging pipe. Both of these cases represent *nonuniform flow*. To obtain this convective acceleration component in the s direction, we have

$$a_s \Big|_{\text{conv}} = \lim_{\Delta t \to 0} \frac{\Delta V_s}{\Delta t} = \lim_{\Delta t \to 0} \frac{\Delta s}{\Delta t} \frac{\Delta V_s}{\Delta s} = V \frac{\partial V}{\partial s}$$

The normal component $\Delta \mathbf{V}_n$ in Fig. 3–19d is due to the *change in the direction* of $\mathbf{V}$, since it indicates how the velocity vector "swings" as the particle moves or is *convected* through the control volume. Since $\mathbf{V}$ is always tangent to the path, the change in angle $\Delta \theta$ between $\mathbf{V}$ and $\mathbf{V} + \Delta \mathbf{V}$, Fig. 3–19d, must be the same angle $\Delta \theta$ shown in Fig. 3–19e. And, because $\Delta \theta$ is very small, then $\Delta \theta = \Delta s/R$ (Fig. 3–19e) and also $\Delta \theta = \Delta V_n/V$ (Fig. 3–19d). Equating these results gives $\Delta V_n = (V/R)\Delta s$. Therefore the convective acceleration component in the n direction becomes

$$a_n \Big|_{\text{conv}} = \lim_{\Delta t \to 0} \frac{\Delta V_n}{\Delta t} = \frac{V}{R} \lim_{\Delta t \to 0} \frac{\Delta s}{\Delta t} = \frac{V^2}{R}$$

A typical example of this acceleration component occurs in a curved pipe, because the direction of the velocity of the particle will change as the particle moves from the entrance to the exit control surface.

Resultant Acceleration. If we now combine both the local and the convective changes using the above results, the streamline and normal acceleration components become

$$a_s = \left(\frac{\partial V}{\partial t} \right)_s + V \frac{\partial V}{\partial s} \tag{3–11}$$

$$a_n = \left(\frac{\partial V}{\partial t} \right)_n + \frac{V^2}{R} \tag{3–12}$$

To summarize, the first terms on the right are the *local changes* in the velocity's magnitude and direction, caused by *unsteady flow*, and the second terms are *convective changes* in the velocity's magnitude and direction, caused by *nonuniform flow*.

Important Points

- The material derivative is used to determine the acceleration of a particle when the velocity field is known. It consists of two parts, the *local* or time change made within the control volume, and the *convective* or position change made as the particle moves into and out of the control surfaces.

- Streamline coordinates are located at a point on a streamline. They consist of an *s coordinate* axis that is tangent to the streamline, positive in the direction of flow, and an *n coordinate* axis that is normal to the *s* axis. It is positive if it acts toward the center of curvature of the streamline at the point.

- The velocity of a fluid particle always acts in the +*s* direction.

- The *s component of acceleration* of a particle consists of the *magnitude change in the velocity*. This is the result of the local time rate of change, $(\partial V / \partial t)_s$, and convective change, $V(\partial V / \partial s)$.

- The *n component of acceleration* of a particle consists of *directional change in the velocity*. This is the result of the local time rate of change, $(\partial V / \partial t)_n$, and convective change, V^2 / R.

Using particles of smoke, streamlines provide a method of visualizing the flow of air over the body of this automobile. (© Frank Herzog/Alamy)

EXAMPLE | 3.7

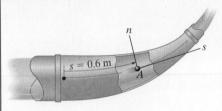

Fig. 3–20

As a fluid flows through the fixed curved conduit in Fig. 3–20, the velocity of particles on the streamline is described by $V = (0.4s^2)e^{-0.4t}$ m/s, where s is in meters and t is in seconds. Determine the magnitude of the acceleration of the fluid particle located at point A, where $s = 0.6$ m, when $t = 1$ s. The radius of curvature of the streamline at A is $R = 0.5$ m.

SOLUTION

Fluid Description.   Since this Eulerian description of the motion is a function of space and time, the flow is nonuniform and unsteady.

Analysis.   The streamline coordinates are established at point A.

Streamline Acceleration Component.   Applying Eq. 3–11, to determine the magnitude change in velocity, we have

$$a_s = \left(\frac{\partial V}{\partial t}\right)_s + V\frac{\partial V}{\partial s}$$

$$= \frac{\partial}{\partial t}\left[(0.4s^2)e^{-0.4t}\right] + \left[(0.4s^2)e^{-0.4t}\right]\frac{\partial}{\partial s}\left[(0.4s^2)e^{-0.4t}\right]$$

$$a_s = 0.4s^2(-0.4e^{-0.4t}) + (0.4s^2)e^{-0.4t}(0.8s\,e^{-0.4t})$$

$$= 0.4(0.6 \text{ m})^2\left[-0.4\,e^{-0.4(1\,\text{s})}\right] + (0.4)(0.6 \text{ m})^2e^{-0.4(1\,\text{s})}\left[0.8(0.6 \text{ m})e^{-0.4(1\,\text{s})}\right]$$

$$= -0.00755 \text{ m/s}^2$$

Normal Acceleration Component.   Because the pipe is not rotating, then the streamline is not rotating, and so the n axis remains in a *fixed direction* at A. Therefore, there is no local change in the *direction* of velocity within the control volume along the n axis. Applying Eq. 3–12, we only have a convective change in the n direction.

$$a_n = \left(\frac{\partial V}{\partial t}\right)_n + \frac{V^2}{R} = 0 + \frac{\left[0.4(0.6 \text{ m})^2e^{-0.4(1\,\text{s})}\right]^2}{0.5}$$

$$= 0.01863 \text{ m/s}^2$$

Acceleration.   The magnitude of acceleration is therefore

$$a = \sqrt{a_s^2 + a_n^2} = \sqrt{(-0.00755 \text{ m/s}^2)^2 + (0.01863 \text{ m/s}^2)^2}$$

$$= 0.0201 \text{ m/s}^2 = 20.1 \text{ mm/s}^2 \qquad\qquad Ans.$$

References

1. D. Halliday, et al, *Fundamentals of Physics*, 7th ed, J. Wiley and Sons, Inc., N.J., 2005

2. W. Merzkirch, *Flow Visualization*. 2nd ed., Academic Press, New York, NY, 1897.

3. R. C. Baker, *Introductory Guide to Flow Measurement*, 2nd ed., John Wiley, New York, NY, 2002.

4. R. W. Miller, *Flow Measurement Engineering Handbook*, 3rd ed., McGraw-Hill, New York, NY, 1997.

5. G. S. Settles, *Schlieren and Shadowgraph Techniques: Visualizing Phenomena in Transport Media,* Springer, Berlin, 2001.

FUNDAMENTAL PROBLEMS

Sec. 3.1–3.3

F3–1. The velocity field of a two-dimensional flow field is defined by $u = \left(\frac{1}{4}x\right)$ m/s and $v = (2t)$ m/s, where x is in meters and t is in seconds. Determine the position (x, y) of a particle when $t = 2$ s, if the particle passes through point $(2 \text{ m}, 6 \text{ m})$ when $t = 0$.

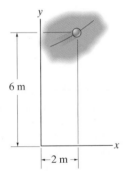

Prob. F3–1

F3–2. A flow field is defined by the velocity components $u = \left(2x^2\right)$ m/s and $v = (8y)$ m/s, where x and y are in meters. Determine the equation of the streamline passing through the point $(2 \text{ m}, 3 \text{ m})$.

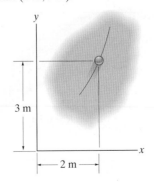

Prob. F3–2

Sec. 3.4

F3–3. The flow of water through the nozzle causes the velocity of particles along the center streamline to be $V = \left(200x^3 + 10t^2\right)$ m/s, where t is in seconds and x is in meters. Determine the acceleration of a particle at $x = 0.1$ m when $t = 0.2$ s.

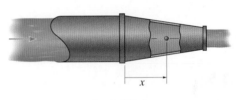

Prob. F3–3

F3–4. The velocity of dioxitol along the x axis is given by $u = 3(x + 4)$ m/s, where x is in meters. Find the acceleration of a particle located at $x = 100$ mm. What is the position of a particle when $t = 0.02$ s if it starts from $x = 0$ when $t = 0$?

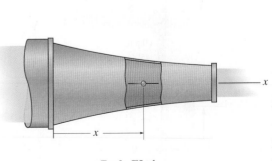

Prob. F3–4

F3–5. A flow field has velocity components of $u = (3x + 2t^2)$ m/s and $v = (2y^3 + 10t)$ m/s, where x and y are in meters and t is in seconds. Determine the magnitudes of the local and convective accelerations of a particle at point $x = 3$ m, $y = 1$ m when $t = 2$ s.

F3–7. A fluid flows through the curved pipe with a constant average speed of 3 m/s. Determine the magnitude of acceleration of water particles on the streamline along the centerline of the pipe.

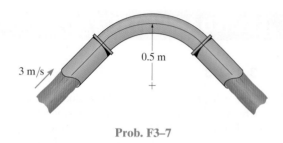

Prob. F3–7

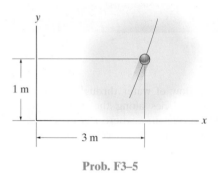

Prob. F3–5

Sec. 3.5

F3–6. Fluid flows through the curved pipe at a steady rate. If the velocity of a particle along the center streamline is $V = (20s^2 + 4)$ m/s, where s is in meters, determine the magnitude of the acceleration of a particle at point A.

F3–8. Fluid flows through the curved pipe such that along the center streamline $V = (20s^2 + 1000t^{3/2} + 4)$ m/s, where s is in meters and t is in seconds. Determine the magnitude of the acceleration of a particle at point A, where $s = 0.3$ m, when $t = 0.02$ s.

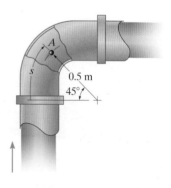

Prob. F3–6

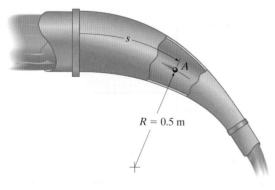

Prob. F3–8

PROBLEMS

Sec. 3.1–3.3

3–1. A marked particle is released into a flow when $t = 0$, and the pathline for the particle is shown. Draw the streakline and the streamline for the particle when $t = 2$ s and $t = 4$ s.

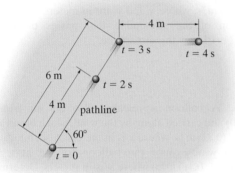

Prob. 3–1

3–2. A flow field for a fluid is described by $u = (2 + y)$ m/s and $v = (2y)$ m/s, where y is in meters. Determine the equation of the streamline that passes through point $(3$ m, 2 m$)$, and find the velocity of a particle located at this point. Draw this streamline.

3–3. A fluid has velocity components of $u = [30/(2x + 1)]$ m/s and $v = (2ty)$ m/s, where x and y are in meters and t is in seconds. Determine the streamlines that passes through point $(1$ m, 4 m$)$ at times $t = 1$ s, $t = 2$ s, and $t = 3$ s.

*3–4. The flow of a liquid is originally along the positive y axis at 3 m/s for 4 s. If it then suddenly changes to 2 m/s along the positive x axis for $t > 4$ s, draw the pathline and streamline for the first marked particle when $t = 2$ s and $t = 6$ s. Also draw the streakline at these two times.

3–5. The soap bubble is released in the air and rises with a velocity of $\mathbf{V} = \{(0.8x)\mathbf{i} + (0.06t^2)\mathbf{j}\}$ m/s, where x is in meters and t is in seconds. Determine the magnitude of the bubble's velocity, and its direction measured counterclockwise from the x axis, when $t = 5$ s, at which time $x = 2$ m and $y = 3$ m. Draw its streamline at this instant.

Prob. 3–5

3–6. A flow of water is defined by $u = 5$ m/s and $v = 8$ m/s. If metal flakes are released into the flow at the origin $(0, 0)$, draw the streamline and pathline for these particles.

3–7. A velocity field is defined by $u = (4x)$ m/s and $v = (2t)$ m/s, where t is in seconds and x is in meters. Determine the pathline that passes through point $(2$ m, 6 m$)$ when $t = 1$ s. Plot this pathline for 0.25 m $\leq x \leq 4$ m.

*3–8. The flow of a liquid is originally along the positive x axis at 2 m/s for 3 s. If it then suddenly changes to 4 m/s along the positive y axis for $t > 3$ s, draw the pathline and streamline for the first marked particle when $t = 1$ s and $t = 4$ s. Also draw the streakline at these two times.

3–9. A two-dimensional flow field for a fluid can be described by $\mathbf{V} = \{(2x + 1)\mathbf{i} - (y + 3x)\mathbf{j}\}$ m/s, where x and y are in meters. Determine the magnitude of the velocity of a particle located at (2 m, 3 m), and its direction measured counterclockwise from the x axis.

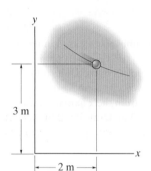

y

3 m

2 m

x

Prob. 3–9

3–10. A two-dimensional flow field for a liquid can be described by $\mathbf{V} = \{(5y^2 - x)\mathbf{i} + (3x + y)\mathbf{j}\}$ m/s, where x and y are in meters. Determine the magnitude of the velocity of a particle located at (5 m, −2 m), and its direction measured counterclockwise from the x axis.

3–11. A two-dimensional flow field for a liquid can be described by $\mathbf{V} = \{(6y^2 - 1)\mathbf{i} + (3x + 2)\mathbf{j}\}$ m/s, where x and y are in meters. Find the streamline that passes through point (6 m, 2 m) and determine the velocity at this point. Sketch the velocity on the streamline.

*3–12. A stream of water has velocity components of $u = -2$ m/s, $v = 3$ m/s for $0 \le t < 10$ s; and $u = 5$ m/s, $v = -2$ m/s for 10 s $< t \le 15$ s. Plot the pathline and streamline for a particle released at point (0, 0) when $t = 0$ s.

3–13. A flow field is defined by $u = (0.8t)$ m/s and $v = 0.4$ m/s, where t is in seconds. Plot the pathline for a particle that passes through the origin when $t = 0$. Also, draw the streamline for the particle when $t = 4$ s.

3–14. A flow field is defined by $u = \left[8x/(x^2 + y^2)\right]$ m/s and $v = \left[8y/(x^2 + y^2)\right]$ m/s, where x and y are in meters. Determine the equation of the streamline passing through point (1 m, 1 m). Draw this streamline.

3–15. A flow field is defined by $u = (8y)$ m/s and $v = (6x)$ m/s, where x and y are in meters. Determine the equation of the streamline that passes through point (1 m, 2 m). Draw this streamline.

*3–16. A balloon is released into the air from point (1 m, 0) and carried along by the wind, which blows at a rate of $u = (0.8x)$ m/s, where x is in meters. Also, buoyancy and thermal winds cause the balloon to rise at a rate of $v = (1.6 + 0.4y)$ m/s, where y is in meters. Determine the equation of the streamline for the balloon, and draw this streamline.

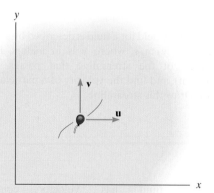

y

$\mathbf{v}$

$\mathbf{u}$

x

Prob. 3–16

3–17. A balloon is released into the air from the origin and carried along by the wind, which blows at a constant rate of $u = 0.5$ m/s. Also, buoyancy and thermal winds cause the balloon to rise at a rate of $v = (0.8 + 0.6y)$ m/s, where y is in meters. Determine the equation of the streamline for the balloon, and draw this streamline.

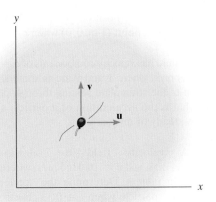

Prob. 3–17

3–18. A fluid has velocity components of $u = [30/(2x + 1)]$ m/s and $v = (2ty)$ m/s, where x and y are in meters and t is in seconds. Determine the pathline that passes through point (2 m, 6 m) at time $t = 2$ s. Plot this pathline for $0 \leq x \leq 4$ m.

3–19. The velocity field of a fluid is defined by $u = \left(\frac{1}{2}x\right)$ m/s, $v = \left(\frac{1}{8}y^2\right)$ m/s for $0 \leq t < 5$ s and by $u = \left(-\frac{1}{4}x^2\right)$ m/s, $v = \left(\frac{1}{4}y\right)$ m/s for 5 s $< t \leq 10$ s, where x and y are in meters. Plot the streamline and pathline for a particle released at point (1 m, 1 m) when $t = 0$ s.

*3–20. A fluid has velocity components of $u = [30/(2x + 1)]$ m/s and $v = (2ty)$ m/s, where x and y are in meters and t is in seconds. Determine the streamlines that pass through point (2 m, 6 m) at times $t = 2$ s and $t = 5$ s. Plot these streamlines for $0 \leq x \leq 4$ m.

3–21. The velocity for an oil flow is defined by $\mathbf{V} = \left\{3y^2\mathbf{i} + 8\mathbf{j}\right\}$ m/s, where y is in meters. What is the equation of the streamline that passes through point (2 m, 1 m)? If a particle is at this point when $t = 0$, at what point is it located when $t = 1$ s?

3–22. A flow field for a liquid is described by $\mathbf{V} = \left\{(2x + 1)\mathbf{i} - y\mathbf{j}\right\}$ m/s, where x and y are in meters. Determine the magnitude of the velocity of a particle located at point (3 m, 1 m). Sketch the velocity on the streamline.

3–23. The circulation of a fluid is defined by the velocity field $u = (6 - 3x)$ m/s and $v = 2$ m/s, where x is in meters. Plot the streamline that passes through the origin for $0 \leq x < 2$ m.

*3–24. Particles travel within a flow field defined by $\mathbf{V} = \left\{2y^2\mathbf{i} + 4\mathbf{j}\right\}$ m/s, where x and y are in meters. Determine the equation of the streamline passing through point (1 m, 2 m), and find the velocity of a particle located at this point. Draw this streamline.

3–25. A flow field is described by $u = \left(x^2 + 5\right)$ m/s and $v = (-6xy)$ m/s, where x and y are in meters. Determine the equation of the streamline that passes through point (5 m, 1 m), and find the velocity of a particle located at this point. Draw this streamline.

3–26. A particle travels along a streamline defined by $y^3 = 8x - 12$. If its speed is 5 m/s when it is at $x = 1$ m, determine the two components of its velocity at this point. Sketch the velocity on the streamline.

3–27. A velocity field is defined by $u = (4x)$ m/s and $v = (2t)$ m/s, where t is in seconds and x is in meters. Determine the equation of the streamline that passes through point (2 m, 6 m) when $t = 1$ s. Plot this streamline for 0.25 m $\leq x \leq 4$ m.

Sec. 3.4

*3–28. Air flows uniformly through the center of a horizontal duct with a velocity of $V = (6t^2 + 5)$ m/s, where t is in seconds. Determine the acceleration of the flow when $t = 2$ s.

3–29. A fluid has velocity components of $u = (5y^2)$ m/s and $v = (4x - 1)$ m/s, where x and y are in meters. Determine the equation of the streamline passing through point $(1 \text{ m}, 1 \text{ m})$. Find the components of the acceleration of a particle located at this point, and sketch the acceleration on the streamline.

3–30. Air flowing through the center of the duct decreases in speed from $V_A = 8$ m/s to $V_B = 2$ m/s in a linear manner. Determine the velocity and acceleration of a particle moving horizontally through the duct as a function of its position x. Also, find the position of the particle as a function of time if $x = 0$ when $t = 0$.

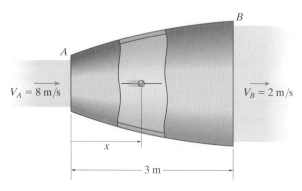

Prob. 3–30

3–31. The velocity for the flow of a gas along the center streamline of the pipe is defined by $u = (10x^2 + 200t + 6)$ m/s, where x is in meters and t is in seconds. Determine the acceleration of a particle when $t = 0.01$ s and it is at A, just before leaving the nozzle.

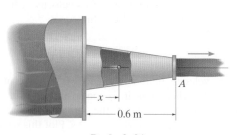

Prob. 3–31

*3–32. A fluid has velocity components of $u = (2x^2 - 2y^2 + y)$ m/s and $v = (y + xy)$ m/s, where x and y are in meters. Determine the magnitude of the velocity and acceleration of a particle located at point $(2 \text{ m}, 4 \text{ m})$.

3–33. The velocity field for a flow of water is defined by $u = (2x)$ m/s, $v = (6tx)$ m/s, and $w = (3y)$ m/s, where t is in seconds and x, y, z are in meters. Determine the acceleration and the position of a particle when $t = 0.5$ s if this particle is at point $(1 \text{ m}, 0, 0)$ when $t = 0$.

3–34. A fluid has velocity components of $u = (2y^2)$ m/s and $v = (8xy)$ m/s, where x and y are in meters. Determine the equation of the streamline passing through point $(1 \text{ m}, 2 \text{ m})$. Also, what is the acceleration of a particle at this point? Is the flow steady or unsteady?

3–35. A velocity field for oil is defined by $u = (100y)$ m/s, and $v = (0.03t^2)$ m/s, where t is in seconds and y is in meters. Determine the acceleration and the position of a particle when $t = 0.5$ s. The particle is at the origin when $t = 0$.

*3–36. If $u = (2x^2)$ m/s and $v = (-y)$ m/s, where x and y are in meters, determine the equation of the streamline that passes through point $(2 \text{ m}, 6 \text{ m})$, and find the acceleration of a particle at this point. Sketch this streamline for $x > 0$, and find the equations that define the x and y components of acceleration of the particle as a function of time if $x = 2$ m and $y = 6$ m when $t = 0$.

3–37. As the valve is closed, oil flows through the nozzle such that along the center streamline it has a velocity of $V = [6(1 + 0.4x^2)(1 - 0.5t)]$ m/s, where x is in meters and t is in seconds. Determine the acceleration of an oil particle at $x = 0.25$ m when $t = 1$ s.

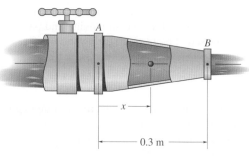

Prob. 3–37

3–38. The velocity field for a fluid is defined by $u = \left[y/(x^2 + y^2)\right]$ m/s and $v = \left[4x/(x^2 + y^2)\right]$ m/s, where x and y are in meters. Determine the acceleration of a particle located at point (2 m, 0) and a particle located at point (4 m, 0). Sketch the equations that define the streamlines that pass through these points.

3–39. The velocity of gasoline, along the centerline of a tapered pipe, is given by $u = (4tx)$ m/s, where t is in seconds and x is in meters. Determine the acceleration of a particle when $t = 0.8$ s if $u = 0.8$ m/s when $t = 0.1$ s.

*3–40. A fluid has velocity components of $u = (5y^2 - x)$ m/s and $v = (4x^2)$ m/s, where x and y are in meters. Determine the velocity and acceleration of a particle passing through point (2 m, 1 m).

3–41. A fluid has velocity components of $u = (8t^2)$ m/s and $v = (7y + 3x)$ m/s, where x and y are in meters and t is in seconds. Determine the velocity and acceleration of a particle passing through point (1 m, 1 m) when $t = 2$ s.

3–42. The velocity of a flow field is defined by $u = (-y/4)$ m/s and $v = (x/9)$ m/s, where x and y are in meters. Determine the magnitude of the velocity and acceleration of a particle that passes through point (3 m, 2 m). Find the equation of the streamline passing through this point, and sketch the velocity and acceleration at the point on this streamline.

3–43. The velocity of a flow field is defined by $V = \{4x\mathbf{i} + 2\mathbf{j}\}$ m/s, where x is in meters. Determine the magnitude of the velocity and acceleration of a particle that passes through point (1 m, 2 m). Find the equation of the streamline passing through this point, and sketch the velocity and acceleration at the point on this streamline.

*3–44. A flow field has velocity components of $u = -(4x + 6)$ m/s and $v = (10y + 3)$ m/s, where x and y are in meters. Determine the equation for the streamline that passes through point (1 m, 1 m), and find the acceleration of a particle at this point.

3–45. The velocity of a flow field is defined by $V = \{4y\mathbf{i} + 2x\mathbf{j}\}$ m/s, where x and y are in meters. Determine the magnitude of the velocity and acceleration of a particle that passes through point (2 m, 1 m). Find the equation of the streamline passing through this point, and sketch the velocity and acceleration at the point on this streamline.

3–46. The velocity of a flow field is defined by $u = (2x^2 - y^2)$ m/s and $v = (-4xy)$ m/s, where x and y are in meters. Determine the magnitude of the velocity and acceleration of a particle that passes through point (1 m, 1 m). Find the equation of the streamline passing through this point, and sketch the velocity and acceleration at the point on this streamline.

3–47. Airflow through the duct is defined by the velocity field $u = (2x^2 + 8)$ m/s and $v = (-8x)$ m/s, where x is in meters. Determine the acceleration of a fluid particle at the origin (0,0) and at point (1 m,0). Also, sketch the streamlines that pass through these points.

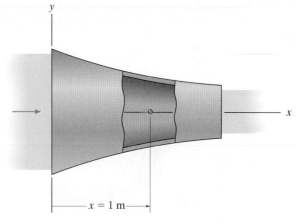

Prob. 3–47

3

Sec. 3.5

***3–48.** Fluid particles have velocity components of $u = (8y)$ m/s and $v = (6x)$ m/s, where x and y are in meters. Determine the acceleration of a particle located at point $(1\text{ m}, 1\text{ m})$. Determine the equation of the streamline passing through this point.

3–49. Water flows into the drainpipe such that it only has a radial velocity component $V = (-3/r)$ m/s, where r is in meters. Determine the acceleration of a particle located at point $r = 0.5$ m, $\theta = 20°$. At $s = 0, r = 1$ m.

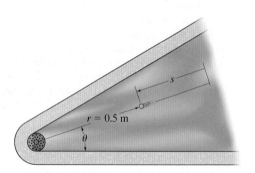

Prob. 3–49

3–50. As water flows steadily over the spillway, one of its particles follows a streamline that has a radius of curvature of 16 m. If its speed at point A is 5 m/s, which is increasing at 3 m/s², determine the magnitude of acceleration of the particle.

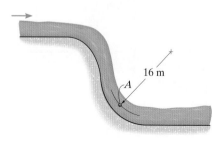

Prob. 3–50

3–51. The motion of a tornado can, in part, be described by a free vortex, $V = k/r$, where k is a constant. Consider the steady motion at the radial distance $r = 3$ m, where $V = 18$ m/s. Determine the magnitude of the acceleration of a particle traveling on the streamline having a radius of $r = 9$ m.

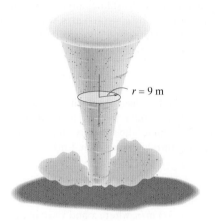

Prob. 3–51

***3–52.** Air flows around the front circular surface. If the steady-stream velocity is 4 m/s upstream from the surface, and the velocity along the surface is defined by $V = (16 \sin \theta)$ m/s, determine the magnitude of the streamline and normal components of acceleration of a particle located at $\theta = 30°$.

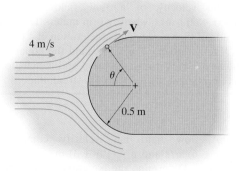

Prob. 3–52

3–53. A particle located at a point within a fluid flow has velocity components of $u = 4$ m/s and $v = -3$ m/s, and acceleration components of $a_x = 2$ m/s² and $a_y = 8$ m/s². Determine the magnitude of the streamline and normal components of acceleration of the particle.

3–54. A particle moves along the circular streamline, such that it has a velocity of 3 m/s, which is increasing at 3 m/s². Determine the acceleration of the particle, and show the acceleration on the streamline.

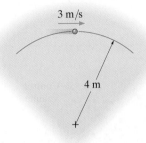

Prob. 3–54

3–55. A fluid has velocity components of $u = (2y^2)$ m/s and $v = (8xy)$ m/s, where x and y are in meters. Determine the magnitude of the streamline and normal components of acceleration of a particle located at point $(1\,\text{m}, 2\,\text{m})$.

*3–56. A fluid has velocity components of $u = (2y^2)$ m/s and $v = (8xy)$ m/s, where x and y are in meters. Determine the magnitude of the streamline and normal components of acceleration of a particle located at point $(1\,\text{m}, 1\,\text{m})$. Find the equation of the streamline passing through this point, and sketch the streamline and normal components of acceleration at this point.

3–57. Fluid particles have velocity components of $u = (8y)$ m/s and $v = (6x)$ m/s, where x and y are in meters. Determine the magnitude of the streamline and the normal components of acceleration of a particle located at point $(1\,\text{m}, 2\,\text{m})$.

CHAPTER REVIEW

A Lagrangian description follows the motion of a *single particle* as the particle moves through the flow field.

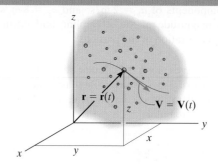

A Eulerian description considers a specific region (or control volume) in the flow, and it measures the motion or a fluid property of all the particles that pass through this region.

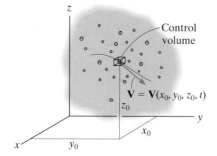

Laminar flow occurs when the fluid flows in thin layers so that the fluid particles follow smooth paths.

Turbulent flow is very erratic, causing mixing of the fluid particles, and therefore more internal friction than laminar flow.

Steady flow occurs when the flow does not change with time.

Uniform flow occurs when the flow does not change with location.

A *streamline* is a curve that indicates the direction of the velocity of particles located on it at a particular *instant of time*.

$$\frac{dy}{dx} = \frac{v}{u}$$

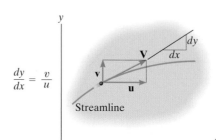

A *pathline* shows the path of a particle during a specified time period. It is determined using a time-lapse photograph.

Streaklines are formed when smoke or colored dye is released into the flow from the same point, and an instantaneous photograph is taken of the trace of all the marked particles.

If the flow is steady, the streamlines, pathlines, and streaklines will all coincide.

Fluid motion around objects having a high speed, or those producing heat, can be visualized using a shadowgraph or a schlieren photograph. Also, complex flows can be visualized using a computational fluid dynamics program.

If a Eulerian description is used to define the velocity field $\mathbf{V} = \mathbf{V}(x, y, z, t)$, then the acceleration will have both local and convective components. The *local acceleration* accounts for the *time rate of change* in velocity within the control volume, and the *convective acceleration* accounts for its *spatial change*, from the point where the particle enters a control surface to the point where it exits another surface.

$$\mathbf{a} = \frac{D\mathbf{V}}{Dt} = \frac{\partial \mathbf{V}}{\partial t} + \left(u\frac{\partial \mathbf{V}}{\partial x} + v\frac{\partial \mathbf{V}}{\partial y} + w\frac{\partial \mathbf{V}}{\partial z} \right)$$

Total   Local   Convective accel.
accel.   accel.

Streamline coordinates s, n have their origin at a point on a streamline. The s coordinate is tangent to the streamline and it is positive in the direction of flow. The normal coordinate n is positive toward the streamline's center of curvature at the point.

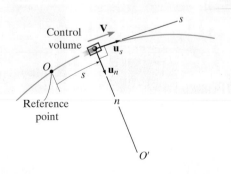

Chapter 4

(©wu kailiang/Alamy)

The analysis of flow through the various ducts and vessels in this chemical processing plant depends upon the conservation of mass.

Conservation of Mass

CHAPTER OBJECTIVES

■ To define the concept of a finite control volume, and then show how a Lagrangian and a Eulerian description of fluid behavior can be related using the Reynolds transport theorem.

■ To show how to determine the volume and mass flow, and the average velocity in a conduit.

■ To use the Reynolds transport theorem to derive the continuity equation, which represents the conservation of mass.

■ To demonstrate application of the continuity equation to problems involving finite fixed, moving, and deformable control volumes.

4.1 Finite Control Volumes

In Sec. 3.1 we defined a control volume as a selected differential volume of space within a system of particles through which some of the fluid particles flow. Recall that the boundary of this volume is the *control surface*. A portion of the surface of this volume may be *open*, where the fluid particles flow into or out of the control volume, and the remaining portion is a *closed* surface. Although in the previous sections we considered a fixed *differential size* control volume, we can also consider control volumes that have a *finite size*, such as the one shown in Fig. 4–1.

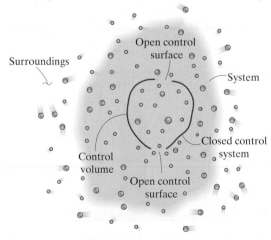

Fig. 4–1

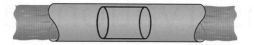

Fixed control volume

(a)

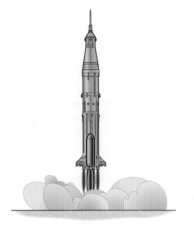

Moving control volume

(b)

Changing control volume

(c)

Fig. 4–2

Actually, depending upon the problem, a control volume can be *fixed*, it can *move*, or it can *change shape*. Also, it may include solid parts of an object within its boundary. For example, a fixed control volume within the pipe in Fig. 4–2a is indicated by the red boundary. The control volume that outlines the rocket engine and the fluid within it, Fig. 4–2b, moves with the rocket as it travels upward. Finally, the control volume within the inflatable structure in Fig. 4–2c changes its shape while air is pumped into its open control surface.

As we have seen, using a control volume is fundamental to applying a Eulerian description of the flow. It is for this reason that a control volume approach will be used in this chapter to solve problems that involve the continuity of flow. Then later we will extend its application to problems that involve energy, Chapter 5, and momentum, Chapter 6. In all these cases, it will be necessary to clearly define the boundaries of a selected control volume, and to specify the size and orientation of its open control surfaces.

Open Control Surfaces.

The open control surfaces of a control volume will have an area that either lets fluid flow into the control volume, A_{in}, or lets fluid flow out of it, A_{out}. In order to properly identify these areas, we will express each as a vector, where its direction is normal to the area and *always* directed *outward* from the control volume. For example, if the fixed control volume in Fig. 4–3 is used to study the flow into and out of the tee connection, then the direction of each open control surface *area* is defined by its outward normal, and is represented by the vectors $\mathbf{A}_A$, $\mathbf{A}_B$, and $\mathbf{A}_C$.

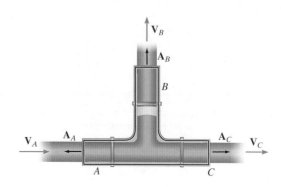

Fig. 4–3

Velocity.   When using a control volume, we will also have to specify the velocity of flow both into and out of each control surface. Since the *outward* normal for each control surface area is positive, then flow *into* a control surface will be *negative*, and flow *out* of a control surface will be *positive*. By this convention, $\mathbf{V}_A$ in Fig. 4–4 is in the negative direction, and $\mathbf{V}_B$ and $\mathbf{V}_C$ are in the positive direction.

Steady Flow.   In some problems it will be advantageous to select a moving control volume in order to observe steady flow, and thereby simplify the analysis. For example, consider the blade moving with a velocity $\mathbf{V}_b$, shown in Fig. 4–4a. For a *fixed observer*, the flow at A will appear to be $\mathbf{V}_f$ at time t; however, when the blade advances, the velocity at A then becomes $\mathbf{V}_f'$ at time $t + \Delta t$. This is a case of *unsteady flow* since it changes with time. If we select a control volume that contains the fluid on the blade, and then move this control volume and observer *with the blade* at $\mathbf{V}_{\text{cv}} = \mathbf{V}_b$, then the flow at A will appear as *steady flow*, Fig. 4–4b. Since the velocity of the fluid stream from the nozzle is $\mathbf{V}_f$, the steady-flow velocity of the fluid relative to the open control surface, $\mathbf{V}_{f/\text{cs}}$, is determined from the *relative-velocity equation*, $\mathbf{V}_f = \mathbf{V}_{\text{cs}} + \mathbf{V}_{f/\text{cs}}$, or

$$\mathbf{V}_{f/\text{cs}} = \mathbf{V}_f - \mathbf{V}_{\text{cs}}$$

where $\mathbf{V}_{\text{cs}} = \mathbf{V}_b$. In scalar form this equation becomes

$$(\overset{+}{\rightarrow}) \qquad V_{f/\text{cs}} = V_f - (-V_{\text{cs}}) = V_f + V_{\text{cs}}$$

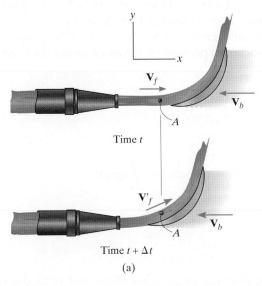

(a)

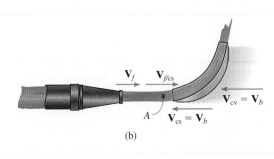

(b)

Fig. 4–4

4.2   The Reynolds Transport Theorem

Much of fluid behavior is based on the conservation of mass, the principle of work energy, and the principle of impulse and momentum. These laws were originally formulated for a *particle*, and they were described using a Lagrangian approach. However, for application in fluid mechanics, we must have a means to *convert* these laws from their Lagrangian description to a Eulerian description. This conversion for a system of particles is done using the Reynolds transport theorem. In this section we will formalize this theorem, and then we will use it in the next section to develop the continuity equation, and then later in Chapters 5 and 6 use it to develop the energy and momentum equations. Before we establish this theorem, however, we will first discuss how to best describe each fluid property in terms of the fluid's mass and volume.

Fluid Property Description.   Any fluid property that *depends on the amount of volume or mass* in a system is called an **extensive property**, N, because the volume or mass "extends" throughout the system. For example, momentum is an extensive property since it represents mass times velocity, $N = mV$. Fluid properties that are independent of the system's mass are called **intensive properties**, η (eta). Examples include temperature and pressure.

We can represent an extensive property N as an intensive property η simply by expressing it per unit mass, that is, $\eta = N/m$. Since momentum is $N = mV$, then $\eta = V$. Likewise, kinetic energy is $N = (1/2)mV^2$, and so $\eta = (1/2)V^2$. Since mass is related to volume by $m = \rho V$, then in general, the relation between an extensive and intensive property for a system of fluid particles, expressed in terms of either its mass or its volume, is

$$N = \int_m \eta \, dm = \int_V \eta \rho \, dV \tag{4–1}$$

The integrations are over the entire mass of the system or the volume it covers.

The Reynolds Transport Theorem.   We are now ready to relate the *time rate of change* of any extensive property N for a fluid system to its time rate of change as seen from the control volume. To do this, we will consider the control volume as fixed within a conduit, and outlined by the red boundary in Fig. 4–5a. At time t, the entire *system* of fluid particles is considered to be within, and coincident with the control volume (CV). At time $t + \Delta t$, a portion of this system of particles exits the open control surface and is now in region R_{out}, *outside* the control volume, Fig. 4–5b. This will leave a void R_{in} within the control volume. In other words, the system of fluid particles went from occupying CV at time t, to occupying $[CV + (R_{out} - R_{in})]$ at time $t + \Delta t$.

Since this change in the location of the system of particles occurs during the time Δt, then the fluid's extensive property N within this system will also change. By definition of the derivative, we can write this change as

$$\left(\frac{dN}{dt}\right)_{syst} = \lim_{\Delta t \to 0} \frac{(N_{syst})_{t+\Delta t} - (N_{syst})_t}{\Delta t}$$

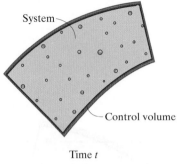

System

Control volume

Time t

(a)

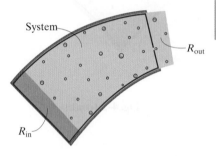

System

R_{out}

R_{in}

Time $t + \Delta t$

(b)

Fig. 4–5

If we represent these changes from the point of view of the control volume, then from the above discussion, we have

$$\left(\frac{dN}{dt}\right)_{syst} = \lim_{\Delta t \to 0}\left[\frac{(N_{cv})_{t+\Delta t} + (\Delta N_{out} - \Delta N_{in}) - (N_{cv})_t}{\Delta t}\right]$$

$$= \lim_{\Delta t \to 0}\left[\frac{(N_{cv})_{t+\Delta t} - (N_{cv})_t}{\Delta t}\right] + \lim_{\Delta t \to 0}\left[\frac{\Delta N_{out}}{\Delta t}\right] - \lim_{\Delta t \to 0}\left[\frac{\Delta N_{in}}{\Delta t}\right] \quad (4-2)$$

The first term on the right represents the *local derivative* since it is the change of N *within the control volume with respect to time*. Using Eq. 4–1, to express this result in terms of the corresponding intensive property η, we have

$$\lim_{\Delta t \to 0}\left[\frac{(N_{cv})_{t+\Delta t} - (N_{cv})_t}{\Delta t}\right] = \frac{\partial N_{cv}}{\partial t} = \frac{\partial}{\partial t}\int_{cv} \eta\rho\, dV \quad (4-3)$$

The second term on the right side of Eq. 4–2 is a *convective derivative* of the extensive property as the system exits the control surface. Since, in general, $\Delta N / \Delta t = \eta\Delta m / \Delta t$, and $\Delta m = \rho\Delta V$, then

$$\frac{\Delta N}{\Delta t} = \eta\rho\frac{\Delta V}{\Delta t}$$

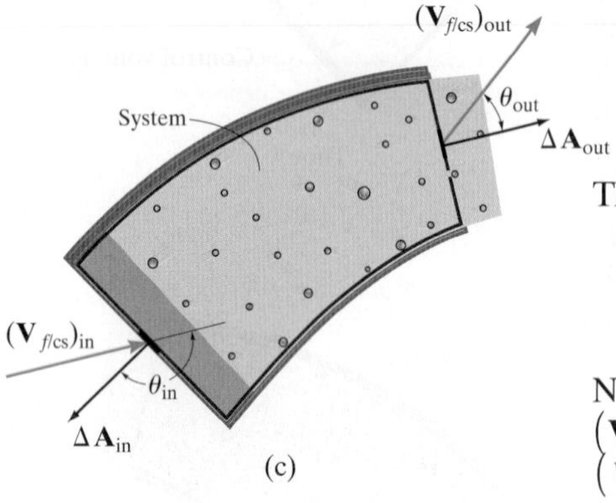

As shown in Fig. 4–5c, the rate at which a small *volume* of fluid particles will flow out of the control surface, having an area ΔA_{out} is $(\Delta V)_{out}/\Delta t = \Delta A_{out}\left[\left(V_{f/cs}\right)_{out}\cos\theta_{out}\right]$. Since ΔA_{out} can be expressed as a vector, then using the dot product,* we can also write $\Delta V_{out}/\Delta t = \left(\mathbf{V}_{f/cs}\right)_{out}\cdot\Delta\mathbf{A}_{out}$. Therefore, the above equation becomes $(\Delta N/\Delta t)_{out} = \eta\rho\left(\mathbf{V}_{f/cs}\right)_{out}\cdot\Delta\mathbf{A}_{out}$. For the *entire* exit control surface,

(c)

Fig. 4–5 (cont.)

$$\lim_{\Delta t\to 0}\left(\frac{\Delta N_{out}}{\Delta t}\right) = \int \eta\rho\left(\mathbf{V}_{f/cs}\right)_{out}\cdot d\mathbf{A}_{out}$$

The same arguments apply for the last term in Eq. 4–9, so

$$\lim_{\Delta t\to 0}\left(\frac{\Delta N_{in}}{\Delta t}\right) = \int \eta\rho\left(\mathbf{V}_{f/cs}\right)_{in}\cdot d\mathbf{A}_{in}$$

Notice that here the dot product will produce a negative result, since $\left(\mathbf{V}_{f/cs}\right)_{in}$ is inward and $d\mathbf{A}_{in}$ is outward. In other words, $\left(\mathbf{V}_{f/cs}\right)_{in}\cdot d\mathbf{A}_{in} = \left(V_{f/cs}\right)_{in} dA_{in}\cos\theta$, where $\theta > 90°$, Fig. 4–5c.

If we combine the above two terms, and express it as a "net" flow through the control volume, then with Eq. 4–3, Eq. 4–2 becomes

$$\boxed{\left(\frac{DN}{Dt}\right)_{syst} = \frac{\partial}{\partial t}\int_{cv}\eta\rho\, dV + \int_{cs}\eta\rho\mathbf{V}\cdot d\mathbf{A}} \qquad (4\text{–}4)$$

Local   Convective
change   change

This result is referred to as the **Reynolds transport theorem** since it was first developed by the British scientist Osborne Reynolds. In summary, it relates the time rate of change of *any* extensive property N of a system of fluid particles, defined from a Lagrangian description, to the changes of the same property from the viewpoint of the control volume, that is, as defined from a Eulerian description. The first term on the right side is the *local change*, since it represents the time rate of change in the intensive property *within* the control volume. The second term on the right is the *convective change*, since it represents the *net flow* of the intensive property through the control surfaces.

The two terms on the right side of Eq. 4–4 form the material derivative of N, and that is why we have symbolized the left side by this operator. In Sec. 3.4 we discussed how the material derivative was used to determine the time rate of change of the velocity for a single fluid particle, when the velocity is expressed as a velocity field (Eulerian description). The Reynolds transport theorem does the same thing; however, here it relates these changes for a fluid property having a finite continuous number of particles within the system.

---

*Recall that the dot product $\mathbf{V}_{f/cs}\cdot\Delta\mathbf{A} = V_{f/cs}\,\Delta A\cos\theta$, where the angle θ $(0° \le \theta \le 180°)$ is measured between the *tails* of the vectors.

Applications.

When applying the Reynolds transport theorem, it is *first necessary* to specify the control volume that contains a selected portion of the fluid system. Once this is done, the local changes of the fluid property *within* the control volume can be determined, as well as the convective changes that occur through its open control surfaces. A few examples will illustrate how this is done.

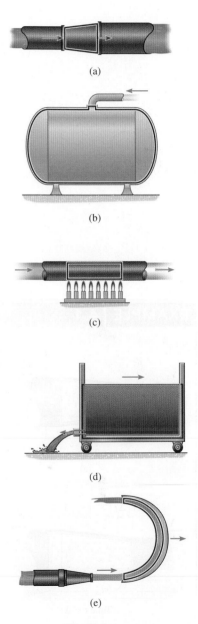

(a)

(b)

(c)

(d)

(e)

Fig. 4–6

- As shown in Fig. 4–6a, an incompressible fluid flows through the pipe transition at a *steady rate*. If we take the volume outlined in red as the fixed control volume, then there will be *no* local changes of the fluid within it because the mass flow is steady and the fluid mass within the control volume remains constant. Convective changes occur at each of the two open control surfaces because there is a flow of fluid mass across these surfaces.

- Air is being pumped into the tank in Fig. 4–6b. If the control volume is taken as the entire volume within the tank, then local changes occur because the mass of the air in the tank is *increasing* with time. Also, convective changes occur at the open control surface or pipe connection.

- As air flows at a steady rate through the pipe in Fig. 4–6c, it is being heated. If we consider the region outlined in red as the control volume, then even though the heating will affect the density of the air, the mass within the control volume is constant, so *no local time rate of change* in mass will occur within the control volume. The density will change, however, this will cause the air to expand, and so the velocity of the air will increase at the exit. We have nonuniform flow. Convective changes occur at the inlet and outlet because the air is moving through these surfaces.

- An incompressible liquid leaks out of the moving cart in Fig. 4–6d. The control volume that contains this liquid in the cart is both moving and deformable. Local changes occur because the mass in the control volume is decreasing with time. Convective change occurs at the open control surface (outlet).

- The liquid on the moving blade in Fig. 4–6e is taken as a control volume. If we observe the motion from the blade, the flow will be steady, and so no local changes occur to the fluid mass within this control volume. Convective changes occur through the open control surfaces.

Further examples of how to select a proper control volume, and specify the local and convective changes that occur, will be presented throughout the text whenever we use the Reynolds transport theorem as it relates to the conservation of mass, the principle of work and energy, and the principle of impulse and momentum.

Important Points

- A Lagrangian description is used to describe the motion of a fluid particle within a *system* of particles, whereas a Eulerian description uses a fixed, moving, or deformable control volume to describe the flow of particles that enter and exit its open control surfaces.

- Fluid properties that depend upon volume or mass, such as energy and momentum, are called *extensive properties, N*. Those properties that are independent of mass, such as temperature and pressure, are called *intensive properties, η*. Any extensive property can be made intensive by dividing it by the fluid's mass, $\eta = N/m$.

- The Reynolds transport theorem provides a means for relating the time rate of change of an extensive property N of a fluid, as measured for a system, to the time rate of change as measured from a control volume. The control volume change consists of two parts: a *local change* that measures the change of the extensive property *within* the control volume, and a *convective change* that measures the change in the net amount of this property that enters and exits the open control surfaces. This net amount must be measured relative to the control surfaces if they are moving.

EXAMPLE | 4.1

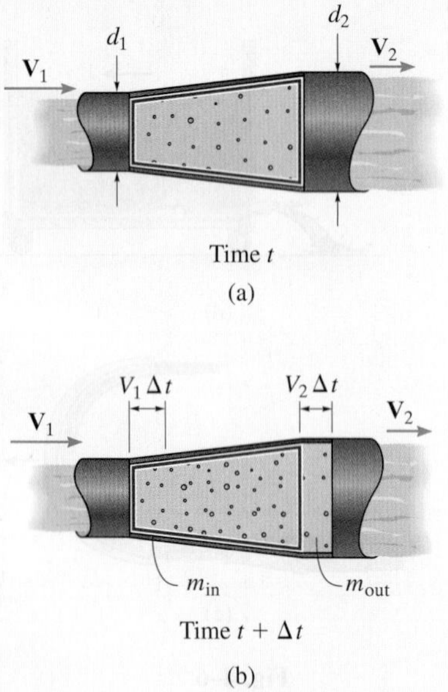

Time t

(a)

Time $t + \Delta t$

(b)

Fig. 4–7

An ideal fluid flows through the divergent section of pipe in Fig. 4–7a such that it enters with a velocity V_1. If the flow is steady, determine the velocity V_2 at which it exits.

SOLUTION

Fluid Description.   This is a case of one-dimensional, steady, nonuniform flow. It is nonuniform because the velocities are different at each location. Because the fluid is assumed to be ideal, its density is constant and the viscosity is zero. For this reason, the velocity profile will be uniform over each cross section.

Analysis I.   To analyze the flow, we will consider a fixed control volume to represent the fluid system within the divergent section, Fig. 4–7a. At time $t + \Delta t$ this system will move to the position shown in Fig. 4–7b. Therefore, the amount of mass moving out into the pipe of diameter d_2 during the time Δt is $m_{out} = \rho \Delta V_{out} = \rho(V_2\Delta t)(\frac{1}{4} \pi d_2^2)$, and the amount of mass lost at the entrance is equivalent to an amount of mass that would move through the pipe of diameter d_1. It is $m_{in} = \rho \Delta V_{in} = \rho(V_1\Delta t)\left(\frac{1}{4} \pi d_1^2\right)$.

Due to steady flow, the mass of fluid *within* the control volume remains constant. Therefore, because the total amount of the system's mass is constant, we require

$$m_{\text{out}} - m_{\text{in}} = 0$$

$$\rho(V_2 \Delta t)\left(\frac{1}{4}\pi d_2^2\right) - \rho(V_1 \Delta t)\left(\frac{1}{4}\pi d_1^2\right) = 0$$

$$V_2 = V_1\left(\frac{d_1}{d_2}\right)^2 \qquad\qquad \textit{Ans.}$$

The result indicates the velocity decreases, $V_2 < V_1$, something to be expected, since the flow is out of a larger area.

Analysis II. Now let's see how we can obtain this same result by applying the Reynolds transport theorem. Here the extensive property is the mass $N = m$, so $\eta = m/m = 1$. Applying Eq. 4–4, we have

$$\left(\frac{Dm}{Dt}\right)_{\text{syst}} = \frac{\partial}{\partial t}\int_{cv} \rho \, dV + \int_{cs} \rho \mathbf{V}_{f/cs} \cdot d\mathbf{A}$$

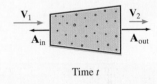

V_1 V_2

A_{in} A_{out}

Time t

(c)

The term on the left is zero because the system mass does not change with time. Also, the first term on the right is zero because the flow is *steady*, that is, there is no local change of the mass within the control volume, Fig. 4–7c. Because both the density and the velocity at each open control surface are *constant*, they can be factored out of the integral in the last term, and integration then yields the in and out control surface areas, A_{in} and A_{out}. These areas have positive outward directions, as shown in Fig. 4–7c, and so the dot product evaluation of the last term in the above equation reduces to

$$0 = 0 + \rho V_2 A_{\text{out}} - \rho V_1 A_{\text{in}}$$

$$0 = 0 + V_2\left(\frac{1}{4}\pi d_2^2\right) - V_1\left(\frac{1}{4}\pi d_1^2\right)$$

$$V_2 = V_1\left(\frac{d_1}{d_2}\right)^2 \qquad\qquad \textit{Ans.}$$

We will extend this application of the Reynolds transport theorem further in Sec. 4.4.

4.3   Volumetric Flow, Mass Flow, and Average Velocity

Due to viscosity, the *velocity* of individual fluid particles flowing through a conduit can vary substantially. To simplify an analysis, especially for problems involving one-dimensional flow, we can sometimes consider the fluid as having an average velocity, or describe the flow in terms of its volume or mass per unit time. We will now formally define these terms.

Volumetric Flow.   The rate at which a *volume* of fluid flows through a cross-sectional area A is called the **volumetric flow**, or simply the **flow** or **discharge**. It can be determined provided we know the velocity profile for the flow across the area. For example, consider the flow of a viscous fluid through a pipe, such that its velocity profile has the axisymmetric shape shown in Fig. 4–8. If particles passing through the differential area dA have a velocity v, then during the time dt, a volume element of fluid of length $v\, dt$ will pass through the area. Since this volume is $d\Psi = (v\, dt)(dA)$, then the *volumetric flow* dQ through the area is determined by dividing the volume by dt, which gives $dQ = d\Psi/dt = v\, dA$. If we integrate this over the entire cross-sectional area A, we have

$$Q = \int_A v\, dA$$

Here Q can be measured in m^3/s.

Integration is possible if the velocity can be expressed as a function of the coordinates describing the area. For example, we will show in Chapter 9 that the velocity profile in Fig. 4–8 is a paraboloid, provided the flow is laminar. Sometimes, however, the velocity profile must be determined experimentally, as for turbulent flows, in which case a graphical integration of the velocity profile may be performed. In either case, the integral in the equation *geometrically* represents the *volume* within the velocity diagram shown in Fig. 4–8.

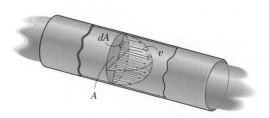

Fig. 4–8

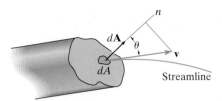

Fig. 4–9

When calculating Q, it is important to remember that the velocity must be *normal* to the cross-sectional area through which the fluid flows. If this is not the case, as in Fig. 4–9, then we must consider the velocity's *normal component* $v \cos \theta$ for the calculation. By considering the area as a vector, $d\mathbf{A}$, where its normal is *positive outward*, we can use the dot product, $\mathbf{v} \cdot d\mathbf{A} = v \cos \theta \, d\mathbf{A}$, to express the integral in the previous equation in a more general form, namely

$$Q = \int_A \mathbf{v} \cdot d\mathbf{A} \qquad (4\text{–}5)$$

Average Velocity.   When the fluid is considered ideal, then its viscous or frictional effects can be neglected. As a result, the velocity profile of the fluid over the cross section will be *uniform* as shown in Fig. 4–10. This type of profile also closely resembles the case of turbulent flow, where we have seen, in Fig. 3–3 how turbulent mixing of the fluid tends to *flatten* the velocity profile to be somewhat uniformly distributed. For the case of, $\mathbf{v} = \mathbf{V}$, Eq. 4–5 gives

$$Q = \mathbf{V} \cdot \mathbf{A} \qquad (4\text{–}6)$$

Here $\mathbf{V}$ is the *average velocity* and $\mathbf{A}$ is the area of the cross section.

For any real fluid, the average velocity can be determined by requiring the flow to be equivalent for both the actual and the average velocity distributions, Fig. 4–8 and Fig. 4–10, so that

$$Q = VA = \int_A \mathbf{v} \cdot d\mathbf{A}$$

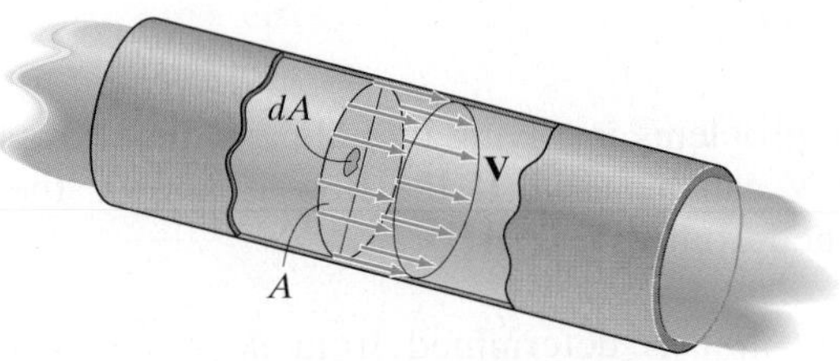

Inviscid or ideal fluids produce an
average velocity distribution

Fig. 4–10

The mass flow of air through this duct must be determined by using the open area of the duct and the velocity component that is perpendicular to this area.

Therefore, the average velocity is

$$V = \frac{\int_A \mathbf{v} \cdot d\mathbf{A}}{A} \tag{4–7}$$

Most often, however, the *average velocity* of the flow is determined if we know Q through the cross-sectional area A. Combining Eqs. 4–5 and 4–7, it is

$$V = \frac{Q}{A} \tag{4–8}$$

Mass Flow. Since the mass of the element in Fig. 4–8 is $dm = \rho \, d\mathbb{V} = \rho(v \, dt)dA$, the **mass flow** or **mass discharge** of the fluid through the entire cross section becomes

$$\dot{m} = \frac{dm}{dt} = \int_A \rho \mathbf{v} \cdot d\mathbf{A} \tag{4–9}$$

Measurements can be made in kg/s.

If the fluid is incompressible, then ρ is constant, and for the special case of a uniform velocity profile Eq. 4–9 gives

$$\dot{m} = \rho \mathbf{V} \cdot \mathbf{A} \tag{4–10}$$

Important Points

- The *volumetric flow* or *discharge* through an area is determined from $Q = \int_A \mathbf{v} \cdot d\mathbf{A}$, where $\mathbf{v}$ is the velocity of each fluid particle passing through the area. The dot product is used because the calculation requires the velocity to be perpendicular to the area. Here Q can have units of m^3/s.

- In many problems involving one-dimensional flow, the average velocity $\mathbf{V}$ can be used. If the flow is known, then it can be determined from $V = Q/A$.

- The *mass flow* is determined from $\dot{m} = \int \rho \mathbf{v} \cdot d\mathbf{A}$, or for an incompressible fluid having an average velocity, $\dot{m} = \rho \mathbf{V} \cdot \mathbf{A}$. Here $\dot{m}$ can have units of kg/s.

EXAMPLE | 4.2

The velocity profile for the steady laminar flow of water through a 0.4-m-diameter pipe is defined by $v = 3(1 - 25r^2)$ m/s, where r is in meters, Fig. 4–11a. Determine the volumetric flow through the pipe and the average velocity of the flow.

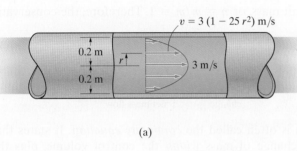
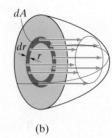

(a)   (b)

Fig. 4–11

SOLUTION

Fluid Description. Here one-dimensional steady flow occurs.

Analysis. The volumetric flow is determined using Eq. 4–5. A differential ring element of thickness dr is selected, Fig. 4–11b, so that $dA = 2\pi r \, dr$. Thus

$$Q = \int_A \mathbf{v} \cdot d\mathbf{A} = \int_0^{0.2 \text{ m}} 3(1 - 25r^2) \, 2\pi r \, dr$$

$$= 6\pi \left[\left(\frac{r^2}{2} \right) - \left(\frac{25r^4}{4} \right) \right]_0^{0.2 \text{ m}}$$

$$= 0.1885 \text{ m}^3/\text{s} = 0.188 \text{ m}^3/\text{s} \qquad Ans.$$

To avoid this integration, Q can also be determined by calculating the *volume* under the "velocity paraboloid" of radius 0.2 m and height 3 m/s. It is

$$Q = \frac{1}{2} \pi r^2 h = \frac{1}{2} \pi (0.2 \text{ m})^2 (3 \text{ m/s}) = 0.188 \text{ m}^3/\text{s} \qquad Ans.$$

The average velocity is determined from Eq. 4–4.

$$V = \frac{Q}{A} = \frac{0.1885 \text{ m}^3/\text{s}}{\pi (0.2 \text{ m})^2} = 1.5 \text{ m/s} \qquad Ans.$$

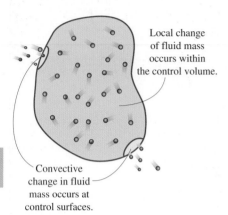

Local change of fluid mass occurs within the control volume.

Convective change in fluid mass occurs at control surfaces.

Fig. 4–12

4.4   Conservation of Mass

The conservation of mass states that within a region, apart from any nuclear process, matter can neither be created nor destroyed. From a Lagrangian point of view, the mass of all the particles in a system of particles must be *constant* over time, and so we require the change in the mass to be $(dm/dt)_{sys} = 0$. In order to develop a similar statement that relates to a control volume, we must use the Reynolds transport theorem, Eq. 4–4. Here the extensive property $N = m$, and so the corresponding intensive property is mass per unit mass, or $\eta = m/m = 1$. Therefore, the conservation of mass requires

$$\frac{\partial}{\partial t}\int_{cv} \rho\, d\forall \;+\; \int_{cs} \rho\, \mathbf{V}\cdot d\mathbf{A} = 0 \qquad (4\text{–}11)$$

Local mass change   Convective net mass flow

This equation is often called the *continuity equation*. It states that the *local rate* of change of mass *within* the control volume, plus the *net convective rate* at which mass enters and exits the open control surfaces, must equal zero, Fig. 4–12.

Special Cases.   Provided we have a control volume with a fixed size that is completely filled with an incompressible fluid, then there will be no local change of the fluid mass within the control volume. In this case, the first term in Eq. 4–11 is zero, and so the *net* mass flow into and out of the open control surfaces must be zero. In other words, "what flows in must flow out." Thus for both steady and unsteady flow,

$$\int_{cs} \rho\mathbf{V}\cdot d\mathbf{A} = \Sigma \dot{m}_{out} - \Sigma \dot{m}_{in} = 0 \qquad (4\text{–}12)$$

Incompressible flow

Assuming the *average velocity* occurs through each control surface, then V will be constant, and integration yields

$$\Sigma \rho\mathbf{V}\cdot\mathbf{A} = \Sigma \dot{m}_{out} - \Sigma \dot{m}_{in} = 0 \qquad (4\text{–}13)$$

Incompressible flow

Finally, if the *same fluid* is flowing at a *steady rate* into and out of the control volume, then the density can be factored out, and we have

$$\Sigma \mathbf{V}\cdot\mathbf{A} = \Sigma Q_{out} - \Sigma Q_{in} = 0 \qquad (4\text{–}14)$$

Incompressible steady flow

A conceptual application of this equation is shown in Fig. 4–13. By our sign convention, notice that whenever fluid *exits* a control surface, $\mathbf{V}$ and $\mathbf{A}_{out}$ are *both* directed outward, and by the dot product this term is *positive*. If the fluid *enters* a control surface, $\mathbf{V}$ is directed inward and $\mathbf{A}_{in}$ is directed outward, and so their dot product will be *negative*.

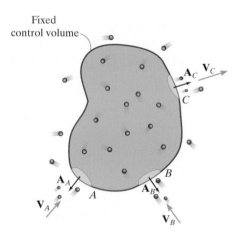

Fixed control volume

$$\Sigma\mathbf{V}\cdot\mathbf{A} = 0$$
$$-V_A A_A - V_B A_B + V_C A_C = 0$$

Steady flow of an incompressible fluid

Fig. 4–13

Important Points

- The continuity equation is based on the conservation of mass, which states that the mass of all the particles within a system remains constant.

- If the control volume is entirely filled with an incompressible fluid, then there will be no local change in the control volume as the fluid flows through its open control surface. Instead, only convective changes occur.

The mass flow of air through this duct assembly must be calculated in order to balance the airflow out of each of the room vents.

Procedure for Analysis

The following procedure can be used when applying the continuity equation.

Fluid Description

- Classify the flow as steady or unsteady, uniform or nonuniform. Also, determine if the fluid can be assumed inviscid and/or incompressible.

Control Volume

- Establish the control volume and determine what type it should be. Fixed control volumes are generally used for objects that are at rest and have a fixed amount of fluid passing through them, like pipes. Moving control volumes are used for objects, such as pump or turbine blades, and changing control volumes can be used for containers that have a variable amount of fluid within them. Care should be taken to orient the open control surfaces so that their planar areas are perpendicular to the flow. Also, locate these surfaces in a region where the flow is uniform and well established.

Continuity Equation

- Consideration must be given both to the rate of change of mass within the control volume and to the rate at which mass enters and exits each open control surface. For application, we will always write the fundamental equation, Eq. 4–11, and then show how this equation reduces to a specific case. For example, if the control volume does not deform and is completely filled with an incompressible fluid, then the local change of the mass within the control volume will be zero, so "what flows in must flow out," as indicated by Eqs. 4–12 through 4–14.

  Remember that planar areas **A** of open control surfaces are defined by vectors that are always directed *outward*, normal to the control surface. Thus, flow *into* a control surface will be *negative* since **V** and **A** will be in opposite directions, whereas flow *out* of a control surface is *positive* since both vectors are outward. If a control surface is moving, then for steady flow, the velocity of flow into or out of a control surface must be measured *relative* to the moving surface, that is, $\mathbf{V} = \mathbf{V}_{f/cs}$ in the above equations.

EXAMPLE 4.3

Water flows into the 250-mm-diameter barrel pipe of the fire hydrant at $Q_C = 0.15 \text{ m}^3/\text{s}$, Fig. 4–14a. If the velocity out the 75-mm-diameter nozzle at A is 12 m/s, determine the discharge out the 100-mm-diameter nozzle at B.

SOLUTION

Fluid Description.   This is a condition of *steady flow*, where the water will be considered an ideal fluid. Therefore, average velocities will be used.

Control Volume.   We will take the control volume to be *fixed* and have it enclose the volume within the fire hydrant and extended portions of hose as shown in the figure. There is no local change within the control volume because the flow is steady. Convective changes occur through the three open control surfaces.

Since the flow at C is known, the average velocity there is

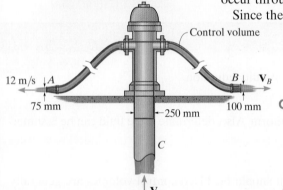

(a)

$$V_C = \frac{Q_C}{A_C} = \frac{0.15 \text{ m}^3/\text{s}}{\pi \, (0.125 \text{ m})^2} = \frac{9.6}{\pi} \text{ m/s}$$

Continuity Equation.   For steady, incompressible flow,

$$\frac{\partial}{\partial t} \int_{cv} \rho \, d\!\!\forall + \int_{cs} \rho \, \mathbf{V} \cdot d\mathbf{A} = 0 \tag{1}$$

$$0 - V_C A_C + V_A A_A + V_B A_B = 0 \tag{2}$$

$$0 - \left(\frac{9.6}{\pi} \text{ m/s}\right)\!\left[\pi(0.125 \text{ m})^2\right] + (12 \text{ m/s})\!\left[\pi(0.0375 \text{ m})^2\right] +$$
$$V_B\!\left[\pi(0.05 \text{ m})^2\right] = 0$$

$$V_B = 12.35 \text{ m/s}$$

The discharge at B is therefore

$$Q_B = V_B A_B = (12.35 \text{ m/s})\!\left[\pi(0.05 \text{ m})^2\right] = 0.0970 \text{ m}^3/\text{s} \quad \textit{Ans.}$$

This solution is actually a direct application of Eq. 4–14, represented here by Eq. 2. Also, notice that since $Q_C = V_C A_C$ and $Q_B = V_B A_B$, it was really not necessary to calculate V_C and V_B as an intermediate step. But be careful to recognize that if Q_C is used in Eq. 2, then it is a negative term, i.e., Eq. 2 becomes $0 - Q_C + V_A A_A + Q_B = 0$.

NOTE: If the viscosity of the water was considered, and a specific velocity profile, like the one shown in Fig. 4–14b, was established, then the solution would require integration of the second term in Eq. 1. (See the first step in Example 4.1.)

(b)

Fig. 4–14

EXAMPLE | 4.4

Air flows into the gas heater in Fig. 4–15 at a steady rate, such that at A its absolute pressure is 203 kPa, its temperature is 20°C, and its velocity is 15 m/s. When it exits at B, it is at an absolute pressure of 150 kPa and a temperature of 75°C. Determine its velocity at B.

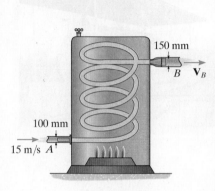

SOLUTION

Fluid Description.   As stated, we have steady flow. We will neglect viscosity and use average velocities through the pipe. Due to the pressure and temperature changes within the heater, the air changes its density, and so we must include the effects of compressibility.

Fig. 4–15

Control Volume.   As shown, the control volume is fixed and consists of the region of pipe within the heater, along with short extensions from it. Here there are no local changes *of the mass* within the control volume because the flow remains *steady*. Convective changes occur, however, due to the flow through the open control surfaces at A and B.

Continuity Equation.   The pressure and the temperature affect the air density at the open control surfaces. The flow of air *in* at A is negative.

$$\frac{\partial}{\partial t}\int_{cv} \rho \, d\forall + \int_{cs} \rho \, \mathbf{V} \cdot d\mathbf{A} = 0$$

$$0 - \rho_A V_A A_A + \rho_B V_B A_B = 0$$

$$-\rho_A(15 \text{ m/s})\left[\pi(0.05 \text{ m})^2\right] + \rho_B V_B\left[\pi(0.075 \text{ m})^2\right] = 0$$

$$V_B = 6.667\left(\frac{\rho_A}{\rho_B}\right) \tag{1}$$

Ideal Gas Law.   The densities at A and B are obtained from the ideal gas law. We have

$$p_A = \rho_A R T_A; \quad 203(10^3) \text{ N/m}^2 = \rho_A R(20 + 273) \text{ K}$$

$$p_B = \rho_B R T_B; \quad 150(10^3) \text{ N/m}^2 = \rho_B R(75 + 273) \text{ K}$$

The value of R for air is tabulated in Appendix A, although here it can be eliminated by division.

$$1.607 = \frac{\rho_A}{\rho_B}$$

Substituting this into Eq. 1 yields

$$V_B = 6.667(1.607) = 10.7 \text{ m/s} \qquad Ans.$$

EXAMPLE | 4.5

The tank in Fig. 4–16 has a volume of 1.5 m³ and is being filled with air, which is pumped into it at an average rate of 8 m/s through a hose having a diameter of 10 mm. As the air enters the tank, its temperature is 30°C and its absolute pressure is 500 kPa. Determine the rate at which the density of the air within the tank is changing at this instant.

SOLUTION

Fluid Description. Due to mixing, we will assume the air has a uniform density within the tank. This density is changing because the air is compressible. The flow into the tank is steady.

Control Volume. We will consider the fixed control volume to be the air contained within the tank. Local changes occur within this control volume because the *mass of air* within the tank is *changing with time*. The average velocity of the air stream will be considered at the open control surface at A.

Continuity Equation. Applying the continuity equation, realizing that the control volume (tank) has a constant volume, and assuming that the density within it is changing in a uniform manner, we have

$$\frac{\partial}{\partial t}\int_{cv}\rho\, d\forall + \int_{cs}\rho\,\mathbf{V}\cdot d\mathbf{A} = 0$$

$$\frac{\partial\rho_a}{\partial t}\forall_t - \rho_A V_A A_A = 0 \tag{1}$$

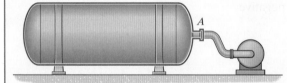

Fig. 4–16

Ideal Gas Law. The density of air flowing into the tank is determined using the ideal gas law. From Appendix A, $R = 286.9$ J/(kg · K), and so

$$p_A = \rho_A R T_A; \qquad 500\left(10^3\right) \text{N/m}^2 = \rho_A\left[286.9 \text{ J/(kg · K)}\right](30 + 273) \text{ K}$$

$$\rho_A = 5.752 \text{ kg/m}^3$$

Therefore,

$$\frac{\partial\rho_a}{\partial t}\left(1.5 \text{ m}^3\right) - \left[\left(5.752 \text{ kg/m}^3\right)\left(8 \text{ m/s}\right)\right]\left[\pi(0.005 \text{ m})^2\right] = 0$$

$$\frac{\partial\rho_a}{\partial t} = 2.41\left(10^{-3}\right) \text{kg/}\left(\text{m}^3 \cdot \text{s}\right) \qquad\qquad Ans.$$

This positive result indicates that the density of the air within the tank is increasing, which is to be expected.

EXAMPLE | 4.6

The rocket sled in Fig. 4–17 is propelled by a jet engine that burns fuel at a rate of 60 kg/s. The air duct at A has an opening of 0.2 m² and takes in air having a density of 1.20 kg/m³. If the engine exhausts the gas relative to the nozzle at B with an average velocity of 300 m/s, determine the density of the exhaust. The sled is moving forward at a constant rate of 80 m/s and the nozzle has a cross-sectional area of 0.35 m².

SOLUTION

Fuel Description.   The air–fuel system is compressible, and so its density will be different at the inlet A and exhaust B. We will use average velocities.

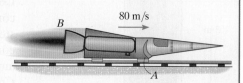

Fig. 4–17

Control Volume.   The control volume is represented by the enclosed region within the engine that accepts the air and fuel, burns it, and exhausts it. We will assume it moves with the rocket. From this viewpoint (as a passenger), the flow is *steady*, so there is no local time rate of change of the air–fuel mass within the control volume. Convective changes occur at the air intake, the fuel line intake, and the nozzle. Also, assuming the outside air is stationary, then the relative velocity of airflow at the intake A is

$$\xrightarrow{+} V_A = V_{cs} + V_{A/cs}$$

$$0 = 80 \text{ m/s} + V_{A/cs}$$

$$V_{A/cs} = -80 \text{ m/s} = 80 \text{ m/s} \leftarrow$$

At B, $V_{B/cs} = 300$ m/s because the velocity of the exhaust gas is measured relative to the control surface.

Continuity Equation.   The mass flow of fuel is $\dot{m}_f = 60$ kg/s, and since there is no local change, we have

$$\frac{\partial}{\partial t} \int_{cv} \rho \, d\forall + \int_{cs} \rho \mathbf{V}_{f/cs} \cdot d\mathbf{A} = 0$$

$$0 - \rho_a V_{A/cs} A_A + \rho_g V_{B/cs} A_B - \dot{m}_f = 0$$

$$-1.20 \text{ kg/m}^3 (80 \text{ m/s})(0.2 \text{ m}^2) + \rho_g (300 \text{ m/s})(0.35 \text{ m}^2) - 60 \text{ kg/s} = 0$$

$$\rho_g = 0.754 \text{ kg/m}^3 \qquad \qquad Ans.$$

Notice that if the selected control volume were in a *fixed location* in space, and the rocket sled were passing this location, then local changes *would occur* within the control volume as the rocket sled passed by. In other words, the flow through the control volume would appear as *unsteady flow*.

EXAMPLE | 4.7

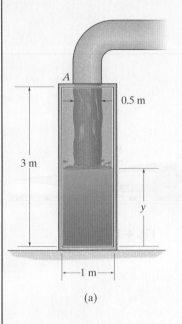

A

0.5 m

3 m

y

←1 m→

(a)

Fig. 4–18

The 1-m-diameter tank in Fig. 4–18a is being filled with water using a 0.5-m-diameter pipe, which has a discharge of 0.15 m³/s. Determine the rate at which the level of water is rising in the tank.

SOLUTION

Fluid Description. This is a case of steady flow. We assume the water to be an incompressible fluid so that ρ_w is constant.

Control Volume I. We will choose a *fixed control volume*, which consists of the volume of the *entire tank*, Fig. 4–18a. Although we have steady flow into this control volume, *local changes* occur because the control volume is not entirely filled with water. In other words, the amount of *mass* within the control volume changes with time. *Convective changes* occur at the open control surface at A. We will exclude the local change of the air mass within the tank, which equals its mass flow out from the top of the tank. To calculate the volume of water within the tank, we will assume that as the water falls it maintains a constant diameter of 0.5 m as shown.*

Continuity Equation. Applying the continuity equation, realizing that $Q_A = V_A A_A$, we have

$$\frac{\partial}{\partial t}\int_{cv} \rho_w \, d\forall + \int_{cs} \rho_w \mathbf{V} \cdot d\mathbf{A} = 0$$

$$\rho_w \frac{d\forall}{dt} - \rho_w Q_A = 0$$

Here $\forall$ is the total volume of water *within* the control volume at the instant the depth is y. Factoring out ρ_w, we have

$$\frac{d}{dt}\Big[\pi(0.5\,\text{m})^2 y + \pi(0.25\,\text{m})^2(3\,\text{m} - y)\Big] - \big(0.15\,\text{m}^3/\text{s}\big) = 0$$

$$\pi \frac{d}{dt}(0.1875y + 0.1875) = 0.15$$

$$0.1875\frac{dy}{dt} + 0 = \frac{0.15}{\pi}$$

$$\frac{dy}{dt} = \frac{0.8}{\pi}\,\text{m/s} = 0.255\,\text{m/s} \qquad\qquad Ans.$$

*If this water column disperses, the same volume would fall into the tank.

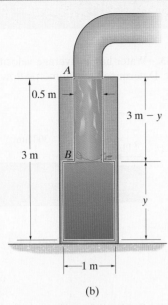

(b)

Fig. 4–18 (cont.)

Control Volume II.   We can also work this problem by considering a fixed control volume consisting only of the water within the tank, Fig. 4–18b. In this case no local changes occur because the water within this control volume is considered incompressible, so that the *mass* remains constant. But convective changes occur because water flows into the control surface of area $\pi(0.25 \text{ m})^2$ at A, and it flows out of the control surface at B, having an area of $[\pi(0.5 \text{ m})^2 - \pi(0.25 \text{ m})^2]$.

Continuity Equation.   Since $Q_A = V_A A_A$, we have

$$\frac{\partial}{\partial t} \int_{cv} \rho_w \, d\mathcal{V} + \int_{cs} \rho_w \mathbf{V} \cdot d\mathbf{A} = 0$$

$$0 - V_A A_A + V_B A_B = 0$$

$$-\left(0.15 \text{ m}^3/\text{s}\right) + V_B \left[\left(\pi(0.5 \text{m})^2 - \pi(0.25 \text{ m})^2\right)\right] = 0$$

$$V_B = \frac{dy}{dt} = \frac{0.8}{\pi} \text{ m/s} = 0.255 \text{ m/s} \qquad\qquad Ans.$$

References

1. ASME, *Flow Meters*, 6th ed., ASME, New York, NY, 1971.

2. S. Vogel, *Comparative Biomechanics*, Princeton University Press, Princeton, NJ, 2003.

3. S. Glasstone and A. Sesonske, *Nuclear Reactor Engineering*, D. van Nostrand, Princeton, NJ, 2001.

FUNDAMENTAL PROBLEMS

Sec. 4.3

F4–1. Water flows into the tank through a rectangular tube. If the average velocity of the flow is 16 m/s, determine the mass flow. Take $\rho_w = 1000 \text{ kg/m}^3$.

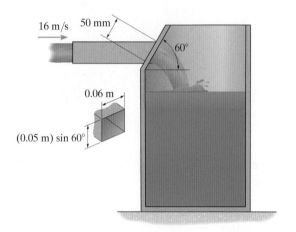

Prob. F4–1

F4–2. Air flows through the triangular duct at 0.7 kg/s when the temperature is 15°C and the gage pressure is 70 kPa. Determine its average velocity. Take $\rho_{atm} = 101$ kPa.

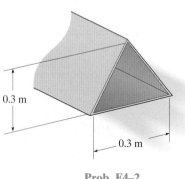

Prob. F4–2

F4–3. Water has an average velocity of 8 m/s through the pipe. Determine the volumetric flow and mass flow.

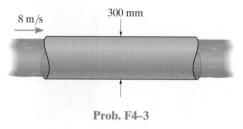

Prob. F4–3

F4–4. Crude oil flows through the pipe at 0.02 m³/s. If the velocity profile is assumed to be as shown, determine the maximum velocity V_0 of the oil and the average velocity.

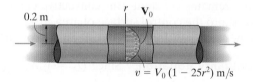

$$v = V_0 (1 - 25r^2) \text{ m/s}$$

Prob. F4–4

F4–5. Determine the mass flow of air having a temperature of 20°C and pressure of 80 kPa as it flows through the circular duct with an average velocity of 3 m/s.

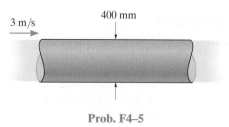

Prob. F4–5

F4–6. If the velocity profile for a very viscous liquid as it flows through a 0.5-m-wide rectangular channel is approximated as $u = (6y^2)$ m/s, where y is in meters, determine the volumetric flow.

F4–8. A liquid flows into the tank at A at 4 m/s. Determine the rate, dy/dt, at which the level of the liquid is rising in the tank. The cross-sectional area of the pipe at A is $A_A = 0.1\,\text{m}^2$.

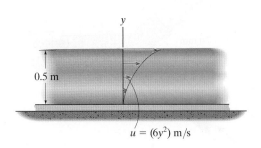

Prob. F4–6

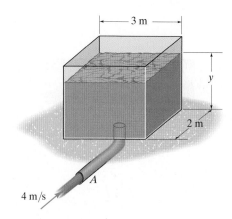

Prob. F4–8

Sec. 4.4

F4–7. The average velocity of the steady flow at A and B is indicated. Determine the average velocity at C. The pipes have cross-sectional areas of $A_A = A_C = 0.1\,\text{m}^2$ and $A_B = 0.2\,\text{m}^2$.

F4–9. As air exits the tank at 0.05 kg/s, it is mixed with water at 0.002 kg/s. Determine the average velocity of the mixture as it exits the 20-mm-diameter pipe if the density of the mixture is 1.45 kg/m³.

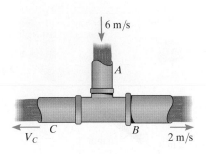

Prob. F4–7

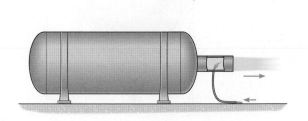

Prob. F4–9

F4–10. Air at a temperature of 16°C and gage pressure of 200 kPa flows through the pipe at 12 m/s when it is at *A*. Determine its average velocity when it exits the pipe at *B* if its temperature there is 70°C.

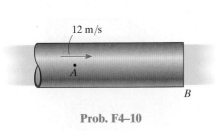

Prob. F4–10

F4–11. Determine the rate at which water is rising in the triangular container when $t = 10$ s if the water flows from the 50-mm-diameter pipe with an average speed of 6 m/s. The container is 1 m long. At $t = 0$ s, $h = 0.1$ m.

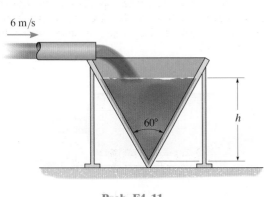

Prob. F4–11

F4–12. Determine the required diameter of the pipe at *C* so that water flows through the pipe at the rates shown.

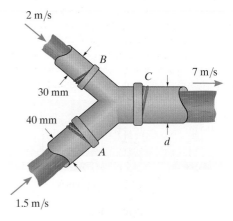

Prob. F4–12

F4–13. Oil flows into the tank with an average velocity of 4 m/s through the 50-mm-diameter pipe at *A*. It flows out of the tank at 2 m/s through the 20-mm-diameter pipe at *B*. Determine the rate at which the depth *y* of the oil in the tank is changing.

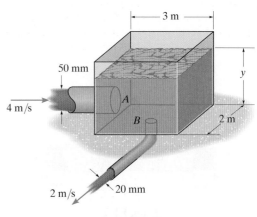

Prob. F4–13

PROBLEMS

Sec. 4.1–4.2

4–1. Water flows steadily through the pipes with the average velocities shown. Outline the control volume that contains the water in the pipe system. Indicate the open control surfaces, and show the positive direction of their areas. Also, indicate the direction of the velocities through these surfaces. Identify the local and convective changes that occur. Assume water to be incompressible.

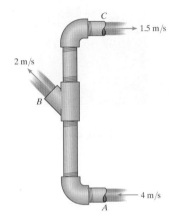

Prob. 4–1

4–2. The toy balloon is filled with air and is released. At the instant shown, the air is escaping at an average velocity of 4 m/s, measured relative to the balloon, while the balloon is accelerating. For an analysis, why is it best to consider the control volume to be moving? Select this control volume so that it contains the air in the balloon. Indicate the open control surface, and show the positive direction of its area. Also, indicate the direction of the velocity through this surface. Identify the local and convective changes that occur. Assume the air to be incompressible.

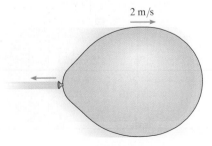

Prob. 4–2

4–3. Water is drawn steadily through the pump. The average velocities are indicated. Select a control volume that contains the water in the pump and extends slightly past it. Indicate the open control surfaces, and show the positive direction of their areas. Also, indicate the direction of the velocities through these surfaces. Identify the local and convective changes that occur. Assume water to be incompressible.

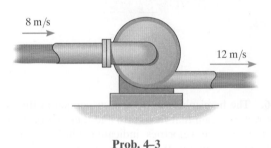

Prob. 4–3

*4–4. The average velocities of water flowing steadily through the nozzle are indicated. If the nozzle is connected onto the end of the hose, outline the control volume to be the entire nozzle and the water inside it. Also, select another control volume to be just the water inside the nozzle. In each case, indicate the open control surfaces, and show the positive direction of their areas. Specify the direction of the velocities through these surfaces. Identify the local and convective changes that occur. Assume water to be incompressible.

Prob. 4–4

4–5. Air flows through the tapered duct, and during this time heat is being added at an increasing rate, changing the density of the air within the duct. The average velocities are indicated. Select a control volume that contains the air in the duct. Indicate the open control surfaces, and show the positive direction of their areas. Also, indicate the direction of the velocities through these surfaces. Identify the local and convective   changes that occur. Assume the air is incompressible.

4–7. The tank is being filled with water at *A* at a rate faster than it is emptied at *B*; the average velocities are shown. Select a control volume that includes only the water in the tank. Indicate the open control surfaces, and show positive direction of their areas. Also, indicate the direction of the velocities through these surfaces. Identify the local and convective changes that occur. Assume water to be incompressible.

2 m/s   7 m/s

Prob. 4–5

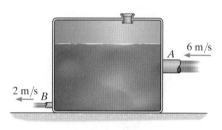

A   6 m/s

2 m/s *B*

Prob. 4–7

4–6. The hemispherical bowl is suspended in the air by the water stream that flows into and then out of the bowl at the average velocities indicated. Outline a control volume that contains the bowl and the water entering and leaving it. Indicate the open control surfaces, and show the positive direction of their areas. Also, indicate the direction of the velocities through these surfaces. Identify the local and convective changes that occur. Assume water to be incompressible.

***4–8.** Compressed air is being released from the tank, and at the instant shown it has a velocity of 3 m/s. Select a control volume that contains the air in the tank. Indicate the open control surface, and show the positive direction of its area. Also, indicate the direction of the velocity through this surface. Identify the local and convective changes that occur. Assume the air to be compressible.

3 m/s   3 m/s

Prob. 4–6

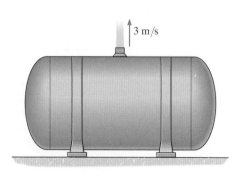

3 m/s

Prob. 4–8

4

4–9. The blade on the turbine is moving to the left at 6 m/s. Water is ejected from the nozzle at *A* at avn average velocity of 2 m/s. For the analysis, why is it best to consider the control volume as moving? Outline this moving control volume that contains the water on the blade. Indicate the open control surfaces, and show the positive direction of their areas through which flow occurs. Also, indicate the magnitudes of the relative velocities and their directions through these surfaces. Identify the local and convective changes that occur. Assume water to be incompressible.

4–11. The balloon is rising at a constant velocity of 3 m/s . Hot air enters from a burner and flows into the balloon at *A* at an average velocity of 1 m/s, measured relative to the balloon. For an analysis, why is it best to consider the control volume as moving? Outline this moving control volume that contains the air in the balloon. Indicate the open control surface, and show the positive direction of its area. Also, indicate the magnitude of the velocity and its direction through this surface. Identify the local and convective changes that occur. Assume the air to be incompressible.

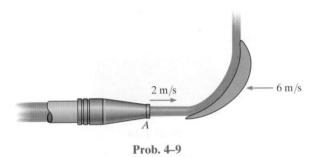

Prob. 4–9

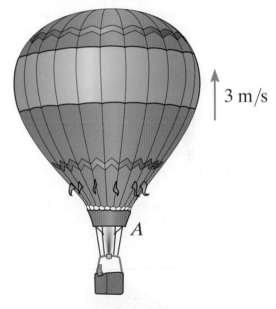

Prob. 4–11

4–10. The jet engine is moving forward with a constant speed of 800 km/h. Fuel from a tank enters the engine and is mixed with the intake air, burned, and exhausted with an average relative velocity of 1200 km/h. Outline the control volume as the jet engine and the air and fuel within it. For an analysis, why is it best to consider this control volume to be moving? Indicate the open control surfaces, and show the positive direction of their areas. Also, indicate the magnitudes of the relative velocities and their directions through these surfaces. Identify the local and convective changes that occur. Assume the fuel is incompressible and the air is compressible.

Prob. 4–10

Sec. 4.3

*4–12. Water flows along a rectangular channel having a width of 0.75 m. If the average velocity is 2 m/s, determine the volumetric discharge.

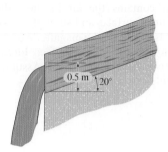

Prob. 4–12

4–13. Air enters the turbine of a jet engine at a rate of 40 kg/s. If it is discharged with an absolute pressure of 750 kPa and temperature of 120°C, determine its average velocity at the exit. The exit has a diameter of 0.3 m.

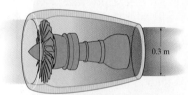

Prob. 4–13

4–14. Water flows along the triangular trough with an average velocity of 5 m/s. Determine the volumetric discharge and the mass flow if the vertical depth of the water is 0.3 m.

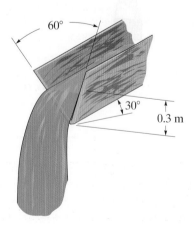

Prob. 4–14

4–15. The velocity profile in a channel carrying a very viscous liquid is approximated by $u = 3(e^{0.2y} - 1)$ m/s, where y is in meters. If the channel is 1 m wide, determine the volumetric discharge from the channel.

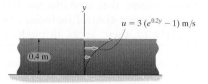

Prob. 4–15

*4–16. A fluid flowing between two plates has a velocity profile that is assumed to be linear as shown. Determine the average velocity and volumetric discharge in terms of U_{max}. The plates have a width of w.

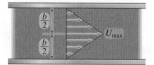

Prob. 4–16

4–17. The velocity field for a flow is defined by $u = (6x)$ m/s and $v = (4y^2)$ m/s, where x and y are in meters. Determine the discharge through the surfaces at A and at B.

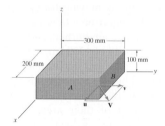

Prob. 4–17

4–18. Water flowing at a constant rate fills the tank to a height of $h = 3$ m in 5 minutes. If the tank has a width of 1.5 m, determine the average velocity of the flow from the 0.2-m-diameter pipe at A.

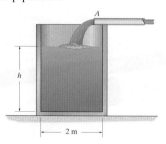

Prob. 4–18

4–19. Water flows through the pipe at a constant average velocity of 0.5 m/s. Determine the relation between the time needed to fill the tank to a depth of $h = 3$ m and the diameter D of the pipe at A. The tank has a width of 1.5 m. Plot the time in minutes (vertical axis) versus the diameter 0.05 m $\le D \le 0.25$ m. Give values for increments of $\Delta D = 0.05$ m.

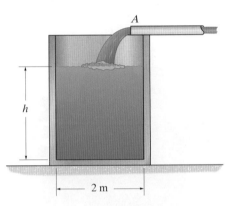

Prob. 4–19

*4–20. Determine the mass flow of air in the duct if it has an average velocity of 15 m/s. The air has a temperature of 30°C, and the (gage) pressure is 50 kPa.

4–21. Air flows through the duct at an average velocity of 20 m/s. If the temperature is maintained at 20°C, plot the variation of the mass flow (vertical axis) versus the (gage) pressure for the range of $0 \le p \le 100$ kPa. Give values for increments of $\Delta p = 20$ kPa. The atmospheric pressure is 101.3 kPa.

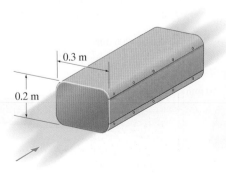

Probs. 4–20/21

4–22. The velocity profile in a channel carrying a very viscous liquid is approximated by $u = 3(e^{0.2y} - 1)$ m/s, where y is in meters. Determine the average velocity of the flow. The channel has a width of 1 m.

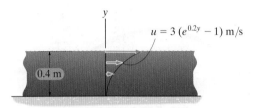

Prob. 4–22

4–23. The velocity profile for a liquid in the channel is determined experimentally and found to fit the equation $u = (8y^{1/3})$ m/s, where y is in meters. Determine the volumetric discharge if the width of the channel is 0.5 m.

*4–24. The velocity profile for a liquid in the channel is determined experimentally and found to fit the equation $u = (8y^{1/3})$ m/s, where y is in meters. Determine the average velocity of the liquid. The channel has a width of 0.5 m.

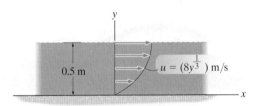

Probs. 4–23/24

4–25. The velocity profile for a fluid within the circular pipe for fully developed turbulent flow is modeled using Prandtl's one-seventh power law $u = U(y/R)^{1/7}$. Determine the average velocity for this case.

4–26. The velocity profile for a fluid within the circular pipe for fully developed turbulent flow is modeled using Prandtl's one-seventh power law $u = U(y/R)^{1/7}$. Determine the mass flow of the fluid if it has a density ρ.

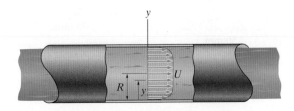

Probs. 4–25/26

4–27. Determine the volumetric flow through the 50-mm-diameter nozzle of the fire boat if the water stream reaches point B, which is $R = 24$ m from the boat. Assume the boat does not move.

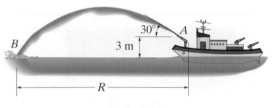

30°/ A

B 3 m

R

Prob. 4–27

***4–28.** Determine the volumetric flow through the 50-mm-diameter nozzle of the fire boat as a function of the distance R of the water stream. Plot this function of flow (vertical axis) versus the distance for $0 \le R \le 25$ m. Give values for increments of $\Delta R = 5$ m. Assume the boat does not move.

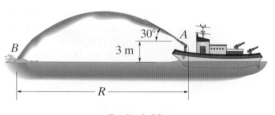

30°/ A

B 3 m

R

Prob. 4–28

4–29. For a short time, the flow of carbon tetrachloride through the circular pipe transition can be expressed as $Q = (0.8t + 5)(10^{-3})$ m³/s, where t is in seconds. Determine the average velocity and average acceleration of a particle located at A and B when $t = 2$ s.

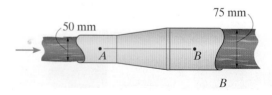

75 mm

50 mm

A B

B

Prob. 4–29

4–30. The human heart has an average discharge of $0.1(10^{-3})$ m³/s, determined from the volume of blood pumped per beat and the rate of beating. Careful measurements have shown that blood cells pass through the capillaries at about 0.5 mm/s. If the average diameter of a capillary is 6 µm, estimate the number of capillaries that must be in the human body.

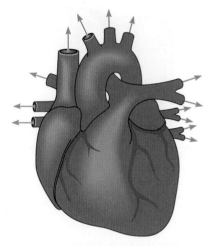

Prob. 4–30

4–31. The radius of the circular duct varies as $r = (0.05e^{-3x})$ m, where x is in meters. The flow of a fluid at A is $Q = 0.004$ m³/s at $t = 0$, and it is increasing at $dQ/dt = 0.002$ m³/s². If a fluid particle is originally located at $x = 0$ when $t = 0$, determine the time for this particle to arrive at $x = 100$ mm.

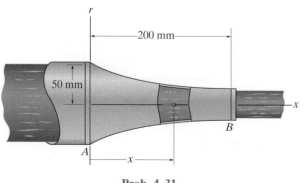

r

200 mm

50 mm

x

A

x B

Prob. 4–31

*4–32. The radius of the circular duct varies as $r = (0.05e^{-3x})$ m, where x is in meters. If the flow of the fluid at A is $Q = 0.004$ m³/s at $t = 0$, and it is increasing at $dQ/dt = 0.002$ m³/s², determine the time for this particle to arrive at $x = 200$ mm.

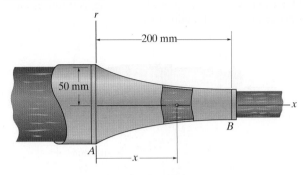

Prob. 4–32

4–33. Air flows through the gap between the vanes at 0.75 m³/s. Determine the average velocity of the air passing through the inlet A and the outlet B. The vanes have a width of 400 mm and the vertical distance between them is 200 mm.

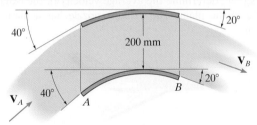

Prob. 4–33

4–34. The tapered pipe transfers ethyl alcohol to a mixing tank such that a particle at A has a velocity of 2 m/s. Determine the velocity and acceleration of a particle at B, where $x = 75$ mm.

4–35. The tapered pipe transfers ethyl alcohol to a mixing tank such that when a valve is opened, a particle at A has a velocity at A of 2 m/s, which is increasing at 4 m/s². Determine the velocity of the same particle when it arrives at B, where $x = 75$ mm.

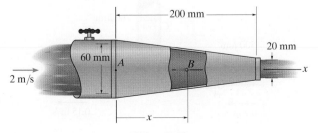

Probs. 4–34/35

Sec. 4.4

*4–36. Water flows through the pipe at A at 300 kg/s, and then out the double wye with an average velocity of 3 m/s through B and an average velocity of 2 m/s through C. Determine the average velocity at which it flows through D.

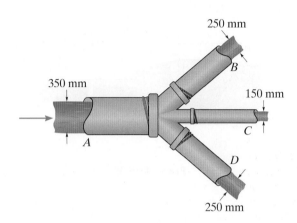

Prob. 4–36

4–37. If water flows at 150 kg/s through the double wye at B, at 50 kg/s through C, and at 150 kg/s through D, determine the average velocity of flow through the pipe at A.

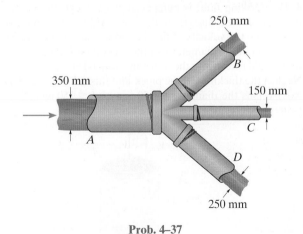

Prob. 4–37

4–38. The unsteady flow of linseed oil is such that at A it has a velocity of $V_A = (0.7t + 4)$ m/s, where t is in seconds. Determine the acceleration of a fluid particle located at $x = 0.2$ m when $t = 1$ s. *Hint:* Determine $V = V(x, t)$, then use Eq. 3–4.

4–39. The unsteady flow of linseed oil is such that at A it has a velocity of $V_A = (0.4t^2)$ m/s, where t is in seconds. Determine the acceleration of a fluid particle located at $x = 0.25$ m when $t = 2$ s. *Hint*: Determine $V = V(x, t)$, then use Eq. 3–4.

4–42. Water flows through the nozzle at a rate of 0.2 m³/s. Determine the velocity V of a particle as it moves along the centerline as a function of x.

4–43. Water flows through the nozzle at a rate of 0.2 m³/s. Determine the acceleration of a particle as it moves along the centerline as a function of x.

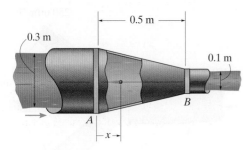

Probs. 4–38/39

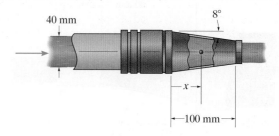

Probs. 4–42/43

****4–44.** A jet engine draws in air at 25 kg/s and jet fuel at 0.2 kg/s. If the density of the expelled air–fuel mixture is 1.356 kg/m³, determine the average velocity of the exhaust relative to the plane. The exhaust nozzle has a diameter of 0.4 m.

****4–40.** Drilling fluid is pumped down through the center pipe of a well and then rises up within the annulus. Determine the diameter d of the inner pipe so that the average velocity of the fluid remains the same in both regions. Also, what is this average velocity if the discharge is 0.02 m³/s? Neglect the thickness of the pipes.

4–41. Drilling fluid is pumped down through the center pipe of a well and then rises up within the annulus. Determine the velocity of the fluid forced out of the well as a function of the diameter d of the inner pipe, if the velocity of the fluid forced into the well is maintained at $V_{in} = 2$ m/s. Neglect the thickness of the pipes. Plot this velocity (vertical axis) versus the diameter for 50 mm $\leq d \leq$ 150 mm. Give values for increments of $\Delta d = 25$ mm.

Prob. 4–44

4–45. Oil flows into the pipe at A with an average velocity of 0.2 m/s and through B with an average velocity of 0.15 m/s. Determine the maximum velocity V_{max} of the oil as it emerges from C if the velocity distribution is parabolic, defined by $v_C = V_{max}(1 - 100r^2)$, where r is in meters measured from the centerline of the pipe.

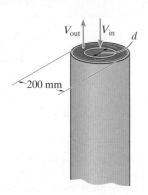

Probs. 4–40/41

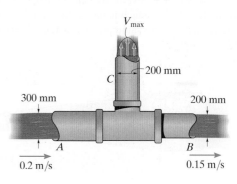

Prob. 4–45

4–46. An oscillating water column (OWC), or gully generator, is a device for producing energy created by ocean waves. As noted, a wave will push water up into the air chamber, forcing the air to pass through a turbine, producing energy. As the wave falls back, the air is drawn into the chamber, reversing the rotational direction of the turbine, but still creating more energy. Assuming a wave will reach an average height of $h = 0.5$ m in the 0.8-m-diameter chamber at B, and it falls back at an average speed of 1.5 m/s, determine the speed of the air as it moves through the turbine at A, which has a net area of 0.26 m². The air temperature at A is $T_A = 20°C$, and at B it is $T_B = 10°C$.

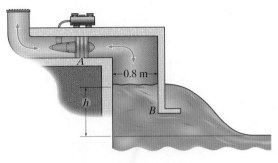

Prob. 4–46

4–47. An oscillating water column (OWC), or gully generator, is a device for producing energy created by ocean waves. As noted, a wave will push water up into the air chamber, forcing the air to pass through a turbine, producing energy. As the wave falls back, the air is drawn into the chamber, reversing the rotational direction of the turbine, but still creating more energy. Determine the speed of the air as it moves through the turbine at A, which has a net open area of 0.26 m², if the speed of the water in the 0.8-m- diameter chamber is 5 m/s. The air temperature at A is $T_A = 20°C$, and at B it is $T_B = 10°C$.

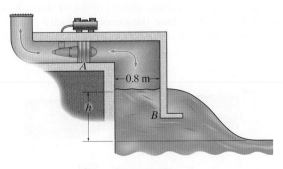

Prob. 4–47

*__4–48.__ The cylindrical syringe is actuated by applying a force on the plunger. If this causes the plunger to move forward at 10 mm/s, determine the average velocity of the fluid passing out of the needle.

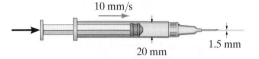

Prob. 4–48

4–49. Carbon dioxide flows into the tank at A at $V_A = 4$ m/s, and nitrogen flows in at B at $V_B = 3$ m/s. Both enter at a gage pressure of 300 kPa and a temperature of 250°C. Determine the steady mass flow of the mixed gas at C.

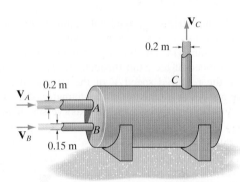

Prob. 4–49

4–50. Carbon dioxide flows into the tank at A at $V_A = 10$ m/s, and nitrogen flows in at B with a velocity of $V_B = 6$ m/s. Both enter at a gage pressure of 300 kPa and a temperature of 250°C. Determine the average velocity of the mixed gas leaving the tank at a steady rate at C. The mixture has a density of $\rho = 1.546$ kg/m³.

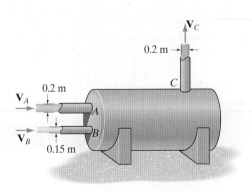

Prob. 4–50

4–51. The flat strip is sprayed with paint using the six nozzles, each having a diameter of 2 mm. They are attached to the 20-mm-diameter pipe. The strip is 50 mm wide, and the paint is to be 1 mm thick. If the average speed of the paint through the pipe is 1.5 m/s, determine the required speed V of the strip as it passes under the nozzles.

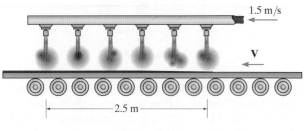

Prob. 4–51

*4–52.** The flat strip is sprayed with paint using the six nozzles, which are attached to the 20-mm-diameter pipe. The strip is 50 mm wide and the paint is to be 1 mm thick. If the average speed of the paint through the pipe is 1.5 m/s, determine the required speed V of the strip as it passes under the nozzles as a function of the diameter of the pipe. Plot this function of speed (vertical axis) versus diameter for $10 \text{ mm} \le D \le 30 \text{ mm}$. Give values for increments of $\Delta D = 5 \text{ mm}$.

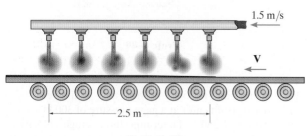

Prob. 4–52

4–53. The unsteady flow of glycerin through the reducer is such that at A its velocity is $V_A = (0.8t^2)$ m/s, where t is in seconds. Determine its average velocity at B, and its average acceleration at A, when $t = 2$ s. The pipes have the diameters shown.

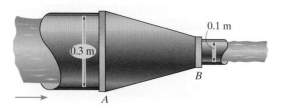

Prob. 4–53

4–54. The cylindrical plunger traveling at $V_p = (0.004t^{1/2})$ m/s, where t is in seconds, injects a liquid plastic into the mold to make a solid ball. If $d = 50$ mm, determine the amount of time needed to do this if the volume of the ball is $V = \frac{4}{3}\pi r^3$.

4–55. The cylindrical plunger traveling at $V_p = (0.004t^{1/2})$ m/s, where t is in seconds, injects a liquid plastic into the mold to make a solid ball. Determine the time needed to fill the mold as a function of the plunger diameter d. Plot the time needed to fill the mold (vertical axis) versus the diameter of the plunger for $10 \text{ mm} \le d \le 50 \text{ mm}$. Give values for increments of $\Delta d = 10 \text{ mm}$. The volume of the ball is $V = \frac{4}{3}\pi r^3$.

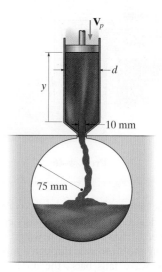

Probs. 4–54/55

*4–56.** The wind tunnel is designed so that the lower pressure outside the testing region draws air out in order to reduce the boundary layer or frictional effects along the wall within the testing tube. Within region B there are 2000 holes, each 3 mm in diameter. If the pressure is adjusted so that the average velocity of the air through each hole is 40 m/s, determine the average velocity of the air exiting the tunnel at C. Assume the air is incompressible.

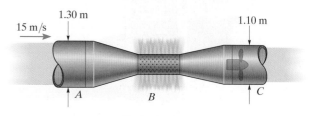

Prob. 4–56

4–57. The pressure vessel of a nuclear reactor is filled with boiling water having a density of $\rho_w = 850$ kg/m³. Its volume is 185 m³. Due to failure of a pump needed for cooling, the pressure release valve A is opened and emits steam having a density of $\rho_s = 35$ kg/m³ and an average speed of $V = 400$ m/s. If it passes through the 40-mm-diameter pipe, determine the time needed for all the water to escape. Assume that the temperature of the water and the velocity at A remain constant.

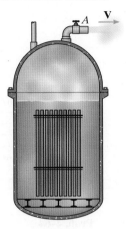

Prob. 4–57

4–58. The pressure vessel of a nuclear reactor is filled with boiling water having a density of $\rho_w = 850$ kg/m³. Its volume is 185 m³. Due to failure of a pump, needed for cooling, the pressure release valve is opened and emits steam having a density of $\rho_s = 35$ kg/m³. If the steam passes through the 40-mm-diameter pipe, determine the average speed through the pipe as a function of the time needed for all the water to escape. Plot the speed (vertical axis) versus the time for $0 \le t \le 3$ h. Give values for increments of $\Delta t = 0.5$ h. Assume that the temperature of the water remains constant.

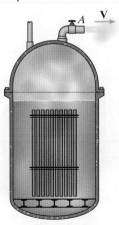

Prob. 4–58

4–59. Water flows through the pipe such that it has a parabolic velocity profile $V = 3(1 - 100r^2)$ m/s, where r is in meters. Determine the time needed to fill the tank to a depth of $h = 1.5$ m if $h = 0$ when $t = 0$. The width of the tank is 3 m.

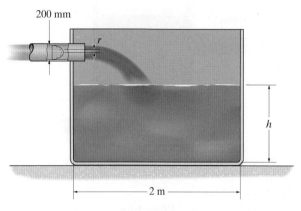

200 mm

2 m

Prob. 4–59

*4–60. The piston is traveling downward at $V_p = 3$ m/s, and as it does, air escapes radially outward through the entire bottom of the cylinder. Determine the average speed of the escaping air. Assume the air is incompressible.

4–61. The piston is travelling downward with a velocity V_p, and as it does, air escapes radially outward through the entire bottom of the cylinder. Determine the average velocity of the air at the bottom as a function of V_p. Plot this average velocity of the escaping air (vertical axis) versus the velocity of the piston for $0 \le V_p \le 5$ m/s. Give values for increments of $\Delta V_p = 1$ m/s. Assume the air is incompressible.

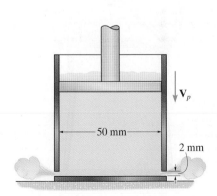

50 mm

2 mm

Probs. 4–60/61

4–62. Water enters the cylindrical tank at A with an average velocity of 2 m/s, and oil exits the tank at B with an average velocity of 1.5 m/s. Determine the rates at which the top level C and interface level D are moving. Take $\rho_o = 900$ kg/m³.

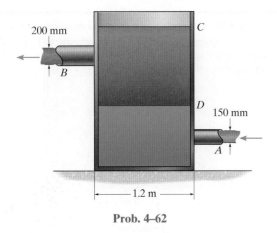

200 mm

C

B

D

150 mm

A

1.2 m

Prob. 4–62

4–63. The 0.5-m-wide lid on the barbecue grill is being closed at a constant angular velocity of $\omega = 0.2$ rad/s, starting at $\theta = 90°$. In the process, the air between A and B will be pushed out in the *radial direction* since the sides of the grill are covered. Determine the average velocity of the air that emerges from the front of the grill at the instant $\theta = 45°$ rad. Assume that the air is incompressible.

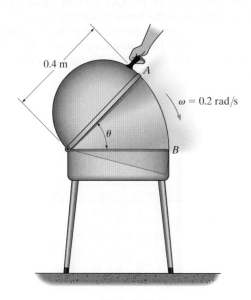

0.4 m

A

$\omega = 0.2$ rad/s

θ

B

Prob. 4–63

*4–64. As air flows over the plate, frictional effects on its surface tend to form a boundary layer in which the velocity profile changes from that of being uniform to one that is parabolic, defined by $u = \left[1000y - 83.33(10^3)y^2\right]$ m/s, where y is in meters, $0 \le y < 6$ mm. If the plate is 0.2 m wide and this change in velocity occurs within the distance of 0.5 m, determine the mass flow through the sections AB and CD. Since these results will not be the same, how do you account for the mass flow difference? Take $\rho = 1.226$ kg/m³.

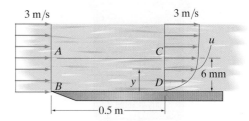

3 m/s   3 m/s

A C u

B y D 6 mm

0.5 m

Prob. 4–64

4–65. The tank originally contains oil. If kerosene having a mass flow of 0.2 kg/s enters the tank at A and mixes with the oil, determine the rate of change of the density of the mixture in the tank if 0.28 kg/s of the mixture exits the tank at the overflow B. The tank is 3 m wide.

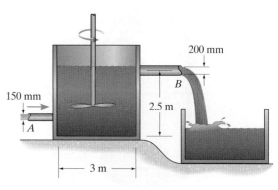

200 mm

B

150 mm   2.5 m

A

3 m

Prob. 4–65

4–66. Air is pumped into the tank using a hose having an inside diameter of 6 mm. If the air enters the tank with an average speed of 6 m/s and has a density of 1.25 kg/m^3, determine the initial rate of change in the density of the air within the tank. The tank has a volume of 0.04 m^3.

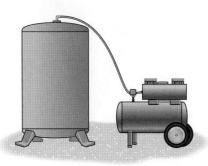

Prob. 4–66

4–67. The tank contains air at a temperature of 20°C and absolute pressure of 500 kPa. Using a valve, the air escapes with an average speed of 120 m/s through a 15-mm-diameter nozzle. If the volume of the tank is 1.25 m^3, determine the rate of change in the density of the air within the tank at this instant. Is the flow steady or unsteady?

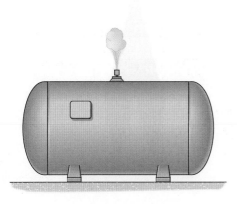

Prob. 4–67

*4–68. The 2-m-diameter cylindrical emulsion tank is being filled at A with cyclohexanol at an average rate of $V_A = 4$ m/s and at B with thiophene at an average rate of $V_B = 2$ m/s. Determine the rate at which the depth increases as a function of depth h.

4–69. The 2-m-diameter cylindrical emulsion tank is being filled at A with cyclohexanol at an average rate of $V_A = 4$ m/s and at B with thiophene at an average rate of $V_B = 2$ m/s. Determine the rate at which the depth of the mixture is increasing when $h = 1$ m. Also, what is the average density of the mixture? Take $\rho_{cy} = 779 \text{ kg/m}^3$, and $\rho_t = 1051 \text{ kg/m}^3$.

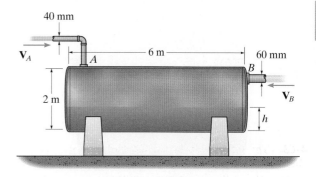

Probs. 4–68/69

4–70. The cylindrical tank in a food-processing plant is filled with a concentrated sugar solution having an initial density of $\rho_s = 1400 \text{ kg/m}^3$. Water is piped into the tank at A at $0.03 \text{ m}^3/\text{s}$ and mixes with the sugar solution. If an equal flow of the diluted solution exits at B, determine the amount of water that must be added to the tank so that the density of the sugar solution is reduced by 10% of its original value.

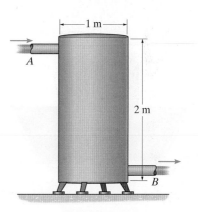

Prob. 4–70

4–71. The cylindrical pressure vessel contains methane at an initial absolute pressure of 2 MPa. If the nozzle is opened, the mass flow depends upon the absolute pressure and is $\dot{m} = 3.5(10^{-6})p$ kg/s, where p is in pascals. Assuming the temperature remains constant at 20°C, determine the time required for the pressure to drop to 1.5 MPa.

*4–72. The cylindrical pressure vessel contains methane at an initial absolute pressure of 2 MPa. If the nozzle is opened, the mass flow depends upon the absolute pressure and is $\dot{m} = 3.5(10^{-6})p$ kg/s, where p is in pascals. Assuming the temperature remains constant at 20°C, determine the pressure in the tank as a function of time. Plot this pressure (vertical axis) versus the time for $0 \leq t \leq 15$ s. Give values for increments of $\Delta t = 3$ s.

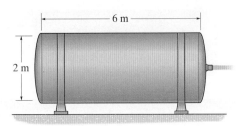

Probs. 4–71/72

4–73. A part is manufactured by placing molten plastic into the trapezoidal container and then moving the cylindrical die down into it at a constant speed of 20 mm/s. Determine the average speed at which the plastic rises in the form as a function of y_c. The container has a width of 150 mm.

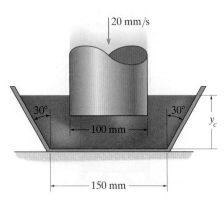

Prob. 4–73

4–74. Water flows out of the stem of the funnel at an average speed of $V = \left(3e^{-0.05t}\right)$ m/s, where t is in seconds. Determine the average speed at which the water level is falling at the instant $y = 100$ mm. At $t = 0$, $y = 200$ mm.

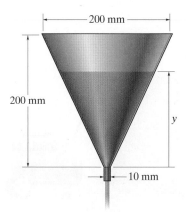

Prob. 4–74

4–75. As part of a manufacturing process, a 0.1-m-wide plate is dipped into hot tar and then lifted out, causing the tar to run down and then off the sides of the plate as shown. The thickness w of the tar at the bottom of the plate decreases with time t, but it still is assumed to maintain a linear variation along the height of the plate as shown. If the velocity profile at the bottom of the plate is approximately parabolic, such that $u = \left[0.5(10^{-3})(x/w)^{1/2}\right]$ m/s, where x and w are in meters, determine w as a function of time. Initially, when $t = 0$, $w = 0.02$ m.

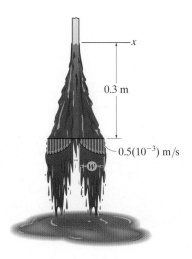

Prob. 4–75

*4–76. Hexylene glycol is flowing into the trapezoidal container at a constant rate of 600 kg/min. Determine the rate at which the level is rising when $y = 0.5$ m. The container has a constant width of 0.5 m. $\rho_{hg} = 924$ kg/m³.

4–78. The cylinder is pushed down into the tube at a rate of $V = 5$ m/s. Determine the average velocity of the liquid as it rises in the tube.

4–79. Determine the speed V at which the cylinder must be pushed down into the tube so that the liquid in the tube rises with an average velocity of 4 m/s.

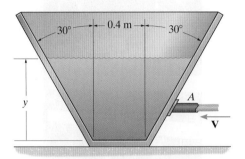

Prob. 4–76

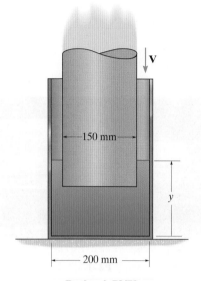

Probs. 4–78/79

4–77. Hexylene glycol is flowing into the container at a constant rate of 600 kg/min. Determine the rate at which the level is rising when $y = 0.5$ m. The container is in the form of a conical frustum. *Hint*: The volume of a cone is $V = \frac{1}{3} \pi r^2 h$. $\rho_{hg} = 924$ kg/m³.

*4–80. The conical shaft is forced into the conical seat at a constant speed of V_0. Determine the average velocity of the liquid as it is ejected from the horizontal section AB as a function of y. *Hint*: The volume of a cone is $V = \frac{1}{3} \pi r^2 h$.

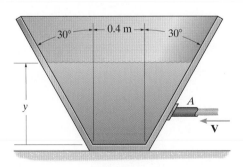

Prob. 4–77

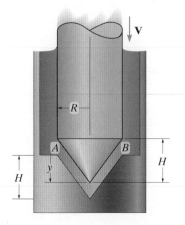

Prob. 4–80

CONCEPTUAL PROBLEMS

P4–1. Air flows to the left through this duct transition. Will the air accelerate or decelerate? Explain why.

P4–2. As water falls from the opening, it narrows as shown and forms what is called a *vena contracta*. Explain why this occurs, and why the water remains together in a stream.

P4–1

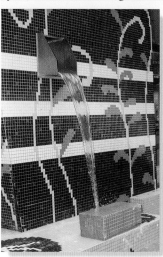

P4–2

CHAPTER REVIEW

A control volume is used for a Eulerian description of the flow. Depending on the problem, this volume can be fixed, be moving, or have a changing shape.

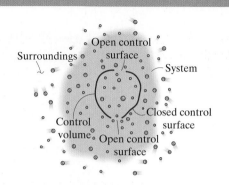

Using the Reynolds transport theorem, we can relate the time rate of change of a fluid property N for a system of particles to its time rate of change as measured from a control volume. Determining the control volume change requires measuring both a *local change* within the control volume, and a *convection change*, as the fluid passes through its open control surfaces.

The volumetric flow or discharge Q through a planar area A is determined by finding the velocity of the flow *perpendicular* to the area. If the velocity profile is known, then integration must be used to determine Q. If the average velocity V is used, then $Q = \mathbf{V} \cdot \mathbf{A}$.

$$Q = \int_A \mathbf{v} \cdot d\mathbf{A}$$

The mass flow depends on the density of the fluid and on the velocity profile passing through the area. If the average velocity V is used, then $\dot{m} = \rho \mathbf{V} \cdot \mathbf{A}$.

$$\dot{m} = \int_A \rho \mathbf{v} \cdot d\mathbf{A}$$

The continuity equation is based on the conservation of mass, which requires that the mass of a fluid system remain constant with respect to time. In other words, its time rate of change is zero.

$$\frac{\partial}{\partial t} \int_{cv} \rho \, d V + \int_{cs} \rho \, \mathbf{V} \cdot d\mathbf{A} = 0$$

We can use a fixed, a moving, or a changing control volume to apply the continuity equation. In particular, if the control volume is completely filled with fluid and the flow is steady, then no local changes will occur within the control volume, so only convective changes need be considered.

If the control volume is attached to a body, moving with constant velocity, then steady flow will occur, and so at the control surface $\mathbf{V} = \mathbf{V}_{f/cs}$ must be used when applying the continuity equation.

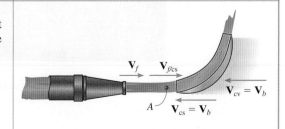

Chapter 5

The design of fountains requires application of the principles of work and energy. Here the velocity of the flow out of the nozzles is transformed into lifting the water to its maximum height. (© Don Andreas/Fotolia)

Work and Energy of Moving Fluids

CHAPTER OBJECTIVES

- To develop Euler's equations of motion and the Bernoulli equation for streamline coordinates, and to illustrate some important applications.

- To show how to establish the energy grade line and hydraulic grade line for a fluid system.

- To develop the energy equation from the first law of thermodynamics, and to show how to solve problems that include pumps, turbines, and frictional loss.

5.1   Euler's Equations of Motion

In this section we will apply Newton's second law of motion to study the motion of a single fluid particle as it travels along a streamline with steady flow, Fig. 5–1a. Here the streamline coordinate s is in the direction of motion and tangent to the streamline. The normal coordinate n is directed positive toward the streamline's center of curvature. The particle has a length Δs, height Δn, and width Δx. Since the flow is *steady*, the streamline will remain fixed, and since it happens to be curved, the particle will have *two* components of acceleration. Recall from Sec. 3.4 the *tangential* or streamline component a_s measures the time rate of change in the *magnitude* of the particle's velocity. It is determined from $a_s = V(dV/ds)$. The *normal* component a_n measures the time rate of change in the *direction* of the velocity. It is determined from $a_n = V^2/R$, where R is the radius of curvature of the streamline at the point where the particle is located.

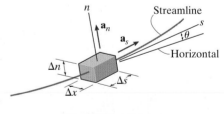

Ideal fluid particle

(a)

Fig. 5–1

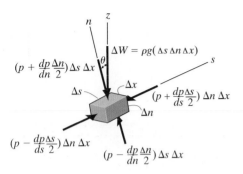

Free-body diagram
(b)

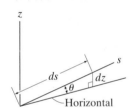

(c)

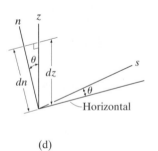

(d)

Fig. 5–1 (cont.)

The free-body diagram of the particle is shown in Fig. 5–1b. If we assume the fluid is *inviscid*, then the shearing forces of viscosity are not present, and only the forces caused by weight and pressure act on the particle. If the pressure at the center of the particle is p, then there will be a change in pressure, as noted on each face of the particle, due to its finite size.* Finally, the weight of the particle is $\Delta W = \rho g \Delta V = \rho g \, (\Delta s \Delta n \Delta x)$, and its mass is $\Delta m = \rho \Delta V = \rho(\Delta s \Delta n \Delta x)$.

s Direction.   Applying the equation of motion $\Sigma F_s = ma_s$ in the s direction, we have

$$\left(p - \frac{dp}{ds}\frac{\Delta s}{2}\right)\Delta n \Delta x - \left(p + \frac{dp}{ds}\frac{\Delta s}{2}\right)\Delta n \Delta x - \rho g(\Delta s \Delta n \Delta x)\sin\theta$$
$$= \rho(\Delta s \Delta n \Delta x)V\left(\frac{dV}{ds}\right)$$

Dividing through by the mass, $\rho(\Delta s \Delta n \Delta x)$, and rearranging the terms yields

$$\frac{1}{\rho}\frac{dp}{ds} + V\left(\frac{dV}{ds}\right) + g\sin\theta = 0 \qquad (5\text{–}1)$$

As noted in Fig. 5–1c, $\sin\theta = dz/ds$. Therefore,

$$\frac{dp}{\rho} + V\,dV + g\,dz = 0 \qquad (5\text{–}2)$$

n Direction.   Applying the equation of motion $\Sigma F_n = ma_n$ in the n direction, Fig. 5–1b, we have

$$\left(p - \frac{dp}{dn}\frac{\Delta n}{2}\right)\Delta s \Delta x - \left(p + \frac{dp}{dn}\frac{\Delta n}{2}\right)\Delta s \Delta x - \rho g(\Delta s \Delta n \Delta x)\cos\theta$$
$$= \rho(\Delta s \Delta n \Delta x)\left(\frac{V^2}{R}\right)$$

Using $\cos\theta = dz/dn$, Fig. 5–1d, and dividing by the volume, $(\Delta s \Delta n \Delta x)$, this equation reduces to

$$-\frac{dp}{dn} - \rho g\frac{dz}{dn} = \frac{\rho V^2}{R} \qquad (5\text{–}3)$$

Equations 5–2 and 5–3 are differential forms of the equations of motion that were originally developed by the Swiss mathematician Leonhard Euler. For this reason they are often called **Euler's differential equations of motion.** They apply only in the s and n directions, to the *steady flow* of an *inviscid fluid* particle that *moves along* a streamline. We will now consider a few important applications.

*For this change, we have considered only the *first term* in a Taylor series expansion at this point, since the higher-order terms will cancel out as the particle's size becomes infinitesimal. (See the footnote on page 52.)

Steady Horizontal Flow of an Ideal Fluid. Shown in Fig. 5–2 are straight horizontal open and closed conduits, where an *ideal fluid* is flowing at *constant velocity*. In both cases, the pressure at A is p_A, and we wish to determine the pressure at points B and C. Since A and B lie on the *same streamline*, we can apply Euler's equation in the s direction, Eq. 5–2, and integrate it for a particle traveling between these two points. Here $V_A = V_B = V$, and because there is no elevation change, $dz = 0$. Also, since the fluid density is constant, we have

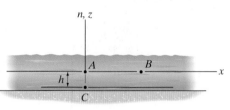

Open conduit

$$\frac{dp}{\rho} + V\,dV + g\,dz = 0$$

$$\frac{1}{\rho}\int_{p_A}^{p_B} dp + \int_{V}^{V} V\,dV + 0 = 0$$

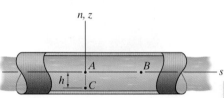

$$\frac{1}{\rho}(p_B - p_A) + 0 + 0 = 0 \quad \text{or} \quad p_B = p_A$$

Thus, for an ideal fluid, the pressure along both the open and closed conduits *remains constant in the horizontal direction.* This result is to be expected because no viscous friction forces have to be overcome by the pressure that pushes the fluid forward.

Closed conduit

Fig. 5–2

To determine the pressure at C, notice that A and C lie on *different streamlines* or s axes, Fig. 5–2. However, C is on the n axis, which has its origin at A. Since the radius of curvature of the horizontal streamline at A is $R \to \infty$, Eq. 5–3 becomes

$$-\frac{dp}{dn} - \rho g\frac{dz}{dn} = \frac{\rho V^2}{R} = 0$$

$$-dp - \rho g\,dz = 0$$

Integrating from A to C, noting that C is at $z = -h$ from A, we have

$$-\int_{p_A}^{p_C} dp - \rho g\int_{0}^{-h} dz = 0$$

$$-p_C + p_A - \rho g(-h - 0) = 0$$

$$p_C = p_A + \rho gh$$

This result indicates that *in the vertical direction, the pressure is the same as if the fluid were static*; that is, it is the same result as Eq. 2–5. The term "static pressure" is often used here, since it is a measure of the pressure *relative to the flow*, in which case the fluid will appear to be at rest.

Notice that if the conduit, and hence the streamlines, were *curved*, then we would not obtain this result, because particles at A and C would move along streamlines having *different* radii of curvature, so they would *change the directions of their velocities* at different rates. In other words, the term $\rho V^2/R \neq 0$, and so the smaller R becomes, the *larger* the pressure required to change the fluid particles' direction and keep them on their streamline. The following example should help to illustrate this point.

EXAMPLE | 5.1

(a)

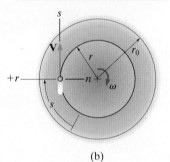

(b)

Fig. 5–3

A tornado has winds that essentially move along horizontal circular streamlines, Fig. 5–3a. Within the eye, $0 \leq r \leq r_0$, the wind velocity is $V = \omega r$, which represents a forced vortex, that is, flow rotating at a constant angular rate ω as described in Sec. 2.14. Determine the pressure distribution within the eye of the tornado as a function of r, if at $r = r_0$ the pressure is $p = p_0$.

SOLUTION

Fluid Description.   We have steady flow, and we will assume the air is an ideal fluid, that is, it is inviscid and has a constant density ρ.

Analysis.   The streamline for a fluid particle having a radius r is shown in Fig. 5–3b. To find the pressure distribution as a function of r (positive outward), we must apply Euler's equation in the n direction (positive inward).

$$-\frac{dp}{dn} - \rho g \frac{dz}{dn} = \frac{\rho V^2}{R}$$

Since the path is horizontal, $dz = 0$. Also, for the chosen streamline, $R = r$ and $dn = -dr$. Since the velocity of the particle is $V = \omega r$, Fig. 5–3b, the above equation becomes

$$\frac{dp}{dr} - 0 = \frac{\rho(\omega r)^2}{r}$$

$$dp = \rho \omega^2 r \, dr$$

Notice that the pressure increases, $+dp$, as we move away $+dr$ from the center. This pressure is needed to change the direction of the velocity of the air particles and keep them on their circular path. By comparison, this same effect occurs in a cord that maintains the motion of a ball swinging in a horizontal circular path. The longer the cord, the greater the force the cord must exert on the ball to maintain the same rotation.

Since $p = p_0$ at $r = r_0$, then

$$\int_p^{p_0} dp = \rho \omega^2 \int_r^{r_0} r \, dr$$

$$p = p_0 - \frac{\rho \omega^2}{2}(r_0^2 - r^2) \qquad\qquad Ans.$$

5.2   The Bernoulli Equation

As we have seen, Euler's equations represent the application of Newton's second law of motion to the steady flow of an inviscid fluid particle, expressed in terms of the streamline coordinates s and n. Since particle motion *only occurs* in the s direction, we can integrate Eq. 5–2 along a streamline and thereby obtain a relationship between the motion of a particle and the pressure and gravitational forces that act upon it. We have

$$\int \frac{dp}{\rho} + \int V\, dV + \int g\, dz = 0$$

Provided the fluid density can be expressed as a function of pressure, integration of the first term can be carried out. The most common case, however, is to consider the fluid to be *ideal*, that is, both inviscid and incompressible. Since we previously assumed the fluid is inviscid, we now have an *ideal fluid*. For this case, integration yields

$$\frac{p}{\rho} + \frac{V^2}{2} + gz = \text{const.} \qquad (5\text{–}4)$$

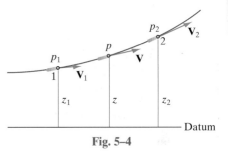

Fig. 5–4

Here z is the *elevation* of the particle, measured from an *arbitrarily* chosen *fixed* horizontal plane or *datum*, Fig. 5–4. For any particle *above* this datum, z is positive, and for any particle *below* it, z is negative. For a particle on the datum, $z = 0$.

   Equation 5–4 is referred to as the **Bernoulli equation**, named after Daniel Bernoulli, who stated it around the mid-18th century. Sometime later, though, it was expressed as a formula by Leonhard Euler. When it is applied between any two points, 1 and 2, located on the *same streamline*, Fig. 5–4, it can be written in the form

$$\boxed{\; \frac{p_1}{\rho} + \frac{V_1^2}{2} + gz_1 = \frac{p_2}{\rho} + \frac{V_2^2}{2} + gz_2 \;} \qquad (5\text{–}5)$$

steady flow, ideal fluid, same streamline

If the fluid is a gas, such as air, the elevation terms can generally be neglected, since the density of a gas is small. In other words, its weight is not considered a significant force compared to that caused by the pressure within the gas.

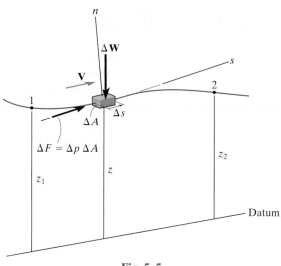

Fig. 5–5

Interpretation of Terms. As noted by its derivation, the Bernoulli equation is actually an integrated form of Newton's second law of motion, written for the steady flow of an ideal fluid in the *s* direction. However, it can be interpreted as a statement of the principle of work and energy, as it applies to a particle representing a *unit mass of fluid.* To show this, we first rewrite it as

$$\left(\frac{V_2^2}{2} - \frac{V_1^2}{2}\right) = \left(\frac{p_1}{\rho} - \frac{p_2}{\rho}\right) + g\,(z_1 - z_2)$$

Here each of the terms in parentheses has units of energy or work per unit mass, J/kg. In the form shown, this equation states that the change in kinetic energy of the particle of unit mass is equal to the work done by the pressure and gravitational forces, as the particle moves from position 1 to position 2. Recall that work is the product of the component of force in the direction of displacement, times the displacement. The work of the pressure force is called **flow work**, and it is always directed along the streamline, Fig. 5–5. The gravitational work is done in the vertical direction (*z*) since the weight is in this direction. We will discuss these concepts further in Sec. 5.5, where we will show that the Bernoulli equation is actually a special case of the energy equation.

Limitations.   It is very important to remember that the Bernoulli equation can only be applied when we have *steady flow* of an *ideal fluid*—one that is incompressible (ρ is constant) and inviscid ($\mu = 0$). As developed here, its application is between any two points lying on the *same streamline*. If these conditions cannot be justified, then application of this equation will produce erroneous results. With reference to Fig. 5–6, the following is a list of some common situations in which the Bernoulli equation *should not be used*.

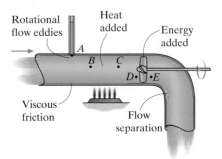

Places where the Bernoulli equation *does not* apply

Fig. 5–6

- Many fluids, such as air and water, have rather low viscosities, so in some situations they may be assumed to be ideal fluids. However, realize that in certain regions of the flow, the effects of viscosity can not be ignored. For example, viscous flow will always predominate near a solid boundary, such as the walls of the pipe in Fig. 5–6. This region is called the *boundary layer*, and we will discuss it in more detail in Chapter 11. Here the velocity gradient is high, and the fluid friction, or shear, which causes boundary layer formation, will produce heat and thereby remove energy from the flow. The Bernoulli equation *cannot* be applied within the boundary layer because of this resulting energy loss.

- Sudden changes in the direction of a solid boundary can cause the boundary layer to thicken and result in flow separation from the boundary. This has occurred in Fig. 5–6 along the inside wall of the pipe at the bend. Here turbulent mixing of the fluid not only produces frictional heat loss, but greatly affects the velocity profile and causes a severe pressure drop. The streamlines within this region are not well defined, and so the Bernoulli equation does not apply. Flow through connections, such as "tees" and valves, can produce similar energy losses due to flow separation. Furthermore, flow separation and the development of turbulent mixing within the flow also occur at the pipe–tube interface A in Fig. 5–6. Hence the Bernoulli equation *does not* apply across this interface.

- Energy changes within the flow occur within regions of heat removal or heat application, such as between B and C. Also, pumps or turbines can supply or remove energy from the flow, such as between D and E. The Bernoulli equation *does not* account for these energy changes, and so it cannot be applied within these regions.

- If the fluid is a gas, then its density will change as the speed of the flow increases. Normally, as a general rule for engineering calculations, a gas can be considered incompressible provided its speed remains below about 30% of the speed at which sound travels within it. For example, at 15°C the sonic speed in air is 340 m/s, so a limiting value would be 102 m/s. For faster flows, compressibility effects will cause heat loss, and this becomes important. For these high speeds, the Bernoulli equation *will not* give acceptable results.

5.3   Applications of the Bernoulli Equation

In this section we will provide some basic applications of the Bernoulli equation to show how to determine either the velocity or the pressure at different points on a streamline.

Flow from a Large Reservoir.

When water flows from a tank or reservoir through a drain, Fig. 5–7, the flow is actually *unsteady*. This happens because when the distance h is large, then due to the greater water pressure at the drain, the water level will drop at a faster rate than when h is small. However, if the reservoir has a *large volume* and the drain has a relatively small diameter, then the movement of water within the reservoir is very slow and so, at its surface, $V_A \approx 0$, Fig. 5–7. Under these circumstances it is reasonable to assume *steady flow* through the drain. Also, for small-diameter openings, the *elevation difference* between C and D is small and so $V_C \approx V_D \approx V_B$. Furthermore, the pressure at the centerline B of the opening is *atmospheric*, as it is at C and D.

If we assume that water is an ideal fluid, then the Bernoulli equation can be applied between points A and B that lie on the selected streamline in Fig. 5–7. Setting the gravitational datum at B, and using gage pressures, where $p_A = p_B = 0$, we have

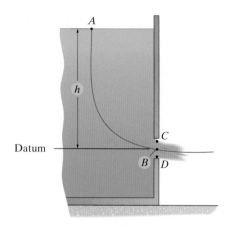

Fig. 5–7

$$\frac{p_A}{\rho} + \frac{V_A^2}{2} + gz_A = \frac{p_B}{\rho} + \frac{V_B^2}{2} + gz_B$$

$$0 + 0 + gh = 0 + \frac{V_B^2}{2} + 0$$

$$V_B = \sqrt{2gh}$$

This result is known as ***Torricelli's law*** since it was first formulated by Evangelista Torricelli in the 17th century. As a point of interest, it can be shown that V_B is the *same* as that obtained by a fluid particle that is simply dropped from rest at the same height h, even though the time of travel for this freely falling particle is much shorter than for the one flowing through the tank.

Flow around a Curved Boundary.   As a fluid flows around a smooth obstacle, the energy of the fluid will be transformed from one form into another. For example, consider the horizontal streamline that intersects the front of the curved surface in Fig. 5–8. Since *B* is a *stagnation point*, fluid particles moving from *A* to *B* must decelerate as they approach the boundary, so that their velocity at *B* momentarily becomes zero, before the flow begins to separate and then move along the sides of the surface. If we apply the Bernoulli equation between points *A* and *B*, we have

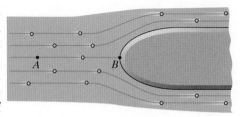

Fig. 5–8

$$\frac{p_A}{\rho} + \frac{V_A^2}{2} + gz_A = \frac{p_B}{\rho} + \frac{V_B^2}{2} + gz_B$$

$$\frac{p_A}{\rho} + \frac{V_A^2}{2} + 0 = \frac{p_B}{\rho} + 0 + 0$$

$$\underset{\substack{\text{Stagnation} \\ \text{pressure}}}{p_B} = \underset{\substack{\text{Static} \\ \text{pressure}}}{p_A} + \underset{\substack{\text{Dynamic} \\ \text{pressure}}}{\rho \frac{V_A^2}{2}}$$

This pressure is referred to as the **stagnation pressure** because it represents the **total pressure** exerted by the fluid at the stagnation point, *B*. As noted in the previous section, the pressure p_A is a **static pressure** because it is measured relative to the flow. Finally, the *increase* in pressure, $\rho V_A^2 / 2$, is called the **dynamic pressure** because it represents the additional pressure required to bring the fluid to rest at *B*.

Flow in an Open Channel.   One method of determining the velocity of a moving liquid in an *open channel*, such as a river, is to immerse a bent tube into the stream and observe the height *h* to which the liquid rises within the tube, Fig. 5–9. Such a device is called a **stagnation tube**, or a **Pitot tube**, named after Henri Pitot who invented it in the early 18th century.

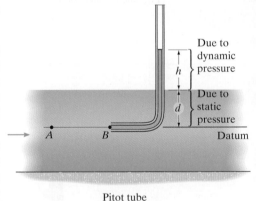

Pitot tube

Fig. 5–9

To show how it works, consider the two points *A* and *B* located on the horizontal streamline. Point *A* is upstream within the fluid, where the velocity of flow is V_A and the pressure is $p_A = \rho g d$. Point *B* is at the opening of the tube. It is the stagnation point, since the velocity of flow has momentarily been reduced to zero due to its impact with the liquid within the tube. The liquid at this point produces *both* a *static pressure*, which causes the liquid in the tube to rise to a level *d*, and a *dynamic pressure*, which forces additional liquid farther up the vertical segment to a height *h* above the liquid surface. Thus the total pressure of the liquid at *B* is $p_B = \rho g(d + h)$. Applying the Bernoulli equation with the gravitational datum on the streamline, we have

$$\frac{p_A}{\rho} + \frac{V_A^2}{2} + gz_A = \frac{p_B}{\rho} + \frac{V_B^2}{2} + gz_B$$

$$\frac{\rho g d}{\rho} + \frac{V_A^2}{2} + 0 = \frac{\rho g(d + h)}{\rho} + 0 + 0$$

$$V_A = \sqrt{2gh}$$

Hence, by measuring *h* on the Pitot tube, the velocity of the flow can be determined.

Flow in a Closed Conduit.   If the liquid is flowing in a *closed conduit* or pipe, Fig. 5–10*a*, then it will be necessary to use both a piezometer and a Pitot tube to determine the velocity of the flow. The *piezometer* measures the *static pressure* at *A*. This pressure is caused by the internal pressure in the pipe, ρgh, and the hydrostatic pressure ρgd, caused by the weight of the fluid. The total pressure at *A* is therefore $\rho g(h + d)$. The total pressure at the stagnation point *B* will be larger than this, due to the dynamic pressure $\rho V_A^2/2$. If we apply the Bernoulli equation at points *A* and *B* on the streamline, using the measurements *h* and $(l + h)$ from these two tubes, the velocity V_A can be obtained.

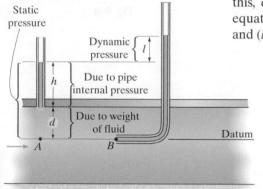

Static pressure

Dynamic pressure l

h Due to pipe internal pressure

d Due to weight of fluid

Datum

A B

Piezometer and pitot tubes

(a)

$$\frac{p_A}{\rho} + \frac{V_A^2}{2} + gz_A = \frac{p_B}{\rho} + \frac{V_B^2}{2} + gz_B$$

$$\frac{\rho g(h + d)}{\rho} + \frac{V_A^2}{2} + 0 = \frac{\rho g(h + d + l)}{\rho} + 0 + 0$$

$$V_A = \sqrt{2gl}$$

Rather than use two separate tubes in the manner just described, a more elaborate single tube called a ***Pitot-static tube*** is often used to determine the velocity of the flow in a closed conduit. It is constructed using two concentric tubes as shown in Fig. 5–10*b*. Like the Pitot tube in Fig. 5–10*a*, the stagnation pressure at *B* can be measured from the pressure tap at *E* in the inner tube. Downstream from *B* there are several open holes *D* on the outer tube. This outer tube acts like the piezometer in Fig. 5–10*a* so that the static pressure can be measured from the pressure tap at *C*. Using these two measured pressures and applying the Bernoulli equation between points *A* and *B*, neglecting any elevation differences between *C* and *E*, we have

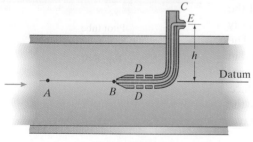

C

E

D

h

Datum

A B D

Pitot-static tube

(b)

Fig. 5–10

$$\frac{p_A}{\rho} + \frac{V_A^2}{2} + gz_A = \frac{p_B}{\rho} + \frac{V_B^2}{2} + gz_B$$

$$\frac{p_C + \rho gh}{\rho} + \frac{V_A^2}{2} + 0 = \frac{p_E + \rho gh}{\rho} + 0 + 0$$

$$V_A = \sqrt{\frac{2}{\rho}(p_E - p_C)}$$

In actual practice, the *difference in pressure* can be determined either by using a manometer, attached to the outlets *C* and *E*, and measuring the differential height of the manometer liquid; or by using a pressure transducer. Corrections are sometimes made to the readings since the flow may be slightly disturbed at the side inlet holes *D*, due to its movement around the front of the tube at *B* and past its vertical segment.

Venturi meter.

A **venturi meter** is a device that can also be used to measure the average velocity or the flow of an incompressible fluid through a pipe, Fig. 5–11. It was conceived of by Giovanni Venturi in 1797, but the principle was not applied until about one hundred years later by the American engineer Clemens Herschel. This device consists of a reducer followed by a *venturi tube* or *throat* of diameter d_2, and then a gradual transition segment back to the original pipe. As the fluid flows through the reducer, the flow accelerates, causing a higher velocity and lower pressure to be developed within the throat. Applying the Bernoulli equation along the center streamline, between point 1 in the pipe and point 2 in the throat, gives

$$\frac{p_1}{\rho} + \frac{V_1^2}{2} + gz_1 = \frac{p_2}{\rho} + \frac{V_2^2}{2} + gz_2$$

$$\frac{p_1}{\rho} + \frac{V_1^2}{2} + 0 = \frac{p_2}{\rho} + \frac{V_2^2}{2} + 0$$

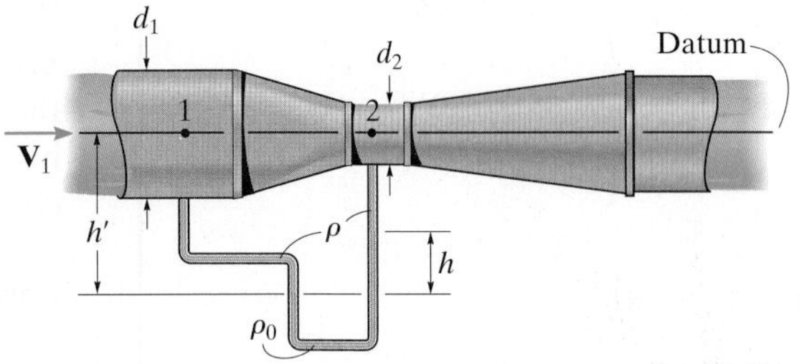

In addition, the continuity equation can be applied at points 1 and 2. For steady flow we have

$$\frac{\partial}{\partial t} \int_{cv} \rho d\Psi + \int_{cs} \rho \mathbf{V} \cdot d\mathbf{A} = 0$$

$$0 - V_1 \pi \left(\frac{d_1^2}{4}\right) + V_2 \pi \left(\frac{d_2^2}{4}\right) = 0$$

Combining these two results and solving for V_1, we get

$$V_1 = \sqrt{\frac{2(p_2 - p_1)/\rho}{1 - (d_1/d_2)^4}}$$

Venturi meter

Fig. 5–11

The static pressure difference $(p_2 - p_1)$ is often measured using a pressure transducer or a manometer. For example, if a manometer is used, as in Fig. 5–11, and ρ is the density of the fluid in the pipe, and ρ_0 is the density for the fluid in the manometer, then applying the manometer rule, we have

$$p_1 + \rho g h' - \rho_0 g h - \rho g (h' - h) = p_2$$

$$p_2 - p_1 = (\rho - \rho_0)g h$$

Once the measurement for h is made, the result $(p_2 - p_1)$ is substituted into the above equation to obtain V_1. The volumetric flow can then be determined from $Q = V_1 A_1$.

Important Points

- The Euler differential equations of motion apply to a fluid particle that moves along a streamline. These equations are based on the *steady flow of an inviscid fluid*. Because viscosity is neglected, the flow is affected only by pressure and gravitational forces. In the *s* direction these forces cause a *change in magnitude* of a fluid particle's velocity, giving it a tangential acceleration; and in the *n* direction these forces cause a *change in direction* of the velocity, thereby producing a normal acceleration.

- When the streamlines for the flow are *straight horizontal lines*, the Euler equations show that for an ideal (frictionless) fluid having steady flow, the pressure p_0 in the *horizontal direction* is *constant*. Furthermore, since the velocity does not change its direction, there is no normal acceleration. Consequently, in a horizontal open or closed conduit, the pressure variation in the vertical direction is hydrostatic. In other words, this is a measure of *static pressure* because it can be made relative to the moving fluid.

- The Bernoulli equation is an integrated form of Euler's equation in the *s* direction. It can be interpreted as a statement of the fluid's work and energy. It is applied at two points located on the *same streamline,* and it requires *steady flow of an ideal fluid.* This equation *cannot* be applied to viscous fluids, or at transitions where the flow separates and becomes turbulent. Also, it cannot be applied between points where fluid energy is added or withdrawn by external sources, such as pumps and turbines, or in regions where heat is applied or removed.

- The Bernoulli equation indicates that if the flow is horizontal, then z is constant, and so $p/\rho + V^2/2 = $ const. Therefore, through a *converging* duct or nozzle, the *velocity* will *increase* and the *pressure* will *decrease*. Likewise, for a *diverging* duct, the velocity will *decrease* and the pressure will *increase*. (See the photo on the next page.)

- A *Pitot tube* can be used to measure a fluid's velocity at a point in an *open channel*. The flow creates a dynamic pressure $\rho V^2/2$ at the tube's stagnation point, which forces the fluid up the tube. For a *closed conduit*, both a piezometer and a Pitot tube must be used to measure velocity. A device that combines both of these is called a Pitot-static tube.

- A *venturi meter* can be used to measure the velocity or volumetric flow of a fluid through a closed duct or pipe.

Procedure for Analysis

The following procedure provides a means for applying the Bernoulli equation.

Fluid Description.

- *Be sure* that the fluid can be assumed to be an ideal fluid, that is, incompressible and inviscid. Also, steady flow must occur.

Bernoulli Equation.

- Select two points on the *same streamline* within the flow, where some values of pressure and velocity are known. The elevation of these points is measured from an arbitrarily established *fixed* datum. At pipe outlets to the atmosphere, and at open surfaces, the pressure can be assumed to be atmospheric, that is, the gage pressure is zero.

- The velocity at each point can be determined if the volumetric flow and cross-sectional area of a conduit are known, $V = Q/A$.

- Tanks or reservoirs that drain slowly have liquid surfaces that are essentially at rest; that is, $V \approx 0$.

- When the ideal fluid is a gas, elevation changes, measured from the datum, can generally be neglected.

- Once the known and unknown values of p, V, and z have been identified at each of the two points, the Bernoulli equation can be applied. When substituting the data, be sure to use a consistent set of units.

- If more than one unknown is to be determined, consider relating the velocities using the continuity equation, or the pressures using the manometer equation if it applies.

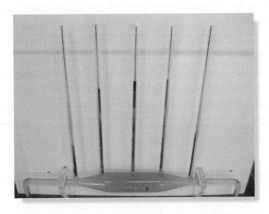

As noted by the water level in the piezometers, the pressure of the water flowing through this pipe will vary in accordance with the Bernoulli equation. Where the diameter is small, the velocity is high and the pressure is low; and where the diameter is large, the velocity is low and the pressure is high.

EXAMPLE | 5.2

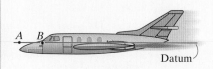

Fig. 5–12

The jet plane in Fig. 5–12 is equipped with a piezometer and a Pitot tube. The piezometer indicates an absolute pressure of 47.2 kPa, while the Pitot tube B reads an absolute pressure of 49.6 kPa. Determine the altitude of the plane and its speed.

SOLUTION

The piezometer measures the static pressure in the air, and so the altitude of the plane can be determined from the table in Appendix A. For an absolute pressure of 47.2 kPa, the altitude is approximately

$$h = 6 \text{ km} \qquad Ans.$$

Fluid Description. We will assume that the speed of the plane is slow enough so that the air can be considered incompressible and inviscid— an ideal fluid. Doing so, we can apply the Bernoulli equation, provided we observe *steady flow*. This can be realized if we view the motion *from the plane.** Thus, the air at A, which is actually at rest, will have the same speed as the plane when observed from the plane, $V_A = V_p$. The air at B, a stagnation point, will appear to be at rest when observed from the plane, $V_B = 0$. From Appendix A, for air at an elevation of 6 km, $\rho_a = 0.6601 \text{ kg/m}^3$.

Bernoulli Equation. Applying the Bernoulli equation at points A and B on the horizontal streamline, we have

$$\frac{p_A}{\rho} + \frac{V_A^2}{2} + gz_A = \frac{p_B}{\rho} + \frac{V_B^2}{2} + gz_B$$

$$\frac{47.2(10^3) \text{ N/m}^2}{0.6601 \text{ kg/m}^3} + \frac{V_p^2}{2} + 0 = \frac{49.6(10^3) \text{ N/m}^2}{0.6601 \text{ kg/m}^3} + 0 + 0$$

$$V_p = 85.3 \text{ m/s} \qquad Ans.$$

NOTE: We will show in Chapter 13 that this speed is about 25% the speed of sound in air, and since 25% < 30%, our assumption of air being incompressible is valid. Most planes are equipped with either a piezometer and a Pitot tube, or a combination Pitot-static tube. The pressure readings are directly converted to altitude and air speed and are shown on the instrument panel. If more accuracy is needed, corrections are made by factoring in a reduced density of the air at high altitudes. Finally, realize that for proper operation, the opening of any Pitot tube must be free of debris, such as caused by nesting insects or ice formation.

---

*If the flow is observed from the ground, then the flow is *unsteady* since the velocity of the air particles will *change with time* as the plane flies past these particles. Remember that the Bernoulli equation *does not* apply for unsteady flow.

EXAMPLE 5.3

Determine the average velocity of the flow of water in the pipe in Fig. 5–13, and the static and dynamic pressure at point B. The water level in each of the tubes is indicated. Take $\rho_w = 1000\,\text{kg/m}^3$.

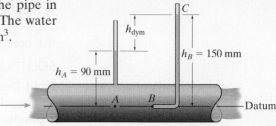

Fig. 5–13

SOLUTION

Fluid Description. We have steady flow. Also, we will consider the water to be an ideal fluid.

Bernoulli Equation. At A the total pressure is the *static pressure* found from $p_A = \rho_w g h_A$, and at B the total (or stagnation) pressure is a *combination* of static and dynamic pressures found from $p_B = \rho_w g h_B$. Knowing these pressures, we can determine the average velocity of flow V_A using the Bernoulli equation, applied at points A and B, where B is a stagnation point on the streamline. We have

$$\frac{p_A}{\rho} + \frac{V_A^2}{2} + gz_A = \frac{p_B}{\rho} + \frac{V_B^2}{2} + gz_B$$

$$\frac{\rho_w g h_A}{\rho_w} + \frac{V_A^2}{2} + 0 = \frac{\rho_w g h_B}{\rho_w} + 0 + 0$$

$$\frac{V_A^2}{2} = g(h_B - h_A) = (9.81\,\text{m/s}^2)(0.150\,\text{m} - 0.090\,\text{m})$$

$$V_A = 1.085\,\text{m/s} = 1.08\,\text{m/s} \qquad \text{Ans.}$$

The static pressure at both A and B is determined from the piezometric head.

$$(p_A)_{\text{static}} = (p_B)_{\text{static}} = \rho_w g h_A = (1000\,\text{kg/m}^3)(9.81\,\text{m/s}^3)(0.09\,\text{m}) = 883\,\text{Pa} \qquad \text{Ans.}$$

The dynamic pressure at B is determined from

$$\rho_w \frac{V_A^2}{2} = (1000\,\text{kg/m}^3)\frac{(1.085\,\text{m/s})^2}{2} = 589\,\text{Pa} \qquad \text{Ans.}$$

This value can also be obtained by noticing that

$$h_{\text{dyn}} = 0.15\,\text{m} - 0.09\,\text{m} = 0.06\,\text{m}$$

so that

$$(p_B)_{\text{dyn}} = \rho_w g h_{\text{dyn}} = (1000\,\text{kg/m}^3)(9.81\,\text{m/s}^2)(0.06\,\text{m}) = 589\,\text{Pa} \qquad \text{Ans.}$$

EXAMPLE 5.4

A transition is placed in a rectangular air duct as shown in Fig. 5–14. If 2 kg/s of air flows steadily through the duct, determine the pressure change that occurs between the ends of the transition. Take $\rho_a = 1.23\,\text{kg/m}^3$.

SOLUTION

Fluid Description.   We have steady flow. At slow velocities the air passing through the duct will be considered an ideal fluid, that is, incompressible and inviscid.

Analysis.   To solve this problem, we will first use the continuity equation to obtain the average velocity of flow at A and at B. Then we will use the Bernoulli equation to determine the pressure difference between A and B.

Continuity Equation.   We will consider a fixed control volume that contains the air within the duct, Fig. 5–14. Thus, for steady flow,

$$\frac{\partial}{\partial t}\int_{\text{cv}} \rho\, d V + \int_{\text{cs}} \rho \mathbf{V}\cdot d\mathbf{A} = 0$$

$$0 - V_A A_A + V_B A_B = 0$$

$$Q = V_A A_A = V_B A_B$$

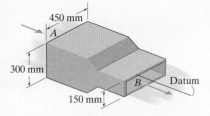

Fig. 5–14

But

$$Q = \frac{\dot{m}}{\rho} = \frac{2\,\text{kg/s}}{1.23\,\text{kg/m}^3} = 1.626\,\text{m}^3/\text{s}$$

so that

$$V_A = \frac{Q}{A_A} = \frac{1.626\,\text{m}^3/\text{s}}{(0.45\,\text{m})(0.3\,\text{m})} = 12.04\,\text{m/s}$$

and

$$V_B = \frac{Q}{A_B} = \frac{1.626\,\text{m}^3/\text{s}}{(0.45\,\text{m})(0.15\,\text{m})} = 24.09\,\text{m/s}$$

Bernoulli Equation.   Selecting points at A and B on the horizontal streamline, we have

$$\frac{p_A}{\rho} + \frac{V_A^2}{2} + gz_A = \frac{p_B}{\rho} + \frac{V_B^2}{2} + gz_B$$

$$\frac{p_A}{1.23\,\text{kg/m}^3} + \frac{(12.04\,\text{m/s})^2}{2} + 0 = \frac{p_B}{1.23\,\text{kg/m}^3} + \frac{(24.09\,\text{m/s})^2}{2} + 0$$

$$p_A - p_B = 267.66\,\text{Pa} = 0.268\,\text{kPa} \qquad\qquad Ans.$$

This small drop in pressure, or the low velocities, will not significantly change the density of the air, and so here it is reasonable to have assumed the air to be incompressible.

EXAMPLE 5.5

Water flows up through the vertical pipe that is connected to the transition, Fig. 5–15. If the volumetric flow is 0.02 m³/s, determine the height h to which the water will rise in the Pitot tube. The piezometer level at A is indicated.

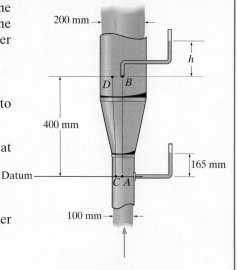

Fig. 5–15

SOLUTION

Fluid Description. The flow is steady, and the water is assumed to be an ideal fluid, where $\rho_w = 1000$ kg/m³.

Bernoulli Equation. From the piezometer reading, the pressure at A is

$$p_A = \rho_w g h_A = \left(1000 \text{ kg/m}^3\right)\left(9.81 \text{ m/s}^2\right)(0.165 \text{ m}) = 1618.65 \text{ Pa}$$

This total pressure is caused by the static pressure in the water. In other words, it is the pressure within the enclosed pipe at this level.

Since the flow is known, the velocity at A can be determined.

$$Q = V_A A_A; \qquad 0.02 \text{ m}^3/\text{s} = V_A \left[\pi(0.05 \text{ m})^2\right]$$

$$V_A = 2.546 \text{ m/s}$$

Also, since B is a stagnation point, $V_B = 0$. We can now determine the pressure at B by applying the Bernoulli equation at points A and B on the vertical streamline in Fig. 5–15. The datum is placed at A, and so

$$\frac{p_A}{\rho_w} + \frac{V_A^2}{2} + gz_A = \frac{p_B}{\rho_w} + \frac{V_B^2}{2} + gz_B$$

$$\frac{1618.65 \text{ N/m}^2}{1000 \text{ kg/m}^3} + \frac{(2.546 \text{ m/s})^2}{2} + 0 = \frac{p_B}{1000 \text{ kg/m}^3} + 0 + \left(9.81 \text{ m/s}^2\right)(0.4 \text{ m})$$

$$p_B = 936.93 \text{ Pa}$$

Since B is a stagnation point, this total pressure is caused by *both* static and dynamic pressures at B. For the Pitot tube we require

$$h = \frac{p_B}{\rho_w g} = \frac{936.93 \text{ Pa}}{(1000 \text{ kg/m}^3)(9.81 \text{ m/s}^2)} = 0.09551 \text{ m} = 95.5 \text{ mm} \qquad Ans.$$

NOTE: Although it is not part of this problem, the pressure at D can be obtained ($p_D = 734$ Pa) by applying the Bernoulli equation along the streamline CD. First, though, the velocity $V_D = 0.6366$ m/s must be obtained by applying $Q = V_D A_D$.

5

EXAMPLE 5.6

After a long time storage, the gas tank contains a 0.6-m depth of gasoline and a 0.2-m depth of water as shown in Fig. 5–16. Determine the time needed to drain the water if the drain hole has a diameter of 25 mm. The tank is 1.8 m wide and 3.6 m long. The density of gasoline is $\rho_g = 726 \text{ kg/m}^3$, and for water, $\rho_w = 1000 \text{ kg/m}^3$.

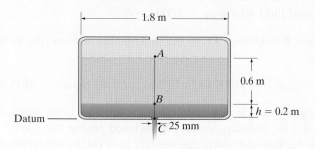

Fig. 5–16

SOLUTION

Fluid Description. The gasoline is on top of the water because its specific gravity is less than that of water. Because the tank is large relative to the drain hole, we will assume steady flow and will consider the two fluids to be ideal.

Bernoulli Equation. Here we will select the vertical streamline containing points B and C, Fig. 5–16. At any instant the level of water is h, as measured from the datum, and so the pressure at B is due to the weight of the gasoline above it, that is

$$p_B = \gamma_g h_{AB} = (726 \text{ kg/m}^3)(9.81 \text{ m/s}^2)(0.6 \text{ m}) = 4.273(10^3) \text{ N/m}^2$$

To simplify the analysis for using the Bernoulli equation, we will neglect the velocity at B since $V_B \approx 0$, and so V_B^2 will be even smaller. Since C is open to the atmosphere, $p_C = 0$. Thus,

$$\frac{p_B}{\rho} + \frac{V_B^2}{2} + gz_B = \frac{p_C}{\rho} + \frac{V_C^2}{2} + gz_C$$

$$\frac{4.273(10^3) \text{ N/m}^2}{1000 \text{ kg/m}^3} + 0 + (9.81 \text{ m/s}^2)h = 0 + \frac{V_C^2}{2} + 0$$

$$V_C = 4.429\sqrt{h + 0.4356} \tag{1}$$

Continuity Equation. The continuity of flow at B and C will allow us to relate the *actual* nonzero V_B to V_C. We will choose a control volume that contains all the water up to the depth h. Since V_B is downward and h is positive upward, then at the top control surface, $V_B = -dh/dt$. Thus,

$$\frac{\partial}{\partial t} \int_{cv} \rho \, dV + \int_{cs} \rho \mathbf{V} \cdot d\mathbf{A} = 0$$

$$0 - V_B A_B + V_C A_C = 0$$

$$0 - \left(-\frac{dh}{dt}\right)\left[(1.8 \text{ m})(3.6 \text{ m})\right] + V_C\left[\pi(0.0125 \text{ m})^2\right] = 0$$

$$\frac{dh}{dt} = -75.752(10^{-6})V_C$$

Now, using Eq. 1,

$$\frac{dh}{dt} = -75.752(10^{-6})(4.429\sqrt{h} + 0.4356)$$

or

$$\frac{dh}{dt} = -0.3355(10^{-3})\sqrt{h + 0.4356} \tag{2}$$

Notice that when $h = 0.2\,\text{m}$, $V_B = -dh/dt = 0.268(10^{-3})\,\text{m/s}$, which is very slow by comparison to $V_C = 3.53\,\text{m/s}$, as determined from Eq. 1.

If t_d is the time needed to drain the tank, then separating the variables in Eq. 2 and integrating, we get

$$\int_{0.2\,\text{m}}^{0} \frac{dh}{\sqrt{h + 0.4356}} = -0.3355(10^{-3})\int_{0}^{t_d} dt$$

$$\left(2\sqrt{h + 0.4356}\right)\Big|_{0.2\,\text{m}}^{0} = -0.3355(10^{-3})t_d$$

Evaluating the limits, we get

$$-0.2745 = -0.3355(10^{-3})t_d$$

$$t_d = 818.06\text{ s} = 13.6\text{ min.} \qquad Ans.$$

5.4   Energy and Hydraulic Grade Lines

For some applications involving liquids, it is convenient to substitute $\gamma = \rho g$ into the Bernoulli equation and rewrite it in the form

$$H = \frac{p}{\gamma} + \frac{V^2}{2g} + z \qquad (5\text{--}6)$$

Here each term is expressed as energy per unit weight, having units of J/N. However, we can also consider these terms as having units of length, m. Then the first term on the right represents the static **pressure head**, which is the height of a fluid column supported by a pressure p acting at its base. The second term is the **kinetic** or **velocity head**, which indicates the vertical distance a fluid particle must fall from rest to attain the velocity V. And finally, the third term is the **gravitational head**, which is the height of a fluid particle placed above (or below) a selected datum. The **total head** H is the sum of these three terms, and a plot of this value along the length of a pipe or channel that contains the fluid is called the **energy grade line** (EGL). Although each of the terms in Eq. 5–6 may change, their sum H will *remain constant* at every point along the same streamline, provided there are no frictional losses and there is no addition or removal of energy by an external source, such as a pump or turbine. Experimentally, H can be obtained at any point using a Pitot tube, as shown in Fig. 5–17.

For problems involving the design of pipe systems or channels, it is often convenient to plot the energy grade line and also its counterpart, the **hydraulic grade line** (HGL). This line shows how the **hydraulic head** $p/\gamma + z$ will vary along the pipe (or channel). Here a piezometer can be used to experimentally obtain its value, Fig. 5–17. By comparison, notice that the EGL will *always lie above* the HGL by a distance of $V^2/2g$.

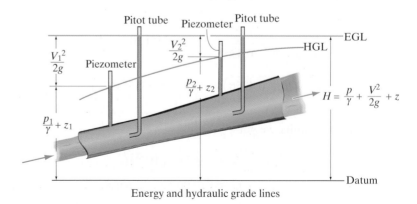

Energy and hydraulic grade lines

Fig. 5–17

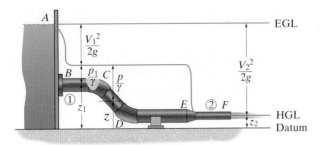

Fig. 5–18

Figure 5–18 shows how the HGL varies along the streamline at the *centerline* of a pipe. With practically no velocity, the HGL and EGL originally coincide at the water surface of the reservoir A. As the water, assumed here to be an ideal fluid, begins to flow through the pipe at B, it accelerates to a velocity V_1. This causes the HGL to drop by $V_1^2/2g$. Due to continuity of flow, this velocity must be maintained throughout the pipe $BCDE$. Consequently, the gravitational head z follows the centerline of the pipe, and the pressure head p/γ is above it. Specifically, within the section BC, the HGL is at $p_1/\gamma + z_1$, and along the inclined section CD, the pressure head will increase in proportion to a drop in the gravitational head. Continuity then requires an *increase* in velocity out of the transition E to V_2, and this speed is then maintained along the pipe EF. This increased velocity causes the pressure head to drop to zero (or atmospheric pressure), both *within* and just outside of the pipe at F, so that the HGL is then defined only by its gravitational head z_2 above the datum*. In other words, as we have discussed in Sec. 5.1, uniform flow of an *ideal fluid* through a straight horizontal pipe does not require a pressure difference along the pipe's length to push the fluid through the pipe since the pressure *does not* have to overcome frictional resistance.

Important Points

- The Bernoulli equation can be expressed in terms of the total head H of fluid. This head is measured in units of length and remains constant along a streamline provided no friction losses occur, and no energy is added to the fluid or withdrawn from it due to external sources. $H = p/\gamma + V^2/2g + z = \text{const.}$

- A plot of the total head H versus the distance in the direction of flow is called the *energy grade line* (EGL). For the cases considered here, this line will always be horizontal, and its value can be calculated from any point along the flow where p, V, and z are known.

- The *hydraulic grade line* (HGL) is a plot of the hydraulic head, $p/\gamma + z$, versus the distance in the direction of flow. If the EGL is known, then the HGL will always be $V^2/2g$ *below* the EGL.

---

*Actually there will be a hydrostatic difference in pressure along the diameter of the pipe as noted on p.225 because the fluid within the pipe is *supported* by the pipe. Once the fluid is ejected then it is in free fall and at atmospheric pressure.

EXAMPLE 5.7

Water flows through the 100-mm-diameter pipe at 0.025 m³/s, Fig. 5–19a. If the pressure at A is 225 kPa, determine the pressure at C, and construct the energy and hydraulic grade lines from A to D. Take $\gamma_w = 9.81$ kN/m³.

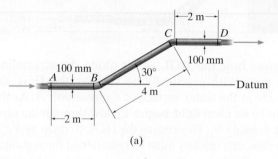

(a)

Fig. 5–19

SOLUTION

Fluid Description.   We have steady flow, and we will assume water is an ideal fluid.

Bernoulli Equation.   The average velocity of flow through the pipe is

$$Q = VA; \qquad 0.025 \text{ m}^3/\text{s} = V\left[\pi(0.05 \text{ m})^2\right]$$

$$V = \frac{10}{\pi} \text{ m/s}$$

Since the pipe has a constant diameter over its length, this velocity remains *constant* in order to satisfy the continuity equation.

The pressure at A and B is the *same* since segment AB is horizontal. We can find the pressure at C (and D) by applying the Bernoulli equation at points B and C, which lie on the same streamline. With the gravitational datum through AB, noting that $V_B = V_C = V$, we have

$$\frac{p_B}{\gamma_w} + \frac{V_B^2}{2g} + z_B = \frac{p_C}{\gamma} + \frac{V_C^2}{2g} + z_C$$

$$\frac{225(10^3) \text{ N/m}^2}{9.81(10^3) \text{ N/m}^3} + \frac{\left(\dfrac{10}{\pi} \text{ m/s}\right)^2}{2(9.81 \text{ m/s}^2)} + 0 =$$

$$\frac{p_C}{9.81(10^3) \text{ N/m}^3} + \frac{\left(\dfrac{10}{\pi} \text{ m/s}\right)^2}{2(9.81 \text{ m/s}^2)} + (4 \text{ m}) \sin 30°$$

$$p_C = p_D = 205.38(10^3) \text{ Pa} = 205 \text{ Pka} \qquad Ans.$$

Notice that the pressure at C has dropped because the pressure at B has to do work to lift the fluid to C.

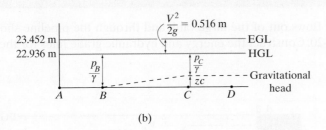

(b)

EGL and HGL.   The total head remains constant because there are no friction losses. This head can be determined from the conditions at any point along the pipe. Using point B, we have

$$H = \frac{p_B}{\gamma} + \frac{V_B^2}{2g} + z_B = \frac{225(10^3) \text{ N/m}^2}{9.81(10^3) \text{ N/m}^3} + \frac{\left(\frac{10}{\pi} \text{ m/s}\right)^2}{2(9.81 \text{ m/s}^2)} + 0$$

$$= 23.5 \text{ m}$$

The EGL is located as shown in Fig. 5–19b.

   The velocity throughout the pipe is constant, and so the velocity head is

$$\frac{V^2}{2g} = \frac{\left(\frac{10}{\pi} \text{ m/s}\right)^2}{2(9.81 \text{ m/s}^2)} = 0.516 \text{ m}$$

Now that this is known, the HGL is plotted 0.516 m *below* the EGL, Fig. 5–19b. Notice that the HGL can also be calculated along AB as

$$\frac{p_B}{\gamma} + z_B = \frac{225(10^3) \text{ N/m}^2}{9.81(10^3) \text{ N/m}^3} + 0 = 22.9 \text{ m}$$

or along CD as

$$\frac{p_C}{\gamma} + z_C = \frac{205.38(10^3) \text{ N/m}^2}{9.81(10^3) \text{ N/m}^3} + (4 \text{ m}) \sin 30° = 22.9 \text{ m}$$

Along BC the gravitational head increases, and the pressure head will correspondingly decrease ($p_C/\gamma < p_B/\gamma$).

EXAMPLE | 5.8

Water flows out of the large tank and through the pipeline shown in Fig. 5–20. Construct the energy and hydraulic grade lines for the pipe.

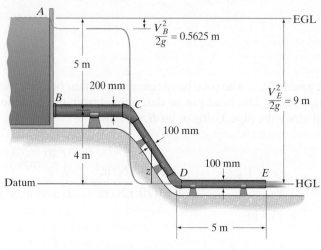

Fig. 5–20

SOLUTION

Fluid Description. We will assume the water level in the tank remains essentially constant so that steady flow will be maintained. The water is assumed to be an ideal fluid.

Energy Grade Line. We will take the gravitational datum through DE. At A the velocity and pressure heads are both zero, and so the total head is equal to the gravitational head, which is at a level of

$$H = \frac{p_A}{\gamma} + \frac{V_A^2}{2g} + z = 0 + 0 + (4 \text{ m} + 5 \text{ m}) = 9 \text{ m}$$

The EGL remains at this level since the fluid is ideal, and so there are no energy losses due to friction as the water flows through the pipe.

Hydraulic Grade Line.   Since the (gage) pressure at both A and E is zero, the velocity of the water exiting the pipe at E can be determined by applying the Bernoulli equation at these points, which lie on the same streamline.

$$\frac{p_A}{\gamma} + \frac{V_A^2}{2g} + z_A = \frac{p_E}{\gamma} + \frac{V_E^2}{2g} + z_E$$

$$0 + 0 + 9 \text{ m} = 0 + \frac{V_E^2}{2(9.81 \text{ m/s}^2)} + 0$$

$$V_E = 13.29 \text{ m/s}$$

The velocity of the water through pipe BC can now be determined from the continuity equation, considering the fixed control volume to contain the water within the entire pipe. We have

$$\frac{\partial}{\partial t} \int_{cv} \rho \, d\mathcal{V} + \int_{cs} \rho \mathbf{V} \cdot d\mathbf{A} = 0$$

$$0 - V_B A_B + V_E A_E = 0$$

$$-V_B \left[\pi (0.1 \text{ m})^2 \right] + 13.29 \text{ m/s} \left[\pi (0.05 \text{ m})^2 \right] = 0$$

$$V_B = 3.322 \text{ m/s}$$

The HGL can now be established. It is located *below* the EGL, a distance defined by the velocity head $V^2/2g$. For pipe segment BC this head is

$$\frac{V_B^2}{2g} = \frac{(3.322 \text{ m/s})^2}{2(9.81 \text{ m/s}^2)} = 0.5625 \text{ m}$$

The HGL is maintained at $9 \text{ m} - 0.5625 \text{ m} = 8.44 \text{ m}$ until the transition at C changes the velocity head within CDE to

$$\frac{V_E^2}{2g} = \frac{(13.29 \text{ m/s})^2}{2(9.81 \text{ m/s}^2)} = 9 \text{ m}$$

This causes the HGL to drop to $9 \text{ m} - 9 \text{ m} = 0$. In other words, along pipe CDE the HGL is *at the gravitational datum.* Along CD, z is always positive, Fig. 5–20, and therefore, a corresponding *negative* pressure head $-p/\gamma$ must be developed within the flow to maintain a zero hydraulic head, i.e., $p/\gamma + z = 0$. If this negative pressure becomes large enough, it can cause cavitation, something we will discuss in the next example. Finally, along DE, $z = 0$ and also $p_D = p_E = 0$.

EXAMPLE **5.9**

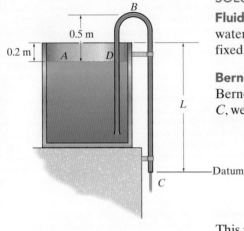

(a)

Fig. 5–21

The siphon in Fig. 5–21a is used to draw water from the large open tank. If the absolute vapor pressure for the water is $p_v = 1.23$ kPa, determine the shortest drop length L of the 50-mm-diameter tube that will cause cavitation in the tube. Draw the energy and hydraulic grade lines for the tube.

SOLUTION

Fluid Description. As in the previous example, we will assume the water is an ideal fluid, and the level in the tank remains essentially fixed, so that we have steady flow. $\gamma = 9810$ N/m^3.

Bernoulli Equation. To obtain the velocity at C, we will apply the Bernoulli equation at points A and C. With the gravitational datum at C, we have

$$\frac{p_A}{\gamma} + \frac{V_A^2}{2g} + z_A = \frac{p_C}{\gamma} + \frac{V_C^2}{2g} + z_C$$

$$0 + 0 + (L - 0.2 \text{ m}) = 0 + \frac{V_C^2}{2(9.81 \text{ m/s}^2)} + 0$$

$$V_C = 4.429\sqrt{(L - 0.2 \text{ m})} \qquad (1)$$

This result is valid provided the pressure at any point within the tube does not drop to or below the vapor pressure. If it does, it will cause the water to boil (cavitate), causing a "hissing" noise and energy loss. Of course, this will invalidate the application of the Bernoulli equation. Since the flow is assumed to be steady, and the tube has a constant diameter, then due to continuity, $V^2/2g$ is constant throughout the tube, and so the hydraulic head $(p/\gamma + z)$ must also be constant.

The *smallest pressure* in the tube occurs at B, where z, measured from the datum, is a maximum. Using standard atmospheric pressure of 101.3 kPa, the gage vapor pressure for the water is 1.23 kPa − 101.3 kPa = −100.07 kPa. Assuming this negative pressure develops at B, then applying the Bernoulli equation at points B and C, realizing that $V_B = V_C$, we have

$$\frac{p_B}{\gamma} + \frac{V_B^2}{2g} + z_B = \frac{p_C}{\gamma} + \frac{V_C^2}{2g} + z_C$$

$$\frac{-100.07(10^3) \text{ N/m}^2}{9810 \text{ N/m}^3} + \frac{V^2}{2g} + (L + 0.3 \text{ m}) = 0 + \frac{V^2}{2g} + 0$$

$$L + 0.3 \text{ m} = 10.20 \text{ m}$$

$$L = 9.90 \text{ m} \qquad \qquad Ans.$$

From Eq. 1 the critical velocity is

$$V_C = 4.429\sqrt{(9.90 \text{ m} - 0.2 \text{ m})} = 13.80 \text{ m/s}$$

If L is equal to or greater than 9.90 m, cavitation will occur in the siphon at B because then the pressure at B will be equal to or lower than −100.07 kPa.

Notice that we can also obtain this result by applying the Bernoulli equation, first between A and B, to obtain V_B, then between B and C to obtain L.

EGL and HGL. Since $V_B = V_C = 13.80$ m/s, throughout the tube PBC the velocity head is*

$$\frac{V^2}{2g} = \frac{(13.80 \text{ m/s})^2}{2(9.81 \text{ m/s}^2)} = 9.70 \text{ m}$$

The total head can be determined from C, it is

$$H = \frac{p_C}{\gamma} + \frac{V_C^2}{2g} + z_C = 0 + 9.70 \text{ m} + 0 = 9.70 \text{ m}$$

Both the EGL and the HGL are shown in Fig. 5–21b. Here the HGL remains at zero. We can determine the pressure head within the tube at D by applying the Bernoulli equation between D and C

$$\frac{p_D}{y} + \frac{V_D^2}{2g} + Z_D = \frac{p_C}{y} + \frac{V_C^2}{2g} + Z_C$$

$$\frac{p_D}{y} + 9.70 \text{ m} + (9.90 \text{ m} - 0.2 \text{ m}) = 0 + 9.70 \text{ m} + 0$$

$$\frac{p_D}{y} = -9.70 \text{ m}$$

Therefore, the pressure decreases from -9.70 m at D to $p/\gamma = -100.07(10^3)$ N/m^2/9810 N/m^3 = -10.2 m at B, while the gravitational head z increases from 9.70 m to 9.70 m + 0.5 m = 10.2 m, Fig. 5–21b. After rounding the top of the pipe at B, the pressure head increases, while the gravitational head decreases by a corresponding amount.

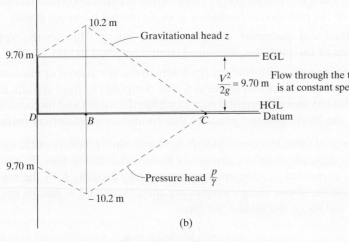

(b)

Fig. 5–21 (cont.)

*Cavitation is prevented when V_C is actually slightly less than this value.

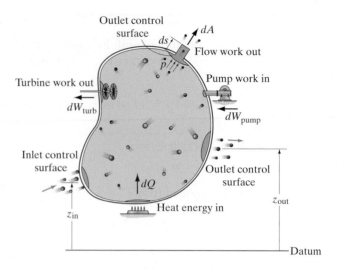

Fig. 5–22

5.5   The Energy Equation

In this section we will expand our application of work and energy methods beyond the limitation of the Bernoulli equation and will include heat and viscous fluid flow, along with work input from a pump and work output to a turbine. Before we begin, however, we will first discuss the various forms of energy that a fluid system can have when it is contained within the control volume shown in Fig. 5–22.

System Energy.   At any instant, the total energy E of the fluid system consists of three parts:

Kinetic energy.   This is energy of motion that depends upon the *macroscopic speed* of the particles, as measured from an inertial reference frame.

Gravitational potential energy.   This is energy due to the vertical position of the particles, measured from a selected datum.

Internal energy.   Internal energy refers to the vibrational or *microscopic motion* of the atoms and molecules that compose the fluid system. It also includes any stored potential energy within the atoms and molecules that cause the binding of the particles due to nuclear or electrical forces.

The total of these three energies, E, is an extensive property of the system, since it depends upon the amount of mass within the system. However, it can be expressed as an intensive property e by dividing E by the mass. In this case the above three energies are then expressed as energy per unit mass, and so for the system, we have

$$e = \frac{1}{2}V^2 + gz + u \qquad (5\text{--}7)$$

Let us now consider the various forms of heat energy and work done by the fluid system in Fig. 5–22.

Heat Energy.

Heat energy dQ can be added or drawn through an open control surface, via the process of conduction, convection, or radiation. It *increases* the total energy of the system within the control volume if it *flows in* (system heated) and *decreases* the total energy if it *flows out* (system cooled).

Work.

Work dW can be done *by the enclosed system* on its surroundings through an open control surface. Work *decreases* the total energy of the system when it is *done by the system*, and it *increases* the total energy of the system when it is done *on the system*. In fluid mechanics we will be interested in three types of work.

Flow Work.

When a fluid is subjected to a pressure, it can push a volume $d\forall$ of the system's mass outward from a control surface opening. This is *flow work*, dW_p. To calculate it, consider the small volume of the system $d\forall = dA\,ds$ in Fig. 5–22 being pushed *outward* by the (gage) pressure p within the system. Since dA is the cross-sectional area of this volume, then the force exerted by the system is $dF = p\,dA$. If the distance the volume moves outward is ds, then the flow work for this small volume is $dW_p = dF\,ds = p(dA\,ds) = p\,d\forall$.

Shaft Work.

If work is performed on a *turbine* by the fluid system within the control volume, then the work will *subtract* energy from the system at an open control surface, Fig. 5–22. However, it is also possible for work to be performed on the system by a *pump*, thereby *adding* external energy to the fluid. In both cases, this type of work is called *shaft work* because a shaft is used to input or extract the work.

Shear Work.

The viscosity of any real fluid will cause shear stress τ to develop tangent to the control volume's inner surface. Because of the no-slip condition on a *fixed* control surface, no work can be done on the surface since the shear stress *does not move* along the surface. Only along an open control surface can this shear stress *move*, thereby creating *shear work* dW_τ. Here, however, we will *neglect* this form of work because any open control surfaces will always be selected *perpendicular* to the flow of the fluid coming into and passing out of the control volume. Because of this, no displacement of fluid occurs tangent to an open control surface, and so no work due to shear stress is done.*

Energy Equation.

The conservation of energy for a fluid system contained within the control volume is formalized by the *first law of thermodynamics*. This law states that the time rate at which heat is *added* or put *into* the system, $\dot{Q}_{in}$, plus the rate of work added to the system, is equal to the time rate of change of the total energy within the system.

$$\dot{Q}_{in} + \dot{W}_{in} = \left(\frac{dE}{dt}\right)_{syst.} \qquad (5\text{–}8)$$

*Normal viscous stress can occur if the flow is nonuniform. However, any work done by this stress will be zero provided the fluid is inviscid; or if it is considered viscous, then the normal stress will equal the pressure if the streamlines are parallel. We will consider this to be the case in this text. Also, see the footnote on page 402.

The term on the right can be converted to the rate of change of energy within the control volume using the Reynolds transport theorem, Eq. 3–17, where $\eta = e$, defined by Eq. 5–7. We now have

$$\dot{Q}_{in} + \dot{W}_{in} = \frac{\partial}{\partial t} \int_{cv} e\rho \, d\mathcal{V} + \int_{cs} e\rho \mathbf{V} \cdot d\mathbf{A}$$

The two terms on the right indicate the local rate of change of energy per unit mass *within* the control volume, plus the net convective amount of energy per unit mass passing through the open control surfaces. Assuming the *flow is steady*, then this first term will be equal to zero. Substituting Eq. 5–7 for e into the last term yields

$$\dot{Q}_{in} + \dot{W}_{in} = 0 + \int_{cs} \left(\frac{1}{2}V^2 + gz + u \right) \rho \mathbf{V} \cdot d\mathbf{A} \qquad (5\text{–}9)$$

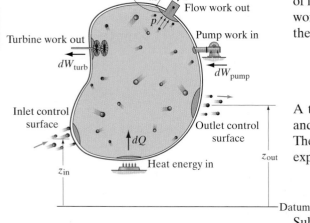

Fig. 5–22

The time rate of output work can be represented by the rates of its flow work and shaft work. As stated previously, the flow work is caused by pressure, where $dW_p = p(dA \, ds)$, and thus the *rate* of flow work out through a control surface is

$$\dot{W}_p = \frac{dW_p}{dt} = -\int_{cs} p\left(\frac{ds}{dt}dA \right) = -\int_{cs} p\mathbf{V} \cdot d\mathbf{A}$$

A turbine will produce output shaft work (positive), dW_{turb}, and a pump will produce input shaft work (negative), dW_{pump}. Therefore, the time rate of total work *out of the system* can be expressed as

$$\dot{W}_{out} = -\int_{cs} p\mathbf{V} \cdot d\mathbf{A} + \dot{W}_{pump} - \dot{W}_{turb}$$

Substituting this result into Eq. 5–9 and rearranging the terms gives

$$\dot{Q}_{in} + \dot{W}_{pump} - \dot{W}_{turb} = \int_{cs}\left[\frac{p}{\rho} + \frac{1}{2}V^2 + gz + u \right]\rho\mathbf{V} \cdot d\mathbf{A} \qquad (5\text{–}10)$$

The integration must be carried out over the outlet (out) and inlet (in) control surfaces. For our case we will assume the flow is uniform and one dimensional, and so average velocities will be used. Also, we will assume that the pressure p and location z at each opening are *constant*, Fig. 5–22. Continuity requires the mass flow in to be equal to the mass flow out, so $\dot{m} = \rho_{in}V_{in}A_{in} = \rho_{out}V_{out}A_{out}$, and therefore, Eq. 5–10 now becomes

$$\dot{Q}_{in} + \dot{W}_{pump} - \dot{W}_{turb} =$$
$$\left[\left(\frac{p_{out}}{\rho_{out}} + \frac{V_{out}^2}{2} + gz_{out} + u_{out} \right) - \left(\frac{p_{in}}{\rho_{in}} + \frac{V_{in}^2}{2} + gz_{in} + u_{in} \right) \right]\dot{m} \quad (5\text{–}11)$$

This is the *energy equation* for one-dimensional *steady flow*, and it applies to both compressible and incompressible fluids.

Incompressible Flow.

If we assume the flow is steady and incompressible, then $\rho_{in} = \rho_{out} = \rho$. Furthermore, if Eq. 5–11 is divided by $\dot{m}$, and the terms are rearranged, we obtain

$$\frac{p_{in}}{\rho} + \frac{V_{in}^2}{2} + gz_{in} + w_{pump} = \frac{p_{out}}{\rho} + \frac{V_{out}^2}{2} + gz_{out} + w_{turb} + (u_{out} - u_{in} - q_{in})$$

Here each term represents energy per unit mass, J/kg. Specifically, w_{pump} and w_{turb} are the shaft work per unit mass, performed by the pump and turbine, respectively; and the term q_{in}, a scalar, is the heat energy per unit mass that goes *into* the system.

Later, in Chapter 7, we will express the friction losses that produce the change in internal energy $(u_{out} - u_{in})$ in terms of velocity coefficients. Here, we will simply state it collectively as fl (friction loss). Finally, if we exclude problems involving heat transfer, then $q_{in} = 0$; and so for our purposes, a general expression of the energy equation becomes

$$\frac{p_{in}}{\rho} + \frac{V_{in}^2}{2} + gz_{in} + w_{pump} = \frac{p_{out}}{\rho} + \frac{V_{out}^2}{2} + gz_{out} + w_{turb} + fl \quad (5\text{–}12)$$

This equation states that the total available energy per unit mass passing through the *inlet* control surfaces, plus the work per unit mass that is *added* to the fluid within the control volume by a pump, is *equal to* the total energy per unit mass that passes through the *outlet* control surfaces, plus the energy *removed* from the fluid within the control volume by a turbine, plus the *energy losses* that occur within the control volume due to fluid friction.

If Eq. 5–12 is divided by g, then the terms represent energy per unit weight or "head of fluid."

Occurs within control volume

$$\frac{p_{in}}{\gamma} + \frac{V_{in}^2}{2g} + z_{in} + h_{pump} = \frac{p_{out}}{\gamma} + \frac{V_{out}^2}{2g} + z_{out} + h_{turb} + h_L \quad (5\text{–}13)$$

Occurs at open control surfaces

The last term is called the **head loss**, $h_L = fl/g$, and the terms h_{pump} and h_{turb} are referred to as the **pump head** and **turbine head**, respectively. Therefore, this form of the energy equation requires that the total input head plus the pump head equals the total output head plus the turbine head plus the head loss. Note that Eq. 5–13 reduces to the Bernoulli equation if there is no shaft work and no change in the fluid's internal energy, $h_L = 0$.

Compressible Fluid.   For compressible gas flow, Eq. 5–11 is generally expressed in terms of the enthalpy of a unit mass of gas. *Enthalpy h* is defined as the sum of the flow work and internal energy. Since the flow work for a volume of fluid is $p \, d\mathcal{V}$, then for a unit mass, $p \, d\mathcal{V}/dm = p/\rho$. Therefore,

$$h = p/\rho + u \tag{5–14}$$

If we substitute this into Eq. 5–11, we get

$$\dot{Q}_{in} - \dot{W}_{turb} + \dot{W}_{pump} = \left[\left(h_{out} + \frac{V_{out}^2}{2} + gz_{out} \right) - \left(h_{in} + \frac{V_{in}^2}{2} + gz_{in} \right) \right] \dot{m} \tag{5–15}$$

Application of this equation will become important in Chapter 13, where we discuss compressible flow.

Power and Efficiency.   The power output of a turbine or power input of a pump is measured in watts (1 W = 1 J/s), and is defined as its *time rate* of doing work, $\dot{W} = dW/dt$. We can express the power in terms of the pump or turbine head, where either is referred to as the **shaft head**, h_s, by noting $w_s = \dot{W}_s/\dot{m}$ or $\dot{W}_s = w_s \dot{m}$. From the derivation of Eq. 5–13, recall that $h_s = w_s/g$, or $w_s = h_s g$, and since $\dot{m} = \rho Q = \gamma Q/g$, then

$$\boxed{\dot{W}_s = \dot{m}gh_s = Q\gamma h_s} \tag{5–16}$$

Since pumps (and turbines) have friction losses, they will never be 100% efficient. For pumps, **mechanical efficiency** e is the ratio of mechanical power delivered to the fluid $(\dot{W}_s)_{out}$ divided by the electrical power required to run the pump, $(\dot{W}_s)_{in}$. Thus,

$$e = \frac{(\dot{W}_s)_{out}}{(\dot{W}_s)_{in}} \qquad 0 < e < 1 \tag{5–17}$$

Nonuniform Velocity.   Integration of the velocity term in the energy equation, Eq. 5–10, was possible because we assumed a *uniform* or *constant flow* occurs over the inlet and outlet control surfaces. However, within pumps and turbines, the flow is never steady as the fluid passes through the machine. Although this is the case under constant rotational speed, the flow is *cyclic*, and most often these cycles are fast. Provided the time considered for observing the flow is larger than that of a single cycle, then the time average of a quantity of flow through the open control surfaces can be reasonably determined using the energy equation. We refer to this flow passing through the open control surfaces as **quasi-steady flow**.

Also, if a velocity profile for the flow at the inlet and outlet control surfaces is nonuniform, as it is in all cases of viscous flow, then the velocity profile must be known in order to carry out the integration in Eq. 5–10. One way of representing this integration is to use a dimensionless **kinetic energy coefficient** α, and express the integration of the velocity profile in terms of the average velocity V of the profile, as determined from Eq. 4–3, that is, $V = \int v\,dA/A$. In other words, the velocity term in Eq. 5–10 can be written as $\int_{cs} \frac{1}{2}V^2\rho\mathbf{V}\cdot d\mathbf{A} = \alpha\frac{1}{2}V^2\dot{m}$, so that

$$\alpha = \frac{1}{\dot{m}V^2}\int_{cs} V^2\rho\mathbf{V}\cdot d\mathbf{A} \qquad (5\text{–}18)$$

Therefore, in cases where it may be necessary to consider the nonuniformity of the velocity at a control surface, we can substitute the terms involving V^2 in the energy equation with αV^2. For example, it will be shown in Chapter 9 that for laminar flow, the velocity profile for the fluid in a pipe is a *paraboloid*, Fig. 5–23a, and so for this case, integration will yield $\alpha = 2$.* For turbulent flow it is usually sufficient to take $\alpha = 1$ since turbulent mixing of the fluid will cause the velocity profile to become approximately uniform, Fig. 5–23b.

The examples that follow illustrate various applications of the energy equation to problems involving heat and head losses due to viscous friction, and the additional work due to turbines and pumps. Further applications of this important equation will be given in other chapters throughout the text.

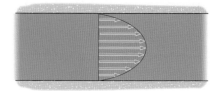

Velocity profile for
laminar flow

(a)

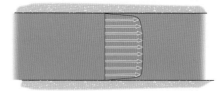

Mean velocity profile for
turbulent flow

(b)

Fig. 5–23

Important Points

- The energy equation is based upon the first law of thermodynamics, which states that the time rate of change of the total energy within a fluid system is equal to the rate at which heat is *added* to the system minus the rate of work done by the system.

- In general, the system's total energy E within a control volume consists of the kinetic energy of all the fluid particles, their potential energy, and their atomic or molecular internal energies.

- The work done by a system can be *flow work* due to pressure, *shaft work* due to a pump or turbine, or *shear work* caused by viscous friction on the sides of an open control surface. Shear work is not considered here, since the flow will always be *perpendicular* to any open control surface.

- The energy equation is identical to the Bernoulli equation when no internal friction losses and heat transfer occur, and no shaft work is done on or by the fluid system.

*See Prob. 5–77.

5

Procedure for Analysis

The following procedure can be used when applying the various forms of the energy equation.

- **Fluid Description.** As developed here, the energy equation applies to one-dimensional steady flow of either compressible or incompressible fluids.

- **Control Volume.** Select the control volume that contains the fluid, and indicate the open control surfaces. Be sure that these surfaces are located in regions where the flow is uniform and well defined.

  Establish a fixed datum to measure the elevation (potential energy) of the fluid moving into and out of each control surface.

  Remember that if the fluid is assumed inviscid or ideal, then its velocity profile is uniform as it passes through an open control surface. The average velocity V can then be used. If a viscid fluid is considered, then the coefficient α can be determined using Eq. 5–17, and αV^2 is used instead of V^2. The average velocities "in" and "out" of the open control surfaces can be determined if the volumetric flow is known, $Q = VA$.

- **Energy Equation.** Write the energy equation, and below it substitute in the numerical data using a consistent set of units. Heat energy dQ_{in} is *positive* if heat flows *into* the control volume, and it is *negative* if heat flows *out*.

  Reservoirs, or large tanks that drain slowly, have liquid surfaces that are essentially at rest, $V \approx 0$.

  The terms $(p/\gamma + z)$ in Eq. 5–13 represent the hydraulic head at the "in" and "out" control surfaces. This head remains constant over each surface, and so it may be calculated at *any point* on the surface. Once this point is selected, its elevation z is *positive* if it is *above* a datum and *negative* if it is *below* a datum.

  If more than one unknown is to be determined, consider relating the velocities using the continuity equation or relating the pressures using the manometer equation if it applies.

EXAMPLE | 5.10

The turbine in Fig. 5–24 is used in a small hydroelectric plant, along with a 0.3-m-diameter pipe. If the discharge at B is 1.7 m³/s, determine the amount of power that is transferred from the water to the turbine blades. The frictional head loss through the pipe and turbine is 4 m.

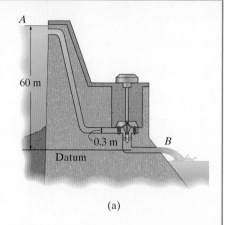

(a)

Fig. 5–24

SOLUTION

Fluid Description. This is a case of steady flow. Here viscous frictional losses occur within the fluid. We consider the water to be incompressible, where $\gamma_w = 9810$ N/m³.

Control Volume. A portion of the reservoir, along with water within the pipe and turbine, is selected to be the fixed control volume. The average velocity at B can be determined from the discharge.

$$Q = V_B A_B; \qquad 1.7 \text{ m}^3/\text{s} = V_B\left[\pi(0.15 \text{ m})^2\right]$$

$$V_B = 24.05 \text{ m/s}$$

Energy Equation. For convenience, vertical measurements z to the datum are made from the *centerline* of the pipe.* Applying the energy equation between A (in) and B (out), with the gravitational datum set at B, we have

$$\frac{p_A}{\gamma} + \frac{V_A^2}{2g} + z_A + h_{\text{pump}} = \frac{p_B}{\gamma} + \frac{V_B^2}{2g} + z_B + h_{\text{turb}} + h_L$$

$$0 + 0 + 60 \text{ m} + 0 = 0 + \frac{(24.05 \text{ m/s})^2}{2(9.81 \text{ m/s}^2)} + 0 + h_{\text{turb}} + 4 \text{ m}$$

$$h_{\text{turb}} = 26.52 \text{ m}$$

As expected, the result is positive, indicating that energy is supplied by the water (system) to the turbine.

Power. Using Eq. 5–16, the power transferred to the turbine is therefore

$$\dot{W}_s = Q\gamma_w h_s = (1.7 \text{ m}^3/\text{s})(9810 \text{ N/m}^3)(26.52 \text{ m})$$

$$= 442 \text{ kW} \qquad \qquad Ans.$$

Notice that the power *lost* due to the effects of friction is

$$\dot{W}_L = Q\gamma_w h_L = (1.7 \text{ m}^3/\text{s})(9810 \text{ N/m}^3)(4 \text{ m}) = 66.7 \text{ kW}$$

As a point of interest, the section of pipe that transfers the water from the reservoir to the turbine is called a *penstock*.

---

*Since the hydraulic head $H = p/\gamma + z$ is constant over the cross section of the horizontal segment of the pipe, measurement to any point on the cross section can be considered. See the following example.

EXAMPLE | 5.11

Water is flowing with an average velocity of 6 m/s when it comes down the spillway of a dam as shown in Fig. 5–25. Within a short distance a hydraulic jump occurs, which causes the water to transition from a depth of 0.8 m to 2.06 m. Determine the energy loss caused by the turbulence within the jump. The spillway has a constant width of 2 m.

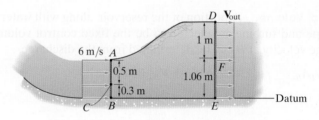

Fig. 5–25

SOLUTION

Fluid Behavior.   Steady flow occurs before and after the jump. The water is assumed to be incompressible.

Control Volume.   Here we will consider a fixed control volume that contains the water within the jump and a short distance from it, Fig. 5–25. Steady flow passes through the open control surfaces because these surfaces are removed from the regions within the jump where the flow is not well defined.

Continuity Equation.   Since the cross sections of AB and DE are known, we can determine the average velocity out of DE using the continuity equation.

$$\frac{\partial}{\partial t} \int_{cv} \rho d\forall + \int_{cs} \rho \mathbf{V} \cdot d\mathbf{A} = 0$$

$$0 - \left(1000 \text{ kg/m}^3\right)(6 \text{ m/s})(0.8 \text{ m})(2 \text{ m}) + \left(1000 \text{ kg/m}^3\right)V_{out}(2.06 \text{ m})(2 \text{ m}) = 0$$

$$V_{out} = 2.3301 \text{ m/s}$$

Energy Equation. We will place the datum at the bottom control surface, Fig. 5–25. From this we can determine the hydraulic head $(p/\gamma + z)$ at each open control surface. If we select points A and D, then since $p_{in} = p_{out} = 0$, we get

$$\frac{p_{in}}{\gamma} + z_{in} = 0 + 0.8 \text{ m} = 0.8 \text{ m}$$

$$\frac{p_{out}}{\gamma} + z_{out} = 0 + 2.06 \text{ m} = 2.06 \text{ m}$$

If instead we take points B and E, then since $p = \gamma h$, we again have

$$\frac{p_{in}}{\gamma} + z_{in} = \frac{\gamma(0.8 \text{ m})}{\gamma} + 0 = 0.8 \text{ m}$$

$$\frac{p_{out}}{\gamma} + z_{out} = \frac{\gamma(2.06 \text{ m})}{\gamma} + 0 = 2.06 \text{ m}$$

Finally, if we use intermediate points C and F, then once again

$$\frac{p_{in}}{\gamma} + z_{in} = \frac{\gamma(0.5 \text{ m})}{\gamma} + 0.3 \text{ m} = 0.8 \text{ m}$$

$$\frac{p_{out}}{\gamma} + z_{out} = \frac{\gamma(1 \text{ m})}{\gamma} + 1.06 \text{ m} = 2.06 \text{ m}$$

In all cases we obtain the *same results*, and so it does not matter which pair of points on the control surface we choose. Here we will choose points A and D. Since no shaft work is done, we have

$$\frac{p_{in}}{\gamma} + \frac{V_{in}^2}{2g} + z_{in} + h_{pump} = \frac{p_{out}}{\gamma} + \frac{V_{out}^2}{2g} + z_{out} + h_{turb} + h_L$$

$$0 + \frac{(6 \text{ m/s})^2}{2(9.81 \text{ m/s}^2)} + 0.8 \text{ m} + 0 = 0 + \frac{(2.3301 \text{ m/s})^2}{2(9.81 \text{ m/s}^2)} + 2.06 \text{ m} + 0 + h_L$$

$$h_L = 0.298 \text{ m} \qquad\qquad Ans.$$

This energy loss produces turbulence and frictional heating within the jump.

EXAMPLE | 5.12

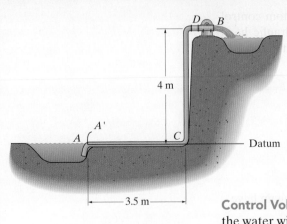

(a)

Fig. 5–26

The irrigation pump in Fig. 5–26a is used to supply water to the pond at B at a rate of $0.09 \text{ m}^3/\text{s}$. If the pipe is 150 mm in diameter, determine the required power of the pump. Assume the frictional head loss per meter length of pipe is 0.1 m/m. Draw the energy and hydraulic grade lines for this system.

SOLUTION

Fluid Description.   Here we have steady flow. The water is assumed to be incompressible, but viscous friction losses occur. $\gamma_w = 9.81 \text{ kN/m}^3$.

Control Volume.   We will select a fixed control volume that contains the water within the reservoir A, along with that in the pipe and pump. For this case, the velocity at A is essentially zero, and the pressure at the inlet and outlet surfaces A and B is zero. Since the volumetric flow is known, the average velocity at the outlet is

$$Q = V_B A_B; \quad 0.09 \text{ m}^3/\text{s} = V_B [\pi (0.075 \text{ m})^2]$$

$$V_B = \frac{16}{\pi} \text{ m/s}$$

Energy Equation.   Establishing the gravitational datum at A, and applying the energy equation between A (in) and B (out), we have

$$\frac{p_A}{\gamma} + \frac{V_A^2}{2g} + z_A + h_{\text{pump}} = \frac{p_B}{\gamma} + \frac{V_B^2}{2g} + z_B + h_{\text{turb}} + h_L$$

$$0 + 0 + 0 + h_{\text{pump}} = 0 + \frac{\left(\dfrac{16}{\pi} \text{ m/s}\right)^2}{2(9.81 \text{ m/s}^2)} + 4 \text{ m} + 0 + (0.1 \text{ m/m})(7.5 \text{ m})$$

$$h_{\text{pump}} = 6.072 \text{ m}$$

This positive result indicates the pump head or energy per unit weight of water that is transferred *into* the system by the pump.

Power.   The pump must therefore have a power of

$$\dot{W}_s = Q\gamma_w h_{\text{pump}} = (0.09 \text{ m}^3/\text{s})[9.81(10^3) \text{ N/m}^3](6.072 \text{ m})$$
$$= 5.361(10^3) \text{ W} = 5.36 \text{ kW} \qquad \textit{Ans.}$$

Of this amount, the power needed to overcome the frictional head loss is

$$\dot{W}_L = Q\gamma_w h_L = (0.09 \text{ m}^3/\text{s})[9.81(10^3) \text{ N/m}^3][(0.1 \text{ m/m})(7.5 \text{ m})]$$
$$= 662.18 \text{ W}$$

EGL and HGL. Recall the EGL is a plot of the *total head* $H = p/\gamma + V^2/2g + z$ along the pipe. The HGL lies $V^2/2g$ below and parallel to the EGL. Before we draw these lines, notice that the velocity head is

$$\frac{V^2}{2g} = \frac{\left(\frac{16}{\pi}\text{ m/s}\right)^2}{2\left(9.81 \text{ m/s}^2\right)} = 1.322 \text{ m}$$

It remains constant since the pipe has the same diameter throughout its length. Before we draw EGL and HGL, the pressure head at $A'*$, C and D must be determined first.

At A',

$$\frac{p_A}{\gamma} + \frac{V_A^2}{2g} + z_A + h_{\text{pump}} = \frac{p_{A'}}{\gamma} + \frac{V_{A'}^2}{2g} + z_{A'} + h_{\text{turb}} + h_L$$

$$0 + 0 + 0 + 0 = \frac{p_{A'}}{\gamma} + 1.322 \text{ m} + 0 + 0 + 0$$

$$\frac{p_{A'}}{\gamma} = -1.322 \text{ m}$$

At C,

$$\frac{p_A}{\gamma} + \frac{V_A^2}{2g} + z_A + h_{\text{pump}} = \frac{p_C}{\gamma} + \frac{V_C^2}{2g} + z_C + h_{\text{turb}} + h_L$$

$$0 + 0 + 0 + 0 = \frac{p_C}{\gamma} + 1.322 \text{ m} + 0 + 0 + (0.1 \text{ m/m})(3.5 \text{ m})$$

$$\frac{p_C}{\gamma} = -1.672 \text{ m}$$

And at D,

$$\frac{p_A}{\gamma} + \frac{V_A^2}{2g} + z_A + h_{\text{pump}} = \frac{p_D}{\gamma} + \frac{V_D^2}{2g} + z_D + h_{\text{turb}} + h_L$$

$$0 + 0 + 0 + 0 = \frac{p_D}{\gamma} + 1.322 \text{ m} + 4 \text{ m} + 0 + (0.1 \text{ m/m})(7.5 \text{ m})$$

$$\frac{p_D}{\gamma} = -6.072 \text{ m}$$

The negative sign indicate the negative pressure caused by the suction of the pump. The total head at A', C, D, and B is therefore

$$H = \frac{p}{\gamma} + \frac{V^2}{2g} + z$$

$$H_{A'} = -1.322 \text{ m} + 1.322 \text{ m} + 0 = 0$$

$$H_C = -1.672 \text{ m} + 1.322 \text{ m} + 0 = -0.35 \text{ m}$$

$$H_D = -6.072 \text{ m} + 1.322 \text{ m} + 4 \text{ m} = -0.75 \text{ m}$$

$$H_B = 0 + 1.322 \text{ m} + 4 \text{ m} = 5.32 \text{ m}$$

Stretching the pipe out, these values are plotted in Fig. 5–26*b* to produce the EGL. The HGL lies 1.322 m below and parallel to the EGL.

*A' is a point within the pipe located at a short distance from A where the water has reached the average velocity V within the pipe, having been accelerated from rest at the entrance of the pipe.

5.32 m

4 m

$h_{\text{pump}} = 6.07$ m

A A' C D B Datum

−0.35 m

EGL

−1.32 m

−0.75 m

HGL

−1.67 m Friction loss along pipe

−2.07 m

(b)

EXAMPLE | 5.13

The pump in Fig. 5–27 discharges water at $80(10^3)$ l/h. The pressure at A is 150 kPa, whereas the exit pressure in the pipe at B is 500 kPa. The pipe filter causes the internal energy of the water to increase by 50 J/kg at its exit due to frictional heating, while there is a heat conduction loss from the water of 250 J/s. Determine the power developed by the pump.

SOLUTION

Fluid Description. We have steady flow into and out of the pump. The water is considered incompressible, but viscous frictional losses occur. $\gamma_w = 9.81(10^3)$ N/m^3.

Control Volume. The fixed control volume contains the water within the pump, filter, and pipe extensions. Since there is 1 liter in 1000 cm^3 of water, the volumetric and mass flows are

$$Q = \left[80(10^3)\frac{\text{liter}}{\text{h}} \right]\left(\frac{1000 \text{ cm}^3}{1 \text{ liter}} \right)\left(\frac{1 \text{ m}}{100 \text{ cm}} \right)^3\left(\frac{1 \text{ h}}{3600 \text{ s}} \right) = 0.02222 \text{ m}^3/\text{s}$$

and

$$\dot{m} = \rho Q = (1000 \text{ kg/m}^3)(0.02222 \text{ m}^3/\text{s}) = 22.22 \text{ kg/s}$$

Therefore, the velocities at A (in) and B (out) are

$$Q = V_A A_A;\quad 0.02222 \text{ m}^3/\text{s} = V_A[\pi(0.075 \text{ m})^2];\; V_A = 1.258 \text{ m/s}$$

$$Q = V_B A_B;\quad 0.02222 \text{ m}^3/\text{s} = V_B[\pi(0.025 \text{ m})^2];\; V_B = 11.32 \text{ m/s}$$

Energy Equation. Since there is heat conduction *loss*, $\dot{Q}_{in}$ is negative; that is, the heat flows out. Also, there is no elevation change ($Z_A = Z_B = Z$) in the flow from A to B. For this problem we will apply Eq. 5–11.

$$\dot{Q}_{in} + \dot{W}_{pump} - \dot{W}_{turb} = \left[\left(\frac{p_B}{\rho} + \frac{V_B^2}{2} + gz_B + u_B \right) \right.$$
$$\left. - \left(\frac{p_A}{\rho} + \frac{V_A^2}{2} + gz_A + u_A \right) \right]\dot{m}$$

$$- 250 \text{ J/s} + \dot{W}_{pump} - 0$$

$$= \left[\left(\frac{500(10^3) \text{ N/m}^2}{1000 \text{ kg/m}^3} + \frac{(11.32 \text{ m/s})^2}{2} + gz + 50 \text{ J/kg} \right) \right.$$
$$\left. - \left(\frac{150(10^3) \text{ N/m}^2}{1000 \text{ kg/m}^3} + \frac{(1.258 \text{ m/s})^2}{2} + gz + 0 \right) \right](22.22 \text{ kg/s})$$

$$\dot{W}_{pump} = 10.54(10^3) \text{ W} = 10.5 \text{ kW} \qquad \textit{Ans.}$$

The positive result indicates that indeed energy is *added* to the water within the control volume using the pump.

150 mm
A

50 mm
B

Datum

Fig. 5–27

5

EXAMPLE | 5.14

The turbine in Fig. 5–28 takes in steam with an enthalpy of $h_A = 2.80$ MJ/kg at 40 m/s. A steam–water mixture exits the turbine with an enthalpy of 1.73 MJ/kg at 15 m/s. If the heat loss to the surroundings during this process is 500 J/s, determine the power the fluid supplies to the turbine. The mass flow through the turbine is 0.8 kg/s.

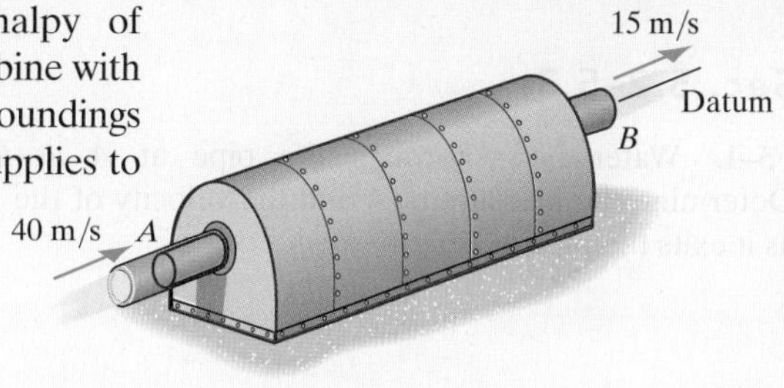

Fig. 5–28

SOLUTION

Fluid Description. Outside the turbine, uniform steady flow occurs because the steam is located away from the turbine's moving blades. Compressibility and frictional effects occur and are reported in terms of the change in enthalpy.

Control Volume. We will consider the control volume to contain the steam within the turbine and within a portion of the pipes at the entrance A (in) and the exit B (out). Uniform steady flow occurs through these open control surfaces since they are located away from the internal moving parts of the turbine.

Energy Equation. Since part of the energy is reported in terms of the steam's enthalpy ($h = p/\rho + u$), we will use Eq. 5–15. Here the elevation change for the steam is zero. Also, heat flows *out* from the control volume, so this is a negative numerical quantity. Thus,

$$\dot{Q}_{in} + \dot{W}_{pump} - \dot{W}_{turb} = \left[\left(h_B + \frac{V_B^2}{2} + gz_B \right) - \left(h_A + \frac{V_A^2}{2} + gz_A \right) \right] \dot{m}$$

$$-500 \text{ J/s} + 0 - \dot{W}_{turb} =$$

$$\left[\left(1.73(10^6) \text{ J/kg} + \frac{(15 \text{ m/s})^2}{2} + 0 \right) - \left(2.80(10^6) \text{ J/kg} + \frac{(40 \text{ m/s})^2}{2} + 0 \right) \right](0.8 \text{ kg/s})$$

$$\dot{W}_{turb} = 856 \text{ kW} \qquad\qquad Ans.$$

The result is positive, which indicates that energy or power is indeed drawn out of the system and transferred to the turbine.

References

1. D. Ghista, *Applied Biomedical Engineering Mechanics*, CRC Press, Boca Raton, FL, 2009.
2. A. Alexandrou, *Principles of Fluid Mechanics*, Prentice Hall, Upper Saddle River, NJ, 2001.
3. I. H. Shames, *Mechanics of Fluids*, McGraw Hill, New York, NY, 1962.
4. L. D. Landau, E. M. Lifshitz, *Fluid Mechanics*, Pergamon Press, Addison-Wesley Pub., Reading, MA, 1959.

FUNDAMENTAL PROBLEMS

Sec. 5.2–5.3

F5–1. Water flows through the pipe at A at 6 m/s. Determine the pressure at A and the velocity of the water as it exits the pipe at B.

F5–3. The fountain is to be designed so that water is ejected from the nozzle and reaches a maximum elevation of 2 m. Determine the required water pressure in the pipe at A, a short distance AB from the nozzle exit.

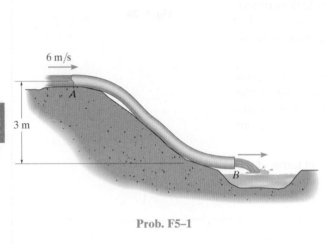

Prob. F5–1

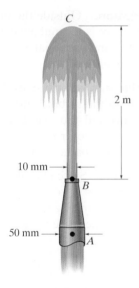

Prob. F5–3

F5–2. Oil is subjected to a pressure of 300 kPa at A, where its velocity is 7 m/s. Determine its velocity and the pressure at B. $\rho_o = 940 \text{ kg/m}^3$.

F5–4. Water flows through the pipe at 8 m/s. Determine the pressure reading on the gage C if the pressure at A is 80 kPa.

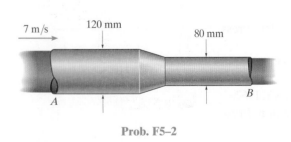

Prob. F5–2

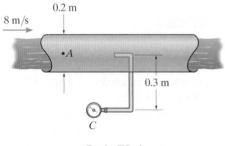

Prob. F5–4

F5–5. The tank has a square base and is filled with water to the depth of $y = 0.4$ m. If the 20-mm-diameter drain pipe is opened, determine the initial volumetric flow of the water and the volumetric flow when $y = 0.2$ m.

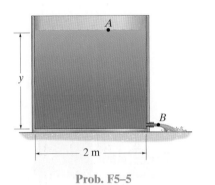

Prob. F5–5

Sec. 5.4–5.5

F5–7. Water flows from the reservoir through the 100-mm-diameter pipe. Determine the discharge at B. Draw the energy grade line and the hydraulic grade line for the flow from A to B.

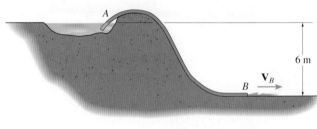

Prob. F5–7

F5–6. Air at a temperature of 80°C flows through the pipe. At A, the pressure is 20 kPa, and the average velocity is 4 m/s. Determine the pressure reading at B. Assume the air is incompressible.

F5–8. Crude oil flows through the 50-mm-diameter pipe such that at A its average velocity is 4 m/s and the pressure is 300 kPa. Determine the pressure of the oil at B. Draw the energy grade line and the hydraulic grade line for the flow from A to B.

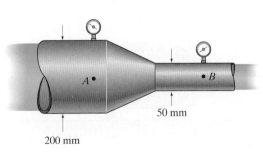

Prob. F5–6

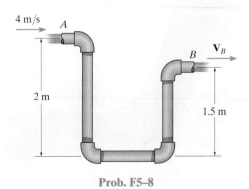

Prob. F5–8

F5–9. At A, water at a pressure of 400 kPa and a velocity of 3 m/s flows through the transitions. Determine the pressure and velocity at B and C. Draw the energy grade line and the hydraulic grade line for the flow from A to C.

F5–11. Water is supplied to the pump at a pressure of 80 kPa and a velocity of $V_A = 2$ m/s. If the discharge is required to be 0.02 m³/s through the 50-mm-diameter pipe, determine the power that the pump must supply to the water to lift it 8 m. The total head loss is 0.75 m.

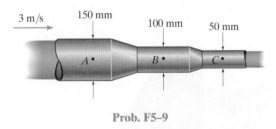

Prob. F5–9

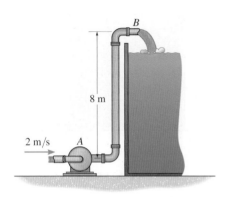

Prob. F5–11

F5–10. Water from the reservoir flows through the 150-m-long, 50-mm-diameter pipe into the turbine at B. If the head loss in the pipe is 1.5 m for every 100-m length of pipe, and the water exits the pipe at C with an average velocity of 8 m/s, determine the power output of the turbine. The turbine operates with 60% efficiency.

F5–12. The jet engine takes in air and fuel having an enthalpy of 600 kJ/kg at 12 m/s. At the exhaust, the enthalpy is 450 kJ/kg and the velocity is 48 m/s. If the mass flow is 2 kg/s, and the rate of heat loss is 1.5 kJ/s, determine the power output of the engine.

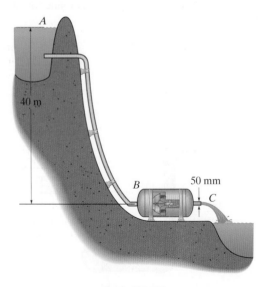

Prob. F5–10

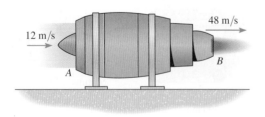

Prob. F5–12

PROBLEMS

Unless otherwise stated, in the following problems, assume the fluid is an ideal fluid, that is, incompressible and frictionless.

Sec. 5.1–5.3

5–1. Water flows in the horizontal tapered pipe. Determine the average decrease in pressure in 4 m along a horizontal streamline so that the water has an acceleration of 0.5 m/s².

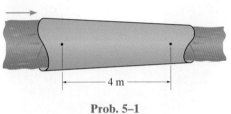

Prob. 5–1

5–2. An ideal fluid having a density ρ flows with a velocity V through the *horizontal* pipe bend. Plot the pressure variation within the fluid as a function of the radius r, where $r_i \leq r \leq r_o$ and $r_o = 2r_i$. For the calculation assume the velocity is constant over the cross section.

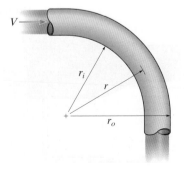

Prob. 5–2

5–3. Piston C moves to the right at a constant speed of 5 m/s, and as it does, outside air at atmospheric pressure flows into the circular cylinder through the opening at B. Determine the pressure within the cylinder and the power required to move the piston. Take $\rho_a = 1.23$ kg/m³. *Hint:* Recall that power is force F times velocity V, where $F = pA$.

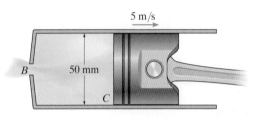

Prob. 5–3

*5–4. An infusion pump produces pressure within the syringe that gives the plunger A a velocity of 20 mm/s. If the saline fluid has a density of $\rho_s = 1050$ kg/m³, determine the pressure developed in the syringe at B.

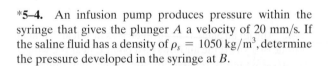

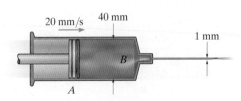

Prob. 5–4

5–5. By applying a force **F**, a saline solution is ejected from the 15-mm-diameter syringe through a 0.6-mm-diameter needle. If the pressure developed within the syringe is 60 kPa, determine the average velocity of the solution through the needle. Take $\rho = 1050 \text{ kg/m}^3$.

5–6. By applying a force **F**, a saline solution is ejected from the 15-mm-diameter syringe through a 0.6-mm-diameter needle. Determine the average velocity of the solution through the needle as a function of the force F applied to the plunger. Plot this velocity (vertical axis) as a function of the force for $0 \le F \le 20 \text{ N}$. Give values for increments of $\Delta F = 5 \text{ N}$. Take $\rho = 1050 \text{ kg/m}^3$.

5–9. Water is discharged through the drain pipe at B from the large basin at $0.03 \text{ m}^3/\text{s}$. If the diameter of the drainpipe is $d = 60 \text{ mm}$, determine the pressure at B just inside the drain when the depth of the water is $h = 2 \text{ m}$.

5–10. Water is discharged through the drain pipe at B from the large basin at $0.03 \text{ m}^3/\text{s}$. Determine the pressure at B just inside the drain as a function of the diameter d of the drainpipe. The height of the water is maintained at $h = 2 \text{ m}$. Plot the pressure (vertical axis) versus the diameter for $60 \text{ mm} < d < 120 \text{ mm}$. Give values for increments of $\Delta d = 20 \text{ mm}$.

15 mm

F

Probs. 5–5/6

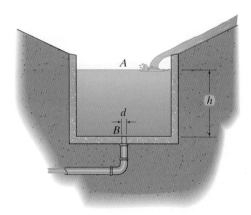

A

d

B

h

Probs. 5–9/10

5–7. The jet airplane is flying at 80 m/s in still air, A, at an altitude of 3 km. Determine the absolute stagnation pressure at the leading edge B of the wing.

*5–8. The jet airplane is flying at 80 m/s in still air, A, at an altitude of 4 km. If the air flows past point C near the wing at 90 m/s, measured relative to the plane, determine the difference in pressure between the air near the leading edge B of the wing and point C.

5–11. The mercury in the manometer has a difference in elevation of $h = 0.15 \text{ m}$. Determine the volumetric discharge of gasoline through the pipe. Take $\rho_{gas} = 726 \text{ kg/m}^3$.

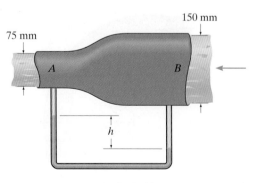

150 mm

75 mm

A B

h

Prob. 5–11

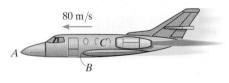

80 m/s

A

B

Probs. 5–7/8

*5–12. The average human lung takes in about 0.6 liter of air with each inhalation, through the mouth and nose, A. This lasts for about 1.5 seconds. Determine the power required to do this if it occurs through the trachea B having a cross-sectional area of 125 mm². Take $\rho_a = 1.23 \text{ kg/m}^3$. *Hint:* Recall that power is force F times velocity V, where $F = pA$.

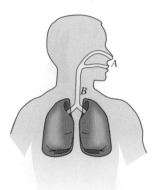

Prob. 5–12

5–13. A fountain is produced by water that flows up the tube at $Q = 0.08 \text{ m}^3/\text{s}$ and then radially through two cylindrical plates before exiting to the atmosphere. Determine the velocity and pressure of the water at point A.

5–14. A fountain is produced by water that flows up the tube at $Q = 0.08 \text{ m}^3/\text{s}$ and then radially through two cylindrical plates before exiting to the atmosphere. Determine the pressure of the water as a function of the radial distance r. Plot the pressure (vertical axis) versus r for 200 mm $\leq r \leq$ 400 mm. Give values for increments of $\Delta r = 50$ mm.

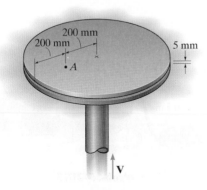

Probs. 5–13/14

5–15. A fountain ejects water through the four nozzles, which have inner diameters of 10 mm. Determine the pressure in the pipe and the required volumetric flow through the supply pipe so that the water stream always reaches a height of $h = 4$ m.

*5–16. A fountain ejects water through the four nozzles, which have inner diameters of 10 mm. Determine the maximum height h of the water stream passing through the nozzles as a function of the volumetric flow rate into the 60-mm-diameter pipe at E. Also, what is the corresponding pressure at E as a function of h?

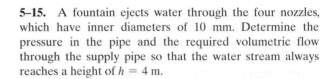

Probs. 5–15/16

5–17. If water flows into the pipe at a constant rate of 30 kg/s, determine the pressure acting at the inlet A when $y = 0.5$ m. Also, what is the rate at which the water surface at B is rising when $y = 0.5$ m? The container is circular.

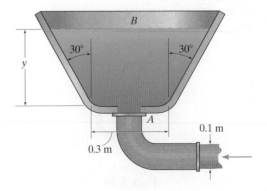

Prob. 5–17

5

5–18. One method of producing energy is to use a tapered channel (TAPCHAN), which diverts sea water into a reservoir as shown in the figure. As a wave approaches the shore through the closed tapered channel at A, its height will begin to increase until it begins to spill over the sides and into the reservoir. The water in the reservoir then passes through a turbine in the building at C to generate power and is returned to the sea at D. If the speed of the water at A is $V_A = 2.5$ m/s, and the water depth is $h_A = 3$ m, determine the minimum height h_B at the back B of the channel to prevent the water from entering the reservoir.

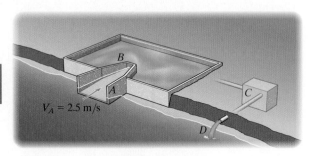

Prob. 5–18

5–19. Blood flows from the left ventricle (LV) of the heart, which has an exit diameter of $d_1 = 16$ mm, through the stenotic aortic valve of diameter $d_2 = 8$ mm, and then into the aorta A having a diameter of $d_3 = 20$ mm. If the cardiac output is 4 liters per minute, the heart rate is 90 beats per minute, and each ejection of blood lasts 0.31 s, determine the pressure change over the valve. Take $\rho_b = 1060$ kg/m^3.

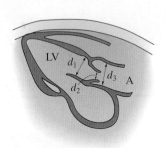

Prob. 5–19

*5–20. Air enters the tepee door at A with an average speed of 2 m/s and exits at the top B. Determine the pressure difference between these two points and find the average speed of the air at B. The areas of the openings are $A_A = 0.3$ m^2 and $A_B = 0.05$ m^2. The density of the air is $\rho_a = 1.20$ kg/m^3.

Prob. 5–20

5–21. Water flows from the hose at B at the rate of 4 m/s when the water level in the large tank is 0.5 m. Determine the pressure of air that has been pumped into the top of the tank at A.

5–22. If the hose at A is used to pump air into the tank with a pressure of 150 kPa, determine the discharge of water at the end of the 15-mm-diameter hose at B when the water level is 0.5 m.

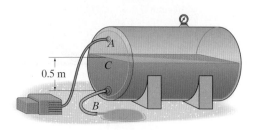

Probs. 5–21/22

5–23. Determine the flow of oil through the pipe if the difference in height of the water column in the manometer is $h = 100$ mm. Take $\rho_o = 875$ kg/m^3.

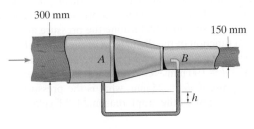

300 mm

150 mm

A B

h

Prob. 5–23

*5–24. Determine the difference in height h of the water column in the manometer if the flow of oil through the pipe is 0.04 m^3/s. Take $\rho_o = 875$ kg/m^3.

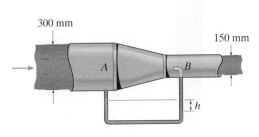

300 mm

150 mm

A B

h

Prob. 5–24

5–25. Air at a temperature of 40°C flows into the nozzle at 6 m/s and then exits to the atmosphere at B, where the temperature is 0°C. Determine the pressure at A.

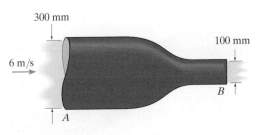

300 mm

100 mm

6 m/s

B

A

Prob. 5–25

5–26. A water-cooled nuclear reactor is made with plate fuel elements that are spaced 3 mm apart and 800 mm long. During an initial test, water enters at the bottom of the reactor (plates) and flows upward at 0.8 m/s. Determine the pressure difference in the water between A and B. Take the average water temperature to be 80°C.

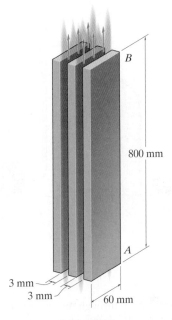

B

800 mm

A

3 mm

3 mm

60 mm

Prob. 5–26

5–27. Determine the velocity of water through the pipe if the manometer contains mercury held in the position shown. Take $\rho_{Hg} = 13\,550$ kg/m^3.

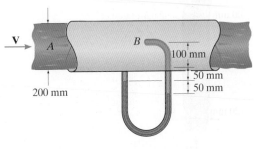

V

A B

100 mm

50 mm

50 mm

200 mm

Prob. 5–27

*5–28. Determine the air pressure that must be exerted at the top of the kerosene in the large tank at B so that the initial discharge through the drainpipe at A is 0.1 m³/s once the valve at A is opened.

5–29. If air pressure at the top of the kerosene in the large tank is 80 kPa, determine the initial discharge through the drainpipe at A once the valve is opened.

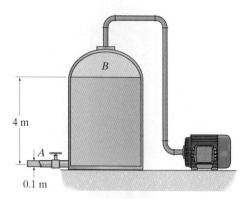

Probs. 5–28/29

5–30. Water flows up through the *vertical pipe*. Determine the pressure at A if the average velocity at B is 4 m/s.

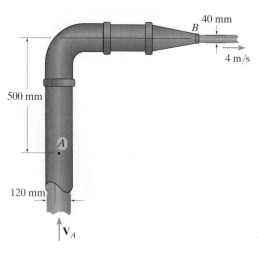

Prob. 5–30

5–31. Water flows up through the *vertical pipe* such that when it is at A, it is subjected to a pressure of 150 kPa and has a velocity of 3 m/s. Determine the pressure and its velocity at B. Set $d = 75$ mm.

*5–32. Water flows through the *vertical pipe* such that when it is at A, it is subjected to a pressure of 150 kPa and has a velocity of 3 m/s. Determine the pressure and velocity at B as a function of the diameter d of the pipe at B. Plot the pressure and velocity (vertical axis) versus the diameter for 25 mm $\leq d \leq$ 100 mm. Give values for increments of $\Delta d = 25$ mm. If $d_B = 25$ mm, what is the pressure at B? Is this lower region of the graph reasonable? Explain.

Probs. 5–31/32

5–33. Water flows in a rectangular channel over the 1-m drop. If the width of the channel is 1.5 m, determine the volumetric flow in the channel.

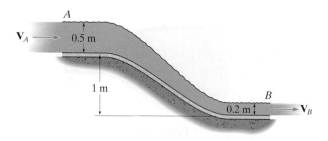

Prob. 5–33

5–34. Water flows through the pipe at A with a velocity of 6 m/s and at a pressure of 280 kPa. Determine the velocity of the water at B and the difference in elevation h of the mercury in the manometer.

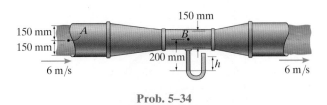

150 mm

150 mm
150 mm

A

6 m/s

200 mm

B

h

6 m/s

Prob. 5–34

5–35. Carbon dioxide at 20°C flows past the Pitot tube B such that mercury within the manometer is displaced 50 mm as shown. Determine the mass flow if the duct has a cross-sectional area of 0.18 m².

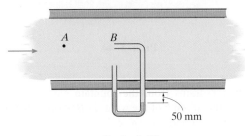

A B

50 mm

Prob. 5–35

*5–36.** Water flows along the rectangular channel such that after it falls to the lower elevation, the depth becomes $h = 0.3$ m. Determine the volumetric discharge through the channel. The channel has a width of 1.5 m.

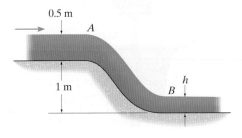

0.5 m

A

1 m

h

B

Prob. 5–36

5–37. Water flows at 3 m/s at A along the rectangular channel that has a width of 1.5 m. If the depth at A is 0.5 m, determine the depth at B.

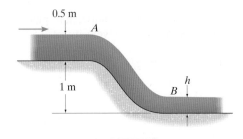

0.5 m

A

1 m

B

h

Prob. 5–37

5–38. If the velocity of water changes uniformly along the transition from $V_A = 10$ m/s to $V_B = 4$ m/s, determine the pressure difference between A and x.

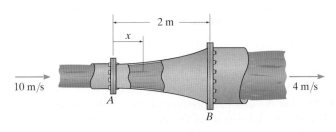

2 m

x

10 m/s

A

B

4 m/s

Probs. 5–38

5–39. If the velocity of water changes uniformly along the transition from $V_A = 10$ m/s to $V_B = 4$ m/s, find the pressure difference between A and $x = 1.5$ m.

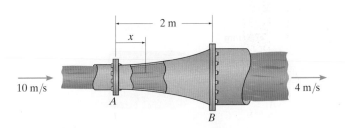

2 m

x

10 m/s

A

B

4 m/s

Probs. 5–39

5

*5–40. Oil flows through the horizontal pipe under a pressure of 400 kPa and at a velocity of 2.5 m/s at A. Determine the pressure in the pipe at B if the pressure at C is 150 kPa. Neglect any elevation difference. Take $\rho_o = 880 \text{ kg/m}^3$.

5–41. Oil flows through the horizontal pipe under a pressure of 100 kPa and at a velocity of 2.5 m/s at A. Determine the pressure in the pipe at C if the pressure at B is 95 kPa. Take $\rho_o = 880 \text{ kg/m}^3$.

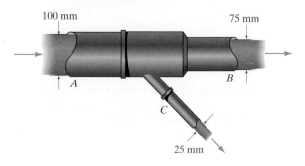

Probs. 5–40/41

5–42. Air at 15°C and an absolute pressure of 275 kPa flows through the 200-mm-diameter duct at $V_A = 4 \text{ m/s}$. Determine the absolute pressure of the air after it passes through the transition and into the 400-mm-diameter duct B. The temperature of the air remains constant.

5–43. Air at 15°C and an absolute pressure of 250 kPa flows through the 200-mm-diameter duct at $V_A = 20 \text{ m/s}$. Determine the rise in pressure, $\Delta p = p_B - p_A$, when the air passes through the transition and into the 400-mm-diameter duct B. The temperature of the air remains constant.

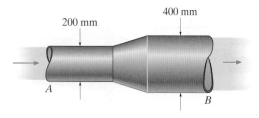

Probs. 5–42/43

*5–44. The open cylindrical tank is filled with linseed oil. A crack having a length of 50 mm and average height of 2 mm occurs at the base of the tank. How many liters of oil will slowly drain from the tank in eight hours? Take $\rho_o = 940 \text{ kg/m}^3$.

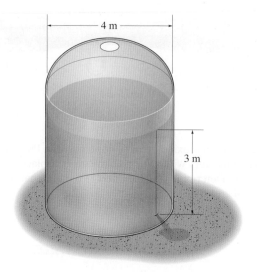

Prob. 5–44

5–45. Water drains from the fountain cup A to cup B. Determine the depth h of the water in B in order for steady flow to be maintained. Take $d = 25 \text{ mm}$.

5–46. Water drains from the fountain cup A to cup B. If the depth in cup B is $h = 50 \text{ mm}$, determine the velocity of the water at C and the diameter d of the opening at D so that steady flow occurs.

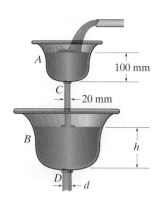

Probs. 5–45/46

5–47. As air flows downward through the venturi constriction, it creates a low pressure A that causes ethyl alcohol to rise in the tube and be drawn into the air stream. If the air is then discharged to the atmosphere at C, determine the smallest volumetric flow of air required to do this. Take $\rho_{ea} = 789 \text{ kg/m}^3$ and $\rho_a = 1.225 \text{ kg/m}^3$.

*5–48. As air flows downward through the venturi constriction, it creates a low pressure at A that causes ethyl alcohol to rise in the tube and be drawn into the air stream. Determine the velocity of the air as it passes through the tube at B in order to do this. The air is discharged to the atmosphere at C. Take $\rho_{ea} = 789 \text{ kg/m}^3$ and $\rho_a = 1.225 \text{ kg/m}^3$.

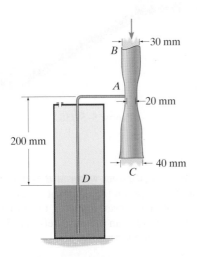

Probs. 5–47/48

5–49. Determine the volumetric flow and the pressure in the pipe at A if the height of the water column in the Pitot tube is 0.3 m and the height in the piezometer is 0.1 m.

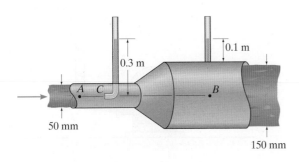

Prob. 5–49

5–50. Determine the velocity of the flow out of the vertical pipes at A and B, if water flows into the Tee at 8 m/s and under a pressure of 40 kPa.

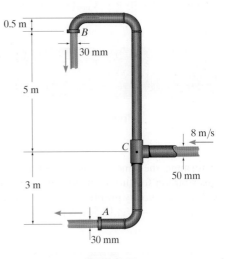

Prob. 5–50

5–51. At the instant shown, the level of water in the conical funnel is $y = 200$ mm. If the stem has an inner diameter of 5 mm, determine the rate at which the surface level of the water is dropping.

*5–52. If the stem of the conical funnel has a diameter of 5 mm, determine the rate at which the surface level of the water is dropping as a function of the depth y. Assume steady flow. *Note*: For a cone, $V = \frac{1}{3}\pi r^2 h$.

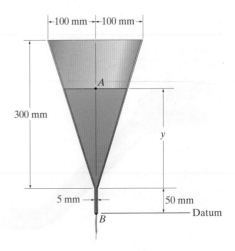

Probs. 5–51/52

Sec. 5.4–5.5

5–53. Determine the kinetic energy coefficient α if the velocity distribution for laminar flow in a smooth pipe has a velocity profile defined by $u = U_{max}(1 - (r/R)^2)$.

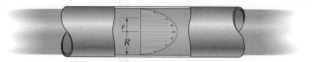

Prob. 5–53

5–54. Determine the kinetic energy coefficient α if the velocity distribution for turbulent flow in a smooth pipe is assumed to have a velocity profile defined by Prandtl's one-seventh power law, $u = U_{max}(1 - r/R)^{1/7}$.

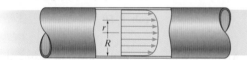

Prob. 5–54

5–55. Water at a pressure of 80 kPa and a velocity of 2 m/s at A flows through the transition. Determine the velocity and the pressure at B. Draw the energy and hydraulic grade lines for the flow from A to B using a datum at B.

*5–56. Water at a pressure of 80 kPa and a velocity of 2 m/s at A flows through the transition. Determine the velocity and the pressure at C. Plot the pressure head and the gravitational head for AB using a datum at B.

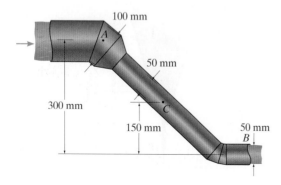

Probs. 5–55/56

5–57. Oil flows through the constant-diameter pipe such that at A the pressure is 50 kPa, and the velocity is 2 m/s. Determine the pressure and velocity at B. Draw the energy and hydraulic grade lines for AB using a datum at B. Take $\rho_o = 900$ kg/m³.

5–58. Oil flows through the constant-diameter pipe such that at A the pressure is 50 kPa, and the velocity is 2 m/s. Plot the pressure head and the gravitational head for AB using a datum at B. Take $\rho_o = 900$ kg/m³.

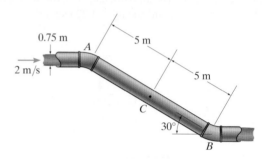

Probs. 5–57/58

5–59. The pump discharges water at B at 0.05 m³/s. If the friction head loss between the intake at A and the outlet at B is 0.9 m, and the power input to the pump is 8 kW, determine the difference in pressure between A and B. The efficiency of the pump is $e = 0.7$.

*5–60. The power input of the pump is 10 kW and the friction head loss between A and B is 1.25 m. If the pump has an efficiency of $e = 0.8$, and the increase in pressure from A to B is 100 kPa, determine the volumetric flow of water through the pump.

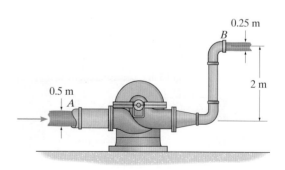

Probs. 5–59/60

5–61. Water is siphoned from the open tank. Determine the volumetric discharge from the 20-mm-diameter hose. Draw the energy and hydraulic grade lines for the hose using a datum at *B*.

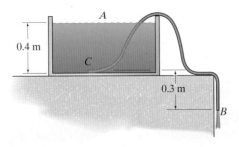

Prob. 5–61

5–62. Water in the reservoir flows through the 0.2-m-diameter pipe at *A* into the turbine. If the discharge at *B* is 0.5 m³/s, determine the power output of the turbine. Assume the turbine runs with an efficiency of 65%. Neglect frictional losses in the pipe.

5–63. Water in the reservoir flows through the 0.2-m-diameter pipe at *A* into the turbine. If the discharge at *B* is 0.5 m³/s, determine the power output of the turbine. Assume the turbine runs with an efficiency of 65%, and there is a head loss of 0.5 m through the pipe.

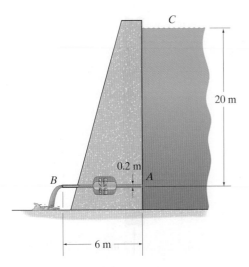

Probs. 5–62/63

*5–64. A 300-mm-diameter horizontal oil pipeline extends 8 km, connecting two large open reservoirs having the same level. If friction in the pipe creates a head loss of 3 m for every 200 m of pipe length, determine the power that must be supplied by a pump to produce a flow of 6 m³/min through the pipe. The ends of the pipe are submerged in the reservoirs. Take $\rho_o = 880$ kg/m³.

5–65. A pump is used to deliver water from a large reservoir to another large reservoir that is 20 m higher. If the friction head loss in the 200-mm-diameter, 4-km-long pipeline is 2.5 m for every 500 m of pipe length, determine the required power output of the pump so the flow is 0.8 m³/s. The ends of the pipe are submerged in the reservoirs.

5–66. The pump draws water from the large reservoir *A* and discharges it at 0.2 m³/s at *C*. If the diameter of the pipe is 200 mm, determine the power the pump delivers to the water. Neglect friction losses. Construct the energy and hydraulic grade lines for the pipe using a datum at *B*.

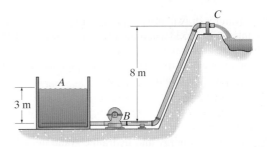

Prob. 5–66

5–67. Solve Prob. 5–66, but include a friction head loss in the pump of 0.5 m, and a friction loss of 1 m for every 5 m length of pipe. The pipe extends 3 m from the reservoir to *B*, then 12 m from *B* to *C*.

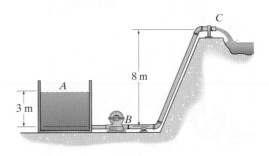

Prob. 5–67

*5–68. As air flows through the duct, its absolute pressure changes from 220 kPa at A to 219.98 kPa at B. If the temperature remains constant at $T = 60°C$, determine the head loss between these points. Assume the air is incompressible.

5–71. The turbine C removes 300 kW of power from the water that passes through it. If the pressure at the intake A is $p_A = 300$ kPa and the velocity is 8 m/s, determine the pressure and velocity of the water at the exit B. Neglect the frictional losses between A and B.

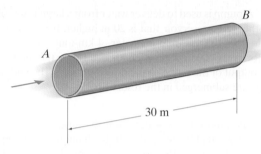

Prob. 5–68

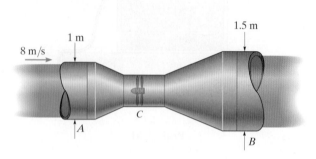

Prob. 5–71

5–69. Water is drawn into the pump, such that the pressure at the inlet A is -35 kPa and the pressure at B is 120 kPa. If the discharge at B is 0.08 m³/s, determine the power output of the pump. Neglect friction losses. The pipe has a constant diameter of 100 mm. Take $h = 2$ m.

*5–72. The vertical pipe is filled with oil. When the valve at A is closed, the pressure at A is 160 kPa, and at B it is 90 kPa. When the valve is open, the oil flows at 2 m/s, and the pressure at A is 150 kPa and at B it is 70 kPa. Determine the head loss in the pipe between A and B. Take $\rho_o = 900$ kg/m³.

5–70. Draw the energy and hydraulic grade lines for the pipe ACB in Prob. 5–99 using a datum at A.

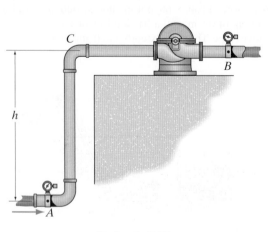

Probs. 5–69/70

Prob. 5–72

5–73. The flow of air through a 200-mm-diameter duct has an absolute inlet pressure of 180 kPa, a temperature of 15°C, and a velocity of 10 m/s. Farther downstream a 2-kW exhaust system increases the outlet velocity to 25 m/s. Determine the density of the air at the outlet, and the change in enthalpy of the air. Neglect heat transfer through the pipe.

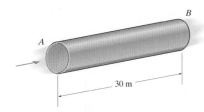

Prob. 5–73

5–74. Nitrogen gas having an enthalpy of 250 J/kg is flowing at 6 m/s into the 10-m-long pipe at A. If the heat loss from the walls of the duct is 60 W, determine the enthalpy of the gas at the exit B. Assume that the gas is incompressible with a density of $\rho = 1.36 \text{ kg/m}^3$.

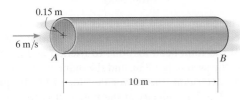

Prob. 5–74

5–75. If the pressure at A is 60 kPa, and the pressure at B is 180 kPa, determine the power output supplied by the pump if the water flows at 0.02 m³/s. Neglect friction losses.

*5–76. The pump supplies a power of 1.5 kW to the water producing a volumetric flow of 0.015 m³/s. If the total frictional head loss within the system is 1.35 m, determine the pressure difference between the inlet A and outlet B of the pipes.

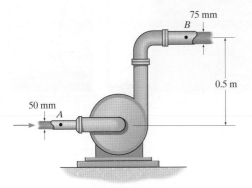

Probs. 5–75/76

5–77. The wave overtopping device consists of a floating reservoir that is continuously filled by waves, so that the water level in the reservoir is always higher than that of the surrounding ocean. As the water drains out at A, the energy is drawn by the low-head hydroturbine, which then generates electricity. Determine the power that can be produced by this system if the water level in the reservoir is always 1.5 m above that of the ocean, The waves add 0.3 m³/s to the reservoir, and the diameter of the tunnel containing the turbine is 600 mm. The head loss through the turbine is 0.2 m. Take $\rho_w = 1050 \text{ kg/m}^3$.

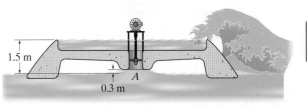

Prob. 5–77

5–78. Air and fuel enter a turbojet engine (turbine) having an enthalpy of 800 kJ/kg and a relative velocity of 15 m/s. The mixture exits with a relative velocity of 60 m/s and an enthalpy of 650 kJ/kg. If the mass flow is 30 kg/s, determine the power output of the jet. Assume no heat transfer occurs.

Prob. 5–78

5–79. The measured water pressure at the inlet and exit portions of the pipe are indicated for the pump. If the flow is 0.1 m³/s, determine the power that the pump supplies to the water. Neglect friction losses.

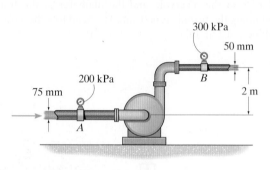

Prob. 5–79

***5–80.** The circular hovercraft draws in air through the fan A and discharges it through the bottom B near the ground, where it produces a pressure of 1.50 kPa on the ground. Determine the average velocity of the air entering at A that is needed to lift the hovercraft 100 mm off the ground. The open area at A is 0.75 m². Neglect friction losses. Take $\rho_a = 1.22$ kg/m³.

5–81. The pump at C produces a discharge of water at B of 0.035 m³/s. If the pipe at B has a diameter of 50 mm and the hose at A has a diameter of 30 mm, determine the power output supplied by the pump. Assume frictional head losses within the pipe system are determined from $3V_B^2/2g$.

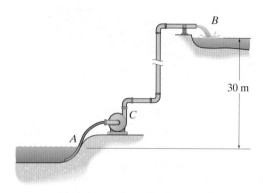

Prob. 5–81

5–82. The pump is used to transfer carbon tetrachloride in a processing plant from a storage tank A to the mixing tank C. If the total head loss due to friction and the pipe fittings in the system is 1.8 m, and the diameter of the pipe is 50 mm, determine the power developed by the pump when $h = 3$ m. The velocity at the pipe exit is 10 m/s, and the storage tank is opened to the atmosphere. Take $\rho_{ct} = 1590$ kg/m³.

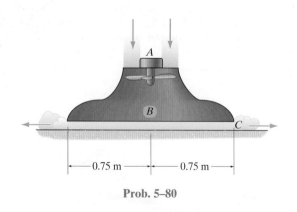

Prob. 5–80

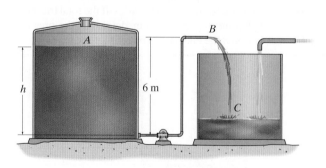

Prob. 5–82

CONCEPTUAL PROBLEMS

P5–1. The level of coffee is measured by the standpipe *A*. If the valve is pushed open and the coffee begins to flow out, will the level of coffee in the standpipe go up, go down, or remain the same? Explain.

P5–1

P5–2. The ball is suspended in the air by the stream of air produced by the fan. Explain why it will return to this position if it is displaced slightly to the right or left.

P5–2

P5–3. When air flows through the hose, it causes the paper to rise. Explain why this happens.

P5–3

CHAPTER REVIEW

For an *inviscid fluid*, the *steady flow* of a particle along a streamline is caused by pressure and gravitational forces. The Euler differential equations describe this motion. Along the streamline or *s* direction, the forces change the *magnitude* of the particle's velocity, and along the normal or *n* direction, they change the *direction* of its velocity.

$$\frac{dp}{\rho} + V\,dV + g\,dz = 0$$

$$-\frac{dp}{dn} - \rho g \frac{dz}{dn} = \frac{\rho V^2}{R}$$

The Bernoulli equation is an integrated form of Euler's equation in the *s* direction. It applies between two points on the *same streamline* for *steady flow* of an *ideal fluid*. It cannot be used at points where energy losses occur, or between points where fluid energy is added or withdrawn by external sources. When applying the Bernoulli equation, remember that points at atmospheric openings have zero gage pressure and that the velocity is zero at a stagnation point or at the top surface of a large reservoir.

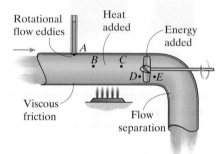

Places where the Bernoulli equation *does not* apply

$$\frac{p_1}{\rho} + \frac{V_1^2}{2} + gz_1 = \frac{p_2}{\rho} + \frac{V_2^2}{2} + gz_2$$

steady flow, ideal fluid, same streamline

Pitot tubes can be used to measure the velocity of a liquid in an *open channel*. To measure the velocity of a liquid in a *closed conduit*, it is necessary to use a Pitot tube along with a piezometer, which measures the static pressure in the liquid. A venturi meter can also be used to measure the average velocity or the volumetric flow.

The Bernoulli equation can be expressed in terms of the total head H of the fluid. A plot of the total head is called the energy grade line, EGL, which will always be a constant horizontal line, provided there are no frictional losses. The hydraulic grade line, HGL, is a plot of the hydraulic head, $p/\gamma + z$. This line will always be *below* the EGL by an amount equal to the kinetic head, $V^2/2g$.

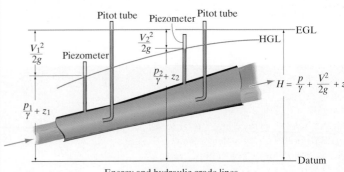

Energy and hydraulic grade lines

$$H = \frac{p}{\gamma} + \frac{V^2}{2g} + z = \text{const.}$$

When the fluid is viscous and/or energy is added to or removed from the fluid, the energy equation should be used. It is based on the first law of thermodynamics, and a control volume must be specified when it is applied. It can be expressed in various forms.

$$\dot{Q}_{in} - \dot{W}_{turb} + \dot{W}_{pump} =$$

$$\left[\left(h_{out} + \frac{V_{out}^2}{2} + gz_{out} \right) - \left(h_{in} + \frac{V_{in}^2}{2} + gz_{in} \right) \right] \dot{m}$$

$$\frac{p_{in}}{\rho} + \frac{V_{in}^2}{2} + gz_{in} + w_{pump} = \frac{p_{out}}{\rho} + \frac{V_{out}^2}{2} + gz_{out} + w_{turb} + fl$$

$$\frac{p_{in}}{\gamma} + \frac{V_{in}^2}{2g} + z_{in} + h_{pump} = \frac{p_{out}}{\gamma} + \frac{V_{out}^2}{2g} + z_{out} + h_{turb} + h_L$$

Power is the rate of doing shaft work.

$$\dot{W}_s = \dot{m} \, g h_s = Q \gamma h_s$$

Chapter 6

(© Sander van der Werf/Shutterstock)

Impulse and momentum principles play an important role in the design of both windmills and wind turbines.

Fluid Momentum

6.1 The Linear Momentum Equation

The design of many hydraulic structures, such as floodgates and flow diversion blades, as well as pumps and turbines, depends upon the forces that a fluid flow exerts on them. In this section we will obtain these forces by using a linear momentum analysis, which is based on Newton's second law of motion, written in the form $\Sigma \mathbf{F} = m\mathbf{a} = d(m\mathbf{V})/dt$. For application of this equation, it is important to measure the time rate of change in the momentum, $m\mathbf{V}$, from an *inertial* or nonaccelerating frame of reference, that is, a reference that either is fixed or moves with constant velocity.

Because of the fluid flow, a control volume approach works best for this type of analysis, and so we will apply the Reynolds transport theorem to determine the time derivative $d(m\mathbf{V})/dt$ before we apply Newton's second law. Linear momentum is an extensive property of a fluid, where $\mathbf{N} = m\mathbf{V}$, and so $\boldsymbol{\eta} = m\mathbf{V}/m = \mathbf{V}$. Therefore, Eq. 4–11 becomes

$$\left(\frac{d\mathbf{N}}{dt} \right)_{\text{syst}} = \frac{\partial}{\partial t} \int_{\text{cv}} \boldsymbol{\eta} \rho \, d\forall + \int_{\text{cs}} \boldsymbol{\eta} \rho \mathbf{V} \cdot d\mathbf{A}$$

$$\left(\frac{d(m\mathbf{V})}{dt} \right)_{\text{syst}} = \frac{\partial}{\partial t} \int_{\text{cv}} \mathbf{V} \rho \, d\forall + \int_{\text{cs}} \mathbf{V} \rho \mathbf{V} \cdot d\mathbf{A}$$

Now, substituting this result into Newton's second law of motion, we obtain our result, the *linear momentum equation*.

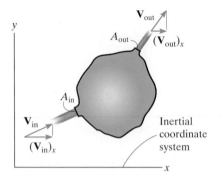

Control Volume

(a)

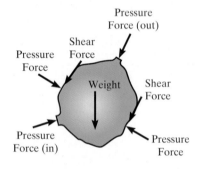

Free-Body Diagram

(b)

Fig. 6–1

$$\Sigma \mathbf{F} = \frac{\partial}{\partial t} \int_{cv} \mathbf{V} \rho \, d\forall + \int_{cs} \mathbf{V} \rho \mathbf{V} \cdot d\mathbf{A} \qquad (6\text{–}1)$$

It is *very important* to realize how the velocity $\mathbf{V}$ is used in the last term of this equation. It stands alone as a *vector quantity* $\mathbf{V}$, and as a result it has *components* along the x, y, z axes. But it is also involved in the dot product operation with $d\mathbf{A}$ in order to define the *mass flow* through an open control surface, that is, $\rho \mathbf{V} \cdot d\mathbf{A}$. This is a *scalar quantity*, and so it *does not have components*. To emphasize this point, consider the flow of an ideal fluid into and out of the two control surfaces in Fig. 6–1a. In the x direction, the last term in Eq. 6–1 becomes $\int_{cs} V_x \rho \, \mathbf{V} \cdot d\mathbf{A} = (V_{in})_x(-\rho V_{in} A_{in}) + (V_{out})_x(\rho V_{out} A_{out})$. Here $(V_{in})_x$ and $(V_{out})_x$ are the x components of $\mathbf{V}_{in}$ and $\mathbf{V}_{out}$. They both act in the $+x$ direction, Fig. 6–1a. When writing the expression for the dot products, we have followed our positive sign convention, that is, A_{in} and A_{out} are *both* positive out, but $\mathbf{V}_{in}$ is negative, since it is directed into the control volume. For this reason, $\rho V_{in} A_{in}$ is a negative quantity.

Steady Flow. If the flow is *steady*, then no local change of momentum will occur within the control volume, and the first term on the right of Eq. 6–1 will be equal to zero. Therefore

$$\Sigma \mathbf{F} = \int_{cs} \mathbf{V} \rho \mathbf{V} \cdot d\mathbf{A} \qquad (6\text{–}2)$$

Steady flow

Furthermore, if we have an *ideal fluid*, then ρ is constant and viscous friction is zero. Thus the velocity will be uniformly distributed over the open control surfaces, and so integration of Eq. 6–2 gives

$$\Sigma \mathbf{F} = \Sigma \mathbf{V} \rho \mathbf{V} \cdot \mathbf{A} \qquad (6\text{–}3)$$

Steady ideal flow

The above equations are often used in engineering, to obtain the fluid forces acting on various types of surfaces that deflect or transport the flow.

Free-Body Diagram. When Eq. 6–1 is applied, there will generally be three types of external forces $\Sigma \mathbf{F}$ that can act on the fluid system contained within the control volume. As shown on its free-body diagram, Fig. 6–1b, they are *shear forces* that act tangent to the closed control surface, *pressure forces* that act normal to the open and closed control surfaces, and *weight* that acts through the center of gravity of the mass of fluid within the control volume. For analysis, we will represent the *resultant* shear and pressure forces by their single resultant force. The opposite of this resultant is the effect of the fluid system on this surface. It is referred to as the ***dynamic force***.

6.2 Applications to Bodies at Rest

On occasion a vane, pipe, or other type of conduit can be subjected to fluid forces because it can change the direction of the flow. For these cases, the following procedure can be used to apply a linear momentum analysis to determine the resultant force caused by pressure and shear that a moving fluid exerts on its fixed surface.

Procedure for Analysis

Fluid Description.

- Identify the type of flow as steady, unsteady, uniform, or nonuniform. Also specify if the fluid is compressible, incompressible, viscous, or inviscid.

Control Volume and Free-Body Diagram.

- The control volume can be fixed, moving, or deformable, and, depending upon the problem, it can include *both* solid and fluid parts. Open control surfaces should be located in a region where the flow is uniform and well established. These surfaces should be oriented so that their planar areas are perpendicular to the flow. For inviscid and ideal fluids, the velocity profile will be uniform over the cross section, and so represents an average velocity.

- The free-body diagram of the control volume should be drawn to identify all of the external forces acting on it. These forces generally include the weight of the contained fluid and the weight of any solid portions of the control volume, the resultant of the frictional shear and pressure forces or their components that act on a *closed control surface*, and the pressure forces acting on the *open control surfaces*. Note that the pressure forces will be zero if the control surface is open to the atmosphere; however, if an open control surface is contained within the fluid, then the pressure on this surface may have to be determined by using the Bernoulli equation.

Linear Momentum.

- If the volumetric flow is known, the average velocities at the open control surfaces can be determined using $V = Q/A$, or by applying the continuity equation, the Bernoulli equation, or the energy equation.

- Establish the inertial x, y coordinate system, and apply the linear momentum equation, using the x and y components of the velocity, shown on the open control surfaces, and the forces shown on the free-body diagram. Remember, the term $\rho \mathbf{V} \cdot d\mathbf{A}$ in the momentum equation is a *scalar quantity* that represents the *mass flow* through the area $\mathbf{A}$ of each open control surface. The product $\rho \mathbf{V} \cdot \mathbf{A}$ is *negative* for mass flow *into* a control surface since $\mathbf{V}$ and $\mathbf{A}$ are in opposite directions, whereas $\rho \mathbf{V} \cdot \mathbf{A}$ is *positive* for mass flow *out* of a control surface, since then $\mathbf{V}$ and $\mathbf{A}$ are in the same direction.

The attachments on this cover plate must resist the change in the momentum of the flow of the water coming out of the opening and striking the plate.

EXAMPLE 6.1

The end of a pipe is capped with a reducer as shown in Fig. 6–2a. If the water pressure within the pipe at A is 200 kPa, determine the shear force that glue along the sides of the pipe exerts on the reducer to hold it in place.

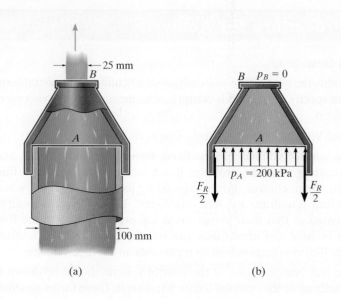

(a)   (b)

Fig. 6–2

SOLUTION

Fluid Description.   We have steady flow, and we will assume water is an ideal fluid, where $\rho_w = 1000\,\text{kg/m}^3$.

Control Volume and Free-Body Diagram.   We will select the control volume to represent the reducer along with the portion of water within it, Fig. 6–2a. The reason for this selection is to show or "expose" the required shear force F_R on the free-body diagram of the control volume, Fig. 6–2b. Also shown is the water pressure p_A, developed at the inlet control surface A. There is no pressure force at the outlet control surface since the (gage) pressure is $p_B = 0$. Excluded is the symmetric horizontal or normal force of the wall of the pipe on the reducer. We have also excluded the *weight* of the reducer, along with the weight of water within it, since here they can be considered negligible.

Continuity Equation. Before applying the momentum equation we must first obtain the velocity of the water at A and B. Applying the continuity equation for steady flow, we have

$$\frac{\partial}{\partial t}\int_{cv}\rho\,d\mathsf{V} + \int_{cs}\rho\mathbf{V}\cdot d\mathbf{A} = 0$$

$$0 - \rho V_A A_A + \rho V_B A_B = 0$$

$$-V_A\big[\pi(0.05\text{ m})^2\big] + V_B\big[\pi(0.0125\text{ m})^2\big] = 0$$

$$V_B = 16V_A \qquad\qquad (1)$$

Bernoulli Equation. Since the pressures at A and B are known, the velocities at these points can be determined by using Eq. 1 and applying the Bernoulli equation to points on a vertical streamline passing through A and B.* Neglecting the elevation difference from A to B, we have

$$\frac{p_A}{\gamma} + \frac{V_A^2}{2g} + z_A = \frac{p_B}{\gamma} + \frac{V_B^2}{2g} + z_B$$

$$\frac{200(10^3)\text{ N}/\text{m}^2}{(1000\text{ kg}/\text{m}^3)(9.81\text{ m}/\text{s}^2)} + \frac{V_A^2}{2(9.81\text{ m}/\text{s}^2)} + 0 = 0 + \frac{(16V_A)^2}{2(9.81\text{ m}/\text{s}^2)} + 0$$

$$V_A = 1.252\text{ m}/\text{s}$$

$$V_B = 16(1.252\text{ m}/\text{s}) = 20.04\text{ m}/\text{s}$$

Linear Momentum. To obtain F_R, we can now apply the momentum equation in the vertical direction.

$$\Sigma\mathbf{F} = \frac{\partial}{\partial t}\int_{cv}\mathbf{V}\rho\,d\mathsf{V} + \int_{cs}\mathbf{V}\rho\mathbf{V}\cdot d\mathbf{A}$$

Since the flow is steady, no local change in momentum occurs, and so the first term on the right is zero. Also, since the fluid is ideal, ρ_w is constant and average velocities can be used. Thus,

$$+\uparrow\Sigma F_y = 0 + V_B(\rho_w V_B A_B) + (V_A)(-\rho_w V_A A_A)$$

$$+\uparrow\Sigma F_y = \rho_w(V_B^2 A_B - V_A^2 A_A)$$

$$\big[200(10^3)\text{ N}/\text{m}^2\big]\big[\pi(0.05\text{ m})^2\big] - F_R = (1000\text{ kg}/\text{m}^3)\big[(20.04\text{ m}/\text{s})^2(\pi)(0.0125\text{ m})^2 - (1.252\text{ m}/\text{s})^2(\pi)(0.05\text{ m})^2\big]$$

$$F_R = 1.39\text{ kN} \qquad\qquad Ans.$$

This positive result indicates that the shear force acts downward on the reducer (control surface), as initially assumed.

*Actually a fitting such as this will generate frictional losses, and so the energy equation should be used. This will be discussed in Chapter 10.

EXAMPLE | 6.2

Water is discharged from the 50-mm diameter nozzle of a fire hose at 15 liters/s, Fig. 6–3a. The flow strikes the fixed surface such that 3/4 flows along B, while the remaining 1/4 flows along C. Determine the x and y components of force exerted on the surface.

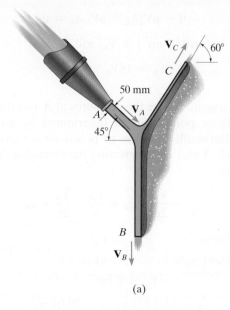

(a)

Fig. 6–3

SOLUTION

Fluid Description.   This is a case of steady flow. We consider the water as an ideal fluid, for which $\rho_w = 1000 \text{ kg/m}^3$.

Control Volume and Free-Body Diagram.   We will select a fixed control volume that contains the water from the nozzle and on the surface, Fig. 6–3a. The pressure distributed over the surface creates resultant horizontal and vertical reaction components, $\mathbf{F}_x$ and $\mathbf{F}_y$, on the closed control surface. We will neglect the weight of the water within the control volume. Also, because the (gage) pressure on all the open control surfaces is zero, no force is acting on these surfaces.

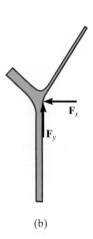

(b)

If we apply the Bernoulli equation to a point on the fluid stream along A, and another point on the fluid stream along B (or C), neglecting elevation changes between these points, and realizing that the (gage) pressure is zero, then $p/\gamma + V^2/2g + z = \text{const.}$ shows that the speed of the water onto and along the fixed surface is the same everywhere. That is, $V_A = V_B = V_C$. The surface only changes the direction of the velocity.

Linear Momentum. The velocity can be determined from the flow rate at A, because here the cross-sectional area is known.

$$Q_A = \left(\frac{15 \text{ liters}}{\text{s}}\right)\left(\frac{1000 \text{ cm}^3}{1 \text{ liter}}\right)\left(\frac{1 \text{ m}}{100 \text{ cm}}\right)^3 = 0.015 \text{ m}^3/\text{s}$$

$$V_A = V_B = V_C = \frac{Q_A}{A_A} = \frac{0.015 \text{ m}^3/\text{s}}{\pi(0.025 \text{ m})^2} = \frac{24}{\pi} \text{ m/s}$$

For this case of steady flow, the linear momentum equation becomes

$$\Sigma \mathbf{F} = \frac{\partial}{\partial t} \int_{cv} \mathbf{V} \rho \, d\mathbb{V} + \int_{cs} \mathbf{V} \rho \mathbf{V} \cdot d\mathbf{A}$$

$$\Sigma \mathbf{F} = 0 + \mathbf{V}_B(\rho V_B A_B) + \mathbf{V}_C(\rho V_C A_C) + \mathbf{V}_A(-\rho V_A A_A)$$

Notice that the last term is negative because the flow is *into* the control volume. Because $Q = VA$, then

$$\Sigma \mathbf{F} = \rho(Q_B \mathbf{V}_B + Q_C \mathbf{V}_C - Q_A \mathbf{V}_A)$$

Now when we resolve this equation in the x and y directions, only the velocities, $\mathbf{V}$, will have *components*. The flows, $Q = VA$, are *scalar quantities*. Thus,

$$\xrightarrow{+} \Sigma F_x = \rho(Q_B V_{Bx} + Q_C V_{Cx} - Q_A V_{Ax})$$

$$-F_x = (1000 \text{ kg/m}^3)\left\{0 + \frac{1}{4}(0.015 \text{ m}^3/\text{s})\left[\left(\frac{24}{\pi} \text{ m/s}\right)\cos 60°\right]\right.$$
$$\left. - (0.015 \text{ m}^3/\text{s})\left[\left(\frac{24}{\pi} \text{ m/s}\right)\cos 45°\right]\right\}$$

$$F_x = 66.70 \text{ N} = 66.7 \text{ N} \leftarrow \qquad\qquad Ans.$$

$$+\uparrow \Sigma F_y = \rho\left[Q_B(-V_{By}) + Q_C V_{Cy} - Q_A(-V_{Ay})\right]$$

$$F_y = (1000 \text{ kg/m}^3)\left\{\frac{3}{4}(0.015 \text{ m}^3/\text{s})\left(-\frac{24}{\pi} \text{ m/s}\right) + \frac{1}{4}(0.015 \text{ m}^3/\text{s})\left[\left(\frac{24}{\pi} \text{ m/s}\right)\sin 60°\right]\right.$$
$$\left. - (0.015 \text{ m}^3/\text{s})\left[\left(-\frac{24}{\pi} \text{ m/s}\right)\sin 45°\right]\right\}$$

$$F_y = 19.89 \text{ N} = 19.9 \text{ N} \uparrow \qquad\qquad Ans.$$

These results produce a resultant force of 69.6 N that acts on the water. The equal but opposite *dynamic force* acts on the surface. Note that if the nozzle is directed *perpendicular* to a flat surface, a similar calculation shows that the force is 115 N. This can become quite dangerous if the stream is directed at someone!

EXAMPLE | 6.3

Air flows through the duct in Fig. 6–4a such that at A it has a temperature of 30°C and an absolute pressure of 300 kPa, whereas due to cooling, at B it has a temperature of 10°C and an absolute pressure of 298.5 kPa. If the average velocity of the air at A is 3 m/s, determine the resultant frictional force acting along the walls of the duct between these two locations.

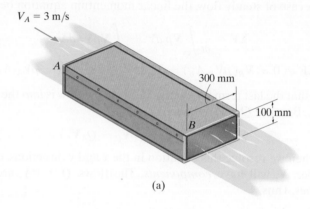

(a)

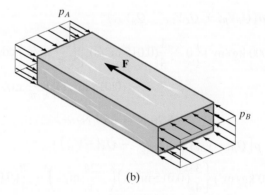

(b)

Fig. 6–4

SOLUTION

Fluid Description. Here, because of a change in density, we have steady flow of a viscous, compressible fluid. We consider the air to be an ideal gas at the temperatures and pressures considered.

Control Volume and Free-Body Diagram. The air within the duct will be taken as the fixed control volume, Fig. 6–4a. The friction force **F** that acts on the control surface along the walls of the duct is the result of the viscous effect of the air, Fig. 6–4b. The weight of the air can be neglected, but the pressure forces at A and B must be considered. (Pressure forces on the closed or side control surfaces are not shown since they produce a zero resultant, and they act perpendicular to the flow.) To obtain $\mathbf{F}$ we must apply the momentum equation, but first we must determine the density of the air at A and B, and the average velocity at B.

Ideal Gas Law. From Appendix A, $R = 286.9\,\text{J}/(\text{kg} \cdot \text{K})$ so that

$$p_A = \rho_A R T_A$$

$$300(10^3)\,\text{N/m}^2 = \rho_A\left[286.9\,\text{J}/(\text{kg} \cdot \text{K})\right](30°\text{C} + 273)\,\text{K}$$

$$\rho_A = 3.451\,\text{kg/m}^3$$

$$p_B = \rho_B R T_B$$

$$298.5(10^3)\,\text{N/m}^2 = \rho_B\left[286.9\,\text{J}/(\text{kg} \cdot \text{K})\right](10°\text{C} + 273)\,\text{K}$$

$$\rho_B = 3.676\,\text{kg/m}^3$$

Continuity Equation. Although the velocity profile is affected by the frictional effects of viscosity, primarily along the inside *surface* of the duct, here we will use average velocities.* We can determine this (average) velocity of the air at B by applying the continuity equation. For steady flow, we have

$$\frac{\partial}{\partial t}\int_{cv} \rho\,d\forall + \int_{cs} \rho\mathbf{V} \cdot d\mathbf{A} = 0$$

$$0 - \rho_A V_A A_A + \rho_B V_B A_B = 0$$

$$0 - (3.451\,\text{kg/m}^3)(3\,\text{m/s})(0.3\,\text{m})(0.1\,\text{m}) + (3.676\,\text{kg/m}^3)(V_B)(0.3\,\text{m})(0.1\,\text{m}) = 0$$

$$V_B = 2.816\,\text{m/s}$$

Linear Momentum. Because steady compressible flow occurs,

$$\Sigma\mathbf{F} = \frac{\partial}{\partial t}\int_{cv} \mathbf{V}\rho\,d\forall + \int_{cs} \mathbf{V}\rho\mathbf{V} \cdot d\mathbf{A}$$

$$\xrightarrow{+} \Sigma F = 0 + V_B(\rho_B V_B A_B) + (V_A)(-\rho_A V_A A_A)$$

$$\left[300(10^3)\,\text{N/m}^2\right](0.3\,\text{m})(0.1\,\text{m}) - \left[298.5(10^3)\,\text{N/m}^2\right](0.3\,\text{m})(0.1\,\text{m}) - F$$
$$= 0 + (2.816\,\text{m/s})\left[(3.676\,\text{kg/m}^3)(2.816\,\text{m/s})(0.3\,\text{m})(0.1\,\text{m})\right]$$
$$- (3\,\text{m/s})\left[(3.451\,\text{kg/m}^3)(3\,\text{m/s})(0.3\,\text{m})(0.1\,\text{m})\right]$$

$$F = 45.1\,\text{N} \qquad\qquad Ans.$$

EXAMPLE 6.4

When the sluice gate G in Fig. 6–5a is in the open position as shown, water flows out from under it at a depth of 1 m. If the gate is 3 m wide, determine the resultant horizontal force that must be applied by its supports to hold the gate in place. Assume the channel behind the gate maintains a constant depth of 5 m.

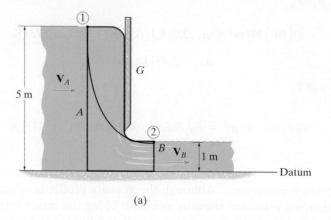

(a)

Fig. 6–5

6

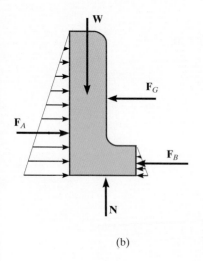

(b)

SOLUTION

Fluid Description. Since the channel depth is assumed constant, the flow will be steady. We will also assume water to be an ideal fluid so that it will flow with an average velocity through the gate. Take $\rho_w = 1000 \text{ kg/m}^3$.

Control Volume and Free-Body Diagram. To determine the force on the gate, the fixed control volume will include a control surface along the face of the gate, and a volume of water on each side of the gate, Fig. 6–5a. There are three forces acting on this control volume in the horizontal direction as shown on the free-body diagram, Fig. 6–5b. They are the unknown pressure force resultant of the gate F_G, and the two hydrostatic pressure force resultants from the water on the control surfaces at A and B. The viscous frictional forces acting on the closed control surfaces on the gate and on the ground are neglected since we have assumed the water to be inviscid. (Actually, this force will be comparatively small.)

Bernoulli and Continuity Equations. We can determine the average velocities at A and B by applying the Bernoulli equation (or the energy equation) and the continuity equation, Fig. 6–5a. When a streamline is chosen through points 1 and 2, the Bernoulli equation gives*

$$\frac{p_1}{\gamma} + \frac{V_1^2}{2g} + z_1 = \frac{p_2}{\gamma} + \frac{V_2^2}{2g} + z_2$$

$$0 + \frac{V_A^2}{2(9.81 \text{ m/s}^2)} + 5\text{ m} = 0 + \frac{V_B^2}{2(9.81 \text{ m/s}^2)} + 1\text{ m}$$

$$V_B^2 - V_A^2 = 78.48 \tag{1}$$

For continuity,

$$\frac{\partial}{\partial t} \int_{cv} \rho \, d\mathcal{V} + \int_{cs} \rho \mathbf{V} \cdot d\mathbf{A} = 0$$

$$0 - V_A(5\text{ m})(3\text{ m}) + V_B(1\text{ m})(3\text{ m}) = 0$$

$$V_B = 5V_A \tag{2}$$

Solving Eqs. 1 and 2 yields

$$V_A = 1.808 \text{ m/s} \quad \text{and} \quad V_B = 9.042 \text{ m/s}$$

Linear Momentum

$$\Sigma \mathbf{F} = \frac{\partial}{\partial t} \int_{cv} \mathbf{V} \rho \, d\mathcal{V} + \int_{cs} \mathbf{V} \rho \mathbf{V} \cdot d\mathbf{A}$$

$$\overset{+}{\rightarrow} \Sigma F_x = 0 + V_B(\rho V_B A_B) + (V_A)(-\rho V_A A_A)$$

$$= 0 + \rho (V_B^2 A_B - V_A^2 A_A)$$

Referring to the free-body diagram, Fig. 6–5b,

$$\tfrac{1}{2}[(1000 \text{ kg/m})(9.81 \text{ m/s}^2)(5 \text{ m})](3 \text{ m})(5 \text{ m})$$

$$- \tfrac{1}{2}[(1000 \text{ kg/m}^2)(9.81 \text{ m/s}^2)(1 \text{ m})](3 \text{ m})(1 \text{ m}) - F_G$$

$$= (1000 \text{ kg/m}^3)[(9.042 \text{ m/s})^2 (3 \text{ m}) (1 \text{ m}) - (1.808 \text{ m/s})^2 (3 \text{ m}) (5 \text{ m})]$$

$$F_G = 156.96 (10^3) \text{ N} = 157 \text{ kN} \qquad \textit{Ans.}$$

As a point of comparison, the hydrostatic force on the gate is

$$(F_G)_{st} = \tfrac{1}{2}(\rho g h b)h = \tfrac{1}{2}[(1000 \text{ kg/m}^3)(9.81 \text{ m/s}^2)(5 \text{ m} - 1 \text{ m})(3 \text{ m})](5 \text{ m} - 1 \text{ m})$$

$$= 235.44(10^3) \text{ N} = 235 \text{ kN}$$

*All the particles on the surface will eventually pass under the gate. Here we have chosen the particle at point 1 that happens to wind up at point 2.

EXAMPLE | 6.5

Oil flows through the open pipe AB in Fig. 6–6a such that at the instant shown, it has a velocity at A of 6 m/s, which is increasing at 0.9 m/s². Determine the pump pressure at B that creates this flow. Take $\rho_o = 900$ kg/m³.

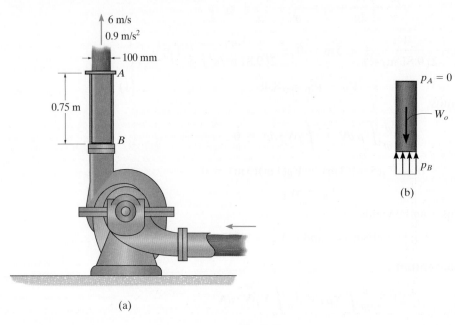

(a)

(b)

Fig. 6–6

SOLUTION

Fluid Description. Because of the acceleration, we have a case of unsteady flow. We will assume the oil is an ideal fluid.

Control Volume and Free-Body Diagram. Here we consider a fixed control volume that contains the oil within the vertical section AB of the pipe, Fig. 6–6a. The forces shown on its free-body diagram are the weight of oil within the control volume, $W_o = \rho_o g \forall_o$, and the pressure at B, Fig. 6–6b.* The (gage) pressure at A is zero.

---

*The horizontal pressure caused by the sides of the pipe on the control volume surface is not included here since it will produce a zero resultant force. Also, friction along the sides is excluded because the oil is assumed to be inviscid.

Linear Momentum. Because an ideal fluid produces average velocities, the momentum equation becomes

$$\Sigma \mathbf{F} = \frac{\partial}{\partial t} \int_{cv} \mathbf{V} \rho \, d\mathbb{V} + \int_{cs} \mathbf{V} \rho \mathbf{V} \cdot d\mathbf{A}$$

$$+\uparrow \Sigma F_y = \frac{\partial}{\partial t} \int_{cv} V \rho \, d\mathbb{V} + V_A(\rho V_A A_A) + (V_B)(-\rho V_B A_B)$$

Since ρ is constant and $A_A = A_B$, the continuity equation requires $V_A = V_B = V = 6$ m/s. As a result, the last two terms will cancel each other. In other words, the flow is uniform, and so there is no net convective effect.

The *unsteady flow* term (local effect) indicates that the momentum changes within the control volume due to the time rate of change in the flow velocity (unsteady flow). Since both this rate of change, 0.9 m/s^2, and the density are constant throughout the control volume, the above equation becomes

$$+\uparrow \Sigma F_y = \frac{dV}{dt} \rho \mathbb{V}; \qquad p_B A_B - \rho_o g \mathbb{V}_o = \frac{dV}{dt} \rho_o \mathbb{V}_o \qquad (1)$$

$$p_B[\pi(0.05 \text{ m})^2] - (900 \text{ kg/m}^3)(9.81 \text{ m/s}^2)[\pi(0.05 \text{ m})^2(0.75 \text{ m})]$$

$$= (0.9 \text{ m/s}^2)(900 \text{ kg/m}^3)[\pi(0.05 \text{ m})^2(0.75 \text{ m})]$$

$$p_B = 7.229(10^3) \text{ Pa} = 7.23 \text{ kPa} \qquad \qquad Ans.$$

Notice that Eq. 1 is actually the application of $\Sigma F_y = ma_y$, Fig. 6–6b.

6.3   Applications to Bodies Having Constant Velocity

For some problems, a blade or vane may be moving with constant velocity, and when this occurs, the forces on the blade can be obtained by selecting the control volume so that it *moves with the body*. Provided this is the case, the velocity and the mass flow terms in the momentum equation are then measured relative to each control surface, so that $V = V_{f/cs}$, and therefore Eq. 6–1 becomes

$$\Sigma \mathbf{F} = \frac{\partial}{\partial t} \int_{cv} \mathbf{V}_{cv} \rho \, d\mathbb{V} + \int_{cs} \mathbf{V}_{f/cs} \rho \mathbf{V}_{f/cs} \cdot d\mathbf{A}$$

With this equation, the solution is simplified, because the flow will appear to be *steady flow* relative to the control volume, and the first term on the right side will be zero.

Using the procedure for analysis outlined in Sec. 6.2, the following examples illustrate application of the momentum equation for a body moving with constant velocity. In Sec. 6.5 we will consider application of this equation to propellers and wind turbines.

EXAMPLE	6.6

The truck in Fig. 6–7a is moving to the left at 5 m/s into a 50-mm-diameter stream of water, which has a discharge of 8 liter/s. Determine the dynamic force the stream exerts on the truck if it is deflected off the windshield as shown.

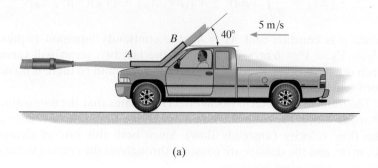

(a)

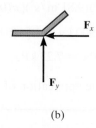

(b)

Fig. 6–7

SOLUTION

Fluid Description. In this problem the driver will observe *steady flow*, and so we will fix the x, y inertial coordinate system to the truck, so that it moves with constant velocity. We will assume water to be an ideal fluid, and so average velocities can be used because friction is negligible. $\rho = 1000 \text{ kg/m}^3$.

Control Volume and Free-Body Diagram. We will consider the moving control volume to include the portion AB of the water stream that is in contact with the truck, Fig. 6–7a. As shown on its free-body diagram, Fig. 6–7b, only the horizontal and vertical components of force *caused by the truck* on the control volume will be considered significant. (The pressure forces at the open control surfaces are atmospheric and the weight of the water is neglected.)

The nozzle velocity of the stream is determined first.

$$Q = VA; \quad \left(\frac{8 \text{ liter}}{\text{s}}\right)\left(\frac{10^{-3} \text{ m}^3}{1 \text{ liter}}\right) = V\left[\pi(0.025 \text{ m})^2\right] \quad V = 4.074 \text{ m/s}$$

Relative to the control volume (or driver), the velocity of the water at A is therefore

$$\overset{+}{\rightarrow}(V_{f/cs})_A = V_f - V_{cv}$$

$$(V_{f/cs})_A = 4.074 \text{ m/s} - (-5 \text{ m/s}) = 9.074 \text{ m/s}$$

Application of the Bernoulli equation will show that this relative average speed is maintained as the water flows off the windshield at B (neglecting the effect of elevation). Also, the size of the cross-sectional area at B (open control surface) must remain the same as at A to maintain continuity, $V_A A_A = V_B A_B = VA$, although its shape will certainly change.

Linear Momentum.   For steady incompressible flow we have

$$\Sigma \mathbf{F} = \frac{\partial}{\partial t} \int_{cv} \mathbf{V}_{f/cv} \rho \, d\Psi + \int_{cs} \mathbf{V}_{f/cs} \, \rho \mathbf{V}_{f/cs} \cdot d\mathbf{A}$$

$$\Sigma \mathbf{F} = 0 + (\mathbf{V}_{f/cs})_B \big[\rho(V_{f/cs})_B A_B\big] + (\mathbf{V}_{f/cs})_A \big[-\rho(V_{f/cs})_A A_A\big]$$

Carefully note that here we only must consider the components of the velocity $\mathbf{V}_{f/cs}$, whereas the mass flow terms $\rho V_{f/cs} A$ are scalars. Applying this equation in the x and y directions yields

$$\overset{+}{\rightarrow} \Sigma F_x = 0 + \big[(V_{f/cs})_B \cos 40°\big]\big[\rho \, (V_{f/cs})_B A_B\big] - (V_{f/cs})_A \big[\rho \, (V_{f/cs})_A A_A\big]$$

$$-F_x = \big[(9.074 \text{ m/s}) \cos 40°\big]\big[\big(1000 \text{ kg/m}^3\big)(9.074 \text{ m/s})\big[\pi(0.025 \text{ m})^2\big]\big]$$

$$- (9.074 \text{ m/s})\big[\big(1000 \text{ kg/m}^3\big)(9.074 \text{ m/s})\big[\pi(0.025 \text{ m})^2\big]\big]$$

$$F_x = 37.83 \text{ N}$$

$$+\uparrow \Sigma F_y = \big[(V_{f/cs})_B \sin 40°\big]\big[\rho \, (V_{f/cs})_B A_B\big] - 0$$

$$F_y = \big[(9.074 \text{ m/s}) \sin 40°\big]\big[(1000 \text{ kg/m}^3)(9.074 \text{ m/s})\big[\pi(0.025 \text{ m})^2\big]\big] - 0$$

$$= 103.9 \text{ N}$$

Thus,

$$F = \sqrt{(37.83 \text{ N})^2 + (103.9 \text{ N})^2} = 111 \text{ N} \qquad \qquad Ans.$$

This (dynamic) force also acts on the truck, but in the opposite direction.

EXAMPLE   6.7

The jet of water having a cross-sectional area of $2(10^{-3})$ m^2 and a velocity of 45 m/s strikes the vane of a turbine, causing it to move at 20 m/s, Fig. 6–8a. Determine the dynamic force of the water on the vane, and the power output caused by the water.

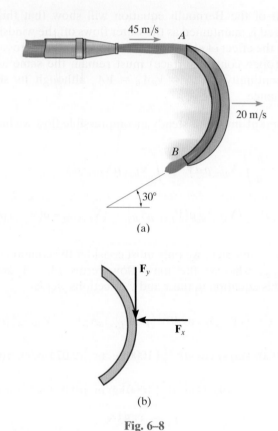

(a)

(b)

Fig. 6–8

SOLUTION

Fluid Description.   To observe steady flow, the x, y reference is fixed to the vane. It can be considered as an inertial coordinate system since it moves with constant velocity. We assume the water is an ideal fluid, where $\rho = 1000$ kg/m^3.

Control Volume and Free-Body Diagram.   We will take the moving control volume to contain the water on the vane from A to B, Fig. 6–8a. As shown on its free-body diagram, Fig. 6–8b, the force components of the vane on the control volume are denoted as $\mathbf{F}_x$ and $\mathbf{F}_y$. The weight of the water is neglected, and the pressure forces on the open control surfaces are atmospheric.

Linear Momentum.   The velocity of the jet relative to the control surface at A is

$$\overset{+}{\rightarrow} V_{f/cs} = V_f - V_{cv}$$

$$V_{f/cs} = 45 \text{ m/s} - 20 \text{ m/s} = 25 \text{ m/s}$$

With negligible elevation change, the Bernoulli equation will show that water leaves the vane at B with this same speed, and so to satisfy continuity, $A_A = A_B = A$. For steady ideal fluid flow, the momentum equation becomes

$$\Sigma \mathbf{F} = \frac{\partial}{\partial t} \int_{cv} \mathbf{V}_{f/cv}\, \rho\, d\mathbb{V} + \int_{cs} \mathbf{V}_{f/cs}\, \rho \mathbf{V}_{f/cs} \cdot d\mathbf{A}$$

$$\Sigma \mathbf{F} = 0 + (\mathbf{V}_{f/cs})_B \left[\rho (V_{f/cs})_B A_B \right] + (\mathbf{V}_{f/cs})_A \left[-\rho (V_{f/cs})_A A_A \right]$$

As in the previous example, the velocity $\mathbf{V}_{f/cs}$ has *components* in this equation. The mass flow terms $\rho V_{f/cs} A$ are scalars. Therefore,

$$\overset{+}{\rightarrow} \Sigma F_x = \left[-(V_{f/cs})_B \cos 30° \right] \left[\rho (V_{f/cs})_B A_B \right] + (V_{f/cs})_A \left[-\rho (V_{f/cs})_A A_A \right]$$

$$-F_x = \left[-(25 \text{ m/s}) \cos 30° \right] \left[(1000 \text{ kg/m}^3)(25 \text{ m/s})(2(10^{-3}) \text{ m}^2) \right]$$

$$-(25 \text{ m/s}) \left[-(1000 \text{ kg/m}^3)(25 \text{ m/s})(2(10^{-3}) \text{ m}^2) \right]$$

$$F_x = 2333 \text{ N}$$

$$+\uparrow \Sigma F_y = \left[-(V_{f/cs})_B \sin 30° \right] \left[\rho (V_{f/cs})_B A_B \right] - 0$$

$$-F_y = \left[-(25 \text{ m/s}) \sin 30° \right] \left[(1000 \text{ kg/m}^3)(25 \text{ m/s})(2(10^{-3}) \text{ m}^2) \right]$$

$$F_y = 625 \text{ N}$$

$$F = \sqrt{(2333 \text{ N})^2 + (625 \text{ N})^2} = 2.41 \text{ kN} \qquad\qquad Ans.$$

The equal but opposite force acts on the vane. This represents the dynamic force.

Power.   By definition, power is work per unit time, or the product of a force and the parallel component of velocity. Here only $\mathbf{F}_x$ produces power since $\mathbf{F}_y$ does not displace up or down, and hence does no work. Since the vane is moving at 20 m/s,

$$\dot{W} = \mathbf{F} \cdot \mathbf{V}; \qquad \dot{W} = (2333 \text{ N})(20 \text{ m/s}) = 46.7 \text{ kW} \qquad\qquad Ans.$$

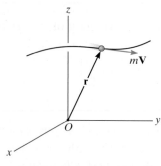

Angular momentum $\mathbf{r} \times m\mathbf{V}$

Fig. 6–9

6.4   The Angular Momentum Equation

The ***angular momentum*** of a particle is the *moment* of the particle's linear momentum $m\mathbf{V}$ about a point or axis. Using the vector cross product to define this moment, the angular momentum becomes $\mathbf{r} \times m\mathbf{V}$, where $\mathbf{r}$ is the position vector extending from the moment point O to the particle, Fig. 6–9. If we perform the cross product operation with $\mathbf{r}$ and Newton's second law of motion, $\mathbf{F} = m\mathbf{a} = d(m\mathbf{V})/dt$, then it follows that the sum of the moments of the external forces acting on the system, $\Sigma(\mathbf{r} \times \mathbf{F})$, is equal to the time rate of change of the system's angular momentum, i.e., $\Sigma\mathbf{M} = \Sigma(\mathbf{r} \times \mathbf{F}) = d(\mathbf{r} \times m\mathbf{V})/dt$. For a Eulerian description, we can obtain the time derivative of $\mathbf{r} \times m\mathbf{V}$, using the Reynolds transport theorem, Eq. 4–11, where $\mathbf{N} = \mathbf{r} \times m\mathbf{V}$, so that $\boldsymbol{\eta} = \mathbf{r} \times \mathbf{V}$. We have

$$\left(\frac{d\mathbf{N}}{dt}\right)_{syst} = \frac{\partial}{\partial t}\int_{cv} \boldsymbol{\eta}\rho \, d\forall + \int_{cs} \boldsymbol{\eta}\rho \, \mathbf{V} \cdot d\mathbf{A}$$

$$\frac{d}{dt}\left(\mathbf{r} \times m\mathbf{V}\right) = \frac{\partial}{\partial t}\int_{cv} (\mathbf{r} \times \mathbf{V})\rho \, d\forall + \int_{cs} (\mathbf{r} \times \mathbf{V})\rho\mathbf{V} \cdot d\mathbf{A}$$

Therefore, the angular momentum equation becomes

$$\boxed{\Sigma\mathbf{M}_O = \frac{\partial}{\partial t}\int_{cv} (\mathbf{r} \times \mathbf{V})\rho \, d\forall + \int_{cs} (\mathbf{r} \times \mathbf{V})\,\rho\mathbf{V} \cdot d\mathbf{A}} \qquad (6\text{–}4)$$

Steady Flow. If we consider steady flow, the first term on the right will be zero. Also, if the average velocity through the open control surfaces is used, and the fluid density remains constant, as in the case of an ideal fluid, then the second term on the right can be integrated, and we obtain

$$\Sigma \mathbf{M} = \Sigma (\mathbf{r} \times \mathbf{V}) \rho \mathbf{V} \cdot \mathbf{A} \qquad (6\text{--}5)$$

Steady flow

This final result is often used to determine the torque on the shaft of a pump or turbine, as will be shown in Example 5.10 and in Chapter 14. It can also be used for finding reactive forces or couple moments on static structures subjected to steady flow.

Procedure for Analysis

Application of the angular momentum equation follows the same procedure as that for linear momentum.

Fluid Description

Define the type of flow, whether it is steady or unsteady and uniform or nonuniform. Also define the type of fluid, whether it is viscous, compressive, or if it can be assumed to be an ideal fluid. For an ideal fluid, the velocity profile will be uniform, and the density will be constant.

Control Volume and Free-Body Diagram

Select the control volume so its free-body diagram includes the unknown forces and couple moments that are to be determined. The forces on the free-body diagram include the weight of the fluid and any solid portion of the body, the pressure forces at the open control surfaces, and the components of the resultant pressure and frictional shear forces acting on closed control surfaces.

Angular Momentum

If the flow is known, the average velocities through the open control surfaces can be determined using $V = Q/A$, or by applying the continuity equation, the Bernoulli equation, or the energy equation. Set up the inertial x, y, z coordinate axes, and apply the angular momentum equation about an axis so that a selected unknown force or moment can be obtained.

The impulse of water flowing onto and off the blades of this waterwheel causes the wheel to turn.

EXAMPLE 6.8

Water flows out of the fire hydrant in Fig. 6–10a at 120 liters/s. Determine the reactions at the fixed support necessary to hold the fire hydrant in place.

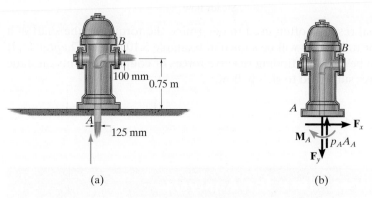

(a)   (b)

Fig. 6–10

SOLUTION

Fluid Description. This is a case of steady flow. The water will be assumed to be an ideal fluid, where $\rho_W = 1000 \text{ kg/m}^3$.

Control Volume and Free-Body Diagram. We will consider the entire fire hydrant and water contained within it as a fixed control volume. Since the support at A is fixed, three reactions act on its free-body diagram, Fig. 6–10b. Also, the pressure force $p_A A_A$ acts on the open control surface at A. Since atmospheric pressure exists at B, there is no pressure force at B. Here we will neglect the weight of the fire hydrant and the water within it.

Bernoulli Equation. We must first determine the pressure at A. Here

$$Q = \left(120 \frac{\text{liters}}{\text{s}}\right)\left(\frac{1000 \text{ cm}^3}{1 \text{ liter}}\right)\left(\frac{1 \text{ m}}{100 \text{ cm}}\right)^3 = 0.12 \text{ m}^3/\text{s}$$

Then, the velocities at A and B are

$$Q = V_A A_A; \quad \left(0.12 \text{ m}^3/\text{s}\right) = V_A \left[\pi(0.0625 \text{ m})^2\right]; \qquad V_A = \frac{30.72}{\pi} \text{ m/s}$$

$$Q = V_B A_B; \quad \left(0.12 \text{ m}^3/\text{s}\right) = V_B \left[\pi(0.05 \text{ m})^2\right]; \qquad V_B = \frac{48}{\pi} \text{ m/s}$$

Thus, with the datum at A,

$$\frac{p_A}{\gamma} + \frac{V_A^2}{2g} + z_A = \frac{p_B}{\gamma} + \frac{V_B^2}{2g} + z_B$$

$$\frac{p_A}{(1000 \text{ kg/m}^3)(9.81 \text{ m/s}^2)} + \frac{\left(\dfrac{30.72}{\pi} \text{ m/s}\right)^2}{2(9.81 \text{ m/s}^2)} + 0 = 0 + \frac{\left(\dfrac{48}{\pi} \text{ m/s}\right)^2}{2(9.81 \text{ m/s}^2)} + 0.75 \text{ m}$$

$$p_A = 76.27(10^3) \text{ Pa}$$

Linear and Angular Momentum.   The force reactions at the support are obtained from the linear momentum equation. For steady flow,

$$\Sigma \mathbf{F} = \frac{\partial}{\partial t} \int_{cv} \mathbf{V}\rho \, d\forall + \int_{cs} \mathbf{V} \, \rho \mathbf{V} \cdot d\mathbf{A}$$

$$\Sigma \mathbf{F} = 0 + \mathbf{V}_B(\rho V_B A_B) + \mathbf{V}_A(-\rho V_A A_A)$$

Considering the x and y components of the velocities, we have

$$\xrightarrow{+} \Sigma F_x = V_{Bx}(\rho V_B A_B) + 0$$

$$F_x = \left(\frac{48}{\pi} \text{ m/s}\right)(1000 \text{ kg/m}^3)\left(\frac{48}{\pi} \text{ m/s}\right)[\pi(0.05 \text{ m})^2]$$

$$F_x = 1833.46 \text{ N} = 1.83 \text{ kN} \qquad\qquad Ans.$$

$$+\uparrow \Sigma F_y = 0 + (V_{Ay})(-\rho V_A A_A)$$

$$[76.27(10^3) \text{ N/m}^2][\pi(0.0625 \text{ m})^2] - F_y = \left(\frac{30.72}{\pi} \text{ m/s}\right)\left\{-(1000 \text{ kg/m}^3)\left(\frac{30.72}{\pi} \text{ m/s}\right)[\pi(0.0625 \text{ m})^2]\right\}$$

$$F_y = 2109.39 \text{ N} = 2.11 \text{ kN} \qquad\qquad Ans.$$

We will apply the angular momentum equation about point A, in order to eliminate the force reactions at this point.

$$\Sigma M_A = \frac{\partial}{\partial t} \int_{cv} (\mathbf{r} \times \mathbf{V})\rho \, d\forall + \int_{cs} (\mathbf{r} \times \mathbf{V}) \, \rho \mathbf{V} \cdot d\mathbf{A}$$

$$\zeta + \Sigma M_A = 0 + (rV_B)(\rho V_B A_B)$$

Here the vector cross product has been evaluated as a scalar moment of $\mathbf{V}_B$ about point A. Therefore,

$$M_A = \left[(0.75 \text{ m})\left(\frac{48}{\pi} \text{ m/s}\right)\right]\left\{(1000 \text{ kg/m}^3)\left(\frac{48}{\pi} \text{ m/s}\right)[\pi(0.05 \text{ m})^2]\right\}$$

$$= 1375.10 \text{ N} \cdot \text{m} = 1.38 \text{ kN} \cdot \text{m} \qquad\qquad Ans.$$

EXAMPLE | 6.9

The arm of the sprinkler in Fig. 6–11a rotates at a constant rate of $\omega = 100$ rev/min. This motion is caused by water that enters the base at 3 liter/s and exits each of the two 20-mm-diameter nozzles. Determine the frictional torque on the shaft of the arm that keeps the rotation rate constant.

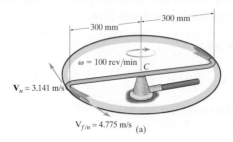

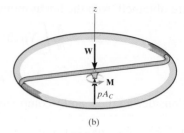

Fig. 6–11

SOLUTION

Fluid Description.   As the arm rotates, the flow will be quasi steady, that is, it is cyclic and repetitive. In other words, *on the average* the flow can be considered steady. We assume the water to be an ideal fluid, where $\rho = 1000$ kg/m³.

Control Volume and Free-Body Diagram.   Here we will choose a control volume that is *fixed* and contains the moving arm and the water within it, Fig. 6–11a.* As shown on the free-body diagram, Fig. 6–11b, the forces acting on this control volume include the weight **W** of the arm and the water, the pressure force $p_C A_C$ from the water supply, and the frictional torque **M** on the base of the arm at the axle. The fluid exiting the nozzles is at atmospheric pressure, and so there are no pressure forces here.

---

*If we chose a *rotating* control volume consisting of the arms and the fluid within it, then the coordinate system, which rotates with it, would *not* be an inertial reference. Its rotation would complicate the analysis since it produces additional acceleration terms that must be accounted for in the momentum analysis. See Ref. [2].

Velocity.   Due to symmetry, the discharge through *each nozzle* is equal to half of the flow. Thus, the velocity of the water passing through the fixed opened control surfaces is

$$Q = VA; \qquad \left(\frac{1}{2}\right)\left[\left(\frac{3 \text{ liter}}{\text{s}}\right)\left(\frac{10^{-3}\,\text{m}^3}{1 \text{ liter}}\right)\right] = V_{f/n}\left[\pi(0.01\,\text{m})^2\right]$$

$$V_{f/n} = (V_B)_{\text{rel}} = (V_A)_{\text{rel}} = 4.775\,\text{m/s}$$

The rotation of the arms causes the nozzle to have a velocity of

$$V_n = \omega r = \left(100\,\frac{\text{rev}}{\text{min}}\right)\left(\frac{2\pi\,\text{rad}}{\text{rev}}\right)\left(\frac{1\,\text{min}}{60\,\text{s}}\right)(0.3\,\text{m}) = 3.141\,\text{m/s}$$

Therefore, the tangential exit velocity of the fluid from the opened control surfaces as seen by a fixed observer looking down on the *fixed control volume*, Fig. 6–11a, is

$$V_f = -V_n + V_{f/n} \qquad\qquad (1)$$

$$V_f = -3.141\text{ m/s} + 4.775\text{ m/s} = 1.633\text{ m/s}$$

Angular Momentum.   If we apply the angular momentum equation about the z axis, then no angular momentum occurs at the control inlet surface C since the velocity of the flow is directed along the z axis. The (tangential) velocity V_f of the fluid stream as it exits the control surfaces will, however, produce angular momentum about the z axis. The vector cross products for the two moments of $\mathbf{V}_f$ can be written in terms of the scalar moment of these velocities. Since they are equal, and the flow is steady, we have

$$\Sigma \mathbf{M} = \frac{\partial}{\partial t}\int_{\text{cv}}(\mathbf{r}\times\mathbf{V})\rho\,d\forall + \int_{\text{cs}}(\mathbf{r}\times\mathbf{V})\rho\mathbf{V}_{f/cs}\cdot d\mathbf{A}$$

$$\Sigma M_z = 0 + 2r_A V_f(\rho V_{f/n}A_n) \qquad\qquad (2)$$

$$M = 2(0.3\,\text{m})(1.633\,\text{m/s})\left\{(1000\text{ kg/m}^3)(4.775\text{ m/s})\left[\pi(0.01\,\text{m})^2\right]\right\}$$

$$M = 1.47\text{ N}\cdot\text{m} \qquad\qquad \textit{Ans.}$$

It is interesting to note that if this frictional torque on the shaft were equal to zero, there would actually be an *upper limit* to the rotation ω of the arm. To determine this, $V_n = \omega(0.3\text{ m})$ and so Eq. 1 then becomes

$$V_f = -\omega(0.3\,\text{m}) + 4.775\text{ m/s}$$

Substituting this result into Eq. 2 yields

$$0 = 0 + 2(0.3\,\text{m})\left[-\omega\,(0.3\,\text{m}) + 4.775\,\text{m/s}\right]\left\{(1000\text{ kg/m}^3)(4.775\text{ m/s})\left[\pi(0.01\,\text{m})^2\right]\right\}$$

$$\omega = 15.92\,\text{rad/s} = 152\text{ rev/min}$$

EXAMPLE 6.10

The axial-flow pump in Fig. 6–12a has an impeller with blades having a mean radius of $r_m = 80$ mm. Determine the average torque **T** that must be applied to the impeller, in order to create a flow of water through the pump of 0.1 m³/s while the impeller rotates at $\omega = 120$ rad/s. The open cross-sectional area through the impeller is 0.025 m². The flow of water is delivered onto each blade along the axis of the pump, and it exits the blade with a *tangential component* of velocity of 5 m/s, as shown in Fig. 6–12b.

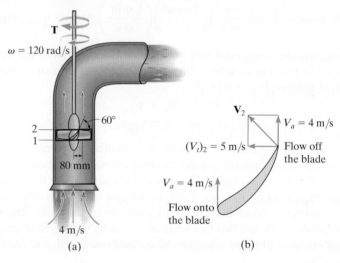

Fig. 6–12

SOLUTION

Fluid Description. Flow through the pump is cyclic unsteady flow, but on a time average basis it may be considered mean steady flow. Here we will consider water as an ideal fluid, where $\rho = 1000$ kg/m³.

Since we are looking for the *torque* developed by the pump on the shaft, we must apply the angular momentum equation.

Control Volume and Free-Body Diagram. As in the previous example, we will consider a fixed control volume that includes the impeller and the water surrounding yet slightly removed from it, Fig. 6–12a. The forces acting on its free-body diagram, Fig. 6–12c, include the pressure of the water acting on the entrance and exit control surfaces, and the torque **T** on the impeller shaft. The weight of the water and blades, along with the pressure distribution around the rim of the closed control surfaces, is not shown since they produce no torque about the shaft.

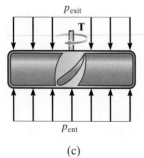

(c)

Continuity Equation. The axial flow through each open control surface is maintained since their areas are equal. That is,

$$\frac{\partial}{\partial t} \int_{cv} \mathbf{V}\rho \, d\mathcal{V} + \int_{cs} \rho \mathbf{V} \cdot d\mathbf{A} = 0$$

$$0 - \rho V_{a1}A + \rho V_{a2}A = 0$$

$$V_{a1} = V_{a2}$$

Therefore, flow in the axial direction is *constant*, and so

$$Q = V_a A; \qquad 0.1 \, \text{m}^3/\text{s} = V_a(0.025 \, \text{m}^2)$$

$$V_a = 4 \, \text{m/s}$$

Angular Momentum. Applying the angular momentum equation about the axis of the shaft, for steady flow we have

$$\Sigma \mathbf{M} = \frac{\partial}{\partial t} \int_{cv} (\mathbf{r} \times \mathbf{V})\rho \, d\mathcal{V} + \int_{cs} (\mathbf{r} \times \mathbf{V})\rho \mathbf{V} \cdot d\mathbf{A}$$

$$T = 0 + \int_{cs} r_m V_t \rho V_a \, dA \tag{1}$$

Here the vector cross product is replaced by the moment of the *tangential component* $\mathbf{V}_t$ of the water's velocity $\mathbf{V}$. Only this component produces a moment about the axis of the shaft, Fig. 6–12b. Also, notice that the *flow* through the open control surfaces is determined *only* from the *axial component* of $\mathbf{V}$, that is, $\rho \mathbf{V} \cdot d\mathbf{A} = \rho V_a dA$. Therefore, integrating over the area, we get

$$T = r_m(V_t)_2(\rho V_a A) + r_m(V_t)_1(-\rho V_a A) \tag{2}$$

The flow through the open control surfaces, onto and off a blade, is shown in Fig. 6–12b. Here it can be seen that there is no initial tangential component of velocity, $V_{t1} = 0$, because the flow is delivered to the blade in the axial direction at the rate of $V_a = 4 \, \text{m/s}$. At the top of the blade, the impeller has given the water a velocity $\mathbf{V}_2 = \mathbf{V}_a + (\mathbf{V}_t)_2$, but as stated, only the tangential component creates angular momentum.

Substituting the data into Eq. 2, we therefore have

$$T = (0.08 \, \text{m})(5 \, \text{m/s})\left[(1000 \, \text{kg/m}^3)(4 \, \text{m/s})(0.025 \, \text{m}^2)\right] - 0$$

$$= 40 \, \text{N} \cdot \text{m} \qquad\qquad\qquad\qquad\qquad Ans.$$

A more thorough analysis of problems involving axial-flow pumps is covered in Chapter 14.

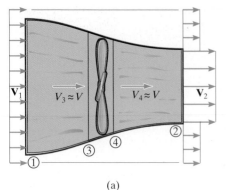

(a)

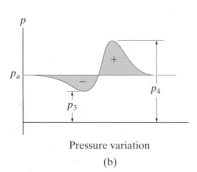

Pressure variation

(b)

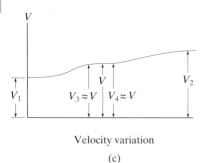

Velocity variation

(c)

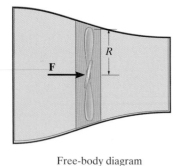

Free-body diagram

(d)

Fig. 6–13

*6.5   Propellers and Wind Turbines

Propellers and wind turbines both act like a screw by using several blades mounted on a rotating shaft. In the case of a boat or airplane propeller, *applying a torque* causes the linear momentum of the fluid in front of the propeller to increase as it flows towards then through the blades. This change in momentum creates a reactive force on the propeller, which then pushes it forward. A wind turbine works the opposite way, extracting fluid energy or *developing a torque* from the wind, as the wind passes through the propeller.

The design of both of these devices is actually based on the same principles used to design airfoils (or airplane wings). See, for example, Refs. [3] and [5]. In this section, however, we will provide some insight into how these devices operate, using a simplified analysis. First we will discuss the propeller and then the wind turbine. In both cases we will assume the fluid to be *ideal*.

Propeller.   So that the flow appears to be *steady flow*, a control volume of fluid, shown in Fig. 6–13a, will be observed *relative to the center of the propeller*, which we will assume to be stationary.* This volume excludes the propeller but includes the outlined portion of the fluid slipstream that passes through it. Fluid at the left control surface, 1, is moving toward the propeller with a velocity V_1. The fluid within the control volume from 1 to 3 is accelerated because of the reduced pressure (suction) within this region of the control volume, Fig. 6–13b. If we assume the propeller has many thin blades, then the velocity V is *essentially constant* as the fluid passes through the propeller from 3 to 4, Figs. 6–13a and 6–13c. The increased pressure that occurs on the right side of the propeller, Fig. 6–13b, pushes the flow and further accelerates it from 4 to 2. Finally, because of continuity of mass flow, the control surface or slipstream boundary *narrows*, such that the velocity at the far right control surface, 2, is increased to V_2, Fig. 6–13a.

Realize that this description of the flow is somewhat simplistic, since it neglects the interactive effects between the fluid and the housing to which the propeller is attached. Also, the boundary at the upper and lower closed control surface, from 1 to 2 in Fig. 6–13a, has a *discontinuity* between the outside still air and the flow within the control volume, when actually there is a smooth transition between the two. Finally, in addition to the axial motion we are considering, the propeller will also impart a rotary motion, or whirl, to the air. In the analysis that follows, we will neglect these effects.

Linear Momentum.   If the linear momentum equation is applied in the horizontal direction to the fluid within the control volume, then the only horizontal force acting on the control volume's free-body diagram is that caused by the force of the propeller on the fluid, Fig. 6–13d. (During operation, the pressure on *all* outside control surfaces remains constant and equal to the atmospheric pressure of the undisturbed fluid. In other words, the gage pressure is zero.) Therefore,

*We can also consider the center of the propeller to be moving to the left at V_1, since the analysis is the same for both cases.

$$\Sigma \mathbf{F} = \frac{\partial}{\partial t} \int_{cv} \mathbf{V} \rho \, d\Psi + \int_{cs} \mathbf{V} \rho \mathbf{V}_{1/cs} \cdot d\mathbf{A}$$

$$\xrightarrow{+} \Sigma F = 0 + V_2(\rho V_2 A_2) + V_1(-\rho V_1 A_1)$$

Since $Q = V_2 A_2 = V_1 A_1 = VA = V\pi R^2$, where R is the radius of the propeller, then

$$F = \rho\big[V(\pi R^2)\big](V_2 - V_1) \qquad (6\text{–}6)$$

As noted in Fig. 6–13c, the velocities $V_3 = V_4 = V$, and so no momentum change occurs between sections 3 and 4. Therefore, the force F of the propeller can also be expressed as the difference in pressure that occurs on sections 3 and 4, that is, $F = (p_4 - p_3)\pi R^2$, Fig. 6–13b. Thus, the above equation becomes

$$p_4 - p_3 = \rho V(V_2 - V_1) \qquad (6\text{–}7)$$

It is now necessary to express V in terms of V_1 and V_2.

Bernoulli Equation. The Bernoulli equation, $p/\gamma + V^2/2g + z$ = const. can be applied along a horizontal streamline between points at 1 and 3 and between points at 4 and 2.* Realizing that the (gage) pressures $p_1 = p_2 = 0$, we have

$$0 + \frac{V_1^2}{2} + 0 = \frac{p_3}{\rho} + \frac{V^2}{2} + 0$$

and

$$\frac{p_4}{\rho} + \frac{V^2}{2} + 0 = 0 + \frac{V_2^2}{2} + 0$$

If we add these equations and solve for $p_4 - p_3$, we obtain

$$p_4 - p_3 = \frac{1}{2}\rho\big(V_2^2 - V_1^2\big)$$

Finally, equating this equation to Eq. 6–7, we get

$$V = \frac{V_1 + V_2}{2} \qquad (6\text{–}8)$$

This result is known as **Froude's theorem**, after William Froude, who first derived it. It indicates that the velocity of the flow through the propeller is actually the average of the upstream and downstream velocities. If it is substituted into Eq. 6–6, then the force or thrust developed by the propeller on the fluid becomes

$$F = \frac{\rho \pi R^2}{2}\big(V_2^2 - V_1^2\big) \qquad (6\text{–}9)$$

*This equation *cannot* be applied between points at 3 and 4 because *energy is added* to the fluid by the propeller within this region. Furthermore, within this region the flow is unsteady.

The rotation ω of the propeller will cause points along its length to have a different velocity in accordance with the equation $v = \omega r$. To maintain a constant angle of attack with the air stream, the blade is given a noticeable angle of twist.

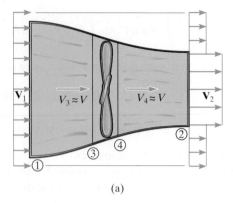

(a)

Fig. 6–13a

Power and Efficiency.   The **power output** of the propeller is caused by the rate of work done by the thrust **F**. If we think of the fluid ahead of the propeller as being at rest, Fig. 6–13a, and the propeller fixed to an airplane and moving forward at V_1, then the power output produced by **F** is

$$\dot{W}_o = FV_1 \tag{6–10}$$

The **power input** is the rate of work needed to maintain the increase of velocity of the slipstream from V_1 to V_2. Since this requires the fluid to have a velocity V through the propeller, then

$$\dot{W}_i = FV \tag{6–11}$$

Finally, the **ideal efficiency** e is the ratio of the power output to the power input. Using Eq. 6–8, we have

$$e_{\text{prop}} = \frac{\dot{W}_o}{\dot{W}_i} = \frac{2V_1}{V_1 + V_2} \tag{6–12}$$

In general, a propeller's *actual efficiency* will increase as the speed of the aircraft or boat to which it is attached is increased, although it can never be equal to 1 (or 100%) because of frictional losses. As the speed increases, though, there is a point at which the efficiency will start to drop off. This can occur on aircraft propellers when the tips of the blades reach or exceed the speed of sound. When this occurs, the drag force on the propeller will significantly increase due to compressibility of the air. Also, in the case of boats, efficiency is reduced because cavitation can occur if the pressure at the tips of the blades reaches the vapor pressure. Experimental tests on airplane propellers show that their *actual efficiencies* are in the range of about 60% to 80%. Boats, with their smaller-diameter propellers, have lower efficiencies, often in the range of 40% to 60%.

Wind Turbine.

Wind Turbine. Wind turbines and windmills extract kinetic energy from the wind. The flow pattern for these devices is opposite to that of a propeller, and so it looks like that shown in Fig. 6–14, where the slipstream gets larger as it passes through the blades. Using a similar analysis as that for propellers, we can show that Froude's theorem also applies, that is,

$$V = \frac{V_1 + V_2}{2} \qquad (6\text{--}13)$$

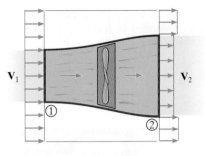

Wind Turbine

Fig. 6–14

Power and Efficiency. Using a derivation that compares with Eq. 6–9 for $\mathbf{F}$, we have

$$F = \frac{\rho \pi R^2}{2}\left(V_1^2 - V_2^2\right)$$

Here the parcel of air drawn into the blades has an area of $A = \pi R^2$ and is given a velocity V, so the *power output* is $\dot{W}_o = FV$ which becomes

$$\dot{W}_o = \frac{1}{2}\rho VA\left(V_1^2 - V_2^2\right) \qquad (6\text{--}14)$$

This result actually represents the time rate of loss in the kinetic energy of the wind once it passes through the blades.

It is customary to measure the *power input* as the time rate of change of the kinetic energy of the wind passing through the area swept by the blades, πR^2, but *without the blades* being present to disturb the velocity V_1. Since $\dot{m} = \rho A V_1$, then $\dot{W}_i = \frac{1}{2}\dot{m}V_1^2 = \frac{1}{2}(\rho A V_1)V_1^2$. Therefore, the efficiency of a wind turbine is

$$e_{\text{turbine}} = \frac{\dot{W}_o}{\dot{W}_i} = \frac{\frac{1}{2}\rho VA\left(V_1^2 - V_2^2\right)}{\frac{1}{2}(\rho A V_1)V_1^2} = \frac{V\left(V_1^2 - V_2^2\right)}{V_1^3}$$

Substituting Eq. 6–13, and simplifying, we get

$$e_{\text{turbine}} = \frac{1}{2}\left[1 - \left(\frac{V_2^2}{V_1^2}\right)\right]\left[1 + \left(\frac{V_2}{V_1}\right)\right] \qquad (6\text{--}15)$$

If e_{turbine} is plotted as a function of V_2/V_1, we can show that the curve passes through a *maximum* of $e_{\text{turbine}} = 0.593$ when $V_2/V_1 = 1/3$.* In other words, a wind turbine can extract a maximum of 59.3% of the total power of the wind. This is known as **Betz's law**, named after the German physicist who derived it in 1919. Wind turbines have a rated power at a specified wind speed, but those currently used in the power industry base their performance on a **capacity factor**. This is a ratio comparing the actual energy output for a year to its rated power output for the year. For modern-day wind turbines, the capacity factor is usually between 0.30 and 0.40. See Ref. [6].

Wind turbines are gaining in popularity as a device for harnessing energy. Like an airplane propeller, each blade acts as an airfoil. The simplified analysis presented here indicates the basic principles; however, a more complete analysis would include treating the propeller as a wing.

*See Prob. 6–64.

EXAMPLE | 6.11

The motor on a small boat has a propeller with a radius of 60 mm, Fig. 6–15a. If the boat is traveling at 2 m/s, determine the thrust on the boat and the ideal efficiency of the propeller if it discharges water through it at 0.04 m³/s.

© Carver Mostardi/Alamy

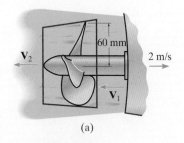

(a)

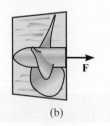

(b)

Fig. 6–15

SOLUTION

Fluid Description.   We consider the flow to be uniform steady flow and the water to be incompressible, where $\rho_W = 1000$ kg/m³.

Analysis.   The average velocity of the water flowing through the propeller is

$$Q = VA; \qquad 0.04 \text{ m}^3/\text{s} = V\left[\pi(0.06 \text{ m})^2\right]$$

$$V = 3.537 \text{ m/s}$$

To achieve steady flow, the control volume moves with the propeller, and so the flow velocity into it is $V_1 = 2$ m/s, Fig. 6–15a. We can now obtain the flow velocity out, V_2, from Eq. 6–8.

$$V = \frac{V_1 + V_2}{2}; \qquad V_2 = 2\,(3.537 \text{ m/s}) - 2 \text{ m/s} = 5.074 \text{ m/s}$$

Applying Eq. 6–9 to find the thrust on the boat, Fig. 6–15b, we have

$$F = \frac{\rho\pi R^2}{2}\,(V_2^2 - V_1^2)$$

$$= \frac{(1000 \text{ kg/m}^3)\,(\pi)\,(0.06 \text{ m})^2}{2}\left[(5.074 \text{ m/s})^2 - (2 \text{ m/s})^2\right]$$

$$= 122.94 \text{ N} = 123 \text{ N} \qquad\qquad\qquad\qquad Ans.$$

The power output and input are determined from Eqs. 6–10 and 6–11, respectively.

$$\dot{W}_o = FV_1 = (122.94 \text{ N})\,(2 \text{ m/s}) = 245.88 \text{ W}$$

$$\dot{W}_i = FV = (122.94 \text{ N})\,(3.537 \text{ m/s}) = 434.82 \text{ W}$$

Thus, the ideal efficiency of the propeller is

$$e = \frac{\dot{W}_o}{\dot{W}_i} = \frac{245.88 \text{ W}}{434.82 \text{ W}} = 0.5654 = 56.5\% \qquad Ans.$$

This same result can also be obtained from Eq. 6–12.

$$e = \frac{2V_1}{V_1 + V_2} = \frac{2(2 \text{ m/s})}{(2 \text{ m/s}) + (5.074 \text{ m/s})} = 0.5654$$

6.6   Applications for Control Volumes Having Accelerated Motion

For some problems, it will be convenient to choose a control volume that is *accelerating*. If Newton's second law is written for the fluid system within this accelerating control volume, we have

$$\Sigma \mathbf{F} = \frac{d(m\mathbf{V})}{dt} \tag{6-16}$$

If the fluid velocity $\mathbf{V}$ is measured from a *fixed* inertial reference, then it is equal to the velocity of the control volume, $\mathbf{V}_{cv}$, plus the velocity of the fluid measured relative to its control surface, $\mathbf{V}'_{f/cs}$, that is

$$\mathbf{V} = \mathbf{V}_{cv} + \mathbf{V}'_{f/cs}$$

Substituting this equation into Eq. 6–16, where the first term becomes the mass times the acceleration of the control volume, we get

$$\Sigma \mathbf{F} = m\frac{d\mathbf{V}_{cv}}{dt} + \frac{d(m\mathbf{V}_{f/cs})}{dt} \tag{6-17}$$

The time rate of change indicated by the last term must be expressed as a   Eulerian description, using the Reynold's transport theorem, with $\eta = \frac{m\mathbf{V}_{rel}}{m} = \mathbf{V}_{rel}$.

$$\left[\frac{d(m\mathbf{V}_{f/cs})}{dt}\right] = \frac{\partial}{\partial t}\int_{cv} \mathbf{V}_{f/cv}\,\rho\,d\forall + \int_{cs} \mathbf{V}_{f/cs}(\rho\mathbf{V}_{f/cs}\cdot d\mathbf{A})$$

Substituting this into Eq. 6–17, we get

$$\Sigma \mathbf{F} = m\frac{d\mathbf{V}_{cv}}{dt} + \frac{\partial}{\partial t}\int_{cv} \mathbf{V}_{f/cv}\,\rho\,d\forall + \int_{cs} \mathbf{V}_{f/cs}(\rho\mathbf{V}_{f/cs}\cdot d\mathbf{A}) \tag{6-18}$$

This result indicates that the sum of the external forces acting on the accelerating control volume is equal to the inertial effect of the *entire mass* contained within the control volume, plus the local rate at which the momentum of the fluid system is changing relative to the control volume, plus the convective rate at which momentum is leaving or entering relative to the control surfaces. We will consider important applications of this equation in the following two sections.

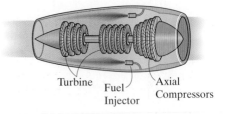

Turbine · Fuel · Axial Compressors
Injector

Turbojet Engine

(a)

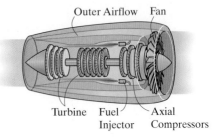

Outer Airflow · Fan

Turbine · Fuel · Axial Compressors
Injector

Turbofan Engine

(b)

Fig. 6–16

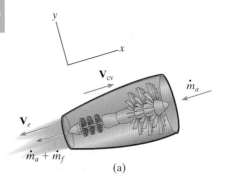

(a)

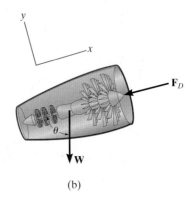

(b)

Fig. 6–17

6.7 Turbojets and Turbofans

A turbojet or turbofan engine is used mainly for the propulsion of aircraft. As shown in Fig. 6–16a, a **turbojet** operates by taking in air at its front, and then increasing its pressure by passing it through a series of fans called a **compressor**. Once the air is under high pressure, fuel is injected and ignited in the combustion chamber. The resulting hot gases *expand* and then move at a high velocity through a turbine. A portion of the kinetic energy of the gas is used to turn the shaft connected to *both* the turbine and the compressor. The remaining energy is used to propel the aircraft, as the gas is ejected through the exhaust nozzle. Like a propeller, the efficiency of a turbojet increases as the speed of the aircraft increases. A **turbofan engine** works on the same principles, except that a fan is added on the front of the turbine and turns with the shaft to supply additional incoming air by propelling some of it through a duct surrounding the turbojet, and thereby developing an increased thrust, Fig. 6–16b.

To analyze the motion and thrust of a turbojet (or turbofan), we will consider the engine represented in Fig. 6–17a.* Here the control volume surrounds the engine and contains the gas (air) and fuel within the engine. If the engine is initially at rest, then the control volume will *accelerate* with the engine, and so we must apply Eq. 6–18, where measurements are made from a *fixed* (inertial) coordinate system. For simplicity, we will consider one-dimensional, incompressible steady flow through the engine. The forces acting near the free-body diagram of the control volume, Fig. 6–17b, consist of the weight of the engine $\mathbf{W}$ and the atmospheric drag $\mathbf{F}_D$. The gage pressure acting near the intake and exit control surfaces is zero since here the air is at atmospheric pressure. Thus, for steady flow, Eq. 6–18, applied along the engine, becomes

$$\left(\overset{+}{\nearrow}\right) -W\cos\theta - F_D = m\frac{dV_{cv}}{dt} + 0 + \int_{cs} \mathbf{V}_{f/cs}(\rho\,\mathbf{V}_{f/cs}\cdot d\mathbf{A})$$

The last term on the right can be evaluated by noting that at the *intake* the air flow has a relative velocity of $(\mathbf{V}_{f/cs})_i = -\mathbf{V}_{cv}$, and at the exhaust the fuel and air mixture has a relative velocity of $(\mathbf{V}_{f/cs})_e = \mathbf{V}_e$. Therefore, using our sign convention,

$$-W\cos\theta - F_D = m\frac{dV_{cv}}{dt} + (-V_{cv})(-\rho_a V_{cv} A_i) + (-V_e)(\rho_{a+f} V_e A_e)$$

At the intake $\dot{m}_a = \rho_a V_{cv} A_i$, and at the exhaust $\dot{m}_a + \dot{m}_f = (\rho_{a+f} V_e A_e)$. Therefore,

$$-W\cos\theta - F_D = m\frac{dV_{cv}}{dt} + \dot{m}_a V_{cv} - (\dot{m}_a + \dot{m}_f)V_e \qquad (6\text{–}19)$$

The thrust of the engine must overcome the two forces on the left side of this equation, along with the inertia caused by the *mass of the engine* and its contents, that is, $m(dV_{cv}/dt)$. The thrust is therefore the result of the last two terms on the right side.

$$T = \dot{m}_a V_{cv} - (\dot{m}_a + \dot{m}_f)V_e \qquad (6\text{–}20)$$

*In this discussion, the engine is assumed to be free of an airplane or a fixed support.

Notice that we have not shown this force on the free-body diagram of the control volume because it is the "effect" caused by the *mass flow*.

6.8   Rockets

A rocket engine burns either solid or liquid fuel that is carried within the rocket. A solid-fueled engine is designed to produce a constant thrust, which is achieved by forming the propellant into a shape that provides *uniform burning* of the fuel. Once it is ignited, it cannot be controlled. A liquid-fueled engine requires a more complex design, involving the use of piping, pumps, and pressurized tanks. It operates by combining a liquid fuel with an oxidizer as the fuel enters a combustion chamber. Control of the thrust is then achieved by adjusting the flow of the fuel.

The performance of either of these types of rockets can be studied by applying the linear momentum equation in a manner similar to that used for the turbojet. If we consider the control volume to be the entire rocket, as shown in Fig. 6–18, then Eq. 6–19 can be applied, except here the weight is vertically downward, and $\dot{m}_a = 0$ since only fuel is used. Therefore, for a rocket we have

$$(+\uparrow)\ -W\ -\ F_D = m\frac{dV_{cv}}{dt}\ -\ \dot{m}_f V_e \qquad (6\text{–}21)$$

A numerical application of this equation is given in Example 6.13.

Turbofan engine for a commercial jet liner.

Fig. 6–18

Important Points

- Propellers are propulsion devices that act as a screw, causing the linear momentum of the fluid in front of the propeller to increase as the fluid flows towards and then past the blades. Wind turbines act in an opposite manner, removing energy from the wind, and thus decreasing the wind's momentum. A simple analysis of the flow can be made for both of these devices using the linear momentum and Bernoulli equations.

- If the control volume is selected to have accelerated motion, then the momentum equation must include an additional term, $m(d\mathbf{V}_{cv}/dt)$, which accounts for the inertia of the mass within the control volume. This term is required because Newton's second law forms the basis of the momentum equation, and measurements must be made from a nonaccelerating or inertial reference frame.

- The resulting momentum equation can be used to analyze the motion of turbojets and rockets. The thrust produced is not shown on the free-body diagram because it is the result of the terms involving the mass flow out of the engine.

© AF archive/Alamy

EXAMPLE | 6.12

The jet plane in Fig. 6–19a is in level flight with a *constant speed* of 140 m/s. Each of its two turbojet engines burns fuel at a rate of 3 kg/s. Air, at a temperature of 15°C, enters the intake, which has a cross-sectional area of 0.15 m². If the exhaust has a velocity of 700 m/s, measured relative to the plane, determine the drag acting on the plane.

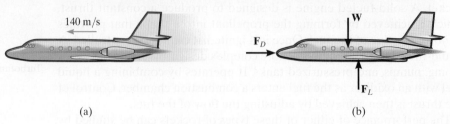

(a)　　　　　　　　　　　　　　　(b)

Fig. 6–19

SOLUTION

Fluid Description.　We consider this a case of steady flow when measured relative to the plane. Within the engine the air is compressible; however, in this problem we do not have to account for this effect since the resulting exhaust velocity is known. From Appendix A, at $T = 15°C$, $\rho_a = 1.23$ kg/m³.

Analysis.　To determine the drag we will consider the entire plane, its two engines, and the air and fuel within the engines as the control volume, Fig. 6–19a. This control volume moves with a constant speed of $V_{cv} = 140$ m/s, so the mass flow of the air delivered to each engine is

$$\dot{m}_a = \rho_a V_{cv} A = \left(1.23 \text{ kg/m}^3\right)\left(140 \text{ m/s}\right)\left(0.15 \text{ m}^2\right)$$
$$= 25.83 \text{ kg/s}$$

Conservation of mass requires that this same mass flow of air pass through the engine. The free-body diagram of the control volume is shown in Fig. 6–19b. Applying Eq. 6–19, realizing there are *two engines*, we have

$$\left(\overset{+}{\leftarrow}\right) -W \cos \theta - F_D = m\frac{dV_{cv}}{dt} + \dot{m}_a V_{cv} - \left(\dot{m}_a + \dot{m}_f\right)V_e$$

$$0 - F_D = 0 + 2\left[(25.83 \text{ kg/s})(140 \text{ m/s}) - (25.83 \text{ kg/s} + 3 \text{ kg/s})(700 \text{ m/s})\right]$$

$$F_D = 33.1\left(10^3\right) \text{ N} = 33.1 \text{ kN} \hspace{2cm} Ans.$$

Because the plane is in equilibrium, this force is equivalent to the thrust provided by the engines.

EXAMPLE 6.13

The rocket and its fuel in Fig. 6–20 have an initial mass of 5 Mg. When it takes off from rest, the 3 Mg of fuel is consumed at the rate of $\dot{m}_f = 80$ kg/s and is expelled at a constant speed of 1200 m/s, relative to the rocket. Determine the maximum velocity of the rocket. Neglect the change in gravity due to altitude and the drag resistance of the air.

SOLUTION

Fluid Description. The flow is steady when measured from the rocket. Compressibility effects of the air on the rocket are not considered here, since the drag resistance is not considered.

Analysis. The rocket and its contents are selected as an accelerating control volume, Fig. 6–20a. The free-body diagram is shown in Fig. 6–20b. For this problem, Eq. 6–19 becomes

$$(+\uparrow) \quad -W = m\frac{dV_{cv}}{dt} - \dot{m}_f V_e$$

Here $V_{cv} = V$. At any instant t during the flight, the mass of the rocket is $m = (5000 - 80t)$ kg, and since $W = mg$, we have

$$-\big[(5000 - 80t)\text{ kg}\big]\big(9.81\text{ m/s}^2\big) = \big[(5000 - 80t)\text{ kg}\big]\frac{dV}{dt} - (80\text{ kg/s})(1200\text{ m/s})$$

Note that the thrust of the engine is represented by the last term. Separating the variables and integrating, where $V = 0$ when $t = 0$, we have

$$\int_0^V dV = \int_0^t \left(\frac{80(1200)}{5000 - 80t} - 9.81\right) dt$$

$$V = -1200 \ln(5000 - 80t) - 9.81t \Big|_0^t$$

$$V = 1200 \ln\left(\frac{5000}{5000 - 80t}\right) - 9.81t$$

Maximum velocity occurs at the instant all the fuel has been exhausted. The time t' needed to do this is

$$m_f = \dot{m}_f t'; \qquad 3(10^3)\text{ kg} = (80\text{ kg/s})t', \qquad t' = 37.5\text{ s}$$

Therefore,

$$V_{\text{max}} = 1200 \ln\left(\frac{5000}{5000 - 80(37.5)}\right) - 9.81(37.5)$$

$$V_{\text{max}} = 732\text{ m/s} \qquad\qquad Ans.$$

Including the effect of air resistance will complicate the solution, and it will be discussed further in Chapter 11.

(a)

(b)

Fig. 6–20

References

1. J. R. Lamarch and A. J. Baratta, *Introduction to Nuclear Engineering*, Prentice Hall, Inc., Upper Saddle River, NJ, 2001.
2. J. A. Fry. *Introduction to Fluid Mechanics*, MIT Press, Cambridge, MA, 1994.
3. National Renewable Energy Laboratory, *Advanced Aerofoil for Wind Turbines (2000)*, DOE/GO-10098-488, Sept. 1998, revised Aug. 2000.
4. M. Fremond et al., "Collision of a solid with an incompressible fluid," *Journal of Theoretical and Computational Fluid Dynamics*, London, UK.
5. D. A. Griffin and T. D. Ashwill "Alternative composite material for megawatt-scale wind turbine blades: design considerations and recommended testing," *J Sol Energy Eng* 125:515–521, 2003.
6. S. M. Hock, R. W. Thresher, and P. Tu, "Potential for far-term advanced wind turbines performance and cost projections," *Sol World Congr Proc Bienn Congr Int Sol Energy Soc* 1:565–570, 1992.
7. B. MacIsaak and R. Langlon, *Gas Turbine Propulsion Systems*, American Institute of Aeronautics and Astronautics, Ruston, VA, 2011.

FUNDAMENTAL PROBLEMS

Sec. 6.1–6.2

F6–1. Water is discharged through the 40-mm-diameter elbow at $0.012 \text{ m}^3/\text{s}$. If the pressure at A is 160 kPa, determine the resultant force the elbow exerts on the water.

F6–2. The 5-kg shield is held at an angle of 60° to deflect the 40-mm-diameter water stream, which is discharged at $0.02 \text{ m}^3/\text{s}$. If measurements show that 30% of the discharge is deflected upwards, determine the resultant force needed to hold the shield in place.

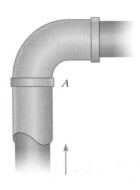

Prob. F6–1

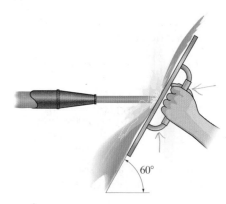

Prob. F6–2

F6–3. Water is flowing at 10 m/s from the 50-mm-diameter open pipe AB. If the flow is increasing at 3 m/s^2, determine the pressure in the pipe at A.

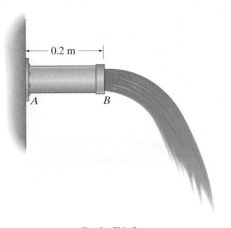

Prob. F6–3

F6–4. Crude oil flows at the same rate out of each branch of the wye fitting. If the pressure at A is 80 kPa, determine the resultant force at A to hold the fitting to the pipe.

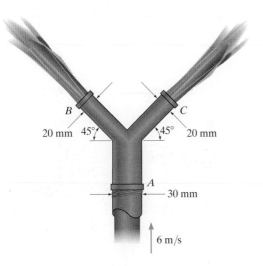

Prob. F6–4

Sec. 6.3

F6–5. The table fan develops a slipstream that has a diameter of 0.25 m. If the air is moving horizontally at 20 m/s as it leaves the blades, determine the horizontal friction force that the table must exert on the fan to hold it in place. Assume that the air has a constant density of 1.22 kg/m^3, and that the air just to the right of the blade is essentially at rest.

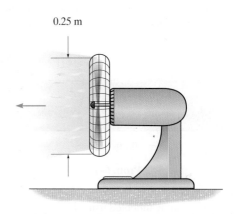

Prob. F6–5

F6–6. As water flows out the 20-mm-diameter pipe, it strikes the vane, which is moving to the left at 1.5 m/s. Determine the resultant force on the vane needed to deflect the water 90° as shown.

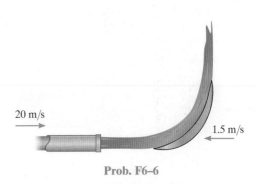

Prob. F6–6

6

PROBLEMS

Unless otherwise stated, in the following problems, assume the fluids are ideal fluids, that is, incompressible and frictionless.

Sec. 6.1–6.2

6–1. Determine the linear momentum of a mass of fluid in a 0.2-m length of pipe if the velocity profile for the fluid is a paraboloid as shown. Compare this result with the linear momentum of the fluid using the average velocity of flow. Take $\rho = 800$ kg/m³.

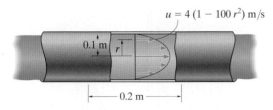

$$u = 4 (1 - 100 \, r^2) \text{ m/s}$$

0.1 m r

0.2 m

Prob. 6–1

6–2. Flow through the circular pipe is turbulent, and the velocity profile can be modeled using Prandtl's one-seventh power law, $v = V_{max}(1 - r/R)^{1/7}$. If ρ is the density, show that the momentum of the fluid per unit time passing through the pipe is $(49/72)\pi R^2 \rho V_{max}^2$. Then show that $V_{max} = (60/49)V$, where V is the average velocity of the flow. Also, show that the momentum per unit time is $(50/49)\pi R^2 \rho V^2$.

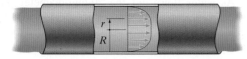

r
R

Prob. 6–2

6–3. Oil flows at 0.05 m³/s through the transition. If the pressure at the transition C is 8 kPa, determine the resultant horizontal shear force acting along the seam AB that holds the cap to the larger pipe. Take $\rho_o = 900$ kg/m³.

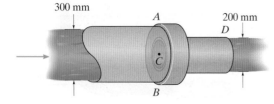

300 mm

A

200 mm
D

$\dot{C}$

B

Prob. 6–3

*6–4. Water flows through the hose with a velocity of 4 m/s. Determine the force that the water exerts on the wall. Assume the water does not splash back off the wall.

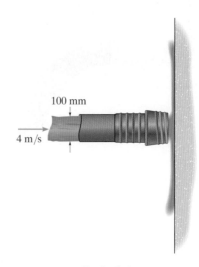

100 mm

4 m/s

Prob. 6–4

6–5. A small marine ascidian called a styela fixes itself on the sea floor and then allows moving water to pass through it in order to feed. If the opening at A has a diameter of 2 mm, and at the exit B the diameter is 1.5 mm, determine the horizontal force needed to keep this organism attached to the rock at C when the water is moving at 0.2 m/s into the opening at A. Take $\rho = 1050$ kg/m³.

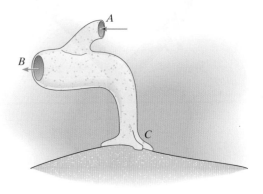

A

B

C

Prob. 6–5

6–6. Oil flows through the 100-mm-diameter pipe with a velocity of 5 m/s. If the pressure in the pipe at A and B is 80 kPa, determine the x and y components of force the flow exerts on the elbow. The flow occurs in the horizontal plane. Take $\rho_o = 900$ kg/m^3.

*6–8. The jet of water flows from the 100-mm-diameter pipe at 4 m/s. If it strikes the fixed vane and is deflected as shown, determine the normal force the jet exerts on the vane.

6–9. The jet of water flows from the 100-mm-diameter pipe at 4 m/s. If it strikes the fixed vane and is deflected as shown, determine the volume flow towards A and towards B if the tangential component of the force that the water exerts on the vane is zero.

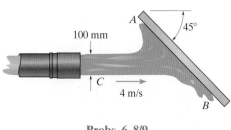

Probs. 6–8/9

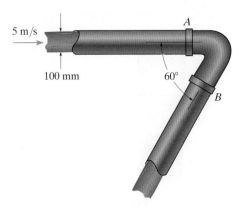

Prob. 6–6

6–7. The apparatus or "jet pump" used in an industrial plant is constructed by placing the tube within the pipe. Determine the increase in pressure $(p_B - p_A)$ that occurs between the back A and front B of the pipe if the velocity of the flow within the 200-mm-diameter pipe is 2 m/s, and the velocity of the flow through the 20-mm-diameter tube is 40 m/s. The fluid is ethyl alcohol having a density of $\rho_{ea} = 790$ kg/m^3. Assume the pressure at each cross section of the pipe is uniform.

6–10. The nozzle has a diameter of 40 mm. If it discharges water with a velocity of 20 m/s against the fixed blade, determine the horizontal force exerted by the water on the blade. The blade divides the water evenly at an angle of $\theta = 45°$.

6–11. The nozzle has a diameter of 40 mm. If it discharges water with a velocity of 20 m/s against the fixed blade, determine the horizontal force exerted by the water on the blade as a function of the blade angle θ. Plot this force (vertical axis) versus θ for $0° \le \theta \le 75°$. Give values for increments of $\Delta\theta = 15°$. The blade divides the water evenly.

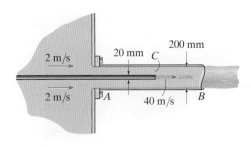

Prob. 6–7

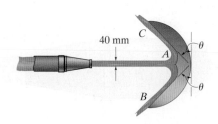

Probs. 6–10/11

6–12. The hemispherical bowl of mass m is held in equilibrium by the vertical jet of water discharged through a nozzle of diameter d. If the volumetric flow is Q, determine the height h at which the bowl is suspended. The water density is ρ_w.

6–13. The 500-g hemispherical bowl is held in equilibrium by the vertical jet of water discharged through the 10-mm-diameter nozzle. Determine the height h of the bowl as a function of the volumetric flow Q of the water through the nozzle. Plot the height h (vertical axis) versus Q for $0.5(10^{-3})\ \text{m}^3/\text{s} \leq Q \leq 1(10^{-3})\ \text{m}^3/\text{s}$. Give values for increments of $\Delta Q = 0.1(10^{-3})\ \text{m}^3/\text{s}$.

Probs. 6–12/13

6–14. Water flows through the 200-mm-diameter pipe at 4 m/s. If it exits into the atmosphere through the nozzle, determine the resultant force the bolts must develop at the connection AB to hold the nozzle onto the pipe.

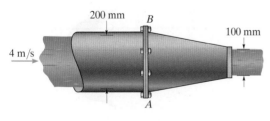

Prob. 6–14

6–15. Water enters A with a velocity of 8 m/s and pressure of 70 kPa. If the velocity at C is 9 m/s, determine the horizontal and vertical components of the resultant force that must act on the transition to hold it in place. Neglect the size of the transition.

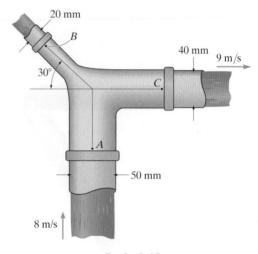

Prob. 6–15

6–16. The barge is being loaded with an industrial waste liquid having a density of 1.2 Mg/m³. If the average velocity of flow out of the 100-mm-diameter pipe is $V_A = 3$ m/s, determine the force in the tie rope needed to hold the barge stationary.

6–17. The barge is being loaded with an industrial waste liquid having a density of 1.2 Mg/m³. Determine the maximum force in the tie rope needed to hold the barge stationary. The waste can enter the barge at any point within the 10-m region. Also, what is the speed of the waste exiting the pipe at A when this occurs? The pipe has a diameter of 100 mm.

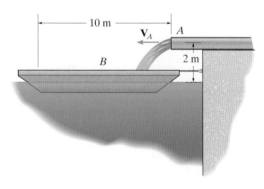

Probs. 6–16/17

6–18. Air flows through the 1-m-wide closed duct with a uniform velocity of 0.3 m/s. Determine the horizontal force **F** that the strap C must exert on the transition to hold it in place. Neglect any force at the slip joints A and B. Take $\rho_a = 1.22 \text{ kg/m}^3$.

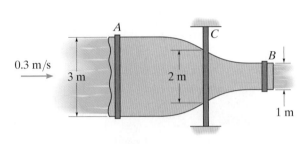

Prob. 6–18

6–19. Crude oil flows through the horizontal tapered 45° elbow at 0.02 m³/s. If the pressure at A is 300 kPa, determine the horizontal and vertical components of the resultant force the oil exerts on the elbow. Neglect the size of the elbow.

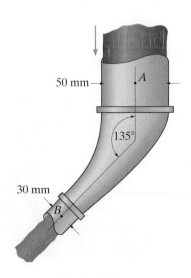

Prob. 6–19

*6–20. As water flows through the pipe at a velocity of 5 m/s, it encounters the orifice plate, which has a hole in its center. If the pressure at A is 230 kPa, and at B it is 180 kPa, determine the force the water exerts on the plate.

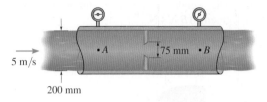

Prob. 6–20

6–21. As oil flows through the 20-m-long, 200-mm-diameter pipeline, it has a constant average velocity of 2 m/s. Friction losses along the pipe cause the pressure at B to be 8 kPa less than the pressure at A. Determine the resultant friction force on this length of pipe. Take $\rho_o = 880 \text{ kg/m}^3$.

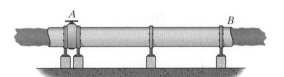

Prob. 6–21

6–22. Water flows into the tank at the rate of 0.05 m³/s from the 100-mm-diameter pipe. If the tank is 500 mm on each side, determine the compression in each of the four springs that support its corners when the water reaches a depth of $h = 1$ m. Each spring has a stiffness of $k = 8$ kN/m. When empty, the tank compresses each spring 30 mm.

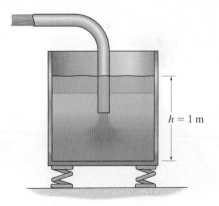

Prob. 6–22

6–23. The toy sprinkler consists of a cap and a rigid tube having a diameter of 20 mm. If water flows through the tube at $0.7(10^{-3})$ m^3/s, determine the vertical force the wall of the tube must support at B. Neglect the weight of the sprinkler head and the water within the curved segment of the tube. The weight of the tube and water within the vertical segment AB is 4 N.

6–25. The disk valve is used to control the flow of 0.008 m^3/s of water through the 40-mm-diameter tube. Determine the force $\mathbf{F}$ required to hold the valve in place for any position x of closure of the valve.

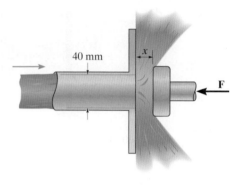

40 mm

Prob. 6–25

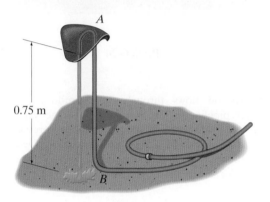

Prob. 6–23

6–26. Water flows through the pipe C at 4 m/s. Determine the horizontal and vertical components of force exerted by elbow D necessary to hold the pipe assembly in equilibrium. Neglect the size and weight of the pipe and the water within it. The pipe has a diameter of 60 mm at C, and at A and B the diameters are 20 mm.

*6–24. The toy sprinkler consists of a cap and a rigid tube having a diameter of 20 mm. Determine the flow through the tube such that it creates a vertical force of 6 N in the wall of the tube at B. Neglect the weight of the sprinkler head and the water within the curved segment of the tube. The weight of the tube and water within the vertical segment AB is 4 N.

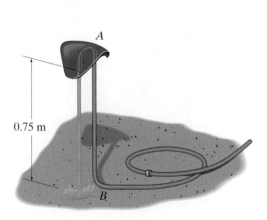

Prob. 6–24

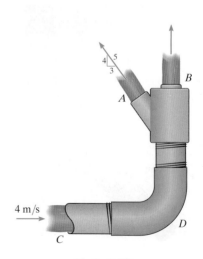

4 m/s

Prob. 6–26

6–27. The 300-kg circular craft is suspended 100 mm from the ground. For this to occur, air is drawn in at 18 m/s through the 200-mm-diameter intake and discharged to the ground as shown. Determine the pressure that the craft exerts on the ground. Take $\rho_a = 1.22 \, \text{kg/m}^3$.

6–29. Oil flows through the 50-mm-diameter vertical pipe assembly such that the pressure at A is 240 kPa and the velocity is 3 m/s. Determine the horizontal and vertical components of force the pipe exerts on the U-section AB of the assembly. The assembly and the oil within it have a combined weight of 60 N. Take $\rho_o = 900 \, \text{kg/m}^3$.

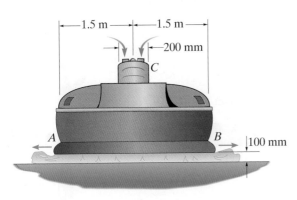

Prob. 6–27

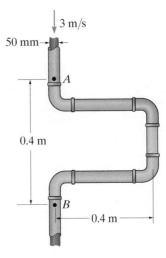

Prob. 6–29

*6–28.** The truck dumps water on the ground such that it flows from the truck through a 100-mm-wide opening at an angle of 60°. The length of the opening is 2 m. Determine the friction force that all the wheels of the truck must exert on the ground to keep the truck from moving at the instant the water depth in the truck is 1.75 m.

6–30. The cylindrical needle valve is used to control the flow of 0.003 m³/s of water through the 20-mm-diameter tube. Determine the force **F** required to hold it in place when $x = 10$ mm.

6–31. The cylindrical needle valve is used to control the flow of 0.003 m³/s of water through the 20-mm-diameter tube. Determine the force **F** required to hold it in place for any position x of closure of the valve.

Prob. 6–28

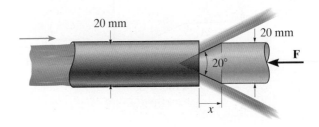

Probs. 6–30/31

*6–32. Air at a temperature of 30°C flows through the expansion fitting such that its velocity at A is 15 m/s and the absolute pressure is 250 kPa. If no heat or frictional loss occurs, determine the resultant force needed to hold the fitting in place.

Sec. 6.3

6–34. Water flows through the hose with a velocity of 2 m/s. Determine the force **F** needed to keep the circular plate moving to the right at 2 m/s.

6–35. Water flows through the hose with a velocity of 2 m/s. Determine the force **F** needed to keep the circular plate moving to the left at 2 m/s.

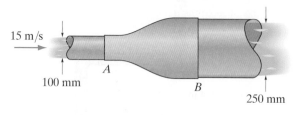

15 m/s

100 mm

A

B

250 mm

Prob. 6–32

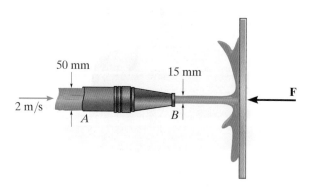

50 mm   15 mm

2 m/s

A B **F**

Probs. 6–34/35

6–33. Air at a temperature of 30°C flows through the expansion fitting such that its velocity at A is 15 m/s and the absolute pressure is 250 kPa. If heat and frictional loss due to the expansion cause the temperature and absolute pressure of the air at B to become 20°C and 7.50 kPa, determine the resultant force needed to hold the fitting in place.

*6–36. Flow from the water stream strikes the inclined surface of the cart. Determine the power produced by the stream if, due to rolling friction, the cart moves to the right with a constant velocity of 2 m/s. The discharge from the 50-mm-diameter nozzle is 0.04 m³/s. One-fourth of the discharge flows down the incline, and three-fourths flows up the incline.

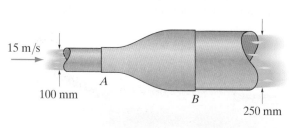

15 m/s

100 mm

A

B

250 mm

Prob. 6–33

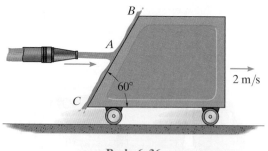

B

A

C

60°

2 m/s

Prob. 6–36

6–37. The boat is powered by the fan, which develops a slipstream having a diameter of 1.25 m. If the fan ejects air with an average velocity of 40 m/s, measured relative to the boat, and the boat is traveling with a constant velocity of 8 m/s, determine the force the fan exerts on the boat. Assume that the air has a constant density of $\rho_a = 1.22$ kg/m^3 and that the entering air at A is essentially at rest relative to the ground.

Prob. 6–37

6–38. The car is used to scoop up water that is lying in a trough at the tracks. Determine the force needed to pull the car forward at constant velocity v for each of the three cases. The scoop has a cross-sectional area A and the density of water is ρ_w.

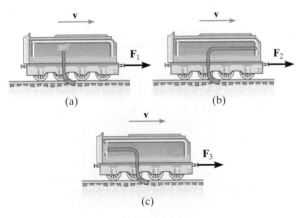

Prob. 6–38

6–39. Water flows at 0.1 m^3/s through the 100-mm-diameter nozzle and strikes the vane on the 150-kg cart, which is originally at rest. Determine the velocity of the cart 3 seconds after the jet strikes the vane.

*6–40. Water flows at 0.1 m^3/s through the 100-mm-diameter nozzle and strikes the vane on the 150-kg cart, which is originally at rest. Determine the acceleration of the cart when it attains a velocity of 2 m/s.

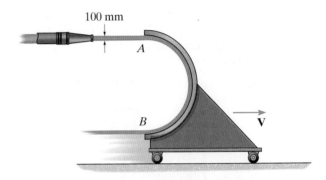

Probs. 6–39/40

6–41. A 25-mm-diameter stream flows at 10 m/s against the blade and is deflected 180° as shown. If the blade is moving to the left at 2 m/s, determine the horizontal force F of the blade on the water.

6–42. Solve Prob. 6–41 if the blade is moving *to the right* at 2 m/s. At what speed must the blade be moving to the right to reduce the force F to zero?

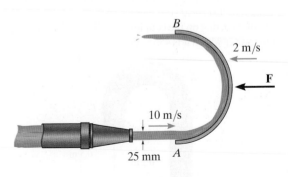

Probs. 6–41/42

Sec. 6.4

6–43. Water flows through the Tee fitting at 0.02 m³/s. If the water exits the fitting at B to the atmosphere, determine the horizontal and vertical components of force, and the moment that must be exerted on the fixed support at A, in order to hold the fitting in equilibrium. Neglect the weight of the fitting and the water within it.

***6–44.** Water flows through the Tee fitting at 0.02 m³/s. If the pipe at B is extended and the pressure in the pipe at B is 75 kPa, determine the horizontal and vertical components of force, and the moment that must be exerted on the fixed support at A, to hold the fitting in equilibrium. Neglect the weight of the fitting and the water within it.

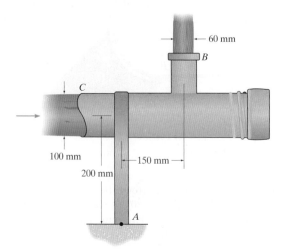

Probs. 6–43/44

6–45. Water flows through the curved pipe at a speed of 5 m/s. If the diameter of the pipe is 150 mm, determine the horizontal and vertical components of the resultant force, and the moment acting on the coupling at A. The weight of the pipe and the water within it is 450 N, having a center of gravity at G.

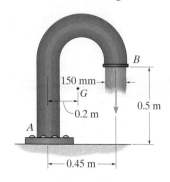

Prob. 6–45

6–46. The chute is used to divert the flow of water. If the flow is 0.4 m³/s and it has a cross-sectional area of 0.03 m², determine the horizontal and vertical force components at the pin A, and the horizontal force at the roller B, necessary for equilibrium. Neglect the weight of the chute and the water on it.

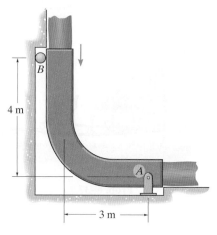

Prob. 6–46

6–47. Air enters into the hollow propeller tube at A with a mass flow of 3 kg/s and exits at the ends B and C with a velocity of 400 m/s, measured relative to the tube. If the tube rotates at 1500 rev/min, determine the frictional torque **M** on the tube.

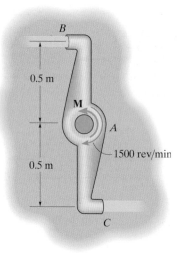

Prob. 6–47

*6–48. Water flows into the bend fitting with a velocity of 3 m/s. If the water exits at *B* into the atmosphere, determine the horizontal and vertical components of force, and the moment at *C*, needed to hold the fitting in place. Neglect the weight of the fitting and the water within it.

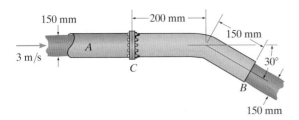

Prob. 6–48

6–49. Water flows into the bend fitting with a velocity of 3 m/s. If the water at *B* exits into a tank having a gage pressure of 10 kPa, determine the horizontal and vertical components of force, and the moment at *C*, needed to hold the fitting in place. Neglect the weight of the fitting and the water within it.

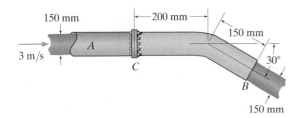

Prob. 6–49

6–50. The 5-mm-diameter arms of a rotating lawn sprinkler have the dimensions shown. Water flows out relative to the arms at 6 m/s, while the arms are rotating at 10 rad/s. Determine the frictional torsional resistance at the bearing *A*, and the speed of the water as it emerges from the nozzles, as measured by a fixed observer.

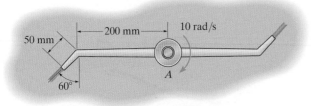

Prob. 6–50

6–51. When operating, the air-jet fan discharges air with a speed of $V = 18$ m/s into a slipstream having a diameter of 0.5 m. If the air has a density of 1.22 kg/m³, determine the horizontal and vertical components of reaction at *C*, and the vertical reaction at each of the two wheels, *D*. The fan and motor have a mass of 25 kg and a center of mass at *G*. Neglect the weight of the frame. Due to symmetry, both of the wheels support an equal load. Assume the air entering the fan at *A* is essentially at rest.

*6–52. If the air has a density of 1.22 kg/m³, determine the maximum speed *V* at which the air-jet fan can discharge air into the slipstream having a diameter of 0.5 m at *B*, so that the fan does not topple over. The fan and motor have a mass of 25 kg and a center of mass at *G*. Neglect the weight of the frame. Due to symmetry, both of the wheels support an equal load. Assume the air entering the fan at *A* is essentially at rest.

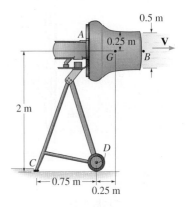

Probs. 6–51/52

6–53. The waterwheel consists of a series of flat plates that have a width *b* and are subjected to the impact of water to a depth *h*, from a stream that has an average velocity of *V*. If the wheel is turning at *ω*, determine the power supplied to the wheel by the water.

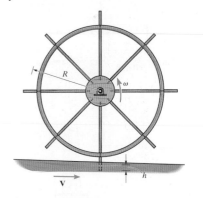

Prob. 6–53

Sec. 6.5–6.8

6–54. A boat has a 250-mm-diameter propeller that discharges 0.6 m³/s of water as the boat travels at 35 km/h in still water. Determine the thrust developed by the propeller on the boat.

6–55. A ship has a 2.5-m-diameter propeller with an ideal efficiency of 40%. If the thrust developed by the propeller is 1.5 MN, determine the constant speed of the ship in still water and the power that must be supplied to the propeller to operate it.

*6–56. The 12-kg fan develops a breeze of 10 m/s using a 0.8-m-diameter blade. Determine the smallest dimension d for the support so that the fan does not tip over. Take $\rho_a = 1.20$ kg/m³.

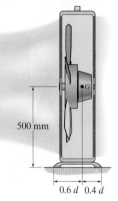

500 mm

0.6 d   0.4 d

Prob. 6–56

6–57. The wind turbine has a rotor diameter of 40 m and an ideal efficiency of 50% in a 12 m/s wind. If the density of the air is $\rho_a = 1.22$ kg/m³, determine the thrust on the blade shaft, and the power withdrawn by the blades.

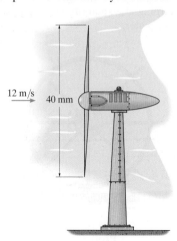

12 m/s   40 mm

Prob. 6–57

6–58. The wind turbine has a rotor diameter of 40 m and an efficiency of 50% in a 12 m/s wind. If the density of the air is $\rho_a = 1.22$ kg/m³, determine the difference between the pressure just in front of and just behind the blades. Also find the mean velocity of the air passing through the blades.

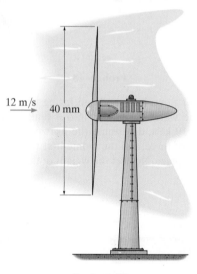

12 m/s   40 mm

Prob. 6–58

6–59. The airplane has a constant speed of 250 km/h in still air. If it has a 2.4-m-diameter propeller, determine the force acting on the plane if the speed of the air behind the propeller, measured relative to the plane, is 750 km/h. Also, what is the ideal efficiency of the propeller, and the power produced by the propeller? Take $\rho_a = 0.910$ kg/m³.

250 km/h

Prob. 6–59

*6–60. The rocket has an initial total mass m_0, including the fuel. When it is fired, it ejects a mass flow of $\dot{m}_e$ with a velocity of v_e measured relative to the rocket. As this occurs, the pressure at the nozzle, which has a cross-sectional area A_e, is p_e. If the drag force on the rocket is $F_D = ct$, where t is the time and c is a constant, determine the velocity of the rocket if the acceleration due to gravity is assumed to be constant.

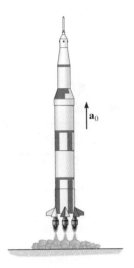

Prob. 6–60

6–61. The jet engine is mounted on the stand while it is being tested with the braking deflector in place. If the exhaust has a velocity of 800 m/s and the pressure just outside the nozzle is assumed to be atmospheric, determine the horizontal force that the supports exert on the engine. The fuel–air mixture has a flow of 11 kg/s.

6–62. If an engine of the type shown in Prob. 6–61 is attached to a jet plane, and it operates the braking deflector with the conditions stated in that problem, determine the speed of the plane in 5 seconds after it lands with a touch-down velocity of 30 m/s. The plane has a mass of 8 Mg. Neglect rolling friction from the landing gear.

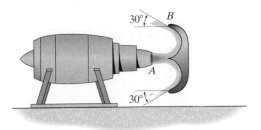

Probs. 6–61/62

6–63. The 12-Mg helicopter is hovering over a lake as the suspended bucket collects 5 m³ of water used to extinguish a fire. Determine the power required by the engine to hold the filled water bucket over the lake. The horizontal blade has a diameter of 14 m. Take $\rho_a = 1.23 \text{ kg/m}^3$.

Prob. 6–63

*6–64. Plot Eq. 6–15 and show that the maximum efficiency of a wind turbine is 59.3% as stated by Betz's law.

6–65. The jet engine on a plane flying at 160 m/s in still air draws in air at standard atmospheric temperature and pressure through a 0.5-m-diameter inlet. If 2 kg/s of fuel is added and the mixture leaves the 0.3-m-diameter nozzle at 600 m/s, measured relative to the engine, determine the thrust provided by the turbojet.

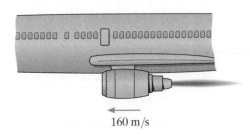

160 m/s

Prob. 6–65

6–66. The jet engine is mounted on the stand while it is being tested. Determine the horizontal force that the engine exerts on the supports, if the fuel–air mixture has a mass flow of 11 kg/s and the exhaust has a velocity of 2000 m/s.

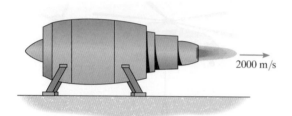

2000 m/s

Prob. 6–66

6–67. The jet plane has a constant velocity of 750 km/h. Air enters its engine nacelle at A having a cross-sectional area of 0.8 m². Fuel is mixed with the air at $\dot{m}_e = 2.5$ kg/s and is exhausted into the ambient air with a velocity of 900 m/s, measured relative to the plane. Determine the force the engine exerts on the wing of the plane. Take $\rho_a = 0.850$ kg/m³.

Prob. 6–67

*6–68. The boat has a mass of 180 kg and is traveling forward on a river with a constant velocity of 70 km/h, measured relative to the river. The river is flowing in the opposite direction at 5 km/h. If a tube is placed in the water, as shown, and it collects 40 kg of water in the boat in 80 s, determine the horizontal thrust T on the tube that is required to overcome the resistance due to the water collection.

$v = 5$ km/h

T

Prob. 6–68

6–69. The jet boat takes in water through its bow at 0.03 m³/s, while traveling in still water with a constant velocity of 10 m/s. If the water is ejected from a pump through the stern at 30 m/s, measured relative to the boat, determine the thrust developed by the engine. What would be the thrust if the 0.03 m³/s of water were taken in along the sides of the boat, perpendicular to the direction of motion? If the efficiency is defined as the work done per unit time divided by the energy supplied per unit time, then determine the efficiency for each case.

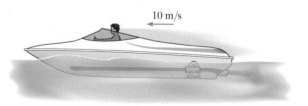

10 m/s

Prob. 6–69

6–70. A commercial jet aircraft has a mass of 150 Mg and is cruising at a constant speed of 850 km/h in level flight ($\theta = 0°$). If each of the two engines draws in air at a rate of 1000 kg/s and ejects it with a velocity of 900 m/s relative to the aircraft, determine the maximum angle θ at which the aircraft can fly with a constant speed of 750 km/h. Assume that air resistance (drag) is proportional to the square of the speed, that is, $F_D = cV^2$, where c is a constant to be determined. The engines are operating with the same power in both cases. Neglect the amount of fuel consumed.

Prob. 6–70

6–71. The balloon has a mass of 20 g (empty) and it is filled with air having a temperature of 20°C. If it is released, it begins to accelerate upward at 8 m/s². Determine the initial mass flow of air from the stem. Assume the balloon is a sphere having a radius of 300 mm.

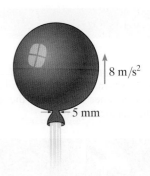

8 m/s²

5 mm

Prob. 6–71

*6–72. The 10-Mg jet plane has a constant speed of 860 km/h when it is flying horizontally. Air enters the intake I at the rate of 40 m³/s. If the engine burns fuel at the rate of 2.2 kg/s, and the gas (air and fuel) is exhausted relative to the plane with a speed of 600 m/s, determine the resultant drag force exerted on the plane by air resistance. Assume that the air has a constant density of $\rho_a = 1.22$ kg/m³.

I

Prob. 6–72

6–73. The 12-Mg jet airplane has a constant speed of 950 km/h when it is flying along a horizontal straight line. Air enters the intake scoops S at the rate of 50 m³/s. If the engine burns fuel at the rate of 0.4 kg/s, and the gas (air and fuel) is exhausted relative to the plane with a speed of 450 m/s, determine the resultant drag force exerted on the plane by air resistance. Assume that air has a constant density of 1.22 kg/m³.

$v = 950$ km/h

S

Prob. 6–73

6–74. The cart has a mass M and is filled with water that has an initial mass m_0. If a pump ejects the water through a nozzle having a cross-sectional area A, at a constant rate of v_0 relative to the cart, determine the velocity of the cart as a function of time. What is the maximum speed of the cart, assuming all the water can be pumped out? The frictional resistance to forward motion is F. The density of the water is ρ.

6–75. The jet is traveling at a constant velocity of 400 m/s in still air, while consuming fuel at the rate of 1.8 kg/s and ejecting it at 1200 m/s relative to the plane. If the engine consumes 1 kg of fuel for every 50 kg of air that passes through the engine, determine the thrust produced by the engine and the efficiency of the engine.

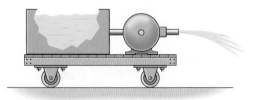

Prob. 6–74

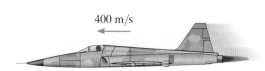

400 m/s

Prob. 6–75

CONCEPTUAL PROBLEMS

P6–1. The water cannon ejects water from this tug in a characteristic parabolic shape. Explain what effect this has on the tug.

P6–2. Water flows onto the buckets of the waterwheel causing the wheel to turn. Have the blades been designed in the most effective manner to produce the greatest angular momentum of the wheel? Explain.

P6–1

P6–2

CHAPTER REVIEW

The linear and angular momentum equations are often used to determine resultant forces and couple moments that a body or surface exerts on a fluid in order to change the fluid's direction.

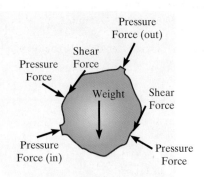

Application of the momentum equations requires identifying a control volume, which can include both solid and fluid parts. The forces and couple moments acting on the control volume are shown on its free-body diagram.

Since the momentum equations are vector equations, they can be resolved into scalar components along x, y, z inertial coordinate axes.

$$\Sigma \mathbf{F} = \frac{\partial}{\partial t} \int_{cv} \mathbf{V} \rho \, d\forall + \int_{cs} \mathbf{V} \, \rho \mathbf{V} \cdot d\mathbf{A}$$

$$\Sigma \mathbf{M}_O = \frac{\partial}{\partial t} \int_{cv} (\mathbf{r} \times \mathbf{V}) \rho \, d\forall + \int_{cs} (\mathbf{r} \times \mathbf{V}) \, \rho \mathbf{V} \cdot d\mathbf{A}$$

If the control volume is in motion, then the velocity entering and exiting each open control surface must be measured *relative* to the control surface.

$$\Sigma \mathbf{F} = \frac{\partial}{\partial t} \int_{cv} \mathbf{V}_{f/cv} \rho \, d\forall + \int_{cs} \mathbf{V}_{f/cs} \, \rho \mathbf{V}_{f/cs} \cdot d\mathbf{A}$$

A propeller acts as a screw, which causes the linear momentum of the fluid to increase as it flows toward, through, then past the blades. Wind turbines decrease the linear momentum of the flow and thereby remove energy from it. A simple analysis of the flow through these two devices can be made using the linear momentum and Bernoulli equations.

If the control volume is chosen to have accelerated motion, as in the case of turbojets and rockets, then Newton's second law of motion, or the momentum equation, must account for the acceleration of the mass of the control volume.

$$\Sigma \mathbf{F} = m \frac{dV_{cv}}{dt} + \frac{\partial}{\partial t} \int_{cv} \mathbf{V}_{f/cv} \rho \, d\forall + \int_{cs} \mathbf{V}_{f/cs} \left(\rho \mathbf{V}_{f/cs} \cdot d\mathbf{A} \right)$$

Chapter 7

(© Worldspec/NASA/Alamy)

Hurricanes are a combination of free and forced vortices, referred to as a combined vortex. Their motion can be analyzed using the equation of differential fluid flow.

Differential Fluid Flow

7.1 Differential Analysis

In the previous chapters, we used a *finite control volume* to apply the conservation of mass, and the energy and momentum equations, to study problems involving fluid flow. There are situations, however, where we may have to determine the pressure and shear-stress *variations* over a surface, or find the fluid's velocity and acceleration profiles within a closed conduit. To do this, we will want to consider a *differential-size element* of fluid, since the variations we are seeking will have to come from the integration of differential equations.

Later in this chapter we will see that for any *real fluid*, this *differential flow analysis*, as it is called, has limited analytical scope. This is because the effects of the fluid's viscosity and its compressibility cause the differential equations that describe the flow to be rather complex. However, if we neglect these effects and consider the fluid to be an *ideal fluid*, then the equations become more manageable, and their solution will provide valuable information for many common types of engineering problems.

323

The techniques used for solving *ideal fluid flow problems* form the basis of the field of **hydrodynamics**. This was the first branch of fluid mechanics since it is a theoretical investigation, without the need for any experimental measurements, apart from knowing the fluid's density. Although this approach *neglects* the effects of viscosity, the results obtained from ideal fluid flow can sometimes provide a *reasonable approximation* for studying the general characteristics of any real fluid flow. Before we analyze differential fluid flow, however, we will first discuss some important aspects of its kinematics.

7.2   Kinematics of Differential Fluid Elements

In general, the forces acting on an element of fluid while it is flowing will tend to cause it to undergo a "rigid-body" displacement, as well as a distortion or change in its shape. The *rigid-body motion* consists of a translation and rotation of the element; the *distortion* causes elongation or contraction of its sides, as well as changes in the angles between them. For example, translation *and* linear distortion can occur when an ideal fluid flows through a converging channel, Fig. 7–1. And translation *and* angular distortion can occur if the fluid is viscous and the flow is steady, Fig. 7–2. In more complex flows, all these motions can occur simultaneously. But before we consider any general motion, we will first analyze each motion and distortion separately, and then show how it is related to the velocity gradient that causes it.

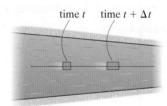

Element translation and linear distortion

Fig. 7–1

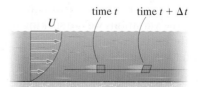

Element translation and angular distortion

Fig. 7–2

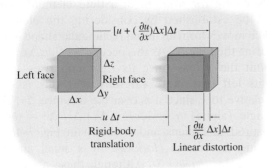

Fig. 7–3

Translation and Linear Distortion.

Consider a differential element of fluid that is moving in an unsteady three-dimensional flow. The rate of translation of the element is defined by its velocity field $\mathbf{V} = \mathbf{V}(x, y, z, t)$. If u is the velocity component in the x direction, Fig. 7–3, then during the time Δt, the element's left face will translate in the x direction by an amount $u\,\Delta t$, whereas the right face will move $\left[u + (\partial u/\partial x)\,\Delta x\right]\Delta t$. The right face moves farther by $\left[(\partial u/\partial x)\,\Delta x\right]\Delta t$ because of an acceleration, that is, an increase ∂u in its velocity.* We have expressed this change as a partial derivative because, in general, u will be a function of both the element's location in the flow and the time; that is, $u = u(x, y, z, t)$.

The result of the movement is therefore a ***rigid-body translation*** of $u\,\Delta t$ and a ***linear distortion*** of $\left[(\partial u/\partial x)\,\Delta x\right]\Delta t$. In the limit, as $\Delta t \to 0$ and $\Delta x \to 0$, the change in the volume of the element due to this distortion becomes $\partial \mathcal{V}_x = \left[(\partial u/\partial x)\,dx\right]dy\,dz\,dt$. If we consider velocity components v in the y direction, and w in the z direction, similar results are obtained. Thus, a general volume change of the element becomes

$$\delta \mathcal{V} = \left[\frac{\partial u}{\partial x} + \frac{\partial v}{\partial y} + \frac{\partial w}{\partial z}\right](dx\,dy\,dz)\,dt$$

The *rate* at which the volume per unit volume changes is called the ***volumetric dilatation rate***. It can be expressed as

$$\frac{\delta \mathcal{V}/d\mathcal{V}}{dt} = \frac{\partial u}{\partial x} + \frac{\partial v}{\partial y} + \frac{\partial w}{\partial z} = \nabla \cdot \mathbf{V} \qquad (7\text{–}1)$$

Here the vector operator "del" is defined as $\nabla = (\partial/\partial x)\mathbf{i} + (\partial/\partial y)\mathbf{j} + (\partial/\partial z)\mathbf{k}$, and the velocity field is $\mathbf{V} = u\mathbf{i} + v\mathbf{j} + w\mathbf{k}$. In vector analysis, this result $\nabla \cdot \mathbf{V}$ is referred to as the ***divergence*** of $\mathbf{V}$, or simply div $\mathbf{V}$.

---

*Throughout this chapter and elsewhere, we will neglect the higher-order terms of this Taylor series expansion, such as $\left(\frac{\partial^2 u}{\partial x^2}\right)\frac{1}{2!}(\Delta x)^2 + \cdots$, since in the limit, as $\Delta t \to 0$, they will all be small compared to the first-order term.

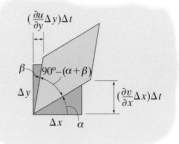

$(u + \frac{\partial u}{\partial y} \Delta y) \Delta t$

$u \Delta t$

Δy

$(v + \frac{\partial v}{\partial x} \Delta x) \Delta t$

$v \Delta t$

Δx

(a)

$(\frac{\partial u}{\partial y} \Delta y) \Delta t$

β $90° - (\alpha + \beta)$

Δy

$(\frac{\partial v}{\partial x} \Delta x) \Delta t$

Δx α

Angular distortion

(b)

Fig. 7–4

Rotation   Criteria for rotation of a fluid element and its angular distortion can be established by considering the element shown in Fig. 7–4a, which moves from an initial rectangular shape into a deformed final shape during the time Δt. As this occurs, the side Δx rotates *counterclockwise* about the z axis since its right end rises $[(\partial v / \partial x) \Delta x] \Delta t$ higher than its left end. Note carefully that this is caused by the change of v relative to x, since it is over the distance Δx. Therefore, the very small angle (alpha) $\alpha = [(\partial v / \partial x) \Delta x] \Delta t / \Delta x = (\partial v / \partial x) \Delta t$. This angle is measured in radians and is shown in Fig. 7–4b. In a similar manner, the side Δy rotates *clockwise* by a small angle (beta) $\beta = [(\partial u / \partial y) \Delta y] \Delta t / \Delta y = (\partial u / \partial y) \Delta t$. Although these are different angles, we will define the ***average angular velocity*** ω_z (omega) of these two adjacent sides as the average time rate of change of α and β as $\Delta t \to 0$. Using the right-hand rule, with the thumb directed along the $+z$ axis (outward), then α is positive and β is negative. Therefore, the average angular velocity, measured in rad/s, is

$$\omega_z = \lim_{\Delta t \to 0} \frac{1}{2} \frac{(\alpha - \beta)}{\Delta t} = \frac{1}{2}(\dot{\alpha} - \dot{\beta})$$

or

$$\omega_z = \frac{1}{2}\left(\frac{\partial v}{\partial x} - \frac{\partial u}{\partial y} \right) \qquad (7\text{–}2)$$

To summarize, the derivatives $\dot{\alpha} = \partial v / \partial x$ and $\dot{\beta} = \partial u / \partial y$ represent the angular velocities of the *sides* of the fluid element, and their average produces the average angular velocity ω_z. Because the bisector between α and β is at the angle $\alpha + \frac{1}{2}(90° - (\beta + \alpha) = 45° + \frac{1}{2}(\alpha - \beta)$, then the time rate of change of this angle is $\frac{1}{2}(\dot{\alpha} - \dot{\beta})$, and so you may also want to think of this as the angular velocity of the *bisector* of α and β.

If the flow is three dimensional, then by the same arguments, we will have x and y components of angular velocity. Therefore, in general,

$$\omega_x = \frac{1}{2}\left(\frac{\partial w}{\partial y} - \frac{\partial v}{\partial z} \right)$$

$$\omega_y = \frac{1}{2}\left(\frac{\partial u}{\partial z} - \frac{\partial w}{\partial x} \right) \qquad (7\text{–}3)$$

$$\omega_z = \frac{1}{2}\left(\frac{\partial v}{\partial x} - \frac{\partial u}{\partial y} \right)$$

These three components can be written in vector form as

$$\boldsymbol{\omega} = \omega_x \mathbf{i} + \omega_y \mathbf{j} + \omega_z \mathbf{k}$$

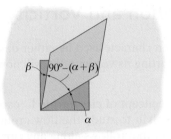

Shear strain

(c)

Fig. 7–4 (cont.)

We can also express Eqs. 7–3 as one-half the **curl** of the velocity. The curl is defined in vector analysis by the vector cross product $\nabla \times \mathbf{V}$. Since $\mathbf{V} = u\mathbf{i} + v\mathbf{j} + w\mathbf{k}$, then

$$\boldsymbol{\omega} = \frac{1}{2} \nabla \times \mathbf{V} = \frac{1}{2} \begin{vmatrix} \mathbf{i} & \mathbf{j} & \mathbf{k} \\ \dfrac{\partial}{\partial x} & \dfrac{\partial}{\partial y} & \dfrac{\partial}{\partial z} \\ u & v & w \end{vmatrix} \qquad (7\text{–}3a)$$

Angular Distortion   Besides using the angles α and β to report the element's angular rotation, they can also be used to define the element's angular distortion, Fig. 7–4c. That is, the 90° angle between the adjacent sides of the element becomes $90° - (\alpha + \beta)$, and so the *change* in this angle is $90° - [90° - (\alpha + \beta)] = \alpha + \beta$. This is called the **shear strain** γ_{xy} (gamma), which is measured in radians. Since *fluids flow*, we will be interested in the *time rate of change in the shear strain*, so in the limit as $\Delta t \to 0$, we have

$$\dot{\gamma}_{xy} = \dot{\alpha} + \dot{\beta} = \frac{\partial v}{\partial x} + \frac{\partial u}{\partial y} \qquad (7\text{–}4)$$

This equation is appropriate for two-dimensional flow, where angular distortion is about the z axis. If the flow is three dimensional, then in a similar manner, shear strains due to angular distortions about the x and y axes give shear strain rates about these two axes as well. In general, then,

$$\dot{\gamma}_{xy} = \frac{\partial v}{\partial x} + \frac{\partial u}{\partial y}$$

$$\dot{\gamma}_{xz} = \frac{\partial w}{\partial x} + \frac{\partial u}{\partial z} \qquad (7\text{–}5)$$

$$\dot{\gamma}_{yz} = \frac{\partial w}{\partial y} + \frac{\partial v}{\partial z}$$

Later in the chapter we will show how these shear-strain rates can be related to the shear stress that causes them, a consequence of the viscosity of the fluid.

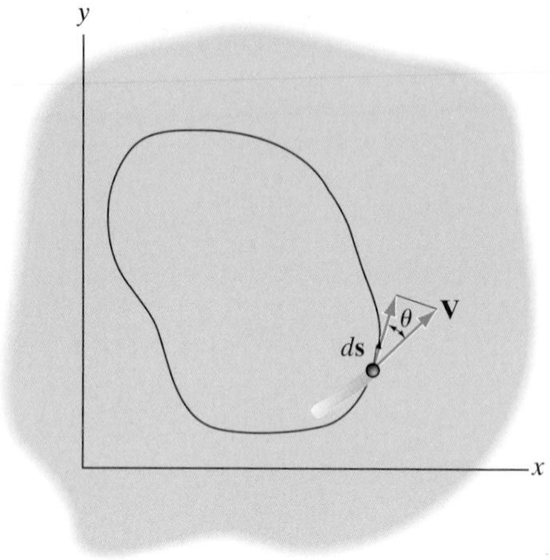

7.3   Circulation and Vorticity

Rotational flow is often characterized by either describing its circulation about a region, or reporting its vorticity. We will now define each of these characteristics.

Circulation.   The concept of *circulation* Γ (capital gamma) was first introduced by Lord Kelvin to study the flow around the *boundary* of a body. It defines the flow that follows along any closed three-dimensional curve. For a unit depth or two-dimensional volumetric flow, the circulation has units of m^2/s. To obtain the circulation, we must integrate around the curve, the component of the velocity that is always *tangent* to the curve, Fig. 7–5. Formally, this is done using a line integral of the dot product $\mathbf{V} \cdot d\mathbf{s} = V\,ds\,\cos\theta$, so that

$$\Gamma = \oint \mathbf{V} \cdot d\mathbf{s} \qquad (7\text{–}6)$$

By convention, the integration is performed counterclockwise, that is, in the $+z$ direction.

To show an application, let's calculate the circulation around a small fluid element located at point (x, y), which is immersed in a general two-dimensional steady flow field $\mathbf{V} = u(x, y)\mathbf{i} + v(x, y)\mathbf{j}$, Fig. 7–6. The average velocities of the flow along each side of the element have the magnitudes and directions shown. Applying Eq. 7–6, we have

$$\Gamma = u\,\Delta x + \left(v + \frac{\partial v}{\partial x}\Delta x\right)\Delta y - \left(u + \frac{\partial u}{\partial y}\Delta y\right)\Delta x - v\,\Delta y$$

which simplifies to

$$\Gamma = \left(\frac{\partial v}{\partial x} - \frac{\partial u}{\partial y}\right)\Delta x \Delta y$$

Note that circulation about this small element, or, for that matter any other size body, *does not* mean that individual fluid particles "circle" around the boundary of the body. After all, the flow on opposite faces of the element is in the same direction, Fig. 7–6. Rather, the circulation is simply the *net result* of the flow around the body as determined from Eq. 7–6.

y

x

ds θ **V**

Circulation

Fig. 7–5

7

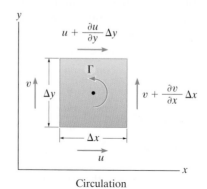

Circulation

Fig. 7–6

Vorticity.

We define the **vorticity** ζ (zeta) at a point (x, y) as the circulation per unit area of an element located at the point. For example, for the element in Fig. 7–7a, dividing its area $\Delta x \Delta y$ into Γ, we get

$$\zeta = \frac{\Gamma}{A} = \frac{\partial v}{\partial x} - \frac{\partial u}{\partial y} \tag{7–7}$$

Comparing this with Eq. 7–2, we see that $\zeta = 2\omega_z$. Perhaps another way to look at this is to imagine a fluid particle rotating at ω_z around a circular path of radius Δr, Fig. 7–7b. Since this particle has a velocity $V = \omega_z \Delta r$, then the circulation is $\Gamma = V(2\pi \Delta r) = 2\pi\omega_z \Delta r^2$. The area of the circle is $\pi \Delta r^2$, and so the vorticity becomes $\zeta = 2\pi\omega_z \Delta r^2 / \pi \Delta r^2 = 2\omega_z$.

The vorticity is actually a vector. In this two-dimensional case, using the right-hand rule, it is directed along the z axis. If we were to consider three-dimensional flow, then by the same development, using Eq. 7–3, we would obtain

$$\boldsymbol{\zeta} = 2\boldsymbol{\omega} = \nabla \times \mathbf{V} \tag{7–8}$$

Irrotational Flow.

The angular rotation, or the vorticity, provides a means of classifying the flow. If $\boldsymbol{\omega} \neq \mathbf{0}$, then the flow is called **rotational flow**; however, if $\boldsymbol{\omega} = \mathbf{0}$ throughout the flow field, then it is termed **irrotational flow**.

Ideal fluids exhibit irrotational flow because no viscous friction forces act on ideal fluid elements, only pressure and gravitational forces. Because these two forces are always *concurrent*, ideal fluid elements cannot be forced to rotate while they are in motion.

The difference between rotational and irrotational flow can be illustrated by a simple example. The velocity profiles for an ideal fluid and a viscous (real) fluid are shown in Fig. 7–8. No rotation occurs in the ideal fluid because the entire element moves with the same velocity, Fig. 7–8a. This is irrotational flow. However, the top and bottom surfaces of the element in the viscous fluid move at *different velocities*, Fig. 7–8b, and this will cause the vertical sides to rotate clockwise at the rate $\dot\beta$. As a result, this produces *rotational* flow $\omega_z = (\dot\alpha - \dot\beta)/2 = (0 - \dot\beta)/2 = -\dot\beta/2$.

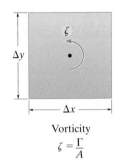

Vorticity

$$\zeta = \frac{\Gamma}{A}$$

(a)

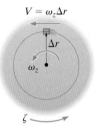

$V = \omega_z \Delta r$

Vorticity

(b)

Fig. 7–7

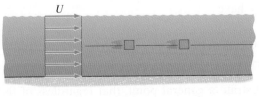

Ideal fluid

(a)

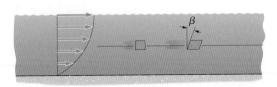

Viscous fluid

(b)

Fig. 7–8

EXAMPLE 7.1

The ideal fluid in Fig. 7–9 has a uniform velocity of $U = 0.2$ m/s. Determine the circulation about the triangular and circular paths.

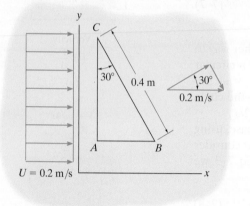

(a)

(b)

Fig. 7–9

SOLUTION

Fluid Behavior. We have steady flow of an ideal fluid within the x–y plane.

Triangular Path. Circulation is defined as positive counterclockwise. In this case we do not need to integrate; rather we determine the length of each side of the triangle, and the component of velocity along each side. From Fig. 7–9a, for CA, AB, BC, we have

$$\Gamma = \oint \mathbf{V} \cdot d\mathbf{s} = (0)(0.4 \text{ m cos } 30°) + (0.2 \text{ m/s})(0.4 \text{ m sin } 30°)$$
$$- (0.2 \sin 30° \text{ m/s})(0.4 \text{ m})$$

$$= 0 \qquad\qquad Ans.$$

Circular Path. The boundary of the cylinder is easily defined using polar coordinates, with θ positive counterclockwise as shown in Fig. 7–14b. Since $ds = (0.1 \text{ m}) d\theta$, we have

$$\Gamma = \oint \mathbf{V} \cdot d\mathbf{s} = \int_0^{2\pi} (-0.2 \sin \theta)(0.1 \text{ m}) \, d\theta = 0.02(\cos \theta)\Big|_0^{2\pi} = 0 \quad Ans.$$

Both of these cases illustrate a general point, that regardless of the shape of the path, for *uniform flow* an ideal fluid will not produce a circulation, and because $\zeta = \Gamma/A$, no vorticity is produced either.

EXAMPLE | 7.2

The velocity of a viscous fluid flowing between the parallel surfaces in Fig. 7–10a is defined by $U = 0.002\left[1 - 10\left(10^3\right)y^2\right]$ m/s, where y is in meters. Determine the vorticity and shear-strain rate of a fluid element located at $y = 5$ mm within the flow.

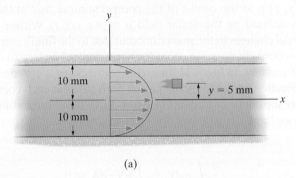

(a)

SOLUTION

Fluid Description.   We have steady one-dimensional flow of a real fluid.

Vorticity.   We must apply Eq. 7–7, where $u = 0.002\left(1 - 10\left(10^3\right)y^2\right)$ m/s and $v = 0$.

$$\zeta = \frac{\partial v}{\partial x} - \frac{\partial u}{\partial y}$$

$$= 0 - 0.002\left[0 - 10\left(10^3\right)(2y)\right] \text{ rad/s}\Big|_{y = 0.005 \text{ m}}$$

$$= 0.200 \text{ rad/s} \qquad\qquad Ans.$$

This vorticity is a consequence of the fluid's viscosity, and since $\zeta \neq 0$, we have rotational flow. Actually, the element has a rotation of $\omega_z = \zeta/2 = 0.1$ rad/s, Fig. 7–10b. It is positive because at $y = 0.005$ m, the velocity profile shows that the top of the fluid element is moving *slower* than its bottom, Fig. 7–10a.

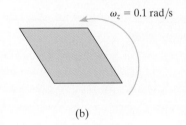

$\omega_z = 0.1$ rad/s

(b)

Shear-Strain Rate.   Applying Eq. 7–4,

$$\dot{\gamma}_{xy} = \dot{\alpha} + \dot{\beta} = \frac{\partial v}{\partial x} + \frac{\partial u}{\partial y}$$

$$= 0 + 0.002\left[0 - 10\left(10^3\right)(2y)\right] \text{ rad/s}\Big|_{y = 0.005 \text{ m}}$$

$$= -0.200 \text{ rad/s} \qquad\qquad Ans.$$

This is a negative rate change. It occurs because β is positive clockwise, and the shear strain is defined as the difference in the angle $90° - (90° + \beta) = -\beta$, Fig. 7–10c.

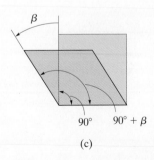

β

$90°$ $90° + \beta$

(c)

Fig. 7–10

7.4   Conservation of Mass

In this section we will derive the continuity equation for an element of fluid flowing through a fixed differential control volume that only has open control surfaces, Fig. 7–11. We will assume three-dimensional flow, where the velocity field has components $u = u(x, y, z, t), v = v(x, y, z, t), w = w(x, y, z, t)$. Point (x, y, z) is at the center of the control volume, and at this point the density is defined by the scalar field $\rho = \rho(x, y, z, t)$. Within the control volume, local changes to the mass can occur due to the fluid's compressibility. Also, convective changes can occur from one control surface to another due to nonuniform flow. In Fig. 7–11 these convective changes are considered only in the x direction, as noted by the partial derivatives at each control surface. If we apply the continuity equation, Eq. 4–12, to the control volume in the x direction, we have

$$\frac{\partial}{\partial t} \int_{cv} \rho \, d\mathcal{V} + \int_{cs} \rho \mathbf{V} \cdot d\mathbf{A} = 0$$

$$\frac{\partial \rho}{\partial t} \Delta x \, \Delta y \, \Delta z + \left(\rho u + \frac{\partial(\rho u)}{\partial x} \frac{\Delta x}{2} \right) \Delta y \, \Delta z - \left(\rho u - \frac{\partial(\rho u)}{\partial x} \frac{\Delta x}{2} \right) \Delta y \, \Delta z = 0$$

Dividing by $\Delta x \, \Delta y \, \Delta z$, and simplifying, we get

$$\frac{\partial \rho}{\partial t} + \frac{\partial(\rho u)}{\partial x} = 0 \tag{7–9}$$

If we include the convective changes in the y and z directions, then the continuity equation becomes

$$\frac{\partial \rho}{\partial t} + \frac{\partial(\rho u)}{\partial x} + \frac{\partial(\rho v)}{\partial y} + \frac{\partial(\rho w)}{\partial z} = 0 \tag{7–10}$$

Finally, using the gradient operator $\nabla = \partial/\partial x \mathbf{i} + \partial/\partial y \mathbf{j} + \partial/\partial z \mathbf{k}$, and expressing the velocity as $\mathbf{V} = u\mathbf{i} + v\mathbf{j} + w\mathbf{k}$, we can write the continuity equation for the differential element in vector form as

$$\frac{\partial \rho}{\partial t} + \nabla \cdot \rho \mathbf{V} = 0 \tag{7–11}$$

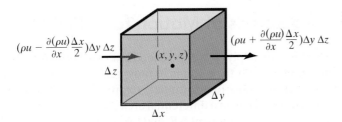

Fig. 7–11

Two-Dimensional Steady Flow of an Ideal Fluid.
Although we have developed the continuity equation in its most general form, often it has applications to two-dimensional steady-state flow of an ideal fluid. For this special case, the fluid is incompressible and so ρ is constant. As a result Eq. 7–10 then becomes

$$\boxed{\frac{\partial u}{\partial x} + \frac{\partial v}{\partial y} = 0} \qquad (7\text{–}12)$$

steady flow
incompressible fluid

Or, from Eq. 7–11, we can write

$$\nabla \cdot \mathbf{V} = 0 \qquad (7\text{–}13)$$

As noted by Eq. 7–1, this is the same as saying the *volumetric dilatation rate must be zero*. In other words, the volume rate of change of the fluid element must be zero since the density of an ideal fluid is constant. For example, if a positive change in size occurs in the x direction ($\partial u/\partial x > 0$), then by Eq. 7–12, a corresponding negative change in size ($\partial v/\partial y < 0$) must occur in the y direction.

Cylindrical Coordinates.
The continuity equation can also be expressed for a differential element in terms of cylindrical coordinates r, θ, z, Fig. 7–12. For the sake of completeness, we will state the result here without proof, and then use it later to describe some important types of symmetrical flow. In the general case,

$$\frac{\partial \rho}{\partial t} + \frac{1}{r}\frac{\partial(r\rho v_r)}{\partial r} + \frac{1}{r}\frac{\partial(\rho v_\theta)}{\partial \theta} + \frac{\partial(\rho v_z)}{\partial z} = 0 \qquad (7\text{–}14)$$

If the fluid is incompressible and the flow is steady, then in two dimensions (r, θ) the continuity equation becomes

$$\boxed{\frac{v_r}{r} + \frac{\partial v_r}{\partial r} + \frac{1}{r}\frac{\partial v_\theta}{\partial \theta} = 0} \qquad (7\text{–}15)$$

steady flow
incompressible fluid

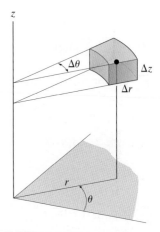

Cylindrical coordinates

Fig. 7–12

7

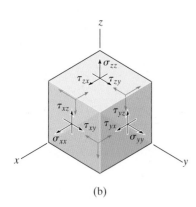

(a)

(b)

7.5   Equations of Motion for a Fluid Particle

In this section we will apply Newton's second law of motion to a differential fluid element and express the result in its most general form. But before we do this, however, we must first formulate expressions that represent the effect of a force $\Delta \mathbf{F}$ acting on a differential area ΔA. As shown in Fig. 7–13a, $\Delta \mathbf{F}$ will have a normal component $\Delta \mathbf{F}_z$ and two shear components $\Delta \mathbf{F}_x$ and $\Delta \mathbf{F}_y$. Stress is the result of these *surface force components*. The normal component creates a **normal stress** on the area, defined as

$$\sigma_{zz} = \lim_{\Delta A \to 0} \frac{\Delta F_z}{\Delta A}$$

and the shear components create **shear stresses**

$$\tau_{zx} = \lim_{\Delta A \to 0} \frac{\Delta F_x}{\Delta A} \qquad \tau_{zy} = \lim_{\Delta A \to 0} \frac{\Delta F_y}{\Delta A}$$

The first letter (z) in this subscript notation represents the *outward* normal direction that defines the direction of the area element ΔA, and the second letter represents the direction of the stress. If we now generalize this idea and consider forces acting on the six faces of a volume element of fluid, then three components of stress will act on each face of the element as shown in Fig. 7–13b.

At each point in the fluid there will be a **stress field** that defines these stresses. And because this field *changes* from one point to the next, the forces these stresses produce on a fluid particle (element) must account for these changes. For example, consider the free-body diagram of the fluid particle in Fig. 7–13c, which shows only the forces of the stress components that act in the x direction. The resultant **surface force** in the x direction is

$$
\begin{aligned}
(\Delta F_x)_{\text{sf}} = {} & \left(\sigma_{xx} + \frac{\partial \sigma_{xx}}{\partial x} \frac{\Delta x}{2} \right) \Delta y\, \Delta z - \left(\sigma_{xx} - \frac{\partial \sigma_{xx}}{\partial x} \frac{\Delta x}{2} \right) \Delta y\, \Delta z \\[6pt]
& + \left(\tau_{yx} + \frac{\partial \tau_{yx}}{\partial y} \frac{\Delta y}{2} \right) \Delta x\, \Delta z - \left(\tau_{yx} - \frac{\partial \tau_{yx}}{\partial y} \frac{\Delta y}{2} \right) \Delta x\, \Delta z \\[6pt]
& + \left(\tau_{zx} + \frac{\partial \tau_{zx}}{\partial z} \frac{\Delta z}{2} \right) \Delta x\, \Delta y - \left(\tau_{zx} - \frac{\partial \tau_{zx}}{\partial z} \frac{\Delta z}{2} \right) \Delta x\, \Delta y
\end{aligned}
$$

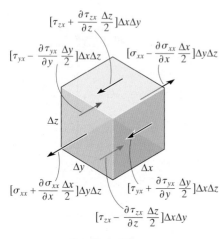

Free-body diagram
(c)

Fig. 7–13

By collecting terms this can be simplified, and in a similar manner, the resultant surface forces produced by the stresses in the y and z directions can also be obtained. We have

$$(\Delta F_x)_{sf} = \left(\frac{\partial \sigma_{xx}}{\partial x} + \frac{\partial \tau_{yx}}{\partial y} + \frac{\partial \tau_{zx}}{\partial z} \right) \Delta x \, \Delta y \, \Delta z$$

$$(\Delta F_y)_{sf} = \left(\frac{\partial \tau_{xy}}{\partial x} + \frac{\partial \sigma_{yy}}{\partial y} + \frac{\partial \tau_{zy}}{\partial z} \right) \Delta x \, \Delta y \, \Delta z$$

$$(\Delta F_z)_{sf} = \left(\frac{\partial \tau_{xz}}{\partial x} + \frac{\partial \tau_{yz}}{\partial y} + \frac{\partial \sigma_{zz}}{\partial z} \right) \Delta x \, \Delta y \, \Delta z$$

Apart from these forces, there is also the **body force** due to the weight of the particle. If Δm is the particle's mass, this force is $\Delta W = (\Delta m)g = \rho g \Delta x \, \Delta y \, \Delta z$. To further generalize this development, we will assume the x, y, z axes have some *arbitrary orientation* so that the weight will have components $\Delta W_x, \Delta W_y, \Delta W_z$ along each axis. Therefore, the sums of all the body and surface force components acting on the fluid particle are

$$\Delta F_x = \left(\rho g_x + \frac{\partial \sigma_{xx}}{\partial x} + \frac{\partial \tau_{yx}}{\partial y} + \frac{\partial \tau_{zx}}{\partial z} \right) \Delta x \, \Delta y \, \Delta z$$

$$\Delta F_y = \left(\rho g_y + \frac{\partial \tau_{xy}}{\partial x} + \frac{\partial \sigma_{yy}}{\partial y} + \frac{\partial \tau_{zy}}{\partial z} \right) \Delta x \, \Delta y \, \Delta z \qquad (7\text{--}16)$$

$$\Delta F_z = \left(\rho g_z + \frac{\partial \tau_{xz}}{\partial x} + \frac{\partial \tau_{yz}}{\partial y} + \frac{\partial \sigma_{zz}}{\partial z} \right) \Delta x \, \Delta y \, \Delta z$$

With these forces established, we can now apply Newton's second law of motion to the particle. Provided the particle's velocity is expressed as a velocity field, $\mathbf{V} = \mathbf{V}(x, y, z, t)$, then the material derivative is used to determine the acceleration, Eq. 3–5. Thus,

$$\Sigma \mathbf{F} = \Delta m \frac{D\mathbf{V}}{Dt} = (\rho \Delta x \, \Delta y \, \Delta z) \left[\frac{\partial \mathbf{V}}{\partial t} + u \frac{\partial \mathbf{V}}{\partial x} + v \frac{\partial \mathbf{V}}{\partial y} + w \frac{\partial \mathbf{V}}{\partial z} \right]$$

When we substitute Eqs. 7–16, factor out the volume $\Delta x \Delta y \Delta z$, and then use $\mathbf{V} = u\,\mathbf{i} + v\,\mathbf{j} + w\,\mathbf{k}$, the x, y, z components of this equation become

$$\rho g_x + \frac{\partial \sigma_{xx}}{\partial x} + \frac{\partial \tau_{yx}}{\partial y} + \frac{\partial \tau_{zx}}{\partial z} = \rho \left(\frac{\partial u}{\partial t} + u \frac{\partial u}{\partial x} + v \frac{\partial u}{\partial y} + w \frac{\partial u}{\partial z} \right)$$

$$\rho g_y + \frac{\partial \tau_{xy}}{\partial x} + \frac{\partial \sigma_{yy}}{\partial y} + \frac{\partial \tau_{zy}}{\partial z} = \rho \left(\frac{\partial v}{\partial t} + u \frac{\partial v}{\partial x} + v \frac{\partial v}{\partial y} + w \frac{\partial v}{\partial z} \right) \qquad (7\text{--}17)$$

$$\rho g_z + \frac{\partial \tau_{xz}}{\partial x} + \frac{\partial \tau_{yz}}{\partial y} + \frac{\partial \sigma_{zz}}{\partial z} = \rho \left(\frac{\partial w}{\partial t} + u \frac{\partial w}{\partial x} + v \frac{\partial w}{\partial y} + w \frac{\partial w}{\partial z} \right)$$

In the next section we will apply these equations to study an ideal fluid. Then later, in Sec. 7.11 we will consider the more general case of a Newtonian fluid.

7

7.6 The Euler and Bernoulli Equations

If we consider the fluid to be an ideal fluid, then the equations of motion will reduce to a simpler form. In particular, there will be no viscous shear stress on the particle (element), and the three normal stress components will represent the pressure. Since these normal stresses have all been defined in Fig. 7–13b as positive outward, and as a convention, positive pressure produces a *compressive stress*, then $\sigma_{xx} = \sigma_{yy} = \sigma_{zz} = -p$. As a result, the general equations of motion for an ideal fluid particle become

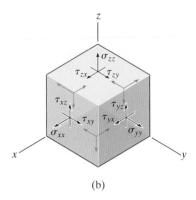

(b)

Fig. 7–13

$$\rho g_x - \frac{\partial p}{\partial x} = \rho\left(\frac{\partial u}{\partial t} + u\frac{\partial u}{\partial x} + v\frac{\partial u}{\partial y} + w\frac{\partial u}{\partial z}\right)$$

$$\rho g_y - \frac{\partial p}{\partial y} = \rho\left(\frac{\partial v}{\partial t} + u\frac{\partial v}{\partial x} + v\frac{\partial v}{\partial y} + w\frac{\partial v}{\partial z}\right) \qquad (7\text{–}18)$$

$$\rho g_z - \frac{\partial p}{\partial z} = \rho\left(\frac{\partial w}{\partial t} + u\frac{\partial w}{\partial x} + v\frac{\partial w}{\partial y} + w\frac{\partial w}{\partial z}\right)$$

These equations are called the **Euler equations of motion**, expressed in x, y, z coordinates. Recall that in Sec. 5.1 we derived them for streamline coordinates, s, n, where they took on a simpler form.

Using the gradient operator, we can also write Eqs. 7–18 in a more compact form, namely

$$\rho \mathbf{g} - \nabla p = \rho\left[\frac{\partial \mathbf{V}}{\partial t} + (\mathbf{V} \cdot \nabla)\mathbf{V}\right] \qquad (7\text{–}19)$$

Two-Dimensional Steady Flow. In many cases we will have steady two-dimensional flow, and the z component of velocity $w = 0$. If we orient the x and y axes so that $\mathbf{g} = -g\mathbf{j}$, then Euler's equations (Eqs. 7–18) become

$$-\frac{1}{\rho}\frac{\partial p}{\partial x} = u\frac{\partial u}{\partial x} + v\frac{\partial u}{\partial y} \qquad (7\text{–}20)$$

$$-\frac{1}{\rho}\frac{\partial p}{\partial y} - g = u\frac{\partial v}{\partial x} + v\frac{\partial v}{\partial y} \qquad (7\text{–}21)$$

The velocity components u and v and the pressure p at any point within the fluid can now be determined, provided we can solve these two partial differential equations along with the continuity equation, Eq. 7–12.

Although we have derived Euler's equations here for an x, y coordinate system, for some problems it will be convenient to express them in polar coordinates r, θ. Without proof, they are

$$-\frac{1}{\rho}\frac{\partial p}{\partial r} = v_r\frac{\partial v_r}{\partial r} + \frac{v_\theta}{r}\frac{\partial v_r}{\partial \theta} - \frac{v\theta^2}{r} \qquad (7\text{–}22)$$

$$-\frac{1}{\rho}\frac{1}{r}\frac{\partial p}{\partial \theta} = v_r\frac{\partial v_\theta}{\partial r} + \frac{v_\theta}{r}\frac{\partial v_\theta}{\partial \theta} + \frac{v_\theta v_r}{r} \qquad (7\text{–}23)$$

The Bernoulli Equation.

In Sec. 5.2 we derived the Bernoulli equation by integrating the streamline component of the Euler equation. There it was shown that the result applies at any two points on the *same streamline*. If a condition of *irrotational flow* exists, meaning $\omega = 0$, then the Bernoulli equation can also be applied between any two points that are on *different streamlines*. To show this, assume we have irrotational two-dimensional flow so that $\omega_z = 0$ or $\partial u/\partial y = \partial v/\partial x$, Eq. 7–2. If we substitute this condition into Eqs. 7–20 and 7–21, we get

$$-\frac{1}{\rho}\frac{\partial p}{\partial x} = u\frac{\partial u}{\partial x} + v\frac{\partial v}{\partial x}$$

$$-\frac{1}{\rho}\frac{\partial p}{\partial y} - g = u\frac{\partial u}{\partial y} + v\frac{\partial v}{\partial y}$$

Since $\partial(u^2)/\partial x = 2u(\partial u/\partial x)$, $\partial(v^2)/\partial x = 2v(\partial v/\partial x)$, $\partial(u^2)/\partial y = 2u(\partial u/\partial y)$, and $\partial(v^2)/\partial y = 2v(\partial v/\partial y)$, the above equations become

$$-\frac{1}{\rho}\frac{\partial p}{\partial x} = \frac{1}{2}\frac{\partial(u^2 + v^2)}{\partial x}$$

$$-\frac{1}{\rho}\frac{\partial p}{\partial y} - g = \frac{1}{2}\frac{\partial(u^2 + v^2)}{\partial y}$$

Integrating with respect to x in the first equation, and with respect to y in the second equation, yields

$$-\frac{p}{\rho} + f(y) = \frac{1}{2}\left(u^2 + v^2\right) = \frac{1}{2}V^2$$

$$-\frac{p}{\rho} - gy + h(x) = \frac{1}{2}(u^2 + v^2) = \frac{1}{2}V^2$$

Here V is the fluid particle's velocity found from its components, $V^2 = u^2 + v^2$. Equating these two results, it is then necessary that $f(y) = -gy + h(x)$. The solution requires that $h(x) = $ Const., since x and y can vary independent of one another. As a result, the unknown function $f(y) = -gy + $ Const. Substituting this and $h(x) = $ Const. into the above two equations, we obtain in either case the Bernoulli equation, that is,

$$\boxed{\frac{p}{\rho} + \frac{V^2}{2} + gy = \text{Const.}} \qquad (7\text{–}24)$$

Steady irrotational flow, ideal fluid

Thus, if the flow is *irrotational*, then the Bernoulli equation may be applied between *any two points* (x_1, y_1) and (x_2, y_2) that are *not* necessarily on the same streamline. Of course, as noted, we must *also* require the *fluid to be ideal* and the *flow to be steady*.

Important Points

- In general, when a differential element of fluid is subjected to forces, it tends to undergo "rigid-body" translation and rotation, as well as linear and angular distortions.

- The rate of translation of a fluid element is determined by the velocity field.

- Linear distortion is measured by the change in volume per unit volume of the fluid element. The rate at which this change occurs is called the volumetric dilatation rate, $\nabla \cdot \mathbf{V}$.

- Rotation of a fluid element is defined by the rotation of the bisector of the fluid element, or the average angular velocities of its two sides. It is expressed as $\boldsymbol{\omega} = \frac{1}{2}\nabla \times \mathbf{V}$. Rotation can also be specified by the vorticity $\boldsymbol{\zeta} = \nabla \times \mathbf{V}$.

- If $\boldsymbol{\omega} = \mathbf{0}$, then the flow is termed *irrotational flow*, that is, no angular motion occurs. This type of flow always occurs in an ideal fluid because viscous shear forces are not present to cause rotation.

- Angular distortion is defined by the rate of change in shear strain, or the rate at which the angle between adjacent sides of the fluid element will change. These strains are caused by shear stress, which is the result of the fluid's viscosity. Ideal or inviscid fluids have no angular distortions.

- Since an ideal fluid is *incompressible*, the continuity equation for steady flow states that the rate of change in volume per unit volume for a fluid element must be zero, $\nabla \cdot \mathbf{V} = 0$.

- Euler's equations relate the pressure and gravitational forces acting on a differential fluid particle of an ideal fluid to its acceleration. If these equations are integrated and combined, then for steady irrotational flow, they produce the Bernoulli equation.

- The Bernoulli equation can be applied between *any two points* not located on the same streamline, provided the fluid is *ideal* and the *steady flow* is *irrotational*, that is, $\boldsymbol{\omega} = \mathbf{0}$.

EXAMPLE 7.3

The velocity field $\mathbf{V} = \{-6x\mathbf{i} + 6y\mathbf{j}\}$ m/s defines the two-dimensional ideal fluid flow in the vertical plane shown in Fig. 7–14. Determine the volumetric dilatation rate and the rotation of a fluid element located at point $B(1\text{ m}, 2\text{ m})$. If the pressure at point $A(1\text{ m}, 1\text{ m})$ is 250 kPa, what is the pressure at point B? Take $\rho = 1200$ kg/m³.

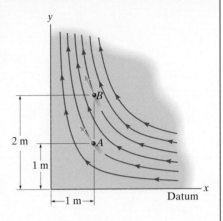

Fig. 7–14

SOLUTION

Fluid Description.   Since the velocity is not a function of time, the flow is steady. The fluid is an ideal fluid.

Volumetric Dilatation.   Applying Eq. 7–1, where $u = (-6x)$ m/s, $v = (6y)$ m/s, and $w = 0$, we have

$$\frac{\delta \forall / d \forall}{\partial t} = \frac{\partial u}{\partial x} + \frac{\partial v}{\partial y} + \frac{\partial w}{\partial z} = -6 + 6 + 0 = 0 \qquad Ans.$$

The result confirms that there is no change in the volume of the fluid element at B as it displaces.

Rotation.   The angular velocity of the fluid element at B is defined by Eq. 7–2.

$$\omega_z = \frac{1}{2}\left(\frac{\partial v}{\partial x} - \frac{\partial u}{\partial y}\right) = \frac{1}{2}(0 - 0) = 0 \qquad Ans.$$

Therefore, the fluid element will not rotate about the z axis. Actually, the above two results apply at *all points* in the fluid, since they are independent of x and y. In other words, an ideal fluid is incompressible and produces irrotational flow.

Pressure.   Since the flow is irrotational and steady, we can apply the Bernoulli equation at two points *not* located on the same streamline, Fig. 7–14. The velocities at A and B are

$$V_A = \sqrt{[-6(1)]^2 + [6(1)]^2} = 8.485 \text{ m/s}$$
$$V_B = \sqrt{[-6(1)]^2 + [6(2)]^2} = 13.42 \text{ m/s}$$

Therefore, with the datum at the x axis, we have

$$\frac{p_A}{\gamma} + \frac{V_A^2}{2g} + y_A = \frac{p_B}{\gamma} + \frac{V_B^2}{2g} + y_B$$

$$\frac{250(10^3)\text{ N/m}^2}{(1200\text{ kg/m}^3)(9.81\text{ m/s}^2)} + \frac{(8.485\text{ m/s})^2}{2(9.81\text{ m/s}^2)} + 1\text{ m} = \frac{p_B}{(1200\text{ kg/m}^3)(9.81\text{ m/s}^2)} + \frac{(13.42\text{ m/s})^2}{2(9.81\text{ m/s}^2)} + 2\text{ m}$$

$$p_B = 173\text{ kPa} \qquad Ans.$$

7.7   The Stream Function

In two dimensions, one method for satisfying the equation of continuity is to replace the *two* unknown velocity components u and v by a *single unknown function*, thus reducing the number of unknowns, and thereby simplifying the analysis of an ideal fluid flow problem. In this section we will use the stream function as a means for doing this, and in the next section we will consider its counterpart, the potential function.

The **stream function** ψ (psi) is the equation that represents *all the equations of the streamlines*. In two dimensions, it is a function of x and y, and for the equation of *each streamline* it is equal to a *specific constant* $\psi(x, y) = C$. You may recall in Sec. 3.3 we developed the technique for finding the equation of a streamline as it relates to the velocity components u and v. Here we will review this procedure, and extend its usefulness.

Velocity Components.   By definition, the velocity of a fluid particle is always tangent to the streamline along which it travels, Fig. 7–15. As a result, we can relate the velocity components **u** and **v** to the slope of the tangent by proportion. As shown in the figure, $dy/dx = v/u$, or

$$u\,dy - v\,dx = 0 \tag{7–25}$$

Now, if we take the total derivative of the streamline equation $\psi(x, y) = C$, which describes the streamline in Fig. 7–15, we have

$$d\psi = \frac{\partial \psi}{\partial x}\,dx + \frac{\partial \psi}{\partial y}\,dy = 0 \tag{7–26}$$

Comparing this with Eq. 7–25, the two components of velocity can be related to ψ. We require

$$\boxed{u = \frac{\partial \psi}{\partial y}, \qquad v = -\frac{\partial \psi}{\partial x}} \tag{7–27}$$

Therefore, if we know the equation of any streamline, $\psi(x, y) = C$, we can obtain the velocity components of a particle that travels along it by using these equations. By obtaining the velocity components in this way, we can show that for steady flow the stream function *automatically satisfies the equation of continuity*. By direct substitution into Eq. 7–12, we find

$$\frac{\partial u}{\partial x} + \frac{\partial v}{\partial y} = 0; \qquad \frac{\partial}{\partial x}\left(\frac{\partial \psi}{\partial y}\right) + \frac{\partial}{\partial y}\left(-\frac{\partial \psi}{\partial x}\right) = 0$$

$$\frac{\partial^2 \psi}{\partial x\,\partial y} - \frac{\partial^2 \psi}{\partial y\,\partial x} = 0$$

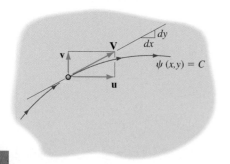

Velocity is tangent to streamline

Fig. 7–15

Later we will see that in some problems, it will be convenient to express the stream function and the velocity components in terms of their polar coordinates, r and θ, Fig. 7–16. Without proof, if $\psi(r, \theta) = C$ is given, then the radial and transverse velocity components are

$$v_r = \frac{1}{r}\frac{\partial \psi}{\partial \theta}, \qquad v_\theta = -\frac{\partial \psi}{\partial r} \qquad (7\text{–}28)$$

Volumetric Flow. The stream function can also be used to determine the volumetric flow between any two streamlines. For example, consider the triangular differential control volume in Fig. 7–17a that is located within the streamtube between the two streamlines ψ and $\psi + d\psi$. Since we have two-dimensional flow, we will consider the flow dq through this element as a measure of flow per unit depth z, that is, as having units of m^2/s. This flow only occurs within this streamtube, because the fluid velocity is always tangent to the streamlines, never perpendicular to them. Continuity requires the flow *into* the control surface AB to be equal to the flow *out of* the control surfaces BC and AC. In the case of BC, since the depth is 1 unit, the flow *out* is $u[dy(1)]$, but in the case of AC, by convention, v is positive *upward*, and so the flow *outward* is $-v[dx(1)]$. Applying the continuity equation, for steady incompressible flow, we have

$$\frac{\partial}{\partial t}\int_{cv} \rho\, d\forall + \int_{cs} \rho \mathbf{V} \cdot d\mathbf{A} = 0$$

$$0 - \rho\, dq + \rho u[dy(1)] - \rho v[dx(1)] = 0$$

$$dq = u\, dy - v\, dx$$

Substituting Eqs. 7–27 into this equation, the right side becomes Eq. 7–26. Therefore,

$$dq = d\psi$$

Thus, the flow dq between the two streamlines is simply found by finding their difference, $(\psi + d\psi) - \psi = d\psi$. The volume flow rate between *any two streamlines* a *finite distance* apart can now be determined by integrating this result. If $\psi_1(x, y) = C_1$ and $\psi_2(x, y) = C_2$, then

$$q = \int_{\psi_1}^{\psi_2} d\psi = \psi_2(x, y) - \psi_1(x, y) = C_2 - C_1 \qquad (7\text{–}29)$$

Let's summarize our results. If the stream function $\psi(x, y)$ is *known*, we can set it equal to various values of the constant $\psi(x, y) = C$ to obtain the streamlines, and thus visualize the flow. We can then use Eqs. 7–27 (or Eqs. 7–28) to determine the velocity components of the flow along a streamline. Also we can determine the volumetric flow between any two streamlines, such as $\psi_1(x, y) = C_1$ and $\psi_2(x, y) = C_2$, by finding the difference in their streamline constants, $q = C_2 - C_1$, Eq. 7–29. As we discussed in Sec. 3.3, once constructed, the distance between the streamlines will also provide an indication of the relative speed of the flow. This is based on the conservation of mass. For example, note how the fluid element in Fig. 7–17b must flatten as it moves through the streamtube in order to preserve its mass (or volume). Thus, at locations where the *streamlines are close together*, the *flow is fast*, and when these *streamlines are farther apart*, the *flow is slow*.

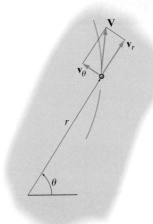

Polar coordinates

Fig. 7–16

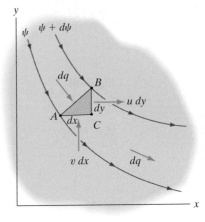

(a)

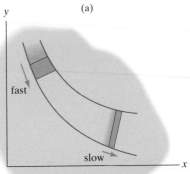

Conservation of mass

(b)

Fig. 7–17

EXAMPLE | 7.4

A flow field is defined by the stream function $\psi(x, y) = y^2 - x$. Draw the streamlines for $\psi_1(x, y) = 0$, $\psi_2(x, y) = 2 \text{ m}^2/\text{s}$, and $\psi_3(x, y) = 4 \text{ m}^2/\text{s}$. What is the velocity of a fluid particle at $y = 1$ m on the streamline $\psi_2(x, y) = 2 \text{ m}^2/\text{s}$?

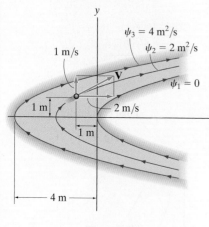

Fig. 7–18

SOLUTION

Fluid Description.   Since time is not involved, this is steady flow of an ideal fluid.

Stream Functions.   The equations for the three streamlines are

$$y^2 - x = 0$$
$$y^2 - x = 2$$
$$y^2 - x = 4$$

These equations are graphed in Fig. 7–18. Each is a parabola and represents a streamline for the constant that defines it.

Velocity.   The velocity components along each streamline are

$$u = \frac{\partial \psi}{\partial y} = \frac{\partial}{\partial y}\left(y^2 - x\right) = (2y) \text{ m/s}$$

$$v = -\frac{\partial \psi}{\partial x} = -\frac{\partial}{\partial x}\left(y^2 - x\right) = -(-1) = 1 \text{ m/s}$$

For the streamline $y^2 - x = 2$, at $y = 1$ m, then $x = -1$ m, so at this point, $u = 2 \text{ m/s}$ and $v = 1 \text{ m/s}$. These two components produce the resultant velocity of a fluid particle at this location, Fig. 7–18. It is

$$V = \sqrt{(2 \text{ m/s})^2 + (1 \text{ m/s})^2} = 2.24 \text{ m/s} \qquad \textit{Ans.}$$

Notice that the directions of the velocity components also provide a means for establishing the *direction* of the flow, as indicated by the small arrows on this streamline, Fig. 7–18.

Although it is not part of this problem, imagine the streamlines for which $\psi_1 = 0$ and $\psi_3 = 4 \text{ m}^2/\text{s}$ represent solid boundaries for a channel, Fig. 7–18. Then from Eq. 7–29, the volumetric flow per unit depth within this channel (or streamtube) would be

$$q = \psi_3 - \psi_1 = 4 \text{ m}^2/\text{s} - 0 = 4 \text{ m}^2/\text{s}$$

EXAMPLE 7.5

Uniform flow occurs at an angle θ with the y axis, as shown in Fig. 7–19. Determine the stream function for this flow.

SOLUTION

Fluid Description. We have steady uniform ideal fluid flow, since **U** is constant.

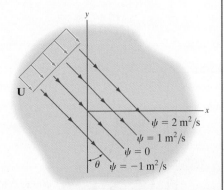

Uniform flow

Fig. 7–19

Velocity. The x and y components of velocity are

$$u = U \sin \theta \quad \text{and} \quad v = -U \cos \theta$$

Stream Function. If we relate the velocity component u to the stream function, we have

$$u = \frac{\partial \psi}{\partial y}; \qquad U \sin \theta = \frac{\partial \psi}{\partial y}$$

Integrating with respect to y, to obtain ψ, yields

$$\psi = (U \sin \theta)y + f(x) \qquad (1)$$

Here $f(x)$ is an unknown function that is to be determined. This can be done by doing the same thing for v, using Eq. 1. We have

$$v = -\frac{\partial \psi}{\partial x}; \qquad -U \cos \theta = -\frac{\partial}{\partial x}\big[(U \sin \theta)y + f(x)\big]$$

$$U \cos \theta = \left(0 + \frac{\partial}{\partial x}\big[f(x)\big]\right)$$

Integrating,

$$(U \cos \theta)x = f(x) + C$$

For convenience, we will set the constant of integration $C = 0$ to produce the stream function. Substituting the result into Eq. 1, we get

$$\psi(x, y) = (U \sin \theta)y + (U \cos \theta)x \qquad \textit{Ans.}$$

We can show that $\psi(x, y)$ produces the velocity U by noting that the velocity components are $u = \partial\psi/\partial y = U \sin \theta$ and $v = -\partial\psi/\partial x = -U \cos \theta$. Therefore, the resultant velocity of fluid particles on each streamline is

$$V = \sqrt{(U \sin \theta)^2 + (-U \cos \theta)^2} = U$$

as noted in Fig. 7–19.

7

EXAMPLE 7.6

The streamlines for steady ideal fluid flow around the 90° corner in Fig. 7–20 are defined by the stream function $\psi(x, y) = (5xy)\ \text{m}^2/\text{s}$. Determine the velocity of the flow at point $x = 2\ \text{m}, y = 3\ \text{m}$. Can the Bernoulli equation be applied between any two points within this flow?

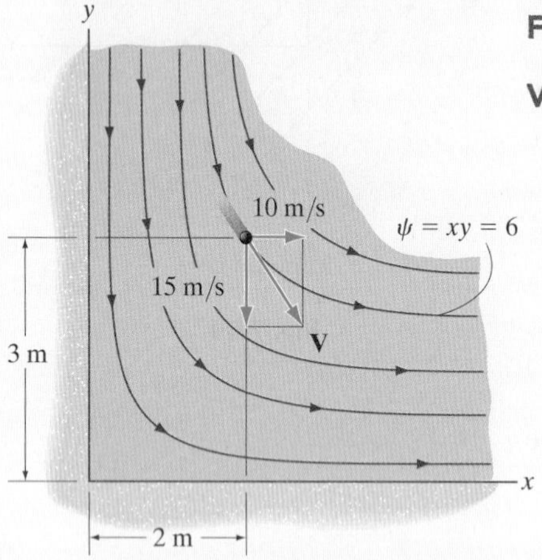

Flow around a 90° corner

Fig. 7–20

SOLUTION

Fluid Description.   As stated, we have steady ideal fluid flow.

Velocity.   The velocity components are determined from Eqs. 7–27.

$$u = \frac{\partial \psi}{\partial y} = \frac{\partial}{\partial y}(5xy) = (5x)\ \text{m/s}$$

$$v = -\frac{\partial \psi}{\partial x} = -\frac{\partial}{\partial x}(5xy) = (-5y)\ \text{m/s}$$

At point $x = 2\,\text{m}, y = 3\,\text{m}$,

$$u = 5(2) = 10\ \text{m/s}$$

$$v = -5(3) = -15\ \text{m/s}$$

The resultant velocity has a magnitude of

$$V = \sqrt{(10\ \text{m/s})^2 + (-15\ \text{m/s})^2} = 18.0\ \text{m/s} \qquad \textit{Ans.}$$

Its direction is tangent to the streamline that passes through point $(2\,\text{m}, 3\,\text{m})$, as shown in Fig. 7–20. To find the equation that defines this streamline, we require $\psi(x, y) = 5(2)(3) = C = 30\ \text{m}^2/\text{s}$. Thus $\psi(x, y) = 5xy = 30$, or $xy = 6$.

The Bernoulli equation can be applied since an ideal fluid has irrotational flow. Applying Eq. 7–2 to check this, we have

$$\omega_z = \frac{1}{2}\left(\frac{\partial v}{\partial x} - \frac{\partial u}{\partial y}\right) = \frac{1}{2}\left(\frac{\partial(-5y)}{\partial x} - \frac{\partial(5x)}{\partial y}\right) = 0$$

Therefore the Bernoulli equation can be used to find the pressure differences between any two points within the fluid.

7.8   The Potential Function

In the previous section we related the velocity components to the stream function, which describes the streamlines for the flow. Another way of relating the velocity components to a single function is to use the *velocity potential* ϕ (phi). It is defined by the potential function $\phi = \phi(x, y)$. The velocity components are determined from $\phi(x, y)$ using the following equations.

$$u = \frac{\partial \phi}{\partial x}, \quad v = \frac{\partial \phi}{\partial y} \qquad (7\text{–}30)$$

The resultant velocity is therefore

$$\mathbf{V} = u\mathbf{i} + v\mathbf{j} = \frac{\partial \phi}{\partial x}\mathbf{i} + \frac{\partial \phi}{\partial y}\mathbf{j} = \nabla \phi \qquad (7\text{–}31)$$

The potential function ϕ only describes *irrotational flow*. To show this, substitute the velocity components, as we have defined them, into Eq. 7–2. This gives

$$\omega_z = \frac{1}{2}\left(\frac{\partial v}{\partial x} - \frac{\partial u}{\partial y}\right)$$

$$= \frac{1}{2}\left[\frac{\partial}{\partial x}\left(\frac{\partial \phi}{\partial y}\right) - \frac{\partial}{\partial y}\left(\frac{\partial \phi}{\partial x}\right)\right] = \frac{1}{2}\left[\frac{\partial^2 \phi}{\partial x\, \partial y} - \frac{\partial^2 \phi}{\partial y\, \partial x}\right] = 0$$

Thus, *if the flow is irrotational flow*, then we can always establish a potential function $\phi(x, y)$, because this function automatically satisfies the condition $\omega_z = 0$.

Another characteristic of $\phi(x, y)$ is that the velocity will always be *perpendicular* to any *equipotential line* $\phi(x, y) = C'$. As a result, any equipotential line will be perpendicular to any intersecting streamline $\psi(x, y) = C$. This can be shown by taking the total derivative of $\phi(x, y) = C'$, which gives

$$d\phi = \frac{\partial \phi}{\partial x}dx + \frac{\partial \phi}{\partial y}dy = 0$$

$$= u\, dx + v\, dy = 0$$

Or,

$$\frac{dy}{dx} = \frac{u}{-v}$$

Graphically, this indicates that the slope θ of the tangent of the streamline $\psi(x, y) = C$ in Fig. 7–21 is the negative reciprocal of the slope of the equipotential line $\phi(x, y) = C'$, Eq. 7–25. Therefore, as shown, *streamlines are always perpendicular to equipotential lines*.

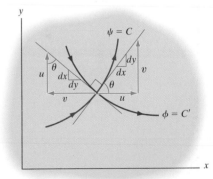

Fig. 7–21

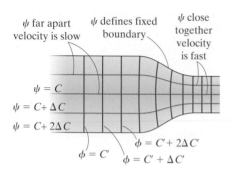

ψ far apart velocity is slow

ψ defines fixed boundary

ψ close together velocity is fast

$\psi = C$

$\psi = C + \Delta C$

$\psi = C + 2\Delta C$

$\phi = C' + 2\Delta C'$

$\phi = C'$ $\phi = C' + \Delta C'$

Flow net for ideal flow through a transition

Fig. 7–22

Finally, if polar coordinates are used to describe the equipotential function, then, without proof, the velocity components v_r and v_θ are related to the potential function by

$$v_r = \frac{\partial \phi}{\partial r}, \qquad v_\theta = \frac{1}{r} \frac{\partial \phi}{\partial \theta} \qquad (7\text{--}32)$$

Flow Net. A family of streamlines and equipotential lines for various values of the constants C and C' makes up a ***flow net***, which can serve as a graphical aid for visualizing the flow. An example of a flow net is shown in Fig. 7–22. Here the streamlines and equipotential lines must be constructed so that they always intersect *perpendicular* to one another and are spaced so that they maintain the same incremental distance ΔC and $\Delta C'$. As noted, where the streamlines are closer together the velocity is high (fast flow), and vice versa. For convenience, a computer can be used to construct a flow net by plotting the equations for $\psi(x, y) = C$ and $\phi(x, y) = C'$, and then incrementally increasing the constants by ΔC and $\Delta C'$.

Important Points

- A stream function $\psi(x, y)$ satisfies the continuity of flow. If $\psi(x, y)$ is known, then it is possible to determine the velocity components at any point within the flow using Eqs. 7–27. Also, the flow between any two streamlines $\psi(x, y) = C_1$ and $\psi(x, y) = C_2$ can be determined by finding the difference between the streamline constants, $q = C_2 - C_1$. The flow can either be rotational or irrotational.

- A potential function $\phi(x, y)$ satisfies the conditions of irrotational flow. If $\phi(x, y)$ is known, the velocity components at any point within the flow can be determined using Eqs. 7–30.

- Equipotential lines are always *perpendicular* to streamlines, and a set of both of these "lines" forms a *flow net*.

- If the velocity components for a flow are known, then the stream function $\psi(x, y)$ or the potentional function $\phi(x, y)$ is determined by integrating Eqs. 7–27 or Eqs. 7–30 and, for convenience, setting the integration constant equal to zero.

- The equations of a streamline and equipotentional line passing through a particular point (x_1, y_1) are determined by first obtaining the constants from $\psi(x_1, y_1) = C_1$ and $\phi(x_1, y_1) = C_1'$, and then writing $\psi(x, y) = C_1$ and $\phi(x, y) = C_1'$.

EXAMPLE | 7.7

A flow has a velocity field defined by $V = \{4xy^2\mathbf{i} + 4x^2y\mathbf{j}\}\,\text{m/s}$. Is it possible to establish a potential function for this flow, and if so, what is the equipotential line passing through point $x = 1\,\text{m}, y = 1\,\text{m}$?

SOLUTION

Fluid Description.   We have steady flow since V is not a function of time.

Analysis.   A potential function can be developed *only if the flow is irrotational*. To find out if it is, we apply Eq. 7–2. Here $u = 4xy^2$ and $v = 4x^2y$, so that

$$\omega_z = \frac{1}{2}\left(\frac{\partial v}{\partial x} - \frac{\partial u}{\partial y}\right) = \frac{1}{2}(8xy - 8xy) = 0$$

Since we have irrotational flow, the potential function can be established. Using the x component of velocity,

$$u = \frac{\partial \phi}{\partial x} = 4xy^2$$

Integrating,

$$\phi = 2x^2y^2 + f(y) \tag{1}$$

The unknown function $f(y)$ must be determined. Using the y component of velocity,

$$v = \frac{\partial \phi}{\partial y}$$

$$4x^2y = \frac{\partial}{\partial y}\left[2x^2y^2 + f(y)\right]$$

$$4x^2y = 4x^2y + \frac{\partial}{\partial y}\left[f(y)\right]$$

$$\frac{\partial}{\partial y}f(y) = 0$$

Therefore, integration gives

$$f(y) = C'$$

The potential function is then determined from Eq. 1, where, for convenience $C' = 0$, so that

$$\phi(x, y) = 2x^2y^2$$

To find the equipotential line passing through point $(1\,\text{m}, 1\,\text{m})$, we require $\phi(x, y) = 2(1)^2(1)^2 = 2$. Thus, $2x^2y^2 = 2$ or

$$xy = 1 \qquad\qquad\qquad Ans.$$

EXAMPLE | 7.8

The potential function for a flow is defined by $\phi(x, y) = 10xy$. Determine the stream function for the flow.

SOLUTION

Fluid Description. This is steady fluid flow, and because it is defined by a potential function, the flow is also irrotational.

Analysis. To solve, we will first determine the velocity components, and then from this, obtain the stream function. Using Eqs. 7–30, we have

$$u = \frac{\partial \phi}{\partial x} = 10y \qquad\qquad v = \frac{\partial \phi}{\partial y} = 10x$$

From the first of Eqs. 7–27 for u, we have

$$u = \frac{\partial \psi}{\partial y}; \qquad\qquad 10y = \frac{\partial \psi}{\partial y}$$

Integrating with respect to y gives

$$\psi = 5y^2 + f(x) \qquad\qquad (1)$$

Here $f(x)$ has to be determined. Using the second of Eqs. 7–27, for v we have

$$v = -\frac{\partial \psi}{\partial x}; \qquad 10x = -\frac{\partial}{\partial x}\left[5y^2 + f(x)\right] = -\left[0 + \frac{\partial}{\partial x}\left[f(x)\right]\right]$$

so that

$$\frac{\partial}{\partial x}\left[f(x)\right] = -10x$$

Integrating yields

$$f(x) = -5x^2 + C$$

Setting $C = 0$ and substituting $f(x)$ into Eq. 1, the stream function becomes

$$\psi(x, y) = 5(y^2 - x^2) \qquad\qquad Ans.$$

The flow net can be plotted by setting $\psi(x, y) = 5(y^2 - x^2) = C$ and $\phi(x, y) = 10xy = C'$ and then plotting these equations for different values of the constants C and C'. When this is done, the flow net will look like that shown in Fig. 7–23a. If we select two streamlines, say $\psi_1 = C_1$ and $\psi_2 = C_2$, to model the sides of a channel, Fig. 7–23b, then our solution can be used to study the flow within the channel, assuming, of course, the fluid is ideal.

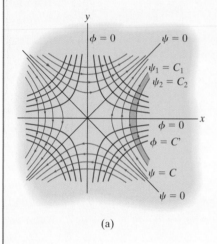

(a)

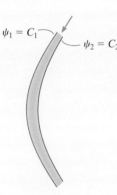

Flow through
a channel

(b)

Fig. 7–23

7.9   Basic Two-Dimensional Flows

The flow of an ideal fluid must satisfy *both* the continuity and irrotational flow conditions. It was stated previously that the stream function ψ automatically satisfies continuity for incompressible flow. However, to ensure that it satisfies irrotationality, we require $\omega_z = \frac{1}{2}(\partial v/\partial x - \partial u/dy) = 0$. When we substitute the velocity components $u = \partial\psi/\partial y$ and $v = -\partial\psi/\partial x$ into this equation, we get

$$\frac{\partial^2 \psi}{\partial x^2} + \frac{\partial^2 \psi}{\partial y^2} = 0 \qquad (7\text{--}33)$$

Or, in vector form,

$$\nabla^2 \psi = 0$$

In a similar manner, since the potential function ϕ automatically satisfies the irrotational flow condition, then to satisfy continuity, $(\partial u/\partial x) + (\partial v/\partial y) = 0$, we substitute the velocity components, $u = \partial\phi/\partial x$ and $v = \partial\phi/\partial y$, into this equation and get

$$\frac{\partial^2 \phi}{\partial x^2} + \frac{\partial^2 \phi}{\partial y^2} = 0$$

or

$$\nabla^2 \phi = 0 \qquad (7\text{--}34)$$

The above two equations are a form of *Laplace's equation*. A solution for the stream function ψ in Eq. 7–33, or the potential function ϕ in the above equation, represents the flow field for an ideal fluid. When one of these equations is solved, the two constants of integration resulting from the solution are evaluated by applying the boundary conditions to the flow. For example, a boundary condition will require that a stream function follow a solid boundary, since no component of velocity can act normal to the surface of the boundary.

Through the years, many investigators have determined ψ or ϕ for various types of ideal flow, either *directly* by solving the above equations, or *indirectly* by knowing the velocity components for the flow. See Refs. [10, 11]. It was this work that formed the basis of the science of *hydrodynamics*, which developed in the late 19th century. As a brief introduction to the methods used in hydrodynamics, we will now present the solutions for ψ and ϕ that involve five basic flow patterns. Once these flows have been introduced, we will then use the results to show how they can be superimposed with one another in order to represent other types of flow.

7

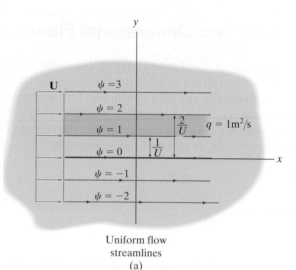

Uniform flow
streamlines
(a)

Fig. 7–24

Uniform Flow.

If the flow is uniform and has a constant velocity U along the x axis, as shown in Fig. 7–24a, then its velocity components are

$$u = U$$
$$v = 0$$

Applying Eqs. 7–27, we can obtain the stream function. Starting with the velocity component u,

$$u = \frac{\partial \psi}{\partial y}; \qquad\qquad U = \frac{\partial \psi}{\partial y}$$

Integrating with respect to y, we get

$$\psi = Uy + f(x)$$

Using this result, we now use the velocity component v.

$$v = -\frac{\partial \psi}{\partial x}; \qquad\qquad 0 = -\frac{\partial}{\partial x}[Uy + f(x)]$$

$$0 = \frac{\partial}{\partial x}[f(x)]$$

Integrating with respect to x yields

$$f(x) = C$$

Thus,

$$\psi = Uy + C$$

We set the integration constant $C = 0$, and so the stream function becomes

$$\psi = Uy$$

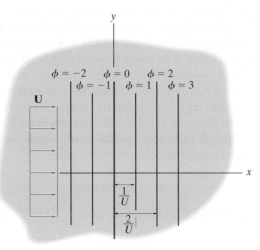

Uniform flow
Equipotential lines
(b)

Fig. 7–24 (cont.)

The streamlines are plotted in Fig. 7–24a by assigning constant values for ψ. For example, when $\psi = 0$, then $Uy = 0$, which represents the streamline passing through the origin. Also, if $\psi = 1$ m^2/s, then $y = 1/U$, and if $\psi = 2$ m^2/s, then $y = 2/U$, etc. Using Eq. 7–29, the flow between $\psi = 1$ m^2/s and $\psi = 2$ m^2/s can be determined, that is, $q = 2$ m^2/s $- 1$ m^2/s $= 1$ m^2/s, Fig. 7–24a.

In a similar manner, using Eqs. 7–30, $u = \partial\phi/\partial x$ and $v = \partial\phi/\partial y$, we can obtain the potential function. By integrating $U = \partial\phi/\partial x$ and $0 = \partial\phi/\partial y$, and obtain

$$\phi = Ux \qquad\qquad (7\text{–}35)$$

Equipotential lines are obtained by assigning constant values to ϕ. For example, $\phi = 0$ corresponds to $x = 0$, and $\phi = 1$ m^2/s corresponds to $x = 1/U$, etc. As expected, these lines will be perpendicular to the streamlines described by ψ, Fig. 7–24b. Together they form the flow net. Also, notice that, as required, both ψ and ϕ will satisfy Laplace's equation, that is, Eqs. 7–33 and 7–34.

Line Source Flow.   In two dimensions, the source of a flow q is defined from a line along the z axis from which the fluid flows radially outward, uniformly in all directions on the x–y plane, Fig. 7–25. Such a flow would approximate water slowly emerging from a pipe connected perpendicular to a horizontal plate, with a second plate just above it. Here q is measured per unit depth, along the z axis (line), so it has units of m^2/s. Due to the angular symmetry, it is convenient to use polar coordinates r, θ to describe this flow. If we consider a circle of radius r, then the flow through the circle having a unit depth passes through an area of $A = 2\pi r(1)$, and since $q = v_r A$, we have

$$q = v_r(2\pi r)(1)$$

The radial component of velocity is therefore

$$v_r = \frac{q}{2\pi r}$$

And due to symmetry, the transverse component is

$$v_\theta = 0$$

The stream function is obtained using Eqs. 7–28. For the radial velocity component, we have

$$v_r = \frac{1}{r}\frac{\partial \psi}{\partial \theta}; \qquad \frac{q}{2\pi r} = \frac{1}{r}\frac{\partial \psi}{\partial \theta}$$

$$\partial \psi = \frac{q}{2\pi}\partial \theta$$

Integrating with respect to θ,

$$\psi = \frac{q}{2\pi}\theta + f(r)$$

Now, considering the transverse velocity component,

$$v_\theta = -\frac{\partial \psi}{\partial r}; \qquad 0 = -\frac{\partial}{\partial r}\left[\frac{q}{2\pi}\theta + f(r)\right]$$

$$0 = \frac{\partial}{\partial r}\left[f(r)\right]$$

Integrating with respect to r,

$$f(r) = C$$

Thus,

$$\psi = \frac{q}{2\pi}\theta + C$$

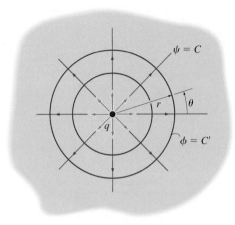

Line source flow

Fig. 7–25

Setting the integration constant $C = 0$, the stream function is

$$\psi = \frac{q}{2\pi} \theta \qquad (7\text{–}36)$$

Hence streamlines for which ψ equals any constant are radial lines at each coordinate θ as expected, Fig. 7–25. For example, when $\psi = 0$, then $(q/2\pi)\theta = 0$ or $\theta = 0$, which represents the horizontal radial line. Likewise, if $\psi = 1$, then $\theta = 2\pi/q$, which defines the angular position of the radial streamline for $\psi = 1$, and so on.

The potential function is determined by integrating Eqs. 7–32.

$$v_r = \frac{\partial \phi}{\partial r}; \qquad\qquad \frac{q}{2\pi r} = \frac{\partial \phi}{\partial r}$$

$$v_\theta = \frac{1}{r}\frac{\partial \phi}{\partial \theta}; \qquad\qquad 0 = \frac{1}{r}\frac{\partial \phi}{\partial \theta}$$

Show that integration of these two equations yields

$$\phi = \frac{q}{2\pi} \ln r \qquad (7\text{–}37)$$

Equipotential lines for which ϕ equals any constant are circles having a center at the source. For example, $\phi = 1$ defines the circle of radius $r = e^{2\pi/q}$, etc., Fig. 7–25. Notice that the source is actually a mathematical singularity since $v_r = q/2\pi r$ approaches infinity as r approaches zero. The flow net we have established, however, is still valid at distances away from the source.

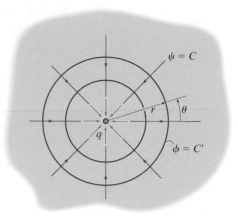

Line sink flow

Fig. 7–26

Line Sink Flow.

When the flow is radially inward toward a line source (z axis), then the strength of the flow q is negative and the flow is termed a line sink flow, Fig. 7–26. This type of flow is similar to the behavior of a shallow constant depth of water in a flat-bottom sink passing through a drain. Here the velocity components are

$$v_r = -\frac{q}{2\pi r}$$

$$v_\theta = 0$$

And the stream and potential functions become

$$\psi = -\frac{q}{2\pi}\theta \tag{7–38}$$

$$\phi = -\frac{q}{2\pi}\ln r \tag{7–39}$$

The flow net for these functions is shown in Fig. 7–26.

Doublet.

When a line source and line sink come close to one another and then combine, they form a *doublet*. To show how to formulate the stream function and potential function for this case, consider the equal-strength source and sink shown in Fig. 7–27a. Using Eqs. 7–36 and 7–38, with θ_1 and θ_2 as the variables for the source and sink, respectively, we have

$$\psi = \frac{q}{2\pi}(\theta_1 - \theta_2)$$

If we rearrange this equation and take the tangent of both sides, using the angle addition formula for the tangent, we get

$$\tan\left(\frac{2\pi\psi}{q}\right) = \tan(\theta_1 - \theta_2) = \frac{\tan\theta_1 - \tan\theta_2}{1 + \tan\theta_1 \tan\theta_2} \qquad (7\text{–}40)$$

From Fig. 7–27a, the tangents of θ_1 and θ_2 can be written as

$$\tan\left(\frac{2\pi\psi}{q}\right) = \frac{\left[y/(x+a)\right] - \left[y/(x-a)\right]}{1 + \left[(y/(x+a))(y/(x-a))\right]}$$

or

$$\psi = \frac{q}{2\pi}\tan^{-1}\left(\frac{-2ay}{x^2 + y^2 - a^2}\right)$$

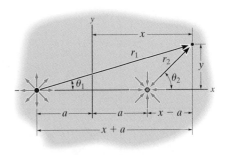

(a)

As the distance a becomes smaller, the angles $\theta_1 \to \theta_2 \to 0$, and so the difference in the angles $(\theta_1 - \theta_2)$ becomes smaller. When this happens, the tangent of the difference will approach the difference itself, that is, $\tan(\theta_1 - \theta_2) \to (\theta_1 - \theta_2)$, so $\tan^{-1}$ in the above equation can be eliminated. If we then convert our result to polar coordinates, where $r^2 = x^2 + y^2$, and $y = r\sin\theta$, we get

$$\psi = -\frac{qa}{\pi}\left(\frac{r\sin\theta}{r^2 - a^2}\right)$$

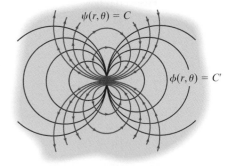

Doublet

(b)

Fig. 7–27

If $a \to 0$, the flows from the source and sink would cancel each other. However, if we consider the *strength q* of both the source and sink as *increasing* as $a \to 0$, then $q \to \infty$ so that the product qa remains *constant*. For convenience, if we define the strength of this *doublet* as $K = qa/\pi$, then in the limit, the stream function becomes

$$\psi = \frac{-K\sin\theta}{r} \qquad (7\text{–}41)$$

We can obtain the potential function in a similar manner. It is

$$\phi = \frac{K\cos\theta}{r} \qquad (7\text{–}42)$$

The flow net for a doublet consists of a series of circles that all intersect at the origin, as shown in Fig. 7–27b. With it, in Sec. 7.10 we will show how it can be superimposed with a uniform flow to represent flow around a cylinder.

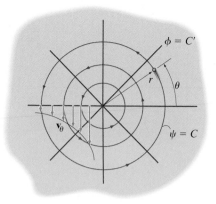

Free-vortex flow

(a)

Free-Vortex Flow.   A free vortex is irrotational flow that is circular. Here the streamlines are circles, and the equipotential lines are radial, Fig. 7–28a. We can represent this by selecting the stream function of a line source, Eq. 7–36, to be the potential function for the vortex. Then, considering the relationship for v_θ from Eqs. 7–28 and 7–32, $-\partial\psi/\partial r = (1/r)(\partial\phi/\partial\theta)$, we can obtain the stream function. The results are

$$\psi = -k \ln r \tag{7–43}$$

$$\phi = k\theta \tag{7–44}$$

where $k = q/(2\pi)$ is a constant. Applying Eqs. 7–28, the velocity components are

$$v_r = \frac{1}{r}\frac{\partial\psi}{\partial\theta}; \qquad\qquad v_r = 0 \tag{7–45}$$

$$v_\theta = -\frac{\partial\psi}{\partial r}; \qquad\qquad v_\theta = \frac{k}{r} \tag{7–46}$$

Note that v_θ becomes larger as r becomes smaller, and the center, $r = 0$, is a singularity since v_θ becomes infinite, Fig. 7–28a. This flow is irrotational since a potential function has been used to describe it. Consequently, fluid elements within the flow will *distort* in such a way that they do not rotate, Fig. 7–28b. Finally, note that this vortex is counterclockwise. To obtain a description of a clockwise vortex, the signs must be changed in Eq. 7–43 and Eq. 7–44.

Circulation.   It is also possible to define the stream and potential functions for a free-vortex flow in terms of its circulation Γ, defined by Eq. 7–6. If we choose circulation about the streamline (circle) at radius r, then

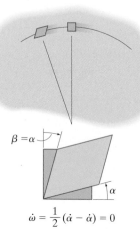

$\beta = \alpha$

$\dot\omega = \frac{1}{2}(\dot\alpha - \dot\alpha) = 0$

Irrotational flow

(b)

Fig. 7–28

$$\Gamma = \oint \mathbf{V} \cdot d\mathbf{s} = \int_0^{2\pi} \frac{k}{r}(r\,d\theta) = 2\pi k$$

Using this result, Eqs. 7–43 and 7–44 become

$$\psi = -\frac{\Gamma}{2\pi}\ln r \tag{7–47}$$

$$\phi = \frac{\Gamma}{2\pi}\theta \tag{7–48}$$

We will use these results in the next section to study the effect of fluid pressure acting on a rotating cylinder.

Forced-Vortex Flow.

A forced vortex is so named because an external torque is required to start or "force" the motion, Fig. 7–29a. Once it is started, the viscous effects of the fluid will eventually cause it to rotate as a rigid body; that is, the fluid elements maintain their shape and *rotate* about a fixed axis, Fig. 7–29b. A typical example was discussed in Sec. 2.14. It is the rotation of a *real fluid* in a container. Since the fluid elements are "rotating," a potential function cannot be established. For this case, like the case discussed in Sec. 2.14, the velocity components are $v_r = 0$ and $v_\theta = \omega r$, where ω is the angular velocity of the fluid, Fig. 7–29a.

$$v_\theta = -\frac{\partial \psi}{\partial r} = \omega r$$

Excluding the constant of integration, the stream function is therefore

$$\psi = -\frac{1}{2} \omega r^2$$

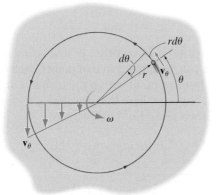

Forced-vortex flow

(a)

Fig. 7–29

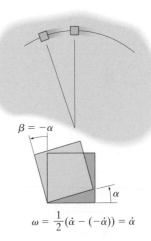

$$\beta = -\alpha$$

$$\omega = \frac{1}{2}(\dot{\alpha} - (-\dot{\alpha})) = \dot{\alpha}$$

Rotational flow

(b)

EXAMPLE | 7.9

A tornado consists of a whirling mass of air such that the winds essentially move along horizontal circular streamlines, Fig. 7–30a. Determine the pressure distribution within the tornado as a function of r.

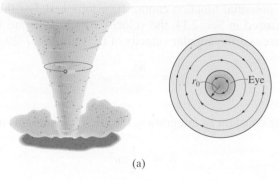

(a)

Fig. 7–30

SOLUTION

Fluid Description. We will assume the air is an ideal fluid that has steady flow. Translational motion of the tornado will be neglected.

Free Vortex. From Eqs. 7–45 and 7–46, the velocity components are

$$v_r = 0 \text{ and } v_\theta = \frac{k}{r} \tag{1}$$

Because the flow within a free vortex is steady *irrotational flow*, the Bernoulli equation can be applied to two points, each lying on *different streamlines*. If we choose a point within the tornado, and another at the same elevation but remote, where the air velocity $V = 0$ and the (gage) pressure is $p = 0$, then using Eq. 7–24, we have

$$\frac{p_1}{\rho} + \frac{V_1^2}{2} + gz_1 = \frac{p_2}{\rho} + \frac{V_2^2}{2} + gz_2$$

$$\frac{p}{\rho} + \frac{k^2}{2r^2} + gz = 0 + 0 + gz$$

$$p = -\frac{\rho k^2}{2r^2} \tag{2}$$

Here k is a constant, as yet to be determined. The negative sign in this equation indicates that a suction pressure develops, and both this pressure and the velocity intensify as r becomes smaller.

   Notice that a free vortex such as this cannot actually exist in a *real fluid* since the velocity and pressure would have to approach infinity as $r \rightarrow 0$. Instead, due to the increasing velocity gradient as r becomes smaller, the viscosity of the air will eventually create enough shear stress to cause the air in the *core*, or "eye," to rotate as a *solid system* having an angular velocity ω. We will assume this transition occurs at a radius of $r = r_0$, Fig. 7–30 *a*.

Forced Vortex. Because of its "rigid-body" motion, the core is a forced vortex, which we have analyzed using Euler's equations of motion in Example 5–1.

In that example, we showed that the pressure distribution is defined by

$$p = p_0 - \frac{\rho\omega^2}{2}\left(r_0^2 - r^2\right) \qquad (3)$$

where p_0 is the pressure at r_0.

In order to fully study the pressure variation, we must now match the two solutions at $r = r_0$. The constant k in Eq. 1 can now be determined since at r_0 the velocity in the forced vortex, $v_\theta = \omega r_0$, must equal that in the free vortex, that is,

$$v_\theta = \omega r_0 = \frac{k}{r_0} \quad \text{so that} \quad k = \omega r_0^2$$

The pressures in Eqs. 2 and 3 must also be equivalent at $r = r_0$, so that

$$-\frac{\rho\left(\omega r_0^2\right)^2}{2r_0^2} = p_0 - \frac{\rho\omega^2}{2}\left(r_0^2 - r_0^2\right)$$

$$p_0 = -\frac{\rho\omega^2 r_0^2}{2}$$

Therefore, after substituting into Eq. 3 and simplifying, we have for the forced vortex, $r \le r_0$,

$$v_\theta = \omega r$$

$$p = \frac{\rho\omega^2}{2}\left(r^2 - 2r_0^2\right)$$

And for the free vortex, $r \ge r_0$,

$$v_\theta = \frac{\omega r_0^2}{r}$$

$$p = -\frac{\rho\omega^2 r_0^4}{2r^2} \qquad \qquad Ans.$$

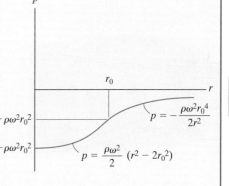

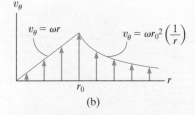

(b)

Fig. 7–30 (cont.)

Using these results, a plot of both the velocity and pressure variations is shown in Fig. 7–30b. Notice that the largest suction (negative pressure) occurs at the *center* of the forced vortex, $r = 0$, and the highest velocity occurs at $r = r_0$. It is the combination of this low pressure and high velocity that makes tornadoes so destructive. In fact tornadoes often reach wind speeds over 322 km/h. From weather reports you may have noticed their effect in leveling well-constructed houses, and lifting cars off the ground.

The type of vortex we have considered here, that is, a combination of a forced vortex surrounded by a free vortex, is sometimes called a **compound vortex**. It not only occurs in tornadoes, but also forms in a kitchen sink as the water drains from the bottom, or in a river as the water flows off a boat oar or around a bridge pier.

7.10   Superposition of Flows

We noted in the previous section that for any ideal flow, both the stream function and the potential function must satisfy Laplace's equation, that is, Eqs. 7–33 and 7–34. Because the second derivatives of ψ and ϕ are of the first power in this equation, that is, they are linear, then several different solutions can be *superimposed*, or added together, to form a new solution. For example, $\psi = \psi_1 + \psi_2$ or $\phi = \phi_1 + \phi_2$. In this way, complex flow patterns can be established from a series of basic flow patterns, such as the ones presented in the previous section. To date, many types of solutions have been produced by this method, oftentimes requiring the application of advanced mathematical analysis. These and other techniques used to find solutions are discussed in books related to hydrodynamics. See Ref. [10] and [11]. What follows are some basic applications using superposition.

Flow Past a Half Body.   If the results for uniform flow and line source flow are added together, the resulting streamline and potential functions are

$$\psi = \frac{q}{2\pi}\theta + Uy = \frac{q}{2\pi}\theta + Ur\sin\theta \qquad (7\text{–}49)$$

$$\phi = \frac{q}{2\pi}\ln r + Ux = \frac{q}{2\pi}\ln r + Ur\cos\theta \qquad (7\text{–}50)$$

Here we have used the coordinate transformation equations, $x = r\cos\theta$ and $y = r\sin\theta$, to represent the results in polar coordinates.

The velocity components can be determined from Eqs. 7–32 (or Eqs. 7–28). We have

$$v_r = \frac{\partial\phi}{\partial r} = \frac{q}{2\pi r} + U\cos\theta \qquad (7\text{–}51)$$

$$v_\theta = \frac{1}{r}\frac{\partial\phi}{\partial\theta} = -U\sin\theta \qquad (7\text{–}52)$$

The resultant flow looks like that shown in Fig. 7–31a. Any one of the streamlines can be selected as the *boundary* for a solid object that fits within this flow pattern. For example, streamlines A and A' form the boundary of an infinitely extended body of the shape shown shaded in Fig. 7–31b. Here, however, we will consider the shape formed by the streamline that passes through the stagnation point P, Fig. 7–31c. This point occurs where the velocity of flow from the source q cancels the uniform flow U, Fig. 7–31a. The stagnation point is located at $r = r_0$, where the two components of velocity must be equal to zero, just before the flow begins to divide equally and pass around the body. For the transverse component, we find

$$0 = -U\sin\theta$$
$$\theta = 0, \pi$$

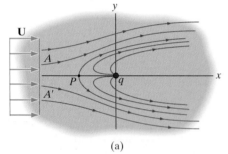

(a)

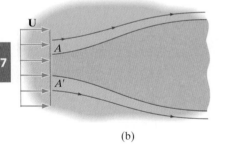

(b)

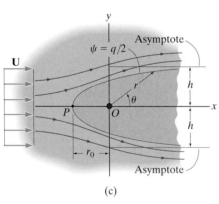

(c)

Flow past a half body

Fig. 7–31

The root $\theta = \pi$ gives the direction of r_0. For the radial component,

$$0 = \frac{q}{2\pi r_0} + U \cos \pi$$

$$r_0 = \frac{q}{2\pi U} \qquad (7\text{--}53)$$

As expected, this radial position of P depends upon the magnitude of the uniform flow velocity U and the strength q of the source.

The boundary of the body can now be specified by the streamline that passes through the point $r = r_0, \theta = \pi$. From Eq. 7–49, the *constant* for this streamline is

$$\psi = \frac{q}{2\pi}\pi + U\left(\frac{q}{2\pi U}\right) \sin \pi = \frac{q}{2}$$

Therefore, the equation of the boundary of the body is

$$\frac{q}{2\pi}\theta + Ur \sin \theta = \frac{q}{2}$$

To simplify, we can solve for q in Eq. 7–53 and substitute it into this equation. This gives

$$r = \frac{r_0(\pi - \theta)}{\sin \theta} \qquad (7\text{--}54)$$

Since the body extends an infinite distance to the right, its top and bottom surfaces approach asymptotes, and so it has no closure. For this reason, it is referred to as a **half body**. The half width h can be determined from Eq. 7–54, by noting that $y = r \sin \theta = r_0(\pi - \theta)$, Fig. 7–31c. As θ approaches 0 or 2π, then $y = \pm h = \pm \pi r_0 = q/(2U)$.

By selecting appropriate values for U and q, we can use the half body to model the front shape of a symmetrical object such as the front surface of an airfoil (wing) subjected to uniform flow U. But this has its limitations. By assuming the fluid to be *ideal*, we then have a *finite value* for velocity at the *boundary* of the body even though all real fluids require a "no-slip" *zero velocity* at the boundary due to their viscosity. In Chapter 11, we will show that this viscous effect is generally *limited* only to a very thin region near the boundary that is formed when fluids of relatively low viscosity, such as air, flow at high velocities. Outside this region the flow can generally be described by the analysis presented here, and indeed, the results have been found to agree rather closely with experimental results.

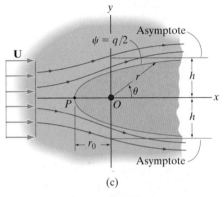

(c)

Flow past a half body

Fig. 7–31 (cont.)

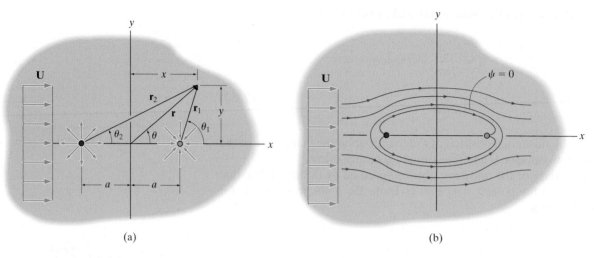

(a)   (b)

Fig. 7–32

Flow around a Rankine Oval.   When uniform flow is superimposed upon a line source and sink of equal strength, each located a distance a away from the origin, Fig. 7–32a, the streamlines so produced will look like those in Fig. 7–32b. Using the established coordinates, we have

$$\psi = Uy + \frac{q}{2\pi}\theta_2 - \frac{q}{2\pi}\theta_1 = Ur\sin\theta + \frac{q}{2\pi}\left(\theta_2 - \theta_1\right) \qquad (7\text{–}55)$$

$$\phi = Ux - \frac{q}{2\pi}\ln r_1 + \frac{q}{2\pi}\ln r_2 = Ur\cos\theta + \frac{q}{2\pi}\ln\frac{r_2}{r_1} \qquad (7\text{–}56)$$

We can also express these functions in Cartesian coordinates, in which case

$$\psi = Uy - \frac{q}{2\pi}\tan^{-1}\frac{2ay}{x^2 + y^2 - a^2} \qquad (7\text{–}57)$$

$$\phi = Ux + \frac{q}{2\pi}\ln\frac{\sqrt{(x + a)^2 + y^2}}{\sqrt{(x - a)^2 + y^2}} \qquad (7\text{–}58)$$

The velocity components are then

$$u = \frac{\partial\phi}{\partial x} = U + \frac{q}{2\pi}\left[\frac{x + a}{(x + a)^2 + y^2} - \frac{x - a}{(x - a)^2 + y^2}\right] \qquad (7\text{–}59)$$

$$v = \frac{\partial\phi}{\partial y} = \frac{q}{2\pi}\left[\frac{y}{(x + a)^2 + y^2} - \frac{y}{(x - a)^2 + y^2}\right] \qquad (7\text{–}60)$$

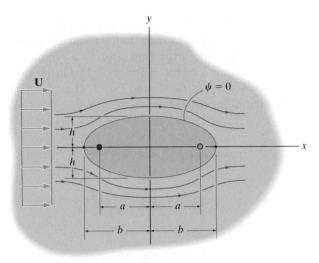

Rankine oval

(c)

Fig. 7–32 (cont.)

Setting $\psi = 0$ in Eq. 7–57, we obtain the shape shown in Fig. 7–32c. It passes through two stagnation points and forms a ***Rankine oval***, named after the hydrodynamist William Rankine, who first developed this idea of combining flow patterns.

To find the location of the stagnation points, we require $u = v = 0$. Thus from Eq. 7–60, with $v = 0$, we get $y = 0$. Then with $u = 0$ at $x = b, y = 0$, Eq. 7–59 gives

$$b = \left(\frac{q}{U\pi} a + a^2 \right)^{1/2} \tag{7–61}$$

This dimension also defines the half-length of the body, Fig. 7–32c. The half-width h is found as the point of intersection of $\psi = 0$ with the y axis $(x = 0)$. From Eq. 7–57,

$$0 = Uh - \frac{q}{2\pi} \tan^{-1} \frac{2ah}{h^2 - a^2}$$

Rearranging the terms yields

$$h = \frac{h^2 - a^2}{2a} \tan\left(\frac{2\pi Uh}{q} \right) \tag{7–62}$$

A specific numerical solution for h in this transcendental equation will require a numerical procedure, as will be demonstrated in Example 7–10. In general, though, when seeking a solution, begin by choosing a number slightly smaller then $q/(2U)$, because the half width of a Rankine oval is somewhat less than the half width of a corresponding half body.

Flow around a Cylinder. If a source and sink of equal strength are placed at the *same point*, producing a doublet, and this is superimposed with a uniform flow, we obtain the flow around a cylinder, Fig. 7–33a. Here a represents the radius of the cylinder, and so using Eqs. 7–34 and 7–35, with $x = r \cos \theta$ and $y = r \sin \theta$, and Eqs. 7–41 and 7–42, the stream functions and potential functions become

$$\psi = Ur \sin \theta - \frac{K \sin \theta}{r}$$

$$\phi = Ur \cos \theta + \frac{K \cos \theta}{r}$$

If we set $\psi = 0$, which passes through the two stagnation points, it will then define the boundary of the cylinder. Thus, $(Ua - K/a) \sin \theta = 0$, so the doublet strength must be $K = Ua^2$. Therefore,

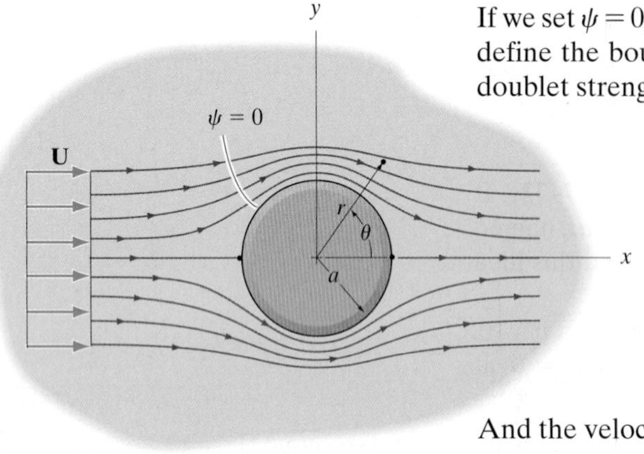

$$\psi = Ur \left(1 - \frac{a^2}{r^2} \right) \sin \theta \tag{7–63}$$

$$\phi = Ur \left(1 + \frac{a^2}{r^2} \right) \cos \theta \tag{7–64}$$

And the velocity components become

$$v_r = \frac{\partial \phi}{\partial r} = U \left(1 - \frac{a^2}{r^2} \right) \cos \theta \tag{7–65}$$

$$v_\theta = \frac{1}{r} \frac{\partial \phi}{\partial \theta} = -U \left(1 + \frac{a^2}{r^2} \right) \sin \theta \tag{7–66}$$

Uniform flow around a cylinder
Ideal flow
(a)

Fig. 7–33

Notice that when $r = a$, then $v_r = 0$ due to the boundary, but $v_\theta = -2U \sin \theta$. The stagnation points occur where $v_\theta = 0$, that is, where $\sin \theta = 0$, or $\theta = 0°$ and $\theta = 180°$. Since the streamlines are close together at the top (or bottom) of the cylinder, $\theta = 90°$, Fig. 7–33a, the maximum velocity occurs here. It is $(v_\theta)_{max} = 2U$. For an ideal fluid, we can have this finite value for v_θ, but remember the viscosity of any real fluid would actually cause this velocity at the boundary to be zero, due to the no-slip condition for viscous flow.

The pressure at a point on or off the cylinder can be determined using the Bernoulli equation, applied at the point and at another one far removed from the cylinder where $p = p_0$ and $V = U$. Neglecting gravitational effects, we have

$$\frac{p}{\rho} + \frac{V^2}{2} = \frac{p_0}{\rho} + \frac{U^2}{2}$$

or

$$p = p_0 + \frac{1}{2}\rho\left(U^2 - V^2\right)$$

Since $V = v_\theta = -2U \sin \theta$ along the surface, $r = a$, then when we substitute this into the above equation, we find that the pressure on the surface is

$$p = p_0 + \frac{1}{2}\rho U^2\left(1 - 4 \sin^2 \theta\right) \tag{7–67}$$

A graph of the result $(p - p_0)$ is shown in Fig. 7–33b. By inspection, this pressure distribution is *symmetrical*, and so it creates no net force on the cylinder. Something that is to be expected for an *ideal fluid*, since the flow does not include the effect of viscous friction. In Chapter 11, however, we will include this effect, and show how this alters the pressure distribution.

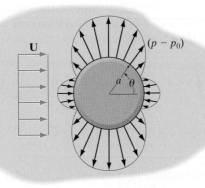

Pressure distribution

(b)

Fig. 7–33 (cont.)

Uniform and Free-Vortex Flow around a Cylinder.

If a counterclockwise *rotating* cylinder is placed in the uniform flow of a *real fluid*, then the fluid particles in contact with the cylinder's surface will stick to the surface, and because of viscosity, move with the cylinder. We can approximately *model* this type of flow using an *ideal fluid* by considering the cylinder as immersed in a uniform flow superimposed with a free vortex, written in terms of its circulation Γ, Eqs. 7–47 and 7–48. Adding these flows results in the stream and potential functions

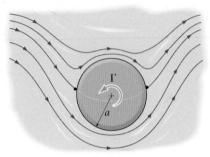

$\Gamma < 4\pi Ua$
Two stagnation points
(a)

$$\psi = Ur\left(1 - \frac{a^2}{r^2}\right)\sin\theta - \frac{\Gamma}{2\pi}\ln r \qquad (7\text{–}68)$$

$$\phi = Ur\left(1 + \frac{a^2}{r^2}\right)\cos\theta + \frac{\Gamma}{2\pi}\theta \qquad (7\text{–}69)$$

The velocity components are therefore

$$v_r = \frac{\partial\phi}{\partial r} = U\left(1 - \frac{a^2}{r^2}\right)\cos\theta \qquad (7\text{–}70)$$

$$v_\theta = \frac{1}{r}\frac{\partial\phi}{\partial\theta} = -U\left(1 + \frac{a^2}{r^2}\right)\sin\theta + \frac{\Gamma}{2\pi r} \qquad (7\text{–}71)$$

From these equations, notice that the velocity distribution *around the surface of the cylinder*, $r = a$, has components of

$$v_r = 0$$

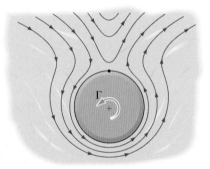

$\Gamma = 4\pi Ua$
One stagnation point
(b)

$$v_\theta = -2U\sin\theta + \frac{\Gamma}{2\pi a} \qquad (7\text{–}72)$$

It may be of interest to show that indeed the circulation around the cylinder is Γ. For this case, $v = v_\theta$ is *always tangent to the cylinder*, and since $ds = a\,d\theta$, we have

$$\Gamma = \oint \mathbf{V}\cdot ds = \int_0^{2\pi}\left(-2U\sin\theta + \frac{\Gamma}{2\pi a}\right)(a\,d\theta)$$

$$= \left(2aU\cos\theta + \frac{\Gamma}{2\pi}\theta\right)\Bigg|_0^{2\pi} = \Gamma$$

The location of the stagnation point on the cylinder is determined by setting $v_\theta = 0$ in Eq. 7–72. We get

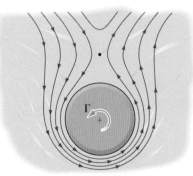

$\Gamma > 4\pi Ua$
No stagnation point
on surface of cylinder
(c)

Fig. 7–34

$$\sin\theta = \frac{\Gamma}{4\pi Ua}$$

As shown by the streamlines in Fig. 7–34a, if $\Gamma < 4\pi Ua$, then two stagnation points will occur on the cylinder since this equation will have two roots for θ. When $\Gamma = 4\pi Ua$, the points will merge and be located at $\theta = 90°$, Fig. 7–34b. Finally, if $\Gamma > 4\pi Ua$, then no root exists, and the flow will not stagnate on the surface of the cylinder; rather, it occurs at a point off the cylinder, Fig. 7–34c.

The pressure distribution around the cylinder is determined by applying the Bernoulli equation in the same manner as the previous case. We get

$$p = p_0 + \frac{1}{2}\rho U^2\left[1 - \left(-2\sin\theta + \frac{\Gamma}{2\pi Ua}\right)^2\right]$$

The general shape of this distribution, $(p - p_0)$, is shown in Fig. 7–34d. Integrating this distribution over the surface of the cylinder in the x and y directions, we can find the components of the resultant force that the (ideal) fluid exerts on the cylinder per unit length. This yields

$$F_x = -\int_0^{2\pi}(p - p_0)\cos\theta\,(a\,d\theta) = 0$$

$$F_y = -\int_0^{2\pi}(p - p_0)\sin\theta(a\,d\theta)$$

$$= -\frac{1}{2}\rho aU^2\int_0^{2\pi}\left[1-\left(-2\sin\theta + \frac{\Gamma}{2\pi Ua}\right)^2\right]\sin\theta\,d\theta$$

$$F_y = -\rho U\Gamma \qquad\qquad (7\text{--}73)$$

Because of the symmetry about the y axis, the result $F_x = 0$ indicates there is no "drag" or retarding force on the cylinder in the x direction.* Only a vertical downward force F_y exists, Fig. 7–34e. Because this force is perpendicular to the uniform flow, it is referred to as "lift." Throwing a spinning ball into the air will also produce lift and cause it to curve. This is called the *Magnus effect*, and we will discuss this effect further in Chapter 11.

Other Applications.
We can extend these ideas of superimposing ideal fluid flows to form other closed bodies having a variety of different shapes. For example, a body having the approximate shape of an airfoil or wing is shown in Fig. 7–35a. It is formed from a streamline that results from the superposition of a uniform flow, in combination with a single line source and a row of sinks having the *same total strength* as the source, but decreasing linearly in intensity.

*Historically, this has been referred to as d'Alembert's paradox, named after Jean le Rond d'Alembert, who in the 1700s was unable to explain why real fluids cause drag on a body. The explanation came later, in 1904, when Ludwig Prandtl developed the concept of the boundary layer, which we will cover in Chapter 11.

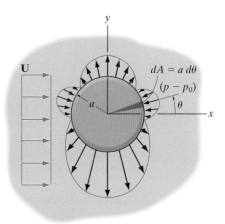

Pressure distribution
(d)

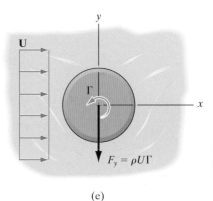

(e)

Fig. 7–34 (cont.)

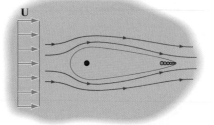

Superposition of a uniform flow
with a source and in-line sinks

Fig. 7–35

Important Points

- The stream function $\psi(x, y)$ or the potential function $\phi(x, y)$ for a given flow can be determined by integrating Laplace's differential equation, which satisfies *both* the continuity and the irrotational flow conditions. The various solutions form the basis of hydrodynamics and are valid *only* for ideal fluids.

- When the viscosity of a fluid is small and the velocity of flow is large, a *thin* boundary layer will form on the surface of a body placed in the flow. Thus, the viscous effects can be limited to the flow within this boundary layer, and the fluid beyond it can often be considered to have ideal flow.

- The basic solutions for $\psi(x, y)$ and $\phi(x, y)$ have been presented for uniform flow, flow from a line source or to a line sink, a doublet, and free-vortex flow. From these solutions, one can obtain the velocity at any point within the flow, and also the pressure at a point can be obtained using the Bernoulli equation.

- Since a forced vortex is *rotational flow*, it has a solution only for $\psi(x, y)$, not for $\phi(x, y)$.

- Laplace's equation is a linear differential equation for $\psi(x, y)$ or $\phi(x, y)$, and so any combination of basic ideal flow solutions to this equation can be superimposed or added together to produce more complicated flows. For example, flow around a half body is formed by superimposing a uniform flow and a line source flow. Flow around a Rankine oval is formed by superimposing a uniform flow and an equal-strength line source and line sink that are not concurrent. And flow around a cylinder is formed by superimposing a uniform flow and a doublet.

- A *drag* is never produced on a symmetric body immersed in an ideal fluid because there are no viscous forces acting on the body that would produce drag.

EXAMPLE | 7.10

A Rankine oval is formed by superimposing a line source and line sink placed 2 m apart, each having a strength of 8 m²/s, Fig. 7–36. If the uniform flow around the body is $U = 10$ m/s, determine the location b of the stagnation points, and find the equation that defines the boundary of the oval and its half width h.

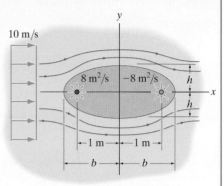

Fig. 7–36

SOLUTION

Fluid Description. We assume steady flow of an ideal fluid having a density ρ.

Analysis. At the stagnation points, the velocity is zero. Using Eq. 7–61, we have

$$b = \left(\frac{q}{U\pi} a + a^2 \right)^{1/2}$$

$$= \left[\frac{8 \text{ m}^2/\text{s}}{(10 \text{ m/s})\pi} (1 \text{ m}) + (1 \text{ m})^2 \right]^{1/2}$$

$$= 1.12 \text{ m} \qquad\qquad\qquad Ans.$$

The stream function for the flow is defined by Eq. 7–57. Here it becomes

$$\psi = Uy - \frac{q}{2\pi} \tan^{-1} \frac{2ay}{x^2 + y^2 - a^2}$$

$$= 10y - \frac{4}{\pi} \tan^{-1} \frac{2y}{x^2 + y^2 - 1}$$

Setting $\psi = 0$ gives the boundary of the body, since it contains the stagnation points.

$$\psi = 10y - \frac{4}{\pi} \tan^{-1} \frac{2y}{x^2 + y^2 - 1} = 0$$

$$\tan (2.5\pi y) = \frac{2y}{x^2 + y^2 - 1} \qquad\qquad Ans.$$

To find the half width of the body, set $x = 0$, $y = h$ in this equation, Fig. 7–36. This gives

$$h = \frac{h^2 - 1}{2} \tan (2.5\pi h)$$

This same result can also be obtained by applying Eq. 7–62. To solve, we note that for a half body, its half width is $q/(2U) = (8 \text{ m}^2/\text{s})/[2(10 \text{ m/s})] = 0.4$ m. And so if we start with a slightly smaller value, say $h = 0.35$ m, then we can adjust it until it satisfies the equation. We find that

$$h = 0.321 \text{ m} \qquad\qquad\qquad Ans.$$

7

7.11   The Navier–Stokes Equations

In the previous sections we considered applications of the equations of motion to *ideal fluids*, where only the forces of gravity and pressure influence the flow. These forces form a concurrent system on each fluid particle or element, and so irrotational flow occurs. Real fluids, however, are viscous, so a more precise set of equations used to describe the flow should include the viscous forces as well.

The general differential equations of motion for a fluid, determined from Newton's second law of motion, were developed in Sec. 7.5 as Eqs. 7–17. For convenience, they are repeated here:

$$\rho g_x + \frac{\partial \sigma_{xx}}{\partial x} + \frac{\partial \tau_{yx}}{\partial y} + \frac{\partial \tau_{zx}}{\partial z} = \rho\left(\frac{\partial u}{\partial t} + u\frac{\partial u}{\partial x} + v\frac{\partial u}{\partial y} + w\frac{\partial u}{\partial z}\right)$$

$$\rho g_y + \frac{\partial \tau_{xy}}{\partial x} + \frac{\partial \sigma_{yy}}{\partial y} + \frac{\partial \tau_{zy}}{\partial z} = \rho\left(\frac{\partial v}{\partial t} + u\frac{\partial v}{\partial x} + v\frac{\partial v}{\partial y} + w\frac{\partial v}{\partial z}\right)$$

$$\rho g_z + \frac{\partial \tau_{xz}}{\partial x} + \frac{\partial \tau_{yz}}{\partial y} + \frac{\partial \sigma_{zz}}{\partial z} = \rho\left(\frac{\partial w}{\partial t} + u\frac{\partial w}{\partial x} + v\frac{\partial w}{\partial y} + w\frac{\partial w}{\partial z}\right)$$

To seek a general solution we will express these equations in terms of the velocity components by relating the stress components to the viscosity of the fluid and the velocity gradients. Recall that for the one-dimensional flow of a *Newtonian fluid*, the shear stress and velocity gradient are related by Eq. 1–14, $\tau = \mu(du/dy)$. However, for three-dimensional flow, similar expressions are more complicated. For the special case where the density is constant and we have a Newtonian fluid, both the normal and shear stresses are linearly related to their associated strain rates. It can be shown, see Ref. [9], that the stress–strain rate relationships then become*

$$\sigma_{xx} = -p + 2\mu\frac{\partial u}{\partial x}$$

$$\sigma_{yy} = -p + 2\mu\frac{\partial v}{\partial y}$$

$$\sigma_{zz} = -p + 2\mu\frac{\partial w}{\partial z}$$

$$\tau_{xy} = \tau_{yx} = \mu\left(\frac{\partial u}{\partial y} + \frac{\partial v}{\partial x}\right) \qquad (7\text{–}74)$$

$$\tau_{yz} = \tau_{zy} = \mu\left(\frac{\partial v}{\partial z} + \frac{\partial w}{\partial y}\right)$$

$$\tau_{zx} = \tau_{xz} = \mu\left(\frac{\partial u}{\partial z} + \frac{\partial w}{\partial x}\right)$$

---

* Notice that the normal stresses are the result of *both* a pressure p, which represents the average normal stress on the fluid element, that is, $p = -\frac{1}{3}(\sigma_{xx} + \sigma_{yy} + \sigma_{zz})$, and a viscosity term that is caused by the motion of the fluid. When the fluid is *at rest* ($u = v = w = 0$) then $\sigma_{xx} = \sigma_{yy} = \sigma_{zz} = -p$, a consequence of Pascal's law. Also, when the *streamlines for the flow are all parallel*, and directed, say, along the x axis (one-dimensional flow), then $v = w = 0$ so $\sigma_{yy} = \sigma_{zz} = -p$. Additionally, with $v = w = 0$, for an *incompressible fluid*, the continuity equation, Eq. 7–9, becomes $\partial u/\partial x = 0$, and so also $\sigma_{xx} = -p$.

If we substitute these equations for the stress components into the equations of motion, and simplify these equations, we obtain

$$\rho\left(\frac{\partial u}{\partial t} + u\frac{\partial u}{\partial x} + v\frac{\partial u}{\partial y} + w\frac{\partial u}{\partial z}\right) = \rho g_x - \frac{\partial p}{\partial x} + \mu\left(\frac{\partial^2 u}{\partial x^2} + \frac{\partial^2 u}{\partial y^2} + \frac{\partial^2 u}{\partial z^2}\right)$$

$$\rho\left(\frac{\partial v}{\partial t} + u\frac{\partial v}{\partial x} + v\frac{\partial v}{\partial y} + w\frac{\partial v}{\partial z}\right) = \rho g_y - \frac{\partial p}{\partial y} + \mu\left(\frac{\partial^2 v}{\partial x^2} + \frac{\partial^2 v}{\partial y^2} + \frac{\partial^2 v}{\partial z^2}\right) \qquad (7\text{--}75)$$

$$\rho\left(\frac{\partial w}{\partial t} + u\frac{\partial w}{\partial x} + v\frac{\partial w}{\partial y} + w\frac{\partial w}{\partial z}\right) = \rho g_z - \frac{\partial p}{\partial z} + \mu\left(\frac{\partial^2 w}{\partial x^2} + \frac{\partial^2 w}{\partial y^2} + \frac{\partial^2 w}{\partial z^2}\right)$$

Here, the terms on the left represent "*ma*," and those on the right represent "ΣF," caused by weight, pressure, and viscosity, respectively. These equations were developed in the early 19th century by the French engineer Louis Navier, and several years later by the British mathematician George Stokes. It is for this reason that they are referred to as the *Navier–Stokes equations*. They apply to uniform, nonuniform, steady, or nonsteady flow of an incompressible Newtonian fluid, for which μ is constant.* Together with the continuity equation, Eq. 7–10,

$$\frac{\partial \rho}{\partial t} + \frac{\partial(\rho u)}{\partial x} + \frac{\partial(\rho v)}{\partial y} + \frac{\partial(\rho w)}{\partial z} = 0 \qquad (7\text{--}76)$$

these *four equations* provide a means of obtaining the velocity components u, v, w and the pressure p within the flow.

Unfortunately, there is no *general solution,* simply because the three unknowns u, v, w appear in *all* the equations, and the first three are nonlinear and of the second order. In spite of this difficulty, for a few problems they all reduce to a simpler form, and thereby produce a solution. This occurs when the boundary and initial conditions are simple, and laminar flow prevails. We will show one of these solutions in the example problem that follows, while others are given as problems. Later, in Chapter 9, we will also show how these equations can be solved for the case of laminar flow between parallel plates, and within a pipe.

*They have also been developed for compressible flow and can be generalized to include a fluid of variable viscosity. See Ref. [9].

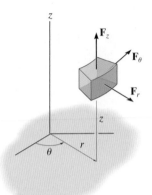

Polar Coodinates

Fig. 7–37

Cylindrical Coordinates.

Although we have presented the Navier–Stokes equations in terms of Cartesian x, y, z coordinates, they can also be developed in terms of cylindrical (or spherical) coordinates. Without proof, and for later use, in cylindrical component form, Fig. 7–37, they are

$$\rho\left(\frac{\partial v_r}{\partial t} + v_r\frac{\partial v_r}{\partial r} + \frac{v_\theta}{r}\frac{\partial v_r}{\partial \theta} - \frac{v_\theta^2}{r} + v_z\frac{\partial v_r}{\partial z}\right)$$

$$= -\frac{\partial p}{\partial r} + \rho g_r + \mu\left[\frac{1}{r}\frac{\partial}{\partial r}\left(r\frac{\partial v_r}{\partial r}\right) - \frac{v_r}{r^2} + \frac{1}{r^2}\frac{\partial^2 v_r}{\partial \theta^2} - \frac{2}{r^2}\frac{\partial v_\theta}{\partial \theta} + \frac{\partial^2 v_r}{\partial z^2}\right]$$

$$\rho\left(\frac{\partial v_\theta}{\partial t} + v_r\frac{\partial v_\theta}{\partial r} + \frac{v_\theta}{r}\frac{\partial v_\theta}{\partial \theta} + \frac{v_r v_\theta}{r} + v_z\frac{\partial v_\theta}{\partial z}\right)$$

$$= -\frac{1}{r}\frac{\partial p}{\partial \theta} + \rho g_\theta + \mu\left[\frac{1}{r}\frac{\partial}{\partial r}\left(r\frac{\partial v_\theta}{\partial r}\right) - \frac{v_\theta}{r^2} + \frac{1}{r^2}\frac{\partial^2 v_\theta}{\partial \theta^2} + \frac{2}{r^2}\frac{\partial v_r}{\partial \theta} + \frac{\partial^2 v_\theta}{\partial z^2}\right] \quad (7\text{–}77)$$

$$\rho\left(\frac{\partial v_z}{\partial t} + v_r\frac{\partial v_z}{\partial r} + \frac{v_\theta}{r}\frac{\partial v_z}{\partial \theta} + v_z\frac{\partial v_z}{\partial z}\right)$$

$$= -\frac{\partial p}{\partial z} + \rho g_z + \mu\left[\frac{1}{r}\frac{\partial}{\partial r}\left(r\frac{\partial v_z}{\partial r}\right) + \frac{1}{r^2}\frac{\partial^2 v_z}{\partial \theta^2} + \frac{\partial^2 v_z}{\partial z^2}\right]$$

And the corresponding continuity equation, Eq. 7–14, is

$$\frac{\partial \rho}{\partial t} + \frac{1}{r}\frac{\partial(r\rho v_r)}{\partial r} + \frac{1}{r}\frac{\partial(\rho v_\theta)}{\partial \theta} + \frac{\partial(\rho v_z)}{\partial z} = 0 \quad (7\text{–}78)$$

EXAMPLE | 7.11

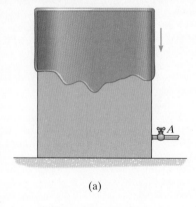

(a)

Fig. 7–38

When the supply valve A is slightly opened, a very viscous Newtonian liquid in the rectangular tank overflows, Fig. 7–38a. Determine the velocity profile of the liquid as it slowly spills over the sides.

SOLUTION

Fluid Description. We will assume the liquid is an incompressible Newtonian fluid that has steady laminar flow. Also, after falling a *short distance* from the top, the liquid along the sides will continue to maintain a constant thickness a.

Analysis. With the coordinate axes established as shown in Fig. 7–38b, there is only a velocity u in the x direction. Furthermore, due to symmetry, u changes only in the y direction, not in the x or z direction, Fig. 7–38c. Because the flow is steady and the liquid is incompressible, the continuity equation becomes

$$\frac{\partial \rho}{\partial t} + \frac{\partial(\rho u)}{\partial x} + \frac{\partial(\rho v)}{\partial y} + \frac{\partial(\rho w)}{\partial z} = 0$$

$$0 + \frac{\partial(\rho u)}{\partial x} + 0 + 0 = 0$$

For constant ρ, integrating yields

$$u = u(y)$$

Using this result, the Navier–Stokes equations in the x and y directions reduce to

$$\rho\left(\frac{\partial u}{\partial t} + u\frac{\partial u}{\partial x} + v\frac{\partial u}{\partial y} + w\frac{\partial u}{\partial z}\right) = \rho g_x - \frac{\partial p}{\partial x} + \mu\left(\frac{\partial^2 u}{\partial x^2} + \frac{\partial^2 u}{\partial y^2} + \frac{\partial^2 u}{\partial z^2}\right)$$

$$0 + 0 + 0 + 0 = \rho g - \frac{\partial p}{\partial x} + 0 + \mu\frac{\partial^2 u}{\partial y^2} + 0 \qquad (1)$$

$$\rho\left(\frac{\partial v}{\partial t} + u\frac{\partial v}{\partial x} + v\frac{\partial v}{\partial y} + w\frac{\partial v}{\partial z}\right) = \rho g_y - \frac{\partial p}{\partial y} + \mu\left(\frac{\partial^2 v}{\partial x^2} + \frac{\partial^2 v}{\partial y^2} + \frac{\partial^2 v}{\partial z^2}\right)$$

$$0 + 0 + 0 + 0 = 0 - \frac{\partial p}{\partial y} + 0 + 0 + 0$$

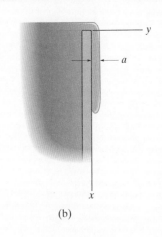

(b)

This last equation shows that the pressure does not change in the y direction; and since p is atmospheric on the surface of the liquid, it remains so within the liquid, that is, $p = 0$ (gage). As a result, Eq. 1 now becomes

$$\frac{\partial^2 u}{\partial y^2} = -\frac{\rho g}{\mu}$$

Integrating twice, we obtain

$$\frac{\partial u}{\partial y} = -\frac{\rho g}{\mu}y + C_1 \qquad (2)$$

$$u = -\frac{\rho g}{2\mu}y^2 + C_1 y + C_2 \qquad (3)$$

To evaluate the constant C_2, we can use the no-slip condition; that is, the velocity of the liquid at $y = 0$ must be $u = 0$, Fig. 7–38c. Therefore, $C_2 = 0$. To obtain C_1, we realize that there is no shear stress τ_{xy} on the *free surface* of the liquid. Applying this condition to Eq. 7–74 yields

$$\tau_{xy} = \mu\left(\frac{\partial u}{\partial y} + \frac{\partial v}{\partial x}\right)$$

$$0 = \mu\left(\frac{\partial u}{\partial y} + 0\right)$$

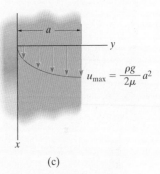

Therefore, $du/dy = 0$ at $y = a$. Substituting this into Eq. 2 yields $C_1 = (\rho g/\mu)a$, so Eq. 3 now becomes

$$u = \frac{\rho g}{2\mu}\left(2ya - y^2\right) \qquad Ans.$$

(c)

Fig. 7–38 (cont.)

Once it is fully developed, this parabolic velocity profile, shown in Fig. 7–38c, is maintained as the liquid flows down the edge of the tank. It is the result of a balance of downward gravity and upward viscous forces.

7

7.12　Computational Fluid Dynamics

In the previous section we have seen that a description of any fluid flow requires satisfying the Navier–Stokes equations and the continuity equation, along with the appropriate boundary or initial conditions for the flow. However, these equations are very complicated, and so their solution has been obtained for only a few special cases involving laminar flow.

Fortunately, during the past decades there has been an exponential growth in the capacity, memory storage, and affordability of high-speed computers, making it possible to use numerical methods to solve these equations. This field of study is referred to as *computational fluid dynamics* (CFD), and it is now widely used to design and analyze many different types of fluid flow problems, such as those involving aircraft, pumps and turbines, heating and ventilation equipment, wind loadings on buildings, chemical processes, and even biomedical implant devices, as well as atmospheric weather modeling.

Several popular CFD computer programs are currently available—for example, FLUENT, FLOW–3D, and ANSYS, to name a few. Other phenomena related to fluid flow, such as heat transfer, chemical reactions, and multiphase changes, have also been incorporated into these programs. As accuracy in predicting the flow improves, the use of these codes offers potential savings by eliminating the need for sophisticated *experimental testing* of models or their prototypes. What follows is a brief introduction and overview of the types of methods that are used to develop CFD software. Further information about this field can be found by consulting one of the many references listed at the end of this chapter, or by taking courses or seminars in this subject, offered at many engineering schools or in the private sector.

The CFD Code.　There are three basic parts to any CFD code. They are the input, the program, and the output. Let's consider each of these in turn.

Input.　The operator must input data related to the fluid properties, specify whether laminar or turbulent flow occurs, and identify the geometry of the boundary for the flow.

Fluid Properties.　Many CFD packages include a list of physical properties, such as density and viscosity, that can be selected to define the fluid. Any data not included in the list must be identified and then entered as separate entries.

Flow Phenomena.　The user must select a physical model provided with the code that relates to the type of flow. In this regard, many commercial packages will have a selection of models that can be used to predict turbulent flow. Needless to say, it takes some experience to select one that is appropriate, since a model that fits one problem may not be suitable for another.

Geometry.　The physical geometry around the flow must also be defined. This is done by creating a grid or mesh system throughout the fluid domain. To make this job user friendly, most CFD packages include various types of boundary and mesh geometries that can be selected to improve computational speed and accuracy.

Program. For many users of a CFD program the various algorithms and numerical techniques used to perform the calculations will be unknown. The process, however, consists of two parts. The first considers the fluid as a system of discrete particles and converts the relevant partial differential equations into a group of *algebraic equations*, and the second uses an iterative procedure to find solutions of these equations that satisfy the initial and boundary conditions for the problem. There are several approaches that can be used to do this. These include the finite difference method, the finite element method, and the finite control volume method.

Finite Difference Method. For unsteady flow, the finite difference method uses a distance–time grid that determines the conditions at a particular point, one time step in the future, based on present conditions at adjacent points. To give some idea of the application of this approach, we will use it later to model unsteady open-channel flow in Chapter 12.

Finite Element Method. As the name implies, this method considers the fluid to be subdivided into small "finite elements," and the equations that describe the flow within each element are then made to satisfy the boundary conditions at the corners, or *nodes*, of adjacent elements. It has the advantage of having a higher degree of accuracy than using the finite difference method; however, the methodologies used are more complex. Also, since the elements or grid can take any irregular form, the finite element method can be made to match any type of boundary.

Finite Control Volume Method. The finite control volume method combines the best attributes of both the finite difference and the finite element methods. It is capable of modeling complex boundary conditions, while expressing the governing differential equations using relatively straightforward finite difference relations. The characteristic feature of this method is that each of the many small control volumes accounts for the local time rate of change in a flow variable such as velocity, density, or temperature within the control volume, and the net flux of the variable convected through its control surfaces. Each of these terms is converted into sets of algebraic equations, which are then solved using an iterative method. As a result of these advantages, the finite control volume method has become well established and is currently used in most CFD software.

Output. The output is usually a graphical form of the flow domain, showing the geometry of the problem, and sometimes the grid or mesh used for the analysis. Superimposed on this, the operator can choose to show contours for the flow variables, including streamlines, or pathlines, and velocity vector plots, etc. Hard-copy printouts of steady flow can be made, or time animation of unsteady flows can be displayed in video format. An example of this output is shown in Fig. 7–39.

General Considerations. If a realistic prediction of a complex flow is to be determined, the operator must have *experience* at running any particular code. Of course, it is important to have a full understanding of the basic principles of fluid mechanics, in order to define a proper model for the flow, and make a reasonable selection of the time-step size and grid layout. Once the solution is obtained, it can be compared with experimental data or with similar existing flow situations. Finally, *always remember* that the responsibility of using any CFD program lies in the hands of the operator (engineer), and for that reason, he or she is ultimately accountable for the results.

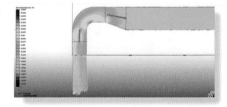

This CFD analysis of flow through the transition and elbow shows how the velocity varies along the cross section as represented by the colors.

Fig. 7–39

References

1. H. Lamb, *Hydrodynamics*, 6th ed., Dover Publications, New York, NY, 1945.

2. J. D. Anderson Jr., *Computational Fluid Dynamics: The Basics with Applications*, McGraw-Hill, New York, NY, 1995.

3. A. Quarteroni, "Mathematical models in science and engineering," *Notices of the AMS*, Vol. 56, No. 1, 2009, pp. 10–19.

4. T. W. Lee, *Thermal and Flow Measurements*, CRC Press, Boca Raton, FL.

5. T. J. Chung, *Computational Fluid Dynamics*, Cambridge, England, 2002.

6. J. Tannechill et al., *Computational Fluid Mechnanics and Heat Transfer*, 2nd ed., Taylor and Francis, Bristol, PA, 1997.

7. C. Chow, *An Introduction to Computational Fluid Mechanics*, John Wiley, New York, NY, 1980.

8. F. White. *Viscous Fluid Flow*, 3rd ed., McGraw-Hill, New York, NY, 2005.

9. H. Rouse, *Advanced Mechanics of Fluids*, John Wiley, New York, NY, 1959.

10. J. M. Robertson, *Hydrodynamics in Theory and Applications*, Prentice Hall, Englewood Cliffs, NJ, 1965.

11. L. Milne-Thomson, *Theoretical Hydrodynamics*, 4th ed., Macmillan, New York, NY, 1960.

12. R. Peyret and T. Taylor, *Computational Methods for Fluid Flow*, Springer-Verlag, New York, NY, 1983.

13. E. Buckingham, "On physically similar system: illustrations of the use of dimensional equations," *Physical Review*, 4, 1914, pp. 345–376.

14. I. H. Shames, *Mechanics of Fluids*, McGraw Hill, New York, NY, 1962.

15. J. Tu et al., *Computational Fluid Dynamics: A Practical Approach*, Butterworth-Heinemann, New York, NY, 2007.

16. A. L. Prasuhn, *Fundamentals of Fluid Mechanics*, Prentice-Hall, Englewood Cliffs, NJ, 1980.

17. L. Larsson, "CFD in ship desgin–prospects and limitations," *Ship Technology Research*, Vol. 44, July 1997, pp. 133–154.

18. J. Piquet, *Turbulent Flow Models and Physics*, Springer, Berlin, 1999.

19. T. Cebui, *Computational Fluid Dynamics for Engineers*, Springer-Verlag, New York, NY, 2005.

20. D. Apsley and W. Hu, "CFD simulation of two- and three-dimensional free-surface flow," *Int J Numer Meth Fluids*, 42, 2003, pp. 465–491.

21. X. Yang and H. Ma, "Cubic eddy-viscosity turbulence models for strongly swirling confined flows with variable density," *Int J Numer Meth Fluids*, 45, 2004, pp. 985–1008.

22. D. Wilcox, *Turbulence Modeling for CFD*, DCW Industries, La Canada, CA, 1993.

PROBLEMS

Sec. 7.1–7.6

7–1. As the top plate is pulled to the right with a constant velocity **U**, the fluid between the plates has a linear velocity distribution as shown. Determine the rate of rotation of a fluid element and the shear-strain rate of the element located at y.

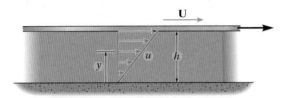

Prob. 7–1

7–2. The velocity within the eye of a tornado is defined by $v_r = 0$, $v_\theta = (0.2r)$ m/s, where r is in meters. Determine the circulation at $r = 60$ m and at $r = 80$ m.

Prob. 7–2

7–3. A uniform flow **V** is directed at an angle θ to the horizontal as shown. Determine the circulation around the rectangular region.

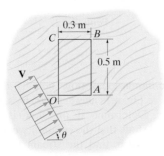

Prob. 7–3

*7–4. A flow is defined by its velocity components $u = (4x^2 + 4y^2)$ m/s and $v = (-8xy)$ m/s, where x and y are in meters. Determine if the flow is irrotational. What is the circulation around the rectangular region?

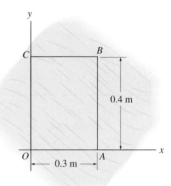

Prob. 7–4

7–5. Consider the fluid element that has dimensions in polar coordinates as shown and whose boundaries are defined by the streamlines with velocities v and $v + dv$. Show that the vorticity for the flow is given by $\zeta = -(v/r + dv/dr)$.

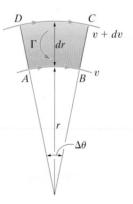

Prob. 7–5

Sec. 7.7–7.8

7–6. Determine the stream and potential functions for the two-dimensional flow field if V_0 and θ are known.

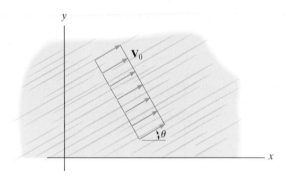

Prob. 7–6

7–7. If the stream function for a flow is $\psi = (3x + 2y)$, where x and y are in meters, determine the potential function and the magnitude of the velocity of a fluid particle at point $(1 \text{ m}, 2 \text{ m})$.

*7–8. The velocity profile of a very thick liquid flowing along the channel of constant width is approximated as $u = (3y^2) \text{ mm/s}$, where y is in millimeters. Determine the stream function for the flow and plot the streamlines for $\psi_0 = 0$, $\psi_1 = 1 \text{ mm}^2/\text{s}$, and $\psi_2 = 2 \text{ mm}^2/\text{s}$.

7–9. The velocity profile of a very thick liquid flowing along the channel of constant width is approximated as $u = (3y^2) \text{ mm/s}$, where y is in millimeters. Is it possible to determine the potential function for the flow? If so, what is it?

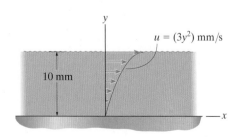

Probs. 7–8/9

7–10. A two-dimensional flow is described by the stream function $\psi = (xy^3 - x^3y) \text{ m}^2/\text{s}$, where x and y are in meters. Show that the continuity condition is satisfied and determine if the flow is rotational or irrotational.

7–11. The liquid confined between two plates is assumed to have a linear velocity distribution as shown. Determine the stream function. Does the potential function exist?

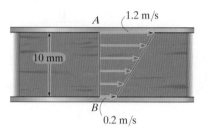

Prob. 7–11

*7–12. The liquid confined between two plates is assumed to have a linear velocity distribution as shown. If the pressure at the top surface of the bottom plate is 600 N/m^2, detemine the pressure at the bottom surface of the top plate. Take $\rho = 1.2 \text{ Mg/m}^3$.

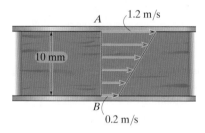

Prob. 7–12

7–13. A two-dimensional flow field is defined by its components $u = (3y) \text{ m/s}$ and $v = (9x) \text{ m/s}$, where x and y are in meters. Determine if the flow is rotational or irrotational, and show that the continuity condition for the flow is satisfied. Also, find the stream function and the equation of the streamline that passes through point $(4 \text{ m}, 3 \text{ m})$. Plot this streamline.

7–14. A fluid has the velocity components shown. Determine the stream and potential functions. Plot the streamlines for $\psi_0 = 0$, $\psi_1 = 1$ m²/s, and $\psi_2 = 2$ m²/s.

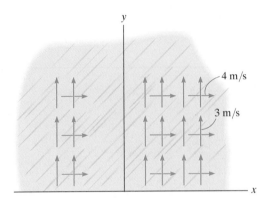

4 m/s

3 m/s

Prob. 7–14

7–15. A flow field is defined by the stream function $\psi = 2(x^2 - y^2)$ m²/s, where x and y are in meters. Determine the flow per unit depth in m²/s that occurs through AB, CB, and AC as shown.

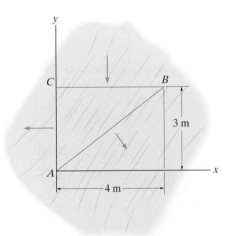

3 m

4 m

Prob. 7–15

*7–16. The stream function for a flow field is defined by $\psi = (4/r^2) \sin 2\theta$. Show that continuity of the flow is satisfied, and determine the r and θ velocity components of fluid particles at point $r = 2$ m, $\theta = (\pi/4)$ rad. Plot the streamline that passes through this point.

7–17. A flow is described by the stream function $\psi = (8x - 4y)$ m²/s, where x and y are in meters. Determine the potential function, and show that the continuity condition is satisfied and that the flow is irrotational.

7–18. The stream function for a flow field is defined by $\psi = 2r^3 \sin 2\theta$. Determine the magnitude of the velocity of fluid particles at point $r = 1$ m, $\theta = (\pi/3)$ rad, and plot the streamlines for $\psi_1 = 1$ m²/s and $\psi_2 = 2$ m²/s.

7–19. Water flow through the horizontal channel is defined by the stream function $\psi = 2(x^2 - y^2)$ m²/s. If the pressure at B is atmospheric, determine the pressure at point $(0.5$ m, $0)$ and the flow per unit depth in m²/s.

7

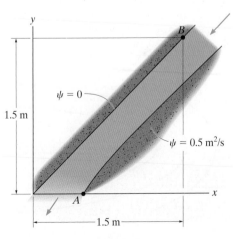

$\psi = 0$

$\psi = 0.5$ m²/s

1.5 m

1.5 m

Prob. 7–19

*7–20. An ideal fluid flows into the corner formed by the two walls. If the stream function for this flow is defined by $\psi = \left(5\,r^4\sin 4\theta\right)$ m²/s, show that continuity for the flow is satisfied. Also, plot the streamline that passes through point $r = 2$ m, $\theta = (\pi/6)$ rad, and find the magnitude of the velocity at this point.

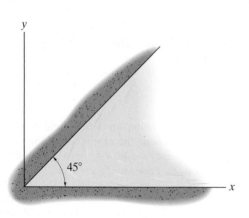

45°

Prob. 7–20

7–21. The flat plate is subjected to the flow defined by the stream function $\psi = \left[8r^{1/2}\sin(\theta/2)\right]$ m²/s. Sketch the streamline that passes through point $r = 4$ m, $\theta = \pi$ rad, and determine the magnitude of the velocity at this point.

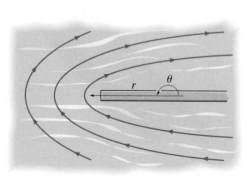

Prob. 7–21

7–22. Determine the potential function for the two-dimensional flow field if $\mathbf{V}_0$ and θ are known.

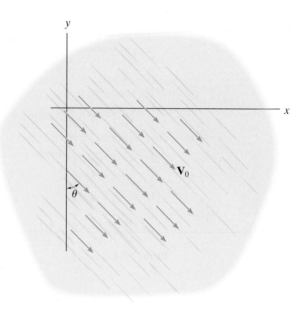

Prob. 7–22

7–23. The stream function for a concentric flow is defined by $\psi = -4r^2$. Determine the velocity components v_r and v_θ, and v_x and v_y. Can the potential function be established? If so, what is it?

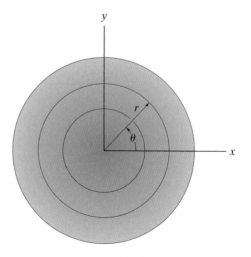

Prob. 7–23

*7–24. If the potential function for a two-dimensional flow is $\phi = (xy)\ \text{m}^2/\text{s}$, where x and y are in meters, determine the stream function, and plot the streamline that passes through the point $(1\ \text{m}, 2\ \text{m})$. What are the x and y components of the velocity and acceleration of fluid particles that pass through this point?

7–25. The horizontal flow confined by the walls is defined by the stream function $\psi = \left[4r^{4/3} \sin\left(\frac{4}{3}\theta\right)\right]\ \text{m}^2/\text{s}$, where r is in meters. Determine the magnitude of the velocity at point $r = 2\ \text{m}$, $\theta = 45°$. Is the flow rotational or irrotational? Can the Bernoulli equation be used to determine the difference in pressure between the two points A and B?

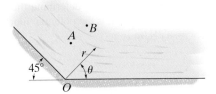

Prob. 7–25

7–26. The horizontal flow between the walls is defined by the stream function $\psi = \left[4r^{4/3} \sin\left(\frac{4}{3}\theta\right)\right]\ \text{m}^2/\text{s}$, where r is in meters. If the pressure at the origin O is 20 kPa, determine the pressure at $r = 2\ \text{m}$, $\theta = 45°$. Take $\rho = 950\ \text{kg/m}^3$.

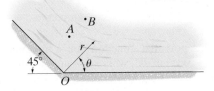

Prob. 7–26

7–27. The flow around the bend in the horizontal channel can be described as a free vortex for which $v_r = 0$, $v_\theta = (8/r)\ \text{m/s}$, where r is in meters. Show that the flow is irrotational. If the pressure at point A is 4 kPa, determine the pressure at point B. Take $\rho = 1100\ \text{kg/m}^3$.

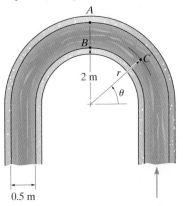

Prob. 7–27

*7–28. A two-dimensional flow is described by the potential function $\phi = \left(8x^2 - 8y^2\right)\ \text{m}^2/\text{s}$, where x and y are in meters. Show that the continuity condition is satisfied, and determine if the flow is rotational or irrotational. Also, establish the stream function for this flow, and plot the streamline that passes through point $(1\ \text{m}, 0.5\ \text{m})$.

7–29. The stream function for a horizontal flow near the corner is $\psi = (8xy)\ \text{m}^2/\text{s}$, where x and y are in meters. Determine the x and y components of the velocity and the acceleration of fluid particles passing through point $(1\ \text{m}, 2\ \text{m})$. Show that it is possible to establish the potential function. Plot the streamlines and equipotential lines that pass through point $(1\ \text{m}, 2\ \text{m})$.

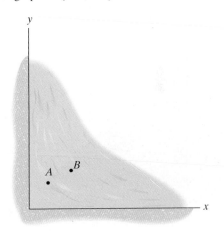

Prob. 7–29

7–30. The stream function for horizontal flow near the corner is defined by $\psi = (8xy)$ m^2/s, where x and y are in meters. Show that the flow is irrotational. If the pressure at point A (1 m, 2 m) is 150 kPa, determine the pressure at point B (2 m, 3 m). Take $\rho = 980$ kg/m^3.

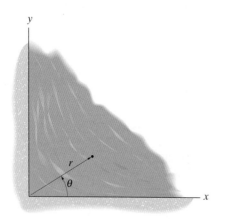

Prob. 7–30

7–31. The stream function for the flow field around the 90° corner is $\psi = 8r^2 \sin 2\theta$. Show that the continuity of flow is satisfied. Determine the r and θ velocity components of a fluid particle located at $r = 0.5$ m, $\theta = 30°$, and plot the streamline that passes through this point. Also, determine the potential function for the flow.

Prob. 7–31

*7–32. A fluid has velocity components $u = 2(x^2 - y^2)$ m/s and $v = (-4xy)$ m/s, where x and y are in meters. Determine the stream function. Also show that the potential function exists, and find this function. Plot the streamlines and equipotential lines that pass through point (1 m, 2 m).

7–33. If the potential function for a two-dimensional flow is $\phi = (xy)$ m^2/s, where x and y are in meters, determine the stream function and plot the streamline that passes through the point (1 m, 2 m). What are the velocity and acceleration of fluid particles that pass through this point?

7–34. A fluid has velocity components of $u = (y^2 - x^2)$ m/s and $v = (2xy)$ m/s, where x and y are in meters. If the pressure at point A (3 m, 2 m) is 600 kPa, determine the pressure at point B (1 m, 3 m). Also what is the potential function for the flow? Take $\gamma = 8$ kN/m^3.

7–35. A fluid has velocity components of $u = (10xy)$ m/s and $v = 5(x^2 - y^2)$ m/s, where x and y are in meters. Determine the stream function, and show that the continuity condition is satisfied and that the flow is irrotational. Plot the streamlines for $\psi_0 = 0$, $\psi_1 = 1$ m^2/s, and $\psi_2 = 2$ m^2/s.

*7–36. A fluid has velocity components of $u = (10xy)$ m/s and $v = 5(x^2 - y^2)$ m/s, where x and y are in meters. Determine the potential function, and show that the continuity condition is satisfied and that the flow is irrotational.

7–37. The potential function for a horizontal flow is $\phi = (x^3 - 5xy^2)$ m^2/s, where x and y are in meters. Determine the magnitude of the velocity at point A (5 m, 2 m). What is the difference in pressure between this point and the origin? Take $\rho = 925$ kg/m^3.

Sec. 7.9–7.10

7–38. Show that the equation that defines a sink will satisfy continuity, which in polar coordinates is written as

$$\frac{\partial(v_r r)}{\partial r} + \frac{\partial(v_\theta)}{\partial \theta} = 0.$$

7–39. Combine a source of strength q with a free counterclockwise vortex, and sketch the resultant streamline for $\psi = 0$.

***7–40.** Pipe A provides a source flow of 5 m²/s, whereas the drain, or sink, at B removes 5 m²/s. Determine the stream function between AB, and show the streamline for $\psi = 0$.

7–41. Pipe A provides a source flow of 5 m²/s, whereas the drain at B removes 5 m²/s. Determine the potential function between AB, and show the equipotential line for $\phi = 0$.

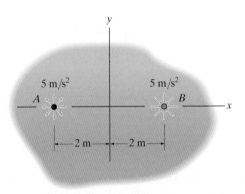

Probs. 7–40/41

7–42. Determine the location of the stagnation point for a combined uniform flow of 8 m/s and a source having a strength of 3 m²/s. Plot the streamline passing through the stagnation point.

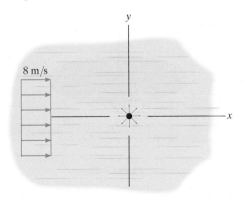

Prob. 7–42

7–43. Two sources, each having a strength of 2 m²/s, are located as shown. Determine the x and y components of the velocity of fluid particles that pass point (x, y). What is the equation of the streamline that passes through point $(0, 8 \text{ m})$ in Cartesian coordinates? Is the flow irrotational?

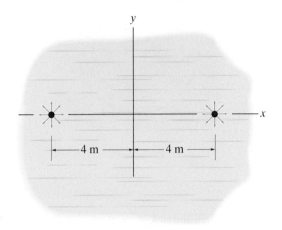

Prob. 7–43

***7–44.** A source at O creates a flow from point O that is described by the potential function $\phi = (8 \ln r)$ m²/s, where r is in meters. Determine the stream function, and specify the velocity at point $r = 5$ m, $\theta = 15°$.

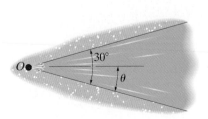

Prob. 7–44

7–45. A free vortex is defined by its stream function $\psi = (-240 \ln r)$ m²/s, where r is in meters. Determine the velocity of a particle at $r = 4$ m and the pressure at points on this streamline. Take $\rho = 1.20$ kg/m³.

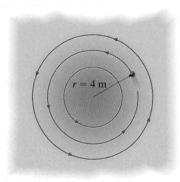

Prob. 7–45

7–46. The source and sink of equal strength q are located a distance d from the origin as indicated. Determine the stream function for the flow, and draw the streamline that passes through the origin.

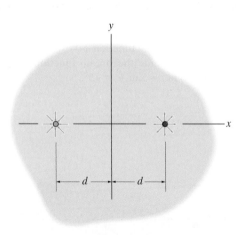

Prob. 7–46

7–47. Two sources, each having a strength q, are located as shown. Determine the stream function, and show that this is the same as having a single source with a wall along the y axis.

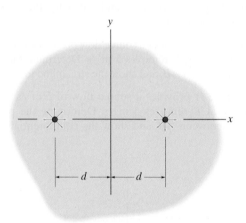

Prob. 7–47

*7–48. The Rankine body is defined by the source and sink, each having a strength of 0.2 m²/s. If the velocity of the uniform flow is 4 m/s, determine the longest and shortest dimensions of the body.

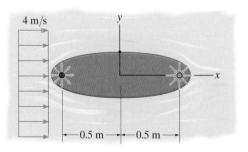

Prob. 7–48

7–49. The Rankine body is defined by the source and sink, each having a strength of 0.2 m²/s. If the velocity of the uniform flow is 4 m/s, determine the equation in Cartesian coordinates that defines the boundary of the body.

7–51. A fluid flows over a half body for which $U = 0.4$ m/s and $q = 1.0$ m²/s. Plot the half body, and determine the magnitudes of the velocity and pressure in the fluid at the point $r = 0.8$ m and $\theta = 90°$. The pressure within the uniform flow is 300 Pa. Take $\rho = 850$ kg/m³.

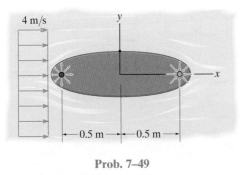

Prob. 7–49

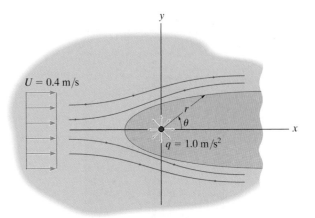

Prob. 7–51

7–50. The half body is defined by a combined uniform flow having a velocity of U and a point source of strength q. Determine the pressure distribution along the top boundary of the half body as a function of θ, if the pressure within the uniform flow is p_0. Neglect the effect of gravity. The density of the fluid is ρ.

*7–52. A source q is emitted from the wall while a flow occurs toward the wall. If the stream function is described as $\psi = (4xy + 8\theta)$ m²/s, where x and y are in meters, determine the distance d from the wall where the stagnation point occurs along the y axis. Plot the streamline that passes through this point.

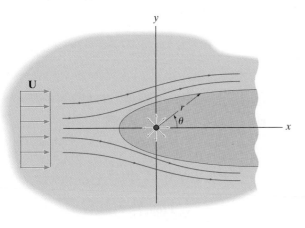

Prob. 7–50

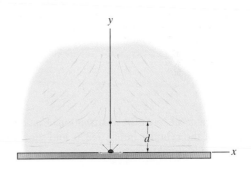

Prob. 7–52

7–53. As water drains from the large cylindrical tank, its surface forms a free vortex having a circulation of Γ. Assuming water to be an ideal fluid, determine the equation $z = f(r)$ that defines the free surface of the vortex. *Hint*: Use the Bernoulli equation applied to two points on the surface.

*7–56. The 0.5-m-diameter bridge pier is subjected to the uniform flow of water at 4 m/s. Determine the maximum and minimum pressures exerted on the pier at a depth of 2 m.

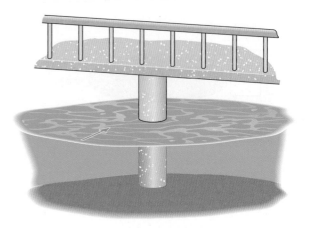

Prob. 7–56

Prob. 7–53

7–57. Air flows around the cylinder such that the pressure, measured at A, is $p_A = -4$ kPa. Determine the velocity U of the flow if $\rho = 1.22$ kg/m³. Can this velocity be determined if instead the pressure at B is measured?

7–54. A fluid has a uniform velocity of $U = 10$ m/s. A source $q = 15$ m²/s is at $x = 2$ m, and a sink $q = -15$ m²/s is at $x = -2$ m. Graph the Rankine body that is formed, and determine the magnitudes of the velocity and the pressure at point (0, 2 m). The pressure within the uniform flow is 40 kPa. Take $\rho = 850$ kg/m³.

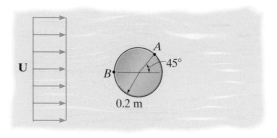

7–55. Integrate the pressure distribution, Eq. 7–67, over the surface of the cylinder in Fig. 7–33b, and show that the resultant force is equal to zero.

Prob. 7–57

7–58. A torque **T** is applied to the cylinder, causing it to rotate counterclockwise with a constant angular velocity of 120 rev/min. If the wind is blowing at a constant speed of 15 m/s, determine the location of the stagnation points on the surface of the cylinder, and find the maximum pressure. The pressure within the uniform flow is 400 Pa. Take $\rho_a = 1.20 \text{ kg/m}^3$.

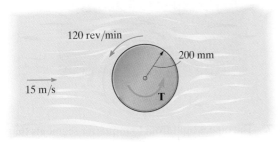

120 rev/min

200 mm

15 m/s

T

Prob. 7–58

7–59. A torque **T** is applied to the cylinder, causing it to rotate counterclockwise with a constant angular velocity of 120 rev/min. If the wind is blowing at a constant speed of 15 m/s, determine the lift per unit length on the cylinder and the minimum pressure on the cylinder. The pressure within the uniform flow is 400 Pa. Take $\rho_a = 1.20 \text{ kg/m}^3$.

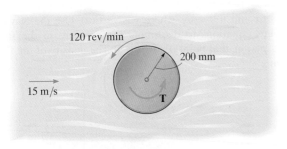

120 rev/min

200 mm

15 m/s

T

Prob. 7–59

*__*7–60.** Air is flowing at $U = 30$ m/s past the Quonset hut of radius $R = 3$ m. Find the velocity and absolute pressure distribution along the y axis for $3 \text{ m} \le y \le \infty$. The absolute pressure within the uniform flow is $p_0 = 100$ kPa. Take $\rho_a = 1.23 \text{ kg/m}^3$.

7–61. The Quonset hut of radius R is subjected to a uniform wind having a velocity U. Determine the resultant vertical force caused by the pressure that acts on the hut if it has a length L. The density of air is ρ.

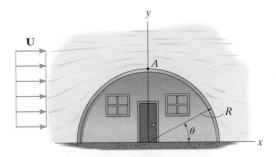

y

U

A

R

θ

x

Probs. 7–60/61

7–62. The Quonset hut of radius R is subjected to a uniform wind having a velocity U. Determine the speed of the wind and the gage pressure at point A. The density of air is ρ.

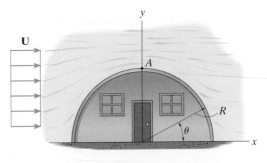

y

U

A

R

θ

x

Prob. 7–62

7

7–63. The 200-mm-diameter cylinder is subjected to a uniform horizontal flow having a velocity of 6 m/s. At a distance far away from the cylinder, the pressure is 150 kPa. Plot the variation of the velocity and pressure along the radial line r, at $\theta = 90°$, and specify their values at $r = 0.1$ m, 0.2 m, 0.3 m, 0.4 m, and 0.5 m. Take $\rho = 1.5$ Mg/m³.

7–64. The 200-mm-diameter cylinder is subjected to a uniform horizontal flow having a velocity of 6 m/s. At a distance far away from the cylinder, the pressure is 150 kPa. Plot the variation of the velocity and pressure along the radial line r, at $\theta = 90°$, and specify their values at $r = 0.1$ m, 0.2 m, 0.3 m, 0.4 m, and 0.5 m. Take $\rho = 1.5$ Mg/m³.

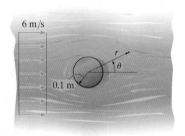

Probs. 7–63/64

7–65. Determine the equation of the boundary of the half body formed by placing a source of 0.5 m²/s in the uniform flow of 8 m/s. Express the result in Cartesian coordinates.

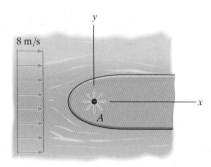

Prob. 7–65

7–66. The half body is defined by a combined uniform flow having a velocity of U and a point source of strength q. Determine the location θ on the boundary of the half body where the pressure p is equal to the pressure p_0 within the uniform flow. Neglect the effect of gravity.

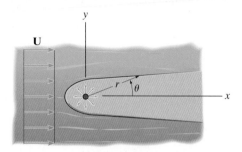

Prob. 7–66

7–67. The pipe is built from four quarter segments that are glued together. If it is exposed to a uniform airflow having a velocity of 8 m/s, determine the resultant force the pressure exerts on the quarter segment AB per unit length of the pipe. Take $\rho = 1.22$ kg/m³.

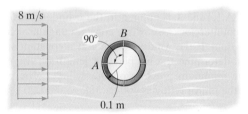

Prob. 7–67

*7–68. The laminar flow of a fluid has velocity components $u = 6x$ and $v = -6y$, where y is vertical. Use the Navier–Stokes equations to determine the pressure in the fluid, $p = p(x, y)$, if at point $(0, 0)$, $p = 0$. The density of the fluid is ρ.

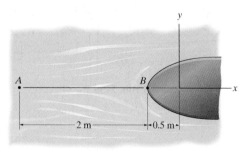

Prob. 7–68

7–69. The steady laminar flow of an ideal fluid toward the fixed surface has a velocity of $u = \left[10\left(1 + 1/(8x^3)\right)\right]$ m/s along the horizontal streamline AB. Use the Navier–Stokes equations and determine the variation of the pressure along this streamline, and plot it for -2.5 m $\leq x \leq -0.5$ m. The pressure at A is 5 kPa, and the density of the fluid is $\rho = 1000$ kg/m^3.

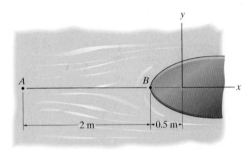

Prob. 7–69

7–70. Liquid is confined between a top plate having an area A and a fixed surface. A force **F** is applied to the plate and gives the plate a velocity **U**. If this causes laminar flow, and the pressure does not vary, show that the Navier–Stokes and continuity equations indicate that the velocity distribution for this flow is defined by $u = U(y/h)$, and that the shear stress within the liquid is $\tau_{xy} = F/A$.

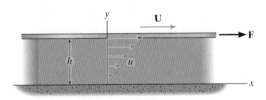

Prob. 7–70

7–71. Fluid having a density ρ and viscosity μ fills the space between the two cylinders. If the outer cylinder is fixed, and the inner one is rotating at ω, apply the Navier–Stokes equations to determine the velocity profile assuming laminar flow.

Prob. 7–71

*7–72. The channel for a liquid is formed by two fixed plates. If laminar flow occurs between the plates, show that the Navier–Stokes and continuity equations reduce to $\partial^2 u/\partial y^2 = (1/\mu)\,\partial p/\partial x$ and $\partial p/\partial y = 0$. Integrate these equations to show that the velocity profile for the flow is $u = (1/(2\mu))\,(dp/dx)\left[y^2 - (d/2)^2\right]$. Neglect the effect of gravity.

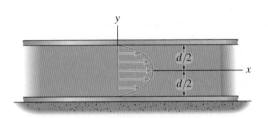

Prob. 7–72

7–73. The sloped open channel has steady laminar flow at a depth h. Show that the Navier–Stokes equations reduce to $\partial^2 u/\partial y^2 = -(\rho\,g\,\sin\theta)/\mu$ and $\partial p/\partial y = -\rho\,g\,\cos\theta$. Integrate these equations to show that the velocity profile is $u = [(\rho\,g\,\sin\theta)/2\mu]\left(2hy - y^2\right)$ and the shear-stress distribution is $\tau_{xy} = \rho\,g\,\sin\theta\,(h - y)$.

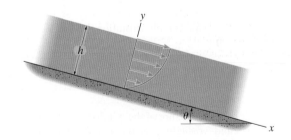

Prob. 7–73

7–74. A horizontal velocity field is defined by $u = 2\left(x^2 - y^2\right)$ m/s and $v = (-4xy)$ m/s. Show that these expressions satisfy the continuity equation. Using the Navier–Stokes equations, show that the pressure distribution is defined by $p = C - \rho V^2/2 - \rho gz$.

7

CHAPTER REVIEW

The rate of *translation* of a fluid element is defined by its velocity field.	$$\mathbf{V} = \mathbf{V}(x, y, z, t)$$
The *linear distortion* of a fluid element is measured as the change in its volume per unit volume. The time rate of this distortion is called the volumetric dilatation.	$$\frac{\delta V / dV}{dt} = \frac{\partial u}{\partial x} + \frac{\partial v}{\partial y} + \frac{\partial w}{\partial z} = \nabla \cdot \mathbf{V}$$
The rate of rotation or *angular velocity* of a fluid element is measured by the average angular velocity of two adjacent sides of a differential element, $\boldsymbol{\omega} = \frac{1}{2}\nabla \times \mathbf{V}$.	$$\omega_z = \frac{1}{2}\left(\frac{\partial v}{\partial x} - \frac{\partial u}{\partial y}\right)$$
The *angular distortion* of a fluid element is defined by the time rate of change of the 90° angle between its adjacent sides. This is the shear-strain rate of the element.	$$\dot{\gamma}_{xy} = \frac{\partial v}{\partial x} + \frac{\partial u}{\partial y}$$
An ideal fluid, which has no viscosity and is incompressible, can exhibit irrotational flow. We require $\boldsymbol{\omega} = \mathbf{0}$.	
Circulation Γ is a measure of the net flow around a boundary, and vorticity ζ is the net circulation about each unit area of fluid.	$$\Gamma = \left(\frac{\partial v}{\partial x} - \frac{\partial u}{\partial y}\right)\Delta x \Delta y \qquad \zeta = \frac{\Gamma}{A} = \frac{\partial v}{\partial x} - \frac{\partial u}{\partial y}$$
The conservation of mass is expressed by the continuity equation, which states $\partial \rho / \partial t + \nabla \cdot \rho \mathbf{V} = 0$, or, for steady incompressible flow, $\nabla \cdot \mathbf{V} = 0$.	$$\frac{\partial u}{\partial x} + \frac{\partial v}{\partial y} = 0$$

Euler's equations of motion apply to an *ideal fluid*. They relate the pressure and gravitational forces acting on a differential fluid element to the components of its acceleration. When integrated, they produce the Bernoulli equation, which can be applied between any two points in the fluid, provided the fluid has steady irrotational flow.	$$\rho g_x + \frac{\partial \sigma_{xx}}{\partial x} + \frac{\partial \tau_{yx}}{\partial y} + \frac{\partial \tau_{zx}}{\partial z} = \rho\left(\frac{\partial u}{\partial t} + u\frac{\partial u}{\partial x} + v\frac{\partial u}{\partial y} + w\frac{\partial u}{\partial z}\right)$$ $$\rho g_y + \frac{\partial \tau_{xy}}{\partial x} + \frac{\partial \sigma_{yy}}{\partial y} + \frac{\partial \tau_{zy}}{\partial z} = \rho\left(\frac{\partial v}{\partial t} + u\frac{\partial v}{\partial x} + v\frac{\partial v}{\partial y} + w\frac{\partial v}{\partial z}\right)$$ $$\rho g_z + \frac{\partial \tau_{xz}}{\partial x} + \frac{\partial \tau_{yz}}{\partial y} + \frac{\partial \sigma_{zz}}{\partial z} = \rho\left(\frac{\partial w}{\partial t} + u\frac{\partial w}{\partial x} + v\frac{\partial w}{\partial y} + w\frac{\partial w}{\partial z}\right)$$
The stream function $\psi(x, y)$ satisfies the continuity equation. If $\psi(x, y)$ is known, then the velocity components of the fluid at any point can be determined from its partial derivatives.	$$u = \frac{\partial \psi}{\partial y} \qquad v = -\frac{\partial \psi}{\partial x}$$
The flow per unit depth, between any two streamlines, $\psi_1(x, y) = C_1$ and $\psi_2(x, y) = C_2$, can be determined from the difference of their stream function constants.	$$q = C_2 - C_1$$
The potential function $\phi(x, y)$ satisfies the conditions of irrotational flow. The velocity components of the fluid at any point can be determined from its partial derivatives.	$$u = \frac{\partial \phi}{\partial x}, \qquad v = \frac{\partial \phi}{\partial y}$$
Solutions for $\psi(x, y)$ and $\phi(x, y)$ have been obtained for a uniform flow, flow from a line source and to a line sink, a doublet, and flow of a free vortex. These solutions can be superimposed (added together or subtracted from one another) to produce more complicated flows, such as flow past a half body, flow around a Rankine oval, or flow around a cylinder.	
When pressure, gravity, and viscous forces are all taken into account, the equations of motion are expressed as the Navier–Stokes equations. Along with the continuity equation, only a limited number of solutions have actually been obtained, and these are for laminar flow. More complex laminar and turbulent flows require a numerical solution of these equations using the methods of computational fluid dynamics (CFD).	$$\rho\left(\frac{\partial u}{\partial t} + u\frac{\partial u}{\partial x} + v\frac{\partial u}{\partial y} + w\frac{\partial u}{\partial z}\right) = \rho g_x - \frac{\partial p}{\partial x} + \mu\left(\frac{\partial^2 u}{\partial x^2} + \frac{\partial^2 u}{\partial y^2} + \frac{\partial^2 u}{\partial z^2}\right)$$ $$\rho\left(\frac{\partial v}{\partial t} + u\frac{\partial v}{\partial x} + v\frac{\partial v}{\partial y} + w\frac{\partial v}{\partial z}\right) = \rho g_y - \frac{\partial p}{\partial y} + \mu\left(\frac{\partial^2 v}{\partial x^2} + \frac{\partial^2 v}{\partial y^2} + \frac{\partial^2 v}{\partial z^2}\right)$$ $$\rho\left(\frac{\partial w}{\partial t} + u\frac{\partial w}{\partial x} + v\frac{\partial w}{\partial y} + w\frac{\partial w}{\partial z}\right) = \rho g_z - \frac{\partial p}{\partial z} + \mu\left(\frac{\partial^2 w}{\partial x^2} + \frac{\partial^2 w}{\partial y^2} + \frac{\partial^2 w}{\partial z^2}\right)$$

7

Chapter 8

(© Georg Gerster/Science Source)

Wind tunnels are often used to test models of aircraft and other vehicles or prototypes. To do this, the model must be properly scaled so that the results correlate to the prototype.

Dimensional Analysis and Similitude

CHAPTER OBJECTIVES

■ To show how to use dimensional analysis to specify the least amount of data needed to experimentally study the behavior of a fluid.

■ To discuss how flow behavior depends on the types of forces that influence the flow, and to present an important set of dimensionless numbers that involve these forces.

■ To formalize a dimensional analysis procedure by obtaining groups of dimensionless numbers using the Buckingham Pi theorem.

■ To show how one can scale a model of a full-size structure or machine, and then use the model to experimentally study the effects of a fluid flow.

8.1 Dimensional Analysis

In the previous chapters we presented many important equations of fluid mechanics, and illustrated their application to the solution of some practical problems. In all these cases we were able to obtain an *algebraic solution* in the form of an equation that describes the flow. In some cases, however, a problem can involve a complicated flow, and the combination of physical variables and fluid properties, such as velocity, pressure, density, viscosity, etc., that describe it may not be fully understood. When this occurs, the flow can then be studied by performing an *experiment*.

Unfortunately, experimental work can be costly and time consuming, and so it becomes important to be able to *minimize the amount of experimental data* that needs to be obtained. The best way to accomplish this is to first perform a dimensional analysis of all the relevant physical variables and fluid properties. Specifically, **dimensional analysis** is a branch of mathematics that is used to organize all these *variables* into sets of **dimensionless groups**. Once these groups are obtained, we can use them to obtain the maximum amount of information from a minimum number of experiments.

TABLE 8–1

Quantity	Symbol	M-L-T	F-L-T
Area	A	L^2	L^2
Volume	$V\!\!\!\!\!-$	L^3	L^3
Velocity	V	LT^{-1}	LT^{-1}
Acceleration	a	LT^{-2}	LT^{-2}
Angular velocity	ω	T^{-1}	T^{-1}
Force	F	MLT^{-2}	F
Mass	m	M	FT^2L^{-1}
Density	ρ	ML^{-3}	FT^2L^{-4}
Specific weight	γ	$ML^{-2}T^{-2}$	FL^{-3}
Pressure	p	$ML^{-1}T^{-2}$	FL^{-2}
Dynamic viscosity	μ	$ML^{-1}T^{-1}$	FTL^{-2}
Kinematic viscosity	ν	L^2T^{-1}	L^2T^{-1}
Power	$\dot{W}_s$	ML^2T^{-3}	FLT^{-1}
Volumetric flow rate	Q	L^3T^{-1}	L^3T^{-1}
Mass flow rate	$\dot{m}$	MT^{-1}	FTL^{-1}
Surface tension	σ	MT^{-2}	FL^{-1}
Weight	W	MLT^{-2}	F
Torque	T	ML^2T^{-2}	FL

The method of dimensional analysis is based on the principle of *dimensional homogeneity*, discussed in Sec. 1.4. It states that *each term in an equation must have the same combination of units.* Except for temperature, the variables in almost all fluid flow problems can be described using the primary dimensions of mass M, length L, and time T; or force F, length L, and time T.* For convenience, Table 8–1 lists the combinations of these dimensions for many of the variables in fluid mechanics.

Although a dimensional analysis will not provide a direct analytical solution for a problem, it will aid in formulating the problem, so that the solution can be obtained experimentally in the simplest way possible. To illustrate this point, and at the same time show the mathematical process involved, let us consider the problem of finding how the power output $\dot{W}_s$ of the pump in Fig. 8–1a *depends* upon the pressure increase Δp it develops from A to B and the flow Q it provides. One way to obtain this unknown relationship experimentally would be to require the pump to produce a specific measured flow Q_1, and then change the power several times and measure each corresponding pressure increase. The data, when

---

*Recall that force depends on mass. They are related by Newton's second law of motion, $F = ma$. Thus in the SI system, force has the dimensions of ML/T^2 (ma).

plotted, would then give the necessary relationship $\dot{W}_s = f(\Delta p, Q_1)$, shown in Fig. 8–1b. Repeating this process for Q_2, etc., we can produce a family of lines or curves as shown. Unfortunately, without a *large number* of such graphs, one for each Q, it would be difficult to obtain the value of $\dot{W}_s$ for any specific Q and Δp.

An easier way to obtain $\dot{W}_s = f(\Delta p, Q)$ is to first perform a dimensional analysis of the variables. Here we must require that Q and Δp be arranged in such a way that their *combined units* are the *same* as those that are used to describe power. Furthermore, since Q and Δp *alone* do not have units of power, these variables are simply not added to or subtracted from one another; rather, they must either be *multiplied or divided*. Hence the unknown functional relationship must have the form

$$\dot{W}_s = CQ^a(\Delta p)^b$$

Here C is some unknown (dimensionless) constant, and a and b are unknown exponents that maintain the dimensional homogeneity for the units of power. Using Table 8–1 and the M-L-T system, the primary dimensions for these variables are $\dot{W}_s(ML^2/T^3)$, $Q(L^3/T)$, and $\Delta p(M/LT^2)$. When the dimensions are substituted into the above equation, it becomes

$$ML^2T^{-3} = (L^3T^{-1})^a(ML^{-1}T^{-2})^b$$
$$= M^bL^{3a-b}T^{-a-2b}$$

Since the dimensions for M, L, and T must be the *same* on each side of this equation,

M:	$1 = b$
L:	$2 = 3a - b$
T:	$-3 = -a - 2b$

Solving yields $a = 1$ and $b = 1$. In other words, the above required functional relationship, shown in Fig 8–1b, is of the form

$$\dot{W}_s = CQ\Delta p$$

Of course, rather than using this formal procedure, we could also obtain this result *by inspection*, realizing that the units of Q and Δp, when multiplied, cancel in such a way so as to produce the units of $\dot{W}_s$.

Now that this relationship has been established, we only need to perform a *single experiment*, measuring the pressure drop Δp_1 and flow Q_1 for a known power $(\dot{W}_s)_1$. Doing this will enable us to determine the *single* unknown constant $C = (\dot{W}_s)_1/(Q_1\Delta p_1)$, and with this known, we will then be able to calculate the power requirement for the pump for *any other* combination of Q and Δp.

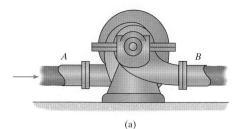

(a)

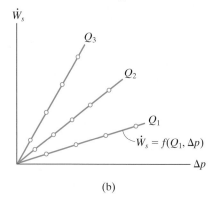

(b)

Fig. 8–1

8

8.2   Important Dimensionless Numbers

The method of applying dimensional analysis, as in the previous example, was originally developed by Lord Rayleigh. It was then later improved upon by Edgar Buckingham, and as we will show in Sec. 8.3, his method requires combining the variables that describe the flow into a set of *dimensionless* ratios or "numbers." Oftentimes these numbers are represented by a ratio of the *forces* that act on the fluid within the flow. And because these ratios appear frequently throughout the study of experimental fluid mechanics, we will introduce some of the more important ones here.

As a matter of convention, each dimensionless ratio is made up of a dynamic or inertia force, and some other force developed by the flow, such as caused by pressure, viscosity, or gravity. The *inertia force* is actually a fictitious force, because it represents the inertia term *ma* in the equation of motion. Specifically, if we write $\Sigma F = ma$ as $\Sigma F - ma = 0$, then the term $-ma$ can be thought of as an *inertia force*, which is needed to produce a zero force resultant on a fluid particle. We will select the inertia force for all these dimensionless ratios because it involves the fluid particle's *acceleration*, and so it plays an important role in almost every problem involving fluid flow. Notice that the inertia force has a magnitude of *ma* or $\rho \forall a$, and so, retaining the fluid property ρ, it has partial dimensions of $\rho(L^3)(L/T^2)$. Since velocity V has dimensions of L/T, we can omit the time dimension and also express this force only in terms of the length dimension, that is, $\rho V^2 L^2$.

We will now briefly define the significance of each dimensionless force ratio, referred to as a "number," and in later sections we will discuss the importance of these ratios for specific applications.

Euler Number.   It is the difference in static pressure between two points in a fluid that causes the fluid to flow, and so the ratio of the force caused by this pressure difference to the inertia force, $\rho V^2 L^2$, is called the *Euler number* or the *pressure coefficient*. The pressure force $\Delta p A$ can be expressed in terms of its length dimensions as $\Delta p L^2$, and so the Euler number is

$$\text{Eu} = \frac{\text{pressure force}}{\text{inertia force}} = \frac{\Delta p}{\rho V^2} \qquad (8\text{--}1)$$

This dimensionless number controls the flow behavior when pressure and inertia forces are dominant, as when a liquid flows through a pipe. It also plays a role in the cavitation of a liquid, and in the study of drag and lift effects produced by a fluid.

Reynolds Number.   The ratio of the inertia force $\rho V^2 L^2$ to the viscous force is called the *Reynolds number*. For a Newtonian fluid, the viscous force is determined from Newton's law of viscosity, $F_v = \tau A = \mu (dV/dy)A$. Retaining the fluid property μ, the viscous force has length dimensions of $\mu(V/L)L^2$, or μVL, so that the Reynolds number becomes

$$\text{Re} = \frac{\text{inertia force}}{\text{viscous force}} = \frac{\rho VL}{\mu} \qquad (8\text{--}2)$$

This dimensionless ratio was first developed by the British engineer Osborne Reynolds while investigating the behavior of laminar and turbulent flow in pipes. When used for this purpose, the "characteristic length" L is chosen as the diameter of the pipe. Notice that when Re is large, the inertia forces within the flow will be greater than the viscous forces, and when this occurs, the flow will become turbulent. Using this idea, in Sec. 9.5 we will show that Reynolds was able to roughly predict when laminar flow begins to transition to turbulent flow. Apart from flow in closed conduits, such as pipes, viscous forces also affect the flow of air around slow-moving aircraft, and water around ships and submarines. For this reason, the Reynolds number has important implications in a number of flow phenomena.

Froude Number.   The ratio of the inertia force $\rho V^2 L^2$ to the fluid's weight $\rho \forall g = (\rho g)L^3$ becomes V^2/gL. If we take the square root of this dimensionless expression, we have another dimensionless ratio called the *Froude number*. It is

$$\text{Fr} = \sqrt{\frac{\text{inertia force}}{\text{gravitational force}}} = \frac{V}{\sqrt{gL}} \qquad (8\text{--}3)$$

This number is named in honor of William Froude, a naval architect who studied surface waves produced by the motion of a ship. Its value indicates the relative importance of inertia and gravity effects on fluid flow. For example, if the Froude number is greater than unity, then inertia effects will outweigh those of gravity. This number is important in the study of any flow having a free surface, as in the case of open channels, or the flow over dams or spillways.

Weber Number.

The ratio of the inertia force $\rho V^2 L^2$ to the surface tension force σL is called the *Weber number*, named after Moritz Weber, who studied the effects of flow in capillary tubes. It is

$$\text{We} = \frac{\text{inertia force}}{\text{surface tension force}} = \frac{\rho V^2 L}{\sigma} \qquad (8\text{–}4)$$

It is important because experiments have shown that capillary flow is controlled by the force of surface tension within a narrow passageway, provided the Weber number is less than one. For most engineering applications, the surface tension force can be neglected; however, it becomes important when studying the flow of thin films of liquid over surfaces, or the flow of small-diameter jets or sprays.

Mach Number.

The square root of the ratio of the inertia force $\rho V^2 L^2$ to the force that causes the compressibility of the fluid is called the *Mach number*. This ratio was developed by Ernst Mach, an Austrian physicist who used it as a reference to study the effects of compressible flow. Recall that fluid compressibility is measured by its bulk modulus, E_V, which is pressure divided by the volumetric strain, Eq. 1–12, $E_V = p/(\Delta V / V)$. Since the units of E_V are the same as those of pressure, and pressure produces a force $F = pA$, the *force* of compressibility has length dimensions of $F = E_V L^2$. The ratio of the inertia force, $\rho V^2 L^2$, to the force of compressibility then becomes $\rho V^2 / E_V$. It will be shown in Chapter 13 that for an ideal gas, $E_V = \rho c^2$, where c is the speed at which a pressure disturbance (sound) will naturally travel within a fluid medium. Substituting this into the ratio, and taking the square root, gives the Mach number.

$$\text{M} = \sqrt{\frac{\text{inertia force}}{\text{compressibility force}}} = \frac{V}{c} \qquad (8\text{–}5)$$

Notice that if M is greater than one, as in the case of supersonic flow, then inertial effects will dominate, and the fluid velocity will be greater than the velocity c of the propagation of any pressure disturbance.

8.3   The Buckingham Pi Theorem

In Sec. 8.1 we discussed a way to perform a dimensional analysis by attempting to combine the variables of a fluid flow as terms that all have a consistent set of units. Since this method is cumbersome when dealing with problems that have a large number of variables, in 1914 the experimentalist Edgar Buckingham developed a more direct method that has since become known as the Buckingham Pi theorem. In this section we will formalize application of this theorem, and show how it applies in cases where four or more physical variables control the flow.

The **Buckingham Pi theorem** states that if a flow phenomenon depends upon n physical variables, such as velocity, pressure, and viscosity, and if present within these physical variables there are m dimensions, such as M, L, and T, then through dimensional analysis, the n variables can be arranged into $(n - m)$ independent dimensionless numbers or groupings. Each of these groupings is called a Π (Pi) term, because in mathematics this symbol is used to symbolize a product. The five dimensionless groupings, or "numbers," discussed in the previous section are typical Π terms. Once the functional relationship among the Π terms is established, it can then be investigated experimentally to see how it relates to the flow behavior using models. The groupings having the most influence are retained, and those having only a slight effect on the flow are rejected. Ultimately, this process will lead to an empirical equation, where any unknown coefficients and exponents are then determined by further experiment.

The proof of the Buckingham Pi theorem is rather long and can be found in books related to dimensional analysis. For example, see Ref. [1]. Here we will be interested only in its application.

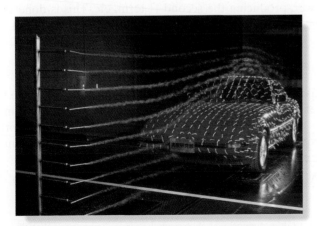

The drag on this car is influenced by the density, viscosity, and velocity of the air, as well as the projected area of the car into the flow. (© Takeshi Takahara/Science Source)

Procedure for Analysis

The Buckingham Pi theorem is used to find the dimensionless groupings among the variables that describe a particular flow phenomenon, and thereby establish a functional relationship between them. The following procedure outlines the steps needed to apply it.

- **Define the Physical Variables.**

  Specify the n physical variables that affect the flow phenomena, and then see if it is possible to form any Π terms *by inspection*. If this cannot be done, then determine the number m of primary dimensions M, L, T or F, L, T that are involved within the *collection* of all these variables.* This will result in $(n - m)\,\Pi$ terms that can be formulated to describe the phenomena. For example, if pressure, velocity, density, and length are the suspected physical variables, then $n = 4$. From Table 8–1, the dimensions of these variables are $ML^{-1}T^{-2}$, LT^{-1}, ML^{-3}, and L, respectively. Since M, L, and T are represented in this collection, $m = 3$ and so this will result in $(4 - 3) = 1\,\Pi$ term.

- **Select the Repeating Variables.**

  From the list of n variables, select m of them so that all these m variables contain the m base dimensions. As a general rule, select the ones with the simplest combination of dimensions. These m variables are called ***repeating variables***. To reduce the amount of work no dimensionless repeating variable should be selected since this in itself is a Π term. Also, no repeating variable should be selected if it is defined in terms of other repeating variables using multiplication or division, such as $Q = VA$. In the above example we can select pressure, velocity, and density ($m = 3$) as the repeating variables, since their collection of dimensions involves M, L, T ($m = 3$).

- **Π Terms.**

  From the remaining list of $(n - m)$ variables, select any one *of them*, called q, and multiply it by the m repeating variables. Raise each of the m variables to an unknown exponent, but keep the q variable to a known power, such as one. This represents the first Π term. Continue the process of selecting any other q variable from the $(n - m)$ variables, and again write the product of q and the *same m repeating variables*, each raised to an unknown exponential power, producing the second Π term, etc. This is done until all $(n - m)\,\Pi$ terms are written. In our example, length L would be selected as q, so that the single Π term is then $\Pi = p^a V^b \rho^c L$.

- **Dimensional Analysis.**

  Express each of the $(n - m)\,\Pi$ terms in terms of its base dimensions (M, L, T or F, L, T), and solve for the unknown exponents by requiring each dimension in the Π term to cancel out, since the Π term must be dimensionless. Once determined, the Π terms are then collected into a functional expression $f(\Pi_1, \Pi_2, \ldots) = 0$, or in the form of an explicit equation, and the numerical values of any remaining unknown constant coefficients or exponents are then determined from experiment.

  The following examples should help clarify these four steps and thereby illustrate application of this procedure.

---

*One can use either the M-L-T or the F-L-T system to do a dimensional analysis; however, it may happen that the variables selected have a *different number* of primary dimensions m in each of these systems. A dimensional matrix approach can be used to identify this situation, but since it does not occur very often, we will not consider its use in this text. See Ref. [5].

EXAMPLE | 8.1

Establish the Reynolds number for a fluid flowing through the pipe in Fig. 8–2 using dimensional analysis, realizing that the flow is a function of the density ρ and viscosity μ of the fluid, along with its velocity V and the pipe's diameter D.

SOLUTION

Define the Physical Variables.   Here $n = 4$. Using the $M\text{-}L\text{-}T$ system and Table 8–1, we have

Density, ρ	ML^{-3}
Viscosity, μ	$ML^{-1}T^{-1}$
Velocity, V	LT^{-1}
Diameter, D	L

Fig. 8–2

Since all three base dimensions (M, L, T) are used here, then $m = 3$. Thus, there is $(n - m) = (4 - 3) = 1\ \Pi$ term.

Select the Repeating Variable.   We will choose ρ, μ, V as the $m = 3$ repeating variables because the collection of their dimensions contains the $m = 3$ primary dimensions. (Of course, another selection, using μ, V, D, can also be made.)

Π Term, D.   Since D has not been selected, it becomes the q variable, which will be set to the first power (exponent one). And so the Π term is $\Pi = \rho^a \mu^b V^c D$.

Dimensional Analysis.   The dimensions for this Π term are

$$\Pi = \rho^a \mu^b V^c D$$
$$= \left(M^a L^{-3a}\right)\left(M^b L^{-b} T^{-b}\right)\left(L^c T^{-c}\right)L = M^{a+b} L^{-3a-b+c+1} T^{-b-c}$$

Π is to be dimensionless, so

For M:	$0 = a + b$
For L:	$0 = -3a - b + c + 1$
For T:	$0 = -b - c$

Solving yields

$$a = 1, b = -1, c = 1$$

so,

$$\Pi_{\text{Re}} = \rho^1 \mu^{-1} V^1 D = \frac{\rho V D}{\mu} \qquad \textit{Ans.}$$

This result is the Reynolds number, Eq. 8–2, where the "characteristic length" L is the pipe diameter D. Later in the text we will give a more thorough discussion of its usefulness and will show how viscous and inertia forces play a predominant role in defining whether the flow through the pipe is laminar or transitions to turbulent flow.

EXAMPLE | 8.2

Fig. 8–3

The wing on the aircraft in Fig. 8–3 is subjected to a drag F_D created by airflow over its surface. It is anticipated that this force is a function of the density ρ and viscosity μ of the air, the "characteristic" length L of the wing, and the velocity V of the flow. Show how the drag force depends on these variables.

SOLUTION

Define the Physical Variables.   Symbolically, the unknown function is $F_D = f(\rho, \mu, L, V)$. In order to collect all the variables as a function, we can rewrite this as follows:* $h(F_D, \rho, \mu, L, V) = 0$. So here, $n = 5$. We will solve the problem using the F-L-T system and Table 8–1. The base dimensions of these five variables are

Drag, F_D	F
Density, ρ	FT^2L^{-4}
Viscosity, μ	FTL^{-2}
Length, L	L
Velocity, V	LT^{-1}

Since all three primary dimensions (F, L, T) are involved in the collection of these variables, $m = 3$, and so we have $(n - m) = (5 - 3) = 2$ Π terms.

Select the Repeating Variables.   The density, length, and velocity will be chosen as the $m = 3$ repeating variables. As required, the collection of their dimensions contains the $m = 3$ primary dimensions.

Π_1 Term F_D.   We will consider F_D as the q variable for the first Π term, $\Pi_1 = \rho^a L^b V^c F_D$.

Dimensional Analysis.   This term has dimensions

$$\Pi_1 = \rho^a L^b V^c F_D$$
$$= \left(F^a T^{2a} L^{-4a}\right)\left(L^b\right)\left(L^c T^{-c}\right) F = F^{a+1} L^{-4a+b+c} T^{2a-c}$$

Therefore,

For F:	$0 = a + 1$
For L:	$0 = -4a + b + c$
For T:	$0 = 2a - c$

Solving yields $a = -1, b = -2, c = -2$, so

$$\Pi_1 = \rho^{-1} L^{-2} V^{-2} F_D = \frac{F_D}{\rho L^2 V^2}$$

*This is like writing $y = 5x + 6$ as $y - 5x - 6 = 0$, where $y = f(x)$ and $h(x, y) = 0$.

Π_2 Term μ.   Finally, we will consider μ as the q variable, thus creating the second Π term, $\Pi_2 = \rho^d L^e V^f \mu$.

Dimensional Analysis.   This term has dimensions

$$\Pi_2 = \rho^d L^e V^f \mu$$
$$= \left(F^d T^{2d} L^{-4d}\right)\left(L^e\right)\left(L^f T^{-f}\right) F T L^{-2} = F^{d+1} L^{-4d+e+f-2} T^{2d-f+1}$$

Therefore,

For F: $0 = d + 1$
For L: $0 = -4d + e + f - 2$
For T: $0 = 2d - f + 1$

Solving yields $d = -1, e = -1, f = -1$, so

$$\Pi_2 = \rho^{-1} L^{-1} V^{-1} \mu = \frac{\mu}{\rho V L}$$

We can also replace Π_2 by Π_2^{-1} since it is *also* a dimensionless ratio—the Reynolds number, Re. The unknown function h between the variables now takes the form

$$h\left(\frac{F_D}{\rho L^2 V^2}, \text{Re}\right) = 0 \qquad\qquad Ans.$$

If we solve for $F_D/\rho L^2 V^2$ in this equation, then we can specify how F_D is related to the Reynolds number. It is

$$\frac{F_D}{\rho L^2 V^2} = f(\text{Re})$$

or

$$F_D = \rho L^2 V^2 [f(\text{Re})] \qquad\qquad (1) \qquad Ans.$$

Later, in Chapter 11, we will show that for experimental purposes, rather than obtain $f(\text{Re})$, it is convenient to express the drag in terms of the fluid's dynamic head, $\rho V^2/2$, and use an *experimentally* determined dimensionless *drag coefficient*, C_D. If we do this, then we require $F_D = \rho L^2 V^2 [f(\text{Re})] = C_D L^2 (\rho V^2/2)$, or $f(\text{Re}) = C_D/2$. Also, replacing L^2 in Eq. 1 by the area A of the wing, we can write Eq. 1 as

$$F_D = C_D A\left(\frac{\rho V^2}{2}\right) \qquad\qquad (2) \qquad Ans.$$

Dimensional analysis did not provide the complete solution to this problem, but as will be shown in Sec. 11.10, once an experiment is performed to determine C_D, then we can use Eq. 2 to obtain F_D.

8

EXAMPLE 8.3

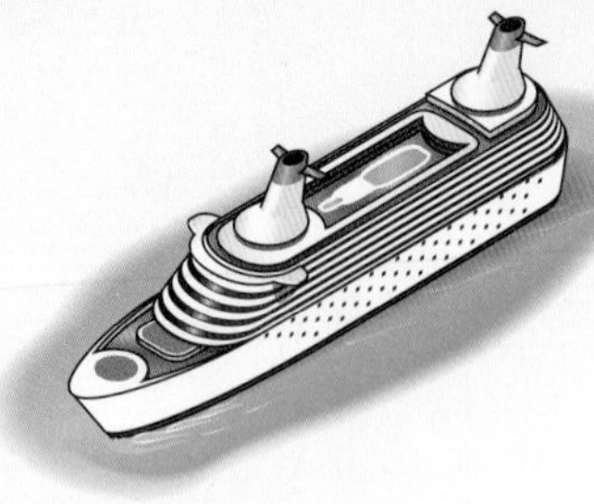

Fig. 8–4

The ship in Fig. 8–4 is subjected to a drag F_D on its hull, created by water passing over and around its surface. It is anticipated that this force is a function of the density ρ and viscosity μ of the water, and since waves are produced, their weight, defined by gravity g, is important. Also, the "characteristic" length of the ship, L, and the velocity of the flow, V, influence the magnitude of the drag. Show how this force depends on all these variables.

SOLUTION

Define the Physical Variables. Here the unknown function is $F_D = f(\rho, \mu, L, V, g)$, which is then expressed as $h(F_D, \rho, \mu, L, V, g) = 0$, where we see $n = 6$. We will solve the problem using the F-L-T system and Table 8–1. The primary dimensions of the variables are

Drag, F_D	F
Density, ρ	FT^2L^{-4}
Viscosity, μ	FTL^{-2}
Length, L	L
Velocity, V	LT^{-1}
Gravity, g	LT^{-2}

Since the three primary dimensions (F, L, T) are involved in the collection of all these variables, $m = 3$, and so there will be $(n - m) = (6 - 3) = 3 \, \Pi$ terms.

Select the Repeating Variables. Density, length, and velocity will be chosen as the repeating variables since the collection of their dimensions contains the $m = 3$ primary dimensions.

Π_1 **Term F_D and Dimensional Analysis.** We will consider F_D as the q variable for the first Π term.

$$\Pi_1 = \rho^a L^b V^c F_D$$
$$= \left(F^a T^{2a} L^{-4a}\right)\left(L^b\right)\left(L^c T^{-c}\right)F = F^{a+1} L^{-4a+b+c} T^{2a-c}$$

Therefore,

For F: $0 = a + 1$
For L: $0 = -4a + b + c$
For T: $0 = 2a - c$

Solving yields $a = -1, b = -2, c = -2$, so

$$\Pi_1 = \rho^{-1}L^{-2}V^{-2}F_D = \frac{F_D}{\rho L^2 V^2}$$

Π_2 Term μ and Dimensional Analysis.   Here we will consider μ as the q variable, thus creating the second Π term.

$$\Pi_2 = \rho^d L^e V^f \mu$$
$$= \left(F^d T^{2d} L^{-4d}\right)\left(L^e\right)\left(L^f T^{-f}\right) F T L^{-2} = F^{d+1} L^{-4d+e+f-2} T^{2d-f+1}$$

For F: $0 = d + 1$

For L: $0 = -4d + e + f - 2$

For T: $0 = 2d - f + 1$

Solving yields $d = -1, e = -1, f = -1$, so

$$\Pi_2 = \rho^{-1} L^{-1} V^{-1} \mu = \frac{\mu}{\rho VL}$$

As in the previous example, we can replace Π_2 by Π_2^{-1} since it represents the Reynolds number, Re.

Π_3 Term g and Dimensional Analysis.   Finally, we consider g as the q variable for the third Π term.

$$\Pi_3 = \rho^h L^i V^j g$$
$$= \left(F^h T^{2h} L^{-4h}\right)\left(L^i\right)\left(L^j T^{-j}\right)\left(L T^{-2}\right)$$

For F: $0 = h$

For L: $0 = -4h + i + j + 1$

For T: $0 = 2h - j - 2$

Solving yields $h = 0, i = 1, j = -2$, so

$$\Pi_3 = \rho^0 L^1 V^{-2} g = gL/V^2$$

Recognizing Π_3^{-1} to be the square of the Froude number, we will consider it instead of Π_3 because both are dimensionless. Thus, the unknown function between the Π terms takes the form

$$h\left(\frac{F_D}{\rho L^2 V^2}, \text{Re}, (\text{Fr})^2\right) = 0 \qquad\qquad Ans.$$

If we solve for $F_D/\rho L^2 V^2$ in this equation, then it can be written symbolically as a function of the Reynolds and Froude numbers.

$$\frac{F_D}{\rho L^2 V^2} = f\left[\text{Re}, (\text{Fr})^2\right]$$

$$F_D = \rho L^2 V^2 f\left[\text{Re}, (\text{Fr})^2\right] \qquad\qquad Ans.$$

Later, in Sec. 8.5, we will discuss how to use this result to determine the actual drag on the ship.

EXAMPLE | 8.4

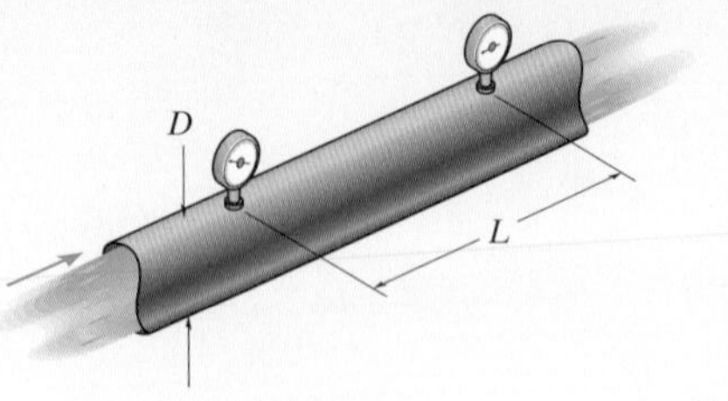

D

L

Fig. 8–5

A pressure drop Δp provides a measure of the frictional losses of a fluid as it flows through a pipe, Fig. 8–5. Determine how Δp is related to the variables that influence it, namely, the pipe diameter D, its length L, the fluid density ρ, viscosity μ, velocity V, and the relative roughness factor ε/D, which is a ratio of the average size of the surface irregularities to the pipe's diameter D.

SOLUTION

Define the Physical Variables.   In this case, $\Delta p = f(D, L, \rho, \mu, V, \varepsilon/D)$ or $h(\Delta p, D, L, \rho, \mu, V, \varepsilon/D) = 0$, and so $n = 7$. Using the M-L-T system and Table 8–1, the primary dimensions of the variables are

Pressure drop, Δp	$ML^{-1}T^{-2}$
Diameter, D	L
Length, L	L
Density, ρ	ML^{-3}
Viscosity, μ	$ML^{-1}T^{-1}$
Velocity, V	LT^{-1}
Relative roughness, ε/D	LL^{-1}

Since all three primary dimensions are involved, $m = 3$ and there will be $(n - m) = (7 - 3) = 4$ Π terms.

Select the Repeating Variables.   Here D, V, ρ will be selected as the $m = 3$ repeating variables. Notice that ε/D cannot be selected because it is *already dimensionless*. Also, we cannot select the length in place of velocity since *both* the diameter and the length have the *same dimensions*.

Π Terms and Dimensional Analysis.   The Π terms are constructed using our repeating variables D, V, ρ, along with Δp for Π_1, L for Π_2, μ for Π_3, and ε/D for Π_4. Thus, for Π_1,

$$\Pi_1 = D^a V^b \rho^c \Delta p$$
$$= (L^a)(L^b T^{-b})(M^c L^{-3c})(ML^{-1}T^{-2}) = M^{c+1}L^{a+b-3c-1}T^{-b-2}$$

For M: $0 = c + 1$
For L: $0 = a + b - 3c - 1$
For T: $0 = -b - 2$

Solving yields $a = 0, b = -2, c = -1$, so

$$\Pi_1 = D^0 V^{-2} \rho^{-1} \Delta p = \frac{\Delta p}{\rho V^2}$$

This is the Euler number.
    Next, for Π_2,

$$\Pi_2 = D^d V^e \rho^f L$$
$$= (L^d)(L^e T^{-e})(M^f L^{-3f})(L) = M^f L^{d+e-3f+1}T^{-e}$$

For M: $0 = f$
For L: $0 = d + e - 3f + 1$
For T: $0 = -e$

Solving yields $d = -1, e = 0, f = 0$, so

$$\Pi_2 = D^{-1} V^0 \rho^0 L = \frac{L}{D}$$

8

Now, for Π_3,

$$\Pi_3 = D^g V^h \rho^i \mu$$
$$= \left(L^g\right)\left(L^h T^{-h}\right)\left(M^i L^{-3i}\right)\left(ML^{-1}T^{-1}\right) = M^{i+1}L^{g+h-3i-1}T^{-h-1}$$

For M: $0 = i + 1$
For L: $0 = g + h - 3i - 1$
For T: $0 = -h - 1$

Solving yields $g = -1, h = -1, i = -1$, so

$$\Pi_3 = D^{-1}V^{-1}\rho^{-1}\mu = \frac{\mu}{DV\rho}$$

This is the inverse of the Reynolds number, so we will consider

$$\Pi_3^{-1} = \frac{\rho VD}{\mu} = \text{Re}$$

Finally, for Π_4,

$$\Pi_4 = D^j V^k \rho^l (\varepsilon/D)$$
$$= \left(L^j\right)\left(L^k T^{-k}\right)\left(M^l L^{-3l}\right)\left(LL^{-1}\right) = M^l L^{j+k-3l+1-1}T^{-k}$$

For M: $0 = l$
For L: $0 = j + k - 3l + 1 - 1$
For T: $0 = -k$

Solving yields $j = 0, k = 0, l = 0$, so

$$\Pi_4 = D^0 V^0 \rho^0 \left(\frac{\varepsilon}{D}\right) = \frac{\varepsilon}{D}$$

Realize that at the start, we could have *saved some time* in solving this problem and determined $\Pi_2 = L/D$ and $\Pi_4 = \varepsilon/D$ simply *by inspection*, because each is a ratio of length over length, *and* each ratio does not contain the *same* two variables. Had this inspection been done at the start, then *only two* Π terms would have had to be determined.

In either case, our results indicate

$$h\left(\frac{\Delta p}{\rho V^2}, \text{Re}, \frac{L}{D}, \frac{\varepsilon}{D}\right) = 0$$

This equation can be solved for the ratio $\Delta p/\rho V^2$ and then written in the form

$$\Delta p = \rho V^2 g\left(\text{Re}, \frac{L}{D}, \frac{\varepsilon}{D}\right) \qquad\qquad Ans.$$

In Chapter 10, we will show how this result has important applications in the design of pipe systems.

8.4   Some General Considerations Related to Dimensional Analysis

The previous four examples illustrate the relatively straightforward method for applying the Buckingham Pi theorem to determine a functional relationship between a dependent variable and a series of dimensionless groupings or Π terms. The most important part of the process, however, is clearly defining the *variables* that influence the flow. This can be done only if one has enough *experience* in fluid mechanics to understand what laws and forces govern the flow. For the selection, these variables include fluid properties such as density and viscosity, the dimensions used to describe the system, and variables such as gravity, pressure, and velocity that are involved in creating the forces within the flow.

If the selection *does not* include an important variable, then a dimensional analysis will produce an incorrect result, leading to an experiment that indicates something is wrong. Also, if irrelevant variables are selected, or the variables are related to one another, such as $Q = VA$, then it will result in too many Π terms, and the experimental work will require additional time and expense to eliminate them.

In summary, if the most important variables that influence the flow are properly selected, then one will be able to minimize the resulting number of Π terms that are involved, and thereby reduce not only the time, but also the cost of any experiment needed to obtain the final result. For example, when studying the motion of a ship, in Example 8–3, we found that the Reynolds and Froude numbers were important. These numbers depend upon the viscous and gravitational forces, respectively. Surface tension, which corresponds to the Weber number, was not considered because this force is negligible for large ships, although for a toy it may become important.

Important Points

- The equations of fluid mechanics are dimensionally homogeneous, which means that each term of an equation must have the same combination of dimensions.

- Dimensional analysis is used to *reduce* the amount of data that must be collected from the variables in an experiment in order to understand the behavior of a flow. This is done by arranging the variables into selected groups that are dimensionless. Once this is done, then it is only necessary to find one relationship among the dimensionless groups, and not to find several relationships among all the separate variables.

- Five important dimensionless ratios of force occur frequently in fluid mechanics. All involve a ratio of the dynamic or inertia force to some other force. These five "numbers" are the Euler number for pressure, the Reynolds number for viscosity, the Froude number for weight, the Weber number for surface tension, and the Mach number for elastic force due to compressibility.

- The Buckingham Pi theorem provides a systematic method for performing a dimensional analysis. The theorem indicates in advance how many unique dimensionless groups of variables (Π terms) to expect, and it provides a way to formulate a relationship among them.

8.5   Similitude

Engineers sometimes resort to using a ***model*** to study the three-dimensional flow around an actual object or ***prototype***, such as a building, automobile, or airplane. They do this because it may be rather difficult to describe the flow using an analytical or computational solution. Even if the flow can be described by a computational analysis, in complicated cases it should be backed up with a corresponding experimental investigation, using a model to verify the results. This is necessary simply because the assumptions made using any computational study may not truly reflect the actual situation involving the complexities of the flow.

If the model and its testing environment are properly proportioned, the experiment will enable the engineer to predict how the flow will affect the prototype. For example, using a model, it is possible to obtain measurements of velocity, depth of liquid flow, pump or turbine efficiencies, and so forth, and with this information, the model can be altered if necessary so that the design of the prototype can be improved.

Usually the model is made smaller than the prototype; however, this may not always be the case. For example, larger models have been constructed to study the flow of gasoline through an injector, or the flow of air through the blades of a turbine used for a dental drill. Whatever its size, it is very important that the model used for an experimental study correspond to the behavior of the prototype when the prototype is subjected to the actual fluid flow. ***Similitude*** is a mathematical process of ensuring that this is the case. It requires the model and the flow around it not only to maintain geometric similarity to the prototype, but also to maintain kinematic and dynamic similarity.

Geometric Similitude. If the model and the fluid flow are geometrically similar to the prototype, then all of the model's linear dimensions must be in the same proportion as those of the prototype, and all its angles must be the same. Take, for example, the prototype (jet plane) shown in Fig. 8–6a. We can express this linear proportion as a ***scale ratio***, which is the ratio of the length L_m of the model to the length L_p of the prototype.

$$\frac{L_m}{L_p}$$

If this ratio is maintained for all dimensions, then it follows that the *areas* of the model and the prototype will be in the proportion L_m^2/L_p^2, and their *volumes* will be in the proportion L_m^3/L_p^3.

The extent to which geometric similitude is achieved depends on the type of problem and on the accuracy required from its solution. For example, *exact* geometric similitude *also* requires the surface roughness of the model to be in proportion to that of the prototype. In some cases, however, this may not be possible, since the reduced size of a model might require its surface to be impossibly smooth. Also, in some modeling, the vertical scale of the model many have to be exaggerated to produce the proper flow characteristics. This would be true in the case of river studies, where the river bottom may be difficult to scale.

Due to the complexity of flow, the effects of wind on high-rise buildings are often studied in wind tunnels.

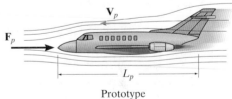

Prototype

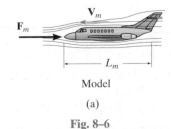

Model

(a)

Fig. 8–6

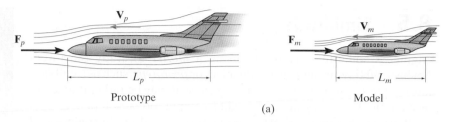

Prototype   Model

(a)

Fig. 8–6

Kinematic Similitude.

The primary dimensions defining *kinematic similitude* are length and time. For example, the velocity of the fluid at corresponding points between the jet model and its prototype in Fig. 8–6a must have a proportional magnitude and be in the same direction. Since velocity depends on length and time, $V = L/T$, then

$$\frac{V_m}{V_p} = \frac{L_m T_p}{L_p T_m}$$

If this requirement holds true, then the length ratio L_m/L_p for geometric similitude must be satisfied, and so must the time ratio T_p/T_m. Satisfying these ratios the acceleration will also be proportional for kinematic similitude. A typical example of having kinematic similitude would be a model of the solar system showing the relative positions of the planets and having the proper time scale for their orbits.

Dynamic Similitude.

To maintain a similar pattern of streamlines around both the prototype and its model, Fig. 8–6b, it is necessary that the forces acting on the fluid particles in both cases be proportional. As stated previously, the inertia force F_i is generally considered the *most important force* that influences fluid flow around an object. For this reason, it is standard convention to use this force, along with each of the other forces F that influence the flow, to produce the ratios for dynamic similitude between model and prototype. We can express each force ratio symbolically as

$$\frac{F_m}{(F_i)_m} = \frac{F_p}{(F_i)_p}$$

The many types of forces to be considered include those due to pressure, viscosity, gravity, surface tension, and elasticity. This means that for *complete* dynamic similitude, the Euler, Reynolds, Froude, Weber, and Mach numbers must be the *same* for both the model and the prototype.

Actually, for complete dynamic similitude, it is not necessary to satisfy the conditions for *all* the forces that affect the flow. Instead, if similitude for *all but one force* is satisfied, then the conditions for this remaining force will *automatically* be satisfied. To understand why this is so, let's consider a case where only the forces of pressure (pr), viscosity (v) and gravity (g) act on two fluid particles having the same mass m and located in the same relative position on the prototype and model, Fig. 8–6b. According to Newton's second law, at any instant the sum of these forces on each particle must be equal to the mass of the particle times its acceleration, $m\mathbf{a}$.[*]

---

[*]The particles have acceleration because the *magnitude* of their velocity can fluctuate, and the streamlines along which they travel can change the *direction* of their velocity.

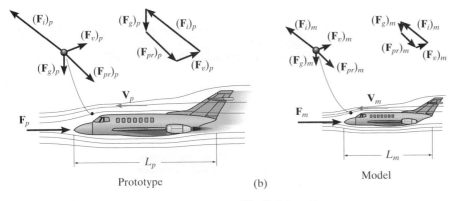

Prototype   (b)   Model

Fig. 8–6 (cont.)

If we express Newton's law as $\Sigma \mathbf{F} - m\mathbf{a} = \mathbf{0}$, and consider the inertia force $\mathbf{F}_i = -m\mathbf{a}$ acting on each particle, Fig. 8–6b, then for each case, the vector addition of these four forces is graphically shown by the vector polygons.

For dynamic similitude the four corresponding forces in these polygons must be proportional, that is, their magnitudes must have the same scaled lengths. However, because the polygons are *closed shapes*, only three of the sides (forces) must satisfy this proportionality, and if this occurs, the fourth side (force) will then *automatically* be proportional. Therefore we can write

$$\frac{\left(F_{pr}\right)_p}{\left(F_{pr}\right)_m} = \frac{\left(F_i\right)_p}{\left(F_i\right)_m}, \qquad \frac{\left(F_v\right)_p}{\left(F_v\right)_m} = \frac{\left(F_i\right)_p}{\left(F_i\right)_m}, \qquad \frac{\left(F_g\right)_p}{\left(F_g\right)_m} = \frac{\left(F_i\right)_p}{\left(F_i\right)_m}$$

Cross multiplying, we can then form the following important force ratios,

$$\text{Eu} = \frac{\left(F_{pr}\right)_p}{\left(F_i\right)_p} = \frac{\left(F_{pr}\right)_m}{\left(F_i\right)_m}, \quad \text{Re} = \frac{\left(F_i\right)_p}{\left(F_v\right)_p} = \frac{\left(F_i\right)_m}{\left(F_v\right)_m}, \quad \text{Fr} = \sqrt{\frac{\left(F_i\right)_p}{\left(F_g\right)_p}} = \sqrt{\frac{\left(F_i\right)_m}{\left(F_i\right)_m}}$$

In other words, the Euler numbers, Reynolds numbers, and Froude numbers, respectively, must be identical between the prototype and the model. From what was previously stated about the force polygons, we can conclude that if the flow satisfies, say, Reynolds and Froude number scaling, then it will *automatically* satisfy Euler number scaling. Furthermore, since the inertia forces $(F_i)_p$ and $(F_i)_m$ have units of ML/T^2, then dynamic similitude *automatically* satisfies both geometric and kinematic similitude, since L and T must be proportional when the inertia forces $\mathbf{F}_i$ are proportional.

Actually, it is rather difficult to obtain *exact similitude* between a prototype and its model, due to the inaccuracies in model construction and in testing procedures. Instead, by using good judgment gained through experience, we can consider only the similitude of the dominating parameters, while those of lesser importance can be safely ignored. The following three cases will illustrate how this is achieved when performing experiments.

Wind turbines are sometimes placed off-shore and the wave loadings on their columns can be substantial. Using similitude, the effect is studied in a wave channel so that these supports can be properly designed.

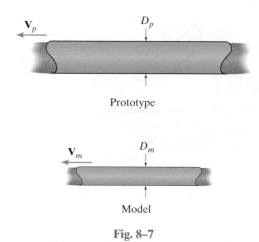

Fig. 8–7

Steady Flow through a Pipe.

If the flow of water through a pipe is to be studied using a model of the pipe, Fig. 8–7, then experience has shown that the inertia and viscous forces are important.* Therefore, for pipe flow, the Reynolds number must be the same for both the model and the prototype to achieve dynamic similitude. For this ratio, the pipe diameter D becomes the "characteristic length" L, so

$$\left(\frac{\rho VD}{\mu}\right)_m = \left(\frac{\rho VD}{\mu}\right)_p$$

Often the same fluid is used for both the model and the prototype, so its properties ρ and μ will be the same. If this is the case, then

$$V_m D_m = V_p D_p$$

Therefore, for given values of V_p, D_p, and V_m, the model of the pipe must have a diameter of $D_m = (VD)_p/V_m$ in order for the flow to have the *same behavior* through both the actual pipe and its model.

---

*The pressure force is also present, but provided the viscous-to-inertia force ratios are equal, the proportionality of pressure-to-inertia force ratios will *automatically* be satisfied, as we have just discussed.

Prototype

Model

Fig. 8–8

Open-Channel Flow.

For open-channel flow, Fig. 8–8, forces caused by surface tension and compressibility can be neglected without introducing appreciable errors, and if the channel length is short, then frictional losses will be negligible, and therefore viscous effects can also be neglected. As a result, the flow through an open channel is primarily governed by the inertia force and gravity. Therefore, the Froude number $\text{Fr} = V/\sqrt{gL}$ can be used to establish similitude between the prototype and its model. For this number, the depth of the fluid, h, in the channel is selected as the "characteristic length" L, and since g is the same, we have

$$\frac{V_m}{\sqrt{h_m}} = \frac{V_p}{\sqrt{h_p}}$$

Besides modeling flow in channels, the Froude number is also used to model the flow over a number of hydraulic structures, such as gates and spillways. However, for rivers this type of scaling can cause problems, because scaling down the model may result in depths that are *too small*, thereby causing the effects of viscosity and surface tension to predominate within the flow around the model. As previously stated, however, such forces are generally neglected, and so only approximate modeling of the flow can be achieved.

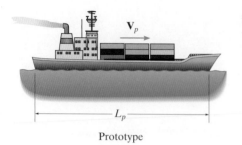

Prototype

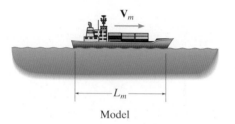

Model

Fig. 8–9

Ships.   As noted in the previous two cases, only one dimensionless ratio had to be satisfied to achieve similitude. However, in the case of a ship, Fig. 8–9, the drag or resistance to forward motion is due to *both* friction along the hull (viscosity) and lifting water against the hull to produce waves (gravity). As a result, the total drag on the ship is a function of *both* the Reynolds and Froude numbers. Similitude between the model and prototype (ship) therefore requires these numbers to be equivalent. Thus,

$$\left(\frac{\rho VL}{\mu}\right)_m = \left(\frac{\rho VL}{\mu}\right)_p$$

$$\left(\frac{V}{\sqrt{gL}}\right)_m = \left(\frac{V}{\sqrt{gL}}\right)_p \tag{8–6}$$

Since the kinematic viscosity is $\nu = \mu/\rho$, and g is the same for both the model and the prototype, these two equations become

$$\frac{V_p}{V_m} = \frac{\nu_p L_m}{\nu_m L_p}$$

$$\frac{V_p}{V_m} = \left(\frac{L_p}{L_m}\right)^{1/2}$$

Equating these ratios to eliminate the velocity ratio yields

$$\frac{\nu_p}{\nu_m} = \left(\frac{L_p}{L_m}\right)^{3/2} \tag{8–7}$$

Since here any model will be *much shorter* than its prototype the ratio L_p/L_m will be *very large*. As a result, to preserve the equality in Eq. 8–7, this would require the model to be tested in a liquid having a *much smaller* kinematic viscosity than water, which is impractical to achieve.*

To get around this difficulty, a method suggested by Froude has been used to solve this problem. In Example 8–3 we showed that a dimensional analysis of all the above variables produces a functional relationship between the total drag F_D and the Reynolds and Froude numbers that has the form

$$F_D = \rho L^2 V^2 f\left[\text{Re}, (\text{Fr})^2\right]$$

---

*Mercury has one of the lowest kinematic viscosities, but even if it were used, it still would require the model to have a length that is too large to be practical.

Realizing this, Froude assumed the total drag on the ship is the *sum of its two components*, namely, skin or viscous friction drag, based only on the Reynolds number, and the drag resistance to wave creation, based only on the Froude number. Thus, the functional dependence for both model and prototype becomes a sum of two separate unknown functions.

$$F_D = \rho L^2 V^2 f_1(\text{Re}) + \rho L^2 V^2 f_2\left[(\text{Fr})^2\right] \qquad (8\text{–}8)$$

The model is then built on the basis of Froude scaling since the effect of wave action (gravity) is the most difficult to predict. Thus, in accordance with Eq. 8–6, the length of the model and its velocity are chosen so that they will produce the same Froude number as the prototype. The *total drag F_D* on the model can then be measured by finding the force needed to pull the model through the water at the required velocity. This represents the action of *both* the wave and viscous forces on the model.

The *viscous forces* alone can be measured by doing a separate test on a fully submerged thin plate that is moving at the same velocity as the model, is made of the same material and roughness as the model, and has the same length and surface area in contact with the water. The wave force on the model can then be obtained by subtracting this viscous drag from the total drag previously measured.

Once the viscous and wave forces *on the model* are known, the total drag *on the prototype* (ship) can then be determined. To do this, from Eq. 8–8, the *gravity forces* on the prototype and model must satisfy

$$\frac{(F_p)_g}{(F_m)_g} = \frac{\rho_p L_p^2 V_p^2 f_2\left[(\text{Fr})^2\right]}{\rho_m L_m^2 V_m^2 f_2\left[(\text{Fr})^2\right]}$$

However, since the unknown function $f_2\left[(\text{Fr})^2\right]$ is to be the same for both the model and the prototype, it will cancel out, and so the wave or gravity force on the prototype (ship) is determined from

$$(F_p)_g = (F_m)_g \left(\frac{\rho_p L_p^2 V_p^2}{\rho_m L_m^2 V_m^2}\right) \qquad (\text{gravity})$$

Finally, the viscous force on the prototype, defined as $\rho L^2 V^2 f_1(\text{Re})$ in Eq. 8–8, is then determined by a similar scaling using a drag coefficient that will be discussed in Chapter 11. With both of these forces calculated, their sum then represents the total drag on the ship.

8

Review. As noted, it is important to have a well-established background in fluid mechanics in order to recognize the primary forces that govern the flow, not only to perform a dimensional analysis, but also to do experimental work. Here are some examples that show which forces are important in certain cases, and the corresponding similitude that must be achieved.

- *Inertia, pressure,* and *viscous forces* predominate within the flow around cars, slow flying airplanes, and the flow through pipes and ducts. Reynolds number similitude.

- *Inertia, pressure,* and *gravity forces* predominate within the flow along open channels, over dams and spillways, or the wave action on structures. Froude number similitude.

- *Inertia, pressure,* and *surface tension forces* predominate within the flow of liquid films, bubble formation, and the passage of liquids through capillarity or small diameter tubes. Weber number similitude.

- *Inertia, pressure,* and *compressibility forces* predominate within the flow around high speed aircraft, or the high speed flow of gas through jet or rocket nozzles, and through pipes. Mach number similitude.

In all these examples and the ones that follow, remember that similitude of the pressure force (Euler number) does not have to be taken into account, because it will be *automatically* satisfied due to Newton's second law.

Important Points

- When building and testing a model, it is important that similitude or similarity between the model and its prototype be achieved. Complete similitude occurs when the model and prototype are geometrically similar, and the flow is kinematically and dynamically similar.

- Geometric similitude occurs when the linear dimensions for the model and prototype are in the same proportion to one another, and all the angles are the same. Kinematic similitude occurs when the velocities and accelerations are proportional. Finally, dynamic similitude occurs when the forces acting on corresponding fluid particles within the flow around the model and the prototype are in specific dimensionless ratios, defined by the Euler, Reynolds, Froude, Weber, and Mach numbers.

- Often it is difficult to achieve *complete* similitude. Instead, engineers will consider only the dominating variables in the flow in order to save cost and time, and yet get reasonable results.

- To satisfy dynamic similitude, it is necessary that *all but one* of the force ratios acting on the fluid particles for both the model and the prototype be equal. Because Newton's second law must be satisfied, the remaining force ratio (generally taken as the Euler number) will automatically be equal.

- It takes good judgment and experience to decide which forces are significant in defining the flow, so that reasonable results can be obtained when testing a model used to predict the performance of its prototype.

EXAMPLE   8.5

Flow through the pipe coupling (union) in Fig. 8–10 is to be studied using a scaled model. The actual pipe is 100-mm in diameter, and the model will use a pipe that is 20-mm in diameter. The model will be made of the same material and transport the same fluid as the prototype. If the velocity of flow through the prototype is estimated to be 1.5 m/s, determine the required velocity through the model.

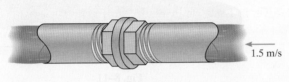

1.5 m/s

Fig. 8–10

SOLUTION

The forces dominating the flow are caused by inertia and viscosity, so Reynolds number similitude must be satisfied. We require

$$\left(\frac{\rho V D}{\mu}\right)_m = \left(\frac{\rho V D}{\mu}\right)_p$$

$$V_m = \left(\frac{\rho_p}{\rho_m}\right)\left(\frac{\mu_m}{\mu_p}\right)\left(\frac{D_p}{D_m}\right)V_p \qquad (1)$$

Since the same fluid is used for both cases,

$$V_m = \left(\frac{D_p}{D_m}\right)V_p = \left(\frac{100\ \text{mm}}{20\ \text{mm}}\right)(1.5\ \text{m/s}) = 7.50\ \text{m/s} \qquad Ans.$$

As noted, Reynolds number similitude leads to high flow velocities for the model due to this scaling factor. To lower this velocity, from Eq. 1, one can use a higher-density or lower-viscosity fluid for the model.

8

EXAMPLE | 8.6

A model of the car in Fig. 8–11 is constructed to a scale of 1/4 and is to be tested at 20°C in a water tunnel. Determine the required velocity of the water if the actual car is traveling at 30 m/s in air at this same temperature.

Fig. 8–11

SOLUTION

Here viscosity creates the predominating force, and so dynamic similitude must satisfy the Reynolds number. Since $\nu = \mu/\rho$, for the Reynolds number, we have

$$\left(\frac{VL}{\nu}\right)_m = \left(\frac{VL}{\nu}\right)_p$$

$$V_m = V_p\left(\frac{\nu_m}{\nu_p}\right)\left(\frac{L_p}{L_m}\right)$$

Using the values of kinematic viscosity of air and water at 20°C, Appendix A, we have

$$V_m = (30 \text{ m/s})\left[\frac{1.00\left(10^{-6}\right) \text{ m}^2/\text{s}}{15.1\left(10^{-6}\right) \text{ m}^2/\text{s}}\right]\left(\frac{4}{1}\right)$$

$$= 7.95 \text{ m/s} \qquad\qquad Ans.$$

EXAMPLE | 8.7

The dam in Fig. 8–12 is to be built such that the estimated average flow over its crest will be $Q = 3000\text{ m}^3/\text{s}$. Determine the required flow over the crest of a model built to a scale of $1/25$.

Fig. 8–12

SOLUTION

Here the weight of the water is the most significant force influencing the flow, and so Froude number similitude must be achieved. Thus,

$$\left(\frac{V}{\sqrt{gL}}\right)_m = \left(\frac{V}{\sqrt{gL}}\right)_p$$

or

$$\frac{V_m}{V_p} = \left(\frac{L_m}{L_p}\right)^{1/2} = \left(\frac{1}{25}\right)^{1/2} \tag{1}$$

We can express the velocity ratio in terms of the flow ratio since $Q = VA$, where A is the product of a height, L_h, and width, L_w. Thus,

$$\frac{V_m}{V_p} = \frac{Q_m A_p}{Q_p A_m} = \frac{Q_m (L_h)_p (L_w)_p}{Q_p (L_h)_m (L_w)_m} = \frac{Q_m}{Q_p}\left(\frac{25}{1}\right)^2$$

Substituting this into Eq. 1, we get

$$\frac{Q_m}{Q_p} = \left(\frac{1}{25}\right)^{5/2}.$$

Thus,

$$Q_m = Q_p\left(\frac{1}{25}\right)^{5/2}$$

$$= \left(3000\text{ m}^3/\text{s}\right)\left(\frac{1}{25}\right)^{5/2} = 0.960\text{ m}^3/\text{s} \qquad Ans.$$

EXAMPLE   8.8

A machine is to be operated in oil having a kinematic viscosity of $\nu = 0.035(10^{-3})\,\text{m}^2/\text{s}$. If viscous and gravity forces predominate during the flow, determine the required viscosity of a liquid used with the model. The model is built to a scale of $1/10$.

SOLUTION

The viscous forces will be similar if the Reynolds numbers are equivalent, and the gravity forces will be similar if the Froude numbers are equivalent. Thus Eq. 8–7 can be used, so

$$
\nu_m = \nu_p \left(\frac{L_m}{L_p} \right)^{3/2} = \left(0.035(10^{-3})\,\text{m}^2/\text{s} \right) \left(\frac{1}{10} \right)^{3/2}
$$

$$
= 1.11(10^{-6})\,\text{m}^2/\text{s} \qquad\qquad Ans.
$$

This value is close to that of water at 15°C, $1.15(10^{-6})\,\text{m}^2/\text{s}$, as noted in Appendix A.

References

1. E. Buckingham, "Model experiments and the form of empirical equations," *Trans ASME*, Vol. 37, 1915, pp. 263–296.

2. S. J. Kline, *Similitude and Approximation Theory*, McGraw-Hill, New York, NY, 1965.

3. P. Bridgman, *Dimensional Analysis*, Yale University Press, New Haven, CN, 1922.

4. E. Buckingham, "On physically similar systems: illustrations of the use of dimensional equations," *Physical Reviews*, Vol. 4, No. 4, 1914, pp. 345–376.

5. T. Szirtes and P. Roza, *Applied Dimensional Analysis and Modeling*, McGraw-Hill, New York, NY, 1997.

6. R. Ettema, *Hydraulic Modeling: Concepts and Practice*, ASCE, Reston, VA, 2000.

PROBLEMS

Sec. 8.1–8.4

8–1. Use inspection to arrange each of the following three variables as a dimensionless ratio: (a) L, t, V, (b) σ, E_V, L, (c) V, g, L.

8–2. Investigate if each ratio is dimensionless. (a) $\rho V^2/p$, (b) $L\rho/\sigma$, (c) $p/V^2 L$, (d) $\rho L^3/V\mu$.

8–3. Determine the F, L, T dimensions of the following terms. (a) $Q/\rho V$, (b) $\rho g/p$, (c) $V^2/2g$, (d) ρgh.

*8–4. Determine the M, L, T dimensions of the following terms. (a) $Q/\rho V$, (b) $\rho g/p$, (c) $V^2/2g$, (d) ρgh.

8–5. The pressure change that occurs in the aortic artery during a short period of time can be modeled by the equation $\Delta p = c_a(\mu V/2R)^{1/2}$, where μ is the viscosity of blood, V is its velocity, and R is the radius of the artery. Determine the M, L, T dimensions for the arterial coefficient c_a.

Prob. 8–5

8–6. The Womersley number is a dimensionless parameter that is used to study transient blood flow through the arteries during heartbeats. It is a ratio of transient to viscous forces and is written as $\text{Wo} = r\sqrt{2\pi f\rho/\mu}$, where r is the vessel radius, f is the frequency of the heartbeat, μ the apparent viscosity, and ρ is the density of blood. Research has shown that the radius r of the aorta of a mammal can be related to its mass m by $r = 0.0024m^{0.34}$, where r is in meters and m is in kilograms. Determine the Womersley number for a horse having a mass of 350 kg and heartbeat rate of 30 beats per minute (bpm), and compare it to that of a rabbit having a mass of 2 kg and heartbeat rate of 180 bpm. The viscosities of blood for the horse and rabbit are $\mu_h = 0.0052$ N·s/m² and $\mu_r = 0.0040$ N·s/m², respectively. The density of blood for both is $\rho_b = 1060$ kg/m³. Plot this variation of Womersley number (vertical axis) with mass for these two animals. The results should show that transient forces increase as the size of the animal increases. Explain why this happens.

8–7. The force of buoyancy F is a function of the volume $V\!\!\!/$ of a body and the specific weight γ of the fluid. Determine how F is related to $V\!\!\!/$ and γ.

*8–8. Express the group of variables p, g, D, ρ as a dimensionless ratio.

8–9. Express the group of variables L, μ, ρ, V as a dimensionless ratio.

8–10. The Womersley number is often used to study blood circulation in biomechanics when there is pulsating flow through a <u>circular</u> tube of diameter d. It is defined as $\text{Wo} = \frac{1}{2}d\sqrt{2\pi f\rho/\mu}$, where f is the frequency of the pressure in cycles per second. Like the Reynolds number, Wo is a ratio of inertia and viscous forces. Show that this number is dimensionless.

8–11. Show that the hydrostatic pressure p of an incompressible fluid can be determined using dimensional analysis by realizing that it depends upon the depth h in the fluid and the fluid's specific weight γ.

*8–12. Establish Newton's law of viscosity using dimensional analysis, realizing that shear stress τ is a function of the fluid viscosity μ and the angular deformation du/dy. *Hint*: Consider the unknown function as $f(\tau, \mu, du, dy)$.

8

8–13. The period of oscillation τ, measured in seconds, of a buoy depends upon its cross-sectional area A, its mass m, and the specific weight γ of the water. Determine the relation between τ and these parameters.

Prob. 8–13

8–14. Laminar flow through a pipe produces a discharge Q that is a function of the pipe's diameter D, the change in pressure Δp per unit length, $\Delta p/\Delta x$, and the fluid viscosity, μ. Determine the relation between Q and these parameters.

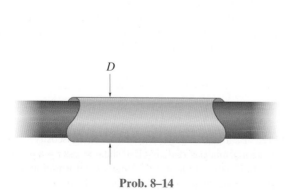

Prob. 8–14

8–15. The flow Q of gas through the pipe is a function of the density ρ of the gas, gravity g, and the diameter D of the pipe. Determine the relation between Q and these parameters.

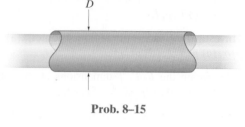

Prob. 8–15

*8–16. The speed of sound V in air is thought to depend on the viscosity μ, the density ρ, and the pressure p. Determine how V is related to these parameters.

8–17. The torsional resistance T of the thrust bearing depends upon the diameter D of the shaft, the axial force F, the shaft rotation ω, and the viscosity μ of the lubricating fluid. Determine the relation between T and these parameters.

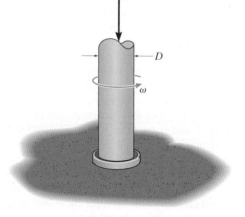

Prob. 8–17

8–18. The capillary rise of a fluid along the walls of the tube causes the fluid to rise a distance h. This effect depends upon the diameter d of the tube, the surface tension σ, the density ρ of the fluid, and the gravitational acceleration g. Determine the relation between h and these parameters.

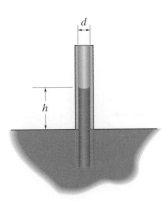

Prob. 8–18

8–19. The velocity c of a wave on the surface of a liquid depends upon the wave length λ, the density ρ, and the surface tension σ of the liquid. Determine the relation between c and these parameters. By what percent will c decrease if the density of the liquid is increased by a factor of 1.5?

*8–20. The discharge Q over the weir A depends upon the width b of the weir, the water head H, and the acceleration of gravity g. If Q is known to be proportional to b, determine the relation between Q and these variables. If H is doubled, how does this affect Q?

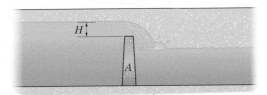

Prob. 8–20

8–21. The head loss h_L in a pipe depends upon its diameter D, the velocity of flow V, and the density ρ and viscosity μ of the fluid. Determine the relation between h_L and these parameters.

8–22. The pressure p within the soap bubble is a function of the bubble's radius r and the surface tension σ of the liquid film. Determine the relation between p and these parameters.

Prob. 8–22

8–23. Fluid flow depends upon the viscosity μ, bulk modulus E_V, gravity g, pressure p, velocity V, density ρ, surface tension σ, and a characteristic length L. Determine the dimensionless groupings for these eight variables.

*8–24. The velocity V of the stream flowing from the side of the tank is thought to depend upon the liquid's density ρ, the depth h, and the acceleration of gravity g. Determine the relation between V and these parameters.

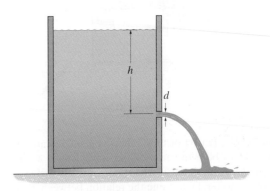

Prob. 8–24

8–25. The time t needed for ethyl ether to drain from the pipette is thought to be a function of the fluid's density ρ and viscosity μ, the nozzle's diameter d, and gravity g. Determine the relation between t and these parameters.

Prob. 8–25

8–26. The pressure difference Δp of air that flows through a fan is a function of the diameter D of the blade, its angular rotation ω, the density ρ of the air, and the flow Q. Determine the relation between Δp and these parameters.

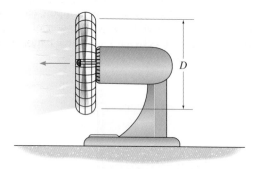

Prob. 8–26

8–27. The period of time τ between small water waves is thought to be a function of the wave length λ, the water depth h, gravitational acceleration g, and the surface tension σ of the water. Determine the relation between τ and these parameters.

*8–28. The drag F_D on the square plate held normal to the wind depends upon the area A of the plate and the air velocity V, density ρ, and viscosity μ. Determine the relation between F_D and these parameters.

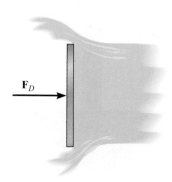

Prob. 8–28

8–29. The speed c of a water wave is a function of the wave length λ, the acceleration of gravity g, and the average depth of the water h. Determine the relation between c and these parameters.

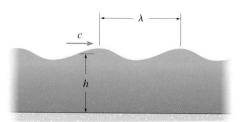

Prob. 8–29

8–30. The discharge Q from a turbine is a function of the generated torque T, the angular rotation ω of the turbine, its diameter D, and the liquid density, ρ. Determine the relation between Q and these parameters. If Q varies linearly with T, how does it vary with the turbine's diameter D?

8–31. The thrust T of the propeller on a boat depends upon the diameter D of the propeller, its angular velocity ω, the speed of the boat V, and the density ρ and viscosity μ of the water. Determine the relation between T and these parameters.

Prob. 8–31

*8–32. The drag force F_D on the jet plane is a function of the speed V, the characteristic length L of the plane, and the density ρ and viscosity μ of the air. Determine the relation between F_D and these parameters.

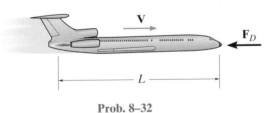

Prob. 8–32

8–33. The torque T developed by a turbine depends upon the depth h of water at the entrance, the density of the water ρ, the discharge Q, and the angular velocity of the turbine ω. Determine the relation between T and these parameters.

8–34. The power P of a blower depends upon the impeller diameter D, its angular velocity ω, the discharge Q, and the fluid density ρ and viscosity μ. Determine the relation between P and these parameters.

8–35. The discharge Q of a pump is a function of the impeller diameter D, its angular velocity ω, the power output P, and the density ρ and viscosity μ of the fluid. Determine the relation between Q and these parameters.

*8–36. As the ball falls through a liquid, its velocity V is a function of the diameter D of the ball, its density ρ_b, the density ρ and viscosity μ of the liquid, and the acceleration due to gravity g. Determine the relation between V and these parameters.

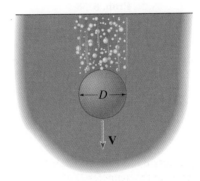

Prob. 8–36

8–37. The thickness δ of the boundary layer for a fluid passing over a flat plate depends upon the distance x from the plate's leading edge, the free-stream velocity U of the flow, and the density ρ and viscosity μ of the fluid. Determine the relation between δ and these parameters.

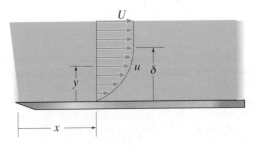

Prob. 8–37

8

8–38. The change in pressure Δp in the pipe is a function of the density ρ and the viscosity μ of the fluid, the pipe diameter D, and the velocity V of the flow. Establish the relation between Δp and these parameters.

Prob. 8–38

8–39. The drag force F_D on the automobile is a function of its velocity V, its projected area A into the wind, and the density ρ and viscosity μ of the air. Determine the relation between F_D and these parameters.

Prob. 8–39

8–40. The drag force F_D acting on a submarine depends upon the characteristic length L of the vessel, the velocity V at which it is traveling, and the density ρ and viscosity μ of the water. Determine the relation between F_D and these parameters.

8–41. The power P supplied by a pump is thought to be a function of the discharge Q, the change in pressure Δp between the inlet and outlet, and the density ρ of the fluid. Use the Buckingham Pi theorem to establish a general relation between these parameters so that an experiment may be performed to determine this relationship.

8–42. When an underwater explosion occurs, the pressure p of the shock wave at any instant is a function of the mass of the explosive m, the initial pressure p_0 formed by the explosion, the spherical radius r of the shock wave, and the density ρ and the bulk modulus E_V of the water. Determine the relation between p and these parameters.

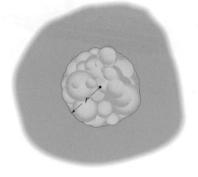

Prob. 8–42

8–43. The diameter D of oil spots made on a sheet of porous paper depends upon the diameter d of the squirting nozzle, the height h of the nozzle from the surface, the velocity V of the oil, and its density ρ, viscosity μ, and surface tension σ. Determine the dimensionless ratios that define this process.

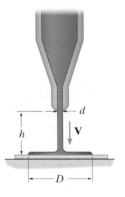

Prob. 8–43

*8–44. Mist from an aerosol produces droplets having a diameter d, which is thought to depend upon the diameter of the nozzle D, the surface tension σ of the droplets, the velocity V at which the droplets are ejected, and the density ρ and viscosity μ of the air. Determine the relation between d and these parameters.

Prob. 8–44

8–45. The discharge Q over a small weir depends upon the water head H, the width b and height h of the weir, the acceleration of gravity g, and the density ρ, viscosity μ, and surface tension σ of the fluid. Determine the relation between Q and these parameters.

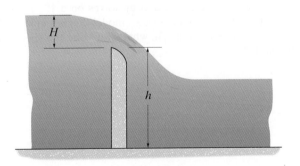

Prob. 8–45

Sec. 8.5

8–46. If water flows through a 50-mm-diameter pipe at 2 m/s, determine the velocity of carbon tetrachloride flowing through a 60-mm-diameter pipe so that they both have the same dynamic characteristics. The temperature of both liquids is 20°C.

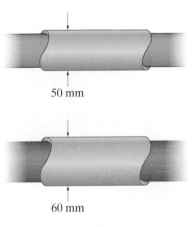

50 mm

60 mm

Prob. 8–46

8–47. Water flowing through a 100-mm-diameter pipe is used to determine the loss in pressure when gasoline flows through a 75-mm-diameter pipe at 3 m/s. If the pressure loss in the pipe transporting the water is 8 Pa, determine the pressure loss in the pipe transporting the gasoline. Take $\nu_g = 0.465(10^{-6})$ m²/s and $\nu_w = 0.890(10^{-6})$ m²/s, $\rho_g = 726$ kg/m³, $\rho_w = 997$ kg/m³.

*8–48. When a 100-mm-diameter sphere travels at 2 m/s in water having a temperature of 15°C, the drag is 2.80 N. Determine the velocity and drag on a 150-mm-diameter sphere traveling through water under similar conditions.

8–49. The model of a river is constructed to a scale of 1/60. If the water in the river is flowing at 6 m/s, how fast must the water flow in the model?

8

8–50. The model of an airplane has a scale of 1/30. If the drag on the prototype is to be determined when the plane is flying at 600 km/h, find the speed of the air in a wind tunnel for the model if the air has the same temperature and pressure. Is it reasonable to do this test?

8–51. The flow of water around the structural support is 1.2 m/s when the temperature is 5°C. If it is to be studied using a model built to a scale of 1/20, and using water at a temperature of 25°C, determine the velocity of the water used with the model.

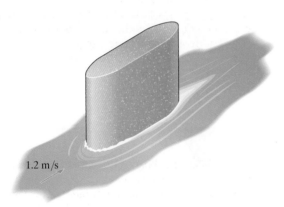

1.2 m/s

Prob. 8–51

*8–52. The optimum performance of mixing blades 0.5 m in diameter is to be tested using a model one-fourth the size of the prototype. If the test of the model in water reveals the optimum speed to be 8 rad/s, determine the optimum angular speed of the prototype when it is used to mix ethyl alcohol. Take $T = 20°C$.

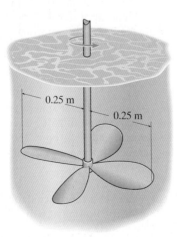

0.25 m

0.25 m

Prob. 8–52

8–53. The flow of water around the bridge pier is to be studied using a model built to a scale of 1/15. If the river flows at 0.8 m/s, determine the corresponding velocity of the water in the model at the same temperature.

0.8 m/s

Prob. 8–53

8–54. The resistance created by waves on a 100-m-long ship is tested in a channel using a model that is 4 m long. If the ship travels at 60 km/h, what should be the speed of the model?

8–55. A model of a ship is built to a scale of 1/20. If the ship is to be designed to travel at 4 m/s, determine the speed of the model in order to maintain the same Froude number.

*8–56. The flow around the airplane flying at an altitude of 10 km is to be studied using a wind tunnel and a model that is built to a 1/15 scale. If the plane has an air speed of 800 km/h, what should the speed of the air be inside the tunnel? Is this reasonable?

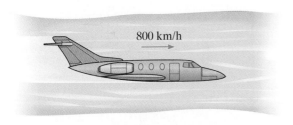

800 km/h

Prob. 8–56

8–57. The velocity of water waves in a channel is studied in a laboratory using a model of the channel one-twelfth its actual size. Determine the velocity of the waves in the channel if they have a velocity of 6 m/s in the model.

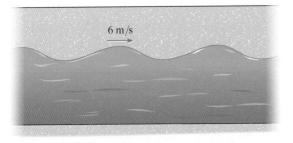

6 m/s

Prob. 8–57

8–58. The motion of water waves in a channel is to be studied in a laboratory using a model one-twelfth the size of the channel. Determine the time for a wave in the channel to travel 10 m if it takes 15 seconds for the wave to travel this distance in the model.

8–59. It is required that a pump be designed for use in a chemical plant such that it delivers 0.8 m^3/s of benzene with a pressure increase of 320 kPa. What is the expected flow and pressure increase produced by a model one-sixth the size of the prototype? If the model produces a power output of 900 kW, what would be the power output of the prototype?

*8–60. The model of a boat is built to a scale of 1/50. Determine the required kinematic viscosity of the water in order to test the model so that the Froude and Reynolds numbers remain the same for the model and the prototype. Is this test practical if the prototype operates in water at $T = 20°C$?

8–61. A model of a plane is built to a scale of 1/15 and is tested in a wind tunnel. If the plane is designed to travel at 800 km/h at an altitude of 5 km, determine the required density of the air in the wind tunnel so that the Reynolds and Mach numbers are the same. Assume the temperature is the same in both cases and the speed of sound in air at this temperature is 340 m/s.

8–62. The model of a hydrofoil boat is to be tested in a channel. The model is built to a scale of 1/20. If the lift produced by the model is 7 kN, determine the lift on the prototype. Assume the water temperature is the same in both cases. This requires Euler number and Reynolds number similarity.

8

8–63. If the jet plane can fly at Mach 2 in air at 2°C, determine the required speed of wind generated in a wind tunnel at 18°C and used on a model built to a scale of 1/25. *Hint*: Use Eq. 13–24, $c = \sqrt{kRT}$, where $k = 1.40$ for air.

Prob. 8–63

*8–64.** A ship has a length of 180 m and travels in the sea where $\rho_s = 1030 \text{ kg/m}^3$. A model of the ship is built to a 1/60 scale, and it displaces 0.06 m³ of water such that its hull has a wetted surface area of 3.6 m². When tested in a towing tank at a speed of 0.5 m/s, the total drag on the model was 2.25 N. Determine the drag on the ship and its corresponding speed. What power is needed to overcome this drag? The drag due to viscous (frictional) forces can be determined using $(F_D)_f = \left(\frac{1}{2}\rho V^2 A\right)C_D$, where C_D is the drag coefficient determined from $C_D = 1.328/\sqrt{\text{Re}}$ for $\text{Re} < 10^6$ and $C_D = 0.455/(\log_{10}\text{Re})^{2.58}$ for $10^6 < \text{Re} < 10^9$. Take $\rho = 1000 \text{ kg/m}^3$ and $\nu = 1.00(10^{-6}) \text{ m}^2/\text{s}$.

8–65. A model of a submarine is built to determine the drag force acting on its prototype. The length scale is 1/100, and the test is run in water at 20°C, with a speed of 8 m/s. If the drag on the model is 20 N, determine the drag on the prototype if it runs in water at the same speed and temperature. This requires that the drag coefficient $C_D = 2F_D/\rho V^2 L^2$ be the same for both the model and the prototype.

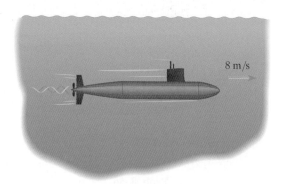

Prob. 8–65

8–66. The drag coefficient on an airplane is defined by $C_D = 2F_D/\rho V^2 L^2$. If the drag acting on the model of a plane tested at sea level is 0.3 N, determine the drag on the prototype, which is 15 times larger and is flying at 20 times the speed of the model at an altitude of 3 km.

8

CHAPTER REVIEW

Dimensional analysis provides a means of combining the variables that influence a flow into groups of dimensionless numbers. This reduces the number of experimental measurements required to describe the flow.

When studying flow behavior, the following important dimensionless ratios of the dynamic or inertia force to some other force often occur.

$$\text{Euler number Eu} = \frac{\text{pressure force}}{\text{inertia force}} = \frac{\Delta p}{\rho V^2}$$

$$\text{Reynolds number Re} = \frac{\text{inertia force}}{\text{viscous force}} = \frac{\rho V L}{\mu}$$

$$\text{Froude number Fr} = \sqrt{\frac{\text{inertia force}}{\text{gravitational force}}} = \frac{V}{\sqrt{gL}}$$

$$\text{Weber number We} = \frac{\text{inertia force}}{\text{surface tension force}} = \frac{\rho V^2 L}{\sigma}$$

$$\text{Mach number M} = \sqrt{\frac{\text{inertia force}}{\text{compressibility force}}} = \frac{V}{c}$$

The Buckingham Pi theorem provides a means for determining the dimensionless groups of variables one can obtain from a set of variables.

Similitude provides a means of making sure that the flow affects the prototype in the same way that it affects its model. The model must be geometrically similar to the prototype, and the flow must be kinematically and dynamically similar.

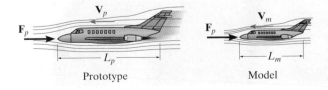

Prototype   Model

Chapter 9

(© Danicek/Shutterstock)

A pressure drop that occurs along a closed conduit such as a pipe is due to frictional losses within the fluid. These losses will be different for laminar and turbulent flows.

Viscous Flow within Enclosed Surfaces

9.1 Steady Laminar Flow between Parallel Plates

In this section we will consider the case of laminar flow of a Newtonian (viscous) fluid confined between two inclined parallel plates. As shown in Fig. 9–1a, the plates are separated by a distance a, and they have a sufficient width and length so that end effects can be neglected. Here, we wish to determine the velocity profile of the fluid, assuming the fluid is *incompressible* and has *steady flow*. To generalize the solution somewhat, we will also assume the top plate is moving with a speed $\mathbf{U}$ relative to the bottom plate. Under these conditions, we have one-dimensional flow since the velocity will vary only in the y direction, while it remains constant in the x and z directions for each value of y.

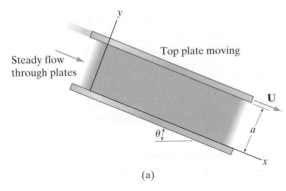

(a)

Fig. 9–1

433

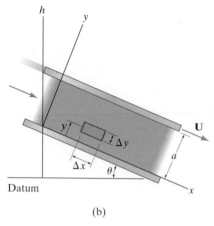

Datum

(b)

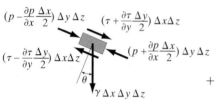

$$\left(p - \frac{\partial p}{\partial x}\frac{\Delta x}{2}\right)\Delta y\,\Delta z \qquad \left(\tau + \frac{\partial \tau}{\partial y}\frac{\Delta y}{2}\right)\Delta x\Delta z$$

$$\left(\tau - \frac{\partial \tau}{\partial y}\frac{\Delta y}{2}\right)\Delta x\Delta z \qquad \left(p + \frac{\partial p}{\partial x}\frac{\Delta x}{2}\right)\Delta y\,\Delta z$$

$$\gamma\,\Delta x\,\Delta y\,\Delta z$$

Free-body diagram

(c)

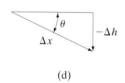

(d)

Fig. 9–1 (cont.)

To analyze the flow, we will apply the momentum equation and select a differential control volume having a length Δx, a thickness Δy, and a width Δz, Fig. 9–1b. The forces acting on the free-body diagram of this control volume along the x direction include the pressure forces on the open control surfaces, Fig. 9–1c, the forces caused by shear stress on the top and bottom closed control surfaces, and the x component of the weight of the fluid within the control volume. The shear forces are different on their opposite surfaces because the movement of the fluid is different on adjacent streamlines. As indicated, both the pressure and the shear stress are assumed to *increase* in the *positive x* and *y* directions, respectively. Since the flow is steady and the fluid incompressible, and $\Delta A_{\text{in}} = \Delta A_{\text{out}}$, then no local and convective changes are made, and so the momentum equation becomes an equilibrium equation, where

$$\Sigma F_x = \frac{\partial}{\partial t}\int_{\text{cv}} \mathbf{V}\rho\,d\forall + \int_{\text{cs}} \mathbf{V}\rho\,\mathbf{V}\cdot d\mathbf{A}$$

$$\left(p - \frac{\partial p}{\partial x}\frac{\Delta x}{2}\right)\Delta y\,\Delta z - \left(p + \frac{\partial p}{\partial x}\frac{\Delta x}{2}\right)\Delta y\,\Delta z$$

$$+ \left(\tau + \frac{\partial \tau}{\partial y}\frac{\Delta y}{2}\right)\Delta x\,\Delta z - \left(\tau - \frac{\partial \tau}{\partial y}\frac{\Delta y}{2}\right)\Delta x\,\Delta z + \gamma\Delta x\,\Delta y\,\Delta z\sin\theta = 0 + 0$$

Factoring out the control volume $\Delta x\,\Delta y\,\Delta z$, noting that $\sin\theta = -\Delta h/\Delta x$, Fig. 9–1d, and taking the limit as Δx and Δh approach zero, we obtain, after simplifying,

$$\frac{\partial \tau}{\partial y} = \frac{\partial}{\partial x}(p + \gamma h)$$

The term on the right is the sum of the pressure gradient and the elevation gradient measured from a datum. Since the flow is steady, this sum is independent of y, and remains the same along each cross section. Therefore, since the pressure is only a function of x, integrating the above equation with respect to y yields

$$\tau = \left[\frac{d}{dx}(p + \gamma h)\right]y + C_1$$

This expression is based *only on a balance of forces*, and so it is valid for *both* laminar and turbulent flows.

If we have a Newtonian fluid, for which *laminar flow* prevails, then we can apply Newton's law of viscosity, $\tau = \mu(du/dy)$, to obtain the velocity profile. Thus,

$$\mu\frac{du}{dy} = \left[\frac{d}{dx}(p + \gamma h)\right]y + C_1$$

Again, integrating with respect to y,

$$u = \frac{1}{\mu}\left[\frac{d}{dx}(p + \gamma h)\right]\frac{y^2}{2} + \frac{C_1}{\mu}y + C_2 \qquad (9\text{–}1)$$

The constants of integration can be evaluated using the "no slip" boundary conditions, namely at $y = 0$, $u = 0$, and at $y = a$, $u = U$. After substituting and simplifying, the above equations for τ and u now become

$$\tau = \frac{U\mu}{a} + \left[\frac{d}{dx}(p + \gamma h)\right]\left(y - \frac{a}{2}\right) \qquad (9\text{--}2)$$

Shear-stress distribution

Laminar and turbulent flow

$$u = \frac{U}{a}y - \frac{1}{2\mu}\left[\frac{d}{dx}(p + \gamma h)\right](ay - y^2) \qquad (9\text{--}3)$$

Velocity profile

Laminar flow

If the pressures and the elevations at *any two points* 1 and 2 on a streamline are known, Fig. 9–1e, then we can obtain the pressure and elevation gradients. For example if points 1 and 2 are selected then*

$$\frac{d}{dx}(p + \gamma h) = \frac{p_2 - p_1}{L} + \gamma\frac{h_2 - h_1}{L}$$

Pressure and elevation gradients

We will now consider a few special cases to better understand how the forces of viscosity, pressure, and gravity influence the flow.

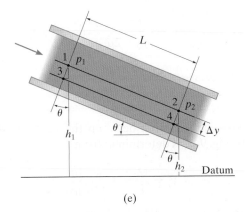

(e)

---

* Notice that if we choose the streamline passing through points 3 and 4 in Fig. 9–1e, then p_1 and p_2 are both *increased* by $\gamma(\Delta y \cos \theta)$, but the elevations h_1 and h_2 are decreased by $(\Delta y \cos \theta)$. Substituting this into the above equation, we obtain the *same* total gradient.

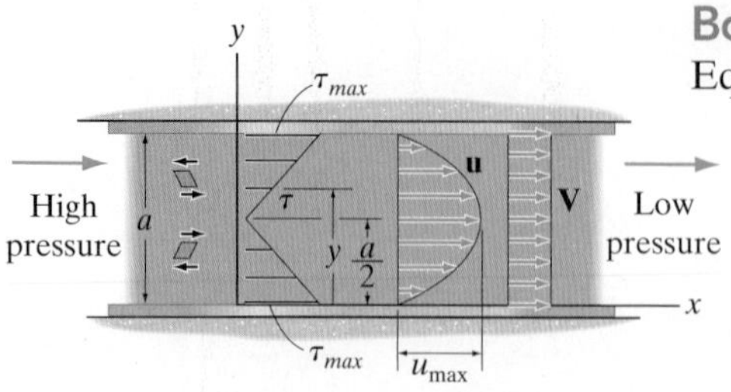

τ_{max}

y

u

τ

High
pressure

a

V   Low
pressure

$y \frac{a}{2}$

x

τ_{max} u_{max}

Shear-stress distribution and actual and average
velocity profiles with *negative* pressure gradient
and no motion of the plates.

Fig. 9–2

Horizontal Flow Caused by a Constant Pressure Gradient—Both Plates Fixed.

In this case $U = 0$ and $dh/dx = 0$, Fig. 9–2, so Eqs. 9–2 and 9–3 become

$$\tau = \frac{dp}{dx}\left(y - \frac{a}{2}\right) \tag{9–4}$$

Shear-stress distribution

$$u = -\frac{1}{2\mu}\frac{dp}{dx}\left(ay - y^2\right) \tag{9–5}$$

Velocity profile
Laminar flow

In order for the flow to move to the right it is necessary that dp/dx be negative. In other words, a higher pressure must be applied to the fluid on the left, causing it to flow to the right. It is the shear stress caused by friction within the fluid that makes the pressure *decrease* in the direction of motion.

A plot of Eq. 9–4 shows that the shear stress varies linearly with y, where τ_{max} occurs at the surface of each plate, Fig. 9–2. Since there is less shear stress on the fluid within the center region between the plates, the velocity there is greater. In fact, Eq. 9–5 shows the resulting velocity distribution is parabolic, and the maximum velocity occurs at the center, where $du/dy = 0$. Substituting $y = a/2$ into Eq. 9–5 reveals that this maximum velocity is

$$u_{max} = -\frac{a^2}{8\mu}\frac{dp}{dx} \tag{9–6}$$

Maximum velocity

The volumetric flow is determined by integrating the velocity distribution over the cross-sectional area between the plates. If the plates have a width b, then $dA = b\,dy$, and we have

$$Q = \int_A u\,dA = \int_0^a -\frac{1}{2\mu}\frac{dp}{dx}\left(ay - y^2\right)(b\,dy)$$

$$Q = -\frac{a^3 b}{12\mu}\frac{dp}{dx} \tag{9–7}$$

Volumetric flow

Finally, the cross-sectional area between the plates is $A = ab$, so the average velocity, Fig. 9–2, is determined from

$$V = \frac{Q}{A} = -\frac{a^2}{12\mu}\frac{dp}{dx} \tag{9–8}$$

Average velocity

If we compare this equation with Eq. 9–6, we find

$$u_{max} = \frac{3}{2}V$$

Horizontal Flow Caused by a Constant Pressure Gradient—Top Plate Moving.

In this case $dh/dx = 0$, so Eqs. 9–2 and 9–3 reduce to

$$\tau = \frac{U\mu}{a} + \frac{dp}{dx}\left(y - \frac{a}{2}\right) \qquad (9\text{–}9)$$

Shear-stress distribution

and

$$u = \frac{U}{a}y - \frac{1}{2\mu}\frac{dp}{dx}(ay - y^2) \qquad (9\text{–}10)$$

Velocity profile
Laminar flow

The volumetric flow is therefore

$$Q = \int_A u\, dA = \int_0^a \left[\frac{U}{a}y - \frac{1}{2\mu}\frac{dp}{dx}(ay - y^2)\right] b\, dy$$

$$= \frac{Uab}{2} - \frac{a^3 b}{12\mu}\frac{dp}{dx} \qquad (9\text{–}11)$$

Volumetric flow

And since $A = ab$, the average velocity is

$$V = \frac{Q}{A} = \frac{U}{2} - \frac{a^2}{12\mu}\frac{dp}{dx} \qquad (9\text{–}12)$$

Average velocity

The location of maximum velocity is determined by setting $du/dy = 0$ in Eq. 9–10. We have

$$\frac{du}{dy} = \frac{U}{a} - \frac{1}{2\mu}\frac{dp}{dx}(a - 2y) = 0$$

$$y = \frac{a}{2} - \frac{U\mu}{a(dp/dx)} \qquad (9\text{–}13)$$

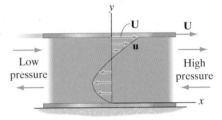

Velocity profile caused by a large
positive pressure gradient
and motion of top plate.

(a)

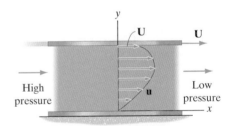

Velocity profile caused by a weak
negative pressure gradient
and motion of top plate.

(b)

Fig. 9–3

If this value is substituted back into Eq. 9–10, the maximum velocity can be obtained. It does not occur at the midpoint; rather, it depends upon the speed of the top plate, U, and the pressure gradient, dp/dx. For example, if a large enough positive or increasing pressure gradient, dp/dx, occurs, then, in accordance with Eq. 9–10, a net *reverse (negative) flow* may take place. A typical velocity profile looks like that shown in Fig. 9–3a, where the top portion of fluid is dragged to the *right*, due to the motion **U** of the plate, and the rest of the fluid moves to the *left*, being pushed by the positive pressure gradient. When the pressure gradient is negative, both this gradient and the plate motion will work together and cause the velocity profile to look something like that shown in Fig. 9–3b.

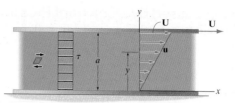

Shear-stress distribution and velocity profile
for zero pressure gradient and
motion of top plate.

Fig. 9–4

Horizontal Flow Caused Only by the Motion of the Top Plate.

If the *pressure gradient dp/dx* and slope *dh/dx* are both zero, then the flow will be caused entirely by the moving plate. In this case, Eqs. 9–2 and 9–3 become

$$\tau = \frac{U\mu}{a}$$

(9–14)

Shear-stress distribution

$$u = \frac{U}{a}y$$

(9–15)

Velocity profile
Laminar flow

These results indicate that the shear stress is constant, while the *velocity profile is linear*, Fig. 9–4.

We discussed this situation in Sec. 1.6, as it relates to Newton's law of viscosity. Fluid motion caused only by movement of the boundary (in this case a plate) is known as **Couette flow**, named after Maurice Couette. In general, however, the term "Couette flow" refers to either laminar or turbulent flow caused only by boundary movement.

Limitations.

It is important to remember that all the velocity-related equations developed in this section apply only for steady-state *laminar flow* of an incompressible Newtonian fluid. Hence, in order to use these equations, we must be sure that laminar flow prevails. In Sec. 9.5 we will discuss how the Reynolds number Re $= \rho V L/\mu$ can be used as a criterion for identifying laminar flow. To do this for parallel plates, the Reynolds number can be calculated using the distance *a* between the plates as the "characteristic length" *L*. Also, using the average velocity to calculate Re, we then have Re $= \rho V a/\mu$. Experiments have shown that laminar flow will occur up to a certain narrow range of values for this Reynolds number. Although it has no particular unique value, in this text we will consider that upper limit to be Re $= 1400$. Therefore,

$$\text{Re} = \frac{\rho V a}{\mu} \leq 1400$$

Laminar flow between plates

Provided this inequality is satisfied, the results of calculating the velocity profiles using the equations presented here have been shown to closely match the velocity profiles obtained from experiments.

9.2 Navier–Stokes Solution for Steady Laminar Flow between Parallel Plates

It is instructive to show that the velocity profile, Eq. 9–3, can *also* be obtained using the continuity equation and application of the Navier–Stokes equations discussed in Sec. 7.11. To do this, we will establish the x, y, z axes as in Fig. 9–5. Since there is steady incompressible flow only in the x direction, then $v = w = 0$, and as a result, the continuity equation, Eq. 7–10, gives

$$\frac{\partial \rho}{\partial t} + \frac{\partial(\rho u)}{\partial x} + \frac{\partial(\rho v)}{\partial y} + \frac{\partial(\rho w)}{\partial z} = 0$$

$$0 + \rho\frac{\partial u}{\partial x} + 0 + 0 = 0$$

so that $\partial u/\partial x = 0$.

Symmetry in the z direction and steady flow indicate that u is not a function of z and x but, rather, only a function of y, that is, $u = u(y)$. Also, from Fig. 9–5, $g_x = g \sin\theta = g(-dh/dx)$ and $g_y = -g\cos\theta$. Using these results, the three Navier–Stokes equations, Eqs. 7–75, reduce to

$$\rho\left(\frac{\partial u}{\partial t} + u\frac{\partial u}{\partial x} + v\frac{\partial u}{\partial y} + w\frac{\partial u}{\partial z}\right) = \rho g_x - \frac{\partial p}{\partial x} + \mu\left(\frac{\partial^2 u}{\partial x^2} + \frac{\partial^2 u}{\partial y^2} + \frac{\partial^2 u}{\partial z^2}\right)$$

$$0 = \rho g\left(-\frac{dh}{dx}\right) - \frac{\partial p}{\partial x} + \mu\frac{d^2 u}{dy^2}$$

$$\rho\left(\frac{\partial v}{\partial t} + u\frac{\partial v}{\partial x} + v\frac{\partial v}{\partial y} + w\frac{\partial v}{\partial z}\right) = \rho g_y - \frac{\partial p}{\partial y} + \mu\left(\frac{\partial^2 v}{\partial x^2} + \frac{\partial^2 v}{\partial y^2} + \frac{\partial^2 v}{\partial z^2}\right)$$

$$0 = -\rho g\cos\theta - \frac{\partial p}{\partial y} + 0$$

$$\rho\left(\frac{\partial w}{\partial t} + u\frac{\partial w}{\partial x} + v\frac{\partial w}{\partial y} + w\frac{\partial w}{\partial z}\right) = \rho g_z - \frac{\partial p}{\partial z} + \mu\left(\frac{\partial^2 w}{\partial x^2} + \frac{\partial^2 w}{\partial y^2} + \frac{\partial^2 w}{\partial z^2}\right)$$

$$0 = 0 - \frac{\partial p}{\partial z} + 0$$

The last equation, when integrated, shows that p is constant in the z direction, which is to be expected. Integrating the second equation gives

$$p = -\rho(g\cos\theta)y + f(x)$$

The first term on the right shows that the pressure varies in a *hydrostatic manner* in the y direction. The second term on the right, $f(x)$, shows the pressure also varies in the x direction. This is due to the viscous shear stress. If we rearrange the first Navier–Stokes equation above, using $\gamma = \rho g$, and integrate it twice, we get

$$\frac{d^2 u}{dy^2} = \frac{1}{\mu}\frac{d}{dx}(p + \gamma h)$$

$$\frac{du}{dy} = \frac{1}{\mu}\frac{d}{dx}(p + \gamma h)y + C_1$$

$$u = \frac{1}{\mu}\left[\frac{d}{dx}(p + \gamma h)\right]\frac{y^2}{2} + C_1 y + C_2$$

This is the same result as Eq. 9–1, so the analysis proceeds as before.

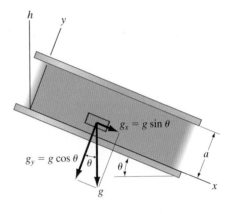

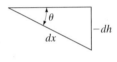

Fig. 9–5

Important Points

- Steady flow between two parallel plates is a balance of the forces of pressure, gravity, and viscosity. The viscous shear stress varies *linearly* along the thickness of the fluid, regardless of whether the flow is laminar or turbulent.

- The velocity profile for *steady laminar flow* of an incompressible Newtonian fluid between two parallel plates is determined using Newton's law of viscosity. In all cases of motion, the fluid at any plate surface has zero velocity *relative to the plate*, since it is assumed to be at rest where it meets the boundary—the "no-slip" condition.

- The formulations in Sec. 9.1 were developed from first principles, and in Sec. 9.2 by solving the continuity and Navier–Stokes equations. These results are in close agreement to experimental measurements. Furthermore, experiments have shown that *laminar flow* between parallel plates will occur up to a critical value of the Reynolds number, which we have taken to be $\text{Re} = \rho V a/\mu \leq 1400$. Here a is the distance between the plates, and V is the average velocity of flow.

Procedure for Analysis

The equations in Sec. 9.1 can be applied using the following procedure.

Fluid Description.

The flow must be steady, and the fluid must be an incompressible Newtonian fluid. Also, *laminar flow must exist*, so be certain to check that the flow conditions produce a Reynolds number $\text{Re} = \rho V a/\mu \leq 1400$.

Analysis.

Establish the coordinates and follow their positive sign convention. Here x is positive in the direction of flow; y is positive measured from the bottom plate, upward and normal to the plate, so it is perpendicular to the flow; and h is positive vertically upward, Fig. 9–1*a*. Finally, be sure to use a consistent set of units when substituting numerical data into any of the equations.

EXAMPLE | 9.1

Glycerin in Fig. 9–6 flows at 0.005 m³/s through the narrow region between the two smooth plates that are 15 mm apart. Determine the pressure gradient acting on the glycerin.

SOLUTION

Fluid Description.   For the analysis, we will assume the plates are wide enough (0.4 m) to neglect end effects. Also, we will assume steady, incompressible, laminar flow. From Appendix A, $\rho_g = 1260 \text{ kg/m}^3$ and $\mu_g = 1.50 \text{ N} \cdot \text{s/m}^2$.

Analysis.   Since the flow is known, we can obtain the pressure gradient by first finding the velocity profile, Eq. 9–3. Here the plates are not moving relative to one another, so $U = 0$, and therefore,

$$u = -\frac{1}{2\mu}\left[\frac{d}{dx}(p + \gamma h)\right](ay - y^2)$$

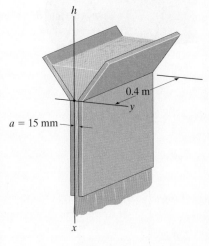

Fig. 9–6

The coordinates are established with x along the left plate edge, positive in the direction of flow (downward), and h positive upward, Fig. 9–6. Thus $dh/dx = -1$, and so the above equation becomes

$$u = -\frac{1}{2\mu}\left(\frac{dp}{dx} - \gamma\right)(ay - y^2) \tag{1}$$

Since Q is known, we can relate it to this velocity profile as follows.

$$Q = \int_A u\, dA = \int_0^a -\frac{1}{2\mu}\left(\frac{dp}{dx} - \gamma\right)(ay - y^2)\, b\, dy$$

$$= -\frac{b}{2\mu}\left(\frac{dp}{dx} - \gamma\right)\int_0^a (ay - y^2)\, dy = -\frac{b}{2\mu}\left(\frac{dp}{dx} - \gamma\right)\left(\frac{a^3}{6}\right)$$

Substituting the data into this equation, we get

$$0.005 \text{ m}^3/\text{s} = \left(-\frac{0.4 \text{ m}}{2(1.50 \text{ N} \cdot \text{s/m}^2)}\right)\left[\frac{dp}{dx} - (1260 \text{ kg/m}^3)(9.81 \text{ m/s}^2)\right]\left(\frac{(0.015 \text{ m})^3}{6}\right)$$

$$\frac{dp}{dx} = -54.3(10^3) \text{ Pa/m} = -54.3 \text{ kPa/m} \qquad Ans.$$

The negative sign indicates the pressure within the glycerin is decreasing in the direction of flow. This is to be expected due to the frictional drag caused by the viscosity.

Finally, we need to check if the flow is indeed laminar by using our Reynolds number criterion. Since $V = Q/A$, we have

$$\text{Re} = \frac{\rho V a}{\mu} = \frac{(1260 \text{ kg/m}^3)\big[(0.005 \text{ m}^3/\text{s})/(0.015 \text{ m})(0.4 \text{ m})\big](0.015 \text{ m})}{1.5 \text{ N} \cdot \text{s/m}^2}$$

$$= 10.5 \le 1400 \qquad \text{(laminar flow)}$$

EXAMPLE | 9.2

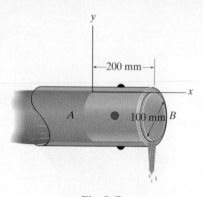

Fig. 9–7

The 100-mm-diameter plug in Fig. 9–7 is placed within a pipe and supported so that oil can flow between it and the walls of the pipe. If the gap between the plug and the pipe is 1.5 mm, and the pressure at A is 400 kPa, determine the discharge of the oil through the gap. Take $\rho_o = 920 \text{ kg/m}^3$ and $\mu_o = 0.2 \text{ N} \cdot \text{s/m}^2$.

SOLUTION

Fluid Description.　We will assume the oil is incompressible and the flow is steady laminar flow. Also, since the gap size is very small as compared to the radius of the plug, we will neglect the curvature of the pipe and any elevation difference, and assume that flow occurs between horizontal "parallel plates" that are at rest.

Analysis.　The discharge is determined from Eq. 9–7. The x coordinate is positive in the direction of flow, so $dp/dx = (p_B - p_A)/L_{AB}$. Since $p_A = 400 \text{ kPa}, p_B = 0$, and $L_{AB} = 0.2 \text{ m}$, we have

$$Q = -\frac{a^3 b}{12\mu_o}\frac{dp}{dx} = -\frac{(0.0015 \text{ m})^3 [2\pi(0.05 \text{ m})]}{12 (0.2 \text{ N} \cdot \text{s/m}^2)}\left[\frac{0 - 400(10^3) \text{ N/m}^2}{0.2 \text{ m}}\right]$$

$$= 0.8836(10^{-3}) \text{ m}^3/\text{s} = 0.884(10^{-3}) \text{ m}^3/\text{s} \qquad\qquad Ans.$$

To check if the flow is laminar, we first obtain the average velocity using Eq. 9–8.

$$V = -\frac{a^2}{12\mu_o}\frac{dp}{dx} = -\frac{(0.0015 \text{ m})^2}{12 (0.2 \text{ N} \cdot \text{s/m}^2)}\left[\frac{0 - 400(10^3) \text{ N/m}^2}{0.2 \text{ m}}\right]$$

$$= 1.875 \text{ m/s}$$

The Reynolds number is therefore

$$\text{Re} = \frac{\rho_o V a}{\mu_o} = \frac{(920 \text{ kg/m}^3)(1.875 \text{ m/s})(0.0015 \text{ m})}{0.2 \text{ N} \cdot \text{s/m}^2}$$

$$= 12.94 < 1400 \qquad \text{(laminar flow)}$$

NOTE: A more exact analysis of this problem can be made by accounting for the curvature of the pipe and plug. It represents steady laminar flow through an annulus, and the relevant equations are developed as part of Prob. 9–38 and 9–43. Also, if the fluid was water, the calculated value of Re would be >1400 and the above analysis would be invalid.

EXAMPLE | 9.3

During a manufacturing process, a 45-mm-wide strip of paper is pulled upward at 0.6 m/s through a narrow channel from a reservoir of glue, as shown in Fig. 9–8a. Determine the force per unit length exerted on the strip when it is in the channel, if the thickness of the glue on each side of the strip is 0.1 mm. Assume the glue is a Newtonian fluid having a viscosity of $\mu = 0.843(10^{-3})$ N·s/m^2 and a density of $\rho = 735$ kg/m^3.

SOLUTION

Fluid Description. Within the channel, steady flow occurs. We will assume the glue is incompressible and the flow is laminar.

Analysis. In this problem, gravity and viscosity predominate. There is no pressure gradient throughout the glue because the pressure at A and B is atmospheric, that is, $p_A = p_B = 0$ and so $\Delta p = 0$ from A to B.

The paper acts as a moving plate, and so to obtain the force per unit length on the paper, we will first apply Eq. 9–2 to obtain the shear stress on the paper, that is

$$\tau = \frac{U\mu}{a} + \left[\frac{\partial}{\partial x}(p + \gamma h) \right]\left(y - \frac{a}{2} \right)$$

The coordinates are established in the usual manner for the glue on the left side, Fig. 9–8b.

As the strip moves upward, the glue adheres to it, but it must overcome the shear stress on its surface at $y = a = 0.1$ mm. Since $dh/dx = 1$ and $\partial p/\partial x = 0$, the above equation becomes

$$
\begin{aligned}
\tau &= \frac{U\mu}{a} + \gamma\left(\frac{a}{2}\right) \\
&= \frac{(0.6 \text{ m/s})(0.843(10^{-3}) \text{ N·s/m}^2)}{0.1(10^{-3}) \text{ m}} + (735 \text{ kg/m}^3)(9.81 \text{ m/s}^2)\left(\frac{0.1(10^{-3}) \text{ m}}{2}\right) \\
&= 5.419 \text{ N/m}^2
\end{aligned}
$$

This stress must be overcome on *each side* of the strip, and because the paper has a width of 45 mm, the force per unit length on the strip is

$$w = 2(5.419 \text{ N/m}^2)(0.045 \text{ m}) = 0.488 \text{ N/m} \qquad Ans.$$

We must now check our assumption of laminar flow. Rather than establishing the actual velocity profile, and then finding the average velocity, here we will consider the maximum velocity, which occurs on the strip at $y = 0.1$ mm. It is $u_{max} = 0.6$ m/s. Since $u_{max} > V$, even at this maximum velocity, we have

$$\text{Re} = \frac{\rho u_{max} a}{\mu} = \frac{(735 \text{ kg/m}^3)(0.6 \text{ m/s})(0.0001 \text{ m})}{0.843(10^{-3}) \text{ N·s/m}^2} = 52.3 \leq 1400 \text{ (laminar flow)}$$

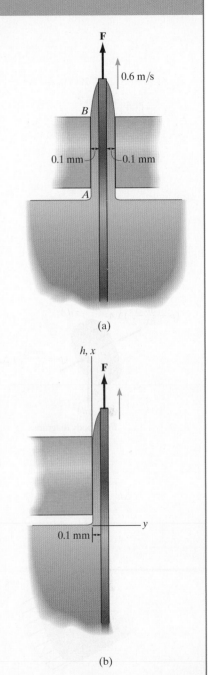

(a)

(b)

Fig. 9–8

9

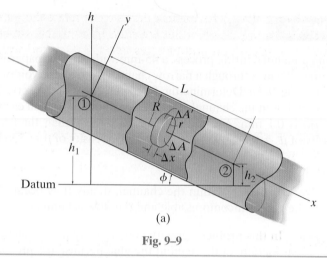

Fig. 9–9

9.3 Steady Laminar Flow within a Smooth Pipe

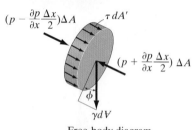

$(p - \dfrac{\partial p}{\partial x}\dfrac{\Delta x}{2})\Delta A$ $\tau\, dA'$

$(p + \dfrac{\partial p}{\partial x}\dfrac{\Delta x}{2})\,\Delta A$

ϕ

γdV

Free-body diagram.

(b)

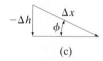

$-\Delta h$ Δx

ϕ

(c)

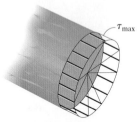

τ_{max}

Shear-stress distribution
for both laminar
and turbulent flow.

(d)

Fig. 9–9 (cont.)

Using an analysis similar to that for the parallel plate, we can also analyze steady laminar flow of an incompressible fluid within a smooth pipe. Here the flow will be axisymmetric, and so it is convenient to consider a control volume element within the fluid to be a differential disk, Fig. 9–9a. The momentum equation applied to this control volume will reduce to a balance of forces (equilibrium), because the flow is steady and the fluid is incompressible. In other words, no convective change occurs between the back and front open control surfaces and no local change occurs within the control volume. As shown on the free-body diagram of the control volume, Fig. 9–9b, the forces to be considered in the x direction are due to pressure, gravity, and viscosity. We have

$$\Sigma F_x = \frac{\partial}{\partial t}\int_{cv} \mathbf{V}\rho\, dV + \int_{cs} \mathbf{V}\,\rho \mathbf{V}\cdot d\mathbf{A}$$

$$\left(p - \frac{\partial p}{\partial x}\frac{\Delta x}{2}\right)\Delta A - \left(p + \frac{\partial p}{\partial x}\frac{\Delta x}{2}\right)\Delta A + \tau\Delta A' + \gamma\Delta V \sin\phi = 0 + 0$$

From Fig. 9–9a, the open control cross-sectional area is $\Delta A = \pi r^2$, the area on the closed control surface is $\Delta A' = 2\pi r\Delta x$, and the control volume is $\Delta V = \pi r^2\Delta x$. Substituting these results into the above equation, noting that $\sin\phi = -\Delta h/\Delta x$, Fig. 9–9c, and taking the limit, we get

$$\tau = \frac{r}{2}\frac{\partial}{\partial x}(p + \gamma h) \qquad (9\text{–}16)$$

Shear-stress distribution
Laminar and turbulent flow

This equation gives the *shear-stress* distribution within the fluid. Notice that it varies directly with r, being largest at the wall, $r = R$, and zero at the center, Fig. 9–9d. Since τ was determined simply from a balance of forces, this distribution is valid for *both laminar and turbulent flows*.

If we consider the flow to be *laminar*, then we can relate the shear stress to the velocity at any point within the fluid by using Newton's law of viscosity, $\tau = \mu(du/dr)$. Substituting this into Eq. 9–16 and rearranging terms, we have

$$\frac{du}{dr} = \frac{r}{2\mu}\frac{\partial}{\partial x}(p + \gamma h)$$

The term $\partial(p + \gamma h)/\partial x$ represents the sum of the pressure and elevation gradients. Since this sum is the hydraulic gradient, it is independent of y, and so integrating the above equation with respect to r, we have

$$u = \frac{r^2}{4\mu}\frac{d}{dx}(p + \gamma h) + C$$

The constant of integration can be determined using the "no-slip" condition that $u = 0$ at $r = R$. Once obtained, the result is

$$u = -\frac{\left(R^2 - r^2\right)}{4\mu}\frac{d}{dx}(p + \gamma h) \qquad (9\text{–}17)$$

Velocity profile
Laminar flow

The *velocity profile* therefore takes the form of a *paraboloid*, Fig. 9–9e. Since τ is small in the central region of the pipe, Fig. 9–9d, the fluid has its greatest velocity there.

The maximum velocity occurs at the center of the pipe, $r = 0$, where $du/dr = 0$. It is

$$u_{max} = -\frac{R^2}{4\mu}\frac{d}{dx}(p + \gamma h) \qquad (9\text{–}18)$$

Maximum velocity

Velocity distribution
for laminar flow

h

(e)

Fig. 9–9 (cont.)

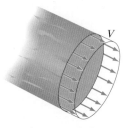

Velocity distribution for laminar flow

(f)

The volumetric flow is determined by integrating the velocity profile over the cross-sectional area. Choosing the differential ring element of area $dA = 2\pi r\, dr$, shown in Fig. 9–9f, we have

$$Q = \int_A u\, dA = \int_0^R u\, 2\pi r\, dr = -\frac{2\pi}{4\mu}\frac{d}{dx}(p + \gamma h)\int_0^R (R^2 - r^2)r\, dr$$

or

$$Q = -\frac{\pi R^4}{8\mu}\frac{d}{dx}(p + \gamma h) \qquad (9\text{–}19)$$

Volumetric flow

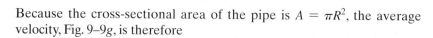

Because the cross-sectional area of the pipe is $A = \pi R^2$, the average velocity, Fig. 9–9g, is therefore

$$V = \frac{Q}{A} = -\frac{R^2}{8\mu}\frac{d}{dx}(p + \gamma h) \qquad (9\text{–}20)$$

Average velocity

Average velocity
distribution

(g)

Fig. 9–9 (cont.)

By comparison with Eq. 9–18, we see that

$$u_{max} = 2V \qquad (9\text{–}21)$$

The negative signs on the right of Eqs. 9–17 through 9–20 result from the sign convention established for the pressure and elevation gradients, $d(p + \gamma h)/dx$. For example, as shown in Fig. 9–9h, if the pressures p_1 and p_2 and the elevations h_1 and h_2 between the two points (cross sections) 1 and 2 on any streamline are known, then these gradients become*

$$\frac{d}{dx}(p + \gamma h) = \frac{p_2 - p_1}{L} + \gamma\frac{h_2 - h_1}{L} \qquad (9\text{–}22)$$

(e)

(h)

*Any streamline can be selected as explained in the footnote on p. 435.

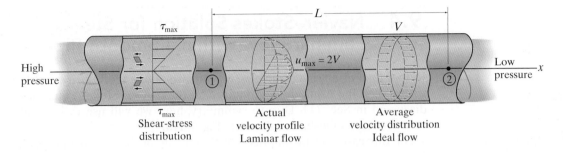

Shear-stress and Velocity Profiles for Negative Pressure Gradient

Fig. 9–10

Horizontal Flow through a Circular Pipe.

If the pipe is horizontal, then the force of gravity will not influence the flow since $dh/dx = 0$. If there is a higher pressure on the left side of the pipe, Fig. 9–10, then over the length L, this pressure will "push" the fluid to the right, but realize the pressure will decrease along the pipe because of fluid friction. This causes a negative pressure gradient ($\Delta p/L < 0$) according to our sign convention. Using this result, and our previous results, the maximum velocity, average velocity, and volumetric flow, expressed in terms of the inner *diameter* of the pipe, $D = 2R$, then become

$$u_{max} = \frac{D^2}{16\mu}\left(\frac{\Delta p}{L}\right) \tag{9–23}$$

$$V = \frac{D^2}{32\mu}\left(\frac{\Delta p}{L}\right) \tag{9–24}$$

$$Q = \frac{\pi D^4}{128\mu}\left(\frac{\Delta p}{L}\right) \tag{9–25}$$

Equation 9–25 is known as the **Hagen–Poiseuille equation**, since it was originally developed from *an experiment* in the mid-1800s by the German engineer Gotthilf Hagen, and independently by a French physician, Jean Louis Poiseuille.* Shortly thereafter the analytical formulation, as developed here, was presented by Gustav Wiedemann.

If we know the flow Q, we can solve the Hagen–Poiseuille equation for the pressure drop that occurs over the length L of the pipe. It is

$$\Delta p = \frac{128\mu L Q}{\pi D^4} \tag{9–26}$$

Notice that the greatest influence on pressure drop comes from the pipe's diameter. For example, a pipe having half the diameter will experience sixteen times the pressure drop due to viscous fluid friction! This effect can have serious consequences on the ability of pumps to provide adequate water flow through pipes that may have narrowed due to the accumulation of corrosion or scale.

*Poiseuille attempted to study the flow of blood using water confined to small-diameter tubes. Actually, however, veins are flexible, and blood is a non-Newtonian fluid, that is, it does not have a constant viscosity.

9.4   Navier–Stokes Solution for Steady Laminar Flow within a Smooth Pipe

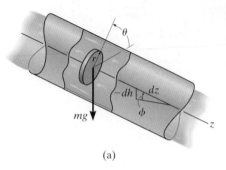

(a)

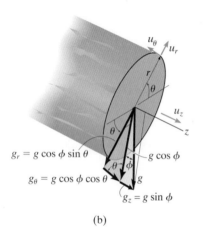

$g_r = g \cos \phi \sin \theta$

$g_\theta = g \cos \phi \cos \theta$

$g_z = g \sin \phi$

(b)

Fig. 9–11

Rather than using first principles, we can also obtain the velocity profile within the pipe using the continuity and the Navier–Stokes equations discussed in Sec. 7.11. Due to symmetry, here we will use cylindrical coordinates, established as shown in Fig. 9–11a.

For this case, the incompressible flow is along the axis of the pipe, Fig. 9–11a, so $v_r = v_\theta = 0$, and the continuity equation, Eq. 7–78, yields

$$\frac{\partial \rho}{\partial t} + \frac{1}{r}\frac{\partial(r\rho v_r)}{\partial r} + \frac{1}{r}\frac{\partial(\rho v_\theta)}{\partial \theta} + \frac{\partial(\rho v_z)}{\partial z} = 0$$

$$0 + 0 + 0 + \rho\frac{\partial v_z}{\partial z} = 0 \quad \text{or} \quad \frac{\partial v_z}{\partial z} = 0$$

Since the flow is steady and symmetrical about the z axis, integration gives $v_z = v_z(r)$.

Carefully note from Fig. 9–11b that the cylindrical components of **g** are $g_r = -g \cos \phi \sin \theta$, $g_\theta = -g \cos \phi \cos \theta$, and $g_z = g \sin \phi$. Using these and the above result, the first Navier–Stokes equation reduces to

$$\rho\left(\frac{\partial v_r}{\partial t} + v_r\frac{\partial v_r}{\partial r} + \frac{v_\theta}{r}\frac{\partial v_r}{\partial \theta} - \frac{v_\theta^2}{r} + v_z\frac{\partial v_r}{\partial z}\right)$$

$$= -\frac{\partial p}{\partial r} + \rho g_r + \mu\left[\frac{1}{r}\frac{\partial}{\partial r}\left(r\frac{\partial v_r}{\partial r}\right) - \frac{v_r}{r^2} + \frac{1}{r^2}\frac{\partial^2 v_r}{\partial \theta^2} - \frac{2}{r^2}\frac{\partial v_\theta}{\partial \theta} + \frac{\partial^2 v_r}{\partial z^2}\right]$$

$$0 = -\frac{\partial p}{\partial r} - \rho g \cos \phi \sin \theta + 0$$

Integrating this equation with respect to r gives

$$p = -\rho g r \cos \phi \sin \theta + f(\theta, z)$$

For the second Navier–Stokes equation,

$$\rho\left(\frac{\partial v_\theta}{\partial t} + v_r\frac{\partial v_\theta}{\partial r} + \frac{v_\theta}{r}\frac{\partial v_\theta}{\partial \theta} + \frac{v_r v_\theta}{r} + v_z\frac{\partial v_\theta}{\partial z}\right)$$

$$= -\frac{1}{r}\frac{\partial p}{\partial \theta} + \rho g_\theta + \mu\left[\frac{1}{r}\frac{\partial}{\partial r}\left(r\frac{\partial v_\theta}{\partial r}\right) - \frac{v_\theta}{r^2} + \frac{1}{r^2}\frac{\partial^2 v_\theta}{\partial \theta^2} + \frac{2}{r^2}\frac{\partial v_r}{\partial \theta} + \frac{\partial^2 v_\theta}{\partial z^2}\right]$$

$$0 = -\frac{1}{r}\frac{\partial p}{\partial \theta} - \rho g \cos \phi \cos \theta + 0$$

Integrating this equation with respect to θ yields

$$p = -\rho g r \cos \phi \sin \theta + f(r, z)$$

Comparing these two results, we require $f(\theta, z) = f(r, z) = f(z)$ since r, θ, z can vary independent of one another. From Fig. 9–11c, the *vertical distance* $h' = r \cos \phi \sin \theta$, so

$$p = -\rho g h' + f(z)$$

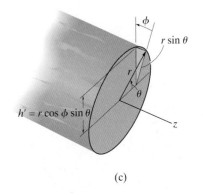

(c)

Fig. 9–11 (cont.)

In other words, the pressure is hydrostatic in the vertical plane since it depends upon the vertical distance h'. The last term, $f(z)$, is the variation in pressure caused by viscosity. Finally, the third Navier–Stokes equation becomes

$$\rho \left(\frac{\partial v_z}{\partial t} + v_r \frac{\partial v_z}{\partial r} + \frac{v_\theta}{r} \frac{\partial v_z}{\partial \theta} + v_z \frac{\partial v_z}{\partial z} \right)$$

$$= -\frac{\partial p}{\partial z} + \rho g_z + \mu \left[\frac{1}{r} \frac{\partial}{\partial r} \left(r \frac{\partial v_z}{\partial r} \right) + \frac{1}{r^2} \frac{\partial^2 v_z}{\partial \theta^2} + \frac{\partial^2 v_z}{\partial z^2} \right]$$

$$0 = -\frac{\partial p}{\partial z} + \rho g \sin \phi + \mu \left[\frac{1}{r} \frac{\partial}{\partial r} \left(r \frac{\partial v_z}{\partial r} \right) \right]$$

From Fig. 9–11a, $\sin \phi = -dh/dz$, so this equation can be rearranged and written as

$$\frac{\partial}{\partial r} \left(r \frac{\partial v_z}{\partial r} \right) = \frac{r}{\mu} \left[\frac{\partial p}{\partial z} + \rho g \left(\frac{\partial h}{\partial z} \right) \right]$$

Integrating twice yields

$$r \frac{\partial v_z}{\partial r} = \frac{r^2}{2\mu} \left[\frac{\partial p}{\partial z} + \rho g \left(\frac{\partial h}{\partial z} \right) \right] + C_1$$

$$v_z = \frac{r^2}{4\mu} \left[\frac{\partial p}{\partial z} + \rho g \left(\frac{\partial h}{\partial z} \right) \right] + C_1 \ln r + C_2$$

The velocity v_z must be finite at the center of the pipe, and since $\ln r \to -\infty$ as $r \to 0$, then $C_1 = 0$. At the wall, $r = R$, $v_z = 0$ because of the "no-slip" condition. Thus,

$$C_2 = -\frac{R^2}{4\mu} \left[\frac{\partial p}{\partial z} + \rho g \left(\frac{\partial h}{\partial z} \right) \right]$$

The final result is therefore

$$v_z = -\frac{R^2 - r^2}{4\mu} \frac{\partial}{\partial z} (p + \gamma h)$$

This is the same as Eq. 9–17, and so a further analysis will produce the rest of the equations in Sec. 9.3.

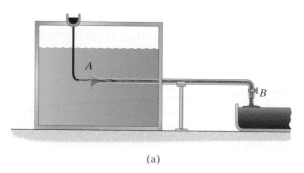

(a)

Fig. 9–12

9.5   The Reynolds Number

In 1883, Osborne Reynolds established a criterion for identifying laminar flow within a pipe. He did this by controlling the flow of water passing through a glass tube, using an apparatus similar to that shown in Fig. 9–12*a*. Here, colored dye was injected within the stream at *A* and the valve at *B* was opened. For slow rates, the flow in the tube was observed to be *laminar* since the dye streak remained straight and uniform, Fig. 9–12*b*. As the flow was increased, by further opening the valve, the streak began to break down as it underwent *transitional* flow, Fig. 9–12*c*. Finally, through a further increase in flow, *turbulence* occurred because the dye became fully dispersed throughout the water in the tube, Fig. 9–12*d*. Experiments using other liquids, as well as gases, showed this same type of behavior. From these experiments, Reynolds *suspected* that this change from laminar to transitional to turbulent flow was dependent on the average velocity V of the fluid, on its density ρ and viscosity μ, and on the tube's diameter D.

For *any two different* experimental setups involving these four variables, Reynolds reasoned that a *similar* flow will occur because then the *forces* acting on the fluid particles in *one flow* would be in the *same ratio* as those acting on the particles in some *other flow*, Fig. 9–13. In other words, as we discussed in Chapter 8, this *dynamic similitude* will ensure both geometric and kinematic similitude for the flow.

Besides the inertia force, the other two significant forces influencing the motion of fluid flow within a pipe are caused by pressure and viscosity, and for similitude, all these forces must satisfy Newton's second law of motion. However, as noted in Sec. 8.4, similitude between the viscous and inertia forces will *automatically* ensure similitude between the pressure and inertia forces. Realizing this, Reynolds chose to study the ratio of the inertia force to the viscous force and, as shown by the dimensional analysis in Example 8–1, so produced the dimensionless "Reynolds number" as the criterion for flow similitude.

Laminar flow

(b)

Transitional flow

(c)

Turbulent flow

(d)

9

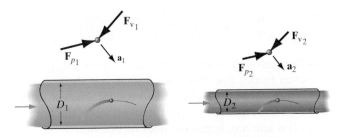

Similitude among pressure, viscous, and inertia
forces for two different experimental setups.

Fig. 9–13

Although *any* velocity and pipe dimension L can be chosen for this ratio, what is important is to realize that for two different flow situations, they must correspond with each other. In practice, the velocity that has been accepted is the *mean* or *average flow velocity* $V = Q/A$, and the dimension, referred to as the "characteristic length," is the *inner diameter D* of the pipe. Using V and D, the Reynolds number then becomes

$$\text{Re} = \frac{\rho V D}{\mu} = \frac{VD}{\nu} \qquad (9\text{–}27)$$

Experiments have confirmed that the higher this number, the greater the chance for laminar flow to break down, since inertia forces will overcome the viscous forces and begin to dominate the flow. And so, from Eq. 9–2, the faster the fluid flows, the more chance the fluid has of becoming unstable. Likewise, the larger the diameter of pipe, the greater the volume of fluid passing through it, and so instability can occur more readily. Finally, a lower kinematic viscosity will cause the flow to become unstable, because there is less chance that any flow disturbances will be dampened out by viscous shear forces.

In practice, it is very difficult to predict at *exactly* what specific value of the Reynolds number the flow in a pipe will suddenly change from laminar to transitional flow. Experiments show that the ***critical velocity*** at which this happens is highly sensitive to any initial vibrations or disturbances to the equipment. Also, the results are affected by initial motion of the fluid, the type of pipe inlet, the pipe's surface roughness, or any slight adjustment that occurs when the valve is opened or closed. For most engineering applications, however, laminar flow begins to change to transitional flow at about Re = 2300.* This value is called the ***critical Reynolds number***, and in this text, unless otherwise stated, we will assume it to be the limiting value for laminar flow in smooth uniform straight pipes. Therefore,

$$\text{Re} \le 2300 \qquad \text{Laminar flow in straight pipes} \qquad (9\text{–}28)$$

This critical value applies only for Newtonian fluids.** By using this as the critical upper bound for laminar flow, we will then be able to estimate the energy loss or pressure drop that occurs within a pipe, and thereby determine the flow through the pipe.

*Some authors use other limits, generally ranging from 2000 to 2400.
**To date, no satisfactory criterion has been established for a non-Newtonian fluid.

Important Points

- Steady flow through a pipe is a balance of pressure, gravitational, and viscous forces. For this case, the viscous shear stress varies linearly from zero at the center and has its greatest value along the wall of the pipe. It does not depend upon the type of flow, whether it is laminar, transitional, or turbulent.

- The *velocity profile* for *steady laminar flow* of an incompressible fluid within a pipe is in the shape of a *paraboloid*. The maximum velocity is $u_{max} = 2V$, and it occurs along the centerline of the pipe. The velocity at the wall is zero since it is a fixed boundary — the "no-slip" condition.

- The formulations for laminar flow in a pipe can be developed from first principles, and also by solving the continuity and Navier–Stokes equations. These results are in close agreement with experimental data.

- Flow within a horizontal pipe is dependent upon both pressure and viscous forces. Reynolds recognized this and formulated the Reynolds number $Re = \rho V D / \mu$ as a criterion for dynamic similitude between two different sets of flow conditions.

- Experiments have indicated that laminar flow in any pipe will occur provided $Re \leq 2300$. This general estimate of an upper limit will be used in this text.

Procedure for Analysis

It is suggested that the equations developed in Sec. 9.3 be applied using the following procedure.

Fluid Description.

Be sure that the fluid is defined as incompressible and the flow is steady. Since laminar flow must prevail, then the flow conditions must not exceed the Reynolds number criterion, $Re \leq 2300$.

Analysis.

Establish the coordinates and follow their positive sign convention. Here the longitudinal axis is positive in the direction of flow, the radial axis is positive outward from the centerline of the pipe, and the vertical axis is positive upward. Finally, be sure to use a consistent set of units when substituting numerical data into any of the equations.

EXAMPLE 9.4

Oil flows through the 100-mm-diameter pipe in Fig. 9–14. If the pressure at A is 34.25 kPa, determine the discharge at B. Take $\rho_o = 870 \text{ kg/m}^2$ and $\mu_o = 0.0360 \text{ N} \cdot \text{s/m}^2$.

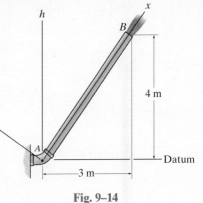

Fig. 9–14

SOLUTION

Fluid Description. We will assume steady flow that is laminar, and that the oil is incompressible.

Analysis. The discharge is determined using Eq. 9–19. The origin of coordinates for x and h is at A, and by convention, the positive x axis is extended in the direction of flow, and the positive h axis is vertically upward. Thus,

$$Q = -\frac{\pi R^4}{8\mu_o} \frac{d}{dx}(p + \gamma h)$$

$$= -\frac{\pi R^4}{8\mu_o}\left(\frac{p_B - p_A}{L} + \frac{\gamma(h_B - h_A)}{L}\right)$$

$$= -\frac{\pi(0.05 \text{ m})^4}{8(0.0360 \text{ N} \cdot \text{s/m}^2)}\left(\frac{0 - 34.25(10^3) \text{ N/m}^2}{5 \text{ m}} + \frac{(870 \text{ kg/m}^3)(9.81 \text{ m/s}^2)(4 \text{ m} - 0)}{5 \text{ m}}\right)$$

$$= 0.001516 \text{ m}^3/\text{s} = 0.00152 \text{ m}^3/\text{s} \qquad\qquad Ans.$$

Since this result is *positive*, the flow is indeed from A to B.

The assumption of laminar flow is checked using the mean velocity and the Reynolds number criterion.

$$V = \frac{Q}{A} = \frac{0.001516 \text{ m}^3/\text{s}}{\pi(0.05 \text{ m})^2} = 0.1931 \text{ m/s}$$

$$\text{Re} = \frac{\rho_o V D}{\mu_o} = \frac{(870 \text{ kg/m}^3)(0.1931 \text{ m/s})(0.1 \text{ m})}{0.0360 \text{ N} \cdot \text{s/m}^2} = 467 < 2300 \qquad \text{(laminar flow)}$$

9

EXAMPLE 9.5

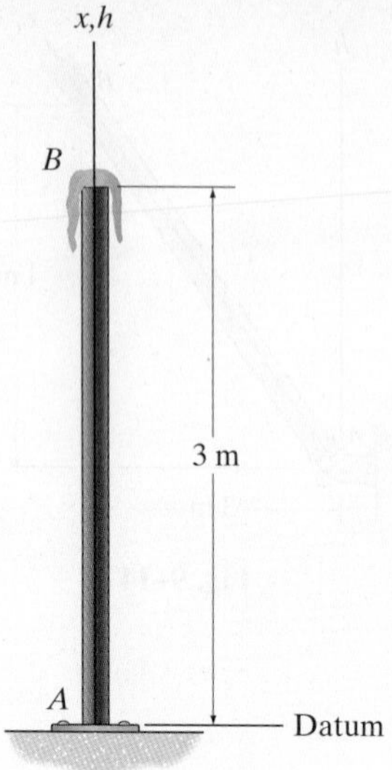

Fig. 9–15

Determine the maximum pressure at A so that the flow of water through the vertical standpipe in Fig. 9–15 remains laminar. The pipe has an inner diameter of 80 mm. Take $\rho_w = 1000 \text{ kg/m}^3$ and $\mu_w = 1.52(10^{-3}) \text{ N} \cdot \text{s/m}^2$.

SOLUTION

Fluid Description.   It is required that laminar flow occur. The flow must also be steady, and we will assume water to be incompressible.

Analysis.   For laminar flow, the maximum average velocity is limited based on the Reynold's number criterion.

$$\text{Re} = \frac{\rho_w V D}{\mu_w}$$

$$2300 = \frac{(1000 \text{ kg/m}^3)(V)(0.08 \text{ m})}{1.52 (10^{-3}) \text{ N} \cdot \text{s/m}^2}$$

$$V = 0.0437 \text{ m/s}$$

We will apply Eq. 9–20 to obtain the pressure at A. Following the sign convention, positive x is in the direction of flow, which is vertically upward, and positive h is also vertically upward, Fig. 9–15. Since $dh/dx = 1$, we have

$$V = -\frac{R^2}{8\mu} \frac{d}{dx}(p + \gamma h)$$

$$0.0437 \text{ m/s} = -\frac{(0.04 \text{ m})^2}{8[1.52(10^{-3}) \text{ N} \cdot \text{s/m}^2]}\left[\left(\frac{0 - P_A}{3 \text{ m}}\right) + (1000 \text{ kg/m}^3)(9.81 \text{ m/s}^2)\left(\frac{3 \text{ m} - 0}{3 \text{ m}}\right)\right]$$

$$P_A = 29.43(10^3) \text{ Pa} = 29.4 \text{ kPa} \qquad Ans.$$

Because the velocity and the viscosity are very small, this pressure is essentially hydrostatic, that is, $p = \gamma h = [(1000 \text{ kg/m}^3)(9.81 \text{ m})](3 \text{ m}) = 29.43(10^3)$ Pa. In other words, the pressure at A is mainly used to *support* the water column, and little is needed to overcome the slight frictional resistance and push the water through the pipe to maintain laminar flow.

9.6   Fully Developed Flow from an Entrance

When fluid flows through the opening of a pipe or duct attached to a reservoir, it will begin to *accelerate* and then transition to either fully laminar or fully turbulent steady flow. We will now consider each of these cases separately.

Flow through long straight pipe will be fully developed. (© Prisma/Heeb Christian/Alamy)

Laminar Flow.   As shown in Fig. 9–16*a*, at the entrance of the pipe the fluid's velocity profile will be nearly uniform. Then, as the fluid travels farther down the pipe, its viscosity will begin to slow down the particles located *near the wall*, since particles at the wall must have zero velocity. With a further advance, the viscous layers that develop near the wall will begin to spread toward the pipe's centerline until the central core of fluid, which originally had uniform velocity, begins to disappear at a distance of L'. Once it does, the flow becomes **fully developed**, that is, the parabolic velocity profile for laminar flow becomes constant.

The transition or **entrance length** L' is actually a function of the pipe diameter D and the Reynolds number. An estimate of this length can be made using an equation formulated by Henry Langhaar. See Ref. [2]. It is

$$L' = 0.06(\text{Re})D \qquad \text{Laminar flow} \qquad (9–29)$$

Using our criterion for laminar flow in pipes, that $\text{Re} \leq 2300$, then as an *upper limit* to the entrance length, $L' = 0.06(2300)D = 138D$. As we will see in the example problem that follows, fully developed laminar flow rarely occurs in pipes, because the velocity will be high, or the flow development will get disrupted by a valve, transition, or bend in the pipe.

Turbulent Flow.   Experiments have shown that the entrance length to fully developed turbulent flow is not very dependent upon the Reynolds number; rather, it depends more upon the shape or type of inlet and upon the actual roughness of the wall of the pipe. For example, a rounded inlet, as shown in Fig. 9–16*b*, produces a shorter transition length to full turbulence than a sharp or 90° inlet. Also, pipes with rough walls produce turbulence at a shorter distance than those with smooth walls. Through experiments, along with a computer analysis, it has been found that fully developed turbulent flow can occur within a relatively short distance. See Ref. [3]. For example, it is on the order of $12D$ for a low Reynolds number, $\text{Re} = 3000$. Although longer transition distances occur at larger Reynolds numbers, for most engineering analysis it is reasonable to assume this transition from unsteady to mean steady turbulent flow is *localized near the entrance*. And as a result, engineers account for the friction or energy loss that occurs at a turbulent entrance length by using a *loss coefficient*, something we will discuss in the next chapter.

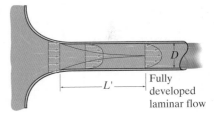

Transition to Laminar Flow

(a)

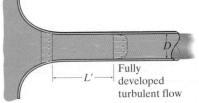

Transition to Turbulent Flow

(b)

Fig. 9–16

EXAMPLE | 9.6

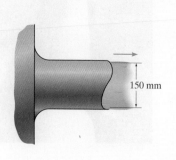

150 mm

Fig. 9–17

If the flow through the 150-mm-diameter drainpipe in Fig. 9–17 is 0.0062 m³/s, classify the flow along the pipe as laminar or turbulent if the fluid is water, and if it is oil. Determine the entrance length to fully developed flow if it is oil. Take $\nu_w = 0.898(10^{-6})$ m²/s and $\nu_o = 0.0353(10^{-3})$ m²/s.

SOLUTION

Fluid Description.   Beyond the entrance length, we consider the flow as steady. The water and oil are both assumed to be incompressible.

Analysis.   The flow is classified on the basis of the Reynolds number. The mean velocity of the flow is

$$V = \frac{Q}{A} = \frac{0.0062 \text{ m}^3/\text{s}}{\pi(0.075 \text{ m})^2} = 0.3508 \text{ m/s}$$

Water.   Here the Reynolds number is

$$\text{Re} = \frac{VD}{\nu_w} = \frac{(0.3508 \text{ m/s})(0.15 \text{ m})}{0.898(10^{-6}) \text{ m}^2/\text{s}} = 58.6(10^3) > 2300 \qquad Ans.$$

The flow is *turbulent*.
    By comparison, note that if Re = 2300, then the average velocity for laminar flow would have to be

$$\text{Re} = \frac{VD}{\nu_w} = 2300; \qquad 2300 = \frac{V(0.15 \text{ m})}{0.898(10^{-6}) \text{ m}^2/\text{s}}$$

$$V = 0.0138 \text{ m/s}$$

This is indeed a very small value, and so practically speaking, due mainly to its relatively low viscosity, *the flow of water through a pipe will almost always be turbulent.*

Oil.   In this case,

$$\text{Re} = \frac{VD}{\nu_o} = \frac{(0.3508 \text{ m/s})(0.15 \text{ m})}{0.0353(10^{-3}) \text{ m}^2/\text{s}} = 1491 < 2300 \qquad Ans.$$

Here *laminar flow* exists in the pipe, although it is not fully developed within the region near the entrance. Applying Eq. 9–29, the transitional length for fully developed laminar flow of the oil is therefore

$$L' = 0.06 \, (\text{Re}) \, D = 0.06(1491)(0.15 \text{ m}) = 13.4 \text{ m} \qquad Ans.$$

This is a rather long distance, but once it occurs, laminar flow is well understood and is defined by a balance of pressure and viscous forces, as we discussed in Sec. 9.3.

9.7   Laminar and Turbulent Shear Stress within a Smooth Pipe

Pipes of circular cross section are by far the most common conduit for a fluid, and for any design or analysis, it is important to be able to specify how shear stress, or frictional resistance, develops within the pipe for both laminar and turbulent flow.

Laminar Flow.   In Sec. 9.3 we obtained the velocity profile for steady laminar flow through a straight pipe, assuming the fluid is viscous, Fig. 9–18*a*. This parabolic profile requires that the fluid surrounding the *smooth wall* have zero velocity, since fluid particles tend to adhere (or stick) to the wall. Layers of fluid a greater distance from the wall have greater velocities, with the maximum velocity occurring at the centerline of the pipe. As discussed in Sec.1.7 the ***viscous shear stress***, or frictional resistance within the fluid, is caused by the continuous exchange of momentum among the *molecules of fluid*, as each layer slides over an adjacent layer.

Turbulent Flow.   If the rate of flow within the pipe is increased, the laminar fluid layers will become unstable and begin to break down as the flow transitions to turbulent flow. As this occurs, the fluid particles move in a *disorderly manner*, causing the formation of eddies or small vortices, and thereby the mixing of fluid throughout the pipe. These effects cause a larger loss of energy, and therefore a greater drop in pressure, compared to laminar flow.

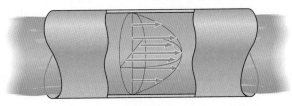

Velocity profile
Laminar flow

(a)

Fig. 9–18

9

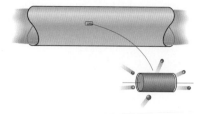

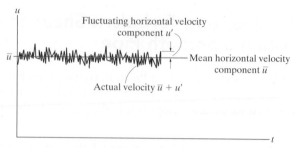

Horizontal velocity components of fluid particles
passing through a control volume.

(b)

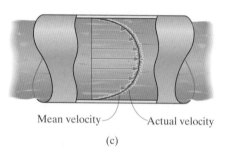

Mean velocity   Actual velocity

(c)

Viscous sublayer near pipe wall

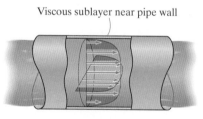

Velocity profile
Turbulent flow

(d)

Fig. 9–18 (cont.)

We can study the effects of turbulent flow by considering a small, fixed control volume located at a point within the pipe, Fig. 9–18b. The velocities of all fluid particles passing through this control volume will have a *random pattern*; however, these velocities can be resolved *horizontally* into a mean value, $\bar{u}$, and a random fluctuating velocity about the mean, u', as shown on the graph. The *fluctuations* will have a very short period, and their magnitude will be small compared to the mean velocity. If this mean velocity component remains *constant*, then we can classify the flow as **steady turbulent flow**, or, more properly, **mean steady flow**. The "turbulent mixing" of the fluid tends to *flatten* the mean horizontal velocity component $\bar{u}$ within a large region around the center of the pipe, and as a result, the velocity profile will be more uniform than laminar flow. The velocity may have an *actual* "wiggly" profile like that shown in Fig. 9–18c, but it will average out to that shown by the dark line.

Although turbulent mixing occurs readily within the central region of the pipe, it tends to *diminish* rapidly near the pipe's inner wall to satisfy the boundary condition of zero velocity at the wall. This region of low velocity produces a **laminar viscous sublayer** near the wall, Fig. 9–18d. The faster the flow, the larger the region of turbulence within the fluid, and the thinner this sublayer. Realize, however, that this sublayer thickness will normally be a very small fraction of the pipe's inner diameter as will be shown in Example 9.9.

Turbulent Shear Stress.   The flow characteristics of the fluid are greatly affected by turbulence. The fluid "particles" in turbulent flow are far greater in size than just the "molecules" being transferred between layers in laminar flow. Think of these particles as formed by the mixing caused by very small eddies or swirling within the flow. For both laminar and turbulent flow, however, the same phenomenon occurs, that is, the slower-moving particles migrate to faster-moving layers, and so they tend to decrease the momentum of the faster layers. Particles migrating from faster- to slower-moving layers will have the opposite effect; they increase the momentum of the slower layer. For turbulent flow, this type of momentum transfer gives rise to an **apparent shear stress** that will be many times greater than the viscous shear stress that is created by *molecular* exchange within the flow.

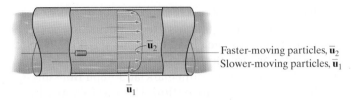

Faster-moving particles, $\bar{\mathbf{u}}_2$
Slower-moving particles, $\bar{\mathbf{u}}_1$

Velocity profile for turbulent flow

(a)

Fig. 9–19

To show conceptually how apparent turbulent shear stress develops, consider the steady turbulent flow along two adjacent fluid layers in Fig. 9–19a. At any instant, the velocity components in the direction of flow can be written in terms of their time-averaged, $\bar{u}$, and mean horizontal fluctuating, u', components, as shown in Fig. 9–18b. That is, for any horizontal velocity,

$$u = \bar{u} + u'$$

Vertical components of velocity only have a fluctuating component, and *no* mean velocity component, since there is no mean flow in this direction. Thus

$$v = v'$$

Now let's consider the movement v_1' of a fluid particle from a slower-moving bottom layer into the faster-moving upper layer, Fig. 9–19b. This transference will *increase* the horizontal velocity component of the transferred fluid particle by u_1'. Since the mass flow through the area dA is $\rho v_1' dA$, then the net change in the momentum, $u_1'(\rho v_1' dA)$, is the result of a force dF that the upper layer produces on the transferred particle. The shear stress is $\tau = dF/dA$, and so the apparent turbulent shear stress produced is then

$$\tau_{\text{turb}} = \rho \overline{u_1' v_1'}$$

where $\overline{u_1' v_1'}$ is the mean product of $u_1' v_1'$.

The same sort of argument can be made for particles transferring down from the faster-moving upper layer to the slower-moving lower one, except here the transference will decrease the horizontal velocity component of the transferred fluid particle by u_2'. This apparent turbulent shear stress, as described here, is sometimes referred to as the **Reynolds stress**, named after Osborne Reynolds who developed these arguments in 1886.

The shear stress within turbulent flow therefore consists of *two components*. The viscous shear stress is due to molecular exchange, $\tau_{\text{visc}} = \mu \, d\bar{u}/dy$, which results from the time average velocity component $\bar{u}$, and the *apparent turbulent shear stress* τ_{turb}, which is based upon the much larger eddy particle exchange between fluid layers. It is the result of the mean horizontal fluctuating component u'. Therefore, we can write

$$\tau = \tau_{\text{visc}} + \tau_{\text{turb}}$$

As noted from Eq. 9–16, τ has a linear variation throughout the flow, as shown in Fig. 9–19c.

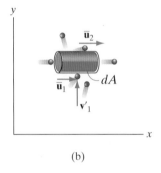

(b)

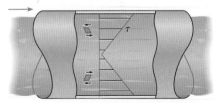

Shear-stress distribution.

(c)

Fig. 9–19 (cont.)

In practice it is difficult to obtain the apparent or turbulent shear-stress component since the vertical and horizontal fluctuating components v' and u' will be different for each location within the flow. In spite of this, empirical formulations for this stress, based on Reynolds's work, were later developed by the French mathematician Joseph Boussinesq, using a concept called the eddy viscosity of flow. This was followed by the work of Ludwig Prandtl, who created a mixing-length hypothesis based on the size of the eddies formed within the flow. Although both of these efforts provide some understanding of the notion of turbulent shear stress, and its relation to velocity, they have very limited application and today are no longer used. Turbulent flow is very complex, due to the erratic motion of the particles, and this has made it practically impossible to obtain a *single accurate mathematical formulation* to describe its behavior. Instead, this difficulty has led to many experimental investigations involving turbulent flow. See Ref. [3], [4]. From this, engineers have produced many different models to predict turbulent behavior, and as we have discussed in Sec. 7.12, some of these have been incorporated into sophisticated computer programs used for computational fluid dynamics (CFD).

9.8   Turbulent Flow within a Smooth Pipe

Careful measurements of the velocity profile for turbulent flow within a pipe have made it possible to identify three different regions of flow within the pipe. These are shown in Fig. 9–20a, and are referred to as the viscous sublayer, the transitional region, and the turbulent flow region.

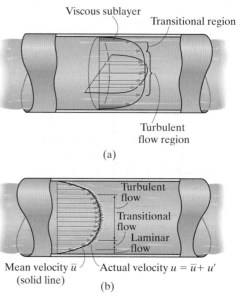

Viscous sublayer

Transitional region

Turbulent flow region

(a)

Turbulent flow

Transitional flow

Laminar flow

Mean velocity $\bar{u}$ (solid line)   Actual velocity $u = \bar{u} + u'$

(b)

Fig. 9–20

Viscous Sublayer.

For almost all fluids, the particles at the wall of the pipe have zero velocity, no matter how great the flow is through the pipe. These particles "stick" to the wall, and the fluid layers near them exhibit laminar flow because of their slow velocity. Consequently, viscous shear stress within the fluid dominates this region, and so if the fluid is a Newtonian fluid, the shear stress can be expressed by $\tau_{\text{visc}} = \mu(du/dy)$. If we integrate this equation, using the boundary condition $u = 0$ at $y = 0$, we can then relate the wall shear stress τ_0 (a constant) to the velocity. Since for laminar flow $u = \bar{u}$, which is the *time-averaged* or *mean velocity*, Fig. 9–20b, we get

$$\tau_0 = \mu \frac{\bar{u}}{y} \qquad (9\text{–}30)$$

We will express this result as a *dimensionless ratio* in order to compare it with experimental results that are normally plotted in terms of "dimensionless" variables. To do this, researchers have used the factor $u^* = \sqrt{\tau_0/\rho}$. This constant has units of velocity, and so it is sometimes referred to as the friction velocity or **shear velocity**. If we divide both sides of the above equation by ρ, and use the kinematic viscosity, $\nu = \mu/\rho$, we get

$$\frac{\bar{u}}{u^*} = \frac{u^* y}{\nu} \qquad (9\text{–}31)$$

Since u^* and ν are constants, then $\bar{u}$ and y form a *linear relationship* that represents the dimensionless velocity profile within the viscous sublayer, Fig. 9–21a. Equation 9–31 is sometimes referred to as the *law of the wall*. On a semi-logarithmic graph it plots as a curve, Fig. 9–21b, and as noted, it fits the experimental data, originally obtained mainly by Johann Nikuradse, for values of $0 \le u^* y/\nu \le 5$. See Ref. [6].

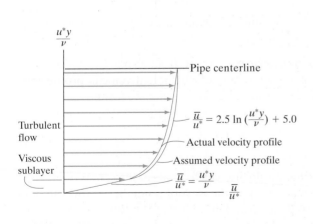

(a)

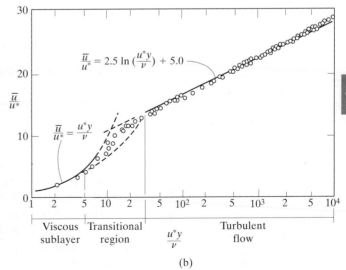

(b)

Fig. 9–21

Transitional and Turbulent Flow Region. Within these two regions, the flow is subjected to *both* viscous and turbulent shear stress, and so here we can express the resultant shear stress as

$$\tau = \tau_{\text{visc}} + \tau_{\text{turb}} = \mu\frac{d\overline{u}}{dy} + |\overline{\rho u' v'}| \qquad (9\text{–}32)$$

Recall from the discussion in the previous section, the turbulent (or Reynolds) shear stress results from the exchanges of large groups of particles between fluid layers. In a sense, this large mass transfer can be thought of as the random distribution of rapid fluctuations and swirling fluid called **eddy currents** within the flow. Of the two components in Eq. 9–32, the turbulent shear stress will predominate within the center of the pipe, but its effect will diminish rapidly as the flow nears the wall and enters the transitional region, where the velocity suddenly drops off, Fig. 9–20a.

The velocity profile for turbulent flow has also been established experimentally, again primarily by J. Nikuradse, Fig. 9–21b. With this Theodore von Kármán and Ludwig Prandtl were able to describe this data by the equation

$$\frac{\overline{u}}{u^*} = 2.5 \ln\left(\frac{u^* y}{\nu}\right) + 5.0 \qquad (9\text{–}33)$$

When plotted, this results in the curve shown in Fig. 9–21a, although on a semi-logarithmic scale, it results in a straight line, Fig. 9–21b.

Notice the scale in Fig. 9–21b. The viscous sublayer and the dashed transition zone extend only a very small distance, $u^* y/\nu \leq 30$, whereas the turbulent flow region extends to $u^* y/\nu = 10^4$. For this reason, and for most engineering applications, flow within the sublayer and transition zone can be neglected. Instead, Eq. 9–33 *alone* can be used to model the velocity profile for the pipe. Here, of course, it is *assumed* that the fluid is incompressible, the flow is fully turbulent and steady in the mean, and the walls of the pipe are *smooth*.

Power Law Approximation. Apart from using Eq. 9–33, other methods have also been used to model a turbulent velocity profile. One involves applying an empirical power law having the form

$$\frac{\overline{u}}{u_{\max}} = \left(1 - \frac{r}{R}\right)^{1/n} \qquad (9\text{–}34)$$

Here $u_{\max}$ is the maximum velocity, which occurs at the center of the pipe, and the exponent n depends upon the Reynolds number. A few values of n for specific values of Re are listed in Table 9–1. See Ref. [4]. The velocity profiles, including the one for laminar flow, are shown in Fig. 9–22. Notice how these profiles flatten out as n gets larger. This is due to the faster flow, or higher Reynolds number. Of these profiles, $n = 7$ is often used for calculations and it provides adequate results for many cases.

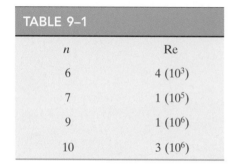

TABLE 9–1

n	Re
6	$4\ (10^3)$
7	$1\ (10^5)$
9	$1\ (10^6)$
10	$3\ (10^6)$

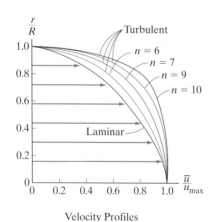

Velocity Profiles

Fig. 9–22

Since the velocity profile in a pipe is axisymmetric, Fig. 9–23, we can integrate Eq. 9–34 and determine the flow for any value of n. We have

$$Q = \int_A \bar{u} \, dA = \int_0^R u_{max} \left(1 - \frac{r}{R}\right)^{1/n} (2\pi r) dr$$

$$= 2\pi R^2 u_{max} \frac{n^2}{(n+1)(2n+1)} \qquad (9\text{–}35)$$

Also, since $Q = V(\pi R^2)$, the average velocity of flow is

$$V = u_{max} \left[\frac{2n^2}{(n+1)(2n+1)}\right] \qquad (9\text{–}36)$$

Besides using Eq. 9–34, there have also been efforts to predict turbulent time-average flow using various "turbulent models" that include chaotic fluctuations within the flow. Research in this important area is ongoing, and hopefully these models will continue to improve through the years.

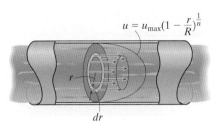

$$u = u_{max}\left(1 - \frac{r}{R}\right)^{\frac{1}{n}}$$

Turbulent Velocity Profile
Approximation

Fig. 9–23

Important Points

- When fluid flows into a pipe from a reservoir, it will accelerate for a certain distance along the pipe before its velocity profile becomes fully developed. For steady laminar flow, this transition or entrance length is a function of the Reynolds number and the pipe diameter. For steady turbulent flow, it depends not only on the type of entrance, but also on the diameter and the surface roughness of the pipe.

- Turbulent flow involves erratic and complex motion of fluid particles. Small eddies form within the flow and cause localized mixing of the fluid. It is for this reason that shear stress and the energy losses for turbulent flow are much greater than those for laminar flow. The shear stress is a combination of viscous shear stress and an "apparent" turbulent shear stress, which is caused by the transfer of groups of fluid particles from one layer of fluid to an adjacent layer.

- The mixing action of turbulent flow tends to "flatten out" the velocity profile and make it more uniform, like an ideal fluid. This profile will always have a narrow viscous sublayer (laminar flow) near the walls of the pipe. Here the fluid must move slowly due to the boundary condition of zero velocity at the wall. The faster the flow, the thinner this sublayer becomes.

- Since turbulent flow is so erratic, an analytical solution for describing the velocity profile cannot be obtained. Instead, it is necessary to rely on experimental methods to define this shape and then to match it with empirical approximations such as Eqs. 9–33 and 9–34.

9

EXAMPLE | 9.7

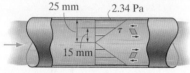

(a)

(b)

1.09 m/s

(c)

Fig. 9–24

The inner wall of the 50-mm-diameter water pipe in Fig. 9–24a is smooth. If turbulent flow exists, and the pressure at A is 10 kPa and at B it is 8.5 kPa, determine the magnitude of the shear stress acting along the wall of the pipe, and also at a distance of 15 mm from its center. What is the velocity at the center of the pipe, and what is the thickness of the viscous sublayer? Take $\rho_w = 1000$ kg/m^3 and $\nu_w = 1.08(10^{-6})$ m^2/s.

SOLUTION

Fluid Description. We have mean steady turbulent flow, and we assume water is incompressible.

Shear Stress. The shear stress along the wall of the pipe is caused by laminar flow, and so it can be determined using Eq. 9–16. Since the pipe is horizontal, $\dfrac{dh}{dx} = 0$, we have

$$\tau_0 = \frac{r}{2} \frac{\Delta p}{L} = \left| \frac{(0.025\ \text{m})}{2} \left(\frac{(8.5 - 10)(10^3)\ \text{N/m}^2}{8\ \text{m}} \right) \right| = 2.344\ \text{Pa} = 2.34\ \text{Pa} \qquad Ans.$$

The shear-stress distribution within the fluid varies *linearly* from the pipe's center as shown in Fig. 9–24b. Recall from Sec. 9.3 that this result comes from a balance of pressure and viscous forces, and it is valid for *both* laminar and turbulent flow. By proportion, we can determine the maximum shear stress at $r = 15$ mm. It is

$$\frac{\tau}{15\ \text{mm}} = \frac{2.344\ \text{Pa}}{25\ \text{mm}}; \qquad \tau = 1.41\ \text{Pa} \qquad Ans.$$

Velocity. Since the flow is turbulent, we will use Eq. 9–33 to determine the centerline velocity, Fig. 9–24c. First,

$$u^* = \sqrt{\tau_0/\rho_w} = \sqrt{(2.344\ \text{N/m}^2)/(1000\ \text{kg/m}^3)} = 0.04841\ \text{m/s}$$

And at the centerline of the pipe, $y = 0.025$ m, and so

$$\frac{u_{max}}{u^*} = 2.5 \ln \left(\frac{u^* y}{\nu_w} \right) + 5.0$$

$$\frac{u_{max}}{0.04841\ \text{m/s}} = 2.5 \ln \left[\frac{(0.04841\ \text{m/s})(0.025\ \text{m})}{1.08(10^{-6})\ \text{m}^2/\text{s}} \right] + 5.0$$

$$u_{max} = 1.09\ \text{m/s} \qquad Ans.$$

Viscous Sublayer. The viscous sublayer extends to $u^* y / \nu = 5$, Fig. 9–21b. Thus,

$$y = \frac{5\nu_w}{u^*} = \frac{5\left[1.08(10^{-6})\ \text{m}^2/\text{s} \right]}{0.04841\ \text{m/s}} = 0.11154(10^{-3})\ \text{m} = 0.112\ \text{mm} \ Ans.$$

Remember that this result is only for a *smooth-walled* pipe. If the pipe has a *rough surface*, then there is a good chance for protuberances to pass through this very thin layer, disrupting the flow and creating added friction. We will discuss this effect in the next chapter.

EXAMPLE 9.8

Kerosene flows through the 100-mm-diameter smooth pipe in Fig. 9–25a, with an average velocity of 20 m/s. Viscous friction causes a pressure drop (gradient) along the pipe to be 0.8 kPa/m. Determine the viscous and turbulent shear-stress components within the kerosene at $r = 10$ mm from the centerline of the pipe. Use a power-law velocity profile, and take $\nu_k = 2(10^{-6})$ m^2/s and $\rho_k = 820$ kg/m^3.

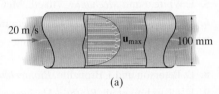

(a)

SOLUTION

Fluid Description. We will assume mean steady turbulent flow. Also, the kerosene can be considered incompressible.

Analysis. To use the power-law velocity profile, we must first determine the exponent n, which depends upon the Reynolds number.

$$\text{Re} = \frac{VD}{\nu_k} = \frac{(20 \text{ m/s})(0.1 \text{ m})}{2(10^{-6}) \text{ m}^2/\text{s}} = 1(10^6)$$

From Table 9–1, for this Reynolds number, $n = 9$.

The shear-stress distribution is linear, as shown in Fig. 9–25b. The maximum shear stress is caused only by viscous effects since it occurs *at the wall* within the viscous sublayer. The magnitude of this stress is determined using Eq. 9–16. With $\dfrac{dh}{dx} = 0$, we have

$$\tau_0 = \frac{r}{2}\frac{dp}{dx} = \left| \frac{(0.05 \text{ m})}{2}\frac{(-800 \text{ N/m}^2)}{1 \text{ m}} \right| = 20 \text{ Pa}$$

By proportion, the *total shear stress* at $r = 10$ mm is then

$$\frac{\tau}{10 \text{ mm}} = \frac{20 \text{ N/m}^2}{50 \text{ mm}}; \qquad \tau = 4 \text{ Pa}$$

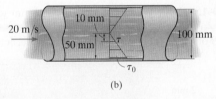

(b)

Fig. 9–25

The *viscous shear-stress component* can be determined at $r = 10$ mm, using Newton's law of viscosity and the power law, Eq. 9–34, since it defines the shape of the velocity profile. First, though, we must determine the maximum velocity $u_{\max}$, Fig. 9–26a. We can do this by applying Eq. 9–36.

$$V = u_{\max}\frac{2n^2}{(n+1)(2n+1)}; \qquad 20 \text{ m/s} = u_{\max}\left[\frac{2(9^2)}{(9+1)[2(9)+1]}\right]$$

$$u_{\max} = 23.46 \text{ m/s}$$

Now, using Eq. 9–34 for u, and $\mu_k = \rho\nu_k$, Newton's law of viscosity becomes

$$\tau_{\text{visc}} = \mu_k\frac{d\bar{u}}{dr} = \mu_k\frac{d}{dr}\left[u_{\max}\left(1 - \frac{r}{R}\right)^{1/n}\right] = \frac{\mu_k u_{\max}}{nR}\left(1 - \frac{r}{R}\right)^{(1-n)/n}$$

$$= \frac{(820 \text{ kg/m}^3)[2(10^{-6}) \text{ m}^2/\text{s}](23.46 \text{ m/s})}{9(0.05 \text{ m})}\left(1 - \frac{0.01 \text{ m}}{0.05 \text{ m}}\right)^{(1-9)/9}$$

$$= 0.1042 \text{ Pa}$$

This is a very small contribution. Instead, the turbulent shear-stress component provides the majority of the shear stress at $r = 10$ mm. It is

$$\tau = \tau_{\text{visc}} + \tau_{\text{turb}}; \qquad 4 \text{ N/m}^2 = 0.1042 \text{ Pa} + \tau_{\text{turb}}$$

$$\tau_{\text{turb}} = 3.90 \text{ Pa} \qquad\qquad Ans.$$

References

1. S. Yarusevych et al., "On vortex shedding from an airfoil in low-Reynolds-number flows," *J Fluid Mechanics*, Vol. 632, 2009, pp. 245–271.

2. H. Langhaar, "Steady flow in the transition length of a straight tube," *J Applied Mechanics*, Vol. 9, 1942, pp. 55–58.

3. J. T. Davies. *Turbulent Phenomena*, Academic Press, New York, NY, 1972.

4. J. Hinze, *Turbulence*, 2nd ed., McGraw-Hill, New York, NY, 1975.

5. F. White, *Fluid Mechanics*, 7th ed., McGraw-Hill, New York, NY, 2008.

6. J. Schetz et. al., *Boundary Layer Analysis*, 2nd ed, American Institute of Aeronautics and Astronautics, 2011.

7. T. Leger and S. L., Celcio, "Examination of the flow near the leading edge of attached cavitation," *J Fluid Mechanics*, Cambridge University Press, UK, Vol. 373, 1998, pp. 61–90.

8. D. Peterson and J. Bronzino, *Biomechanics: Principles and Applications*, CRC Press, Boca Raton, FL, 2008.

9. K. Chandran et al., *Biofluid Mechanics: The Human Circulation*, CRC Press, Boca Raton, FL, 2007.

10. H. Wada, *Biomechanics at Micro and Nanoscale Levels*, Vol. 11, World Scientific Publishing, Singapore, 2006.

11. L. Waite and J. Fine, *Applied Biofluid Mechanics*, McGraw-Hill, New York, NY, 2007.

12. A. Draad and F. Nieuwstadt, "The Earth's rotation and laminar pipe flow," *J Fluid Mechanics*, Vol. 361, 1988, pp. 297–308.

PROBLEMS

Sec. 9.1–9.2

9–1. Crude oil flows through the 2-mm gap between the two fixed parallel plates due to a drop in pressure from A to B of 4 kPa. If the plates are 800 mm wide, determine the flow.

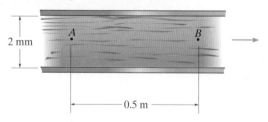

Prob. 9–1

9–2. Glue is applied to the surface of the plastic strip, which has a width of 200 mm, by pulling the strip through the container. Determine the force **F** that must be applied to the tape if the tape moves at 10 mm/s. Take $\rho_g = 730 \text{ kg/m}^3$ and $\mu_g = 0.860 \text{ N} \cdot \text{s/m}^2$.

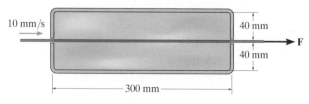

Prob. 9–2

9–3. Crude oil flows through the gap between the two fixed parallel plates due to a drop in pressure from A to B of 4 kPa. Determine the maximum velocity of the oil and the shear stress on each plate.

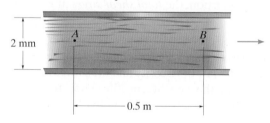

Prob. 9–3

***9–4.** The 20-kg uniform plate is released and slides down the inclined plane. If an oil film under its surface is 0.2 mm thick, determine the terminal velocity of the plate along the plane. The plate has a width of 0.5 m. Take $\rho_o = 880 \text{ kg/m}^3$ and $\mu_o = 0.0670 \text{ N} \cdot \text{s/m}^2$.

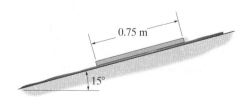

Prob. 9–4

9–5. The two sections of the building wall have a 10-mm-wide crack between them. If the difference in pressure between the inside and outside of the building is 1.5 Pa, determine the flow of air out of the building through the crack. The air temperature is 30°C.

9–7. Using pins, the plug is attached to the cylinder such that there is a gap between the plug and the walls of 0.2 mm. If the pressure within the oil contained in the cylinder is 4 kPa, determine the initial volumetric flow of oil up the sides of the plug. Assume the flow is similar to that between parallel plates since the gap size is very much smaller than the radius of the plug. Take $\rho_o = 880 \text{ kg/m}^3$ and $\mu_o = 30.5(10^{-3}) \text{ N} \cdot \text{s/m}^2$.

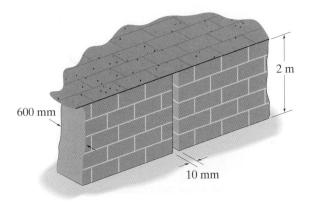

Prob. 9–5

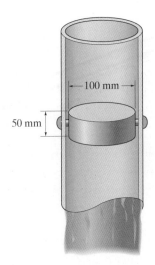

Prob. 9–7

9–6. The boy has a mass of 50 kg and attempts to slide down the inclined plane. If a 0.3-mm-thick oil surface develops between his shoes and the surface, determine his terminal velocity down the incline. Both of his shoes have a total contact area of 0.0165 m². Take $\rho_o = 900 \text{ kg/m}^3$ and $\mu_o = 0.0638 \text{ N} \cdot \text{s/m}^2$.

*9–8. The water tank has a rectangular crack on its side having a width of 100 mm and an average opening of 0.1 mm. If laminar flow occurs through the crack, determine the volumetric flow of water through the crack. The water is at a temperature of $T = 20°C$.

Prob. 9–6

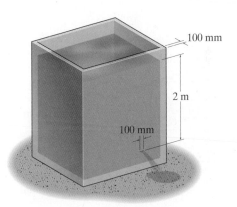

Prob. 9–8

9–9. A solar water heater consists of two flat plates that rest on the roof. Water enters at A and exits at B. If the pressure drop from A to B is 60 Pa, determine the largest gap a between the plates so that the flow remains laminar. For the calculation, assume the water has an average temperature of 40°C.

9–11. The liquid has laminar flow between the two fixed plates due to a pressure gradient dp/dx. Using the coordinate system shown, determine the shear-stress distribution within the liquid and the velocity profile for the liquid. The viscosity is μ.

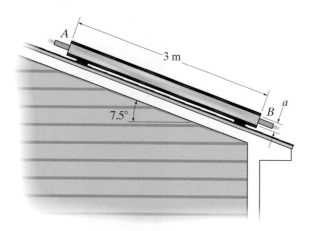

Prob. 9–9

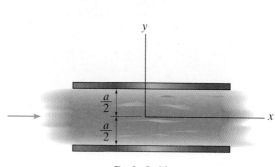

Prob. 9–11

9–10. The 100-mm-diameter shaft is supported by an oil-lubricated bearing. If the gap within the bearing is 2 mm, determine the torque **T** that must be applied to the shaft, so that it rotates at a constant rate of 180 rev/min. Assume no oil leaks out due to sealing, and the flow behavior is similar to that which occurs between parallel plates, since the gap size is very much smaller than the radius of the shaft. Take $\rho_o = 840 \text{ kg/m}^3$ and $\mu_o = 0.22 \text{ N} \cdot \text{s/m}^2$.

***9–12.** The belt is moving at a constant rate of 3 mm/s. The 2-kg plate between the belt and surface is resting on a 0.5-mm-thick film of oil, whereas oil between the top of the plate and the belt is 0.8 mm thick. Determine the plate's terminal velocity as it slides along the surface. Assume the velocity profile is linear. Take $\rho_o = 900 \text{ kg/m}^3$ and $\mu_o = 0.0675 \text{ N} \cdot \text{s/m}^2$.

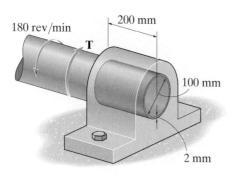

Prob. 9–10

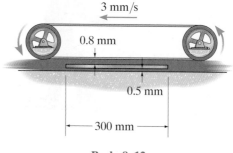

Prob. 9–12

9–13. A thin layer of engine oil is trapped between the belts, which are moving in different directions and at different speeds as shown. Plot the velocity profile within the oil film and the shear-stress distribution. The pressures at A and B are atmospheric. Take $\mu_o = 0.22 \text{ N} \cdot \text{s/m}^2$ and $\rho_o = 876 \text{ kg/m}^3$.

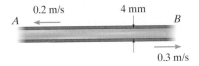

0.2 m/s   4 mm

A B

0.3 m/s

Prob. 9–13

9–14. The water and oil films have the same thickness a and are subjected to the movement of the top plate. Plot the velocity profile and the shear-stress distribution for each fluid. There is no pressure gradient between A and B. The viscosities of water and oil are μ_w and μ_o, respectively.

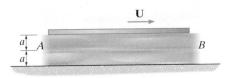

U

a A

B

a

Prob. 9–14

9–15. Use the Navier–Stokes equations to show that the velocity distribution of the steady laminar flow of a fluid flowing down the inclined surface is defined by $u = [\rho g \sin \theta / (2\mu)](2hy - y^2)$, where ρ is the fluid density and μ is its viscosity.

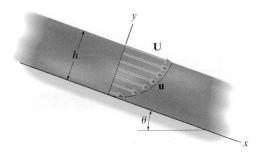

y

h

U

u

θ

x

Prob. 9–15

*9–16. When you inhale, air flows through the turbinate bones of your nasal passages as shown. Assume that for a short length of 15 mm, the flow is passing through parallel plates, the plates having a mean total width of $w = 20$ mm and spacing of $a = 1$ mm. If the lungs produce a pressure drop of $\Delta p = 50$ Pa, and the air has a temperature of 20°C, determine the power needed to inhale air.

1 mm

Prob. 9–16

9–17. A fluid has laminar flow between the two parallel plates, each plate moving in the same direction, but with different velocities, as shown. Use the Navier–Stokes equations, and establish an expression that gives the shear-stress distribution and the velocity profile for the fluid. Plot these results. There is no pressure gradient between A and B.

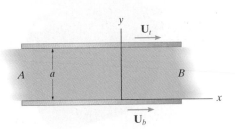

y U_t

A a B

x

U_b

Prob. 9–17

9

Sec. 9.3–9.6

9–18. The retinal arterioles supply the retina of the eye with blood flow. The inner diameter of an arteriole is 0.08 mm, and the mean velocity of flow is 28 mm/s. Determine if this flow is laminar or turbulent. Blood has a density of 1060 kg/m^3 and an apparent viscosity of 0.0036 N·s/m^2.

9–19. A 50-mm-diameter vertical pipe carries oil having a density of $\rho_o = 890$ kg/m^3. If the pressure drop in a 2-m length of the pipe is 500 Pa, determine the shear stress acting along the wall of the pipe. The flow is downward.

*9–20. Lymph is a fluid that is filtered from blood and forms an important part of the immune system. Assuming it is a Newtonian fluid, determine its average velocity if it flows from an artery into a 0.08-μm-diameter precapillary sphincter at a pressure of 120 mm of mercury, then passes vertically upwards through the leg for a length of 1200 mm, and emerges at a pressure of 25 mm of mercury. Take $\rho_l = 1030$ kg/m^3 and $\mu_l = 0.0016$ N·s/m^2.

9–21. Most blood flow in humans is laminar, and apart from pathological conditions, turbulence can occur in the descending portion of the aorta at high flow rates as when exercising. If blood has a density of 1060 kg/m^3, and the diameter of the aorta is 25 mm, determine the largest average velocity blood can have before the flow becomes transitional. Assume that blood is a Newtonian fluid and has a viscosity of $\mu_b = 0.0035$ N·s/m^2. At this velocity, determine if turbulence occurs in an arteriole of the eye, where the diameter is 0.008 mm.

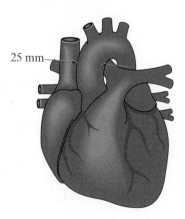

25 mm

Prob. 9–21

9–22. Oil having a density of 900 kg/m^3 and a viscosity of 0.370 N·s/m^2 has a flow of 0.05 m^3/s through the 150-mm-diameter pipe. Determine the drop in pressure caused by viscous friction over the 8-m-long section.

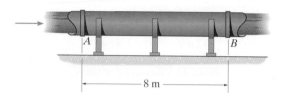

A   B

8 m

Prob. 9–22

9–23. Oil has a flow of 0.004 m^3/s through the 150-mm-diameter pipe. Determine the drop in pressure caused by viscous friction over the 8-m-long section. Take $\rho_o = 900$ kg/m^3 and $\mu_o = 0.370$ N·s/m^2.

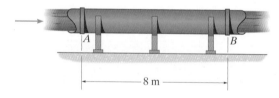

A   B

8 m

Prob. 9–23

*9–24. The 50-mm-diameter smooth pipe drains engine oil out of a large tank at the rate of 0.01 m^3/s. Determine the horizontal force the tank must exert on the pipe to hold it in place. Assume fully developed flow occurs along the pipe. Take $\rho_o = 876$ kg/m^3 and $\mu_o = 0.22$ N·s/m^2.

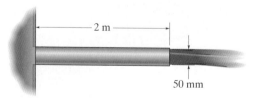

2 m

50 mm

Prob. 9–24

9–25. The 100-mm-diameter horizontal pipe transports castor oil in a processing plant. If the pressure drops 100 kPa in a 10-m length of the pipe, determine the maximum velocity of the oil in the pipe and the maximum shear stress in the oil. Take $\rho_o = 960 \text{ kg/m}^3$ and $\mu_o = 0.985 \text{ N} \cdot \text{s/m}^2$.

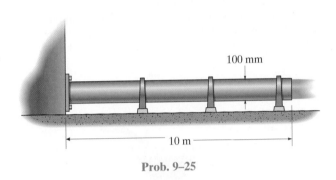

100 mm

10 m

Prob. 9–25

9–26. A smooth 100-mm-diameter pipe transfers kerosene at 20°C with an average velocity of 0.05 m/s. Determine the pressure drop that occurs along the 20-m length. Also, what is the shear stress along the pipe wall?

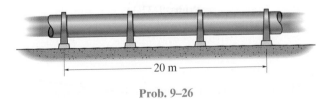

20 m

Prob. 9–26

9–27. Crude oil is flowing vertically upward through a 50-mm-diameter pipe. If the difference in pressure between two points 3 m apart along the pipe is 26.4 kPa, determine the volumetric flow. Take $\rho_o = 880 \text{ kg/m}^3$ and $\mu_o = 30.2(10^{-3}) \text{ N} \cdot \text{s/m}^2$.

9–28. Oil and kerosene are brought together through the wye as shown. Determine if they will mix, that is, create turbulent flow, as they travel along the 60-mm-diameter pipe. Take $\rho_o = 880 \text{ kg/m}^3$, and $\rho_k = 810 \text{ kg/m}^3$. The mixture has a viscosity of $\mu_m = 0.024 \text{ N} \cdot \text{m/s}^2$.

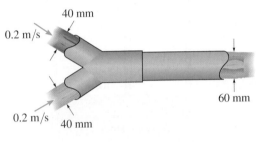

40 mm

0.2 m/s

60 mm

0.2 m/s   40 mm

Prob. 9–28

9–29. Castor oil is subjected to a pressure of 550 kPa at A and to a pressure of 200 kPa at B. If the pipe has a diameter of 30 mm, determine the shear stress acting on the pipe wall and the maximum velocity of the oil. Also, what is the flow Q? Take $\rho_{co} = 960 \text{ kg/m}^3$ and $\mu_{co} = 0.985 \text{ N} \cdot \text{s/m}^2$.

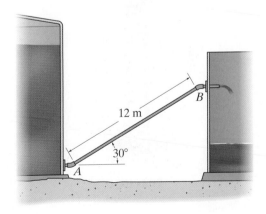

12 m

B

30°

A

Prob. 9–29

9

9–30. Glycerin is at a pressure of 15 kPa at A when it enters the vertical segment of the 100-mm-diameter pipe. Determine the discharge at B.

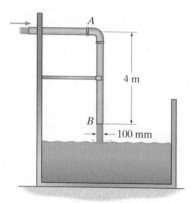

Prob. 9–30

9–31. Crude oil at 20°C is ejected through the 50-mm-diameter smooth pipe. If the pressure drop from A to B is 36.5 kPa, determine the maximum velocity within the flow, and plot the shear-stress distribution within the oil.

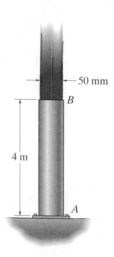

Prob. 9–31

*9–32. Castor oil is poured into the funnel so that the level of 200 mm is maintained. It flows through the stem at a steady rate and accumulates in the cylindrical container. Determine the time needed for the level to reach $h = 50$ mm. Take $\rho_o = 960$ kg/m^3 and $\mu_o = 0.985$ N·m/s^2.

9–33. Castor oil is poured into the funnel so that the level of 200 mm is maintained. It flows through the stem at a steady rate and accumulates in the cylindrical container. If it takes 5 seconds to fill the container to a depth of $h = 80$ mm, determine the viscosity of the oil. Take $\rho_o = 960$ kg/m^3.

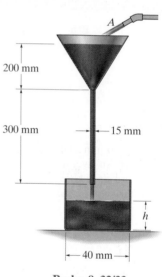

Probs. 9–32/33

9–34. The resistance to breathing can be measured using a spirometer that measures the time to expire a full volume of air from the lungs. About 20% of the resistance occurs in the medium-size bronchi, where laminar flow exists. The resistance to flow R can be thought of as the driving pressure gradient dp/dx divided by the volumetric flow Q. Determine its value as a function of the diameter D of the bronchi, and plot its values for 2 mm $\leq D \leq$ 8 mm. Take $\mu_a = 18.9(10^{-6})$ N·s/m^2.

9–35. It is thought that the shear stress on the endothelial cells that line the walls of an artery may be important for the development of various vascular disorders. If we assume the velocity profile of the blood flow in an arteriole (or very small artery) to be parabolic, and the vessel diameter is 80 μm, determine the wall shear stress as a function of the average velocity. Plot the results for 20 mm/s $\leq V \leq$ 50 mm/s. Assume here that blood is a Newtonian fluid and take $\mu_b = 0.0035$ N·s/m².

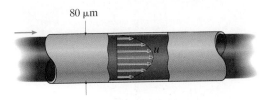

80 μm

Prob. 9–35

9–37. The Reynolds number Re $= \rho V D_h / \mu$ for an annulus is determined using a *hydraulic diameter*, which is defined as $D_h = 4\,A/P$, where A is the open cross-sectional area within the annulus and P is the wetted perimeter. Determine the Reynolds number for water at 30°C if the flow is 0.01 m³/s. Is this flow laminar? Take $r_i = 40$ mm and $r_o = 60$ mm.

9–38. A Newtonian fluid has laminar flow as it passes through the annulus. Use the Navier–Stokes equations to show that the velocity profile for the flow is

$$v_z = \frac{1}{4\mu}\frac{dp}{dz}\left[r^2 - r_0^2 - \left(\frac{r_o^2 - r_i^2}{\ln(r_o/r_i)}\right)\ln\frac{r}{r_o}\right].$$

r_o
r_i

Probs. 9–37/38

*9–36. The cylindrical tank is to be filled with glycerin using the 50-mm-diameter pipe. If the flow is to be laminar, determine the shortest time needed to fill the tank to a depth of 2.5 m. Air escapes through the top of the tank.

9–39. Oil having a density of $\rho_o = 880$ kg/m³ and a viscosity of $\mu_o = 0.0680$ N·s/m² flows through the 20-mm-diameter pipe at 0.001 m³/s. Determine the reading h of the mercury manometer. Take $\rho_{Hg} = 13\,550$ kg/m³.

*9–40. Oil having a density of $\rho_o = 880$ kg/m³ and a viscosity of $\mu_o = 0.0680$ N·s/m² flows through the 20-mm-diameter pipe. If the mercury manometer reads $h = 40$ mm, determine the volumetric flow. Take $\rho_{Hg} = 13\,550$ kg/m³.

9

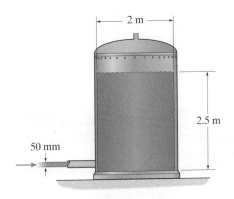

2 m

2.5 m

50 mm

Prob. 9–36

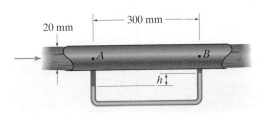

300 mm

20 mm

A B

h

Probs. 9–39/40

9–41. The large flask is filled with a liquid having a density ρ and a viscosity μ. When the valve at A is opened, the liquid begins to flow through the horizontal tube, where $d \ll D$. If $h = h_1$ at $t = 0$, determine the time when $h = h_2$. Assume that laminar flow occurs within the tube.

9–43. A Newtonian fluid has laminar flow as it passes through the annulus. Use the result of Prob. 9–38, and show that the shear-stress distribution for the flow is

$$\tau_{rz} = \frac{1}{4}\frac{dp}{dz}\left(2r - \frac{r_o^2 - r_i^2}{r \ln(r_o/r_i)}\right).$$

Prob. 9–43

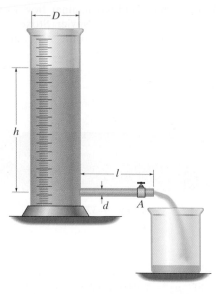

Prob. 9–41

***9–44.** As oil flows from A to B, the pressure drop is 40 kPa through the annular channel. Determine the shear stress it exerts on the walls of the channel, and the maximum velocity of the flow. Use the results of Probs. 9–38 and 9–43. Take $\mu_o = 0.220 \text{ N} \cdot \text{s/m}^2$.

9–42. Use the result of Prob. 9–46 to show that if Poiseuille's equation was used to calculate the apparent viscosity of blood, it could be written as

$$\mu_{app} = \frac{\mu_p}{\left[1 - \left(1 - \dfrac{\delta}{R}\right)^4\left(1 - \dfrac{\mu_p}{\mu_c}\right)\right]}.$$

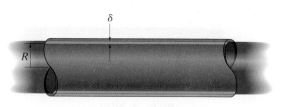

Prob. 9–42

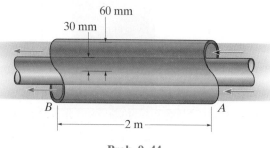

Prob. 9–44

9–45. As oil flows from A to B, the pressure drop is 40 kPa through the annular channel. Determine the volumetric flow. Use the result of Prob. 9–38. Take $\mu_o = 0.220 \text{ N} \cdot \text{s/m}^2$.

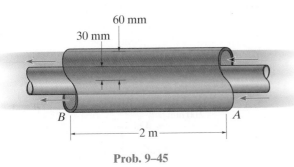

60 mm

30 mm

B A

—2 m—

Prob. 9–45

9–46. When blood flows through a large artery, it tends to separate into a core consisting of red blood cells and an outer annulus called the plasma skimming layer, which is cell free. This phenomenon can be described using the "cell-free marginal layer model," where the artery is considered a circular tube of inner radius R, and the cell-free region has a thickness δ. The equations that govern these regions are $-\dfrac{\Delta p}{L} = \dfrac{1}{r}\dfrac{d}{dr}\left[\mu_c\, r\dfrac{du_c}{dr}\right]$, $0 \le r \le R - \delta$, and $-\dfrac{\Delta p}{L} = \dfrac{1}{r}\dfrac{d}{dr}\left[\mu_p r\dfrac{du_p}{dr}\right]$, $R - \delta \le r \le R$, where μ_c and μ_p are the viscosities (assumed to be Newtonian fluids), and u_c and u_p are the velocities for each region. Integrate these equations and show that the flow is $Q = \dfrac{\pi \Delta p R^4}{8\mu_p L}\left[1 - \left(1 - \dfrac{\delta}{R}\right)^4\left(1 - \dfrac{\mu_p}{\mu_c}\right)\right]$.

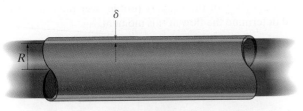

δ

R

Prob. 9–46

9–47. Glycerin flows from a large tank through the smooth 100-mm-diameter pipe. Determine its maximum volumetric flow if the flow is to remain laminar. How far L from the pipe entrance will fully developed laminar flow begin to occur?

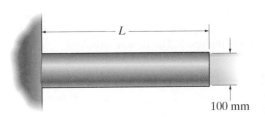

L

100 mm

Prob. 9–47

*9–48. Water flows from a beaker into a 4-mm-diameter tube with an average velocity of 0.45 m/s. Classify the flow as laminar or turbulent if the water temperature is 10°C and if it is 30°C. If the flow is laminar, then find the length of pipe for fully developed flow.

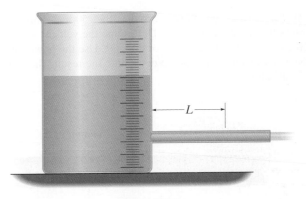

L

Prob. 9–48

9

Sec. 9.7–9.8

9–49. A 70-mm-diameter horizontal pipe has a smooth interior surface and transports crude oil at 20°C. If the pressure decrease over a 5-m-long segment is 180 kPa, determine the thickness of the viscous sublayer and the velocity along the centerline of the pipe. The flow is turbulent.

9–50. Crude oil flows through the 50-mm-diameter smooth pipe at 0.0054 m³/s. If the pressure at A is 16 kPa and at B is 9 kPa, determine the viscous and turbulent shear-stress components within the oil 10 mm from the wall of the pipe. Use the power-law velocity profile, Eq. 9–34, to determine the result. The flow is turbulent.

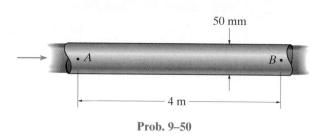

50 mm

Prob. 9–50

9–51. Crude oil flows through the 50-mm-diameter smooth pipe. If the pressure at A is 16 kPa and at B is 9 kPa, determine the thickness of the viscous sublayer, and find the maximum shear stress and the maximum velocity of the oil in the pipe. Use Eq. 9–33 to determine the result. The flow is turbulent.

*9–52. Crude oil flows through the 50-mm-diameter smooth pipe at 0.0054 m³/s. Determine the velocity within the oil 10 mm from the wall of the pipe. Use the power-law velocity profile, Eq. 9–34, to determine the result. The flow is turbulent.

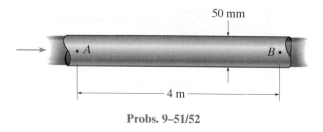

50 mm

4 m

Probs. 9–51/52

9–53. The 100-mm-diameter smooth pipe transports benzene with an average velocity of 7.5 m/s. If the pressure drop from A to B is 400 Pa, determine the viscous and turbulent shear-stress components within the benzene at $r = 25$ mm and $r = 50$ mm from the centerline of the pipe. Use a power-law velocity profile, Eq. 9–34. Take $\rho_{bz} = 880$ kg/m³ and $\nu_{bz} = 0.75(10^{-6})$ m²/s.

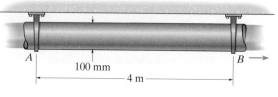

100 mm

4 m

Prob. 9–53

9–54. Crude oil flows through the 50-mm-diameter smooth pipe. If the pressure at A is 16 kPa and at B is 9 kPa, determine the shear stress and the velocity within the oil 10 mm from the wall of the pipe. Use Eq. 9–33 to determine the result. The flow is turbulent.

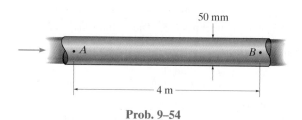

50 mm

4 m

Prob. 9–54

9–55. Experimental testing of artificial grafts placed on the inner wall of the carotid artery indicates that blood flow through the artery at a given moment has a velocity profile that can be approximated by $u = 8.36(1 - r/3.4)^{1/n}$ mm/s, where r is in millimeters and $n = 2.3 \log_{10} \text{Re} - 4.6$. If $\text{Re} = 2(10^9)$, plot the velocity profile over the artery wall, and determine the flow at this moment.

3.4 mm

r

Prob. 9–55

CHAPTER REVIEW

If steady flow occurs between two parallel plates, or within a pipe, then regardless if the flow is laminar or turbulent, the shear stress within the fluid varies in a linear manner such that it will balance the forces of pressure, gravity, and viscosity.

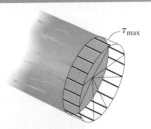

Shear-stress distribution
for both laminar
and turbulent flow.

In this text, laminar flow between two parallel plates requires Re $\leq$ 1400, and for pipes, Re $\leq$ 2300. If this occurs, then for Newtonian fluids, Newton's law of viscosity can be used to determine the velocity profile and the pressure drop along these conduits. When using the relevant equations, for either plates or pipes, be sure to follow the sign convention as it relates to the established coordinates.

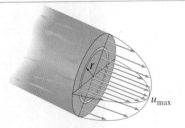

Velocity distribution
for laminar flow

When fluid flows from a large reservoir into a pipe, it will accelerate a certain distance before it becomes fully developed steady laminar or turbulent flow.

Turbulent flow within a pipe causes additional frictional losses due to the erratic mixing of the fluid. This mixing tends to even out the mean velocity profile, making it more uniform; however, along the walls of the pipe there will always be a narrow viscous sublayer having laminar flow.

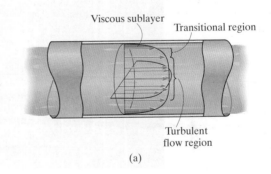

The velocity profile for turbulent flow cannot be studied analytically because it is so unpredictable. Instead, we must use the results of experiments to develop empirical equations that describe this profile.

$$\frac{\bar{u}}{u^*} = 2.5 \ln\left(\frac{u^* y}{v}\right) + 5.0$$

$$\frac{\bar{u}}{u_{max}} = \left(1 - \frac{r}{R}\right)^{1/n}$$

Chapter 10

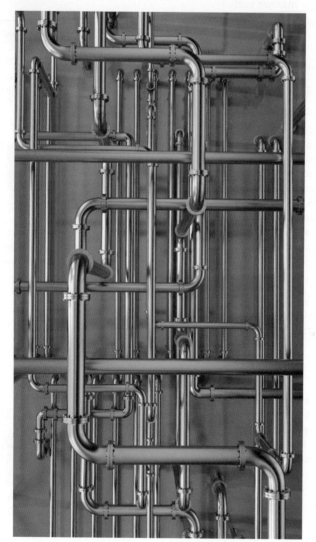

(© whitehoune/Fotolia)

In order to design a system of pipes it is necessary to know the frictional losses within the pipe along with any losses that occur at the connections and fittings.

Analysis and Design for Pipe Flow

CHAPTER OBJECTIVES

■ To discuss frictional losses due to surface roughness within a pipe, and describe how to use experimental data to determine these losses.

■ To show how to analyze and design pipe systems having various fittings and connections.

■ To explain some of the methods engineers use to measure flow through a pipe.

10.1 Resistance to Flow in Rough Pipes

We will now extend our discussion of the previous chapter, and discuss how frictional resistance along the *rough walls* of a pipe contributes to the pressure drop within the pipe. This is important when designing any pipe system, or selecting a pump that is required to maintain a specific flow. Here we will focus on straight pipes that have a circular cross section, since this shape provides the greatest structural strength for resisting pressure, and furthermore a circular cross section will transport the largest amount of fluid with the smallest frictional resistance.

In engineering practice, any frictional or energy loss due to *both* fluid friction and wall roughness is often referred to as a ***major head loss***, h_L, or simply a ***major loss***. We can determine the major loss in a pipe by measuring the pressure at two locations a distance L apart, Fig. 10–1, and then apply the energy equation between these two points. Since no shaft work is done, the pipe is horizontal, $z_{in} = z_{out} = 0$, and $V_{in} = V_{out} = V$, for steady incompressible flow, we have

$$\frac{p_{in}}{\gamma} + \frac{V_{in}^2}{2g} + z_{in} + h_{pump} = \frac{p_{out}}{\gamma} + \frac{V_{out}^2}{2g} + z_{out} + h_{turb} + h_L$$

$$\frac{p_{in}}{\gamma} + \frac{V^2}{2g} + 0 + 0 = \frac{p_{out}}{\gamma} + \frac{V^2}{2g} + 0 + 0 + h_L$$

$$h_L = \frac{p_{in} - p_{out}}{\gamma} = \frac{\Delta p}{\gamma} \qquad (10\text{–}1)$$

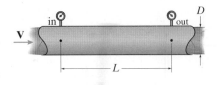

Fig. 10–1

479

Thus the head loss in the pipe, $h_L = \Delta p/L$, results in a pressure drop over the pipe's length L, because the pressure must do work to overcome the frictional resistance that creates this loss. For this reason $p_{in} > p_{out}$. Of course, if the fluid is an *ideal fluid* then $h_L = 0$, since frictional resistance would not occur.

Laminar Flow.   For laminar flow, the major head loss occurs within the fluid. It is due to the frictional resistance or shear stress developed *between layers of fluid* when they slide over one another with different relative velocities. For a Newtonian fluid, this shear stress is related to the velocity gradient by Newton's law of viscosity, $\tau = \mu \, (du/dy)$. In Sec. 9.3, we were able to use this expression to relate the average velocity of flow in the pipe to the pressure gradient $\Delta p/L$. The result is Eq. 9–25, $V = (D^2/32\mu)(\Delta p/L)$. With it, and Eq. 10–1, we can now write the head loss in terms of the average velocity as

$$h_L = \frac{32\mu VL}{D^2\gamma} \tag{10–2}$$

Laminar flow

Notice that the loss *increases* as the internal diameter of the pipe *decreases*, since the losses vary inversely with the *square* of D. This loss is entirely due to the viscosity of the fluid and is produced throughout the flow. Any *mild* surface roughness on the pipe wall will generally not affect laminar flow to any appreciable degree, and so it will have a negligible effect on the loss.

For convenience later on, we will express Eq. 10–2 in terms of the Reynolds number, $\text{Re} = \rho VD/\mu$, and rearrange it in the form

$$h_L = f\frac{L}{D}\frac{V^2}{2g} \tag{10–3}$$

where

$$\boxed{f = \frac{64}{\text{Re}}} \tag{10–4}$$

Laminar flow

This term f is called the **friction factor**. For laminar flow, it is seen to be a function only of the Reynolds number, and again, it does not depend upon whether the inner surface of the pipe's wall is smooth or rough. Here the frictional loss is produced *only* by the viscosity of the fluid.

10

Turbulent Flow.   Since there is no analytical means for determining the head loss in a pipe due to turbulent flow, it then becomes necessary to measure the pressure drop either with two pressure gages, as in Fig. 10–2a, or by using a manometer, Fig. 10–2b. Such experiments have shown that this pressure drop depends upon the pipe diameter D, the pipe length L, the fluid's density ρ, viscosity μ, average velocity V, and the roughness or average height ε of the protuberances from the pipe's inner surface. In Chapter 8, Example 8–4, we showed, using dimensional analysis, that the relationship between these variables and the pressure drop can be written in terms of three dimensionless ratios, namely

$$\Delta p = \rho V^2 g_1 \left(\text{Re}, \frac{L}{D}, \frac{\varepsilon}{D} \right)$$

where g_1 defines this unknown function. Further experiments have shown that the pressure drop is *directly proportional* to the length of the pipe—the longer the pipe, the greater the pressure drop, but *inversely proportional* to the pipe diameter—the smaller the diameter, the greater the pressure drop. As a result, the above relationship then becomes

$$\Delta p = \rho V^2 \frac{L}{D} g_2 \left(\text{Re}, \frac{\varepsilon}{D} \right)$$

Finally, using this result and applying Eq. 10–1 to determine the head loss in the pipe, realizing that $\gamma = \rho g$, we have

$$h_L = \frac{L}{D} \frac{V^2}{2g} g_3 \left(\text{Re}, \frac{\varepsilon}{D} \right)$$

For convenience we have incorporated the factor 2, in order to express h_L in terms of the velocity head $V^2/2g$. In other words, our unknown function is now $g_3(\text{Re}, \varepsilon/D) = 2g_2$. The fact that the head loss is also *directly proportional* to the velocity head is something that has also been confirmed by experiment.

If we compare the above equation with Eq. 10–3, letting the friction factor represent

$$f = g_3 \left(\text{Re}, \frac{\varepsilon}{D} \right)$$

then we can express the head loss for turbulent flow in the *same form* as we did for laminar flow, that is,

$$\boxed{h_L = f \frac{L}{D} \frac{V^2}{2g}} \qquad (10\text{–}5)$$

This important result is called the **Darcy–Weisbach equation**, named after Henry Darcy and Julius Weisbach, who first proposed its use in the late 19th century. It was derived by dimensional analysis, and it applies to fluids having either laminar or turbulent flow. In the case of laminar flow, the friction factor is determined from Eq. 10–4; however, for turbulent flow we must determine the friction-factor relationship $f = g_3(\text{Re}, \varepsilon/D)$ through experiment.

The first attempts at doing this were made by Johann Nikuradse, and then by others, using pipes artificially roughened by uniform sand grains

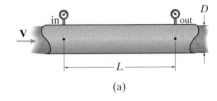

(a)

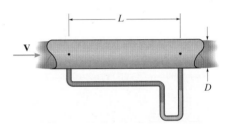

Manometer

(b)

Fig. 10–2

10

of a specific size so that ε is well defined. Unfortunately, for practical applications, commercially available pipes do not have a uniform well-defined roughness. However, using a similar approach, Lewis Moody and Cyril Colebrook were able to extend the work of Nikuradse by performing experiments using commercially available pipes.

Moody Diagram.

Moody presented his data for $f = g_3(\text{Re}, \varepsilon/D)$ in the form of a graph plotted on a log–log scale. It is often called the **Moody diagram**, and for convenience it is shown on the inside back cover. To use this diagram it is necessary to know the average **surface roughness** ε of the pipe's inner wall, Fig. 10–3c. The table above the Moody diagram gives some typical values, provided the pipe is in fairly good condition. However, realize that through use, pipes can become corroded, or scale can build up on their walls. This can significantly alter the value of ε or, in extreme cases, lower the value of D. It is for this reason that engineers must exercise *conservative judgment* for proper selection of ε.

Once ε is known, then the **relative roughness** ε/D and the Reynolds number can be calculated and the friction factor f determined from the Moody diagram. Notice on this diagram that the flow through the pipe is divided into different regions, depending upon the Reynolds number.

Laminar Flow. Experimental evidence indicates that if *laminar flow* is maintained, the friction factor will be *independent* of the roughness of the pipe and, instead, will vary inversely with the Reynolds number in accordance with Eq. 10–4, $f = 64/\text{Re}$. This is to be expected, since here the Reynolds number is low, and the resistance to flow is caused *only* by the laminar shear stress within the fluid, Fig. 10–3a.

Critical Zone and Transitional Flow. If the flow in the pipe is increased just above the Reynolds number of $\text{Re} = 2300$, then the f values are uncertain (critical zone) because the flow becomes unstable. Here and for transitional flow, the flow can switch between laminar and turbulent, or be a combination of both. When this is the case, it is important to be conservative and select a somewhat high value of f. Here, turbulence will begin to occur within regions of the pipe. Along the wall, however, the slower moving fluid will still maintain laminar flow. This *laminar sublayer* will become thinner as the velocity increases, and eventually some of the rough elements on the pipe wall will pass through this sublayer, Fig. 10–3b. Hence the effect of *surface roughness* will begin to become important, and so now the friction factor becomes a function of *both* the Reynolds number and the relative roughness, $f = g_3(\text{Re}, \varepsilon/D)$.

Turbulent Flow. At very large Reynolds numbers, most of the rough elements will penetrate through the laminar sublayer, and so the friction factor then depends primarily on the size ε of these elements, Fig. 10–3c. Here the curves of the Moody diagram tend to *flatten out* and become horizontal. In other words, the values for f become less dependent on the Reynolds number. The turbulent shear stress *near the wall* strongly influences the friction factor, rather than the shear stress within the fluid.

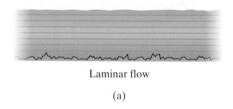

Laminar flow

(a)

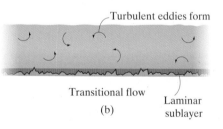

Turbulent eddies form

Transitional flow

(b)

Laminar sublayer

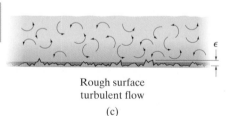

ϵ

Rough surface turbulent flow

(c)

Fig. 10–3

The Moody diagram also shows that very *smooth pipes* (low value of ε/D) have a rapidly decreasing friction factor with increasing Reynolds number, as opposed to those having rough walls. Also, since the surface roughness ε of a particular material is practically the same for all diameters of a pipe made of this material, then the smaller-diameter pipes (large value of ε/D) will have a larger friction factor compared to those that have a larger diameter (small value of ε/D).

Empirical Solutions.   Rather than using the Moody diagram to determine f, we can also obtain this value using an empirical formula. This is particularly helpful when using a computer program or spreadsheet. The Colebrook equation is most often used for this purpose, since it describes the curves of the Moody diagram within the range of complete turbulence [Ref. 2]. It is

$$\frac{1}{\sqrt{f}} = -2 \log \left(\frac{\varepsilon/D}{3.7} + \frac{2.51}{\mathrm{Re}\sqrt{f}} \right) \qquad (10\text{--}6)$$

Unfortunately, this is a transcendental equation that cannot be solved explicitly for f, and therefore it must be solved using an iterative trial-and-error procedure, something that can be done on a pocket calculator or personal computer.

A more direct approximation would be to use the following formulation developed by S. Haaland in 1983, Ref. [5].

$$\frac{1}{\sqrt{f}} = -1.8 \log \left[\left(\frac{\varepsilon/D}{3.7} \right)^{1.11} + \frac{6.9}{\mathrm{Re}} \right] \qquad (10\text{--}7)$$

This equation gives a result that is very close to that obtained using the Colebrook equation.*

Whatever method is used to determine f, remember that realistically, as stated before, the surface roughness of a pipe and its diameter will change with time due to sediment and scale deposits, or corrosion. Thus, calculations based on f have a rather limited reliability. Sufficient allowance should be made for future use, by increasing any value of f using sound judgment.

Noncircular Conduits.   Throughout this discussion we have only considered pipes having a circular cross section; however, the formulations can also be applied to conduits having a noncircular cross section, such as those that are oval or rectangular. In such cases the **hydraulic diameter** for the conduit is normally used as the "characteristic length" when calculating the Reynolds number. This "diameter" is $D_h = 4A/P$, where A is the cross-sectional area of the conduit and P is its perimeter. For example, for a circular pipe, $D_h = \left[4(\pi D^2/4) \right]/(\pi D) = D$. Once D_h is known, then the Reynolds number, relative roughness, and Eq. 10–4, $f = 64/\mathrm{Re}$, and the Moody diagram can be used in the usual manner. The results obtained are generally within a range of accuracy that is acceptable for engineering practice, although they are not very reliable for extremely narrow shapes such as an annulus or elongated opening; Ref. [19].

*Another commonly used formula has been developed by P. K. Swamee and A. K. Jain. See Ref. [10].

Important Points

- Resistance to *laminar flow* in rough pipes is *independent* of the surface roughness of the pipe, since the surface conditions will not severely disrupt the flow. Instead the friction factor is only a function of the Reynolds number, and for this case it can be determined analytically using $f = 64/\text{Re}$.

- Resistance to *turbulent flow* in rough pipes is characterized by a friction factor f that depends upon both the Reynolds number and the relative roughness ε/D of the pipe wall. This relationship, $f = g_3(\text{Re}, \varepsilon/D)$, is expressed graphically by the Moody diagram, or analytically by the empirical Colebrook equation or an alternative form such as Eq. 10–7. As the Moody diagram indicates, for *very high Reynolds numbers*, f depends mostly on the relative roughness of the pipe wall, and not very much on the Reynolds number.

Procedure for Analysis

Many problems involving the head loss within a single pipe require satisfying the conditions of three important equations.

- The head loss in the pipe is related to the variables f, L, D, and V by the Darcy–Weisbach equation,

$$h_L = f\left(\frac{L}{D}\right)\frac{V^2}{2g}$$

- The friction factor f is related to Re and ε/D either by using the Moody diagram, which graphically represents $f = g_3(\text{Re}, \varepsilon/D)$, or analytically by using Eq. 10–6 or 10–7.

- The pressure drop Δp over a length of pipe is related to the head loss h_L using the energy equation.

$$\frac{p_{\text{in}}}{\gamma} + \frac{V_{\text{in}}^2}{2g} + z_{\text{in}} + h_{\text{pump}} = \frac{p_{\text{out}}}{\gamma} + \frac{V_{\text{out}}^2}{2g} + z_{\text{out}} + h_{\text{turb}} + h_L$$

Depending upon the problem, satisfying these three equations may be very direct, as in Examples 10–1 and 10–2, but in cases where f and h_L are *unknown*, the use of the Moody diagram will be required for the solution. Problems of this sort are represented by Examples 10–3 and 10–4.

EXAMPLE | 10.1

The 200-mm-diameter galvanized iron pipe in Fig. 10–4 transports water from a reservoir at a temperature of 20°C. Determine the head loss and pressure drop in 200 m of the pipe if the flow is $Q = 90$ liters/s.

SOLUTION

Fluid Description. We will assume fully developed steady flow, and the water can be considered incompressible. From Appendix A, at $T = 20°C$, $\rho_w = 998.3 \, \text{kg/m}^3$ and $\nu_w = 1.00\left(10^{-6}\right) \text{m}^2/\text{s}$. In order to classify the flow, we must calculate the Reynolds number.

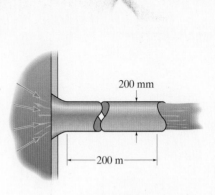

$$V = \frac{Q}{A} = \frac{\left(90 \dfrac{\text{liters}}{\text{s}}\right)\left(\dfrac{1 \, \text{m}^3}{1000 \, \text{liters}}\right)}{\pi(0.1 \, \text{m})^2} = 2.865 \, \text{m/s}$$

Fig. 10–4

$$\text{Re} = \frac{VD}{\nu_w} = \frac{(2.865 \, \text{m/s})(0.2 \, \text{m})}{1.00(10^{-6}) \, \text{m}^2/\text{s}} = 5.73(10^5) > 2300 \; (\text{turbulent})$$

Analysis   The value of ε is taken from the table on top of the Moody diagram for galvanized iron pipe. The relative roughness is then $\varepsilon/D = \dfrac{0.15 \, \text{mm}}{200 \, \text{mm}} = 0.00075$. Using this value and Re, the Moody diagram gives $f = 0.019$. Therefore, from the Darcy–Weisbach equation, the head loss is

$$h_L = f\frac{L}{D}\frac{V^2}{2g} = (0.019)\left(\frac{200 \, \text{m}}{0.2 \, \text{m}}\right)\left[\frac{(2.865 \, \text{m/s})^2}{2(9.81 \, \text{m/s})}\right] = 7.947 \, \text{m} = 7.95 \, \text{m} \quad Ans.$$

This represents the loss of energy along 200 m of the pipe, which results in a pressure drop that can be determined from the energy equation, which in this case becomes Eq. 10–1.

$$\frac{p_{\text{in}}}{\gamma} + \frac{V_{\text{in}}^2}{2g} + z_{\text{in}} + h_{\text{pump}} = \frac{p_{\text{out}}}{\gamma} + \frac{V_{\text{out}}^2}{2g} + z_{\text{out}} + h_{\text{turb}} + h_L$$

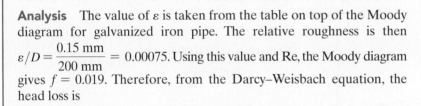

$$\frac{p_{\text{in}}}{\gamma_w} + \frac{V^2}{2g} + 0 + 0 = \frac{p_{\text{out}}}{\gamma_w} + \frac{V^2}{2g} + 0 + 0 + h_L$$

$$h_L = \frac{\Delta p}{\gamma_w}$$

$$7.947 \, \text{m} = \frac{\Delta p}{\left(998.3 \, \text{kg/m}^3\right)\left(9.81 \, \text{m/s}^2\right)}$$

$$\Delta p = 77.83(10^3) \, \text{Pa} = 77.8 \, \text{kPa} \qquad Ans.$$

The "flow work" produced by this pressure drop is needed to overcome the frictional resistance of the fluid within the pipe.

10

EXAMPLE | **10.2**

Heavy fuel oil flows through 3 km of cast iron pipe having a diameter of 250 mm, Fig. 10–5. If the volumetric flow is 40 liter/s, determine the head loss in the pipe. Take $v_o = 0.120(10^{-3})$ m²/s.

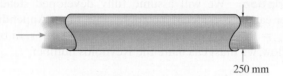

250 mm

Fig. 10–5

SOLUTION

Fluid Description.   We have fully developed steady flow, and we will assume the oil is incompressible. To classify the flow, we must check the Reynolds number.

$$V = \frac{Q}{A} = \frac{(40 \text{ liter/s})\left(1\text{m}^3/1000 \text{ liter}\right)}{\pi(0.125 \text{ m})^2} = 0.8149 \text{ m/s}$$

Then

$$\text{Re} = \frac{VD}{v_o} = \frac{(0.8149 \text{ m/s})(0.250 \text{ m})}{0.120(10^{-3}) \text{ m}^2/\text{s}} = 1698 < 2300 \text{ (laminar)}$$

Analysis.   Rather than use the Moody diagram to obtain f, for laminar flow we can obtain f directly from Eq. 10–4.

$$f = \frac{64}{\text{Re}} = \frac{64}{1698} = 0.0377$$

Thus,

$$h_L = f\frac{L}{D}\frac{V^2}{2g} = (0.0377)\left(\frac{3000 \text{ m}}{0.250 \text{ m}}\right)\left[\frac{(0.8149 \text{ m/s})^2}{2(9.81 \text{ m/s}^2)}\right]$$

$$= 15.3 \text{ m} \qquad\qquad Ans.$$

Here the head loss is a consequence of the oil's viscosity and does not depend upon the surface roughness of the pipe.

10

EXAMPLE | 10.3

The fan in Fig. 10–6 is used to force air having a temperature of 20°C through the 200-mm-diameter galvanized sheet metal duct. Determine the required power output of the fan if the length of the duct is 150 m and the flow is to be 0.15 m³/s. Take $\varepsilon = 0.15$ mm.

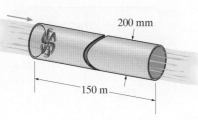

200 mm

150 m

Fig. 10–6

SOLUTION

Fluid Description.   We will assume the air is incompressible and the fan maintains fully developed steady flow. From Appendix A, for 20°C air at atmospheric pressure, $\rho_a = 1.202 \text{ kg/m}^3$ and $\nu_a = 15.1(10^{-6}) \text{ m}^2/\text{s}$. The type of flow is determined from the Reynolds number. Since

$$V = \frac{Q}{A} = \frac{0.15 \text{ m}^3/\text{s}}{\pi(0.1 \text{ m})^2} = \frac{15}{\pi} \text{ m/s}$$

$$\text{Re} = \frac{VD}{\nu_a} = \frac{\left(\dfrac{15}{\pi} \text{ m/s}\right)(0.2 \text{ m})}{15.1(10^{-6}) \text{ m}^2/\text{s}} = 6.32(10^4) > 2300 \text{ (turbulent)}$$

Analysis.   We can determine the shaft head of the fan by applying the energy equation between the inlet and outlet of the duct, but first we must determine the head loss along the duct. Here $\varepsilon/D = \dfrac{0.15 \text{ mm}}{200 \text{ mm}} = 0.00075$. Using this value and Re, the Moody diagram gives $f = 0.023$. Therefore, from the Darcy–Weisbach equation, the head loss through the duct is

$$h_L = f\frac{L}{D}\frac{V^2}{2g} = (0.023)\left(\frac{150 \text{ m}}{0.2 \text{ m}}\right)\left[\frac{\left(\dfrac{15}{\pi} \text{ m/s}\right)^2}{2(9.81 \text{ m/s}^2)}\right] = 20.04 \text{ m}$$

We will select a control volume that includes the fan, a portion of still air just outside the duct, to the left of the fan, and the moving air along the duct. Then the pressure $p_{\text{in}} = p_{\text{out}} = 0$, since it is atmospheric, and $V_{\text{in}} \approx 0$, since the air is still. The fan acts like a pump and will add energy to the air. Therefore, the energy equation becomes

$$\frac{p_{\text{in}}}{\gamma} + \frac{V_{\text{in}}^2}{2g} + z_{\text{in}} + h_{\text{pump}} = \frac{p_{\text{out}}}{\gamma} + \frac{V_{\text{out}}^2}{2g} + z_{\text{out}} + h_{\text{turb}} + h_L$$

$$0 + 0 + 0 + h_{\text{fan}} = 0 + \frac{\left(\dfrac{15}{\pi} \text{ m/s}\right)^2}{2(9.81 \text{ m/s}^2)} + 0 + 0 + 20.04 \text{ m}$$

$$h_{\text{fan}} = 21.205 \text{ m}$$

Notice that, because of the high velocity, most of this head is used to overcome the frictional resistance of the air (20.04 m), and little is used (1.162 m) to provide kinetic energy. The power output of the fan is therefore

$$\dot{W}_s = \gamma_a Q h_{\text{fan}} = (1.202 \text{ kg/m}^3)(9.81 \text{ m/s}^2)(0.15 \text{ m}^3/\text{s})(21.205 \text{ m})$$
$$= 37.5 \text{ W} \qquad\qquad Ans.$$

10

EXAMPLE 10.4

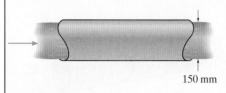

150 mm

Fig. 10–7

Crude oil flows through the 150-mm-diameter steel pipe in Fig. 10–7. Determine its maximum average velocity, if the head loss is not to be greater than $h_L = 1.5$ m in 100 m of pipe. Take $v_o = 40.0(10^{-6})$ m²/s and $\varepsilon = 0.045$ mm.

SOLUTION

Fluid Description. We assume the oil is incompressible and has fully developed steady flow.

Analysis. Since the head loss is given, it can be related to the velocity using the Darcy–Weisbach equation,

$$h_L = f\frac{L}{D}\frac{V^2}{2g}; \qquad 1.5\text{ m} = f\left(\frac{100\text{ m}}{0.15\text{ m}}\right)\left(\frac{V^2}{2(9.81\text{ m/s}^2)}\right)$$

so

$$V = \sqrt{\frac{0.044145}{f}} \qquad (1)$$

In order to obtain the friction factor, we will use the Moody diagram. To do so, we need to calculate the Reynolds number. It can be expressed in terms of velocity as

$$\text{Re} = \frac{VD}{v_o} = \frac{V(0.15\text{ m})}{40.0(10^{-6})\text{ m}^2/\text{s}} = 3750V \qquad (2)$$

If we *assume* Re to be very large, around 10^7, then from the Moody diagram, interpolating for $\varepsilon/D = 0.045$ mm/150 mm = 0.0003, an *estimate* for f would be $f = 0.015$. Thus, from Eq. 1,

$$V = \sqrt{\frac{0.044145}{0.015}} = 1.72\text{ m/s}$$

And from Eq. 2, this produces a Reynolds number of

$$\text{Re} = 3750(1.72\text{ m/s}) = 6.43(10^3)$$

With this value, the Moody diagram gives a new value of $f = 0.034$. With this, using Eqs. 1 and 2, $V = 1.14$ m/s and Re $= 4.27(10^3)$. Using this value of Re, from the Moody diagram, $f = 0.038$, which is close to the previous value of 0.034. ($\leq 10\%$ difference is usually adequate.) Thus, Eq. 1 gives

$$V = 1.08\text{ m/s} \qquad \qquad \textit{Ans.}$$

Realize that we can also solve this problem by expressing Re in terms of f using Eqs. 1 and 2, substituting this result into Eq. 10–6 or 10–7, and then solving for f using a numerical method on a calculator.

10

EXAMPLE | 10.5

The cast iron pipe in Fig. 10–8 is used to transport water at $0.30 \text{ m}^3/\text{s}$. If the head loss is to be no more than 0.006 m for every 1-m length of pipe, determine the smallest diameter D of pipe that can be used. Take $\nu_w = 1.15(10^{-6}) \text{ m}^2/\text{s}$.

SOLUTION

Fluid Description.   We assume water to be incompressible and we have fully developed steady flow.

Fig. 10–8

Analysis.   In this problem, both the friction factor f and pipe diameter D are unknown. However, since the head loss is known, we can relate f and D using the Darcy–Weisbach equation.

$$h_L = f\frac{L}{D}\frac{V^2}{2g}$$

$$0.006 \text{ m} = f\left(\frac{1 \text{ m}}{D}\right)\frac{\left[\dfrac{0.3 \text{ m}^3/\text{s}}{(\pi/4)D^2}\right]^2}{2(9.81 \text{ m/s}^2)}$$

$$D^5 = 1.2394\,f \qquad (1)$$

The Reynolds number can also be expressed in terms of the pipe's diameter as

$$\text{Re} = \frac{VD}{\nu_w} = \frac{\left[\dfrac{0.3 \text{ m}^3/\text{s}}{(\pi/4)D^2}\right]D}{1.15(10^{-6}) \text{ m}^2/\text{s}}$$

$$\text{Re} = \frac{3.3215(10^5)}{D} \qquad (2)$$

(© Prisma/Heeb Christian/Alamy)

The Moody diagram relates Re to f, but we do not know these values, and so we must use a trial-and-error approach. We begin by assuming a value for f. Typically, choose a mid-range value of say $f = 0.025$. Then, from Eqs. 1 and 2, $D = 0.4991$ m and $\text{Re} = 6.65(10^5)$. Since $\varepsilon = 0.00026$ m for cast iron pipes, then $\varepsilon/D = 0.000521$. Using these values of ε/D and Re, we get $f \approx 0.0176$ from the Moody diagram. Substituting this value into Eqs. 1 and 2 gives $D = 0.4653$ m and $\text{Re} = 7.14(10^5)$. Now $\varepsilon/D = 0.000559$, in which case the Moody diagram gives $f \approx 0.0175$, which is close to the previous value. Therefore,

$$D = 0.4653 \text{ m} = 465 \text{ mm} \qquad\qquad Ans.$$

Here we should select a slightly larger diameter pipe based on the available manufactured size. Also, as noted from the Darcy–Weisbach equation, this *larger size D* will produce a slight reduction in the calculated head loss.

10.2 Losses Occurring from Pipe Fittings and Transitions

In the previous sections, we showed how a major head loss occurs along the length of a pipe due to the frictional effects of fully developed flow. In addition to this, head losses also occur at pipe connections, such as bends, fittings, entrances, and transitions. These are called **minor losses**. This term is somewhat of a misnomer, because for many industrial and commercial applications, these losses are often *greater* than the major losses in the pipe system.

Minor losses are the result of *turbulent mixing* of the fluid within the connection as the fluid passes through it. The eddies or swirls that are produced are carried downstream, where they decay and generate heat, before fully developed laminar or turbulent flow is restored. Although a minor loss is not necessarily localized within the connection, we will assume it is and will express this loss in terms of the velocity head as we did in the case for a major loss. Here we will formulate it as

$$h_L = K_L \frac{V^2}{2g} \tag{10–8}$$

Here K_L is called the **resistance** or **loss coefficient**, which is determined from experiment.* Design manuals often provide such data; however, care should be taken when selecting a loss coefficient, since the reported values can vary for a particular fitting that is manufactured from different sources. See Refs. [13] and [19]. Generally, the manufacturer's recommendations should be considered. What follows is a partial list of values for K_L for some common types of fittings encountered in practice. We will use these values for problem solving.

Frictional head losses from valves, elbows, tees, and other fittings have to be considered when choosing a pump to be used with this pipe system. (© Aleksey Stemmer/Fotolia)

Flow coefficients are sometimes used in the valve industry to report minor losses. This is particularly true for control valves. This factor is similar to the resistance coefficient, and can be related to it by the Darcy—Weisbach equation. Further details are given in Ref. [19]. Later in this chapter we will discuss how *discharge coefficients* are used to represent losses incurred in various types of nozzles and flow meters.

Inlet and Exit Transitions.   When fluid enters a pipe from a reservoir, it will cause a minor loss that depends upon the type of transition that is used. Well-rounded transitions, as in Fig. 10–9a, will cause the smallest loss, since they provide a gradual change in flow. The value of K_L depends upon the radius r of the transition; however, as noted in the figure, if $r/D \geq 0.15$ then $K_L = 0.04$ can be used. Entrance transitions that produce greater losses may have a flush entrance, $K_L = 0.5$, Fig. 10–9b, or have a re-entrant pipe, $K_L = 1.0$, Fig. 10–9c. These situations can cause the fluid to separate from the wall of the pipe and form a **vena contracta**, or "necking," near the entrance, because the fluid streamlines cannot bend 90° around the corner. This constricts the flow and causes an increase in velocity near the entrance. This in turn lowers the pressure and creates flow separation producing localized eddies at these locations.

At the discharge end of a pipe into a large reservoir, the loss coefficient is $K_L = 1.0$, regardless of the shape of the transition, Fig. 10–9d. Here the kinetic energy of the fluid is converted into thermal energy as the fluid exits the pipe and eventually comes to rest within the reservoir.

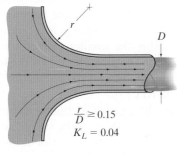

$\dfrac{r}{D} \geq 0.15$
$K_L = 0.04$

Well-rounded entrance
(a)

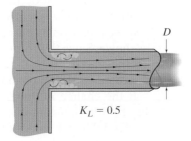

$K_L = 0.5$

Flush entrance
(b)

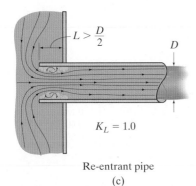

$L > \dfrac{D}{2}$

$K_L = 1.0$

Re-entrant pipe
(c)

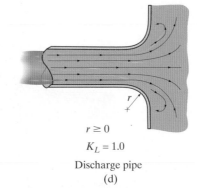

$r \geq 0$
$K_L = 1.0$

Discharge pipe
(d)

Fig. 10–9

10

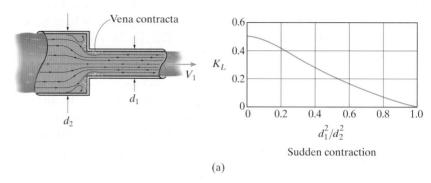

Vena contracta

d_2

d_1

V_1

K_L

Sudden contraction

(a)

Fig. 10–10

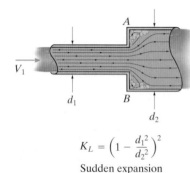

$$K_L = \left(1 - \frac{d_1^2}{d_2^2}\right)^2$$

Sudden expansion

Expansion and Contraction. A sudden expansion or contraction, from one diameter of pipe to another, will cause a minor loss that depends upon the ratios of the cross-sectional diameters, Fig. 10–10a. For a *sudden expansion,* the equation for K_L has been determined from the continuity, energy, and momentum equations. Here the stagnation that occurs within the corners A and B of the larger pipe contributes only slightly to the loss. When the fitting is a *sudden contraction,* a vena contracta will form within the smaller-diameter pipe as shown. Since this formation depends upon the diameter ratios, the loss coefficient has to be determined experimentally, and so the results are indicated on the graph in Fig. 10–10a. See Ref. [3].

If the change in flow is gradual, as in the case of a conical diffuser, Fig. 10–10b, then for rather small angles, $\theta < 8°$, losses can be significantly reduced. Higher values of θ will produce not only wall frictional losses but also flow separation, and the formation of eddies. For this case, values of K_L may be larger than those for a sudden expansion. See Ref. [17]. Some typical values of K_L for these fittings are given in Fig. 10–10b. All these coefficients apply to the velocity head $\left(V^2/2g\right)$ calculated for the pipe having the *smaller* cross-sectional diameter.

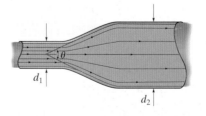

d_1

d_2

θ

$\theta \left(\dfrac{d_2}{d_1} = 4\right)$	K_L
10°	0.13
20°	0.40
30°	0.80

Conical diffuser

(b)

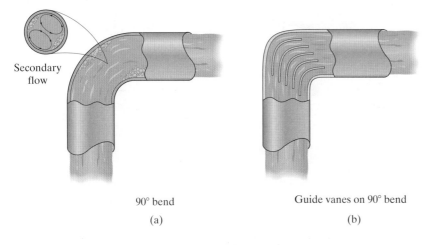

90° bend

(a)

Guide vanes on 90° bend

(b)

Fig. 10–11

Bends.

Changes in the direction of the flow can cause fluid separation from the inner wall of the pipe due to normal or radial accelerations along the streamlines, Fig. 10–11a. The loss created can be increased when **secondary flow** occurs within the bend. It is produced by the radial pressure variation and the frictional resistance both within the fluid and along the pipe wall. The result is the formation of twin eddies and a double swirling motion through the bend. To avoid these effects, a larger-radius bend or "long sweep" can be used, or guide vanes can be placed in sharp bends of larger pipes in order to reduce the head loss, Fig. 10–11b.

Threaded Fittings.

A representative list of the loss coefficients for some threaded pipe fittings, such as valves, elbows, bends, and tees, is given in Table 10–1.

TABLE 10–1	
Loss coefficients for pipe fittings	K_L
Gate valve—fully opened	0.19
Globe valve—fully opened	10
90° elbow	0.90
45° bend	0.40
Tee along pipe run	0.40
Tee along branch	1.8

Even for large-diameter pipe systems such as this, an assessment must be made of the head loss from filters, elbows, and tees.

10

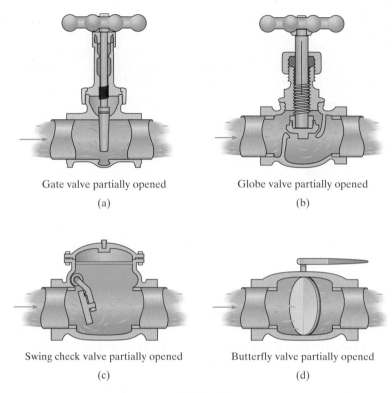

Gate valve partially opened

(a)

Globe valve partially opened

(b)

Swing check valve partially opened

(c)

Butterfly valve partially opened

(d)

Fig. 10–12

Typical butterfly valve

Valves. There are many types of valves used to control the flow of fluids for industrial and commercial applications. In particular the *gate valve* works by blocking the flow with a "gate" or plate that is perpendicular to the flow, as shown in Fig. 10–12a. Gate valves are mainly used to either permit or prevent the flow of liquids, and as result they are either fully open or closed. In the open position they allow little or no obstruction to the flow, and so have a very low resistance. The *globe valve*, shown in Fig. 10–12b, is designed to regulate flow. It consists of a stopper disc that goes up and down from a stationary ring seat. The name "globe" refers to the spherical shape of the outer housing, although modern designs are generally not fully spherical. As shown in Table 10–1, losses are greater for this valve, since the flow is more disruptive. Two additional valves are also shown in Fig. 10–12, namely the *swing check valve*, which prevents flow reversal, Fig. 10–12c, and the *butterfly valve*, Fig. 10–12d, which provides a low-cost, quick shut-off means of regulating the flow. For all these cases, the loss is greatly increased when the valves are partially opened, as shown in each figure, as opposed to when they are fully opened.

Pipe Connections.   Consideration should also be given to the possibility of head losses at pipe-to-pipe connections. For example, small-diameter pipes are generally threaded, so if any burrs remain on cut sections, then they can disturb the flow and cause additional losses through the connection. Likewise, larger-diameter pipes are welded, flanged, or glued together, and these joints can also produce further head loss unless they are properly fabricated and connected.

For any flow analysis, all the minor losses should be carefully accounted for, so that sufficient accuracy in predicting the total head loss is achieved. This is particularly true if a pipe system is made up of many short lengths and has several fittings and transitions.

Equivalent Length.   Another way to describe the hydraulic resistance of valves and fittings is to use an *equivalent-length ratio*, L_{eq}/D. This requires converting the frictional loss within a fitting or valve to an equivalent length of pipe, L_{eq}, that would produce the same loss due to its loss coefficient K_L. Since the head loss through a straight pipe is determined from the Darcy–Weisbach equation, $h_L = f(L_{eq}/D)V^2/2g$, and the head loss through a valve or fitting is expressed as $h_L = K_L V^2/2g$, then by comparison,

$$K_L = f\left(\frac{L_{eq}}{D}\right)$$

Therefore, the equivalent length of pipe producing the loss is

$$L_{eq} = \frac{K_L D}{f} \tag{10–9}$$

The overall head loss or pressure drop for the system is then calculated from the total length of pipe plus the equivalent lengths determined for each fitting.

Important Points

- The frictional loss in a straight pipe is expressed as a "major" head loss, which is determined from the Darcy–Weisbach equation, $h_L = f(L/D)V^2/2g$.

- "Minor" head losses occur in pipe fittings, entrances, transitions, and connections. They are expressed as $h_L = K_L(V^2/2g)$, where the loss coefficient K_L is determined from tabular data found by experiment and published in design handbooks or manufacturers' catalogs.

- Both major and minor head losses are responsible for a pressure drop along the pipe.

10

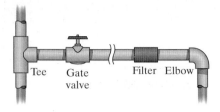

Tee   Gate   Filter   Elbow
       valve

Fig. 10–13

10.3   Single-Pipeline Flow

Many pipelines consist of a single-diameter pipe with bends, valves, filters, and transitions, as in Fig. 10–13. These systems are often used to transport water for industrial and residential usage, and for hydroelectric power. They may also be used to transport crude oil or lubricants through mechanical equipment. The following procedure can be used to properly design such a system.

Procedure for Analysis

Problems that involve flow through a *single pipeline* must satisfy both the *energy equation and the continuity equation,* and account for all the major and minor head losses throughout the system. For incompressible, steady flow, these two equations, referenced to points where the flow is "in" and "out," are

$$\frac{p_{in}}{\gamma} + \frac{V_{in}^2}{2g} + z_{in} + h_{pump} = \frac{p_{out}}{\gamma} + \frac{V_{out}^2}{2g} + z_{out} + h_{turb}$$

$$+ f\frac{L}{D}\frac{V^2}{2g} + \Sigma K_L\left(\frac{V^2}{2g}\right)$$

and

$$Q = V_{in}A_{in} = V_{out}A_{out}$$

Depending upon what is known and unknown, this results in three basic types of problems.

Determine the Pressure Drop.

- The pressure drop for a pipe having a known length, diameter, elevation, roughness, and discharge can be determined directly using the energy equation.

Determine the Flow.

- When the pipe length, diameter, roughness, elevation, and pressure drop are all known, then a trial-and-error solution is necessary to determine the flow (or average velocity V), since the Reynolds number, $Re = VD/\nu$, is unknown, and therefore the friction factor cannot be directly determined from the Moody diagram.

Determine the Length or Diameter of the Pipe.

- The design of a pipe generally requires specifying the length of the pipe and its diameter. Either of these parameters can be found provided the other is known, along with the flow (or average velocity) and the allowed pressure drop or head loss. With the Moody diagram, the solution requires a trial-and-error procedure because, as in the previous case, the Reynolds number and friction factor must be obtained.

The examples that follow illustrate application of each of these types of problems.

10

EXAMPLE | 10.6

When the globe valve at B in Fig. 10–14 is fully opened, it is observed that water flows out through the 65-mm-diameter cast iron pipe with an average velocity of 2 m/s. Determine the pressure in the pipe at A. Take $\rho_w = 998 \text{ kg/m}^3$ and $\nu_w = 0.8(10^{-6}) \text{ m}^2/\text{s}$.

SOLUTION

Fluid Description.   We assume steady incompressible flow and use average velocities.

Analysis.   The pressure drop can be determined from the energy equation, but first we must determine the major and minor head losses.

For the major head loss, the friction factor is determined from the Moody diagram. For cast iron pipe, $\varepsilon/D = 0.26 \text{ mm}/65 \text{ mm} = 0.004$. Also,

$$\text{Re} = \frac{VD}{\nu_w} = \frac{(2 \text{ m/s})(0.065 \text{ m})}{0.8(10^{-6}) \text{ m}^2/\text{s}} = 1.625(10^5)$$

Thus, $f = 0.0290$.

Minor head loss for the elbow is $0.9(V^2/2g)$, and for the fully opened globe valve, it is $10(V^2/2g)$. Thus the total head loss is

$$h_L = f\frac{L}{D}\frac{V^2}{2g} + 0.9\left(\frac{V^2}{2g}\right) + 10\left(\frac{V^2}{2g}\right)$$

$$= 0.0290\left(\frac{10 \text{ m}}{0.065 \text{ m}}\right)\left[\frac{(2 \text{ m/s})^2}{2(9.81 \text{ m/s}^2)}\right] + (0.9 + 10)\left[\frac{(2 \text{ m/s})^2}{2(9.81 \text{ m/s}^2)}\right]$$

$$= 0.9096 \text{ m} + 2.222 \text{ m} = 3.132 \text{ m}$$

Comparing the two terms, notice that the minor losses provide the *largest* contribution (2.222 m) to the total loss, even though this loss is referred to as "minor."

We will consider the control volume to contain the water in the pipe from A to C. Since the pipe has the same diameter throughout, continuity requires $V_A A = V_C A$ or $V_A = V_C = 2 \text{ m/s}$. With the gravitational datum through C, the energy equation becomes

$$\frac{p_A}{\gamma_w} + \frac{V_A{}^2}{2g} + z_A + h_{\text{pump}} = \frac{p_C}{\gamma_w} + \frac{V_C{}^2}{2g} + z_C + h_{\text{turb}} + h_L$$

$$\frac{p_A}{(998 \text{ kg/m}^3)(9.81 \text{ m/s}^2)} + \frac{(2 \text{ m/s})^2}{2(9.81 \text{ m/s}^2)} + 6\text{m} + 0 = 0 + \frac{(2 \text{ m/s})^2}{2(9.81 \text{ m/s}^2)} + 0 + 0 + 3.132 \text{ m}$$

Solving yields

$$p_A = -28.08(10^3) \text{ Pa} = -28.1 \text{ kPa} \qquad \textit{Ans.}$$

The result indicates a suction in the pipe occurs, but this pressure will not cause cavitation since it is greater than the vapor pressure.

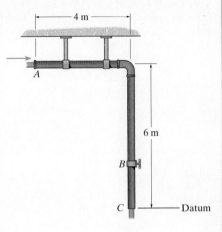

Fig. 10–14

10

EXAMPLE | 10.7

The commercial steel pipe in Fig. 10–15 has a diameter of 80 mm and transfers glycerin at $T = 20°C$ from the large tank to the outlet at B. If the tank is open at the top, determine the initial discharge at B when the gate valve at C is fully opened.

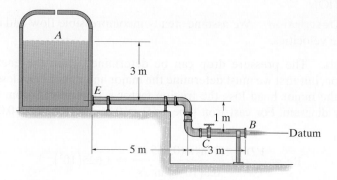

Fig. 10–15

SOLUTION

Fluid Description. We assume steady incompressible flow with $V_A \approx 0$. Using Appendix A, for glycerin $\rho_g = 1260 \text{ kg/m}^3$ and $\nu_g = 1.19(10^{-3}) \text{ m}^2/\text{s}$.

Analysis. The control volume contains the glycerin in the reservoir and pipe. The energy equation will be applied between A and B, with the gravitational datum at B. The major head loss in the pipe is determined from the Darcy–Weisbach equation. Minor head losses are calculated for the flush entrance at E, $0.5(V^2/2g)$, the two elbows, $2[0.9(V^2/2g)]$, and the fully opened gate valve at C, $0.19(V^2/2g)$. We have

$$\frac{p_A}{\gamma} + \frac{V_A^2}{2g} + z_A + h_{\text{pump}} = \frac{p_B}{\gamma} + \frac{V_B^2}{2g} + z_B + h_{\text{turb}} + h_L$$

$$0 + 0 + 4 \text{ m} + 0 = 0 + \frac{V^2}{2g} + 0 + 0 + f\left(\frac{5 \text{ m} + 1 \text{ m} + 3 \text{ m}}{0.08 \text{ m}}\right)\left(\frac{V^2}{2g}\right)$$

$$+ \, 0.5\left(\frac{V^2}{2g}\right) + 2\left[0.9\left(\frac{V^2}{2g}\right)\right] + 0.19\left(\frac{V^2}{2g}\right)$$

$$4 = (112.5f + 3.49)\left[\frac{V^2}{2(9.81 \text{ m/s}^2)}\right] \tag{1}$$

10

A second relationship between f and V can be obtained using the Moody diagram and a trial-and-error procedure. To do this, we assume a value of f or V and then solve for the other one using Eq. 1. Then we calculate the Reynolds number and check the value of f using the Moody diagram.

Rather than doing this, however, we will assume the flow is laminar, since glycerin has a high kinematic viscosity. Then Eq. 10–4 can be used to relate f to V.

$$f = \frac{64}{\text{Re}} = \frac{64\nu_g}{VD} = \frac{64\left[1.19\left(10^{-3}\right) \text{ m}^2/\text{s}\right]}{V(0.08 \text{ m})} = \frac{0.952}{V}$$

Substituting this relation into Eq. 1,

$$4 = \left[112.5\left(\frac{0.952}{V}\right) + 3.49\right]\left[\frac{V^2}{2(9.81)}\right]$$

or

$$3.49V^2 + 107.1V - 78.48 = 0$$

Solving for the positive root, we get

$$V = 0.7161 \text{ m/s}$$

Checking the Reynolds number, we find that

$$\text{Re} = \frac{VD}{\nu_g} = \frac{(0.7161 \text{ m/s})(0.08 \text{ m})}{1.19\left(10^{-3}\right) \text{ m}^2/\text{s}} = 48.1 < 2300 \quad \text{(Laminar flow)}$$

Thus,

$$Q = VA = (0.7161 \text{ m/s})\left[\pi(0.04 \text{ m})^2\right] = 0.00360 \text{ m}^3/\text{s} \quad \textit{Ans.}$$

10

EXAMPLE 10.8

Determine the required diameter of the galvanized iron pipe in Fig. 10–16 if the discharge at C is to be 0.475 m³/s when the gate valve at A is fully opened. The reservoir is filled with water to the depth shown. Take $\nu_w = 1(10^{-6})$ m²/s.

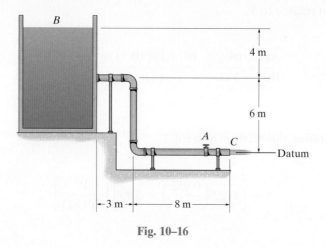

Fig. 10–16

SOLUTION

Fluid Description.   We will assume the reservoir is large so that $V_B \approx 0$, and therefore the flow will be steady. Also, the water is assumed to be incompressible.

Analysis.   Continuity requires the velocity through the pipe to be the same at all points since the pipe has the same diameter throughout. Therefore,

$$Q = VA; \qquad 0.475 \text{ m}^3/\text{s} = V\left(\frac{\pi}{4}D^2\right)$$

$$V = \frac{0.6048}{D^2} \qquad\qquad (1)$$

To obtain a second equation relating V and D, the energy equation will be applied between B and C, with the gravitational datum through C, Fig. 10–16. The control volume for this case contains the water in the reservoir and pipe.

The major loss is determined from the Darcy–Weisbach equation. The minor losses through the pipe come from the flush entrance, $0.5(V^2/2g)$, the two elbows, $2[0.9(V^2/2g)]$, and the fully opened gate valve, $0.19(V^2/2g)$. Thus,

$$\frac{p_B}{\gamma_w} + \frac{V_B^2}{2g} + z_B + h_{\text{pump}} = \frac{p_C}{\gamma_w} + \frac{V_C^2}{2g} + z_C + h_{\text{turb}} + h_L$$

$$0 + 0 + (4\text{ m} + 6\text{ m}) + 0 = 0 + \frac{V^2}{2g} + 0 + 0 +$$

$$\left[f\left(\frac{17\text{ m}}{D}\right)\left(\frac{V^2}{2g}\right) + 0.5\left(\frac{V^2}{2g}\right) + 2\left[0.9\left(\frac{V^2}{2g}\right)\right] + 0.19\left(\frac{V^2}{2g}\right) \right]$$

or

$$10 = \left[f\left(\frac{17}{D}\right) + 3.49 \right]\left[\frac{V^2}{2(9.81\text{ m/s}^2)}\right] \qquad (2)$$

Combining Eqs. 1 and 2 by eliminating V, we obtain

$$536.40D^5 - 3.49D - 17f = 0 \qquad (3)$$

To avoid assuming a value of f and then solving this fifth-order equation for D, it is easier to assume a value of D, calculate f, and then verify this result using the Moody diagram. For example, if we assume $D = 0.350$ m, then from Eq. 3, $f = 0.0939$. From Eq. 1, $V = 4.937$ m/s, and so

$$\text{Re} = \frac{VD}{\nu_w} = \frac{4.937\text{ m/s }(0.350\text{ m})}{1(10^{-6})\text{m}^2/\text{s}} = 1.73(10^6)$$

For galvanized iron pipe $\varepsilon/D = 0.15\,\text{mm}/350\,\text{mm} = 0.000429$. Therefore, from the Moody diagram with these values of ε/D and Re, we obtain $f = 0.0165 \neq 0.0939$.

On the next iteration, choose a value of D that gives an f smaller than $f = 0.0939$ in Eq. 3. Say $D = 0.3$ m, then $f = 0.01508$. Then $V = 6.72$ m/s, $\text{Re} = 2.02(10^6)$, and $\varepsilon/D = 0.15\,\text{mm}/300\,\text{mm} = 0.0005$. With these new values, $f = 0.017$ from the Moody diagram, which is fairly close to the previous value (0.01508). Therefore, we will use

$$D = 300\,\text{mm} \qquad \qquad Ans.$$

10

10.4   Pipe Systems

If several pipes, having different diameters and lengths, are connected together, they form a *pipe system*. In particular, if the pipes are connected successively, as shown in Fig. 10–17a, the system is in **series**, whereas if the pipes cause the flow to divide into different branches, Fig. 10–17b, the system is in **parallel**. We will now give each of these cases separate treatment.

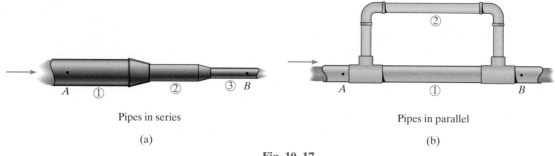

Pipes in series

(a)

Pipes in parallel

(b)

Fig. 10–17

Pipes in Series.   The analysis of a pipe system in series is similar to that used to analyze a single pipe. In this case, however, to satisfy continuity, the flow through each pipe must be the same, so that for the three-pipe system in Fig. 10–17a, we require

$$Q = Q_1 = Q_2 = Q_3$$

Also, the total head loss for the system is equal to the sum of the major head loss along each length of pipe, plus all the minor head losses for the system. Therefore, the energy equation between A (in) and B (out) becomes

$$\frac{p_A}{\gamma} + \frac{V_A^2}{2g} + z_A = \frac{p_B}{\gamma} + \frac{V_B^2}{2g} + z_B + h_L$$

where

$$h_L = h_{L1} + h_{L2} + h_{L3} + h_{minor}$$

Compared to having a single pipe, as in the last section, the problem here is more complex, because the friction factor and the Reynolds number will be different for each pipe.

10

Pipes in Parallel.

Although it is possible for a parallel system to have several branches, here we will consider a system having only two, as shown in Fig. 10–17b. If the problem requires finding the pressure drop between A and B and the flow in each of the pipes, then for continuity of flow, we require

$$Q_A = Q_B = Q_1 + Q_2$$

If the energy equation is applied between A (in) and B (out), then

$$\frac{p_A}{\gamma} + \frac{V_A^2}{2g} + z_A = \frac{p_B}{\gamma} + \frac{V_B^2}{2g} + z_B + h_L$$

Since the fluid will always take the path of least resistance, the amount of flow through each branch pipe will automatically adjust, to maintain the *same head loss* or resistance to flow in each branch. Therefore, for each length we require $h_{L1} = h_{L2}$. Using this, the analysis of a two-branch system is straightforward and is based on the above two equations.

Of course, if a parallel system has more than two branches, then it becomes more difficult to analyze. For example, consider the case of a network of pipes shown in Fig. 10–18. Such a system forms loops and which is representative of the type used for large buildings, industrial processes, or municipal water supply systems. Due to its complexity, the direction of flow and its rate within each loop may not be certain, and so a trial-and-error analysis will be required for the solution. The most efficient method for doing this is based upon matrix algebra, using a computer. This method has widespread use for industrial and commercial applications. The details for applying it will not be covered here; rather, it is discussed in articles or books related to analyzing the flow in pipe networks. See, for example, Ref. [14].

Pipe system used in a chemical processing plant.

Procedure for Analysis

- The solution of problems involving pipe systems in series or in parallel follows the same procedure outlined in the previous section. In general, the flow must satisfy both the continuity equation and the energy equation, and the order in which these equations are applied depends on the type of problem that is to be solved.

- For pipes in *series*, the *flow* through each pipe must be the *same*, and the *head loss* is the *total* for all the pipes. For pipes in *parallel*, the *total flow* is the accumulation of flow from each branch in the system. Also, since the flow takes the path of least resistance, the *head loss* for each branch will be the *same*.

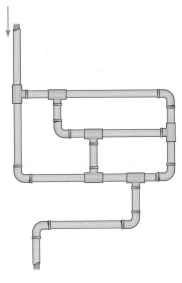

Pipe network

Fig. 10–18

EXAMPLE   10.9

Pipes BC and CE in Fig. 10–19 are made of galvanized iron and have diameters of 200 mm and 100 mm, respectively. If the gate valve at F is fully opened, determine the discharge of water at E in liters/min. The reducer at C has a $K_L = 0.7$. Take $\nu_w = 1.15(10^{-6})$ m²/s.

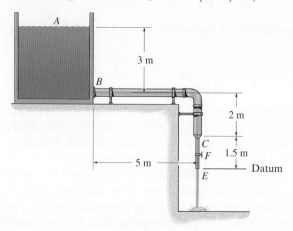

Fig. 10–19

SOLUTION

Fluid Description.   We assume steady incompressible flow occurs, where $V_A \approx 0$.

Continuity Equation.   If the average velocity through the larger-diameter pipe is V, and through the smaller-diameter pipe is V', then choosing a localized control volume of the water within the reducer at C, and applying the continuity equation, we have

$$\frac{\partial}{\partial t}\int_{\text{cv}}\rho\, d\forall + \int_{\text{cs}}\rho \mathbf{V}\cdot d\mathbf{A} = 0$$

$$0 - V\left[(\pi\, 0.1\text{ m})^2\right] + V'\left[(\pi\, 0.05\text{ m})^2\right] = 0$$

$$V' = 4V$$

Energy Equation.   Using this result, we will now apply the energy equation between A and E, in order to obtain a relationship between the velocity and the friction factors. The control volume for this case contains the water throughout the reservoir and pipe system.

$$\frac{p_A}{\gamma_w} + \frac{V_A^2}{2g} + z_A + h_{\text{pump}} = \frac{p_E}{\gamma_w} + \frac{V_E^2}{2g} + z_E + h_{\text{turb}} + h_L$$

$$0 + 0 + 6.5\text{m} + 0 = 0 + \frac{(4V)^2}{2g} + 0 + 0 + h_L \qquad (1)$$

The minor losses in the system come from the flush entrance at B, $0.5(V^2/2g)$, the elbow, $0.9(V^2/2g)$, the reducer, $0.7(V'^2/2g)$, and the fully opened gate valve, $0.19(V'^2/2g)$. Using the Darcy–Weisbach equation for the major head loss in each pipe, and expressing the total head loss in the pipe system in terms of V, we have

$$h_L = f\left(\frac{7 \text{ m}}{0.2 \text{ m}}\right)\frac{V^2}{2g} + f'\left(\frac{1.5 \text{ m}}{0.1 \text{ m}}\right)\left[\frac{(4V)^2}{2g}\right] + 0.5\left(\frac{V^2}{2g}\right) + 0.9\left(\frac{V^2}{2g}\right) + 0.7\left[\frac{(4V)^2}{2g}\right] + 0.19\left[\frac{(4V)^2}{2g}\right]$$

$$h_L = (35f + 240f' + 15.64)\left(\frac{V^2}{2g}\right)$$

Here f and f' are friction factors for the large and small diameter pipes, respectively. Substituting into Eq. 1 and simplifying, we get

$$127.53 = (35f + 240f' + 31.64)V^2 \qquad (2)$$

Moody Diagram.   For galvanized iron pipe, $\varepsilon = 0.15$ mm, so

$$\frac{\varepsilon}{D} = \frac{0.15 \text{ mm}}{200 \text{ mm}} = 0.00075$$

$$\frac{\varepsilon}{D'} = \frac{0.15 \text{ mm}}{100 \text{ mm}} = 0.0015$$

Thus,

$$\text{Re} = \frac{VD}{\nu_w} = \frac{V(0.2 \text{ m})}{1.15(10^{-6}) \text{ m}^2/\text{s}} = 1.739(10^5)V \qquad (3)$$

$$\text{Re}' = \frac{V'D}{\nu_w} = \frac{4V(0.1 \text{ m})}{1.15(10^{-6}) \text{ m}^2/\text{s}} = 3.478(10^5)V \qquad (4)$$

To satisfy the conditions of the Moody diagram, we will assume intermediate values for f and f', say $f = 0.0195$ and $f' = 0.022$. Thus, from Eqs. 2, 3, and 4, we get $V = 1.842$ m/s, $\text{Re} = 3.20(10^5)$, and $\text{Re}' = 6.41(10^5)$. Using these results and checking the Moody diagram, we get $f = 0.0195$ and $f' = 0.022$. Since these values are the same as those we have assumed, then indeed $V = 1.842$ m/s, and so the discharge can be determined by considering, say, the 6-in.-diameter pipe. It is

$$Q = VA = (1.842 \text{ m/s})[\pi(0.1 \text{ m})^2] = 0.05787 \text{ m}^3/\text{s}$$

Or, since there is 1000 liters/m³, then

$$Q = \left(0.05787 \frac{\text{m}^3}{\text{s}}\right)\left(\frac{1000 \text{ liters}}{1 \text{ m}^3}\right)\left(\frac{60 \text{ s}}{1 \text{ min}}\right) = 3472 \text{ liters/min} \quad Ans.$$

EXAMPLE 10.10

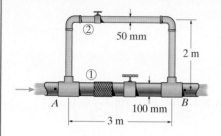

Fig. 10–20

Water flows at a rate of 0.03 m³/s through the branch piping system shown in Fig. 10–20. The 100-mm-diameter pipe has a filter and globe valve on it, and the 50-mm-diameter diverter pipe has a gate valve. The pipes are made of galvanized iron. Determine the flow through each pipe, and the pressure drop between A and B when both valves are fully opened. The head loss due to the filter is $K_L = 1.6\left(V^2/2g\right)$. Take $\gamma_w = 9810 \text{ N/m}^3$ and $\nu_w = 1\left(10^{-6}\right) \text{ m}^2/\text{s}$.

SOLUTION

Fluid Description. We will assume fully developed incompressible steady flow.

Continuity Equation. If we consider the water in the tee at A as the control volume, then continuity requires

$$Q = V_1 A_1 + V_2 A_2$$

$$0.03 \text{ m}^3/\text{s} = V_1\left[\pi(0.05 \text{ m})^2\right] + V_2\left[\pi(0.025 \text{ m})^2\right]$$

$$15.279 = 4V_1 + V_2 \tag{1}$$

Moody Diagram. With this relationship, we will now use the Moody diagram to obtain the velocities through each branch. Branch 1 has flow through two tees, the filter, and the fully opened globe valve. Using Table 10–1, the total head loss is

$$(h_L)_1 = f_1\left(\frac{3 \text{ m}}{0.1 \text{ m}}\right)\left(\frac{V_1^2}{2g}\right) + 2(0.4)\left(\frac{V_1^2}{2g}\right) + 1.6\left(\frac{V_1^2}{2g}\right) + 10\left(\frac{V_1^2}{2g}\right)$$

$$= (30f_1 + 12.4)\frac{V_1^2}{2g} \tag{2}$$

Branch 2 has flow through the branch of two tees, the two elbows, and the fully opened gate valve. Therefore,

$$(h_L)_2 = f_2\left(\frac{7 \text{ m}}{0.05 \text{ m}}\right)\left(\frac{V_2^2}{2g}\right) + 2(1.8)\left(\frac{V_2^2}{2g}\right) + 2(0.9)\left(\frac{V_2^2}{2g}\right) + 0.19\left(\frac{V_2^2}{2g}\right)$$

$$= (140f_2 + 5.59)\frac{V_2^2}{2g} \tag{3}$$

We require the head losses to be the same in each branch, $(h_L)_1 = (h_L)_2$, so that

$$(30f_1 + 12.4)V_1^2 = (140f_2 + 5.59)V_2^2 \tag{4}$$

Equations 1 and 4 contain four unknowns. Since the conditions of the Moody diagram must also be satisfied, we will *assume* intermediate values for the friction factors, say $f_1 = 0.02$ and $f_2 = 0.025$. Therefore, Eqs. 1 and 4 yield

$$V_1 = 2.941 \text{ m/s}$$

$$V_2 = 3.517 \text{ m/s}$$

so that

$$(Re)_1 = \frac{V_1 D_1}{\nu_w} = \frac{(2.941 \text{ m/s})(0.1 \text{ m})}{1(10^{-6}) \text{ m}^2/\text{s}} = 2.94(10^5)$$

$$(Re)_2 = \frac{V_2 D_2}{\nu_w} = \frac{(3.517 \text{ m/s})(0.05 \text{ m})}{1(10^{-6}) \text{ m}^2/\text{s}} = 1.76(10^5)$$

Since $(\varepsilon/D)_1 = 0.15 \text{ mm}/100 \text{ mm} = 0.0015$ and $(\varepsilon/D)_2 = 0.15 \text{ mm}/50 \text{ mm} = 0.003$, then using the Moody diagram, we find $f_1 = 0.022$ and $f_2 = 0.027$. Repeating the calculations with these values, from Eqs. 1 and 4 we get $V_1 = 2.95 \text{ m/s}$ and $V_2 = 3.48 \text{ m/s}$, which are very close to the previous values. Therefore, the flow through each pipe is

$$Q_1 = V_1 A_1 = (2.95 \text{ m/s})[\pi(0.05 \text{ m})^2] = 0.0232 \text{ m}^3/\text{s} \qquad Ans.$$

$$Q_2 = V_2 A_2 = (3.48 \text{ m/s})[\pi(0.025 \text{ m})^2] = 0.0068 \text{ m}^3/\text{s} \qquad Ans.$$

Note that $Q = Q_1 + Q_2 = 0.03 \text{ m}^3/\text{s}$, as required.

Energy Equation. The pressure drop between A and B is determined from the energy equation. The control volume contains all the water in the system from A to B. With the datum through A (in) and B (out), $z_A = z_B = 0$, and $V_A = V_B = V$. We have

$$\frac{p_A}{\gamma_w} + \frac{V_A^2}{2g} + z_A + h_{\text{pump}} = \frac{p_B}{\gamma_w} + \frac{V_B^2}{2g} + z_B + h_{\text{turb}} + h_L$$

$$\frac{p_A}{\gamma_w} + \frac{V^2}{2g} + 0 + 0 = \frac{p_B}{\gamma_w} + \frac{V^2}{2g} + 0 + 0 + h_L$$

or

$$p_A - p_B = \gamma_w h_L$$

Using Eq. 2, we have

$$p_A - p_B = (9810 \text{ N/m}^3)[30(0.022) + 12.4]\left[\frac{(2.95 \text{ m/s})^2}{2(9.81 \text{ m/s}^2)}\right]$$

$$= 56.8(10^3) \text{ Pa} = 56.8 \text{ kPa} \qquad Ans.$$

Since $(h_L)_1 = (h_L)_2$ for this parallel system, we can also obtain this same result using Eq. 3.

10

10.5   Flow Measurement

Through the years, many devices have been developed that measure the volumetric flow or the velocity of a fluid passing through a pipe or a closed conduit. Each method has specific applications, and a choice depends upon the required accuracy, the cost, the size of the flow, and the ease of use. In this section, we will describe some of the more common methods used for this purpose. Greater details can be found in the references listed at the end of the chapter, or on a specific manufacturer's website.

Venturi Meter.   The venturi meter was discussed in Sec. 5.3, and here its principles will be briefly reviewed. As shown in Fig. 10–21, this device provides a converging transition in flow from a pipe to a throat, and then a *gradual* diverging transition back to the pipe. This design prevents flow separation from the walls of the conduit, and thereby minimizes frictional losses within the fluid. It was shown in Sec. 5.3 that by applying the Bernoulli and continuity equations, we can obtain the mean velocity of flow at the throat using the equation

$$V_2 = \sqrt{\frac{2(p_1 - p_2)/\rho}{1 - (D_2/D_1)^4}} \qquad (10\text{–}10)$$

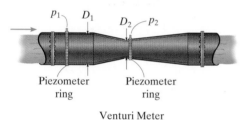

Piezometer
ring

Piezometer
ring

Venturi Meter

Fig. 10–21

For accuracy, a venturi meter is often fitted with two **piezometer rings**, one located at the upstream side of the meter, and the other at its throat, Fig. 10–21. Each ring surrounds a series of annular holes in the pipe so that the average pressure is produced within the ring. A manometer or pressure transducer is then connected to these rings, to measure the average differential static pressure $(p_1 - p_2)$ between them.

Since the Bernoulli equation *does not* account for any frictional losses within the flow, in practice engineers modify the above equation by multiplying it by an experimentally determined **venturi discharge coefficient**, C_v. This coefficient represents the ratio of the actual average velocity in the throat to its theoretical velocity, that is,

$$C_v = \frac{(V_2)_{\text{act}}}{(V_2)_{\text{theo}}}$$

Specific values of C_v are generally reported by the manufacturer as a function of the Reynolds number. Once C_v is obtained, the actual velocity within the throat is then

$$(V_2)_{\text{act}} = C_v \sqrt{\frac{2(p_1 - p_2)/\rho}{1 - (D_2/D_1)^4}}$$

Noting that $Q = V_2 A_2$, the volumetric flow can then be determined from

$$Q = C_v \left(\frac{\pi}{4} D_2^2\right) \sqrt{\frac{2(p_1 - p_2)/\rho}{1 - (D_2/D_1)^4}}$$

Nozzle Meter.

A nozzle meter works basically the same way as a venturi meter. When this device is inserted into the path of the flow, as shown in Fig. 10–22, the flow is constricted at the front of the nozzle, passes through its throat, and then leaves the nozzle without diverting the flow. This causes localized turbulence due to the acceleration of the flow through the nozzle, and then deceleration as the flow adjusts farther downstream. As a result, the frictional losses through the nozzle will be greater than flow through a venturi meter. Measurements of the pressure drop from ports 1 and 2 are used to determine the theoretical velocity, V_2, by applying Eq. 10–10. Here engineers use an experimentally determined **nozzle discharge coefficient**, C_n, to account for any frictional losses. Therefore, the flow becomes

$$Q = C_n\left(\frac{\pi}{4}D_2^2\right)\sqrt{\frac{2(p_1 - p_2)/\rho}{1 - (D_2/D_1)^4}}$$

Values for C_n as a function of the upstream Reynolds number are provided by the manufacturer for various area ratios of the nozzle and pipe.

Orifice Meter.

Another way to measure the flow in a pipe is to constrict the flow with an *orifice meter*, Fig. 10–23. It simply consists of a flat plate with a hole in it. The pressure is measured upstream and at the vena contracta, where the streamlines are parallel and the static pressure is constant. As before, the Bernoulli and continuity equations, applied at these points, result in a theoretical mean velocity defined by Eq. 10–10. The actual flow through the meter is determined by using an **orifice discharge coefficient**, C_o, supplied by the manufacturer, which accounts for both the frictional losses in the flow and the effect of the vena contracta. Thus,

$$Q = C_o\left(\frac{\pi}{4}D_2^2\right)\sqrt{\frac{2(p_1 - p_2)/\rho}{1 - (D_2/D_1)^4}}$$

Of these three meters, the venturi meter is the most expensive, but it will give the most accurate measurement, because the losses within it are minimized. The orifice meter is the least expensive and is easy to install, but it is the most inaccurate since the size of the vena contracta is not well defined. Also, this meter subjects the flow to the largest head loss, or pressure drop. Regardless of which meter is selected, however, it is important that it be installed along a straight section of pipe that is long enough to establish a fully developed flow. In this way, the results should correlate well with those obtained by experiment.

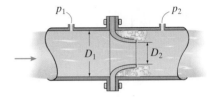

Nozzle Meter

Fig. 10–22

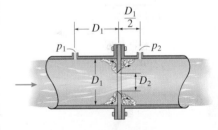

Orifice Meter

Fig. 10–23

10

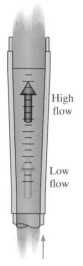

Rotometer

Fig. 10–24

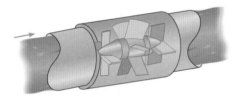

Turbine Flow Meter

Fig. 10–25

Anemometer

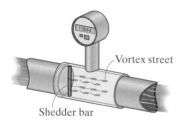

Vortex Flow Meter

Fig. 10–26

Rotometer.

A *rotometer* can be attached to a vertical pipe as shown in Fig. 10–24. The fluid flows in at the bottom, passes up a *tapered* glass tube, and is returned to the pipe after leaving the top. Within the tube there is a weighted float that is pushed upward by the flow. Since the cross-sectional area of the tube becomes larger as the float rises, the velocity of the flow becomes smaller, and the float eventually reaches an equilibrium level, indicated by gradations on the tube. This level is directly related to the flow in the pipe, and so the reading at the level of the float will indicate the flow. For horizontal pipes, a similar device will cause an obstruction to compress a spring a measured distance, and its position can be viewed through a glass tube. Both of these meters can be made to measure flow to an accuracy of about 99%, but they are somewhat limited, since they cannot be used to measure the flow of a very opaque fluid, such as oil.

Turbine Flow Meter.

For large-diameter pipes, such as from 40 mm to 300 mm, a turbine rotor can be installed within a section of the pipe so that the flow of fluid through the pipe will cause the rotor blades to turn, Fig. 10–25. For liquids these devices normally have only a few blades, but for gases more are required in order to generate enough torque to turn the blades. The greater the flow, the faster the blades will turn. One of them is marked, so as it turns, the rotation is detected by an electrical impulse, which is produced as the blade passes by a sensor. Turbine meters are often used to measure the flow of natural gas or water through municipal distribution systems. They can also be designed to be held in the hand, so that the blades can be turned into the wind, for example, to measure its speed. An **anemometer**, shown in the photo, works in a similar manner. It uses cups mounted on an axle, where measuring the rotation of the axle correlates to the wind speed.

Vortex Flow Meter.

If a cylindrical obstruction, called a **shedder bar**, is placed within the flow, as shown in Fig. 10–26, then as the fluid passes around the bar, the disturbance it produces will generate a trail of vortices, called a **Von Kármán vortex street**. The frequency f at which each vortex alternates off each side of the bar can be *measured* using a piezoelectric crystal, which produces a small voltage pulse for every fluctuation. This frequency f is proportional to the fluid velocity V and is related to it by the **Strouhal number**, $St = fD/V$, where the "characteristic length" is D, the diameter of the shedder bar. Since the Strouhal number will have a known constant value, within specific operating limits of the meter, the average velocity V can be determined, i.e., $V = fD/St$. The flow is then $Q = VA$, where A is the cross-sectional area of the meter. The advantage of using a vortex flow meter is that it has no moving parts, and it has an accuracy of about 99%. One disadvantage of its use is the head loss created by the disruption of the flow.

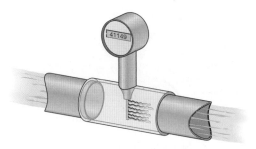

Fig. 10–27

Thermal Mass Flow Meter.
As the name implies, this device measures temperature to determine the velocity of a gas at a very localized region within the flow. One of the most popular types is called a ***constant-temperature anemometer***. It consists of a very small thin wire, usually made of tungsten and having, for example, a diameter of 0.5 μm and a length of 1 mm, Fig. 10–27. When it is placed in the flow, it is heated to a constant temperature, which is maintained electrically as the flow stream tends to cool it. The velocity of the flow can be correlated to the voltage that must be applied to the wire to maintain its temperature. Several of these sensors can even be arranged within a small region to measure the flow in two and three directions.

Since the wire is very fragile, care must be taken so that particulate matter within the gas will not damage or break it. Higher-velocity flow, or gases having a large number of contaminates, can be measured with less sensitivity using a ***hot-film anemometer***, which works on the same principle but consists of a sensor made of a thin metallic film attached to a much thicker ceramic support.

Positive Displacement Flow Meter.
One type of flow meter that can be used to determine the quantity of a liquid that flows past it is called a ***positive displacement flow meter***. It consists of a measuring chamber, such as the volume between the lobes of two gears within the meter, Fig. 10–28. By ensuring close tolerances between the lobes and casing, each revolution allows a measured amount of liquid to pass through. By counting these revolutions, either mechanically or through electrical pulses, the total amount of liquid can be measured.

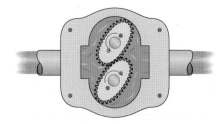

Fig. 10–28

Nutating Disk Meter.
This meter is commonly used to measure the supply of household water or the amount of gasoline drawn from a pump. It has an accuracy of about 99%. As shown in Fig. 10–29, it consists of an inclined disk that isolates a measured volume of the liquid within the chamber of the meter. The pressure from the liquid forces the disk to nutate, or turn about the vertical axis, since the center of the disk is fixed to a ball and spindle. The contained volume of liquid thereby passes through the chamber for every revolution about the axis. Each of these nutations can be recorded, either by a magnetic fluctuation caused by a magnet attached to the rotating disk, or by a gear-and-register arrangement attached to the spindle.

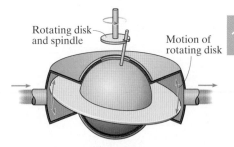

Rotating disk and spindle

Motion of rotating disk

Fig. 10–29

10

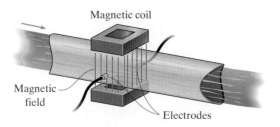

Magnetic flow meter

Fig. 10–30

Magnetic Flow Meter.

This type of flow meter requires very little maintenance and measures the average velocity of a liquid that can conduct electricity, such as seawater, wastewater, liquid sodium, and many types of acidic solutions. The principle of operation is based upon Michael Faraday's law, which states that the *voltage* induced across any conductor (liquid), as it moves at right angles through a magnetic field, is proportional to the velocity of the conductor. To measure the flow, two electrodes are placed on opposite sides of the inner pipe wall and attached to a volt meter. For a ***wafer-style magnetic flow meter***, Fig. 10–30, a magnetic field is established over the entire flow cross section by subjecting the coil within the wafers to an electric current. The volt meter then measures the electric potential or voltage between the electrodes, and this is directly proportional to the velocity of the flow.

Magnetic flow meters can have an accuracy of 99% to 99.5%, and they have been used on pipes up to 12 in. in diameter. The readings are very sensitive to entrained air bubbles at the electrodes and to any static electricity present within the fluid and pipe. For this reason, the pipe must be properly grounded for best performance.

Other Types.

There are other types of meters that can also be used to accurately measure the velocity at a small region within the flow. A ***laser Doppler flow meter*** is based on directing a laser beam toward a targeted area, and measuring the change in frequency of the beam after it is reflected off small particles that pass through the region.* This data is then converted to obtain the velocity in a particular direction. This technique offers high accuracy, and though expensive, it can be set up to determine the velocity components of particles within a region in all three directions. ***Ultrasonic flow meters*** also work on the Doppler principle. They send sound waves through the fluid, and the changes in frequency of any waves that are reflected back are measured using a piezoelectric transducer and then converted to determine the velocity. Finally, ***particle image velocimetry (PIV)*** is a method where very small particles are released into the fluid. Using a camera and laser strobe light, the speed and direction of the flow can then be measured by tracking the illuminated particles.

*The Doppler principle states that a *higher frequency* of a light or sound wave is produced when the source moves *toward* the observer, and a *lower frequency* is produced when it moves *away*. The effect is quite noticeable when one hears the siren on a moving police car or fire truck.

References

1. L. F. Moody, "Friction Factors for Pipe Flow," *Trans ASME*, Vol. 66, 1944, pp. 671–684.

2. F. Colebrook, "Turbulent flow in pipes with particular reference to the transition region between the smooth and rough pipe laws," *J Inst Civil Engineers*, London, Vol. 11, 1939, pp. 133–156.

3. V. Streeter, *Handbook of Fluid Dynamics*, McGraw-Hill, New York, NY.

4. V. Streeter and E. Wylie, *Fluid Mechanics*, 8th ed., McGraw-Hill, New York, NY, 1985.

5. S. E. Haaland, "Simple and explicit formulas for the friction-factor in turbulent pipe flow," *Trans ASME, J Fluids Engineering*, Vol. 105, 1983.

6. H. Ito, "Pressure losses in smooth pipe bends," *J Basic Engineering*, 82 D, 1960, pp. 131–134.

7. C. F. Lam and M. L Wolla, "Computer analysis of water distribution systems," *Proceedings of the ASCE, J Hydraulics Division*, Vol. 98, 1972, pp. 335–344.

8. H. S. Bean, *Fluid Meters: Their Theory and Application*, ASME, New York, NY, 1971.

9. R. J. Goldstein, *Fluid Mechanics Measurements*, 2nd ed., Taylor and Francis, New York, NY, 1996.

10. P. K. Swamee and A. K. Jain, "Explicit equations for pipe-flow problems," *Proceedings of the ASCSE, J Hydraulics Division*, Vol. 102, May 1976, pp. 657–664.

11. H. H. Brunn, *Hot-Wire Anemometry—Principles and Signal Analysis*, Oxford University Press, New York, NY, 1995.

12. R. W. Miller, *Flow Measurement Engineering Handbook*, 3rd ed., McGraw-Hill, New York, NY, 1996.

13. E. F. Brater et al. *Handbook of Hydraulics*, 7th ed., McGraw-Hill, New York, NY, 1996.

14. R. W. Jeppson, *Analysis of Flow in Pipe Networks*, Butterworth-Heinemann, Woburn, MA, 1976.

15. R. J. S. Pigott, "Pressure Losses in Tubing, Pipe, and Fittings," *Trans. ASME*, Vol. 73, 1950, pp. 679–688.

16. *Measurement of Fluid Flow on Pipes Using Orifice, Nozzle, and Venturi*, ASME MFC-3M-2004.

17. J. Vennard and R. Street, *Elementary Fluid Mechanics*, 5th ed., John Wiley and Sons, New York, NY.

18. *Fluid Meters*, ASME, 6th ed., New York, NY.

19. *Flow of Fluid through Valves, Fittings and Pipe*, Technical Paper A10, Crane Co., 2011.

PROBLEMS

Sec. 10.1

10–1. If air flows through the circular duct at 4 m/s, determine the pressure drop that occurs over a 6-m length of the duct. The friction factor is $f = 0.0022$. Take $\rho_a = 1.092 \text{ kg/m}^3$.

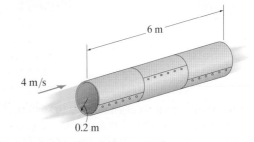

6 m

4 m/s

0.2 m

Prob. 10–1

10–2. A 45-mm-diameter commercial steel pipe is used to carry water at $T = 20°C$. If the head loss in a length of 2 m is 5.60 m, determine the volumetric flow in liters per second.

10–3. Oil flows through a 100-mm-diameter horizontal pipe at 4 m/s. If the pipe is made of cast iron, determine the friction factor. Take $\nu_o = 0.0344(10^{-3})$ m²/s.

*10–4. Air is forced through the circular duct. If the flow is 0.3 m³/s and the pressure drops 0.5 Pa for every 1 m of length, determine the friction factor for the duct. Take $\rho_a = 1.202$ kg/m³.

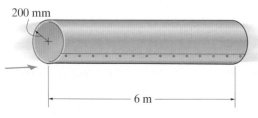

200 mm

6 m

Prob. 10–4

10–5. A nail gun operates using pressurized air, which is supplied through the 10-mm-diameter hose. The gun requires 680 kPa to operate with a 0.003 m³/s airflow. If the air compressor develops 700 kPa, determine the maximum allowable length of hose that can be used for its operation. Assume incompressible flow and a smooth hose. Take $\rho_a = 1.202$ kg/m³, $\nu_a = 15.1(10^{-6})$ m²/s.

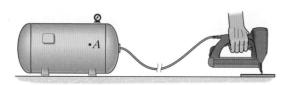

•A

Prob. 10–5

10–6. Determine the diameter of a horizontal 100-m-long PVC pipe that must transport 125 liter/s of turpentine oil so that the pressure drop does not exceed 500 kPa. Take $\varepsilon = 0.0015$ mm, $\rho_t = 860$ kg/m³, and $\mu_t = 1.49(10^{-3})$ N·s/m².

10–7. A pipe has a diameter of 60 mm. When water at 20°C flows through it at 6 m/s, it produces a head loss of 0.3 m when it is smooth. Determine the friction factor of the pipe if, years later, the same flow produces a head loss of 0.8 m.

*10–8. Air flows through the galvanized steel duct, with a velocity of 4 m/s. Determine the pressure drop along a 2-m length of the duct. Take $\rho_a = 1.202$ kg/m³, $\nu_a = 15.1(10^{-6})$ m²/s.

10–9. Determine the greatest air flow Q through the galvanized steel duct so that the flow remains laminar. What is the pressure drop along a 200-m-long section of the duct for this case? Take $\rho_a = 1.202$ kg/m³, $\nu_a = 15.1(10^{-6})$ m²/s.

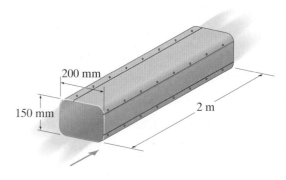

200 mm

150 mm

2 m

Probs. 10–8/9

10–10. Water at 20°C is required to flow through a horizontal commercial steel pipe so that it discharges at 0.013 m³/s. If the maximum pressure drop over a 5-m length is to be no more than 15 kPa, determine the smallest allowable diameter D of the pipe.

10–11. A 75-mm-diameter galvanized iron pipe, having a roughness of $\varepsilon = 0.2$ mm, is used to carry water at a temperature of 60°C and with a velocity of 3 m/s. Determine the pressure drop over its 12-m length if it is horizontal.

*10–12. Water at 20°C passes through the turbine T using a 150-mm-diameter commercial steel pipe. If the length of the pipe is 50 m and the discharge is 0.02 m³/s, determine the power extracted from the water by the turbine.

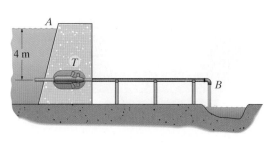

Prob. 10–12

10–13. Determine the power output required to pump 30 liter/s of crude oil through a 200-m-long horizontal cast iron pipe having a diameter of 100 mm. The pipe is open to the atmosphere at its end. Compare this power requirement with pumping water through the same pipe. The temperature is $T = 20$°C for both cases.

10–14. The 20-mm-diameter copper coil is used for a solar hot water heater. If water at an average temperature of $T = 50$°C passes through the coil at 9 liter/min, determine the major head loss that occurs within the coil. Neglect the length of each bend. Take $\varepsilon = 0.03$ mm for the coil.

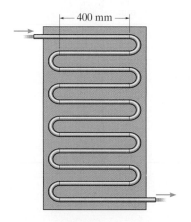

Prob. 10–14

10–15. Water flows through the 50-mm-diameter pipe. If the pressure at A and B is the same, determine the flow. Take $f = 0.035$.

Prob. 10–15

*10–16. Oil flows through the 50-mm-diameter pipe at 0.009 m³/s. If the friction factor is $f = 0.026$, determine the pressure drop that occurs over the 80-m length. Take $\rho_o = 900$ kg/m³.

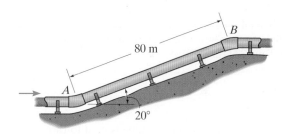

Prob. 10–16

10–17. Oil flows through a 50-mm-diameter cast iron pipe. If the pressure drop over a 10-m-long horizontal segment is 18 kPa, determine the mass flow through the pipe. Take $\rho_o = 900$ kg/m³, $\nu_o = 0.430(10^{-3})$ m²/s.

10–18. If a pipe has a diameter D and a friction factor f, by what percent will the pressure drop in the pipe increase if the volumetric flow is doubled? Assume that f is constant due to a very large Reynolds number.

10

10–19. Oil flows at 2 m/s through a horizontal 50-mm-diameter galvanized iron pipe. Determine if the flow is laminar or turbulent. Also, find the pressure drop that occurs over a 10-m length of the pipe. Take $\rho_o = 850 \text{ kg/m}^3$, $\mu_o = 0.0678 \text{ N} \cdot \text{s/m}^2$.

*10–20. A galvanized steel pipe is required to carry water at 20°C with a velocity of 3 m/s. If the pressure drop over its 200-m horizontal length is not to exceed 15 kPa, determine the required diameter of the pipe.

10–21. A 75-mm-diameter galvanized iron pipe, having a roughness of $\varepsilon = 0.2$ mm, is to be used to carry water at a temperature of 60°C and with a velocity of 3 m/s. Determine the pressure drop over its 12-m length if the pipe is vertical and the flow is upward.

10–22. Methane at 20°C flows through a 30-mm-diameter horizontal pipe at 8 m/s. If the pipe is 200 m long and the roughness is $\varepsilon = 0.4$ mm, determine the pressure drop over the length of the pipe.

10–23. The galvanized iron pipe is used to carry water at 20°C with a velocity of 3 m/s. Determine the pressure drop that occurs over a 4-m length of the pipe.

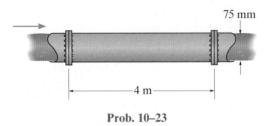

75 mm

4 m

Prob. 10–23

*10–24. Water is pumped from the river through a 40-mm-diameter hose having a length of 3 m. Determine the maximum volumetric discharge from the hose at C so that cavitation will not occur within the hose. The friction factor is $f = 0.028$ for the hose, and the gage vapor pressure for water is -98.7 kPa.

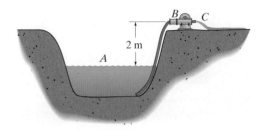

B C

2 m

A

Prob. 10–24

10–25. For a given volumetric flow, the pressure drop is 5 kPa in a horizontal pipe. Determine the pressure drop if the flow is doubled. The flow remains laminar.

10–26. The 50-mm-diameter pipe has a roughness of $\varepsilon = 0.01$ mm. If the discharge of 20°C water is 0.006 m³/s, determine the pressure at A.

10–27. The 50-mm-diameter pipe has a roughness of $\varepsilon = 0.01$ mm. If the water has a temperature of 20°C and the pressure at A is 50 kPa, determine the discharge at B.

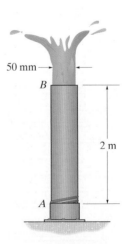

50 mm

B

2 m

A

Probs. 10–26/27

*10–28. Water at 20°C is pumped from the reservoir at A and flows through the 250-mm-diameter smooth pipe. If the discharge at B is 0.3 m³/s, determine the required power output for the pump that is connected to a 200-m length of the pipe. Draw the energy line and the hydraulic grade line for the pipe. Neglect any elevation differences.

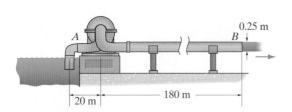

Prob. 10–28

10–29. Sewage, assumed to be water at 20°C, is pumped from the wet well using a pump and a 50-mm-diameter pipe. Determine the maximum discharge from the pump without causing cavitation. The friction factor is $f = 0.026$. The (gage) vapor pressure for water at 20°C is -98.7 kPa. Neglect the loss in the submerged segment of the pipe.

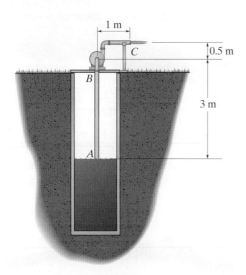

Prob. 10–29

10–30. Sewage, assumed to be water at 20°C, is pumped from the wet well using a pump and a 50-mm-diameter pipe having a friction factor of $f = 0.026$. If the pump delivers 500 W of power to the water, determine the discharge from the pump.

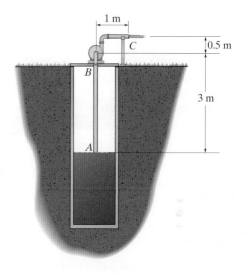

Prob. 10–30

10–31. Water is delivered into the truck using a pump that creates a flow of 300 liter/min through a 40-mm-diameter hose. If the total length of the hose is 8 m, the friction factor is $f = 0.018$, and the tank is open to the atmosphere, determine the power that must be supplied by the pump.

*10–32. Water is delivered at 0.003 m³/s into the truck using a pump and a 40-mm-diameter hose. If the length of the hose from C to A is 10 m, and the friction factor is $f = 0.018$, determine the power output of the pump.

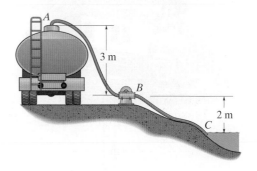

Probs. 10–31/32

Sec. 10.2–10.3

10–33. The 30-mm-diameter 20-m-long commercial steel pipe transports water at 20°C. If the pressure at A is 200 kPa, determine the volumetric flow through the pipe.

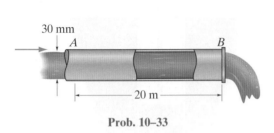

Prob. 10–33

10–34. The geothermal heat pump works on a closed-loop system. The loop consists of plastic pipe having a diameter of 40 mm, a total length of 40 m, and a roughness factor of $\varepsilon = 0.003$ mm. If water at 40°C is to have a flow of $0.002 \text{ m}^3/\text{s}$, determine the power output of the pump. Include the minor losses for the 180° bend, $K_L = 0.6$, and for each of the two 90° elbows, $K_L = 0.4$.

Prob. 10–34

10–35. Determine the power extracted from the water by the turbine at C if the discharge from the pipe at B is $0.02 \text{ m}^3/\text{s}$. The pipe is 38 m long, it has a diameter of 100 mm, and the friction factor is $f = 0.026$. Also, draw the energy line and the hydraulic grade line for the pipe. Neglect minor losses.

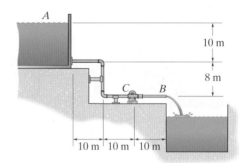

Prob. 10–35

*__10–36.__ Determine the power the pump must supply in order to discharge $0.02 \text{ m}^3/\text{s}$ of water at B from the 100-mm-diameter hose. The friction factor is $f = 0.028$, and the hose is 95 m long. Neglect minor losses.

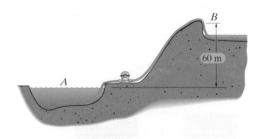

Prob. 10–36

10–37. Water is to be delivered at $0.04 \text{ m}^3/\text{s}$ to point B on the ground, 500 m away from the reservoir. Determine the smallest-diameter pipe that can be used if the pump supplies 40 kW of power. Neglect any elevation changes, and take $f = 0.02$.

Prob. 10–37

10–38. The large tank is filled with water at 20°C to the depth shown. If the gate valve at C is fully opened, determine the power of the water flowing from the end of the nozzle at B. Also, what is the head loss in the system? The galvanized iron pipe is 36 m long and has a diameter of 75 mm. Neglect the minor loss through the nozzle, but include the minor losses at the flush entrance, the two elbows, and the gate valve.

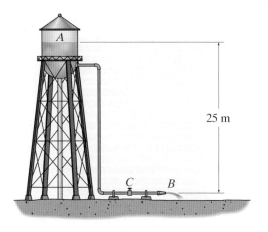

25 m

Prob. 10–38

10–39. Gasoline from the tank of an automobile A is pumped at B through the fuel filter C, and then to the fuel injectors on the engine. The fuel line is stainless steel 4-mm-diameter tubing. Each fuel injector has a nozzle diameter of 0.5 mm. If the loss coefficient is $K_L = 0.5$ at the flush entrance of the fuel tank, $K_L = 1.5$ through the filter, and $K_L = 4.0$ at each nozzle, find the required power output of the pump to deliver fuel at the rate of 0.15 liter/min to four of the cylinders that are under an average pressure of 300 kPa. The fuel line is 2.5 m long. Take $\varepsilon = 0.006$ mm. Assume that the nozzles and the tank are on the same level.

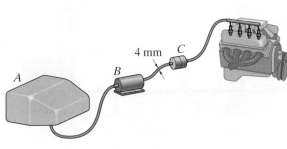

4 mm C

B

A

Prob. 10–39

*10–40. Air at a temperature of 40°C flows through the duct at A with a velocity of 2 m/s. Determine the change in pressure between A and B. Account for the minor loss caused by the sudden change in diameter of the duct.

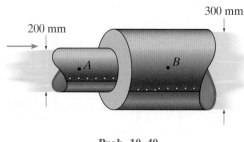

300 mm

200 mm

Prob. 10–40

10–41. Water is stored in the tank using a pump at A that supplies 300 W of power. If the 50-mm-diameter pipe has a friction factor of $f = 0.022$, determine the flow into the tank at the instant shown. The tank is open at the top. Include the minor losses of the 90° elbow and the discharge into the tank at C.

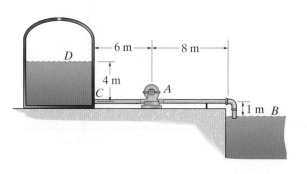

D — 6 m — — 8 m —

4 m

C | A

1 m B

Prob. 10–41

10

10–42. The pump A and pipe system are used to transport oil to the tank. If the pressure developed by the pump is 400 kPa, and the filter at B has a loss coefficient of $K_L = 2.30$, determine the discharge from the pipe at C. The 50-mm-diameter pipe is made of cast iron. Include the minor losses of the filter and three elbows. Take $\rho_o = 890 \text{ kg/m}^3$ and $\nu_o = 52.0\left(10^{-6}\right) \text{ m}^2/\text{s}$.

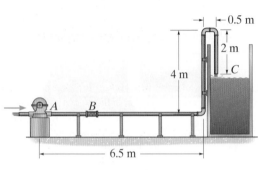

Prob. 10–42

10–43. Water at 20°C flows through the 20-mm-diameter galvanized iron pipe such that it discharges at C from the fully opened gate valve B at 0.003 m³/s. Determine the required pressure at A. Include the minor losses of the four elbows and gate valve.

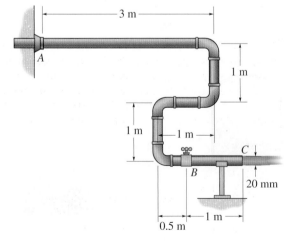

Prob. 10–43

*10–44. Water at 80°C enters the radiator at A with an average velocity of 1.5 m/s and a pressure of 400 kPa. If each 180° bend has a minor loss coefficient of $K_L = 1.03$, determine the pressure at the exit B. The copper pipe has a diameter of 8 mm. Take $\varepsilon = 0.0015$ mm. The radiator is in the vertical plane.

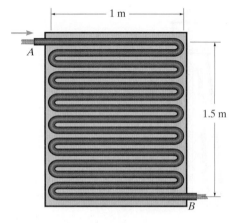

Prob. 10–44

10–45. The pump A and pipe system are used to transport oil to the tank. Determine the required pressure developed by the pump in order to provide a discharge into the tank of 0.003 m³/s. The filter at B has a loss coefficient of $K_L = 2.30$. The 50-mm-diameter pipe is made of cast iron. Include the minor losses of the filter and three elbows. Take $\rho_o = 890 \text{ kg/m}^3$ and $\nu_o = 52.0\left(10^{-6}\right) \text{ m}^2/\text{s}$.

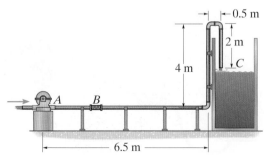

Prob. 10–45

10–46. Water at 20°C is pumped from the reservoir A using a pump that supplies 3 kW of power. Determine the discharge at C if the pipe is made of galvanized iron and has a diameter of 50 mm. Neglect minor losses.

10–47. Water at 20°C is pumped from the reservoir A using a pump that supplies 3 kW of power. Determine the discharge at C if the pipe is made of galvanized iron and has a diameter of 50 mm. Include the minor losses of the four elbows.

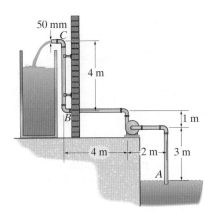

Probs. 10–46/47

*10–48.** Water from the reservoir A is pumped into the large tank B. If the top of the tank is open, and the power output of the pump is 500 W, determine the volumetric flow into the tank when $h = 2$ m. The cast iron pipe has a total length of 6 m and a diameter of 50 mm. Include the minor losses for the elbow and sudden expansion. The water has a temperature of 20°C.

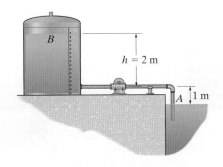

Prob. 10–48

10–49. Water at $T = 20$°C flows from the open tank through the 50-mm-diameter galvanized iron pipe. Determine the discharge at the end B if the globe valve is fully opened. The length of the pipe is 50 m. Include the minor losses of the flush entrance, the four elbows, and the globe valve.

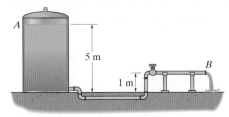

Prob. 10–49

10–50. Water flows down through the vertical pipe at a rate of 3 m/s. If the differential elevation of the mercury manometer is 30 mm as shown, determine the loss coefficient K_L for the filter C contained within the pipe. $\rho_{Hg} = 13\,550$ kg/m^3.

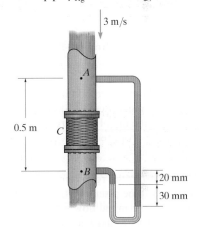

Prob. 10–50

10–51. If the faucet (gate valve) at E is fully opened and the pump produces a pressure of 350 kPa at A, determine the pressure just to the right of the tee connection C. The valve at B remains closed. The pipe and the faucet both have an inner diameter of 30 mm, and $f = 0.04$. Include the minor losses of the tee, the two elbows, and the gate valve.

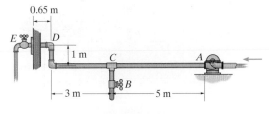

Prob. 10–51

10

Sec. 10.4

*10–52. When the globe valve is fully opened, water at 20°C is discharged at 0.003 m³/s from C. Determine the pressure at A. The galvanized iron pipes AB and BC have diameters of 60 mm and 30 mm, respectively. Consider the minor losses only from the elbows and the globe valve.

10–55. The horizontal galvanized iron pipe system is used for irrigation purposes and delivers water to two different outlets. If the pump delivers a flow of 0.01 m³/s in the pipe at A, determine the discharge at each outlet, C and D. Neglect minor losses. Each pipe has a diameter of 30 mm. Also, what is the pressure at A?

*10–56. Determine the discharge at each outlet, C and D, for the pipe network in Prob. 10–55 by considering the minor losses of the elbow and tee.

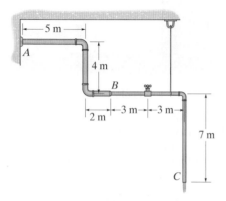

Prob. 10–52

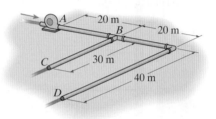

Probs. 10–55/56

10–53. Water from the reservoir at A drains through the 30-mm-diameter pipe assembly. If commercial steel pipe is used, determine the initial discharge into B when the valve E is closed and F is opened. Neglect any minor losses. Take $\nu_w = 1.00(10^{-6})$ m²/s.

10–54. Water from the reservoir at A drains through the 30-mm-diameter pipe assembly. If commercial steel pipe is used, determine the initial flow into pipe D from reservoir A when both valves E and F are fully opened. Neglect any minor losses. Take $\nu_w = 1.00(10^{-6})$ m²/s.

10–57. The two water tanks are connected together using the 100-mm-diameter pipes. If the friction factor for each pipe is $f = 0.024$, determine the flow out of tank C when the valve at A is opened, while the valve at B remains closed. Neglect any minor losses.

10–58. The two water tanks are connected together using the 100-mm-diameter galvanized iron pipes. Determine the flow out of tank C when both valves A and B are opened. Neglect any minor losses. Take $\nu_w = 1.00(10^{-6})$ m²/s.

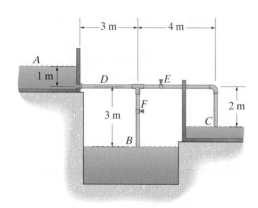

Probs. 10–53/54

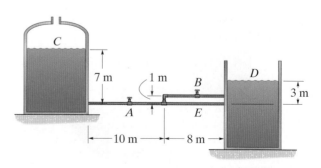

Probs. 10–57/58

CHAPTER REVIEW

The frictional loss within a pipe can be determined analytically for laminar flow. It is defined by the friction factor $f = 64/\text{Re}$. For transitional and turbulent flow, we can determine f either from the Moody diagram or by using an empirical equation that fits the curves of the Moody diagram.	
Once f is obtained, then the head loss, called a "major loss," can be determined using the Darcy–Weisbach equation.	$$h_L = f \frac{L}{D} \frac{V^2}{2g}$$
If a pipe network has fittings and transitions, then the head loss produced by these connections must be taken into account. These losses are called "minor losses."	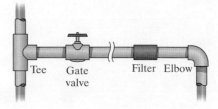 $$h_L = K_L \frac{V^2}{2g}$$
Pipe systems can be arranged in *series*, in which case the flow through each pipe must be the same, and the head loss is the total loss for all the pipes.	$Q = Q_1 = Q_2 = Q_3 \qquad h_L = h_{L1} + h_{L2} + h_{L3} + h_{\text{minor}}$
Pipes can also be arranged in *parallel*, in which case the total flow is the accumulation of flow from each branch pipe in the system, and the head loss for each branch pipe is the same.	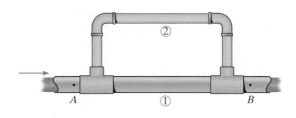 $Q_A = Q_B = Q_1 + Q_2 \qquad (h_L)_1 = (h_L)_2$
Flow through a pipe can be measured in several different ways. These include using a venturi meter, nozzle meter, or orifice meter. Also, a rotometer, turbine flow meter, as well as several other meters can be used.	

10

Chapter 11

(© B.A.E. Inc./Alamy)

Airflow over the surface of this airplane creates both drag and lift. Analysis of these forces requires an experimental investigation.

Viscous Flow over External Surfaces

11.1 The Concept of the Boundary Layer

When a fluid flows over a flat surface, the layer of fluid particles adjacent to the surface will have *zero velocity*, and farther from the surface, the velocity of each layer will increase until they reach the free-stream velocity **U** as shown in Fig. 11–1. This behavior is caused by the shear stress that acts between the layers of the fluid, and for a Newtonian fluid this stress is directly proportional to the velocity gradient, $\tau = \mu(du/dy)$. Notice that this gradient and the shear stress are *largest at the surface*, but both taper off, until farther from the surface, the gradient and the shear stress approach zero. Here the flow behaves as if it is nonviscous, since it is *uniform*, resulting in little or no shear or sliding between adjacent layers of fluid. In 1904, Ludwig Prandtl recognized this difference in flow behavior, and termed the localized region where the velocity is variable the *boundary layer*.

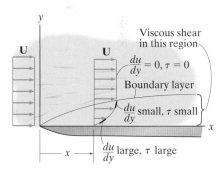

Shear is proportional to the velocity gradient

Fig. 11–1

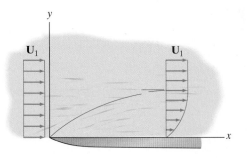

Slow-moving fluids, or fluids with high viscosity, produce a thick boundary layer

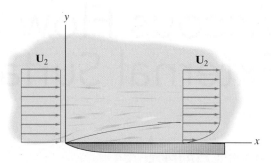

Fast-moving fluids, $U_2 > U_1$, or fluids with low viscosity, produce a thin boundary layer

Fig. 11–2

Understanding boundary layer formation is very important when it becomes necessary to determine a body's *resistance to movement* through a fluid. The design of propellers, wings, turbine blades, and other mechanical and structural elements that interact with moving fluids depends on an analysis of the flow acting within the boundary layer. In this chapter we will study only the effects created by a thin boundary layer, which occurs when the fluid has a low viscosity and the flow over the surface is *relatively fast*. As shown in Fig. 11–2, *slow*-moving fluids, or those having a high viscosity, produce a thick boundary layer, and understanding its effect on the flow requires a specialized experimental analysis or numerical modeling using a computer. See Ref. [5].

Boundary Layer Description.   The development or growth of the boundary layer can best be illustrated by considering the steady uniform flow of a fluid over a long flat plate, which is similar to what occurs along the hull of a ship, a flat section on an airplane, or the side of a building. The basic features of the boundary layer can be divided into three regions. These are shown in Fig. 11–3.

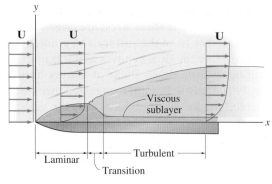

Boundary layer over a flat plate
(greatly exaggerated y scale)

Fig. 11–3

Laminar Flow.   As the fluid flows over the leading edge of the plate with a uniform free-stream velocity **U**, the particles on its surface stick to the surface, while those just above it *slow down* and build up in *smooth layers* farther along the plate's length. In this initial region the flow is laminar. Farther along the plate, the thickness of the boundary layer increases as more and more layers of fluid are influenced by the effect of viscous shear.

Transitional Flow.   A point along the plate is reached where the laminar flow *buildup becomes unstable* and then tends to break down. This is a *transition region*, where some of the fluid particles begin to undergo turbulence, characterized by the random mixing of large groups of particles moving from one layer of fluid to another.

Turbulent Flow.   The mixing of fluid that occurs causes the thickness of the boundary layer to grow rapidly, eventually forming a *turbulent boundary layer*. In spite of this transformation from laminar to turbulent flow, there always remains, below the turbulent boundary layer, a very thin **laminar** or **viscous sublayer** of fluid that is "slow moving" since the fluid must cling along the surface of the plate.

Boundary Layer Thickness.   *At each location* along the plate, the velocity profile within the boundary layer's thickness will *asymptotically* approach the free-stream velocity. Since this thickness is not well defined, engineers use three methods for specifying its value.

Disturbance Thickness.   The simplest way to report the boundary layer's thickness at each location x is to define it as the height δ where the maximum velocity reached is equal to a certain percentage of the free-stream velocity. The agreed-upon value is $u = 0.99U$, as shown in Fig. 11–4.

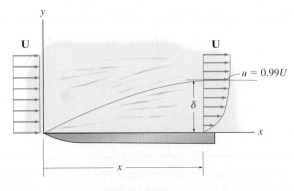

Disturbance thickness

Fig. 11–4

11

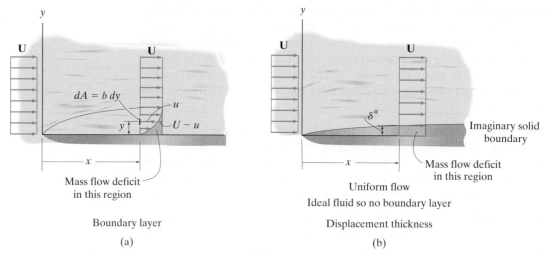

Fig. 11–5

Displacement Thickness.

The boundary layer thickness can also be specified as a *displacement thickness*, δ^*. This refers to the distance the actual surface must be *displaced*, so that if we had an *ideal fluid*, then the mass flow with this new boundary, Fig. 11–5b, would be the same as it is for the real fluid, Fig. 11–5a. This concept is often used to design wind tunnels and the intake of a jet engine.

To determine the distance δ^*, we must find the decrease in mass flow or the *mass flow deficit* for each case. If the surface (or plate) has a width b, then in the case of the real fluid, Fig. 11–5a, the mass flow at y through the differential area $dA = b\,dy$ is $d\dot{m} = \rho u\,dA = \rho u(b\,dy)$. If an ideal fluid were present, then viscous effects would not occur, and so $u = U$, and the mass flow at y would then be $d\dot{m}_0 = \rho U(b\,dy)$. The *mass flow deficit* due to the viscosity is therefore $d\dot{m}_0 - d\dot{m} = \rho(U - u)(b\,dy)$. For the entire boundary layer, integration over its height is necessary to determine this total deficit, shown in dark blue shade in Fig. 11–5a. This deficit must be *the same* for the ideal fluid in Fig. 11–5b. Since it is $\rho U(b\delta^*)$, then,

$$\underbrace{\rho U\left(b\delta^*\right)}_{\substack{\text{Loss in} \\ \text{uniform flow}}} = \underbrace{\int_0^\infty \rho\left(U - u\right)(b\,dy)}_{\substack{\text{Loss in} \\ \text{boundary layer}}}$$

Since ρ, U, and b are constant, we can write this equation as

$$\boxed{\delta^* = \int_0^\infty \left(1 - \frac{u}{U}\right)dy} \qquad (11\text{–}1)$$

Therefore, to determine the displacement thickness, the velocity profile $u = u(y)$ of the boundary layer must be known. If it is, then this integral can be evaluated, either analytically or numerically, at each location x along the plate.

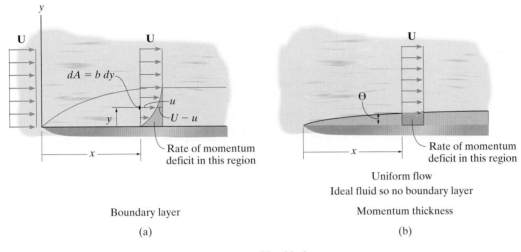

Rate of momentum
deficit in this region

Uniform flow
Ideal fluid so no boundary layer

Boundary layer

Momentum thickness

(a)

(b)

Fig. 11–6

Momentum Thickness. Another way to treat the velocity disturbance brought about by the boundary layer is to consider how the actual surface should be displaced, so that the *rate of momentum* of the flow would be the *same* as if the fluid were ideal. This change in height of the surface is called the *momentum thickness* Θ, Fig. 11–6b. It represents the loss of momentum in the boundary layer compared to having ideal flow. To find it we must determine the rate of momentum flow deficit in each case. If the plate has a width b, then for the real fluid, shown in Fig. 11–6a, at height y the fluid passing through the area dA has a rate of momentum of $d\dot{m}\, u = \rho(dQ)u = \rho(u\, dA)u$. Since $dA = b\, dy$, then $d\dot{m}\, u = \rho(ub\, dy)u$. However, if the mass flow $d\dot{m}$ had a velocity U, then the rate of momentum flow *deficit* would be $\rho[ub\, dy](U - u)$. For the case of the ideal fluid, Fig. 11–6b, the rate of momentum flow deficit is $\rho dQU = \rho(U\Theta b)U$. Therefore, we require

$$\rho(U\Theta b)U = \int_0^\infty \rho u\,(U - u)b\, dy$$

or

$$\boxed{\Theta = \int_0^\infty \frac{u}{U}\left(1 - \frac{u}{U}\right)dy} \qquad (11\text{–}2)$$

In summary, we now have three definitions for the boundary layer thickness: δ refers to the height to which the boundary layer disturbs the flow, up to where the velocity becomes $0.99U$; and δ^* and Θ define the heights to which the surface must be displaced or repositioned so that if the fluid were *ideal* and flowing with the free-stream velocity $\mathbf{U}$ it would produce the same rates of mass and momentum flow, respectively, as in the case of the *real* fluid.

11

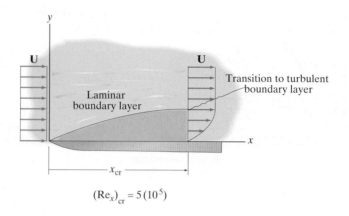

$$(\mathrm{Re}_x)_{\mathrm{cr}} = 5(10^5)$$

Fig. 11–7

Boundary Layer Classification. The magnitude of shear stress a fluid develops on the surface of a plate depends upon the *type of flow* within the boundary layer, and so it is important to find the point where laminar flow begins to transit to turbulent flow. The Reynolds number can be used to do this, since both inertia and viscous forces play a role in boundary layer development. For flow along a flat plate, we will define the Reynolds number on the basis of the "characteristic length" x, which is the distance downstream from the front of the plate, Fig. 11–7. Therefore,

$$\mathrm{Re}_x = \frac{Ux}{\nu} = \frac{\rho Ux}{\mu} \qquad (11\text{–}3)$$

From experiments, it has been found that laminar flow begins to break down at about $\mathrm{Re}_x = 1(10^5)$, but it can be as high as $3(10^6)$. The specific value when this happens is rather sensitive to the surface roughness of the plate, the uniformity of the flow, and any temperature or pressure changes occurring along the plate's surface. See Ref. [11]. In this text, to establish a *consistent value*, we will take this critical value for the Reynolds number to be

$$\boxed{(\mathrm{Re}_x)_{\mathrm{cr}} = 5(10^5)}$$
$$\text{Flat plate}$$

For example, for air at a temperature of 20°C and standard pressure, flowing at 25 m/s, the boundary layer will maintain laminar flow up to a critical distance of $x_{\mathrm{cr}} = (\mathrm{Re}_x)_{\mathrm{cr}}\nu/U = 5(10^5)\left[15.1(10^{-6})\ \mathrm{m^2/s}\right]/(25\ \mathrm{m/s})$ $= 0.302$ m from the front of the plate.

11.2   Laminar Boundary Layers

If $\text{Re}_x \leq 5(10^5)$, then only a laminar boundary layer will form on the surface. In this section, we will discuss how the velocity and the shear stress vary within this type of boundary layer. To do this, it will be necessary to satisfy both the continuity and momentum equations, while accounting for the *viscous flow* within the boundary layer. In Sec. 7.11, we showed that the components of the momentum equation, written for a differential fluid element, became the Navier–Stokes equations. These equations form a complex set of partial differential equations that have no known general solution, although, when they are applied within the region of the boundary layer, with certain assumptions, they can be simplified to produce a usable solution.

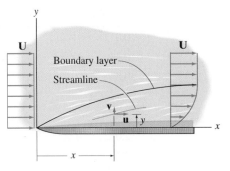

Fig. 11–8

Here we will present this solution and discuss its application to the flow over the flat plate in Fig. 11–8, where the fluid is incompressible and has steady *laminar flow*. Experiments have shown that as the fluid moves over the plate, the streamlines for the flow gradually begin to curve upward, so that a particle located at (x, y) has velocity components u and v. For *high* Reynolds numbers, the boundary layer is very thin, and so the vertical component v will be much smaller than the horizontal component u. Also, due to viscosity, the changes of u and v in the y direction, that is, $\partial u/\partial y$, $\partial v/\partial y$, and $\partial^2 u/\partial y^2$, will be *much greater* than changes $\partial u/\partial x$, $\partial v/\partial x$, and $\partial^2 u/\partial x^2$ in the x direction. Furthermore, because the streamlines within the boundary layer only *slightly* curve upward, the pressure variation in the y direction that causes this curvature is *practically constant*, so that $\partial p/\partial y \approx 0$. Finally, since the pressure above the boundary layer is constant, then within the boundary layer, due to its small height, $\partial p/\partial x \approx 0$. With these assumptions, Prandtl was able to reduce the three Navier–Stokes equations, Eq. 7–75, to just one in the x direction, and it, along with the continuity equation, becomes

$$u \frac{\partial u}{\partial x} + v \frac{\partial u}{\partial y} = \nu \frac{\partial^2 u}{\partial y^2}$$

$$\frac{\partial u}{\partial x} + \frac{\partial v}{\partial y} = 0$$

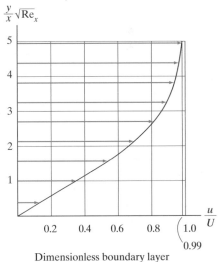

Dimensionless boundary layer velocity profile for laminar flow

Fig. 11–9

To obtain the velocity distribution within the boundary layer it is necessary to solve these equations simultaneously for u and v, using the boundary conditions $u = v = 0$ at $y = 0$ and $u = U$ at $y = \infty$. In 1908 Paul Blasius, who was one of Prandtl's graduate students, did this using a numerical analysis. See Ref. [16]. He presented his results in the form of a curve, shown in Fig. 11–9, that is plotted on axes of dimensionless velocity u/U versus the dimensionless parameter $(y/x)\sqrt{\text{Re}_x}$. Here Re_x is defined by Eq. 11–3. For convenience, numerical values for this curve are listed in Table 11–1. Thus, for a specified point (x, y) and free-stream velocity U, Fig. 11–8, the velocity u for a particle within the boundary layer can be determined from the curve or the table. As expected, the solution indicates that the velocity of the boundary layer approaches the free-stream velocity asymptotically, so that $u/U \rightarrow 1$ as $y \rightarrow \infty$.

11

TABLE 11–1 The Blasius Solution — Laminar Boundary Layer

$\dfrac{y}{x}\sqrt{\mathrm{Re}_x}$	u/U	$\dfrac{y}{x}\sqrt{\mathrm{Re}_x}$	u/U
0.0	0.0	2.8	0.81152
0.4	0.13277	3.2	0.87609
0.8	0.26471	3.6	0.92333
1.2	0.39378	4.0	0.95552
1.6	0.51676	4.4	0.97587
2.0	0.62977	4.8	0.98779
2.4	0.72899	∞	1.00000

Disturbance Thickness. This thickness, $y = \delta$, is the point where the velocity of flow u is at 99% of the free-stream velocity, that is, $u/U = 0.99$. From Blasius's solution in Fig. 11–9, it occurs when

$$\frac{y}{x}\sqrt{\mathrm{Re}_x} = 5.0$$

Thus,

$$\delta = \frac{5.0}{\sqrt{\mathrm{Re}_x}}\,x \qquad (11\text{–}4)$$

Laminar boundary layer
thickness

Using this result, it is worth noting how thin a laminar boundary layer can actually be. For example, if the velocity U is great enough so that the Reynolds number reaches its critical value of $(\mathrm{Re}_x)_{\mathrm{cr}} = 5(10^5)$ at $x_{\mathrm{cr}} = 100\,\mathrm{mm}$, then the laminar boundary layer thickness at this distance is only 0.707 mm.

Displacement Thickness. If Blasius's solution for u/U, Fig. 11–9, is substituted into Eq. 11–1, it can be shown that after numerical integration, the displacement thickness for laminar boundary layers becomes

$$\delta^* = \frac{1.721}{\sqrt{\mathrm{Re}_x}}\,x \qquad (11\text{–}5)$$

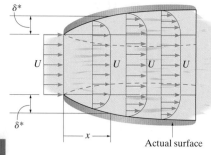

Fig. 11–10

Once obtained, this thickness can then be used to *simulate* the new location of the solid boundary over which laminar flow is considered inviscid or ideal. For example, one can *increase* the opening of a flow chamber to accommodate δ^*, Fig. 11–10. This way, the mass flow through the chamber will be *uniform over a constant cross section*.

Momentum Thickness.

To obtain the momentum thickness of the boundary layer, we must integrate Eq. 11–2, using Blasius's solution for u/U. After a numerical integration, the result becomes

$$\Theta = \frac{0.664}{\sqrt{Re_x}}x \qquad (11\text{--}6)$$

Because $Re_x = Ux/\nu$, notice that in each of these cases the thickness δ, δ^*, and Θ all *decrease* as the Reynolds number or the free-stream velocity U *increases*.

Shear Stress.

For the laminar boundary layer shown in Fig. 11–11a, a *Newtonian fluid* exerts a shear stress on the plate's surface of

$$\tau_0 = \mu\left(\frac{du}{dy}\right)_{y=0} \qquad (11\text{--}7)$$

The velocity gradient at $y = 0$ can be obtained by measuring it from the graph of Blasius's solution, Fig. 11–9. It is shown to be

$$\frac{d\left(\dfrac{u}{U}\right)}{d\left(\dfrac{y}{x}\sqrt{Re_x}\right)}\Bigg|_{y=0} = 0.332$$

For a specific location x and constant values of U, and ν, the Reynolds number $Re_x = Ux/\nu$ will also be constant, and so $d(u/U) = du/U$ and $d\left(y\sqrt{Re_x}/x\right) = dy\sqrt{Re_x}/x$. After rearrangement, the derivative du/dy then becomes

$$\frac{du}{dy} = 0.332\left(\frac{U}{x}\right)\sqrt{Re_x}$$

Substituting this into Eq. 11–7 gives the result

$$\tau_0 = 0.332\mu\left(\frac{U}{x}\right)\sqrt{Re_x} \qquad (11\text{--}8)$$

With this equation, we can now calculate the shear stress on the plate at any position x from the plate's leading edge. Notice that this stress will become smaller, as the distance x increases, Fig. 11–11a.

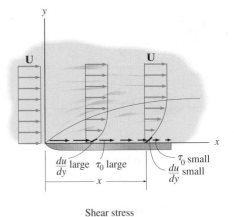

$\dfrac{du}{dy}$ large τ_0 large

τ_0 small
$\dfrac{du}{dy}$ small

Shear stress

(a)

Fig. 11–11

Since τ_0 causes a drag on the plate, in fluid mechanics we often express this effect by writing it in terms of the product of the fluid's dynamic head, that is,

$$\tau_0 = c_f\left(\frac{1}{2}\rho U^2\right) \qquad (11\text{–}9)$$

Substituting Eq. 11–8 for τ_0, and using $\text{Re}_x = \rho U x/\mu$, the **skin friction coefficient** c_f is determined from

$$c_f = \frac{0.664}{\sqrt{\text{Re}_x}} \qquad (11\text{–}10)$$

Like the shear stress, c_f becomes smaller as the distance x from the leading edge of the plate increases.

Friction Drag. If the resultant force acting on the surface of the plate is to be determined, then it is necessary to integrate Eq. 11–8 over the surface. This force is called the **friction drag**, and if the width of the plate is b and its length is L, Fig. 11–11b, we have

$$F_{Df} = \int_A \tau_0\, dA = \int_0^L 0.332\mu\left(\frac{U}{x}\right)\left(\sqrt{\frac{\rho U x}{\mu}}\right)(b\,dx) = 0.332b\ U^{3/2}\sqrt{\mu\rho}\int_0^L \frac{dx}{\sqrt{x}}$$

$$F_{Df} = \frac{0.664b\rho U^2 L}{\sqrt{\text{Re}_L}} \qquad (11\text{–}11)$$

where

$$\text{Re}_L = \frac{\rho U L}{\mu} \qquad (11\text{–}12)$$

Experiments have been carried out to measure the friction drag on a plate caused by laminar boundary layers, and the results have been in close agreement with those obtained from Eq. 11–11.

A dimensionless **friction drag coefficient** for a plate of length L can be defined in terms of the fluid's dynamic head in a manner similar to Eq. 11–9. Since the friction drag acts over the area bL, we have

$$F_{Df} = C_{Df}bL\left(\frac{1}{2}\rho U^2\right) \qquad (11\text{–}13)$$

Substituting Eq. 11–11 for F_{Df} and solving for C_{Df}, we get

$$C_{Df} = \frac{1.328}{\text{Re}_L < 5\left(10^5\right)} \qquad (11\text{–}14)$$

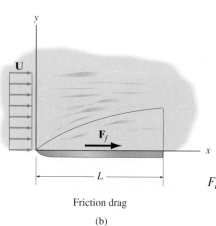

Friction drag

(b)

Fig. 11–11 (cont.)

Important Points

- A very thin boundary layer will form over the surface of a flat plate when a low-viscosity, fast-moving fluid flows over it. Within this layer, the shear stress changes the fluid's velocity profile such that it is zero at the plate's surface and then asymptotically approaches the free-stream velocity U of the flow above the surface.

- Due to viscosity, the thickness of the boundary layer will grow along the length of the plate. As it does, the flow within the boundary layer can change from laminar, to transitional, to turbulent, provided the plate over which it is flowing is long enough. As a convention in this text, we have defined the maximum length x_{cr} of a *laminar* boundary layer as having a Reynolds number of $(Re_x)_{cr} = 5(10^5)$.

- One of the main purposes of using boundary layer theory is to determine the shear-stress distribution that the flow exerts on a plate's surface and, from this, to determine the friction drag on the surface.

- At any location along the plate, we define the *disturbance thickness* δ of the boundary layer as the height at which the flow reaches a velocity of $0.99U$. The *displacement thickness* δ^* and the *momentum thickness* Θ are the heights to which the solid boundary surface must be displaced, or repositioned so that if the fluid were ideal and flowing with the uniform speed U, it would produce the same rates of mass and momentum flow, respectively, as in the case of the real fluid.

- A numerical solution for the thickness, velocity profile, and shear-stress distribution for a *laminar* boundary layer along the surface of a flat plate was determined by Blasius. His results are presented in both graphical and tabular form.

- The friction drag F_{Df} caused by a boundary layer is generally reported using a dimensionless friction drag coefficient C_{Df}, along with the product of the plate area bL and the fluid's dynamic head, $F_{Df} = C_{Df} b L\left(\frac{1}{2}\rho U^2\right)$. Values of C_{Df} are a function of the Reynolds number.

11

EXAMPLE | 11.1

Water flows around the plate in Fig. 11–12a with a mean velocity of 0.25 m/s. Determine the shear-stress distribution and boundary layer thickness along one of its sides, and sketch the boundary layer for 1 m of its length. Take $\rho_w = 1000$ kg/m³ and $\mu_w = 0.001$ N·s/m².

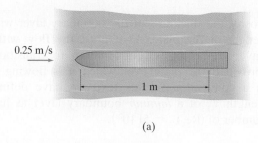

0.25 m/s

1 m

(a)

Fig. 11–12

SOLUTION

Fluid Description. We assume steady incompressible flow along the plate.

Analysis. Using Eq. 11–3, the Reynolds number for the flow in terms of x is

$$\text{Re}_x = \frac{\rho_w U x}{\mu_w} = \frac{(1000 \text{ kg/m}^3)(0.25 \text{ m/s})x}{0.001 \text{ N·s/m}^2} = 2.5(10^5)x$$

When $x = 1$ m, $\text{Re}_x = 2.5(10^5) < 5(10^5)$, so the boundary layer remains laminar. Thus, the shear-stress distribution can be determined from Eq. 11–8.

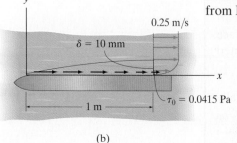

y

0.25 m/s

$\delta = 10$ mm

x

$\tau_0 = 0.0415$ Pa

1 m

(b)

$$\tau_0 = 0.332 \, \mu_w \frac{U}{x} \sqrt{\text{Re}_x}$$

$$= 0.332(0.001 \text{ N·s/m}^2)\left(\frac{0.25 \text{ m/s}}{x}\right)\sqrt{2.5(10^5)x}$$

$$= \left(\frac{0.0415}{\sqrt{x}}\right) \text{Pa} \qquad \qquad Ans.$$

The boundary layer thickness can be determined by applying Eq. 11–4.

$$\delta = \frac{5.0}{\sqrt{\text{Re}_x}} x = \frac{5.0}{\sqrt{2.5(10^5)x}} x = 0.010\sqrt{x} \text{ m} \qquad Ans.$$

These results are shown in Fig. 11–12b for $0 \le x \le 1$ m. Notice that when x increases, τ_0 decreases and δ increases. In particular, when $x = 1$ m, $\tau_0 = 0.0415$ Pa and $\delta = 10$ mm.

EXAMPLE 11.2

The ship in Fig. 11–13a is moving slowly at 0.2 m/s through still water. Determine the thickness δ of the boundary layer at a point $x = 1$ m from the bow. Also, at this location, find the velocity of the water within the boundary layer at $y = \delta$ and $y = \delta/2$. Take $\nu_w = 1.10(10^{-6})$ m²/s.

SOLUTION

Fluid Description. We will assume the ship's hull is a flat plate, and relative to the ship the water exhibits steady incompressible flow.

Disturbance Thickness. First we will check to see if the boundary layer remains laminar at $x = 1$ m.

$$\text{Re}_x = \frac{Ux}{\nu_w} = \frac{(0.2 \text{ m/s})(1 \text{ m})}{1.10(10^{-6}) \text{ m}^2/\text{s}} = 1.818(10^5) < (\text{Re}_x)_{\text{cr}} = 5(10^5) \text{ OK}$$

0.2 m/s

We can now use Blasius's solution, Eq. 11–4, to determine the boundary layer thickness at $x = 1$ m.

$$\delta = \frac{5.0}{\sqrt{\text{Re}_x}} x = \frac{5.0}{\sqrt{1.818(10^5)}}(1 \text{ m}) = 0.01173 \text{ m} = 11.7 \text{ mm} \qquad Ans.$$

Velocity. At $x = 1$ m and $y = \delta = 0.01173$ m, by definition, $u/U = 0.99$. Thus, the velocity of the water at this point, Fig. 11–13b, is

$$u = 0.99(0.2 \text{ m/s}) = 0.198 \text{ m/s} \qquad Ans.$$

To determine the velocity of the water at $x = 1$ m and $y = \delta/2 = 5.86(10^{-3})$ m, we must first find the value of u/U, either from the graph in Fig. 11–9a, or from Table 11–1. Here,

$$\frac{y}{x}\sqrt{\text{Re}_x} = \frac{5.86(10^{-3}) \text{ m}}{1 \text{ m}}\sqrt{1.818(10^5)} = 2.5$$

Using linear interpolation between the values 2.4 and 2.8 given in the table, for 2.5 we obtain

$$\frac{u/U - 0.72899}{2.5 - 2.4} = \frac{0.81152 - 0.72899}{2.8 - 2.4}; \qquad u/U = 0.7496$$

$$u = 0.7496(0.2 \text{ m/s}) = 0.150 \text{ m/s} \qquad Ans.$$

Other points along the velocity profile in Fig. 11–13b can be obtained in a similar manner. Also, this method can be used to find u for other values of x as well, but realize that the maximum value for x cannot exceed x_{cr}, that is,

$$(\text{Re}_x)_{\text{cr}} = \frac{Ux_{\text{cr}}}{\nu_w}; \qquad 5(10^5) = \frac{(0.2 \text{ m/s})x_{\text{cr}}}{1.10(10^{-6}) \text{ m/s}}; \qquad x_{\text{cr}} = 2.75 \text{ m}$$

Beyond this distance the boundary layer begins to transition, and then farther along, it becomes turbulent.

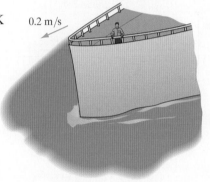

(a)

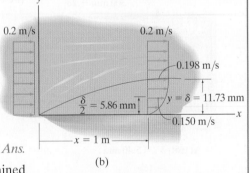

(b)

Fig. 11–13

11

EXAMPLE | 11.3

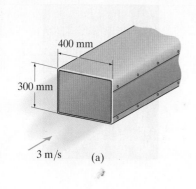

3 m/s   (a)

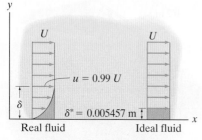

At x = 2 m

(b)

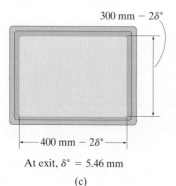

At exit, $\delta^* = 5.46$ mm

(c)

Fig. 11–14

Air flows into the rectangular duct in Fig. 11–14a at 3 m/s. Determine the displacement thickness at the end of its 2-m length, and the uniform velocity of the air as it flows out of the duct. Take $\rho_a = 1.20 \text{ kg/m}^3$ and $\mu_a = 18.1(10^{-6}) \text{ N} \cdot \text{s/m}^2$.

SOLUTION

Fluid Description. We will assume the air is incompressible and has steady flow.

Displacement Thickness. Using Eq. 11–3, the Reynolds number for the air as it moves through the duct is

$$\text{Re}_x = \frac{\rho_a U x}{\mu_a} = \frac{(1.20 \text{ kg/m}^3)(3 \text{ m/s})x}{18.1(10^{-6}) \text{ N} \cdot \text{s/m}^2} = 0.1989(10^6)x$$

When $x = 2$ m, then $\text{Re}_x = 3.978(10^5) < 5(10^5)$, so we have a laminar boundary layer along the duct. Therefore, we can use Eq. 11–5 to determine the displacement thickness.

$$\delta^* = \frac{1.721}{\sqrt{\text{Re}_x}} x = \frac{1.721}{\sqrt{0.1989(10^6)x}} x = 3.859(10^{-3})\sqrt{x} \text{ m} \quad (1)$$

Where $x = 2$ m, $\delta^* = 0.005457$ m $= 5.46$ mm.   *Ans.*

Velocity. If the air were an ideal fluid, then the dimensions of the cross section of the duct at $x = 2$ m would have to be shortened, as shown on the right in Fig. 11–14b, in order to produce the *same* mass flow through the duct as in the case of the actual flow. In other words, the exit cross-sectional area, shown in light shade in Fig. 11–14c, must be

$$A_{\text{out}} = [0.3 \text{ m} - 2(0.005457 \text{ m})][0.4 \text{ m} - 2(0.005457 \text{ m})] = 0.1125 \text{ m}^2$$

Although a constant mass flow passes by each cross section, the uniform portion of the air exiting the duct will have a greater velocity than the uniform airstream entering the duct. To determine its value, we require continuity of flow.

$$\frac{\partial}{\partial t} \int_{\text{cv}} \rho \, d\mathcal{V} + \int_{\text{cs}} \rho \mathbf{V} \cdot d\mathbf{A} = 0$$

$$0 - \rho U_{\text{in}} A_{\text{in}} + \rho U_{\text{out}} A_{\text{out}} = 0$$

$$-(3 \text{ m/s})(0.3 \text{ m})(0.4 \text{ m}) + U_{\text{out}}(0.1125 \text{ m}^2) = 0$$

$$U_{\text{out}} = 3.20 \text{ m/s} \qquad \textit{Ans.}$$

This increase in U has occurred because the boundary layer constricts the flow. In other words, the cross section is decreased by the displacement thickness. If, instead, we wanted to maintain a constant uniform speed of 3 m/s through the duct, it would then be necessary for the cross section to become divergent, so that its dimensions would increase by $2\delta^*$ along its length, in accordance with Eq. 1.

11

EXAMPLE | 11.4

A small submarine has a triangular stabilizing fin on its stern with the dimensions shown in Fig. 11–15a. If the water temperature is 15°C, determine the drag on the fin when the submarine is traveling at 0.5 m/s.

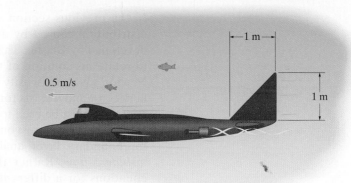

(a)

SOLUTION

Fluid Description.   Relative to the submarine, this is a case of steady incompressible flow. From Appendix A, for water at 15°C, $\rho = 999.2$ kg/m³ and $\nu = 1.15(10^{-6})$ m²/s.

Analysis.   First we will determine if the flow within the boundary layer remains laminar. Since the flow creates the largest Reynolds number at the base of the fin, where $x = 1$ in Fig. 11–15b, is the largest, then

$$(\text{Re})_{max} = \frac{Ux}{\nu} = \frac{(0.5 \text{ m/s})(1 \text{ m})}{1.15(10^{-6}) \text{ m}^2/\text{s}} = 4.35(10^5) < 5(10^5) \quad \text{Laminar}$$

Integration is necessary here since the length x of the fin changes with y. If we establish the x and y axes as shown in Fig. 11–15b, then an arbitrary differential strip of the fin will have a length of $1 - x$ and width of dy. Applying Eq. 11–11 with $b = dy$ and $L = 1 - x$, the total drag on *both sides* of the strip is therefore

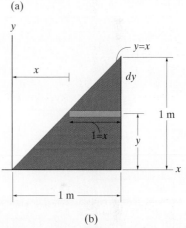

(b)

Fig. 11–15

$$d\mathbf{F}_{Df} = 2\left[\frac{0.664(dy)\rho U^2(1 - x)}{\sqrt{\text{Re}_L}}\right] = 2\left[\frac{0.664(dy)\left(999.2 \text{ kg/m}^3\right)(0.5 \text{ m/s})^2(1 - x)}{\sqrt{\dfrac{(0.5 \text{ m/s})(1 - x)}{1.15(10^{-6}) \text{ m}^2/\text{s}}}}\right]$$

$$= 0.5031(1 - x)^{1/2} \, dy$$

However, $x = y$. Then

$$d\mathbf{F}_{Df} = 0.5031 \left(1 - y\right)^{1/2} dy$$

And so, the total drag acting on all the strips composing the area of the fin is given by

$$F_{Df} = 0.5031 \int_0^{1 \text{ m}} (1 - y)^{1/2} \, dy$$

$$= 0.5031\left[-\frac{2}{3}(1 - y)^{3/2}\right]_0^{1 \text{ m}} = 0.335 \text{ N} \qquad Ans.$$

This is indeed a very small value, the result of the small velocity and low kinematic viscosity.

11

11.3   The Momentum Integral Equation

In the previous section we were able to determine the shear stress distribution caused by a laminar boundary layer using the solution developed by Blasius. This analysis was possible because we were able to relate his solution of the velocity profile to the shear stress using Newton's law of viscosity, $\tau = \mu(du/dy)$. However, no such relationship between τ and u exists for turbulent boundary layers, and so a different approach must be used to study the effects for turbulent flow.

In 1921, Theodore von Kármán proposed an *approximate method* for boundary layer analysis that is suitable for *both laminar and turbulent flow*. Rather than writing the continuity and momentum equations for a differential control volume at a *point*, von Kármán considered doing this for a differential control volume having a thickness dx and extending from the plate's surface up to a *streamline* that intersects the boundary layer, Fig. 11–16a. Through this element the flow is steady, and due to its small height, the pressure within it is virtually constant. Over time, the x component of flow enters the open control surface at the left side, ABC, and exits through the open control surface DE on the right side. No flow can cross over the fixed control surface AE or the streamline boundary CD, since the velocity is always tangent along the streamline.

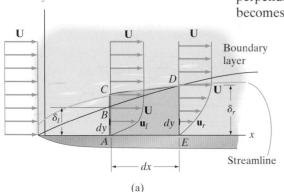

(a)

$\tau_0(1dx)$

Free-body diagram

(b)

Fig. 11–16

Continuity Equation.   If we consider a *unit width* of the plate perpendicular to the page, then for steady flow, the continuity equation becomes

$$\frac{\partial}{\partial t}\int_{cv}\rho \, dV + \int_{cs}\rho \mathbf{V}\cdot d\mathbf{A} = 0$$

$$0 - \int_0^{\delta_l}\rho u_l \, dy - \dot{m}_{BC} + \int_0^{\delta_r}\rho u_r \, dy = 0 \qquad (11\text{–}15)$$

Here $\dot{m}_{BC}$ is due to the constant flow U into the region between the top of the boundary layer and the streamline.

Momentum Equation.   The free-body diagram of the control volume is shown in Fig. 11–16b. Since the pressure p within the control volume is essentially constant, and the height $h_{AC} \approx h_{ED}$, because $AE = dx$, then the resultant forces caused by pressure on each open control surface will cancel. The only external force acting on the closed control surfaces is due to shear stress from the plate. This force is $\tau_0(1dx)$. Applying the momentum equation, we have

$$\rightarrow \Sigma F_x = \frac{\partial}{\partial t} \int_{cv} V \rho \, d\Psi + \int_{cs} V \rho \, \mathbf{V} \cdot d\mathbf{A}$$

$$-\tau_0(1dx) = 0 + \int_0^{\delta_r} \rho u_r^2 \, dy - \int_0^{\delta_l} \rho u_l^2 \, dy - U \dot{m}_{BC}$$

Solving for $\dot{m}_{BC}$ in Eq. 11–15, and substituting into the above equation, realizing that ρ is constant for an incompressible fluid, we have

$$-\tau_0 \, dx = \rho \int_0^{\delta_r} u_r^2 \, dy - \rho \int_0^{\delta_l} u_l^2 \, dy - U\rho \left[\int_0^{\delta_r} u_r \, dy - \int_0^{\delta_l} u_l \, dy \right]$$

$$-\tau_0 \, dx = \rho \int_0^{\delta_r} \left(u_r^2 - U u_r \right) dy - \rho \int_0^{\delta_l} \left(u_l^2 - U u_l \right) dy$$

Since the vertical sides AC and ED are a differential distance dx apart, the terms on the right represent the differential *difference* of the integrals, that is,

$$-\tau_0 \, dx = \rho \, d \left[\int_0^{\delta} \left(u^2 - Uu \right) dy \right]$$

The free-stream velocity U is constant, and so we can also write this equation in terms of the dimensionless velocity ratio, u/U, and express the shear stress as

$$\tau_0 = \rho U^2 \frac{d}{dx} \int_0^{\delta} \frac{u}{U}\left(1 - \frac{u}{U} \right) dy \qquad (11\text{–}16)$$

Recognizing that the integral represents the momentum thickness, Eq. 11–2, we can also write

$$\tau_0 = \rho U^2 \frac{d\Theta}{dx} \qquad (11\text{–}17)$$

Either of the above two equations is called the **momentum integral equation** for a flat plate. To apply it, we must either know the velocity profile $u = u(y)$ at each position x, or approximate it by an equation having the form $u/U = f(y/\delta)$. Some of these profiles are listed in Table 11–2. With any one of these we can evaluate the integral in Eq. 11–16, but due to the upper integration limit, the result will be in terms of δ. In order to determine δ as a function of x, τ_0 must also be related to δ. For laminar boundary layers we can do this using Newton's law of viscosity, which will be done in the following example. It illustrates application of this process to show how the results of δ, c_f, and C_{Df} reported in Table 11–2 were determined. In the next section, we will show how to apply this method to turbulent boundary layer flow.

11

TABLE 11–2

Velocity profile		δ	c_f	C_{Df}
Blasius		$5.00\dfrac{x}{\sqrt{\text{Re}_x}}$	$0.664/\sqrt{\text{Re}_x}$	$1.328/\sqrt{\text{Re}_x}$
Linear	$\dfrac{u}{U} = \dfrac{y}{\delta}$	$3.46\dfrac{x}{\sqrt{\text{Re}_x}}$	$0.578/\sqrt{\text{Re}_x}$	$1.156/\sqrt{\text{Re}_x}$
Parabolic	$\dfrac{u}{U} = -\left(\dfrac{y}{\delta}\right)^2 + 2\left(\dfrac{y}{\delta}\right)$	$5.48\dfrac{x}{\sqrt{\text{Re}_x}}$	$0.730/\sqrt{\text{Re}_x}$	$1.460/\sqrt{\text{Re}_x}$
Cubic	$\dfrac{u}{U} = -\dfrac{1}{2}\left(\dfrac{y}{\delta}\right)^3 + \dfrac{3}{2}\left(\dfrac{y}{\delta}\right)$	$4.64\dfrac{x}{\sqrt{\text{Re}_x}}$	$0.646/\sqrt{\text{Re}_x}$	$1.292/\sqrt{\text{Re}_x}$

EXAMPLE 11.5

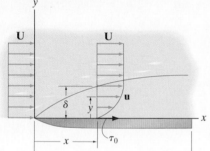

Fig. 11–17

The velocity profile for a laminar boundary layer developed over the plate of width b and length L is approximated by the parabola $u/U = -(y/\delta)^2 + 2(y/\delta)$, as shown in Fig. 11–17. Determine, as a function of x, the thickness δ of the boundary layer, the skin friction coefficient c_f, and the friction drag coefficient C_{Df}.

SOLUTION

Fluid Description.   Here we have steady incompressible laminar flow over the surface.

Boundary Layer Thickness.   Substituting the function u/U into Eq. 11–16 and integrating, we get

$$\tau_0 = \rho U^2 \frac{d}{dx} \int_0^\delta \frac{u}{U}\left(1 - \frac{u}{U}\right) dy$$

$$\tau_0 = \rho U^2 \frac{d}{dx} \int_0^\delta \left(-\left(\frac{y}{\delta}\right)^2 + 2\frac{y}{\delta}\right)\left(1 + \left(\frac{y}{\delta}\right)^2 - 2\frac{y}{\delta}\right) dy = \rho U^2 \frac{d}{dx}\left[\frac{2}{15}\delta\right]$$

$$\tau_0 = \rho U^2 \left[\frac{2}{15}\frac{d\delta}{dx}\right] \tag{1}$$

To obtain δ we must now express τ_0, which is at the plate's surface ($y = 0$), in terms of δ. Since laminar flow exists, this can be done using Newton's law of viscosity, evaluated at $y = 0$.

$$\tau_0 = \mu \frac{du}{dy}\bigg|_{y=0} = \mu U \frac{d}{dy}\left[-\left(\frac{y}{\delta}\right)^2 + 2\left(\frac{y}{\delta}\right)\right]_{y=0} = \frac{2\mu U}{\delta} \tag{2}$$

11

Therefore, Eq. 1 becomes

$$\frac{2\mu U}{\delta} = \rho U^2 \left[\frac{2}{15}\frac{d\delta}{dx}\right] \qquad \text{or} \qquad \rho U \delta \, d\delta = 15\mu \, dx$$

Since $\delta = 0$ at $x = 0$, that is, where the fluid initially contacts the plate, then integration gives

$$\rho U \int_0^\delta \delta \, d\delta = \int_0^x 15\mu \, dx; \qquad \delta = \sqrt{\frac{30\mu x}{\rho U}} \qquad\qquad (3)$$

Since $\text{Re}_x = \rho U x/\mu$, we can also write this expression as

$$\delta = \frac{5.48x}{\sqrt{\text{Re}_x}} \qquad\qquad\qquad Ans.$$

Skin Friction Coefficient.   Substituting Eq. 3 into Eq. 2, the shear stress on the plate as a function of x is therefore

$$\tau_0 = \frac{2\mu U}{\sqrt{\dfrac{30\mu x}{\rho U}}} = 0.365\sqrt{\frac{\mu\rho U^3}{x}} = 0.365\frac{\mu U}{x}\sqrt{\text{Re}_x}$$

Using Eq. 11–9, we get

$$c_f = \frac{\tau_0}{(1/2)\rho U^2} = \frac{0.365\dfrac{\mu U}{x}\sqrt{\text{Re}_x}}{(1/2)\rho U^2} = \frac{0.730}{\sqrt{\text{Re}_x}} \qquad\qquad Ans.$$

Friction Drag Coefficient.   To determine C_{Df}, we must first find F_{Df}.

$$F_{Df} = \int_A \tau_0 \, dA = \int_0^x 0.365\sqrt{\frac{\mu\rho U^3}{x}}(b\,dx) = 0.365b\sqrt{\mu\rho U^3}\left(2x^{1/2}\right)\Big|_0^x = 0.730b\sqrt{\mu\rho U^3 x}$$

Therefore, using Eq. 11–13,

$$C_{Df} = \frac{F_{Df}/bx}{(1/2)\rho U^2} = \frac{1.460b\sqrt{\mu\rho U^3 x}}{\rho U^2 bx}$$

$$C_{Df} = \frac{1.460}{\sqrt{\text{Re}_x}} \qquad\qquad\qquad Ans.$$

The three results obtained here are listed in Table 11–2.

11

11.4 Turbulent Boundary Layers

Turbulent boundary layers are thicker than laminar ones, and the velocity profile within them will be more uniform due to the erratic mixing of the fluid. We can use the momentum integral equation to determine the drag caused by a turbulent boundary layer; however, it is first necessary to express the velocity profile as a function of y. The accuracy of the result, of course, depends upon how close this function approximates the true velocity profile. Although many different formulas have been proposed, one of the simplest that works well is Prandtl's one-seventh power law. It is

$$\frac{u}{U} = \left(\frac{y}{\delta}\right)^{1/7} \tag{11–18}$$

This equation is shown in Fig. 11–18a. Notice that here the profile is *flatter* than the one developed for a laminar boundary layer. The flatness is necessary because, as previously stated, there is a large degree of fluid mixing and momentum transfer within turbulent flow. Also, because of this flatness, there is a *larger* velocity gradient near the plate's surface. As a result, the shear stress developed on the surface will be much *larger* than that caused by a laminar boundary layer.

Prandtl's equation does not apply within the laminar sublayer because the velocity gradient, $du/dy = (U/7\delta)(y/\delta)^{-6/7}$, at $y = 0$ becomes infinite, something that cannot happen. Therefore, as in all cases of turbulent boundary layers, the surface shear stress τ_0 must be *related to δ by experiment*. An empirical formula that agrees well with the data was developed by Prandtl and Blasius. It is

$$\tau_0 = 0.0225\rho U^2 \left(\frac{\nu}{U\delta}\right)^{1/4} \tag{11–19}$$

We will now apply the momentum integral equation with these two equations, and thereby obtain the height δ of the *turbulent boundary layer* as a function of its position x. Applying Eq. 11–16, we have

$$\tau_0 = \rho U^2 \frac{d}{dx} \int_0^\delta \frac{u}{U}\left(1 - \frac{u}{U}\right)dy$$

$$0.0225\rho U^2 \left(\frac{\nu}{U\delta}\right)^{1/4} = \rho U^2 \frac{d}{dx} \int_0^\delta \left(\frac{y}{\delta}\right)^{1/7}\left(1 - \left(\frac{y}{\delta}\right)^{1/7}\right)dy$$

$$0.0225\left(\frac{\nu}{U\delta}\right)^{1/4} = \frac{d}{dx}\left(\frac{7}{8}\delta - \frac{7}{9}\delta\right) = \frac{7}{72}\frac{d\delta}{dx}$$

$$\delta^{1/4}d\delta = 0.231\left(\frac{\nu}{U}\right)^{1/4}dx$$

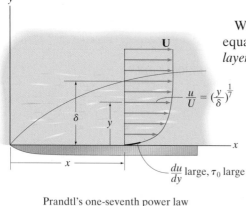

$$\frac{u}{U} = \left(\frac{y}{\delta}\right)^{\frac{1}{7}}$$

$$\frac{du}{dy} \text{ large, } \tau_0 \text{ large}$$

Prandtl's one-seventh power law

(a)

Fig. 11–18

Although all boundary layers are initially laminar, Fig. 11–3, here we will *assume* the front surface of the plate is rather rough, and so the boundary layer will be forced to become turbulent, practically at the outset. Therefore, integrating from $x = 0$, where $\delta = 0$, we have

$$\int_0^\delta \delta^{1/4} d\delta = 0.231\left(\frac{\nu}{U}\right)^{1/4} \int_0^x dx \quad \text{or} \quad \delta = 0.371\left(\frac{\nu}{U}\right)^{1/5} x^{4/5}$$

Using Eq. 11–3 to express the thickness in terms of the Reynolds number, we have

$$\delta = \frac{0.371}{(\text{Re}_x)^{1/5}} x \tag{11–20}$$

Shear Stress along Plate. Substituting Eq. 11–20 into Eq. 11–19, we obtain the shear stress along the plate as a function of x, Fig. 11–18b.

$$\tau_0 = 0.0225\rho U^2 \left[\frac{\nu(\text{Re}_x)^{1/5}}{U(0.371x)}\right]^{1/4}$$

$$= \frac{0.0288\rho U^2}{(\text{Re}_x)^{1/5}} \tag{11–21}$$

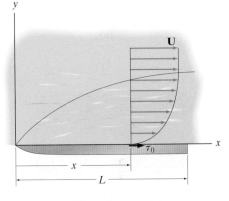

Shear stress on plate

(b)

Drag on Plate. If the plate has a length L and a width b, then the drag can be found by integrating the shear stress over the plate's area, Fig. 11–18c.

$$F_{Df} = \int_A \tau_0 \, dA = \int_0^L \frac{0.0288\rho U^2}{\left(\frac{Ux}{\nu}\right)^{1/5}} (b\,dx) = 0.0360\rho U^2 \frac{bL}{(\text{Re}_L)^{1/5}} \tag{11–22}$$

The frictional drag coefficient, Eq. 11–13, is therefore

$$C_{Df} = \frac{F_{Df}/bL}{(1/2)\rho U^2} = \frac{0.0721}{(\text{Re}_L)^{1/5}} \tag{11–23}$$

This result has been checked by numerous experiments, and it has been found that *a slightly more accurate value* for C_{Df} can be obtained by replacing the constant 0.0721 with 0.0740. The result works well within the following range of Reynolds numbers:

$$C_{Df} = \frac{0.0740}{(\text{Re}_L)^{1/5}} \quad 5(10^5) < \text{Re}_L < 10^7 \tag{11–24}$$

The lower bound represents our limit for laminar boundary layers. For higher values of the Reynolds number, another empirical equation that fits relevant experimental data has been proposed by Hermann Schlichting. See Ref. [16]. It is

$$C_{Df} = \frac{0.455}{(\log_{10} \text{Re}_L)^{2.58}} \quad 10^7 \leq \text{Re}_L < 10^9 \tag{11–25}$$

Keep in mind that all these equations are valid only if a *turbulent boundary layer* extends over the *entire length* of the plate.

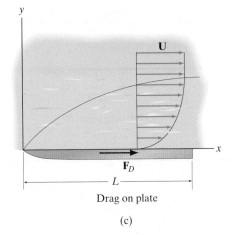

Drag on plate

(c)

Fig. 11–18 (cont.)

11

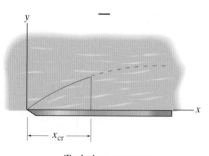

Laminar

Turbulent

x_{cr}

Model

(a)

||

All turbulent

(b)

—

x_{cr}

Turbulent segment

(c)

+

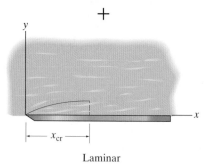

x_{cr}

Laminar

(d)

Fig. 11–19

11.5   Laminar and Turbulent Boundary Layers

As stated in Sec. 11.1, a real boundary layer on a smooth flat plate develops a laminar region, grows in height to become unstable at about $(\mathrm{Re}_x)_{cr} = 5(10^5)$, then eventually transit to become turbulent. Therefore, to develop a more exact approach when calculating the friction drag on the plate, it is necessary to consider both the laminar and turbulent portions of the boundary layer. Through experiment, Prandtl discovered that this could be done by neglecting the friction drag within the transition region, since for fast flows it covers a very short distance along the plate. In other words, we can model the boundary layer like that shown in Fig. 11–19a.

To calculate the friction drag, Prandtl first assumed that the boundary layer is completely turbulent over the plate's entire length L, Fig. 11–19b, and so the drag can be found using Eq. 11–25 for the higher Reynolds numbers or greater distance x. Then an adjustment to this result is made by subtracting a turbulent segment of the drag up to the start of the transition zone, $x = x_{cr}$, Eq. 11–24, Fig. 11–19c, and finally, adding the drag caused by the laminar portion up to this point (Blasius's solution), Eq. 11–14, Fig. 11–19d. The friction drag on the plate is therefore

$$F_{Df} = \frac{0.455}{(\log_{10}\mathrm{Re}_L)^{2.58}}(1/2)\rho U^2(bL) - \frac{0.0740}{(\mathrm{Re}_x)_{cr}^{1/5}}(1/2)\rho U^2(bx_{cr})$$

$$+ \frac{1.328}{\sqrt{(\mathrm{Re}_x)_{cr}}}(1/2)\rho U^2(bx_{cr})$$

Since C_{Df} for the plate is defined by Eq. 11–13, the friction drag coefficient is

$$C_{Df} = \frac{0.455}{(\log_{10}\mathrm{Re}_L)^{2.58}} - \frac{0.0740}{(\mathrm{Re}_x)_{cr}^{1/5}}\frac{x_{cr}}{L} + \frac{1.328}{\sqrt{(\mathrm{Re}_x)_{cr}}}\frac{x_{cr}}{L}$$

Finally, since by proportion $x_{cr}/L = (\mathrm{Re}_x)_{cr}/\mathrm{Re}_L$, and if $(\mathrm{Re}_x)_{cr} = 5(10^5)$, then to fit experimental data for values of the Reynolds number between $5(10^5) \le \mathrm{Re}_L < 10^9$, we have

$$\boxed{C_{Df} = \frac{0.455}{(\log_{10}\mathrm{Re}_L)^{2.58}} - \frac{1700}{\mathrm{Re}_L}} \qquad 5(10^5) \le \mathrm{Re}_L < 10^9 \qquad (11\text{–}26)$$

A plot of the equations used to determine the friction drag coefficient for a range of common Reynolds numbers is shown in Fig. 11–20. Notice that the experimental data points, obtained from several different researchers, indicate close agreement with this theory. Realize that the curves are valid only when the transition in boundary layer flow occurs at a Reynolds number of $5(10^5)$. If the free-stream velocity is subjected to any sudden turbulence, or the surface of the plate is somewhat rough, then the transition in flow will occur at a lower Reynolds number. When

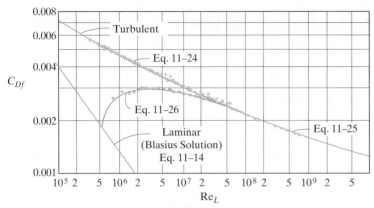

Friction drag coefficient for a flat plate

Fig. 11–20

this occurs, the constant in the second term (1700) in Eq. 11–26 will have to be modified to account for this change. The calculations accounting for these effects will not be outlined here; rather, they are covered in the literature related to this subject. See Ref. [16].

Important Points

- The momentum integral equation provides an approximate method for obtaining the thickness and surface shear-stress distribution caused by either a laminar or a turbulent boundary layer. To apply this equation, it is necessary to know the velocity profile $u = u(y)$ and to be able to express the shear stress in terms of the boundary layer thickness δ, $\tau_0 = f(\delta)$.

- Blasius's solution for $u = u(y)$ can be used for laminar boundary layers, and the shear stress τ_0 can be related to δ using Newton's law of viscosity. For turbulent boundary layers the flow is erratic, and so the needed relationships must be approximated from results obtained from experiments. For example, for a fully turbulent boundary layer, Prandtl's one-seventh power law along with a formulation by Prandtl and Blasius seems to work well.

- A laminar boundary layer exerts a much *smaller* shear stress on a surface than a turbulent boundary layer. A high degree of fluid mixing occurs within turbulent boundary layers, and this creates a larger velocity gradient at the surface, and therefore a higher shear stress on the surface.

- The frictional drag coefficient on a flat surface, caused by a combination laminar and turbulent boundary layer, can be determined by superposition, neglecting the region where transitional flow occurs.

11

EXAMPLE | 11.6

Oil flows over the top of the flat plate in Fig. 11–21 with a free-stream velocity of 20 m/s. If the plate is 2 m long and 1 m wide, determine the friction drag on the plate due to the formation of a combination laminar and turbulent boundary layer. Take $\rho_o = 890 \text{ kg/m}^3$ and $\mu_o = 3.40(10^{-3}) \text{ N} \cdot \text{s/m}^2$.

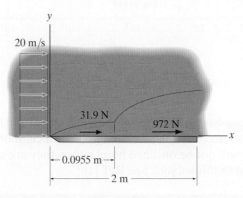

Fig. 11–21

SOLUTION

Fluid Description. We have steady flow and assume the oil is incompressible.

Analysis. We will first determine the position x_{cr} where the laminar boundary layer begins to transition to turbulent. Here $(\text{Re}_x)_{\text{cr}} = 5(10^5)$, so

$$(\text{Re}_x)_{\text{cr}} = \frac{\rho_o U x}{\mu_o}$$

$$5(10^5) = \frac{(890 \text{ kg/m}^3)(20 \text{ m/s})x_{\text{cr}}}{3.40(10^{-3}) \text{ N} \cdot \text{s/m}^2}$$

$$x_{\text{cr}} = 0.0955 \text{ m} < 2 \text{ m}$$

Also, at the end of the plate,

$$\text{Re}_L = \frac{\rho_o U L}{\mu_o} = \frac{(890 \text{ kg/m}^3)(20 \text{ m/s})(2 \text{ m})}{3.40(10^{-3}) \text{ N} \cdot \text{s/m}^2} = 1.047(10^7)$$

11

Using Eq. 11–26 to determine the drag coefficient, since this equation applies for laminar–turbulent boundary layers within the range $5(10^5) \leq \text{Re}_L < 10^9$, we have

$$C_{Df} = \frac{0.455}{(\log_{10} \text{Re}_L)^{2.58}} - \frac{1700}{\text{Re}_L}$$

$$= \frac{0.455}{\left[\log_{10} 1.047(10^7)\right]^{2.58}} - \frac{1700}{1.047(10^7)}$$

$$= 0.002819$$

The *total friction drag* on the plate can now be determined using Eq. 11–13.

$$F_{Df} = C_{Df}\left(\frac{1}{2}\right)\rho U^2 bL$$

$$= 0.002819\left(\frac{1}{2}\right)(890 \text{ kg/m}^3)(20 \text{ m/s})^2(1 \text{ m})(2 \text{ m})$$

$$= 1004 \text{ N} \qquad\qquad\qquad\qquad\qquad Ans.$$

The portion of this force created by the *laminar boundary layer*, which extends along the first 0.0955 m of the plate, can be determined from Eq. 11–11.

$$(F_{Df})_{\text{lam}} = \frac{0.664 b\rho U^2 L}{\sqrt{(\text{Re}_x)_{\text{cr}}}}$$

$$= \frac{0.664(1 \text{ m})(890 \text{ kg/m}^3)(20 \text{ m/s})^2(0.0955 \text{ m})}{\sqrt{5(10^5)}}$$

$$= 31.9 \text{ N}$$

By comparison, the turbulent boundary layer contributes the most friction drag. It is

$$(F_{Df})_{\text{tur}} = 1004 \text{ N} - 31.9 \text{ N} = 972 \text{ N}$$

EXAMPLE 11.7

Water has a free-stream velocity of 10 m/s over the *rough* surface of the plate in Fig. 11–22, causing the boundary layer to suddenly become turbulent. Determine the shear stress on the surface at $x = 2$ m, and the thickness of the boundary layer at this position. Take $\rho_w = 1000$ kg/m^3 and $\mu_w = 1.00(10^{-3})$ N·s/m^2.

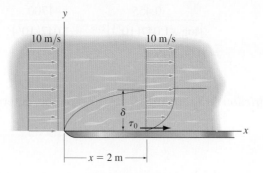

Fig. 11–22

SOLUTION

Fluid Description. We have steady flow and we assume water is incompressible.

Analysis. For this case, the Reynolds number is

$$\mathrm{Re}_x = \frac{\rho_w U x}{\mu_w} = \frac{(1000 \text{ kg/m}^3)(10 \text{ m/s})(2 \text{ m})}{1.00(10^{-3}) \text{ N·s/m}^2} = 20(10^6)$$

Using Eq. 11–21, since the boundary layer has been defined as fully turbulent, the shear stress on the surface at $x = 2$ m is

$$\tau_0 = \frac{0.0288\rho_w U^2}{(\mathrm{Re}_x)^{1/5}} = \frac{0.0288(1000 \text{ kg/m}^3)(10 \text{ m/s})^2}{[20(10^6)]^{1/5}}$$

$$= 99.8 \text{ Pa} \hspace{3cm} Ans.$$

And from Eq. 11–20, the thickness of the boundary layer at $x = 2$ m is

$$\delta = \frac{0.3701x}{(\mathrm{Re}_x)^{1/5}} = \frac{0.371(2 \text{ m})}{[20(10^6)]^{1/5}} = 0.02572 \text{ m} = 25.7 \text{ mm} \quad Ans.$$

Although this is a small thickness, it is much thicker than a laminar boundary layer. Also, the calculated shear stress at $x = 2$ m will be higher than that caused by a laminar boundary layer.

11

EXAMPLE 11.8

Estimate the friction drag on each airplane wing in Fig. 11–23 if it is assumed to be a flat plate, having an average width of 0.9 m and length of 4 m. The plane is flying at 125 m/s. Assume the air is incompressible. Take $\rho_a = 0.819$ kg/m^3 and $\mu_a = 16.6(10^{-6})$ N·s/m^2.

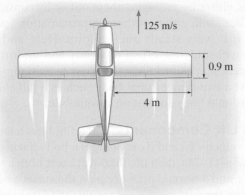

Fig. 11–23

SOLUTION

Fluid Description. We have steady incompressible flow relative to the airplane.

Analysis. At the trailing or back edge of the wing, the Reynolds number is

$$\text{Re}_L = \frac{\rho_a U L}{\mu_a} = \frac{(0.819 \text{ kg/m}^3)(125 \text{ m/s})(0.9 \text{ m})}{16.6(10^{-6}) \text{ N·s/m}^2} = 5.550(10^6) > 5(10^5)$$

Therefore, the wing is subjected to a combined laminar and turbulent boundary layer. Applying Eq. 11–26, since it applies for $5(10^5) < \text{Re}_L < 10^9$, we have

$$C_{Df} = \frac{0.455}{(\log_{10} \text{Re}_L)^{2.58}} - \frac{1700}{\text{Re}_L}$$

$$= \frac{0.455}{\{\log_{10}[5.550(10^6)]\}^{2.58}} - \frac{1700}{5.550(10^6)}$$

$$= 0.003001$$

Since the friction drag acts on *both* the top and bottom wing surfaces, then using Eq. 11–13,

$$F_{Df} = 2\left[C_{Df} b L\left(\frac{1}{2}\rho_a U^2\right)\right]$$

$$= 2\left\{(0.003001)(4 \text{ m})(0.9 \text{ m})\left[\frac{1}{2}(0.819 \text{ kg/m}^3)(125 \text{ m/s})^2\right]\right\}$$

$$= 138.21 \text{ N} = 138 \text{ N} \qquad \qquad Ans.$$

11

11.6   Drag and Lift

In almost all cases, the natural flow of a fluid will be both unsteady and nonuniform. For example, the speed of the wind varies with time and elevation, and so does the speed of water in a river or stream. Even so, for engineering applications we can often approximate the effect of these irregularities, either by averaging them out or by considering their worst condition. Then, with either of these approximations, we can investigate the flow as if it were steady and uniform. Here and in the following sections, we will study the effect of steady and uniform flow on bodies having different shapes, including flow past an axisymmetric body such as a rocket, past a two-dimensional body such as a tall chimney, and past a three-dimensional body such as an automobile.

Drag and Lift Components.   If a fluid has a steady and uniform free-stream velocity **U**, and it encounters a body having a curved surface as shown in Fig. 11–24a, then the fluid will exert both a viscous tangential shear stress τ and a normal pressure p on the surface of the body. For an element dA on the surface, we can resolve the forces produced by τ and p into their horizontal (x) and vertical (y) components, Fig. 11–24b. The **drag** is in the *direction* of **U**. When integrated over the entire surface of the body, this force becomes

$$F_D = \int_A \tau \cos \theta \, dA \; + \; \int_A p \sin \theta \, dA \qquad (11\text{–}27)$$

The **lift** is the force that acts *perpendicular* to **U**. It is

$$F_L = \int_A \tau \sin \theta \, dA \; - \; \int_A p \cos \theta \, dA \qquad (11\text{–}28)$$

Provided the distributions of τ and p are known over the surface, these integrations can be carried out. The following is an example of how this can be done.

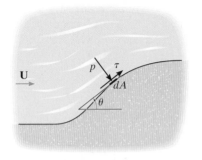

Pressure and shear stress
on a fixed surface

(a)

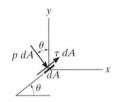

Forces on an element
of surface area

(b)

Fig. 11–24

Since the loading ramps are held in the vertical position, they put a large pressure drag on the truck and reduce its fuel efficiency.

EXAMPLE | 11.9

The semicircular building in Fig. 11–25*a* is 12 m long and is subjected to a steady uniform wind having a velocity of 18 m/s. Assuming the air is an ideal fluid, determine the lift and drag on the building. Take $\rho = 1.23\,\text{kg/m}^3$.

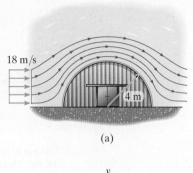

SOLUTION

Fluid Description. Since the air is assumed to be an ideal fluid, there is no viscosity, and so no boundary layer or viscous shear stress acts on the building, only a pressure distribution. We will also assume the building is long enough so end effects do not disturb this two-dimensional steady flow.

Analysis. Using ideal flow theory, as outlined in Sec. 7.10, we have determined that the pressure distribution over the surface of a cylinder is symmetrical, and so for a semicylinder, as shown in Fig. 7–25*b*, it can be described by Eq. 7–67, namely,

$$p = p_0 + \frac{1}{2}\rho U^2\left(1 - 4\sin^2\theta\right)$$

Since $p_0 = 0$ (gage pressure), we have

$$p = 0 + \frac{1}{2}\left(1.23\ \text{kg/m}^3\right)\left(18\ \text{m/s}\right)^2\left(1 - 4\sin^2\theta\right)$$

$$= 199.26\left(1 - 4\sin^2\theta\right)$$

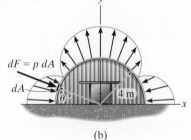

(b)

Fig. 11–25

The drag is the horizontal or *x* component of *d***F**, shown in Fig. 11–25*b*. For the entire building it is

$$F_D = \int_A (p\,dA)\cos\theta = \int_0^{\pi}(199.26)\left(1 - 4\sin^2\theta\right)\cos\theta\left[(12\ \text{m})(4\ \text{m})\,d\theta\right]$$

$$= 9564.48\int_0^{\pi}\left(1 - 4\sin^2\theta\right)\cos\theta\,d\theta$$

$$= 9564.48\left[\sin\theta - \frac{4}{3}\sin^3\theta\right]_0^{\pi} = 0 \qquad\qquad Ans.$$

This result is to be expected since the pressure distribution is *symmetrical* about the *y* axis.

The lift is the vertical or *y* component of *d***F**. Here $d\mathbf{F}_y$ is negative since it is directed in the −*y* direction, so that

$$F_L = \int_A (p\,dA)\sin\theta = -\int_0^{\pi}(199.26)\left(1 - 4\sin^2\theta\right)\sin\theta\left[(12\ \text{m})(4\ \text{m})d\theta\right]$$

$$= -9564.48\int_0^{\pi}\left(1 - 4\sin^2\theta\right)\sin\theta\,d\theta$$

$$= -9564.48\left[3\cos\theta - \frac{4}{3}\cos^3\theta\right]_0^{\pi}$$

$$= 31.88\left(10^3\right)\,\text{N} = 31.9\ \text{kN} \qquad\qquad Ans.$$

The positive sign indicates that the air stream tends to *pull the building up*, as the word "lift" would suggest.

11

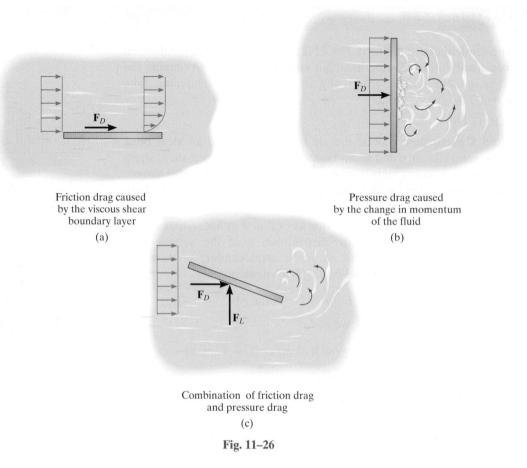

Friction drag caused
by the viscous shear
boundary layer

(a)

Pressure drag caused
by the change in momentum
of the fluid

(b)

Combination of friction drag
and pressure drag

(c)

Fig. 11–26

11.7   Pressure Gradient Effects

In the previous section we noted that the forces of drag and lift are the result of a combination of viscous shear stress and pressure acting on the surface of a body. For example, consider the case of a flat plate. When its surface is aligned with the flow, Fig. 11–26a, only viscous **shear drag** will be produced on the plate. However, when the flow is perpendicular to the plate, as in Fig. 11–26b, then the plate acts as a **bluff body**. Here only **pressure drag** is created. Pressure drag is caused by the change in the momentum of the fluid, as discussed in Chapter 6. The resultant shear drag is zero in this case because the shear stresses act equally up and down the plate's front surface. Notice that in both of these cases, the lift is zero because neither effect produces a resultant force in the vertical direction, perpendicular to the flow. To achieve both lift and drag, the plate must be oriented at an angle to the flow, as shown in Fig. 11–26c. A body that has a curved or irregular surface can also be subjected to the forces of lift and drag, and to better understand how these forces are created, let us consider the uniform flow over the surface of a long cylinder.

11

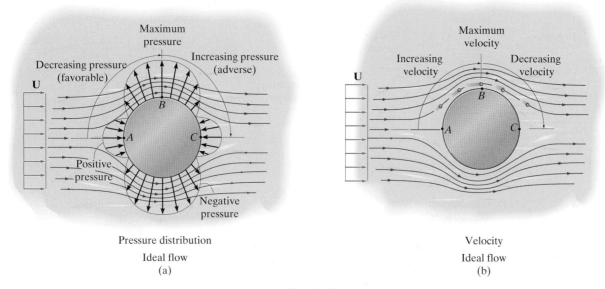

Pressure distribution
Ideal flow
(a)

Velocity
Ideal flow
(b)

Fig. 11–27

Ideal Flow around a Cylinder.   In Sec. 7.10 we discussed how the uniform flow of an *ideal fluid* around a cylinder creates a *variable pressure distribution* over its surface. The result is shown in Fig. 11–27a. Although some portions of the surface are subjected to a positive pressure (pushing), other portions are subjected to a negative pressure (suction). Here we will be interested only in how this pressure *changes*. This change is the **pressure gradient**. The Bernoulli equation $\left(p/\gamma + V^2/2g = \text{const.}\right)$ indicates that the *decreasing pressure* or *negative pressure gradient,* which occurs from A to B, will cause an *increase* in the *velocity* within this region, Fig. 11–27b. This is referred to as a **favorable pressure gradient**, because the flow velocity is *increased*, in this case from zero at the stagnation point A to a maximum at B. Likewise, the *increasing pressure* or *positive pressure gradient*, which occurs from B to C, causes a *decrease* in the *velocity*. This is referred to as an **adverse pressure gradient**, because it slows down the fluid, here from its maximum at B to zero at its stagnation point C. Since we have an ideal fluid, the pressure distribution is *symmetrical* around the cylinder, and so the *resultant pressure drag* (horizontal force) on the cylinder will be equal to zero. Furthermore, since an ideal fluid has no viscosity, there also is no *viscous shear drag* on the cylinder.

11

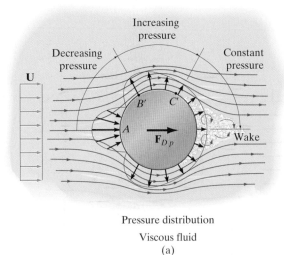

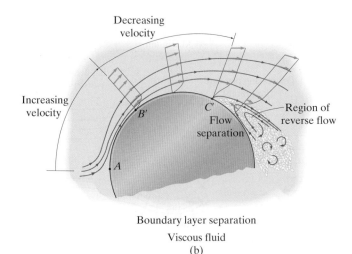

Pressure distribution

Viscous fluid

(a)

Boundary layer separation

Viscous fluid

(b)

Fig. 11–28

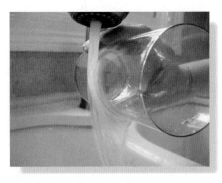

The Coanda effect, indicated by the bending of the waterstream as the water clings to the glass.

Flow separation from the sides of the post is clearly evident in this photo.

Real Flow around a Cylinder.   Unlike an ideal fluid that can slide freely around a cylinder, a real fluid has viscosity, and as a result, the fluid will tend to form a boundary layer and *cling* to the cylinder's surface as it flows around it. This phenomenon was studied in the early 1900s by the Romanian engineer Henri Coanda and is called the ***Coanda effect***.

To understand this viscous behavior, we will consider the long cylinder shown in Fig. 11–28a. The flow starts from the stagnation point at A, and thereafter it forms a laminar boundary layer on the cylinder as the fluid begins to travel around the surface. The favorable *pressure gradient* (pressure decrease) within this initial region increases the velocity, Fig. 11–28b. Because the flow must *overcome* the drag effect of viscous friction within the boundary layer, the minimum pressure and maximum velocity will occur at point B'. This is *sooner* than in the case of ideal fluid flow.

Although the boundary layer continues to grow in thickness farther downstream of point B', the velocity decreases here because of the adverse pressure gradient (increasing pressure) acting within this region, Fig. 11–28a. Point C' marks a separation of flow from the cylinder since the velocity of the slower-moving particles near the surface is finally reduced to zero at this point. Beyond C', within the boundary layer, the flow will begin to back up and move in the opposite direction to the free-stream flow. This ultimately will form a vortex, which will shed from the cylinder, as shown in Fig. 11–28b. A series of these vortices or eddies produce a ***wake***, whose energy will eventually dissipate as heat. The pressure within the wake is relatively constant, and the resultant of the entire pressure distribution around the cylinder produces the *pressure drag* $\mathbf{F}_{Dp}$, Fig. 11–28a. Notice that the magnitude of this force depends to some degree on the location of point C', where the flow separates from the cylinder.

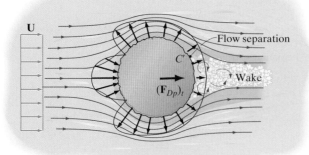

Rough cylinder
Turbulent boundary layer
Smaller drag force
(a)

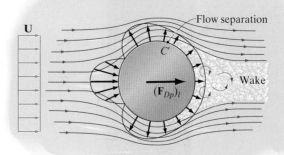

Smooth cylinder
Laminar boundary layer
Larger drag force
(b)

Fig. 11–29

If the flow within the boundary layer is completely *turbulent*, Fig. 11–29*a*, then separation will occur *later* than if the boundary layer flow is *laminar*, Fig. 11–29*b*. This happens because within a turbulent boundary layer, the fluid has *more kinetic energy* than in the laminar case. As a result, the adverse pressure gradient will take *longer* to arrest the flow, and so the point of separation is *farther back* on the surface. Consequently, the resultant of the pressure distribution, Fig. 9–29*a*, will create a *smaller* pressure drag, $(\mathbf{F}_{Dp})_t$, than in the case of laminar flow, where it is $(\mathbf{F}_{Dp})$, Fig. 9–29*b*. Since surface roughness has the effect of producing turbulent boundary layers, it may seem counterintuitive, but one way to *reduce pressure drag* is to roughen the cylinder's front surface.

Unfortunately, the actual point C' of flow separation for either laminar or turbulent boundary layers cannot be determined analytically, except through approximate methods. However, experiments have shown that, as expected, the point of transition from laminar to turbulent flow is a function of the Reynolds number, and therefore the pressure drag, like viscous or friction drag, will be a function of this parameter. For a cylinder, the "characteristic length" for finding the Reynolds number is its diameter D, so $\text{Re} = \rho VD/\mu$.

Vortex Shedding.

When the flow around a cylinder is at a low Reynolds number, *laminar flow* prevails, and the boundary layer will separate from the surface symmetrically from each side, forming eddies that rotate in opposite directions as shown in Fig. 11–30. As the Reynolds number increases, these eddies will elongate, and one will begin to break away from one side of the surface before another breaks away from the other side. This stream of alternating vortex shedding causes the pressure to fluctuate on each side of the cylinder, which in turn tends to vibrate the cylinder *perpendicular* to the flow. Theodore von Kármán was

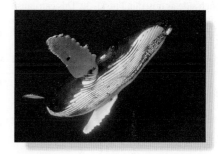

Whale flippers have tubercles on their leading edge as noted in this photo. Wind tunnel tests have revealed that the flow of water between each pair of tubercles produces clockwise and counterclockwise vortices, which energizes the turbulent flow within the boundary layer, thus preventing the boundary layer from separating from the flipper. This gives the whale more maneuverability and less drag. (© MASA USHIODA/Alamy)

11

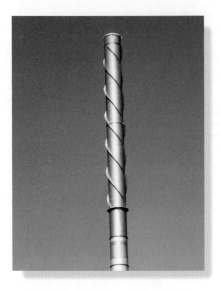

This tall thin-walled metal chimney can be subjected to severe wind loadings. At a critical wind speed, its cylindrical shape will produce a von Kármán vortex street that will shed off each side as noted in Fig. 11–30. This can cause the chimney to oscillate perpendicular to the direction of the wind. To prevent this, the spiral winding, referred to as a fence or "strake," disturbs the flow and prevents the formation of the vortices.

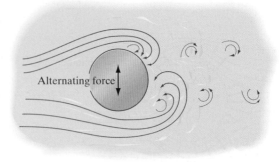

von Kármán vortex street

Fig. 11–30

one of the first to investigate this effect, and the vortex stream so formed is often called a **von Kármán vortex trail** or **vortex street**, so named because the alternate vortices are placed like houses along a street.

The frequency f at which the vortices are shed from each side of a cylinder of diameter D is a function of the **Strouhal number**, which is defined as

$$\text{St} = \frac{fD}{V}$$

This dimensionless number is named after the Czech scientist Vincenc Strouhal, who studied this phenomenon as it pertains to the "singing" noise developed by wires suspended in an airstream. Empirical values for the Strouhal number, as it relates to the Reynolds number for flow around a cylinder, have been developed and can be found in the literature. See Ref. [29]. At very high Reynolds numbers, however, the vortex shedding tends to disintegrate into uniform turbulence off both sides of the cylinder, and so the vibrations will tend to die out. In spite of this, conditions can occur where the oscillating force caused by vortex shedding produces vibrations that must be taken into account when designing structures such as tall chimneys, antennas, submarine periscopes, and even the cables of suspension bridges.

11.8   The Drag Coefficient

As previously stated, the *drag* and *lift* on a body are a combination of the effects of *both* viscous friction and pressure, and if the shear and pressure distributions over the surface can be established, then these forces can be determined, as demonstrated by Example 11.9. Unfortunately, the distributions of p and especially τ are generally difficult to obtain, either through experiment or by some analytical procedure. A simpler method for finding the drag and lift is to *directly* measure them through experiment. In this section we will consider how this is done for drag, and later, in Sec. 11.11, we will consider lift.

It has become a standard procedure in fluid mechanics to express the drag in terms of the fluid's dynamic head, the *projected area A_p* of the body into the fluid stream, and a dimensionless **drag coefficient** C_D. The relationship is

$$F_D = C_D A_p \left(\frac{\rho V^2}{2} \right) \qquad\qquad (11\text{--}29)$$

The value of C_D is determined from experiments, normally performed on a prototype or model placed in a wind or water tunnel or in a channel. For example, C_D for a car is determined by placing it in a wind tunnel, and then, for each wind velocity V, a measurement is made of the horizontal force F_D required to prevent the car from moving. The value of C_D for each velocity is then determined from

$$C_D = \frac{F_D}{(1/2)\rho V^2 A_p}$$

Specific values of C_D are generally available in engineering handbooks and industrial catalogs for a wide variety of shapes. Its value is not constant; rather, experiments indicate that this coefficient depends on a number of factors, and we will now discuss some of these for several different cases.

Reynolds Number.

In general, the drag coefficient is highly dependent on the Reynolds number. In particular, objects or particles that have a very small size and weight, such as powder falling through air, or silt falling through water, have a very low Reynolds number (Re $\ll$ 1). Here no flow separation occurs from the sides of the particle, and the drag is due only to viscous friction caused by laminar flow.

If the object has a spherical shape, then for laminar flow the drag can be determined analytically using a solution developed in 1851 by George Stokes. He obtained his result by solving the Navier–Stokes and continuity equations, and his result has been confirmed by experiment. See Ref. [7]. It is[*]

$$F_D = 3\pi\mu VD$$

Using the definition of the drag coefficient in Eq. 11–29, where the *projected area* of the sphere into the flow is $A_p = (\pi/4)D^2$, we can now express C_D in terms of the Reynolds number, Re $= \rho VD/\mu$. It is

$$C_D = 24/\text{Re} \qquad\qquad (11\text{--}30)$$

Experiments have shown that C_D for other shapes also has this reciprocal dependence on Re, provided Re $\leq$ 1. For example, a circular disk held normal to the flow has a $C_D = 20.4/\text{Re}$. See Ref. [19].

Acceleration occurs when the weight of these people is greater than the drag force caused by the air. As their speed increases, the drag increases until at terminal velocity, equilibrium is achieved at about 200 kph or 120 mi/h.

[*]This equation is sometimes used to measure the viscosity of a fluid having a *very high viscosity*. The experiment is done by dropping a small sphere of known weight and diameter into the fluid contained in a long cylinder and measuring its terminal velocity. Then $\mu = F_D/(3\pi VD)$. Further details are given in Example 11.12 as to how to obtain F_D.

11

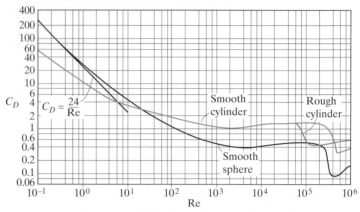

Drag coefficients for a sphere and long cylinder

Fig. 11–31

Cylinder. Experimental values of C_D for flow around the sides of a smooth and a rough cylinder as a function of the Reynolds number are shown on the graph in Fig. 11–31. (From Ref. [18].) Notice that as the Reynolds number increases, C_D begins to *decrease*, until it reaches a constant value at about Re $\approx 10^3$. From this point until Re $\approx 10^5$, laminar flow remains around the cylinder. The sudden *drop* in C_D, at Reynolds numbers within the high range $10^5 < \text{Re} < 10^6$, occurs because the boundary layer over the surface changes from laminar to turbulent, thereby causing the separation point for the flow to occur farther downstream on the back of the cylinder. As mentioned in the previous section, this results in a smaller region for the wake and a *lower pressure drag*. Although there is a *slight increase in the viscous drag*, the net effect will still cause the total drag to be reduced. Also, notice that the drop in C_D occurs sooner for the cylinder with the rough surface, Re $\approx 6(10^4)$, because a rough surface disturbs the boundary layer into becoming turbulent sooner than a smooth surface.

Sphere. When a fluid flows around a three-dimensional body, it behaves in much the same way as it does in a two-dimensional case. However, here the flow can go around the ends of the body as well as around its sides. This has the effect of *extending* the point of flow separation and thereby lessens the drag on the body. For example, the drag force acting on a smooth sphere is also shown in Fig. 11–31, Ref. [18]. Note that it initially follows Stokes' equation for Re < 1. The shape of this curve is similar to that for the cylinder; however, for high Reynolds numbers, the values of C_D for the sphere are about half of those obtained for the cylinder. As before, the characteristic *drop* in C_D at high Reynolds numbers is attributed to the early transition from a laminar to a turbulent boundary layer; and as with the cylinder, a *rough surface* on a sphere will also contribute a further drop in C_D. This is why manufacturers put dimples on golf balls and roughen the surface of tennis balls. Such objects move at high speeds, and so, within the range of the Reynolds number where this occurs, the rough surface will produce a lower C_D, and the ball will travel farther than it would if it had a smooth surface.

11

Froude Number.

When gravity plays a dominant role on drag, then the drag coefficient will be a function of the Froude number, $Fr = V/\sqrt{gl}$. Recall from Sec. 8.4 that both the Froude and Reynolds numbers are used for similitude studies of ships. These numbers are important, since the drag is produced by both viscous friction on the hull and the lifting of water to create waves. We showed how models can be tested to study these effects *independently*. In particular, an experimental relationship between the wave drag coefficient $(C_D)_{wave}$ and the Froude number can be established, and by plotting the data of Fr versus $(C_D)_{wave}$ for different hull geometries of the ship's model, it is then possible to make comparisons before choosing a proper shape for design.

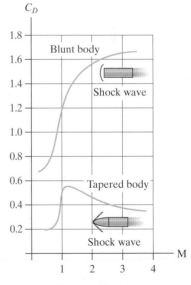

The bulb on the lower bow of this ship significantly reduces the height of the bow wave and the turbulent flow around the bow. The reduced drag coefficient thereby saves fuel costs.

Mach Number.

When the fluid is a gas, such as air, compressibility effects may have to be considered when determining the drag on the body. As a result, the drag coefficient will then be a function of both the Reynolds number and the Mach number, since, apart from pressure, the dominant forces on the body will be caused by inertia, viscosity, and compressibility.[*]

Although the drag is affected by *both* viscous friction and the pressure on the body, when compressibility becomes significant, the effect of these two components will be quite different than when the flow is incompressible. To understand why this is so, consider the variation in C_D versus M for a blunt body (cylinder) and one with a tapered or conical nose, Fig. 11–32, Ref. [6]. At low Mach numbers (M ≪ 1), or velocities below the speed of sound, the drag is primarily affected by the Reynolds number. As a result, in both cases C_D increases only slightly. As sonic flow is approached there is a sharp increase in C_D. At the point M = 1, a shock wave is formed on or in front of the body. Shock waves are very thin, on the order of 0.3 μm, and cause abrupt changes in the flow characteristics, as we will discuss in Chapter 13. What is important here is to realize that a sudden *increase in pressure* occurs across the wave, and this increase causes an additional drag on the body. In aeronautics it is referred to as **wave drag**. At such high speeds, the drag will be independent of the Reynolds number. Instead, only the wave drag, created by the pressure change over the shock wave, produces the predominant amount of the drag.

Notice that the greatest reduction in drag occurs when the nose is tapered, as on a jet aircraft, since then the wave drag is confined to the small "front" region on the nose. As the Mach number increases, for both the blunt and tapered bodies, the shock wave becomes more inclined farther downstream, and the width of the wake behind the wave is thereby reduced, causing C_D to decrease. Incidentally, the shape of the

Drag coefficient for a
Blunt body and a tapered body

Fig. 11–32

[*]Recall that the Mach number is M = V/c, were c is the speed of sound as measured in the fluid.

back end of both the blunt and tapered bodies has very little effect on reducing C_D for supersonic flow, because the predominant drag is caused by the shock wave. For subsonic flow, however, no shock wave is formed, and so the viscous drag is reduced if the back end is tapered, since this discourages flow separation.

11.9   Drag Coefficients for Bodies Having Various Shapes

The flow around bodies having a variety of different shapes follows the same behavior as flow around a cylinder and a sphere. Experiments have shown that for large Reynolds numbers, generally on the order of $\mathrm{Re} > 10^4$, the drag coefficient C_D is essentially *constant* for many of these shapes, because the flow will separate at the sharp edges of the body and thereby become well defined. Typical values of drag coefficients for some common shapes are given in Table 11–3, for $\mathrm{Re} > 10^4$. Many other examples can be found in the literature. See Ref. [19]. Realize that each value of C_D depends not only on the body's shape and the Reynolds number, but also on the angle at which the body is oriented within the flow and the body's surface roughness.

Applications.   If a *composite body* is constructed from the shapes listed in Table 11–3, and each is fully exposed to the flow, then the drag, as determined from Eq. 11–30, $F_D = C_D A_p (\rho V^2/2)$, becomes a superposition of forces calculated for each shape. When using the table, or other available data for C_D, some care should be exercised by considering the conditions under which it is applied. In actual practice, nearby bodies can influence the pattern of flow around the object being studied, and this can greatly affect the actual value of C_D. For example, in a large city, structural engineers construct models of existing buildings surrounding the building to be designed, and then test this entire system in a wind tunnel. Occasionally, the complicated pattern of wind flow will lead to an *increased* pressure loading on the building being considered, and so this increase must be taken into account as part of the design loading. Finally, remember that this data was prepared for steady uniform flow, something that actually never occurs in nature. Consequently, one should exercise good engineering judgment, based on experience and intuition about the flow, whenever any value of C_D is selected.

11

TABLE 11–3 Drag Coefficients for Re $> 10^4$

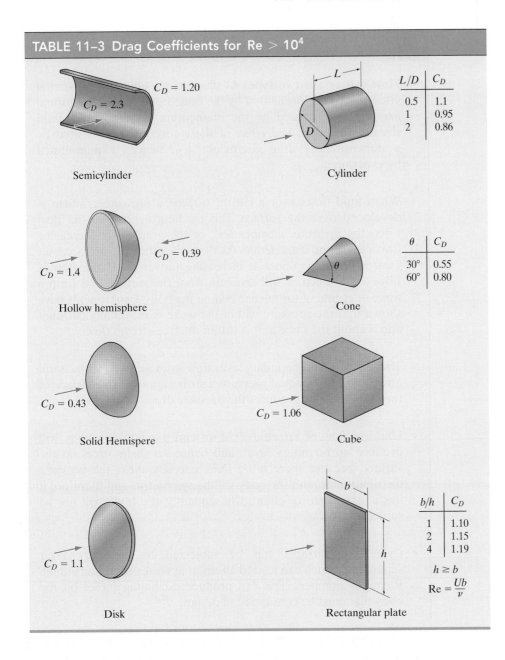

$C_D = 1.20$

$C_D = 2.3$

Semicylinder

L/D	C_D
0.5	1.1
1	0.95
2	0.86

Cylinder

$C_D = 0.39$

$C_D = 1.4$

Hollow hemisphere

θ	C_D
30°	0.55
60°	0.80

Cone

$C_D = 0.43$

Solid Hemispere

$C_D = 1.06$

Cube

$C_D = 1.1$

Disk

b/h	C_D
1	1.10
2	1.15
4	1.19

$h \geq b$

$\text{Re} = \dfrac{Ub}{\nu}$

Rectangular plate

11

Important Points

- Flow can create two types of drag on an object: a tangential viscous *friction drag* caused by the boundary layer, and a normal *pressure drag* caused by the momentum change of the fluid stream. The *combined effect* of these forces is represented by a dimensionless drag coefficient C_D, which is determined by experiment.

- When fluid flows over a *curved surface*, a pressure gradient is developed over the surface. This gradient *accelerates* the flow when the pressure becomes *less*, or the gradient is decreasing (favorable), and it *decelerates* the flow when the pressure becomes *greater*, or the gradient is increasing (adverse). If the flow is slowed too much, it will separate from the surface and form a wake or region of turbulence behind the body. Experiments have shown that the pressure within the wake is essentially constant and is about the same as it is within the free-stream flow.

- The separation of boundary layer flow from a curved surface will give rise to an unequal pressure distribution over the surface of the body, and this produces the pressure drag.

- Uniform flow of an *ideal fluid* around a symmetric body will produce no boundary layer, and hence no shear stress on the surface, because there is no fluid viscosity. Also, the pressure distribution around the body will be symmetric, and therefore it has a zero force resultant. Consequently, the body will not be subjected to drag.

- Cylindrical surfaces can be subjected to boundary layer separation, which can lead to alternating vortex shedding at high Reynolds numbers. This can produce oscillating forces on the body that must be considered in design.

- Experimentally determined drag coefficients, C_D, for cylinders, spheres, and many other simple shapes are available in the literature. Specific values either are tabulated or are presented in graphical form. In either case, C_D is a function of the Reynolds number, the body's shape, its orientation within the flow, and its surface roughness. For some applications, forces other than viscosity become important, and so C_D may, for example, also depend upon the Froude number or the Mach number.

11

EXAMPLE | 11.10

The hemispherical dish in Fig. 11–33 is subjected to a direct uniform wind speed of 15 m/s. Determine the moment of the force created by the wind on the dish about the base of the post at A. For air, take $\rho_a = 1.20 \text{ kg/m}^3$ and $\mu_a = 18.1(10^{-6}) \text{ N} \cdot \text{s/m}^2$.

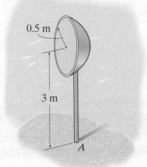

0.5 m

3 m

A

Fig. 11–33

SOLUTION

Fluid Description. Because of the relatively slow velocity, we will assume the air is incompressible and the flow is steady.

Analysis. The "characteristic length" for the dish is its diameter, 1 m. Therefore, the Reynolds number for the flow is

$$\text{Re} = \frac{\rho_a V D}{\mu_a} = \frac{(1.20 \text{ kg/m}^3)(15 \text{ m/s})(1 \text{ m})}{18.1(10^{-6}) \text{ N} \cdot \text{s/m}^2} = 9.94(10^5) > (10^4)$$

From Table 11–3, the drag coefficient is $C_D = 1.4$, so applying Eq. 11–29 yields

$$F_D = C_D A_p \left(\frac{\rho_a V^2}{2} \right)$$

$$= 1.4 \left[\pi (0.5 \text{ m})^2 \right] \left[\frac{(1.20 \text{ kg/m}^3)(15 \text{ m/s})^2}{2} \right]$$

$$= 148.44 \text{ N}$$

Notice the significance of how much the diameter and velocity affect this result, since each is squared. Due to the uniformity of the wind distribution, F_D acts through the geometric center of the dish. The moment of this force about A is therefore

$$M = (148.44 \text{ N})(3 \text{ m}) = 445 \text{ N} \cdot \text{m} \qquad \qquad \textit{Ans.}$$

An additional moment of the wind loading on the post (cylinder) can also be included.

11

EXAMPLE | 11.11

The sports car and driver in Fig. 11–34a have a total mass of 2.3 Mg. The car is traveling at 11 m/s when the driver puts the transmission into neutral and allows the car to freely coast until after 165 s its speed reaches 10 m/s. Determine the drag coefficient for the car, assuming its value is constant. Neglect rolling and other mechanical resistance effects. The projected front area of the car is 0.75 m².

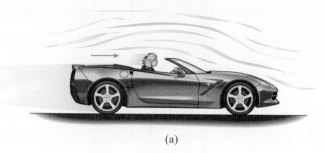

(a)

Fig. 11–34

SOLUTION

Fluid Description.   Relative to the car we have uniform unsteady flow because the car is slowing down. We will assume the air is incompressible. At standard temperature, $\rho_a = 1.23 \text{ kg/m}^3$.

Analysis.   Since the car can be considered as a rigid body, we can apply the equation of motion, $\Sigma F = ma$, and then use $a = dV/dt$ to relate the car's velocity to the time. Alternatively, although it amounts to the same thing, we can choose the car as a "control volume," draw its free-body diagram, Fig. 11–34b, and then apply the momentum equation,

$$\xrightarrow{+} \Sigma F_x = \frac{\partial}{\partial t} \int_{cv} V\rho \, d\forall + \int_{cs} V\rho V \, dA$$

Since there are no open control surfaces, the last term is zero. The first term on the right can be simplified, noting that $V\rho$ is independent of the volume of the car, and so $\int_{cv} d\forall = \forall$. But $\rho\forall = m$, the mass of the car and driver, and therefore, the above equation becomes

$$\xrightarrow{+} \Sigma F_x = \frac{d(mV)}{dt} = m\frac{dV}{dt}$$

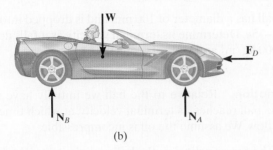

(b)

Finally, since F_D is the drag on the car, we have

$$-C_D A_p \left(\frac{\rho_a V^2}{2} \right) = m \frac{dV}{dt}$$

Separating the variables V and t, and integrating, we get

$$\frac{1}{2} C_D A_p \rho_a \int_0^t dt = -m \int_{V_0}^V \frac{dV}{V^2}$$

$$\frac{1}{2} C_D A_p \rho_a t \bigg|_0^t = m \frac{1}{V} \bigg|_{V_0}^V$$

$$\frac{1}{2} C_D A_p \rho_a t = m \left(\frac{1}{V} - \frac{1}{V_0} \right)$$

Substituting the data,

$$\frac{1}{2} C_D (0.75 \text{ m}^2)(1.23 \text{ kg/m}^3)(165 \text{ s}) = (2.3(10^3)\text{kg}) \left(\frac{1}{10 \text{ m/s}} - \frac{1}{11 \text{ m/s}} \right)$$

$$C_D = 0.275 \qquad\qquad Ans.$$

Aerodynamic design of the 2014 C7 corvette used in this example was based on both a CFD analysis and over 700 hours of wind tunnel testing. The purpose was to achieve an optimum balance of zero lift and required airflow for mechanical cooling, while maintaining a low C_D. (© General Motors, LLC)

11

EXAMPLE | 11.12

(a)

F_b

F_D

mg

(b)

Fig. 11–35

The 0.5-kg ball has a diameter of 100 mm and is dropped into the tank of oil, Fig. 11–35a. Determine its terminal velocity as it falls downward. Take $\rho_o = 900 \text{ kg/m}^3$ and $\mu_o = 0.0360 \text{ N} \cdot \text{s/m}^2$.

SOLUTION

Fluid Description.   Relative to the ball we initially have unsteady flow, until the ball reaches its terminal velocity, at which time we then have steady flow. We assume the oil is incompressible.

Analysis.   The forces acting on the ball include its weight, buoyancy, and drag, Fig. 11–35b. Since equilibrium occurs when the ball reaches terminal velocity, we have

$$+\uparrow \Sigma F_y = 0; \qquad\qquad F_b + F_D - mg = 0$$

The buoyancy force is $F_b = \rho_o g V$, and the drag is expressed by Eq. 11–29. Thus

$$\rho_o g V + C_D A_p \left(\frac{\rho_o V_t^2}{2} \right) - mg = 0$$

$$\left(900 \text{ kg/m}^3\right)\left(9.81 \text{ m/s}^2\right)\left(\frac{4}{3}\right)\pi(0.05 \text{ m})^3 + C_D \pi (0.05 \text{ m})^2 \left(\frac{\left(900 \text{ kg/m}^3\right)V_t^2}{2} \right)$$

$$- (0.5 \text{ kg})\left(9.81 \text{ m/s}^2\right) = 0$$

$$C_D V_t^2 = 0.07983 \text{ m}^2/\text{s}^2 \qquad (1)$$

The value for C_D is found from Fig. 11–31, but it depends upon the Reynolds number.

$$\text{Re} = \frac{\rho_o V_t D}{\mu_o} = \frac{\left(900 \text{ kg/m}^3\right)\left(V_t\right)(0.1 \text{ m})}{0.0360 \text{ N} \cdot \text{s/m}^2} = 2500 V_t \qquad (2)$$

The solution will proceed using an iteration process. First, we will assume a value of C_D and then calculate V_t using Eq. 1. This result will then be used in Eq. 2 to calculate the Reynolds number. Using this value in Fig. 11–31, the corresponding value of C_D is obtained. If it is not close to the assumed value, we must repeat the same procedure until they are approximately equal. The iterations are tabulated.

Iteration	C_D (Assumed)	V_t (m/s) (Eq. 1)	Re (Eq. 2)	C_D (Fig. 11–31)
1	1	0.2825	706	0.55
2	0.55	0.3810	952	0.50
3	0.50	0.3996	999	0.48 (okay)

Therefore, the terminal velocity is

$$V_t = 0.3996 \text{ m/s} = 0.400 \text{ m/s} \qquad\qquad Ans.$$

11.10 Methods for Reducing Drag

In Sec. 11.7 it was shown that when the front of a cylinder is artificially roughened, turbulence occurs sooner in the boundary layer, thereby moving the point of flow separation farther back on the cylinder, Fig. 11–29a. As a result, pressure drag is reduced. Another way of moving the point of separation back is to *streamline* the body, such that it takes the form of a teardrop, as shown in Fig. 11–36. Although *pressure drag* is *reduced*, more of the surface is in contact with the fluid stream, and so the *friction drag* is *increased*. The optimal shape occurs when the *total drag*, which is a combination of both the pressure drag and the friction drag, is a minimum.

The flow around an irregularly shaped body can be complex, so the optimal shape for any streamlined body must be determined by experiment. Also, a design that works well within one range of Reynolds numbers may not be as effective within other ranges. As a general rule, for *low* Reynolds numbers the viscous shear will create the maximum component of drag, and for *high* numbers the pressure drag component will dominate.

Airfoils.

A common streamlined shape is an airfoil, Fig. 11–37a, and the drag acting on it depends upon its **angle of attack** α with the free-flow airstream, Fig. 11–37b. As shown, this angle is defined from the horizontal to the **cord** of the wing, that is, the length measured from the leading edge to the trailing edge. Notice that as α increases, the point of flow separation moves *toward* the leading edge, and this causes the pressure drag to increase, because the inclination of the wing projects a larger area into the airstream and the pressure on the back side is diminished.

To properly design an airfoil to reduce pressure drag, the separation point should therefore be as far back from the leading edge as possible. To accomplish this, modern wings have a smooth surface on the front side of the wing, to maintain a laminar boundary layer, and then, at the point where the transition to turbulence occurs, the boundary layer is energized, either by a rough surface or by using vortex generators, which are small protruding fins on the top of the wing. This allows the boundary layer to *cling* farther back onto the wing's surface, and although this increases friction drag, it will also, as we have seen for the cylinder, decrease the pressure drag.

Apart from defining an appropriate shape for an airfoil, aeronautical engineers have devised other methods for boundary layer control. Delayed separation for large angles of attack can be achieved by using slotted flaps or leading-edge slots, as shown in Fig. 11–38. These devices are designed to transfer fast-moving air from the bottom of the wing to its upper surface in an effort to energize the boundary layer. Another method produces suction of the slow-moving air within the boundary layer, either through slots or through the use of a porous surface. Both of these methods will increase the speed of the flow within the boundary layer and thus delay its separation. They also have the advantage of thinning the boundary layer, and thus delaying its transition from laminar to turbulent. Achieving these

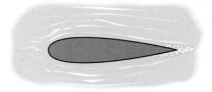

Streamlined body

Fig. 11–36

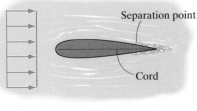

Separation point

Cord

(a)

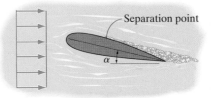

Separation point

α

Flow separation at angle of attack α

(b)

Fig. 11–37

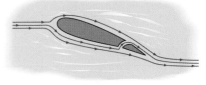

Slotted flap

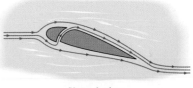

Slotted wing

Fig. 11–38

11

designs can be difficult, however, since it requires some ingenuity in addressing the structural and mechanical problems that arise.

Airfoil Drag Coefficients.   The drag on airfoils has been studied extensively by the National Advisory Committee for Aeronautics (NACA).[*] They, and others, have published graphs that have been used by aeronautical engineers to determine the **section drag coefficient**, $(C_D)_\infty$, that applies to aircraft wings of various shapes. As stated this coefficient is for **section drag**; that is, it assumes the wing has an *infinite length* so that flow around the wing tip is *not considered*. A typical example for a 2409 wing profile is shown in Fig. 11–39. The additional flow around the wing tip produces an *induced drag* on the wing, and we will study its effects in the next section. Once $(C_D)_\infty$ and the induced drag coefficient $(C_D)_i$ are known, then the "total" drag coefficient $C_D = (C_D)_\infty + (C_D)_i$ can be determined. The drag on the airfoil is then

$$F_D = C_D A_{pl}\left(\frac{\rho V^2}{2}\right)$$

(11–31)

Here A_{pl} represents the *planform* of the wing, that is, the projection of its top or bottom area.

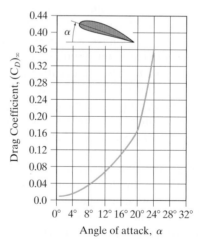

Drag coefficient $(C_D)_\infty$ for NACA airfoil 2409 wing of infinite span

Fig. 11–39

*In 1958, this agency was incorporated into the newly created National Aeronautics and Space Administration (NASA).

Road Vehicles. Through the years it has become important to reduce the aerodynamic drag on cars, buses, and trucks to save on fuel consumption. Although streamlining is limited for these cases by length restrictions on the vehicle, it is possible to decrease C_D for cars by redesigning their front and back profiles, rounding the forward surfaces of sideview mirrors, recessing the door handles, eliminating outside antennas, and rounding the corners of the body. By doing this, automotive engineers have been able to reduce drag coefficients from values as high as 0.60 to around 0.30. For freight trucks, drag coefficients can be as high as 1.35. However, a reduction of about 20% can be achieved by adding shrouds or wind deflectors that direct the airstream smoothly around the cab and along the bottom sides of the trailer. See the accompanying photos.

The drag coefficient for any type of vehicle is a function of the Reynolds number; however, within the typical range of highway speeds, the value of C_D is practically constant. Table 11–4 lists values for some modern-day vehicles, although specific values for any particular vehicle can be obtained from published literature. See Ref. [23]. Once C_D is obtained, the drag can then be determined using Eq. 11–29, that is

$$F_D = C_D A_p \left(\frac{\rho V^2}{2} \right)$$

Here A_p is the *projected area* of the vehicle into the flow. For utility trucks and SUVs this area is about 2.32 m², and for average-size passenger cars it is about 0.790 m².

Modern trucks have rounded grills in their front, skirts along the bottom of their sides, and a roof fairing, which if properly fitted will reduce drag and produce fuel savings of about 6%. Also, the fenders on the back side will further reduce turbulence and decrease drag.

TABLE 11–4	
Vehicle	C_D
Utility truck	0.6–0.8
Sport Utility Vehicle (SUV)	0.35–0.4
Lamborghini Countach	0.42
VW Beetle	0.38
Toyota Celica convertible	0.36
Chevrolet Corvette C5	0.29
Toyota Prius	0.25
Bicycle	1–1.5

Racing cars must maximize downward force on the car to maintain stability and yet minimize drag. (© Alexey Kuznetsov/Fotolia)

11

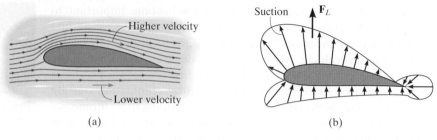

Fig. 11–40

11.11   Lift and Drag on an Airfoil

The effect of a fluid flowing over the surface of a body not only creates drag on the body but can also create lift. We have mentioned that the drag acts in the direction of motion of the airstream, but the lift acts perpendicular to it.

Airfoil Lift.   The phenomenon of lift produced by an airfoil or wing can be explained in different ways. Most commonly, the Bernoulli equation is used to do this.* Basically, the argument states that the higher the velocity, the lower the pressure, and vice versa $\left(p/\gamma + V^2/2g = \text{const.}\right)$. Since the flow over the longer top of an airfoil, Fig. 11–40a, is faster than that beneath its shorter bottom, the pressure on top will be lower than on the bottom. This creates a pressure distribution like that shown in Fig. 11–40b, and its resultant force F_L produces the lift.

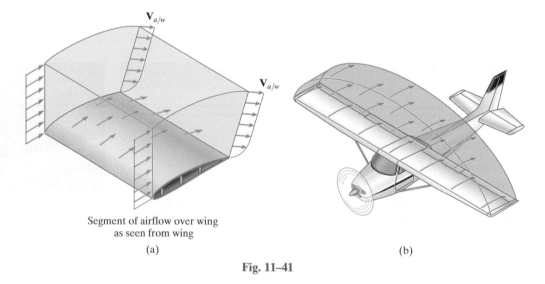

Segment of airflow over wing
as seen from wing

(a)

(b)

Fig. 11–41

* This equation gives a conceptual reason why lift can occur, but it cannot be used to actually calculate the lift because the flow over the wing's top surface will not directly correspond to the flow over the bottom surface.

Lift, however, can be more fully explained only by using the Coanda effect, which was discussed in Sec. 11.7. For example, consider the airstream that passes over the wing's surface in Fig. 11–41a. Because viscosity causes the air to *cling* to the surface, the layers of air just above the surface begin to move faster and faster as they form the boundary layer, until the speed of the air eventually matches the uniform speed of the airflow relative to the wing, $\mathbf{V}_{a/w}$. The shear and pressure forces between these moving layers cause the flow to *bend* in the direction of each slower-moving layer. In other words, the air is forced to follow the surface of the wing. This effect of *redirecting* the airstream propagates upward from the surface of the wing at the speed of sound, since the pressure within the airstream tends to prevent the formation of voids between fluid layers. As a result, a very large volume of air above the wing will be redirected downward and ultimately produce a "downwash" behind the wing, Fig. 11–41b. If the wing moves through calm air with a velocity $\mathbf{V}_w$, as in Fig. 11–41c, then the velocity of the air coming off the trailing edge of the wing, as observed from the wing (or by the pilot), will be $\mathbf{V}_{a/w}$. By vector addition, Fig. 11–41d, the "downwash" velocity of the air as seen by an *observer on the ground* will almost be *vertical* since $\mathbf{V}_a = \mathbf{V}_w + \mathbf{V}_{a/w}$. In other words, when a plane flies close overhead, a ground observer will feel the airstream directed somewhat vertically downward.

The bending of the air around the wing in the manner just described in effect gives the air an (almost) vertical momentum. To do this, the *wing must produce a downward force on the airstream*, and by Newton's third law, the airstream must produce an *equal but opposite upward force on the wing*. It is this force that produces *lift*. Notice from the pressure distribution, Fig. 11–42a, that the greatest lift (greatest negative or suction pressure) is produced on the front top third of the wing, because here the airstream must bend the greatest amount to follow the wing's surface. On the bottom surface there is also a component of lift, caused by a redirection of flow, although here the pressure is positive. Of course, this resultant lift will be even larger if the wing profile is somewhat curved or cambered, as in Fig. 11–42b.

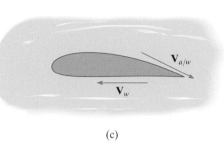

(c)

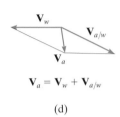

$$\mathbf{V}_a = \mathbf{V}_w + \mathbf{V}_{a/w}$$

(d)

Fig. 11–41 (cont.)

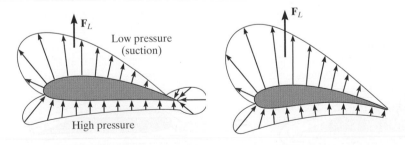

Pressure distribution

(a)

Pressure distribution–cambered airfoil

(b)

Fig. 11–42

11

Circulation.   In Sec. 7.10 we showed that an *ideal fluid* will cause lift on a rotating cylinder when we superimpose a circulation Γ about the cylinder, while the cylinder is subjected to a uniform flow. Martin Kutta and Nikolai Joukowski have independently shown that the lift, as calculated from Eq. 7–73, $L = \rho U \Gamma$, also holds for *any* closed-shaped body subjected to two-dimensional flow. This important result is known as the **Kutta–Joukowski theorem**, and in aerodynamics it is often used to estimate the lift on both airfoils and hydrofoils.

To show how this works, consider the airfoil in Fig. 11–43a when it is subjected to the uniform flow U. Here *stagnation points* develop on the leading edge A and back at the top B of the trailing edge. However, ideal fluids cannot support this flow, because an ideal fluid would have to round the bottom of the trailing edge, and then move up to the stagnation point at B. Physically, this is not possible, since it would take an infinite normal acceleration to change the *direction* of the velocity to round a sharp edge. Therefore, in order to bring the flow in line, so that it leaves the tail smoothly, in 1902 Kutta proposed adding a clockwise circulation Γ about the airfoil, Fig. 11–43b. In this way, when the two flows, Fig. 11–43a and 11–43b, are superimposed, air on top of the airfoil moves faster than air on the bottom, and the stagnation point B moves to the trailing edge, Fig. 11–43c. The faster-moving air on top then has a lower pressure, whereas the slower-moving air on the bottom has a higher pressure, and this pressure difference gives rise to the lift as calculated by $L = \rho U \Gamma$. Aeronautical engineers have used this equation, and it has been shown to be in close agreement with the lift measured experimentally for airfoils having low angles of attack.

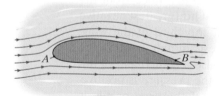

Flow without circulation

(a)

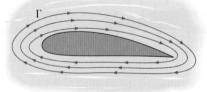

Flow with circulation

(b)

Fig. 11–43

Superposition of flow

(c)

Experimental Data.   Although the *lift coefficient* C_L can be calculated analytically using the circulation for small angles of attack, for larger angles the values of C_L must be determined by experiment. Values of C_L are usually plotted as a function of the attack angle α, and they look something like Fig. 11–44, which, again, is for the 2409 NACA wing section. With this data, and the planform area A_{pl} of the wing, the lift force is then calculated using the equation

$$F_L = C_L A_{pl}\left(\frac{\rho V^2}{2}\right) \qquad (11\text{–}32)$$

It is interesting to note what happens to the lift as the angle of attack of an airfoil increases. As shown in Fig. 11–45a, for a properly designed airfoil, the point of separation of the boundary layer will be near the trailing edge of the wing when the attack angle is zero. As α *increases*, however, this forces the air within the boundary layer to move faster over the top of the leading edge, causing the point of separation to move forward. When the angle of attack reaches a critical value, a large turbulent wake over the top surface of the wing develops, which increases the drag and causes the lift to suddenly drop off. This is a condition of *stall*, Fig. 11–45b. In Fig. 11–44, it is the point having the maximum lift coefficient, $C_L = 1.5$, which occurs at $\alpha \approx 20°$. Obviously, a stall is dangerous for any low-flying aircraft that may not have enough altitude to recover level flight.

Apart from changing the angle of attack to generate lift, modern aircraft also have moveable flaps on their leading and trailing edges to increase curvature of the wing, Fig. 11–46. They are used during takeoffs and landings, when the velocity is low and the need for controlling lift relative to drag is most important.

Race Cars.   Airfoils, such as those described here, are also on racing cars. See photo on p. 615. These devices, along with a unique body shape, are designed to enhance the *braking* of the car by creating a *downward force* due to the aerodynamic effects of the foil. Also, without an airfoil, lifting forces developed underneath the car may cause the tires to lose contact with the road, resulting in a loss of stability and control. Unfortunately, an airfoil used for this purpose can result in a higher-than-normal drag coefficient. For example, a Formula One race car can have a drag coefficient in the range of 0.7–1.1.

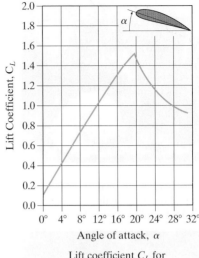

Lift coefficient C_L for NACA airfoil 2409

Fig. 11–44

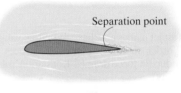

(a)

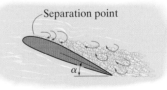

Stall

(b)

Fig. 11–45

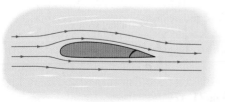

Flap up

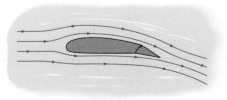

Flap down

Fig. 11–46

The wing vortex trail is clearly evident off the tip of the wing on this agricultural plane. (© NASA Archive/Alamy)

Most jet planes now have turned-up wing tips in order to mitigate trailing vortices that produce induced drag. (© Konstantin Yolshin/Alamy)

Trailing Vortices and Induced Drag.

Our previous discussion of drag produced by an airfoil (wing) was in reference to flow over the wing with *no consideration* for the variation in the length of the wing or for conditions at the wing tips. In other words, the wing was thought to have an *infinite span*. If we account for flow over an *actual wing*, then the air directed down off the trailing edge and tip will produce a swirl, called a ***wing vortex trail***, that contributes an additional drag on the aircraft.

To show how this occurs, consider the wing in Fig. 11–47. As shown, the *higher pressure* on the *bottom* of the wing will cause the flow to travel *up and around* the trailing edge, and *also* around the tip of the wing. This will pull the flow to the left under the wing, and as it comes around off the tip, it pushes the flow on top of the wing to the right. As a result, the *cross flow* off the trailing edge will form a multitude of vortices—a *vortex trail*. From observations, these vortices become unstable, and so they roll up at the edges and form two strong "trailing" vortices at the wing tips. The production of this disturbance requires energy, and so it places an extra burden on the lift, resulting in an ***induced drag***, which must be taken into account when calculating the total drag and strength of the wing. Actually, the turbulence created in this manner by large aircraft can be substantial and may persist for several minutes, creating a hazard for lighter planes flying behind them.

The induced drag on a typical jet aircraft generally amounts to 30% to 50% of the total drag, and at low speeds, such as takeoffs and landings, this percentage is even higher. To reduce this force component, modern aircraft add ***winglets***, or small turned-up airfoils on the wing tips, as shown in the adjacent photo. Experiments in wind tunnels have shown that when winglets are used, the trailing vortices lose strength, leading to about a 5% reduction in total aircraft drag at cruising speed, and to an even greater reduction during takeoffs and landings.

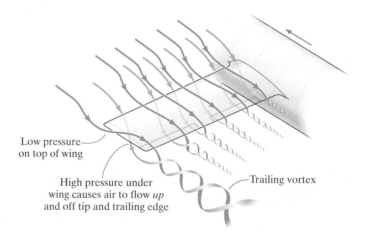

Low pressure on top of wing

High pressure under wing causes air to flow *up* and off tip and trailing edge

Trailing vortex

Fig. 11–47

Induced Drag Coefficient.   An airfoil of *infinite length* and traveling at V_0 must only overcome its section drag, and so in flight it will be directed at the *effective angle of attack* α_0 to maintain the lift $(F_L)_0$, Fig. 11–48a. However, any airfoil of *finite length* must overcome not only the section drag but also the induced drag. If the air underneath the tip of the wing rolls up and then down with an *induced* velocity V', then by vector addition, V_0 becomes V, Fig. 11–48b. To provide the necessary lift in this case, this will change the angle of attack from α_0 to a greater angle α. The difference in these angles, $\alpha_i = \alpha - \alpha_0$, is very small, and so the actual lift $\mathbf{F}_L$ is only slightly greater than $(\mathbf{F}_L)_0$. By vector addition, Fig. 11–48c, we see that the horizontal component $(\mathbf{F}_D)_i$ is the induced drag, and so for small angles its magnitude can be related to the lift by $(F_D)_i = F_L \alpha_i$.

Through both experiment and analysis, Prandtl has shown that if the air that is disturbed over the wing has an *elliptical shape*, as in Fig. 11–41b, which closely approximates many actual cases, then α_i becomes a function of the lift coefficient C_L, the length b of the wing, and its planform area A_{pl}. His result is

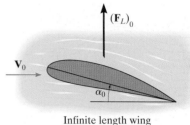

$(F_L)_0$

V_0

α_0

Infinite length wing

(a)

$$\alpha_i = \frac{C_L}{\pi b^2 / A_{pl}} \qquad (11\text{–}33)$$

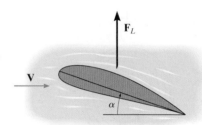

F_L

V

α

Finite length wing

(b)

Since the induced drag and the lift coefficients are proportional to their respective forces, then from Fig. 11–48c,

$$\alpha_i = \frac{(F_D)_i}{F_L} = \frac{(C_D)_i}{C_L}$$

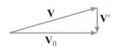

V

V'

V_0

And so, using Eq. 11–33, the ***induced drag coefficient*** is determined from

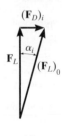

$(F_D)_i$

α_i

F_L

$(F_L)_0$

(c)

Fig. 11–48

$$(C_D)_i = \frac{C_L^2}{\pi b^2 / A_{pl}}$$

Notice that if $b \rightarrow \infty$, then $(C_D)_i \rightarrow 0$, as expected.

The above equation represents the *minimum* induced drag coefficient for any wing shape, and if it is used, then the total drag coefficient for the wing is

$$C_D = (C_D)_\infty + \frac{C_L^2}{\pi b^2 / A_{pl}} \qquad (11\text{–}34)$$

11

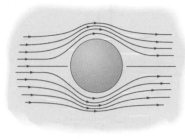

Ideal uniform flow
(a)

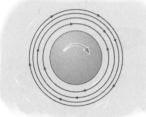

Circulation
(b)

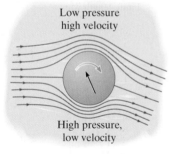

Low pressure
high velocity

High pressure,
low velocity

Combined flow
(c)

Fig. 11–49

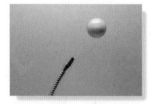

This rotating ball is suspended in the airstream due to the Magnus effect, that is, the lift provided by the rotation and redirection of the airstream balances the ball's weight.

Spinning Ball. Lift can also greatly affect the trajectory of a spinning ball, because the spin will alter the pressure distribution around the ball, thereby changing the direction of the momentum of the air. To show this, consider the ball moving to the left, without spin, Fig. 11–49a. Here the air flows symmetrically around the ball, and so the ball experiences a horizontal drag. If we consider what happens only when the ball spins, we see that its rough surface will pull air around it, forming a boundary layer in the direction of spin, Fig. 11–49b. Adding these two effects produces the condition shown in Fig. 11–49c. That is, in both Fig. 11–49a and Fig. 11–49b, the air is flowing on *top* of the ball in the same direction. This increases its energy and allows the boundary layer to remain attached to the surface for a longer time. Air passing *under* the ball goes in opposite directions for both cases, thereby losing energy and causing early boundary layer separation. Both these effects create a net downwash behind the ball. As a reaction to this, like an airfoil, the air in turn pushes or lifts the ball upward at an angle, Fig. 11–49c.

Experimental data for a *smooth* spinning ball is shown in Fig. 11–50. It is valid for $\text{Re} = VD/v = 6(10^4)$. See Ref. [20]. Notice how, up to a point, the lift coefficient is highly dependent upon the ball's angular velocity. Any increase in ω after this point will hardly affect the lift. Further lift is possible by *roughing* the surface, since this causes turbulence and an increased circulation around the ball, providing an even larger pressure difference between its top and bottom. In addition, the rough surface will reduce the drag because boundary layer separation is delayed. This tendency of a rotating ball to produce lift as described here is called the **Magnus effect**, named after the German scientist Heinrich Magnus, who discovered it. Most people who play baseball, tennis, or ping-pong instinctively notice this phenomenon, and take advantage of it when throwing or striking a ball.

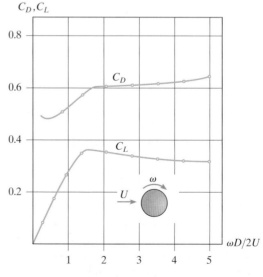

Drag and lift coefficients
for a smooth spinning ball

Fig. 11–50

Important Points

- Streamlining a body tends to decrease the pressure drag on the body, but it has the effect of increasing friction drag. For proper design both of these effects should be minimized.

- Pressure drag is reduced on a streamlined body by extending the laminar boundary layer over the surface, or preventing the boundary layer from separating from the surface. For airfoils, the methods include using wing slots, vortex generators, roughing the surface, and using a porous surface to draw air toward the boundary layer.

- Both drag and lift are important when designing an airfoil. These forces are related to their drag and lift coefficients, C_D and C_L, which are experimentally determined, and are represented graphically as a function of the angle of attack.

- Lift is produced by an airfoil because the airstream is redirected as it passes over the wing. The wing shape generates a force that changes the air's momentum, so the air flows downward. This results in an opposite force reaction of the air on the wing, which produces the lift.

- Trailing vortices are produced off the wing tips of aircraft due to differences in pressure acting on the top and bottom surfaces of the wing. As the air migrates and cross-flows off the wing tips, these vortices produce an induced drag, which must be accounted for in design.

- The trajectory of a spinning ball will curve in a steady airstream because the spinning causes an unequal pressure distribution on the ball's surface, and this results in a change in the direction of the air's momentum, thereby producing lift.

11

EXAMPLE | 11.13

The airplane in Fig. 11–51 has a mass of 1.20 Mg and is flying horizontally at an altitude of 7 km. If each wing is classified as a NACA 2409 section, with a span of 6 m and a cord length of 1.5 m, determine the angle of attack when the plane has an airspeed of 70 m/s. Also, what is the drag on the plane due to the wings, and what is the angle of attack and the speed that will cause the plane to stall?

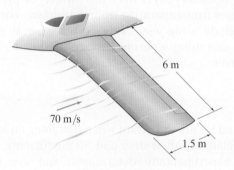

6 m

70 m/s

1.5 m

Fig. 11–51

SOLUTION

Fluid Description.   We have steady flow when measured relative to the plane. Also, the air is considered incompressible. Using Appendix A, at 7-km altitude $\rho_a = 0.590 \text{ kg/m}^3$.

Angle of Attack.   For vertical equilibrium, the lift must be equal to the weight of the plane, and so using Eq. 11–32, we can determine the required lift coefficient. Since we have two wings,

$$F_L = 2C_L A_{pl}\left(\frac{\rho V^2}{2}\right)$$

$$\left(1.20\left(10^3\right) \text{ kg}\right)\left(9.81 \text{ m/s}^2\right) = 2C_L(6 \text{ m})(1.5 \text{ m})\left(\frac{\left(0.590 \text{ kg/m}^3\right)\left(70 \text{ m/s}^2\right)}{2}\right)$$

$$C_L = 0.452$$

From Fig. 11–44, the angle of attack must be approximately

$$\alpha = 5° \qquad\qquad\qquad Ans.$$

11

Drag. The section drag coefficient at this angle of attack is determined from Fig. 11–39. It is for a wing of infinite length. We have, approximately,

$$(C_D)_\infty = 0.02$$

The total drag coefficient for the wings is determined from Eq. 11–34.

$$C_D = (C_D)_\infty + \frac{C_L^2}{\pi b^2 / A_{pl}}$$

$$= 0.02 + \frac{(0.452)^2}{\pi (6 \text{ m})^2 / [(6 \text{ m})(1.5 \text{ m})]}$$

$$= 0.02 + 0.0163 = 0.0363$$

Therefore, the drag on both wings is

$$F_D = 2 C_D A_{pl} \left(\frac{\rho V^2}{2} \right)$$

$$= 2(0.0363) [(6 \text{ m})(1.5 \text{ m})] \left(\frac{(0.590 \text{ kg/m}^3)(70 \text{ m/s})^2}{2} \right)$$

$$F_D = 944 \text{ N} \hspace{4cm} Ans.$$

Stall. From Fig. 11–44, *stall* will occur when the angle of attack is approximately 20° so that $C_L = 1.5$. The speed of the plane when this occurs is then

$$F_L = 2 C_L A_{pl} \left(\frac{\rho V^2}{2} \right)$$

$$\left(1.20(10^3) \text{ kg} \right) \left(9.81 \text{ m/s}^2 \right) = 2(1.5)(6 \text{ m})(1.5 \text{ m}) \left(\frac{(0.590 \text{ kg/m}^3) V_s^2}{2} \right)$$

$$V_s = 38.4 \text{ m/s} \hspace{4cm} Ans.$$

11

References

1. T. von Kármán, "Turbulence and skin friction," *J Aeronautics and Science*, Vol. 1, No. 1. 1934, p. 1.
2. W. P. Graebel, *Engineering Fluid Mechanics*, Taylor Francis, NY, 2001.
3. W. Wolansky et al., *Fundamentals of Fluid Power*, Houghton Mifflin, Boston, MA., 1985.
4. A. Azuma, *The Biokinetics of Flying and Swimming*, 2nd ed., American Institute of Aeronautics and Astronautics, Reston, VA, 2006.
5. J. Schetz, et al., *Boundary Layer Analysis*, 2nd ed., American Institute of Aeronautics and Astronautics, Reston, VA, 2011.
6. J. Vennard and R. Street, *Elementary Fluid Mechanics*, 5th ed., John Wiley, 1975.
7. G. Tokaty, *A History and Philosophy of Fluid Mechanics*, Dover Publications, New York, NY, 1994.
8. E. Torenbeek and H. Wittenberg, *Flight Physics*, Springer-Verlag, New York, NY, 2002.
9. T. von Kármán, *Aerodynamics*, McGraw-Hill, New York, NY, 1963.
10. L. Prandtl and O. G. Tietjens, *Applied Hydro- and Aeromechanics*, Dover Publications, New York, NY, 1957.
11. D. F. Anderson and S. Eberhardt, *Understanding Flight*, McGraw-Hill, New York, NY, 2000.
12. H. Blasius, "The boundary layers in fluids with little friction," NACA. T. M. 1256, 2/1950.
13. L. Prandtl, "Fluid motion with very small friction," NACA. T. M. 452, 3/1928.
14. P. T. Bradshaw et al., *Engineering Calculation Methods for Turbulent Flow*, Academic Press, New York, NY, 1981.
15. O. M. Griffin and S. E. Ramberg, "The vortex street wakes of vibrating cylinders," *J Fluid Mechanics*, Vol. 66, 1974, pp. 553–576.
16. H. Schlichting, *Boundary-Layer Theory*, 8th ed., Springer-Verlag, New York, NY, 2000.
17. F. M. White, *Viscous Fluid Flow*, 3rd ed., McGraw-Hill, New York, NY, 2005.
18. L. Prandtl, *Ergebnisse der aerodynamischen Versuchsanstalt zu Göttingen*, Vol. II, p. 29, R. Oldenbourg, 1923.
19. *CRC Handbook of Tables for Applied Engineering Science*, 2nd ed., CRC Press, Boca Raton, Fl, 1973.
20. S. Goldstein, *Modern Developments in Fluid Dynamics*, Oxford University Press, London, 1938.
21. J. D. Anderson, *Fundamentals of Aerodynamics*, 4th ed., McGraw-Hill, New York, NY, 2007.
22. E. Jacobs, et al., The Characteristics of 78 Related Airfoil Sections from Tests in the Variable-Density Wind Tunnel. National Advisory Committee for Aeronautics, Report 460, U.S. Government Printing Office, Washington, DC.
23. A. Roshko, "Experiments on the flow past a circular cylinder at very high Reynolds numbers," *J Fluid Mechanics*, Vol. 10, 1961, pp. 345–356.
24. W. H. Huchs, *Aerodynamics of Road Vehicles*, 4th ed., Society of Automotive Engineers, Warrendable, PA, 1998.
25. S. T. Wereley, and C. D. Meinhort, "Recent Advances in Micro-Particle Image Velocimetry," *Annual Review of Fluid Mechanics*, 42(1): 557–576, 2010.

PROBLEMS

Sec. 11.1–11.5

11–1. Oil flows with a free-stream velocity of $U = 1.5$ m/s over the flat plate. Determine the distance x_{cr} to where the boundary layer begins to transition from laminar to turbulent flow. Take $\mu_o = 0.0671(10^{-3})$ N·s/m² and $\rho_o = 920$ kg/m³.

11–2. Water at 15°C flows with a free-stream velocity of $U = 2$ m/s over the flat plate. Determine the shear stress on the surface of the plate at point A.

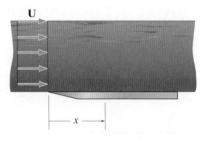

Prob. 11–1

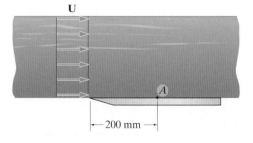

Prob. 11–2

11–3. Air at 60°C flows through the very wide duct. Determine the required dimension a of the duct at $x = 4$ m so that the central 200-mm core flow velocity maintains the constant free-stream velocity of 0.5 m/s.

11–6. The boundary layer for wind blowing over rough terrain can be approximated by the equation $u/U = (y/(y + 0.01))$, where y is in meters. If the free-stream velocity of the wind is 15 m/s, determine the velocity at an elevation $y = 0.1$ m and at $y = 0.3$ m from the ground surface.

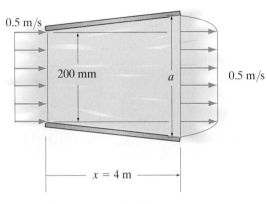

Prob. 11–3

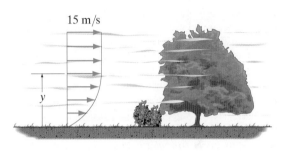

Prob. 11–6

*11–4. An oil–gas mixture flows over the top surface of the plate that is contained in a separator used to process these two fluids. If the free-stream velocity is 0.8 m/s, determine the maximum boundary layer thickness over the plate's surface. Take $\nu = 42(10^{-6})$ m²/s.

11–5. An oil–gas mixture flows over the top surface of the plate that is contained in a separator used to process these two fluids. If the free-stream velocity is 0.8 m/s, determine the friction drag acting on the surface of the plate. Take $\nu = 42(10^{-6})$ m²/s and $\rho = 910$ kg/m³.

11–7. A flat plate is to be coated with a polymer. If the thickness of the laminar boundary layer that occurs during the coating process at a distance of 0.5 m from the plate's front edge is 10 mm, determine the free-stream velocity of this fluid. Take $\nu = 4.68(10^{-6})$ m²/s.

*11–8. A fluid has laminar flow and passes over the flat plate. If the thickness of the boundary layer at a distance of 0.5 m from the plate's edge is 10 mm, determine the boundary layer thickness at a distance of 1 m.

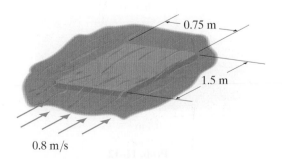

Probs. 11–4/5

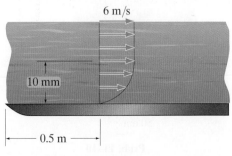

Prob. 11–8

11

11–9. Compare the thickness of the boundary layer of water with air at the end of the 0.4-m-long flat plate. Both fluids are at 20°C and have a free-stream velocity of $U = 0.8$ m/s.

11–11. Oil confined in a channel flows past the diverter fin at $U = 6$ m/s. Determine the friction drag acting on both sides of the fin. Take $\nu_o = 40(10^{-6})$ m^2/s and $\rho_o = 900$ kg/m^3. Neglect end effects.

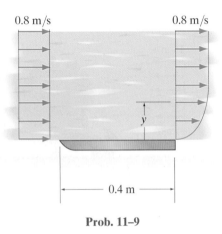

Prob. 11–9

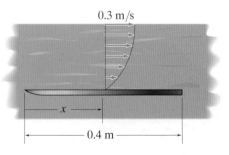

Prob. 11–11

11–10. Determine the friction drag on the bar required to overcome the resistance of the paint if the force **F** lifts the bar at 3 m/s. Take $\rho = 920$ kg/m^3 and $\nu = 42(10^{-6})$ m^2/s.

*11–12. Water at 40°C has a free-stream velocity of 0.3 m/s. Determine the shear stress on the plate's surface at $x = 0.2$ m and at $x = 0.4$ m.

Prob. 11–10

Prob. 11–12

11

11–13. A liquid having a viscosity μ, a density ρ, and a free-stream velocity U flows over the plate. Determine the distance x where the boundary layer has a disturbance thickness that is one-half the depth a of the liquid. Assume laminar flow.

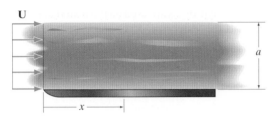

Prob. 11–13

11–14. Water at 40°C has a free-stream velocity of 0.3 m/s. Determine the boundary layer thickness at $x = 0.2$ m and at $x = 0.4$ m on the flat plate.

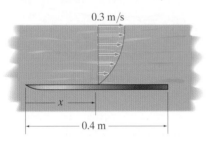

Prob. 11–14

11–15. Assume the boundary layer has a velocity profile that is linear and defined by $u = U(y/\delta)$. Use the momentum integral equation to determine τ_0 for the fluid passing over the flat plate.

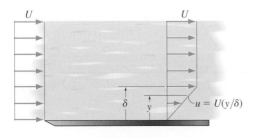

Prob. 11–15

*11–16.** Castor oil flows over the surface of the flat plate at a free-stream speed of 2 m/s. The plate is 0.5 m wide and 2 m long. Plot the boundary layer and the shear stress versus x. Give values for every 0.5 m. Also calculate the friction drag on the plate. Take $\rho_{co} = 960$ kg/m^3 and $\mu_{co} = 985(10^{-3})$ N·s/m^2.

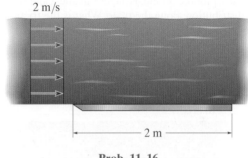

Prob. 11–16

11–17. The wind tunnel operates using air at a temperature of 20°C with a free-stream velocity of 40 m/s. If this velocity is to be maintained at the central 1-m core throughout the tunnel, determine the dimension a at the exit in order to accommodate the growing boundary layer. Show that the boundary layer is turbulent, and use $\delta^* = 0.0463x/(\text{Re}_x)^{1/5}$ to calculate the displacement thickness.

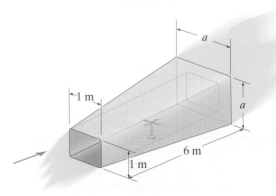

Prob. 11–17

11

11–18. A ship is traveling forward at 10 m/s on a lake. If it is 100 m long and the side of the ship can be assumed to be a flat plate, determine the drag force on a 1-m-wide strip along the entire length of the ship. The water is still and has a temperature of 15°C. Assume the boundary layer is completely turbulent.

11–19. Crude oil at 20°C flows over the surface of the flat plate that has a width of 0.7 m. If the free-stream velocity is $U = 10$ m/s, plot the boundary layer thickness and the shear-stress distribution along the plate. What is the friction drag on the plate?

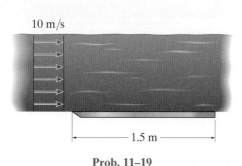

Prob. 11–19

11–21. Assume the turbulent boundary layer for a fluid has a velocity profile that can be approximated by $u = U(y/\delta)^{1/6}$. Use the momentum integral equation to determine the boundary layer thickness as a function of x. Use the empirical formula, Eq. 11–19, developed by Prandtl and Blasius.

11–22. Two hydrofoils are used on the boat that is traveling at 20 m/s. If the water is at 15°C, and if each blade can be considered as a flat plate, 4 m long and 0.25 m wide, determine the thickness of the boundary layer at the trailing or back edge of each blade. What is the drag on each blade? Assume the flow is completely turbulent.

Prob. 11–22

*11–20. Assume the turbulent boundary layer for a fluid has a velocity profile that can be approximated by $u = U(y/\delta)^{1/6}$. Use the momentum integral equation to determine the displacement thickness as a function of x and Re_x. Use the empirical formula, Eq. 11–19, developed by Prandtl and Blasius.

11–23. Two hydrofoils are used on the boat that is traveling at 20 m/s. If the water is at 15°C and if each blade can be considered as a flat plate, 4 m long and 0.25 m wide, determine the drag on each blade. Consider both laminar and turbulent boundary layers.

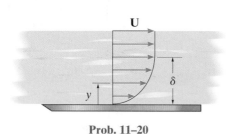

Prob. 11–20

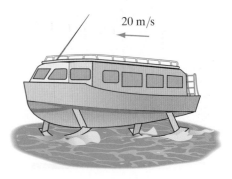

Prob. 11–23

*11–24. The laminar boundary layer for a fluid is assumed to be parabolic, such that $u/U = C_1 + C_2(y/\delta) + C_3(y/\delta)^2$. If the free-stream velocity U starts at $y = \delta$, determine the constants C_1, C_2, and C_3.

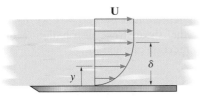

Probs. 11–24

11–25. The laminar boundary layer for a fluid is assumed to be cubic, such that $u/U = C_1 + C_2(y/\delta) + C_3(y/\delta)^3$. If the free-stream velocity U starts at $y = \delta$, determine the constants C_1, C_2, and C_3.

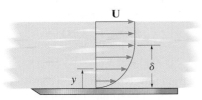

Probs. 11–25

11–26. Air enters the square plenum of an air-handling system with a velocity of 6 m/s and a temperature of 10°C. Determine the thickness of the boundary layer and the momentum thickness of the boundary layer, at $x = 1$ m.

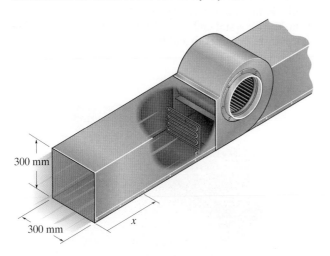

Prob. 11–26

11–27. Air enters the square plenum of an air-handling system with a velocity of 6 m/s and a temperature of 10°C. Determine the displacement thickness δ^* of the boundary layer at a point $x = 1$ m downstream. Also, what is the uniform speed of the air at this location?

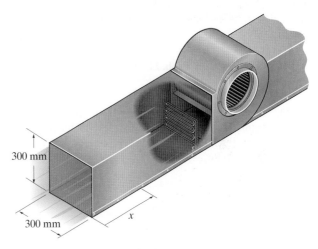

Prob. 11–27

*11–28. Assume a laminar boundary layer for a fluid can be approximated by $u/U = y/\delta$. Determine the thickness of the boundary layer as a function of x and Re_x.

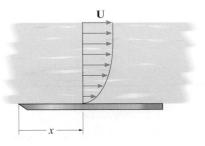

Probs. 11–28

11

11–29. Assume a laminar boundary layer for a fluid can be approximated by $u/U = \sin(\pi y/2\delta)$. Determine the thickness of the boundary layer as a function of x and Re_x.

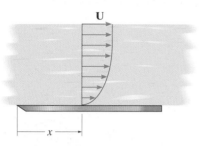

Prob. 11–29

11–30. Assume a laminar boundary layer for a fluid can be approximated by $u/U = \sin(\pi y/2\delta)$. Determine the displacement thickness δ^* for the boundary layer as a function of x and Re_x.

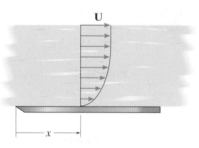

Prob. 11–30

11–31. A boundary layer for laminar flow of a fluid over the plate is to be approximated by the equation $u/U = C_1(y/\delta) + C_2(y/\delta)^2 + C_3(y/\delta)^3$. Determine the constants C_1, C_2, and C_3 using the boundary conditions when $y = \delta, u = U$; when $y = \delta$, $du/dy = 0$; and when $y = 0$, $d^2u/dy^2 = 0$. Find the thickness of the boundary layer as a function of x and Re_x using the momentum integral equation.

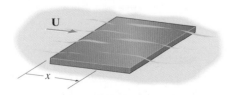

Prob. 11–31

*11–32.** The velocity profile for a laminar boundary layer of a fluid is represented by $u/U = 1.5(y/\delta) - 0.5(y/\delta)^3$. Determine the thickness of the boundary layer as a function of x and Re_x.

11–33. The velocity profile for a laminar boundary layer of a fluid is represented by $u/U = 1.5(y/\delta) - 0.5(y/\delta)^3$. Determine the shear-stress distribution acting on the surface as a function of x and Re_x.

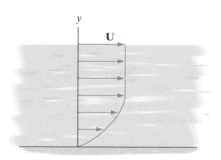

Probs. 11–32/33

11–34. The train travels at 30 m/s and consists of an engine and a series of cars. Determine the approximate thickness of the boundary layer at the top of a car, $x = 18$ m from the front of the train. The air is still and has a temperature of 20°C. Assume the surfaces are smooth and flat, and the boundary layer is completely turbulent.

11–35. The train travels at 30 m/s and consists of an engine and a series of cars. Determine the approximate shear stress acting on the top of a car at $x = 18$ m from the front of the train. The air is still and has a temperature of 20°C. Assume the surfaces are smooth and flat, and the boundary layer is completely turbulent.

30 m/s

Probs. 11–34/35

*11–36. An airplane is flying at a speed of 90 m/s. If the wings are assumed to have a flat surface of width 2.5 m, determine the boundary layer thickness δ and the shear stress at the trailing or back edge. Assume the boundary layer is fully turbulent. The airplane flies at an altitude of 1 km.

11–37. An airplane is flying at an altitude of 1 km and a speed of 90 m/s. If the wings are assumed to have a flat surface of width 2.5 m and length 7 m, determine the friction drag on each wing. Assume the boundary layer is fully turbulent.

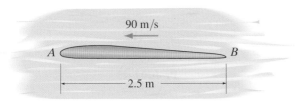

90 m/s

A   B

2.5 m

Probs. 11–36/37

11–38. Wind is blowing at 2 m/s as the truck moves forward into the wind at 8 m/s. If the air has a temperature of 20°C, determine the friction drag acting on the flat side *ABCD* of the truck. Assume the boundary layer is completely turbulent.

11–39. The wind is blowing at 2 m/s as the truck moves forward into the wind at 8 m/s. If the air has a temperature of 20°C, determine the friction drag acting on the top surface *BCFE* of the truck. Assume the boundary layer is completely turbulent.

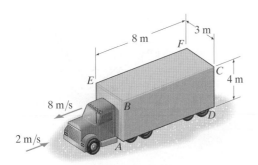

8 m

F

3 m

E

C

4 m

8 m/s

B

D

2 m/s

A

Probs. 11–38/39

*11–40. The oil tanker has a smooth surface area of $4.5(10^3)$ m² in contact with the sea. Determine the friction drag on its hull and the power required to overcome this force if the velocity of the ship is 2 m/s. Take $\rho = 1030$ kg/m³ and $\mu = 1.14(10^{-3})$ N·s/m².

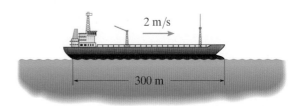

2 m/s

300 m

Prob. 11–40

11–41. The flat-bottom boat is traveling at 4 m/s on a lake for which the water temperature is 15°C. Determine the approximate drag acting on the bottom of the boat if the length of the boat is 10 m and its width is 2.5 m. Assume the boundary layer is completely turbulent.

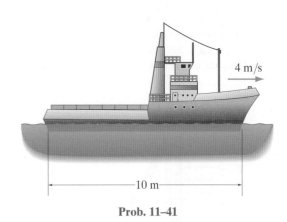

4 m/s

10 m

Prob. 11–41

11–42. An airplane has wings that are, on average, each 5 m long and 3 m wide. Determine the friction drag on the wings when the plane is flying at 600 km/h in still air at an altitude of 2 km. Assume the wings are flat plates and the boundary layer is completely turbulent.

11

Sec. 11.6–11.10

11–43. Wind blows over the inclined surface and produces the approximate pressure distribution shown. Determine the pressure drag acting over the surface if the surface is 3 m wide.

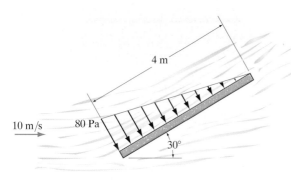

Prob. 11–43

11–45. The plate is 2 m wide and is held at an angle of 12° with the wind as shown. If the average pressure under the plate is 40 kPa, and on the top it is 60 kPa, determine the pressure drag on the plate.

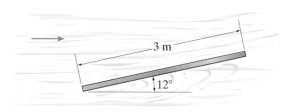

Prob. 11–45

*11–44.** The sign is subjected to a wind profile that produces a pressure distribution that can be approximated by $p = (112.5\, \rho y^{0.6})$ Pa, where y is in meters. Determine the resultant pressure force on the sign due to the wind. The air is at a temperature of 20°C, and the sign is 0.5 m wide.

11–46. The air pressure acting on the inclined surfaces is approximated by the linear distributions shown. Determine the resultant horizontal force acting on the surface if it is 3 m wide.

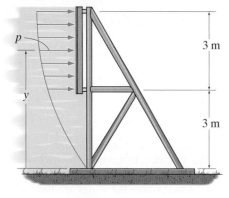

Prob. 11–44

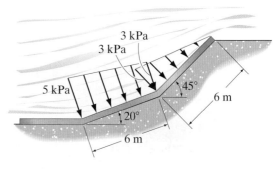

Prob. 11–46

11

11–47. The air pressure acting from A to B on the surface of a curved body can be approximated as $p = (5 - 1.5\theta)$ kPa, where θ is in radians. Determine the pressure drag acting on the body from $0° \leq \theta \leq 90°$. The body has a width of 300 mm.

11–49. Each of the smooth bridge piers (cylinders) has a diameter of 0.75 m. If the river maintains an average speed of 0.08 m/s, determine the drag the water exerts on each pier. The water temperature is 20°C.

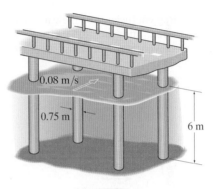

Prob. 11–49

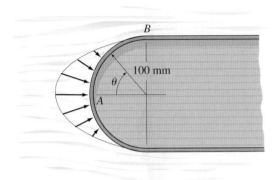

Prob. 11–47

11–50. Determine the moment developed at the base A of the square sign due to wind drag if the front of the signboard is subjected to a 16 m/s wind. The air is at 20°C. Neglect the drag on the pole.

*11–48. Wind at 20°C blows against the 100-mm-diameter flagpole with a speed of 1.20 m/s. Determine the drag on the pole if it has a height of 8 m. Consider the pole to be a smooth cylinder. Would you consider this a significant force?

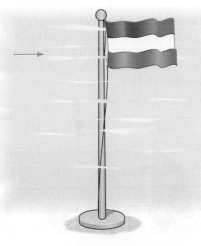

Prob. 11–48

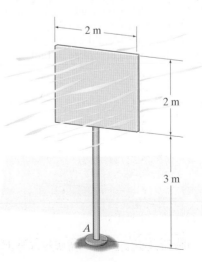

Prob. 11–50

11

11–51. A periscope on a submarine has a submerged length of 2.5 m and a diameter of 50 mm. If the submarine is traveling at 8 m/s, determine the moment developed at the base of the periscope. The water is at a temperature of 15°C. Consider the periscope to be a smooth cylinder.

11–54. The rocket has a nose cone that is 60° and a base diameter of 1.25 m. Determine the drag of the air on the cone when the rocket is traveling at 60 m/s in air having a temperature of 10°C. Use Table 11–3 for the cone, but explain why this may not be an accurate assumption.

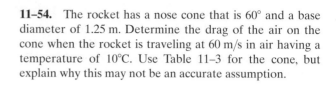

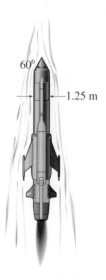

60°

—1.25 m

Prob. 11–54

*11–52. The drag coefficient for the car is $C_D = 0.28$, and the projected area into the 20°C airstream is 2.5 m². Determine the power the engine must supply to maintain a constant speed of 160 km/h.

160 km/h

Prob. 11–52

11–55. The truck has a drag coefficient of $C_D = 1.12$ when it is moving with a constant velocity of 80 km/h. Determine the power needed to drive the truck at this speed if the average front projected area of the truck is 10.5 m². The air is at a temperature of 10°C.

11–53. A rectangular plate is immersed in a stream of oil flowing at 0.5 m/s. Compare the drag acting on the plate if it is oriented so that AB is the leading edge and then when it is rotated 90° counterclockwise so that BC is the leading edge. The plate is 0.8 m wide. Take $\rho_o = 880 \text{ kg/m}^3$. For the calculation use the rectangular plate in Table 11–3.

*11–56. The truck has a drag coefficient of $C_D = 0.86$ when it is moving with a constant velocity of 60 km/h. Determine the power needed to drive the truck at this speed if the average front projected area of the truck is 10.5 m². The air is at a temperature of 10°C.

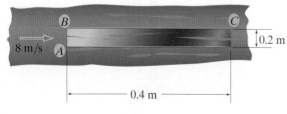

B

C

0.2 m

8 m/s A

0.4 m

Prob. 11–53

Probs. 11–55/56

11–57. Wind at 10°C blows against the 30-m-high chimney at 2.5 m/s. If the diameter of the chimney is 2 m, determine the moment that must be developed at its base to hold it in place. Consider the chimney to be a rough cylinder.

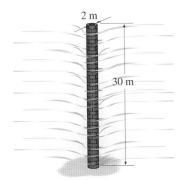

Prob. 11–57

11–58. The parachute has a drag coefficient of $C_D = 1.36$ and an open diameter of 4 m. Determine the terminal velocity as the man parachutes downward. The air is at 20°C. The total mass of the parachute and man is 90 kg. Neglect the drag on the man.

11–59. The parachute has a drag coefficient of $C_D = 1.36$. Determine the required open diameter of the parachute so the man attains a terminal velocity of 10 m/s. The air is at 20°C. The total mass of the parachute and man is 90 kg. Neglect the drag on the man.

*11–60. The man and the parachute have a total mass of 90 kg. If the parachute has an open diameter of 6 m and the man attains a terminal velocity of 5 m/s, determine the drag coefficient of the parachute. The air is at 20°C. Neglect the drag on the man.

Probs. 11–58/59/60

11–61. A 5-m-diameter balloon is at an altitude of 2 km. If it is moving with a terminal velocity of 12 km/h, determine the drag on the balloon.

11–62. The smooth empty drum has a mass of 8 kg and rests on a surface having a coefficient of static friction of $\mu_s = 0.3$. Determine the speed of the wind needed to cause it to either tip over or slide. The air temperature is 30°C.

11–63. The smooth empty drum has a mass of 8 kg and rests on a surface having a coefficient of static friction of $\mu_s = 0.6$. Determine the speed of the wind needed to cause it to either tip over or slide. The air temperature is 30°C.

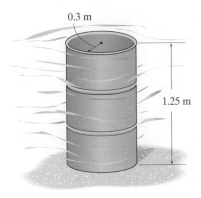

Probs. 11–62/63

*11–64. A ball has a diameter of 60 mm and falls in oil with a terminal velocity of 0.8 m/s. Determine the density of the ball. For oil, take $\rho_o = 880 \text{ kg/m}^3$ and $\nu_0 = 40(10^{-6}) \text{ m}^2/\text{s}$. *Note:* The volume of a sphere is $V = \frac{4}{3}\pi r^3$.

11–65. The smooth cylinder is suspended from the rail and is partially submerged in the water. If the wind blows at 8 m/s, determine the terminal velocity of the cylinder. The water and air are both at 20°C.

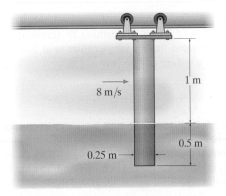

Prob. 11–65

11

11–66. The blades of a mixer are used to stir a liquid having a density ρ and viscosity μ. If each blade has a length L and width w, determine the torque **T** needed to rotate the blades at a constant angular rate ω. Take the drag coefficient of the blade's cross section to be C_D.

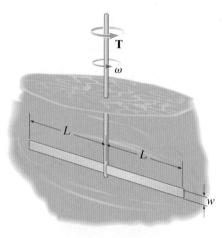

Prob. 11–66

11–67. A solid ball has a diameter of 20 mm and a density of 3.00 Mg/m^3. Determine its terminal velocity if it is dropped into a liquid having a density of $\rho = 2.30 \text{ Mg/m}^3$ and a viscosity of $\nu = 0.052 \text{ m}^2/\text{s}$. *Note*: The volume of a sphere is $V = \frac{4}{3}\pi r^3$.

*11–68. A 5-m-diameter balloon and the gas within it have a mass of 80 kg. Determine its terminal velocity of descent. Assume the air temperature is at 20°C. *Note*: The volume of a sphere is $V = \frac{4}{3}\pi r^3$.

11–69. A boat traveling with a constant velocity of 2 m/s tows a half submerged log having an approximate diameter of 0.35 m. If the drag coefficient is $C_D = 0.85$, determine the tension in the tow rope if it is horizontal. The log is oriented so that the flow is along the length of the log.

11–70. Particulate matter at an altitude of 8 km in the upper atmosphere has an average diameter of 3 μm. If a particle has a mass of $42.5(10^{-12})$ g, determine the time needed for it to settle to the earth. Assume gravity is constant, and for air, $\rho = 1.202 \text{ kg/m}^3$ and $\mu = 18.1(10^{-6}) \text{ N·s/m}^2$.

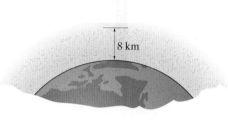

Prob. 11–70

11–71. A ball having a diameter of 0.6 m and a mass of 0.35 kg is falling in the atmosphere at 10°C. Determine its terminal velocity. *Note*: The volume of a sphere is $V = \frac{4}{3}\pi r^3$.

*11–72. A rock is released from rest at the surface of the lake, where the average water temperature is 15°C. If $C_D = 0.5$, determine its speed when it reaches a depth of 600 mm. The rock can be considered a sphere having a diameter of 50 mm and a density of $\rho_r = 2400 \text{ kg/m}^3$. *Note*: The volume of a sphere is $V = \frac{4}{3}\pi r^3$.

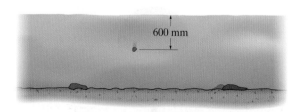

Prob. 11–72

11–73. Determine the velocity of the aerosol solid particles when $t = 10$ μs, if when $t = 0$ they leave the can with a horizontal velocity of 30 m/s. Assume the average diameter of the particles is 0.4 μm and each has a mass of $0.4(10^{-12})$ g. The air is at 20°C. Neglect the vertical component of the velocity. *Note:* The volume of a sphere is $V = \frac{4}{3}\pi r^3$.

Prob. 11–73

11–74. A raindrop has a diameter of 1 mm. Determine its approximate terminal velocity as it falls. Assume that the air has a constant $\rho_a = 1.247$ kg/m³ and $\nu_a = 14.2(10^{-6})$ m²/s. Neglect buoyancy. *Note:* The volume of a sphere is $V = \frac{1}{6}\pi D^3$.

11–75. The 2-Mg race car has a projected front area of 1.35 m² and a drag coefficient of $(C_D)_c = 0.28$. If the car is traveling at 60 m/s, determine the diameter of the parachute needed to reduce the car's speed to 20 m/s in 4 s. Take $(C_D)_p = 1.15$ for the parachute. The air is at 20°C. The wheels are free to roll.

Prob. 11–75

*****11–76.** A 2-mm-diameter sand particle having a density of 2.40 Mg/m³ is released from rest at the surface of oil that is contained in the tube. As the particle falls downward, "creeping flow" will be established around it. Determine the velocity of the particle and the time at which Stokes' law becomes invalid, at about Re = 1. The oil has a density of $\rho_o = 900$ kg/m³ and a viscosity of $\mu_o = 30.2(10^{-3})$ N·s/m². Assume the particle is a sphere, where its volume is $V = \frac{4}{3}\pi r^3$.

Prob. 11–76

11–77. Impure water at 20°C enters the retention tank and rises to a level of 2 m when it stops flowing in. Determine the shortest time needed for all sediment particles having a diameter of 0.05 mm or greater to settle to the bottom. Assume the density of the particles is $\rho = 1.6$ Mg/m³ or greater. *Note:* The volume of a sphere is $V = \frac{4}{3}\pi r^3$.

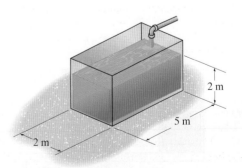

Prob. 11–77

11

11–78. The parachutist has a total mass of 90 kg and is in free fall at 6 m/s when she opens her 3-m-diameter parachute. Determine the time for her speed to be increased to 10 m/s. Also, what is her terminal velocity? For the calculation, assume the parachute to be similar to a hollow hemisphere. The air has a density of $\rho_a = 1.25$ kg/m³.

Prob. 11–78

11–79. A smooth ball has a diameter of 43 mm and a mass of 45 g. When it is thrown vertically upward with a speed of 20 m/s, determine the initial deceleration of the ball. The temperature is 20°C.

*11–80. Dust particles having an average diameter of 0.05 mm and an average density of 450 kg/m³ are stirred up by an airstream and vblown off the edge of the 600-mm-high desk into a horizontal steady wind of 0.5 m/s. Determine the distance d from the edge of the desk where most of them will strike the ground. The air is at a temperature of 20°C. *Note:* The volume of a sphere is $V = \frac{4}{3}\pi r^3$.

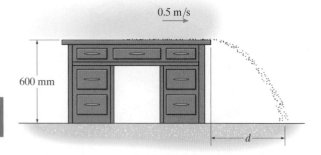

Prob. 11–80

Sec. 11.11

11–81. If it takes 80 kW of power to fly an airplane at 20 m/s, how much power does it take to fly the plane at 25 m/s at the same altitude? Assume C_D remains constant.

11–82. A 4-Mg airplane is flying at a speed of 70 m/s. If each wing can be assumed rectangular of length 5 m and width 1.75 m, determine the drag on each wing when it is flying at the proper angle of attack α. Assume each wing is a NACA 2409 section. The density of air is $\rho_a = 1.225$ kg/m³.

11–83. The plane can take off at 250 km/h when it is at an airport located at an elevation of 2 km. Determine the takeoff speed from an airport at sea level.

Prob. 11–83

*11–84. The 5-Mg airplane has wings that are each 5 m long and 1.75 m wide. Determine its speed in order to generate the same lift when flying horizontally at an altitude of 5 km as it does when flying horizontally at 3 km with a speed of 150 m/s.

Prob. 11–84

11–85. A 3-Mg airplane is flying at a speed of 70 m/s. If each wing can be assumed rectangular of length 5 m and width 1.75 m, determine the smallest angle of attack α to provide lift assuming the wing is a NACA 2409 section. The density of air is $\rho = 1.225 \text{ kg/m}^3$.

11–86. The 5-Mg airplane has wings that are each 5 m long and 1.75 m wide. It is flying horizontally at an altitude of 3 km with a speed of 150 m/s. Determine the lift coefficient.

Probs. 11–85/86

11–87. A 0.5-kg ball having a diameter of 50 mm is thrown with a speed of 10 m/s and has an angular velocity of 400 rad/s. Determine its horizontal deviation d from striking a target a distance of 10 m away. Take $\rho_a = 1.20 \text{ kg/m}^3$ and $\nu_a = 15.0(10^{-6}) \text{ m}^2/\text{s}$, and use Fig. 11–50. Neglect the effect on lift caused by the vertical component of velocity.

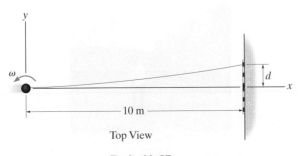

Top View

Prob. 11–87

11–88. The plane of mass 4 mg begins take off from an airport when it attains an airspeed of 200 km/h. If it carries an additional load of mass 350 kg, what must be its airspeed before beginning of takeoff at the same angle of attack?

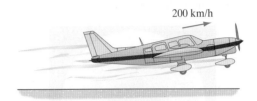

200 km/h

Prob. 11–88

11–89. A baseball has a diameter of 73 mm. If it is thrown with a speed of 5 m/s and an angular velocity of 60 rad/s, determine the lift on the ball. Assume the surface of the ball is smooth. Take $\rho_a = 1.20 \text{ kg/m}^3$ and $\nu_a = 15.0(10^{-6}) \text{ m}^2/\text{s}$, and use Fig. 11–50.

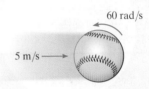

60 rad/s

5 m/s

Prob. 11–89

11–90. The glider has a constant speed of 8 m/s through still air. Determine the angle of descent θ if it has a lift coefficient of $C_L = 0.70$ and a wing drag coefficient of $C_D = 0.04$. The drag on the fuselage is considered negligible compared to that on the wings, since the glider has a very long wingspan.

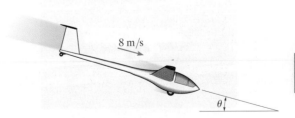

8 m/s

Prob. 11–90

11

CONCEPTUAL PROBLEMS

P11–1. When the cup of hot tea is stirred, top photo, the "leaves" seem to eventually settle at the bottom in the *center* of the cup. Why do they do this, instead of accumulating along the rim? Explain.

P11–1

P11–2. Which structure will best survive in a hurricane, the triangular-shaped building or the dome-shaped building? Draw the pressure distribution and streamlines for the flow in each case to explain your answer.

P11–2

P11–3. Lift, drag, thrust, and weight act on the plane. When the plane is climbing during takeoff is the lift force less than, greater than, or equal to the weight of the plane? Explain.

P11–3

P11–4. A baseball is thrown straight up into the air. Will the time to travel to its highest point be longer, shorter, or the same as the time for it to fall to the same height from which it was thrown?

P11–4

11

CHAPTER REVIEW

The boundary layer is a very thin layer of fluid located in a region just above the surface of a body. Within it, the velocity changes from zero at the surface to the free-stream velocity of the fluid.

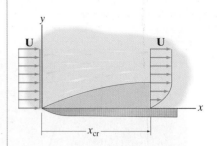

The fluid within the boundary layer formed over the surface of a flat plate will be laminar up to the critical distance x_{cr}. In this text, this distance is determined from $(\text{Re}_x)_{cr} = Ux_{cr}/\nu = 5(10^5)$.

The velocity profile for a *laminar flow* boundary layer has been solved by Blasius. The solution is given in both graphical and tabular form. Knowing this velocity profile, one can find the thickness of the boundary layer and the friction drag that the flow exerts over a flat plate.

The friction drag caused by *turbulent flow* boundary layers is determined by experiment. For both laminar and turbulent cases, this force is reported by using a dimensionless friction drag coefficient C_{Df}, which is a function of the Reynolds number.

$$F_{Df} = C_{Df}A\left(\frac{1}{2}\rho U^2\right)$$

The thickness and shear-stress distribution for both laminar and turbulent boundary layers can be determined by an approximate method, using the momentum integral equation.

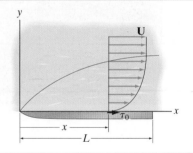

Since turbulent flow creates a larger shear stress on a surface, compared to laminar flow, turbulent boundary layers create a larger friction drag on the surface.

Experimental drag coefficients C_D, caused by both viscous friction and pressure, have been determined for the cylinder, the sphere, and bodies of many other shapes. In general, C_D is a function of the Reynolds number, the body's shape, its orientation within the flow, and its surface roughness. For some cases, this coefficient may also depend on the Froude number or the Mach number.

$$F_D = C_D A_p\left(\frac{\rho V^2}{2}\right)$$

When designing an airfoil, both drag and lift are important. Their coefficients, C_D and C_L, are determined by experiment and are represented graphically as a function of the angle of attack.

11

Chapter 12

(© Tim Roberts Photography/Shutterstock)

Open channels are often used for drainage and irrigation. It is important that they be designed properly so that adequate flow through them is maintained.

Open-Channel Flow

12.1   Types of Flow in Open Channels

An *open channel* is any conduit that has an open or free surface. Examples include rivers, canals, culverts, and flumes. Of these, rivers and streams have variable cross sections, which change over time due to erosion and deposition of earth. *Canals* are generally very long and straight and are used for drainage, irrigation, or navigation. *Culverts* normally do not flow full, and are usually made of concrete or masonry. They are often used to carry drainage under roadways. Finally, a *flume* is a conduit that is supported above ground and is designed to carry drainage over a depression.

If the channel has a constant cross section, it is known as a *prismatic channel*. For example, canals are typically constructed with rectangular or trapezoidal cross sections, whereas culverts and flumes often have circular or elliptical shapes. Rivers and streams have nonprismatic cross sections; however, for approximate analysis they are sometimes modeled by a series of different-size prismatic sections, such as trapezoids and semi-ellipses.

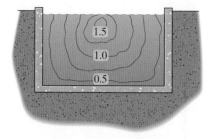

Typical velocity contours,
in m/s, for a rectangular channel

Fig. 12–1

Uniform steady flow,
constant depth

(a)

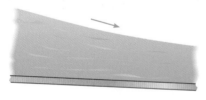

Accelerated
nonuniform flow

(b)

Retarded
nonuniform flow

(c)

Fig. 12–2

Laminar and Turbulent Flows.

Although laminar flow can occur in an open channel, in engineering practice it is rarely encountered. This is because the flow must be very slow to meet any Reynolds number criteria for laminar flow. Instead, open-channel flows are predominantly turbulent. The mixing of the liquid that actually takes place can be caused by the friction of the wind blowing over its surface and friction along the sides of the channel. These effects will cause the velocity profile to become highly irregular; and as a result, the maximum velocity will be somewhere near the top of the liquid surface, but normally not at the surface. A typical velocity profile for water flowing through an open rectangular channel may look something like that shown in Fig. 12–1. Hence, even though the surface may appear calm, as if the flow is laminar, beneath the surface it will be turbulent. In spite of this irregularity, we can often *approximate* the actual flow as being uniform one-dimensional flow, and still achieve reasonable results in estimating the flow.

Uniform and Steady Flow.

Besides being either laminar or turbulent, open-channel flow can also be classified in another way.

Uniform flow occurs when the *depth* of the liquid *remains the same* along the *length* of the channel, because then the velocity of the liquid will not change from one location to the next. An example occurs when the channel has a small slope, so that the gravity force causing the flow is balanced by the friction force that retards it, Fig. 12–2a. If the depth varies along the length, then the flow is *nonuniform*. This can happen if there is a change in slope, or where there is a change in the channel's cross-sectional area. *Accelerated nonuniform flow* occurs when the depth of flow decreases downstream, Fig. 12–2b. An example would be water flowing down a chute or spillway. *Retarded nonuniform flow* occurs if the depth is increasing, Fig. 12–2c, such as when water in a downward-sloping channel backs up to meet the crest of a dam.

Steady flow in a channel occurs when the flow rate *remains constant over time* as in Fig. 12–2a, and so its depth at a specific location remains constant. This is the case for most problems involving open-channel flow. However, if a wave passes by a specific location, the depth and hence the flow will change with time, and this is classified as *unsteady flow*.

Hydraulic Jump.

Apart from the types of flow mentioned above, there is one other phenomenon that can occur in open channels. A **hydraulic jump** is a localized turbulence that rapidly dissipates kinetic energy from the flow. It generally occurs at the bottom of a chute or spillway, Fig. 12–3.

Hydraulic jump

Fig. 12–3

12.2   Open-Channel Flow Classifications

Later in this section it will be shown that the type of flow that occurs in an open channel can be classified by comparing the speed of the liquid in the channel to the speed of a wave on its surface. To make this comparison, however, it is first necessary to formulate a way to obtain the wave speed. Specifically, the speed of the wave relative to the speed of the liquid in the channel is called the **wave celerity**, c.

To determine c we will consider the wave height Δy to be small compared to the depth y of the liquid, Fig. 12–4a. Excluding the effects of surface tension, the propagation of the wave along the channel will be caused by gravity.* To a fixed observer, unsteady flow will occur, since initially the liquid is at rest and then as the wave passes, the liquid under the wave is disturbed and will have a velocity **V**. Realize that the *wave profile* only moves the fluid up and down as it passes by, although it creates the *illusion* that the liquid composing the wave is actually moving over the surface with the velocity c, which is not the case.

For the analysis that follows, it is easier to fix a reference to a control volume that *moves with the wave*, so that for an observer of the wave, the flow will appear as steady flow, Fig. 12–4b. In other words, for one-dimensional flow, the liquid at the open control surface 2 appears to move to the left at c, and the liquid at the open control surface 1 appears to move to the left at **V**. If the liquid is assumed to be ideal, and the channel has a constant width b, then the continuity equation gives

$$\frac{\partial}{\partial t}\int_{cv}\rho\, d\forall + \int_{cs}\rho\,\mathbf{V}_{f/cs}\cdot d\mathbf{A} = 0$$

$$0 + \rho(-c)(yb) + \rho[V(y + dy)b] = 0$$

$$V = \frac{cy}{y + \Delta y}$$

The Bernoulli equation can be applied to points 1 and 2 on the surface streamline.[†] We have

$$\frac{p_1}{\gamma} + \frac{V_1^2}{2g} + z_1 = \frac{p_2}{\gamma} + \frac{V_2^2}{2g} + z_2$$

$$0 + \frac{V^2}{2g} + (y + \Delta y) = 0 + \frac{c^2}{2g} + y$$

Substituting for V, using the result from the continuity equation, and solving for c, we obtain

$$c = \left[\frac{2g\left(y^2 + 2y\Delta y + (\Delta y)^2\right)}{2y + \Delta y}\right]^{1/2}$$

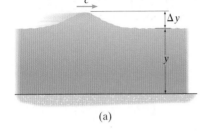

(a)

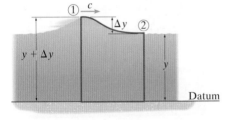

(b)

Fig. 12–4

12

Since the wave has a small height Δy compared to the liquid depth y, then Δy can be neglected, and so the result becomes

$$c = \sqrt{gy} \qquad\qquad (12\text{--}1)$$

The wave speed is only a function of the liquid depth and is independent of any of the fluid's physical properties. It is interesting to note that in the ocean, wave speeds can reach very high values. For example, if the ocean depth is 3 km, any tsunami waves generated by something like an earthquake will travel at 172 m/s!

Froude Number.   The driving force for all open-channel flow is due to *gravity*, and so in 1871, William Froude formulated the Froude number and showed how it can be used for describing this flow. Recall that in Chapter 8 we defined the Froude number as the square root of the ratio of the inertia force to the gravitational force. The result is expressed as

$$\text{Fr} = \frac{V}{\sqrt{gy}} = \frac{V}{c} \qquad\qquad (12\text{--}2)$$

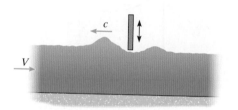

Tranquil flow: Wave moves upstream $c > V$
Critical flow: Standing wave $V = c$
Rapid flow: Wave moves downstream $V > c$

(c)

Fig. 12-4 (cont.)

where V is the average velocity of the liquid in the channel, and y is the depth. To show why the Froude number is important, consider, for example, the case shown in Fig. 12–4c, where a plate momentarily obstructs the steady stream, producing two waves. If Fr $= 1$, then from Eq. 12–2, the liquid must have a velocity $V = c$. When this occurs, the wave on the left will *stand still*. This is referred to as ***critical flow***. If Fr < 1, then $c > V$, and this wave will propagate *upstream*, which is a condition of ***tranquil flow***. In other words, the gravitational force or the weight of the wave overcomes the inertia force caused by its movement. Finally, if Fr > 1, then $V > c$, and the wave will be washed *downstream*. This is termed ***rapid flow*** and is the result of the gravitational force being overcome by the wave's inertia force.

12.3   Specific Energy

The actual behavior of the flow at each location along an open channel depends upon the *total energy* of the flow at that location. To find this energy, we will apply the Bernoulli equation, which assumes steady flow of an ideal liquid. If we establish the datum at the bottom of the channel, as shown in Fig. 12–5, and choose the streamline on the liquid's surface, where the pressure is atmospheric, then $p_1 = p_2 = 0$, and the Bernoulli equation becomes*

$$\frac{p_1}{\gamma} + \frac{V_1^2}{2g} + y_1 = \frac{p_2}{\gamma} + \frac{V_2^2}{2g} + y_2$$

$$y_1 + \frac{V_1^2}{2g} = y_2 + \frac{V_2^2}{2g} \qquad (12\text{-}3)$$

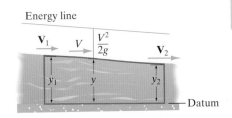

Fig. 12–5

Therefore, at *any intermediate location*, Fig. 12–5, we can also write the total energy for the flow. It is

$$E = \frac{V^2}{2g} + y \qquad (12\text{-}4)$$

This sum is referred to as the **specific energy**, E, since it indicates the amount of the kinetic and potential energy per unit weight of the liquid at a specific location. Stated another way, it is also called the **specific head**, because it has units of length and thereby represents the vertical distance from the *channel bottom* to the energy line, Fig. 12–5.

The specific energy can also be expressed in terms of the volumetric flow by using $Q = VA$. Thus,

$$E = \frac{Q^2}{2gA^2} + y \qquad (12\text{-}5)$$

We can further express E only as a function of y by considering a unique cross section.

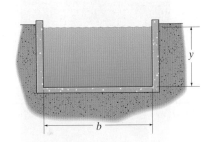

Rectangular cross section
(a)

Rectangular Cross Section. If the cross section is rectangular, Fig. 12–6a, then $A = by$ and so

$$\boxed{E = \frac{Q^2}{2gb^2y^2} + y} \qquad (12\text{-}6)$$

Specific energy

There are two independent variables in this equation, Q and y. However, if Q is held *constant*, then a plot of Eq. 12–6 has the shape shown in Fig. 12–6b. It is called a **specific energy diagram**. In particular, if $Q = 0$, then $E = y$, as shown by the 45° sloped line. This represents a condition of the liquid having no motion or kinetic energy, only potential energy. Notice, however, that when the liquid has a flow Q, then there will be *two* possible depths, y_1 and y_2, that produce the same specific energy $E = E'$. Here the smaller value y_1 represents low potential energy and high kinetic energy. This is rapid or supercritical flow. Likewise, the larger value y_2 represents high potential energy and low kinetic energy. It is referred to as tranquil or subcritical flow.

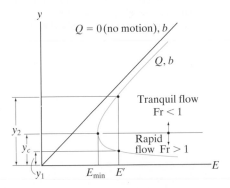

Specific energy diagram
(b)

Fig. 12–6

* Any points on the open control surface can be selected as noted in Example 5.11.

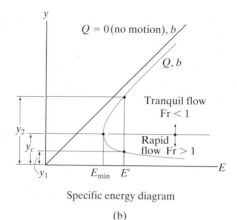

Specific energy diagram

(b)

Fig. 12–6

As shown in the graph, the *minimum value* for the specific energy, E_{min}, occurs at the critical depth. It can be found by setting the derivative of Eq. 12–6 equal to zero and evaluating the result at $y = y_c$. Since Q is constant,

$$\frac{dE}{dy} = \frac{-Q^2}{gb^2 y_c^3} + 1 = 0$$

$$y_c = \left(\frac{Q^2}{gb^2}\right)^{1/3} \tag{12–7}$$

Substituting this into Eq. 12–6, the value of E_{min} is therefore

$$E_{min} = \frac{y_c^3}{2y_c^2} + y_c = \frac{3}{2}y_c \tag{12–8}$$

To summarize, this is the smallest amount of specific energy the liquid can have and yet still maintain the required flow Q. It occurs at the nose of the curve in Fig. 12–6b, where the flow Q is at the critical depth y_c.

To find the critical velocity at this depth, substitute $Q = V_c(by_c)$ into Eq. 12–7, so that

$$y_c = \left(\frac{V_c^2 b^2 y_c^2}{gb^2}\right)^{1/3}$$

and so

$$V_c = \sqrt{gy_c} \tag{12–9}$$

Notice that when the flow is at this critical velocity, the Froude number becomes

$$Fr = \frac{V_c}{\sqrt{gy_c}} = 1$$

Canals having rectangular cross sections are often used for small flows or within confined spaces through crowded neighborhoods.

12

Therefore, for any point on the *upper branch* of the curve in Fig. 12–6b, the depth of flow will exceed the critical depth, $y = y_2 > y_c$. When this happens, $V < V_c$, and so Fr < 1. This is subcritical, or tranquil, flow. Likewise, for any point on the *lower branch* of the curve, the depth of flow will be less than the critical depth, $y = y_1 < y_c$, and then $V > V_c$ and Fr > 1. This is supercritical, or rapid, flow. These three classifications are therefore

Fr $< 1, y > y_c$ or $V < V_c$	Subcritical (tranquil) flow
Fr $= 1, y = y_c$ or $V = V_c$	Critical flow
Fr $> 1, y < y_c$ or $V > V_c$	Supercritical (rapid) flow

(12–10)

Large drainage channels are often made in a trapezoidal shape since this shape is relatively easy to construct. Notice that engineers have placed "weep holes" along the sloped sides in order to reduce the hydrostatic pressure on the inside wall, caused by groundwater absorption.

Because of the branching of the specific energy diagram that takes place at y_c, engineers do not design channels to flow at the critical depth. If they did, then **standing waves** or **undulations** would develop on the surface of the liquid, and any slight disturbance in the depth of flow would cause the liquid to constantly adjust between subcritical and supercritical flow. An unstable condition.

Nonrectangular Cross Section. When the channel cross section is nonrectangular, Fig. 12–7, then the minimum specific energy must be obtained by taking the derivative of Eq. 12–5 and setting it equal to zero, and requiring $A = A_c$. This gives

$$\frac{dE}{dy} = \frac{-Q^2}{gA_c^3}\frac{dA}{dy} + 1 = 0$$

At the top of the channel, the elemental area strip $dA = b_{top}dy$, and so we get

$$\frac{gA_c^3}{Q^2 b_{top}} = 1 \qquad (12–11)$$

Provided b_{top} and A_c can both be related by the geometry of the cross section to the critical depth y_c, Fig. 12–7, then a solution for y_c can be determined from this equation. See Example 12.3.

To find the critical velocity of flow, substitute $Q = V_c A_c$ into the above equation and solve for V_c. We get

$$V_c = \sqrt{\frac{gA_c}{b_{top}}} \qquad (12–12)$$

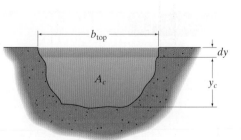

Critical depth in channel with arbitrary cross section

Fig. 12–7

At this speed Fr $= 1$, and so for any other V, the flow can be classified as supercritical or subcritical in accordance with Eqs. 12–10.

12

Important Points

- Open-channel flow is predominantly a turbulent phenomenon due to the mixing action that takes place within the liquid. Although the velocity profile is highly irregular, for a reasonable analysis we can assume the liquid to be an ideal fluid, so that the flow is one dimensional and has an average velocity over the cross section.

- Steady flow requires the velocity profile at a particular cross section to remain constant over time. Uniform flow requires that this profile remain the same at all cross sections. When uniform flow occurs in an open channel, the depth of flow will remain constant throughout the channel.

- Open-channel flow is classified according to the Froude number, which is the square root of the ratio of the inertia force to the gravitational force. When Fr $= 1$, *critical flow* occurs at depth y_c. A wave produced on the surface will remain *stationary*. When Fr < 1, *tranquil flow* occurs at depth $y > y_c$. Any waves on the surface will move *upstream*. Finally, when Fr > 1, *rapid flow* occurs at depth $y < y_c$. Waves will be washed *downstream*.

- The specific energy or specific head E of the flow in an open channel is the sum of its kinetic and potential energies, as measured from a datum *located at the bottom of the channel*. The specific energy diagram is a plot of $E = f(y)$ for a given flow Q. It indicates that when the flow has a specific energy E, it will either move rapidly, $V > V_c$, at a shallow depth, $y < y_c$ (high kinetic energy and low potential energy), or move slowly as tranquil flow, $V < V_c$, at a deeper depth, $y > y_c$ (low kinetic energy and high potential energy).

- The specific energy of a given flow Q is a *minimum* when the flow is at the *critical depth* y_c.

EXAMPLE | 12.1

Water has an average velocity of 4 m/s as it flows in the rectangular channel shown in Fig. 12–8a. If the depth of the flow is 3 m, classify the flow. What is the velocity of the flow at the alternate depth that provides the same specific energy for the flow?

SOLUTION

Fluid Description.   The flow is steady, and water is assumed to be an ideal fluid.

Analysis.   To classify the flow, we must first determine the critical depth from Eq. 12–7. Since the flow is $Q = VA = (4 \text{ m/s})(3 \text{ m})(2 \text{ m}) = 24 \text{ m}^3/\text{s}$,

$$y_c = \left(\frac{Q^2}{gb^2}\right)^{1/3} = \left(\frac{(24 \text{ m}^3/\text{s})^2}{(9.81 \text{ m/s}^2)(2 \text{ m})^2}\right)^{1/3} = 2.45 \text{ m}$$

Here $y_c < y = 3$ m, so the flow will be subcritical, or tranquil.   *Ans.*

For the arbitrary depth y, the specific energy for the flow is determined using Eq. 12–6.

$$E = \frac{(24 \text{ m}^3/\text{s})^2}{2(9.81 \text{ m/s}^2)(2 \text{ m})^2 y^2} + y \tag{1}$$

A plot of this equation is shown in Fig. 12–8c. At $y = 3$ m,

$$E = \frac{Q^2}{2gb^2y^2} + y = \frac{(24 \text{ m}^3/\text{s})^2}{2(9.81 \text{ m/s}^2)(2 \text{ m})^2(3 \text{ m})^2} + 3 \text{ m} = 3.815 \text{ m}$$

To find the alternate depth that provides this *same specific energy* of $E = 3.815$ m, we must substitute this value into Eq. 1, which, after simplification, yields

$$y^3 - 3.815y^2 + 7.339 = 0$$

Solving for the three roots, we obtain

$y = 3.00 \text{ m} > 2.45 \text{ m}$ Subcritical (as before)

$y = 2.02 \text{ m} < 2.45 \text{ m}$ Supercritical   *Ans.*

$y = -1.21 \text{ m}$ Not realistic

For the case of supercritical or rapid flow, Fig. 12–8b, when the depth is $y = 2.02$ m, the velocity must be

$Q = VA;$ $24 \text{ m}^3/\text{s} = V(2.02 \text{ m})(2 \text{ m})$

$V = 5.93 \text{ m/s}$ *Ans.*

The specific energy at critical flow can be determined from Eq. 12–8, $E_{min} = \left(\frac{3}{2}\right)y_c$, or from Eq. 1, using $y_c = 2.45$ m. Its value, 3.67 m, is also shown in Fig. 12–8c.

To summarize, a flow having a specific energy or specific head of $E = 3.815$ m can be rapid at a depth of 2.02 m or tranquil at a depth of 3.00 m. If this same flow has some other specific energy E, then it will occur at two other depths, found from the roots of Eq. 1.

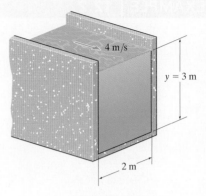

Subcritical flow
(a)

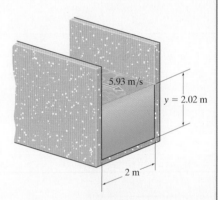

Supercritical flow
(b)

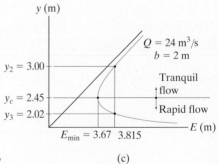

(c)

Fig. 12–8

12

EXAMPLE | 12.2

The horizontal rectangular channel in Fig. 12–9a is 2 m wide and gradually tapers so that it becomes 1 m wide. If water is flowing at 8.75 m³/s and has a depth of 0.6 m while in the 2-m section, determine the depth when it is in the 1 m section.

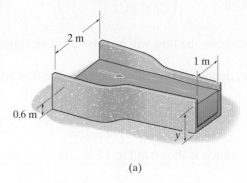

(a)

Fig. 12–9

SOLUTION

Fluid Description. We have steady flow in each region, even though within the transition there is nonuniform flow. The water is assumed to be an ideal fluid.

Analysis. The specific energy for the flow must be the *same* in each section, since no friction losses occur. Within the wide section, the critical depth is

$$y_c = \left(\frac{Q^2}{gb^2} \right)^{1/3} = \left[\frac{(8.75 \text{ m}^3/\text{s})^2}{(9.81 \text{ m/s}^2)(2 \text{ m})^2} \right]^{1/3} = 1.25 \text{ m}$$

Since the depth $y = 0.6 \text{ m} < 1.25 \text{ m}$, the flow is *supercritical*, or *rapid*, in the wide section.

We can find the specific energy for the flow in this section, $b = 2 \text{ m}$, using Eq. 12–6.

$$E = \frac{Q^2}{2gb^2y^2} + y = \frac{(8.75 \text{ m}^3/\text{s})^2}{2(9.81 \text{ m/s}^2)(2 \text{ m})^2y^2} + y \qquad (1)$$

This value of E must remain *constant* through the channel since the channel bottom remains horizontal, and there are no frictional losses.

When the width is 1 m, Eq. 12–6 becomes

$$E = \frac{Q^2}{2gb^2y^2} + y = \frac{(8.75 \text{ m}^3/\text{s})^2}{2(9.81 \text{ m/s}^2)(1 \text{ m})^2y^2} + y \qquad (2)$$

When $y = 0.6$ m, Eq. (1) gives $E = 3.310$ m. Using this value when $b = 1$ m, we have

$$3.310 \text{ m} = \frac{(8.75 \text{ m}^3/\text{s})^2}{2(9.81 \text{ m/s}^2)(1 \text{ m})^2y^2} + y$$

$$y^3 - 3.310y^2 + 3.902 = 0 \qquad (3)$$

The critical depth in this section is

$$y_c = \left(\frac{Q^2}{gb^2}\right)^{1/3} = \left[\frac{(8.75 \text{ m}^3/\text{s})^2}{(9.81 \text{ m/s}^2)(1 \text{ m})^2}\right]^{1/3} = 1.98 \text{ m}$$

Solving Eq. (3) for the depths we obtain

$$y = 2.82 \text{ m} > 1.98 \text{ m} \qquad \text{Subcritical}$$

$$y = 1.45 \text{ m} < 1.98 \text{ m} \qquad \text{Supercritical}$$

$$y = -0.956 \text{ m} \qquad \text{Not realistic}$$

Since the flow was *originally supercritical*, it remains in this state, and so the depth in the 1 m section will be

$$y = 1.45 \text{ m} \qquad \qquad \textit{Ans.}$$

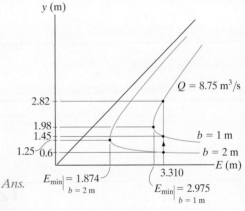

(b)

If we plot Eqs. 1 and 2, Fig. 12–9b, it will help to understand why the flow rate of 8.75 m³/s *remains supercritical* throughout the channel. In the figure, the values of E_{min} were found from Eq. 12–8. As the width of the channel gradually narrows from 2 m to 1 m, the water rises upward from $y = 0.6$ m on the curve for $b = 2$ m, until it reaches the point on the curve for $b = 1$ m. This occurs at the depth of $y = 1.45$ m. It is not possible for the water to reach the greater depth of $y = 2.82$ m in this section, because the specific energy must remain constant, and so it cannot *decrease* to $E_{min} = 2.975$ m and then *increase* again to the required $E = 3.310$ m.

12

EXAMPLE | **12.3**

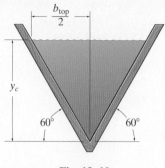

Fig. 12–10

The channel has a triangular cross section as shown in Fig. 12–10. Determine the critical depth if the flow is 12 m³/s.

SOLUTION

Fluid Description.   We assume steady flow of an ideal fluid.

Analysis.   For critical flow, we require the specific energy to be a minimum, which means we must satisfy Eq. 12–11. From Fig. 12–10,

$$b_{top} = 2(y_c \cot 60°) = 1.1547y_c$$

$$A_c = 2\left(\frac{1}{2}(y_c \cot 60°)(y_c)\right) = 0.5774y_c^2$$

Thus,

$$\frac{gA_c^3}{Q^2 b_{top}} = 1$$

$$\frac{\left(9.81 \text{m/s}^2\right)\left(0.5774y_c^2\right)^3}{\left(12 \text{ m}^3/\text{s}\right)^2(1.1547y_c)} = 1$$

Solving yields

$$y_c = 2.45 \text{ m} \qquad\qquad Ans.$$

12.4   Open-Channel Flow over a Rise or Bump

When liquid flows over a rise in a channel bed, as shown in Fig. 12–11a, it will change the depth of flow, since the *increased elevation* of the channel bed will increase the potential energy of the liquid mass. To investigate this effect, we will consider the flow to be one-dimensional, that is, horizontal, by assuming that the change in elevation is small and occurs gradually. We will also neglect any frictional effects, since the change occurs over a short distance.

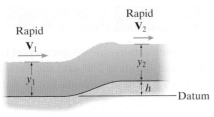

Rapid flow over a rise

(a)

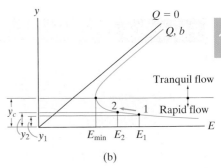

(b)

Rise.

Let's first consider the case where $y_1 < y_c$, so the approach flow is rapid, Fig. 12–11a. As the flow passes over the rise, the liquid is *lifted* a distance h, and so with the datum at the bottom of the lower portion of the channel, the liquid's specific energy will *decrease* from E_1 to E_2, Fig. 12–11b, while the depth of flow will *increase* from y_1 to y_2. In other words, as energy is used to lift the liquid by h (increase in potential energy), because of the continuity of flow, the liquid will slow down (decrease in kinetic energy), although it still remains rapid flow.*

If we now consider $y_1 > y_c$, as in the case of tranquil flow, Fig. 12–11c, then after passing the rise, the specific energy will *decrease* from E_1 to E_2, Fig. 12–11d. Here a *decrease* in the depth of flow (decrease in potential energy) from y_1 to y_2 causes the velocity of flow to increase (increase in kinetic energy); however, tranquil flow will prevail.

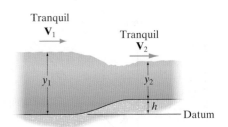

Tranquil flow over a rise

(c)

Bump.

If a bump or hill is placed on the channel bed, Fig. 12–11e, then there will be an upper *limit* by which E can decrease in lifting the fluid. As shown in Fig. 12–11f, it is $(E_1 - E_{min})$. This gives a *maximum rise* of the bump of $(y_c - y_1) = h_c$. If this rise occurs, and the bump is designed properly, specific energy can follow around the nose of the curve, and the flow will thereby change from rapid to tranquil.[†] In other words, just when reaching the top of the bump, the flow will be at the critical depth. Then as the bump begins to slope *downward*, kinetic energy will be added back to the flow. This gets converted to potential energy and raises the depth to y_2 as the specific energy is returned to E_1.

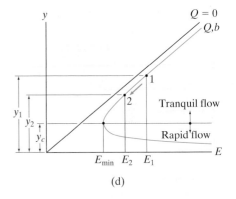

(d)

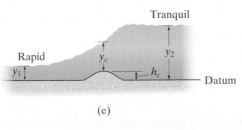

(e)

Fig. 12–11

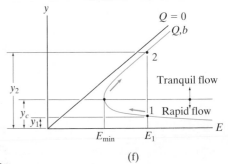

(f)

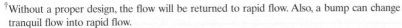

*The continuity of flow requires $Q = V_1(y_1b) = V_2(y_2b)$, or $V_2 = V_1(y_1/y_2)$. But $y_1/y_2 < 1$, so $V_2 < V_1$.

[†]Without a proper design, the flow will be returned to rapid flow. Also, a bump can change tranquil flow into rapid flow.

12

EXAMPLE | 12.4

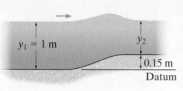

$y_1 = 1$ m

y_2

0.15 m

Datum

(a)

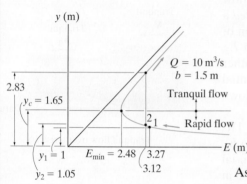

y (m)

$Q = 10$ m³/s
$b = 1.5$ m

Tranquil flow

2.83

$y_c = 1.65$

2

1 ← Rapid flow

$y_1 = 1$ $E_{min} = 2.48$ 3.27

$y_2 = 1.05$

3.12

— E (m)

(b)

Fig. 12–12

Water flows through the rectangular open channel in Fig. 12–12a that is 1.5 m wide. The initial depth is 1 m and the flow is 10 m³/s. If the bed of the channel rises 0.15 m, determine the new depth.

SOLUTION

Fluid Description. We have steady flow. The water is assumed to be an ideal fluid.

Analysis. The critical depth for the flow is determined from Eq. 12–7. We have

$$y_c = \left(\frac{Q^2}{gb^2}\right)^{1/3} = \left(\frac{(10 \text{ m}^3/\text{s})^2}{(9.81 \text{ m/s}^2)(1.5 \text{ m})^2}\right)^{1/3} = 1.65 \text{ m}$$

And the minimum specific energy, Eq. 12–8, is

$$E_{min} = \frac{3}{2}y_c = \frac{3}{2}(1.65 \text{ m}) = 2.48 \text{ m}$$

Since originally $y = 1$ m < 1.65 m, the flow is supercritical, or rapid.

As the water passes over the rise, it gives up some of its kinetic energy because the water is being lifted to increase its potential energy. For the flow, $Q = 10$ m³/s, the specific energy on each side of the rise must be the *same if* we use the *same datum*. Using Eq. 12–6, at $y_1 = 1$ m, we have

$$E = \frac{Q^2}{2gb^2y_1^2} + y_1$$

$$\frac{(10 \text{ m}^3/\text{s})^2}{2(9.81 \text{ m/s}^2)(1.5 \text{ m})^2(1 \text{ m})^2} + 1 \text{ m} = 3.27 \text{ m}$$

and at y_2, for the *same datum*, we require

$$\frac{(10 \text{ m}^3/\text{s})^2}{2(9.81 \text{ m/s}^2)(1.5 \text{ m})^2y_2^2} + (y_2 + 0.15 \text{ m}) = 3.27 \text{ m}$$

$$y_2^3 - 3.115y_2^2 + 2.265 = 0$$

Solving for the three roots yields

$$y_2 = 2.83 \text{ m} > 1.65 \text{ m} \qquad \text{Subcritical}$$
$$y_2 = 1.05 \text{ m} < 1.65 \text{ m} \qquad \text{Supercritical}$$
$$y_2 = -0.764 \text{ m} \qquad \text{Unrealistic}$$

The flow must remain supercritical because it cannot have less specific energy than $E_{min} = 2.48$ m as it is being lifted. In other words, the specific energy E is confined between $(3.27 \text{ m} - 0.15 \text{ m}) = 3.12$ m and 3.27 m, Fig. 12–12b. Consequently,

$$y_2 = 1.05 \text{ m} \qquad \qquad Ans.$$

EXAMPLE 12.5

Water flows through the channel in Fig. 12–13a that has a width of 0.9 m. If the flow is 2.5 m³/s and the depth is originally 0.75 m, show that the upstream flow is rapid, and determine the required height h of a bump placed on the bed of the channel so that the flow is able to change to tranquil downstream from the bump.

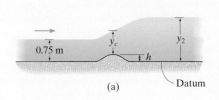

(a)

SOLUTION

Fluid Description.   Steady flow occurs. The water will be considered an ideal fluid.

Analysis.   The critical depth for the flow is

$$y_c = \left(\frac{Q^2}{gb^2}\right)^{1/3} = \left[\frac{(2.5 \text{ m}^3/\text{s})^2}{(9.81 \text{ m/s}^2)(0.9 \text{ m})^2}\right]^{1/3} = 0.9231 \text{ m}$$

Since $y = 0.75 \text{ m} < 0.9231 \text{ m}$, then before the bump the flow is rapid. Also, the minimum specific energy is

$$E_{min} = \frac{3}{2}y_c = 1.384 \text{ m}$$

In general, the specific energy for the flow is

$$E = \frac{Q^2}{2gb^2y^2} + y \quad E = \frac{(2.5 \text{ m}^3/\text{s})^2}{2(9.81 \text{ m/s}^2)(0.9 \text{ m})^2 y^2} + y$$

$$E = \frac{0.3933}{y^2} + y \qquad (1)$$

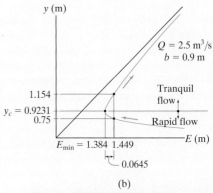

(b)

Fig. 12–13

At the depth $y = 0.75 \text{ m}$, then $E = 1.449 \text{ m}$, Fig. 12–13b. The depth for tranquil flow at this *same specific energy* can be determined by solving

$$1.449 = \frac{0.3933}{y^2} + y$$

$$y^3 - 1.449 y^2 + 0.3933 = 0$$

The three roots are

$$y_2 = 1.154 \text{ m} > 0.9231 \text{ m} \qquad \text{Subcritical}$$
$$y_2 = 0.75 \text{ m} < 0.9231 \text{ m} \qquad \text{Supercritical (as before)}$$
$$y_2 = -0.455 \text{ m} \qquad \text{Unrealistic}$$

As shown in Fig. 12–13b, to produce tranquil flow after the bump, the rise must first *remove* $1.449 \text{ m} - 1.384 \text{ m} = 0.0645 \text{ m}$ of the specific energy from the water. Then, just after the top of the bump, if the bump is designed properly, the same amount of the energy will be returned, so that tranquil flow occurs at the new depth of $y_2 = 1.154 \text{ m}$. Thus, the required height of the bump is

$$h = 0.0645 \text{ m} \qquad \qquad \textit{Ans.}$$

12

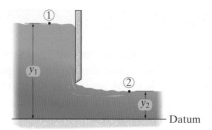

Flow under a sluice gate

Fig. 12–14

12.5 Open-Channel Flow under a Sluice Gate

A *sluice gate* is a structure that is often used to regulate the discharge of a liquid from a reservoir into a channel. An example is shown in Fig. 12–14, where the gate is partially opened. If we neglect frictional losses through the gate and assume we have steady flow of an ideal fluid, then we can apply the Bernoulli equation on the streamline between point 1, where there is essentially no flow, and point 2, where the flow is steady and has an average velocity V_2. This gives

$$\frac{p_1}{\gamma} + \frac{V_1^2}{2g} + y_1 = \frac{p_2}{\gamma} + \frac{V_2^2}{2g} + y_2$$

$$0 + 0 + y_1 = 0 + \frac{V_2^2}{2g} + y_2$$

To obtain the flow as a function of depth, we can use $Q = V_2 b y_2$, where b is the width of the (rectangular) channel. Substituting for V_2 and solving for Q, we get

$$y_1 = \frac{Q^2}{2gb^2 y_2^2} + y_2$$

$$Q = \sqrt{2gb^2}\left(y_2^2 y_1 - y_2^3\right)^{1/2} \tag{12–13}$$

Opening and closing the gate will vary y_2 and also the flow through the gate. Maximum flow can be determined by taking the derivative of the above equation and setting it equal to zero.

$$\frac{dQ}{dy_2} = \sqrt{2gb^2}\left(\frac{1}{2}\right)\left[\left(y_2^2 y_1 - y_2^3\right)^{-1/2}\left(2y_2 y_1 - 3y_2^2\right)\right] = 0$$

The solution requires

$$y_2 = \frac{2}{3}y_1 \tag{12–14}$$

Sluice gates are placed at the crest of this dam in order to maintain control of the water elevation in the reservoir behind the dam.

When y_2 is this critical depth, then the maximum flow is determined from Eq. 12–13. We have

$$Q_{max} = \sqrt{\frac{8}{27} g b^2 y_1^3} \qquad (12\text{–}15)$$

The Froude number at maximum flow, that is, when $y_2 = \frac{2}{3} y_1$ and $V = Q_{max} / b y_2$, is

$$Fr = \frac{V}{\sqrt{g y_2}} = \frac{Q_{max} / b y_2}{\sqrt{g y_2}} = \frac{1}{b y_2 \sqrt{g y_2}} \sqrt{\frac{8 g b^2 y_1^3}{27}}$$

$$= \sqrt{\frac{8}{27} \left(\frac{y_1}{\frac{2}{3} y_1} \right)^3} = 1$$

Therefore, the classification of flow just past the sluice gate is

$$Fr < 1 \qquad \text{Subcritical (tranquil) flow}$$
$$Fr = 1 \qquad \text{Critical flow}$$
$$Fr > 1 \qquad \text{Supercritical (rapid) flow}$$

These results can be interpreted as follows: When the gate is initially opened, the flow *increases* so that (Fr > 1). The flow reaches a maximum discharge when the depth $y_2 = \frac{2}{3} y_1$, (Fr = 1). A further opening of the gate will now cause the flow to *decrease* (Fr < 1). Here the gravitational force becomes greater than the inertia force. In other words, it is more difficult for the liquid to pass under the gate, since on the other side y_2 is high enough so the weight of fluid restrains any increase in the flow. In practice, the results of this analysis are modified somewhat to account for the frictional losses under the gate. This is usually done using an experimentally determined *discharge coefficient* that is multiplied by the flow. We will discuss this in Sec. 12.9, where it also applies to weirs.

Important Points

- A flow will *remain* tranquil or rapid both before and after the channel experiences either a change in width or a change in elevation (rise or drop).

- A bump can be designed to change the flow from one type to another; for example, from rapid to tranquil. This occurs because the bump can first remove specific energy from the flow, bringing the flow to the critical depth, and then return it to its original specific energy. Doing so allows the specific energy to transfer around the "nose" of the specific energy diagram.

- The steady flow over a channel rise, over a bump, or under a sluice gate can be assumed to be one-dimensional flow, with no frictional losses, since the length of the transition is normally short. The flow on each side of the transition is classified according to the Froude number.

- Maximum flow under a sluice gate will occur if the gate is opened so the exit flow depth is $y_2 = \frac{2}{3} y_1$.

EXAMPLE | 12.6

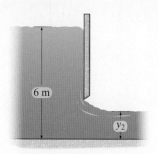

6 m

y_2

Fig. 12–15

The sluice gate in Fig. 12–15 is used to control the flow of water from a large reservoir that has a depth of 6 m. If it opens into a channel that has a width of 4 m, determine the maximum flow that can occur through the channel and the associated depth of flow.

SOLUTION

Fluid Description.   The water surface in the reservoir is assumed to be at a constant elevation so we have steady flow. Also, the water is assumed to be an ideal fluid.

Analysis.   Maximum flow occurs at the critical depth, which is determined from Eq. 12–14.

$$y_2 = \frac{2}{3}y_1 = \frac{2}{3}(6 \text{ m}) = 4 \text{ m} \qquad\qquad Ans.$$

The flow at this depth is determined from Eq. 12–15.

$$Q_{max} = \sqrt{\frac{8}{27}gb^2y_1^3}$$

$$= \sqrt{\frac{8}{27}(9.81 \text{ m/s}^2)(4 \text{ m})^2(6 \text{ m})^3}$$

$$= 100.23 \text{ m}^3/\text{s} = 100 \text{ m}^3/\text{s} \qquad\qquad Ans.$$

As expected, the Froude number for this flow is

$$\text{Fr} = \frac{V_c}{\sqrt{gy_c}} = \frac{(100.23 \text{ m}^3/\text{s})/[4 \text{ m}(4 \text{ m})]}{\sqrt{(9.81 \text{ m/s}^2)(4 \text{ m})}} = 1$$

EXAMPLE | 12.7

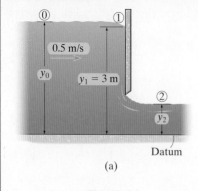

⓪   ①

0.5 m/s

y_0 $y_1 = 3$ m

②

y_2

Datum

(a)

Fig. 12–16

The flow in the 2-m-wide channel in Fig. 12–16a is controlled by the sluice gate, which is partially opened so that it causes the depth of the water near the gate to be 3 m and the mean velocity there to be 0.5 m/s. Determine the depth of the water far upstream, where it is essentially at rest, and also find its depth downstream from the gate.

SOLUTION

Fluid Description.   The water a distance far from the sluice gate is assumed to have a constant depth so that the flow at points 1 and 2 will be steady. Also, the water is assumed to be an ideal fluid.

Analysis.   If the Bernoulli equation is applied between points 0 and 1, located on a streamline at the water surface, we have

$$\frac{p_0}{\gamma} + \frac{V_0^2}{2g} + y_0 = \frac{p_1}{\gamma} + \frac{V_1^2}{2g} + y_1$$

$$0 + 0 + y_0 = 0 + \frac{(0.5 \text{ m/s})^2}{2(9.81 \text{ m/s}^2)} + 3 \text{ m}$$

$$y_0 = 3.01274 \text{ m} = 3.0127 \text{ m} \qquad \textit{Ans.}$$

The Bernoulli equation can be applied between points 1 and 2, but we can *also* apply it between points 0 and 2. If we do this, we have

$$\frac{p_0}{\gamma} + \frac{V_0^2}{2g} + y_0 = \frac{p_2}{\gamma} + \frac{V_2^2}{2g} + y_2$$

$$0 + 0 + 3.01274 \text{ m} = 0 + \frac{V_2^2}{2(9.81 \text{ m/s}^2)} + y_2 \qquad (1)$$

Continuity requires the flow at 1 and 2 to be the same.

$$Q = V_1 A_1 = V_2 A_2$$

$$(0.5 \text{ m/s})[3 \text{ m} (2 \text{ m})] = V_2 y_2 (2 \text{ m})$$

$$V_2 y_2 = 1.5$$

Substituting $V_2 = 1.5/y_2$ into Eq. 1 yields

$$y_2^3 - 3.01274 y_2^2 + 0.11468 = 0$$

Solving for the three roots, we get

$$\begin{array}{lll} y_1 = 3 \text{ m} & \text{Subcritical (as before)} \\ y_2 = 0.2020 \text{ m} & \text{Supercritical} \\ y_2 = -0.1892 \text{ m} & \text{Not realistic} \end{array}$$

The first root indicates the depth, $y_1 = 3$ m, and the second root is the depth downstream from the gate. Thus,

$$y_2 = 0.202 \text{ m} \qquad \textit{Ans.}$$

The specific energy for the flow can be determined from point 0, 1, or 2 since the bed elevation of the channel is constant and frictional losses through the gate have been neglected (ideal fluid). Using point 1,

$$E = \frac{Q^2}{2gb^2 y_1^2} + y_1 = \frac{[(0.5 \text{ m/s})(3 \text{ m})(2 \text{ m})]^2}{2(9.81 \text{ m/s}^2)(2 \text{ m})^2(3 \text{ m})^2} + 3 \text{ m} = 3.01274 \text{ m}$$

A plot of the specific energy is shown in Fig. 12–16b. Here

$$y_c = \frac{2}{3} y_0 = \frac{2}{3}(3.01274 \text{ m}) = 2.008 \text{ m}$$

$$E_{min} = \frac{Q^2}{2gb^2 y_c^2} + y_c = \frac{[(0.5 \text{ m/s})(3 \text{ m})(2 \text{ m})]^2}{2(9.81 \text{ m/s}^2)(2 \text{ m})^2(2.008 \text{ m})^2} + 2.008 \text{ m} = 2.037 \text{ m}$$

As noted, tranquil flow occurs upstream near the gate, and rapid flow occurs downstream.

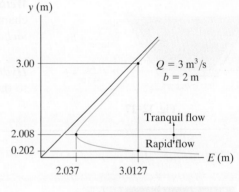

y (m)

3.00

$Q = 3 \text{ m}^3/\text{s}$
$b = 2 \text{ m}$

Tranquil flow

2.008
0.202

Rapid flow

E (m)

2.037   3.0127

(b)

Fig. 12–16 (cont.)

12

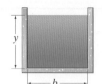

$A = yb$
$P = 2y + b$

$A = \dfrac{R^2}{2}(\alpha - \sin\alpha)$
$P = \alpha R$

$A = \dfrac{y^2}{\tan\alpha}$
$P = \dfrac{2y}{\sin\alpha}$

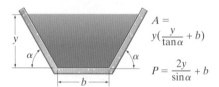

$A = y\left(\dfrac{y}{\tan\alpha} + b\right)$
$P = \dfrac{2y}{\sin\alpha} + b$

Fig. 12–17

Drainage canal having a trapezoidal
cross section

12.6   Steady Uniform Channel Flow

Since all open channels have a *rough surface*, then in order to *maintain* steady uniform flow in the channel, it is essential that it have a *constant slope* and a *constant cross section and surface roughness* along its length. Although these conditions rarely occur in practice, an analysis based on these assumptions is often used to design many types of channels for drainage and irrigation systems. Furthermore, this analysis is sometimes used to approximate the constant flow characteristics of natural channels such as streams and rivers.

Typical prismatic cross sections of open channels that are commonly used in engineering practice are shown in Fig. 12–17. Important geometric properties that are related to these shapes are defined as follows:

Area of Flow, A. The area of the flow cross section.

Wetted Perimeter, P. The distance around the cross section of the channel where the channel is in contact with the liquid. This *does not include* the distance over the free liquid surface.

Hydraulic Radius, R_h. The ratio of the area of the flow cross section to the wetted perimeter.

$$R_h = \frac{A}{P} \tag{12–16}$$

Reynolds Number.   For open-channel flow, the Reynolds number is generally defined as $\text{Re} = VR_h/\nu$, where the hydraulic radius R_h is the "characteristic length." Experiments have shown that laminar flow depends upon the cross-sectional geometry, but in many cases it can be specified as $\text{Re} \le 500$. For example, a channel having a rectangular cross section of width 1 m and flow depth 0.5 m has a hydraulic radius of $R_h = A/P = \left[1\text{ m}(0.5\text{ m})\right]/\left[2(0.5\text{ m}) + 1\text{ m}\right] = 0.25$ m. If the channel is carrying water at standard temperature, it would require the mean velocity for laminar flow to be at most

$$\text{Re} = \frac{VR_h}{\nu}; \qquad 500 = \frac{V(0.25\text{ m})}{1.12\left(10^{-6}\right)\text{ m}^2/\text{s}} \qquad V = 2.24\text{ mm/s}$$

This is extremely slow, and so, as mentioned previously, practically all open-channel flow is turbulent. In fact, nearly all flow occurs at very high Reynolds numbers.

Chézy Equation. To analyze the steady uniform flow along a sloped channel, we will apply the energy equation since the channel has a surface roughness, and so a head loss will occur over its horizontal length L. If we consider the control volume of liquid shown in Fig. 12–18, then the vertical control surfaces have the same depth y. Also, $V_{in} = V_{out} = V$. To calculate the hydraulic head $p/\gamma + z$, we will make reference to points on the liquid surface. Here $p_{in} = p_{out} = 0$.*

$$\frac{p_{in}}{\gamma} + \frac{V_{in}^2}{2g} + z_{in} + h_{pump} = \frac{p_{out}}{\gamma} + \frac{V_{out}^2}{2g} + z_{out} + h_{turb} + h_L$$

$$0 + \frac{V^2}{2g} + y + \Delta y + 0 = 0 + \frac{V^2}{2g} + y + 0 + h_L$$

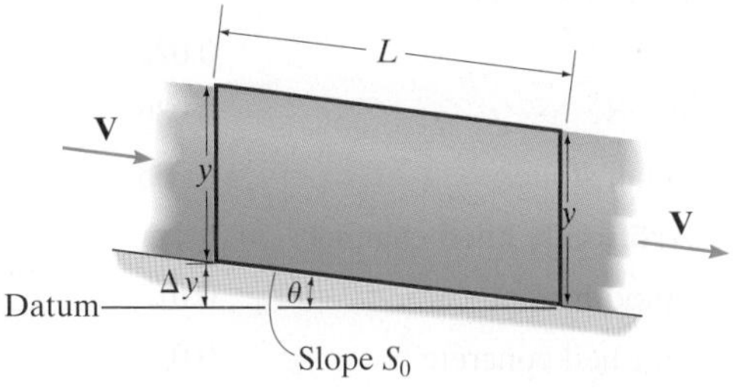

For small slopes, $\Delta y = L \tan \theta \approx LS_0$, and so

$$h_L = LS_0$$

We can also express this head loss using the Darcy–Weisbach equation.

$$h_L = f\left(\frac{L}{D_h}\right)\frac{V^2}{2g}$$

Since channels have a variety of cross sections, here we have used the hydraulic diameter $D_h = 4R_h$, something that we discussed in Sec. 10.1. If we now equate the above two equations and solve for V, we get

$$V = C\sqrt{R_h S_0} \qquad (12\text{–}17)$$

where $C = \sqrt{8g/f}$.

This result is known as the **Chézy formula**, named after Antoine de Chézy, a French engineer who deduced it experimentally in 1775. The coefficient C was originally thought to be constant; however, through later experiments it was found to depend on the shape of the channel's cross section and on the roughness of its surface.

Steady uniform flow

Fig. 12–18

*This head remains constant over the surface and can be calculated at any point on the surface, as shown in Example 5.11.

TABLE 12–1 Surface Roughness Coefficient	
Perimeter	n $(s/m^{1/3})$
Earth channel	
Clean	0.022
Weedy	0.030
Small stones	0.035
Artificially lined channel	
Smooth steel	0.012
Finished concrete	0.012
Unfinished concrete	0.014
Wood	0.012
Brick-lined	0.015

Manning Equation.   In 1891 Robert Manning, an Irish engineer, established experimental values for C by expressing it in terms of the hydraulic radius and a dimensional ***surface roughness coefficient***, n, which has units of $s/m^{1/3}$. He found $C = R_h^{1/6}/n$. Typical values of n, which are in SI units, for some common conditions encountered in engineering are listed in Table 12–1. Expressing the average velocity in terms of n, the ***Manning equation*** is then

$$V = \frac{R_h^{2/3}S_0^{1/2}}{n} \qquad (12\text{–}18)$$

Since $Q = VA$, and the hydraulic radius is $R_h = A/P$, Eq. 12–18 can also be expressed in terms of the volumetric flow. It is

$$Q = \frac{A^{5/3}S_0^{1/2}}{nP^{2/3}} \qquad (12\text{–}19)$$

Best Hydraulic Cross Section.   It can be seen that for a given slope S_0 and surface roughness n, the flow Q in Eq. 12–19 will *increase* if the wetted perimeter P *decreases*. Therefore, we can obtain the maximum Q by minimizing the wetted perimeter P. Such a cross section is called the ***best hydraulic cross section*** since it will both minimize the amount of material needed to construct the channel and maximize the flow. For example, if the channel has a rectangular cross section of given width b, Fig. 12–19, then $A = by$ and $P = 2y + b = 2y + A/y$. Thus, for a constant value of A, although the relationship between b and y is not as yet known, we have

$$\frac{dP}{dy} = \frac{d}{dy}\left(2y + \frac{A}{y}\right) = 2 - \frac{A}{y^2} = 0$$

$$A = 2y^2 = by \quad \text{or} \quad y = \frac{b}{2}$$

Therefore, a rectangular channel flowing at a depth of $y = b/2$ will require the smallest amount of material used for its construction. Since this design size will also provide the maximum amount of uniform flow, it is the best cross section for the rectangle.

In a strict sense, a *semicircular cross section*, flowing full, is the best design shape; however, for very large flows this form is generally difficult to excavate in soil and costly to construct. Instead, large channels have trapezoidal cross sections, or for shallow depths, their cross sections may be rectangular. In any case, the best hydraulic cross section for a selected cross-sectional shape can always be determined as demonstrated here, that is, by expressing the wetted perimeter in terms of the cross-sectional area, and then setting its derivative equal to zero. Once the best cross section is found, then the uniform velocity within it can be determined using Manning's equation.

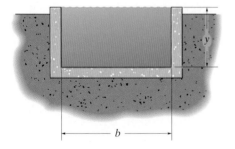

Fig. 12–19

Critical Slope.   If Eq. 12–19 is solved for the slope of the channel, and we express it in terms of the hydraulic radius $R_h = A/P$, then we get

$$S_0 = \frac{Q^2 n^2}{R_h^{4/3} A^2}$$

(12–20)

Channel slope

The *critical slope* for a channel of *any cross section* requires the depth of flow to be at the critical depth, y_c, and since the critical depth is determined from Eq. 12–11, the above equation for critical slope becomes

$$S_c = \frac{n^2 g A_c}{b_{\text{top}} R_{hc}^{4/3}}$$

(12–21)

Critical slope

Flow through a concrete channel having a trapezoidal cross section

Here the critical area A_c and hydraulic radius R_{hc} are determined by setting $y = y_c$ for the section. With this equation, like the depth y, we can compare the actual slope of a channel S_0 with its critical slope S_c, and thereby classify the flow.

$$\begin{aligned}
S_0 < S_c & \qquad \text{Subcritical (tranquil) flow} \\
S_0 = S_c & \qquad \text{Critical flow} \\
S_0 > S_c & \qquad \text{Supercritical (rapid) flow}
\end{aligned}$$

The examples that follow illustrate a few applications of these concepts.

Important Points

- Steady uniform open-channel flow can occur provided the channel has a *constant* slope and surface roughness and the liquid in the channel has a *constant* cross section. When this occurs, the gravitational force on the liquid will be in balance with the frictional forces along the bottom and sides of the channel.

- The Manning equation can be used to determine the average velocity within an open channel having steady uniform flow.

- The best hydraulic cross section for a channel of any given shape having a constant flow, slope, and surface roughness can be determined by minimizing its wetted perimeter.

- The type of steady uniform flow in a channel can be determined by comparing its slope S_0 with the critical slope S_c.

EXAMPLE | 12.8

The channel in Fig. 12–20 is made of finished concrete, and the bed drops 2 m in elevation for a horizontal stretch or reach of 1000 m. Determine the steady uniform volumetric flow rate when the depth of the water is 2 m.

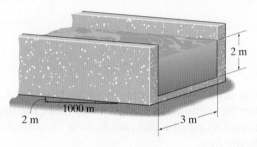

Fig. 12–20

SOLUTION

Fluid Description.   The flow is steady and uniform, and the water is assumed incompressible and flows with an average velocity V.

Analysis.   The slope of the channel is $S_0 = 2 \text{ m}/1000 \text{ m} = 0.002$, and from Table 12–1, for finished concrete $n = 0.012$. Also, for a 2 m depth of water, the hydraulic radius is

$$R_h = \frac{A}{P} = \frac{(3 \text{ m})(2 \text{ m})}{2(2 \text{ m}) + 3 \text{ m}} = 0.8571 \text{ m}$$

Therefore, applying Eq. 12–18

$$V = \frac{R_h^{2/3} S_0^{1/2}}{n}$$

$$\frac{Q}{(3 \text{ m})(2 \text{ m})} = \frac{(0.8571 \text{ m})^{2/3}(0.002)^{1/2}}{0.012}$$

$$Q = 20.18 \text{ m}^3/\text{s} = 20.2 \text{ m}^3/\text{s} \qquad \textit{Ans.}$$

EXAMPLE 12.9

The channel in Fig. 12–21 consists of an unfinished concrete section and overflow regions on each side that contain light brush ($n = 0.050$). If the bottom of the channel has a slope of 0.0015, determine the volumetric flow when the depth is 2.5 m, as shown.

SOLUTION

Fluid Description. We have steady uniform flow, and we will assume the water is an incompressible fluid.

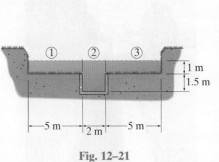

Fig. 12–21

Analysis. The cross section is divided into three composite rectangles, Fig. 12–21. The flow through the entire cross section is thus the sum of the flows through each composite shape. For the calculation, note that the wetted perimeter does not include the liquid boundary between the shapes because n does not act on these surfaces. Since $n = 0.014$ for unfinished concrete, and for light brush $n = 0.050$, as stated, the Manning equation, written in the form of Eq. 12–19, becomes

$$Q = \Sigma \frac{A^{5/3}S_0^{1/2}}{nP^{2/3}} = S_0^{1/2}\left(\frac{A_1^{5/3}}{n_1 P_1^{2/3}} + \frac{A_2^{5/3}}{n_2 P_2^{2/3}} + \frac{A_3^{5/3}}{n_3 P_3^{2/3}}\right)$$

$$= (0.0015)^{1/2}\left[\frac{[(1\text{ m})(5\text{ m})]^{5/3}}{0.050(1\text{ m} + 5\text{ m})^{2/3}} + \frac{[(2\text{ m})(2.5\text{ m})]^{5/3}}{0.014(1.5\text{ m} + 2\text{ m} + 1.5\text{ m})^{2/3}} + \frac{[(5\text{ m})(1\text{ m})]^{5/3}}{0.050(1\text{ m} + 5\text{ m})^{2/3}}\right]$$

$$= 20.7 \text{ m}^3/\text{s} \qquad\qquad Ans.$$

EXAMPLE 12.10

The triangular flume in Fig. 12–22a is used to carry water over a ravine. It is made of wood and has a slope of $S_0 = 0.001$. If the intended flow is to be $Q = 3 \text{ m}^3/\text{s}$, determine the depth of the flow.

SOLUTION

Fluid Description. We assume steady uniform flow of an incompressible fluid.

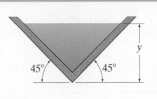

Flume

(a)

Analysis. If y is the depth of the flow, Fig. 12–22b, then

$$P = 2\sqrt{2}y \quad \text{and} \quad A = 2\left[\frac{1}{2}(y)(y)\right] = y^2$$

From Table 12–1, $n = 0.012$ for a wood surface. For SI units,

$$Q = \frac{A^{5/3}S_0^{1/2}}{nP^{2/3}}; \qquad 3 \text{ m}^3/\text{s} = \frac{(y^2)^{5/3}(0.001)^{1/2}}{0.012(2\sqrt{2}y)^{2/3}}$$

$$y = 1.36 \text{ m} \qquad\qquad Ans.$$

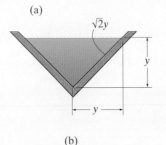

(b)

Fig. 12–22

12

EXAMPLE | 12.11

(© Andrew Orlemann/Shutterstock)

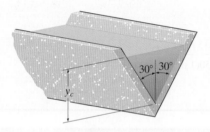

Fig. 12–23

The channel in Fig. 12–23 has a triangular cross section and is made of unfinished concrete. If the flow is 1.5 m³/s, determine the slope that produces critical flow.

SOLUTION

Fluid Description. We assume steady uniform flow of an incompressible fluid.

Analysis. The critical slope is determined using Eq. 12–20, but first we need to determine the critical depth. Since

$$A_c = 2\left[(1/2)(y_c \tan 30°)y_c\right] = y_c^2 \tan 30°$$

$$b_{\text{top}} = 2y_c \tan 30°$$

Then applying Eq. 12–11,

$$\frac{gA_c^3}{Q^2 b_{\text{top}}} = 1$$

$$\frac{(9.81 \text{ m/s}^2)(y_c^2 \tan 30°)^3}{(1.5 \text{ m}^3/\text{s})^2(2y_c \tan 30°)} = 1$$

$$y_c = 1.066 \text{ m}$$

Using this value and $n = 0.014$ for unfinished concrete,

$$b_{\text{top}} = 2(1.066 \text{ m}) \tan 30° = 1.231 \text{ m}$$

$$A_c = (1.066 \text{ m})^2 \tan 30° = 0.6560 \text{ m}^2$$

$$P_c = 2\left(\frac{1.066 \text{ m}}{\cos 30°}\right) = 2.462 \text{ m}$$

$$R_{hc} = \frac{A_c}{P_c} = \frac{0.6560 \text{ m}^2}{2.462 \text{ m}} = 0.2665 \text{ m}$$

Thus,

$$S_c = \frac{n^2 g A_c}{b_{\text{top}} R_{hc}^{4/3}} = \frac{(0.014)^2(9.81 \text{ m/s}^2)(0.6560 \text{ m}^2)}{(1.231 \text{ m})(0.2665 \text{ m})^{4/3}} = 0.00598 \quad Ans.$$

Therefore, when the channel has a flow of $Q = 1.5$ m³/s, then any slope *less* than S_c will produce *subcritical* (tranquil) flow, and any slope *greater* than S_c will produce *supercritical* (rapid) flow.

12.7   Gradual Flow with Varying Depth

In the previous section we considered steady uniform flow in an open channel having a constant slope. This required that the flow occurs at a unique depth, so that this depth remains *constant* over the length of the channel. However, when the slope or cross-sectional area of the channel gradually *changes*, or there is a *change* in surface roughness within the channel, then the depth of the liquid will vary along its length, and steady nonuniform flow will result.

To analyze this case, we will apply the energy equation between the "in" and "out" sections on the differential control volume shown in Fig. 12–24. We will select the points at the top of the liquid surface to calculate the hydraulic head terms,* $p/\gamma + z$. Here $p_{\text{in}} = p_{\text{out}} = 0$, and so

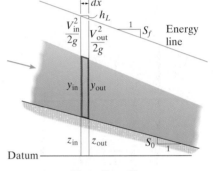

Nonuniform flow

Fig. 12–24

$$\frac{p_{\text{in}}}{\gamma} + \frac{V_{\text{in}}^2}{2g} + z_{\text{in}} = \frac{p_{\text{out}}}{\gamma} + \frac{V_{\text{out}}^2}{2g} + z_{\text{out}} + h_L$$

$$0 + \frac{V_{\text{in}}^2}{2g} + (z_{\text{in}} + y_{\text{in}}) = 0 + \frac{V_{\text{out}}^2}{2g} + (z_{\text{out}} + y_{\text{out}}) + h_L$$

$$\frac{V_{\text{in}}^2}{2g} - \frac{V_{\text{out}}^2}{2g} = (y_{\text{out}} - y_{\text{in}}) + (z_{\text{out}} - z_{\text{in}}) + h_L$$

The two terms on the left represent the *change* in the velocity head over the length dx. Also, $y_{\text{out}} - y_{\text{in}} = dy$, and $z_{\text{out}} - z_{\text{in}} = -S_0\, dx$, where S_0 is the slope of the channel bed, which is positive when it slopes downward to the right. Therefore,

$$-\frac{d}{dx}\left(\frac{V^2}{2g}\right) dx = dy - S_0\, dx + h_L$$

If we define the *friction slope* S_f as the slope of the energy line, then $h_L = S_f dx$, Fig. 12–24. Substituting this for h_L into the above equation, we have, after simplifying,

$$\frac{dy}{dx} + \frac{d}{dx}\left(\frac{V^2}{2g}\right) = S_0 - S_f \qquad (12\text{–}22)$$

---

*Any point on the open control surfaces can be selected, as noted in Example 5.10.

Rectangular Cross Section.
If the channel has a rectangular cross section, then $V = Q/by$, and so

$$\frac{d}{dx}\left(\frac{V^2}{2g}\right) = \frac{d}{dx}\left(\frac{Q^2}{2gb^2y^2}\right) = -2\left(\frac{Q^2}{2gb^2y^3}\right)\frac{dy}{dx} = -\left(\frac{V^2}{gy}\right)\left(\frac{dy}{dx}\right)$$

Finally, we can express this result in terms of the Froude number. Since $\text{Fr} = V/\sqrt{gy}$, then $V^2/gy = \text{Fr}^2$. Therefore,

$$\frac{d}{dx}\left(\frac{V^2}{2g}\right) = -\text{Fr}^2\frac{dy}{dx}$$

For a rectangular cross section, Eq. 12–22 can now be expressed as

$$\frac{dy}{dx} = \frac{S_0 - S_f}{1 - \text{Fr}^2} \tag{12–23}$$

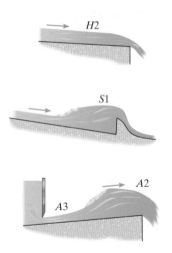

Examples of typical surface profiles

(a)

Fig. 12–25

Surface Profiles.
Since Fr^2 is a function of y, the above equation is a nonlinear first-order differential equation that requires an integration to obtain the depth y of the liquid surface as a function of x along the channel. It is important to be able to determine the *shape of this surface and its depth*, especially when the channel bed changes slope or the flow meets an obstruction such as a dam or a sluice gate. Here there is a possibility that flooding, spillover, or some other unforeseen effect can occur.

As shown in Table 12–2, there are twelve possible shapes the liquid surface can have. Each group of shapes is classified by the channel's slope, that is, horizontal (H), mild (M), critical (C), steep (S), or adverse (A). Also, each shape is classified by a *zone*, indicated by a number that depends upon the *depth* of the *actual flow*, y, compared to the depth for *uniform or normal flow*, y_n, and the depth for *critical flow*, y_c. Zone 1 is for high values of y, zone 2 is for intermediate values, and zone 3 is for low values. Typical examples of *how* these shapes or profiles can occur in a channel are illustrated in Fig. 12–25a.

The shape of the liquid surface for any of these twelve shapes can be sketched by studying the characteristics of its slope as defined by Eq. 12–23. For example, in the case of the *H2* profile shown in Fig. 12–25*b*, because $y > y_c$, the flow will be subcritical or tranquil, so that Fr < 1. Then from Eq. 12–23, when $S_0 = 0$ and Fr < 1, the initial slope of the water surface, dy/dx, will be *negative* as shown, and so indeed the depth will *decrease* as *x* increases.

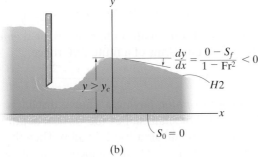

$$\frac{dy}{dx} = \frac{0 - S_f}{1 - \text{Fr}^2} < 0$$

(b)

Fig. 12–25 (cont.)

TABLE 12–2 Surface Profile Classification

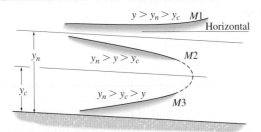

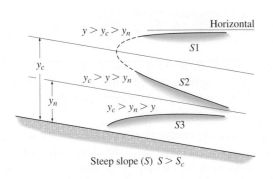

Mild slope (*M*) $S < S_c$

Steep slope (*S*) $S > S_c$

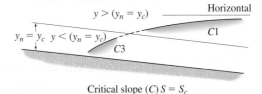

Critical slope (*C*) $S = S_c$

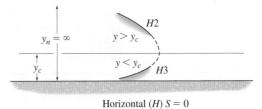

Horizontal (*H*) $S = 0$

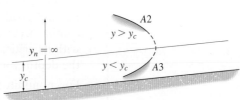

Adverse slope (*A*) $S < 0$

Calculating the Surface Profile.   Once the water surface profile has been classified, then the actual profile can be determined by integrating Eq. 12–22. Over the years, several procedures have been developed for doing this; however, here we will consider using a finite difference method to perform a numerical integration. To do this, we first write Eq. 12–22 in the form

$$\frac{d}{dx}\left(y + \frac{V^2}{2g}\right) = S_0 - S_f \quad \text{or} \quad dx = \frac{d(y + V^2/2g)}{S_0 - S_f}$$

If the channel is divided into small finite reaches, or *segments*, then this equation can be written in terms of a finite difference,

$$\Delta x = \frac{(y_2 - y_1) + (V_2^2 - V_1^2)/2g}{S_0 - S_f} \tag{12–24}$$

We start from a *control point*, having a known discharge Q and water elevation y_1, Fig. 12–26. For small slopes, the vertical depth y_1 can be used to calculate the cross-sectional area A_1 of the flow. Then the mean velocity V_1 can be calculated using $V_1 = Q/A_1$. An increase in the water depth Δy is *assumed*, and the area A_2 at $y_2 = y_1 + \Delta y$ is then calculated. And finally the mean velocity is found from $V_2 = Q/A_2$. If we *further assume* the head loss over the segment is the *same* as that over the same segment having uniform flow, then we can use the Manning equation, Eq. 12–18, to determine the friction slope.

$$S_f = \frac{n^2 V_m^2}{k^2 R_{hm}^{4/3}} \tag{12–25}$$

The values V_m and R_{hm} are average values of the mean velocity and the mean hydraulic radius for the segment. Substituting all the relevant values into Eq. 12–24, we can then calculate Δx. The iteration is continued for the next segment until one reaches the end of the channel. Although this process is tedious to do by hand, it is rather straightforward to program into a pocket calculator or computer. Example 12.13 demonstrates how it is done numerically.

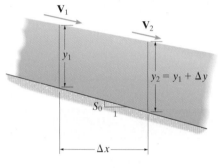

Fig. 12–26

Important Points

- When steady nonuniform flow exists, the liquid depth will gradually vary along the length of the channel. The slope of the liquid surface depends upon the Froude number, Fr, the slope of the channel bed, S_0, and the friction slope S_f.

- For nonuniform flow there are twelve classifications for the liquid surface profile, as shown in Table 12–2. To determine which profile occurs, it is necessary to compare the channel slope S_0 with its critical slope S_c, and to compare the liquid depth y with the depths for uniform or normal flow, y_n, and critical flow, y_c.

- For nonuniform flow, it is possible to use a finite difference method to numerically integrate the differential equation $dy/dx = (S_0 - S_f)/(1 - \text{Fr}^2)$, and thereby obtain a plot of the liquid surface.

EXAMPLE | 12.12

The rectangular channel in Fig. 12–27a is made of unfinished concrete and has a slope of $S_0 = 0.035$. At a specific location, the water has a depth of 1.25 m and the flow is $Q = 0.75 \text{ m}^3/\text{s}$. Classify the surface profile for the flow.

SOLUTION

Fluid Description.   The flow is steady, and here the flow occurs at a depth that is assumed to produce nonuniform flow. The water is considered incompressible.

Analysis.   To classify the surface profile, we must determine the critical depth y_c, the normal flow depth y_n, and the critical slope S_c. From Eq. 12–7,

$$y_c = \left(\frac{Q^2}{gb^2}\right)^{1/3} = \left(\frac{(0.75 \text{ m}^3/\text{s})^2}{9.81 \text{ m/s}^2(2 \text{ m})^2}\right)^{1/3} = 0.2429 \text{ m}$$

(a)

Since $y = 1.25 \text{ m} > y_c = 0.2429 \text{ m}$, the flow is tranquil. For unfinished concrete, $n = 0.014$. The depth y_n that will produce normal or *uniform flow* for $Q = 0.75 \text{ m}^3/\text{s}$ is determined from the Manning equation. Since

$$R_h = \frac{A}{P} = \frac{(2 \text{ m})y_n}{(2y_n + 2 \text{ m})} = \frac{y_n}{(y_n + 1)}$$

then Eq. 12–19 becomes

$$Q = \frac{A^{5/3}S_0^{1/2}}{nP^{2/3}}; \qquad 0.75 \text{ m}^3/\text{s} = \frac{[(2 \text{ m})y_n]^{5/3}(0.035)^{1/2}}{0.014(2y_n + 2 \text{ m})^{2/3}}$$

$$\frac{y_n^{5/3}}{(2y_n + 2 \text{ m})^{2/3}} = 0.017678$$

Solving by trial and error, or using a numerical procedure, we get

$$y_n = 0.1227 \text{ m}$$

The critical slope is now determined from Eq. 12–21,

$$S_c = \frac{n^2gA_c}{b_{top}R_{hc}^{4/3}} = \frac{(0.014)^2(9.81 \text{ m/s}^2)(2 \text{ m})(0.2429 \text{ m})}{(2 \text{ m})\left[\dfrac{2 \text{ m}(0.2429 \text{ m})}{2(0.2429 \text{ m}) + 2 \text{ m}}\right]^{4/3}}$$

$$= 0.004118$$

Since $y = 1.25 \text{ m} > y_c > y_n$ and $S_0 = 0.035 > S_c$, we have nonuniform flow, where the surface is defined by an $S1$ profile. According to Table 12-2 the water surface will look like that shown in Fig. 12–27b.

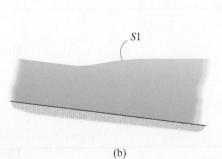

(b)

Fig. 12–27

EXAMPLE | 12.13

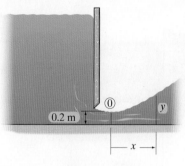

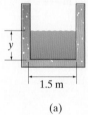

1.5 m

(a)

Fig. 12–28

Water flows from a reservoir under a sluice gate and into the 1.5-m-wide unfinished concrete rectangular channel, which is horizontal, Fig. 12–28a. At the control point 0, the measured flow is 2 m³/s and the depth is 0.2 m. Show how to determine the variation of the water depth along the channel, measured downstream from this point.

SOLUTION

Fluid Description.   Assuming the reservoir maintains a constant level, then the flow will be steady. Just after point 0, it will be non-uniform because its depth is changing. As usual, we assume water to be incompressible and use a mean velocity profile.

Analysis.   First we will classify the water surface profile. The critical depth is

$$y_c = \left(\frac{Q^2}{gb^2}\right)^{1/3} = \left(\frac{\left(2 \text{ m}^3/\text{s}\right)^2}{\left(9.81 \text{ m/s}^2\right)(1.5 \text{ m})^2}\right)^{1/3} = 0.5659 \text{ m}$$

Here $y = 0.2$ m $< y_c = 0.5659$ m, so the flow is rapid. Since the channel is horizontal, $S_0 = 0$. Using Table 12–2, the water surface profile is *H*3, as shown in Fig. 12–28a. This indicates that the water depth will increase as x increases from the control point. (Note that if the channel had a slope, then the Manning equation would have to be used to find y_n since this value is also needed for a profile classification.)

To determine the downstream depths, we will divide the calculations into increments of depth, choosing $\Delta y = 0.01$ m. At the initial station 0, $y = 0.2$ m, and the velocity is

$$V_0 = \frac{Q}{A_0} = \frac{2 \text{ m}^3/\text{s}}{(1.5 \text{ m})(0.2 \text{ m})} = 6.667 \text{ m/s}$$

The hydraulic radius is therefore

$$R_{h0} = \frac{A_0}{P_0} = \frac{(1.5 \text{ m})(0.2 \text{ m})}{(1.5 \text{ m} + 2)(0.2 \text{ m})} = 0.1579 \text{ m}$$

These results are entered into the first line of the table shown in Fig. 12–28b.

Station	y (m)	V (m/s)	V_m (m/s)	R_h (m)	R_{hm} (m)	S_{fm}	Δx (m)	x (m)
0	0.2	6.667		0.1579				0
			6.508		0.1610	0.09479	2.116	
1	0.21	6.349		0.1641				2.12
			6.205		0.1671	0.08200	2.104	
2	0.22	6.061		0.1701				4.22
			5.929		0.1731	0.07144	2.089	
3	0.23	5.797		0.1760				6.31

(b)

At station 1, $y_1 = 0.2$ m $+ 0.01$ m $= 0.21$ m, thus, $V_1 = Q/A_1 = 6.349$ m/s and $R_{h1} = A_1/P_1 = 0.1641$ m (third line of table). The intermediate second line gives the values of

$$V_m = \frac{V_0 + V_1}{2} = \frac{6.667 \text{ m/s} + 6.349 \text{ m/s}}{2} = 6.508 \text{ m/s}$$

$$R_{hm} = \frac{R_{h0} + R_{h1}}{2} = \frac{0.1579 \text{ m} + 0.1641 \text{ m}}{2} = 0.1610 \text{ m}$$

$$S_{fm} = \frac{n^2 V_m^2}{R_{hm}^{4/3}} = \frac{(0.014)^2 (6.508 \text{ m/s})^2}{(0.1610 \text{ m})^{4/3}} = 0.09479$$

$$\Delta x = \frac{(y_1 - y_0) + (V_1^2 - V_0^2)/2g}{S - S_{fm}}$$

$$= \frac{(0.21 \text{ m} - 0.2 \text{ m}) + \left[(6.349 \text{ m/s})^2 - (6.667 \text{ m/s})^2\right]/[2(9.81 \text{ m/s}^2)]}{(0 - 0.09479)}$$

$$= 2.116 \text{ m}$$

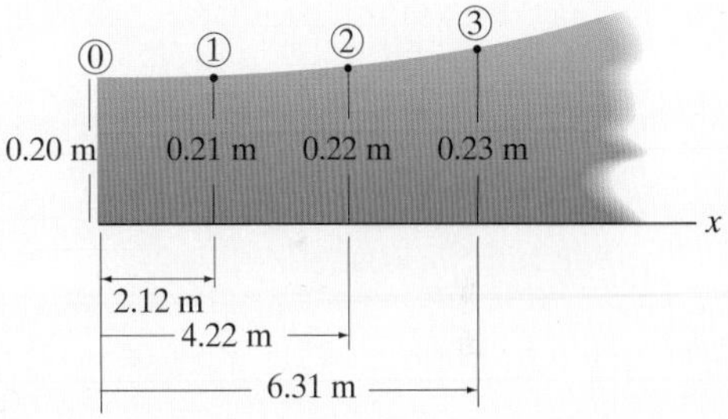

(c)

Fig. 12–28 (cont.)

The calculations are repeated for the next two stations as shown in the table, and the results are plotted in Fig. 12–28c. They appear satisfactory, since each increase $\Delta y = 0.01$ m produces somewhat uniform changes Δx. Had larger changes occurred, then smaller increments of Δy should be selected to improve the accuracy of the finite-difference method.

12

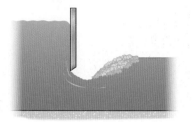

Hydraulic jump

Fig. 12–29

12.8   The Hydraulic Jump

When water passes down the spillway of a dam or underneath a sluice gate, it is typically rapid flow, Fig. 12–29. The transition to much slower tranquil flow downstream can occur, and when it does, it happens in a rather abrupt manner resulting in a ***hydraulic jump***. This is turbulent mixing that releases some of the water's kinetic energy, and in doing so brings the water surface up to the depth needed for tranquil flow.

Regardless of how the jump is formed, it is possible to determine the energy loss and the difference in water depth across the jump. To do this we must apply the continuity, momentum, and energy equations to a control volume containing the region of the jump, Fig. 12–30a. For the analysis, we will consider the jump to occur along the horizontal bed of a rectangular channel having a width b.

Continuity Equation.   Since the control volume extends just beyond the jump into regions of steady flow, we have

$$\frac{\partial}{\partial t} \int_{cv} \rho \, d\Psi + \int_{cs} \rho \, \mathbf{V} \cdot d\mathbf{A} = 0$$

$$0 - \rho V_1(y_1 b) + \rho V_2(y_2 b) = 0$$

$$V_1 y_1 = V_2 y_2 \qquad (12\text{–}26)$$

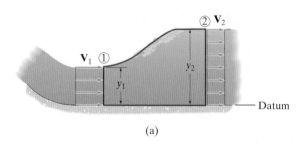

(a)

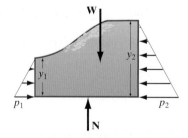

Free-body diagram

(b)

Fig. 12–30

Momentum Equation.   Experiments have shown that the jump happens within a short distance, and so frictional forces acting on the fixed control surfaces at the channel's bed and sides are negligible compared to the forces due to pressure, Fig. 12–30b. Applying the momentum equation in the horizontal direction, we have

$$\Sigma \mathbf{F} = \frac{\partial}{\partial t}\int_{cv} \mathbf{V}\rho\, dV + \int_{cs} \mathbf{V}\rho\, \mathbf{V} \cdot d\mathbf{A}$$

$$\frac{1}{2}(\rho g y_1 b)y_1 - \frac{1}{2}(\rho g y_2 b)y_2 = 0 + V_2\rho\big[V_2(by_2)\big] + V_1\rho\big[-V_1(by_1)\big]$$

Typical formation of a hydraulic jump

Or, after simplifying,

$$\frac{y_1^2}{2} - \frac{y_2^2}{2} = \frac{V_2^2}{g}y_2 - \frac{V_1^2}{g}y_1$$

Using the continuity equation to eliminate V_2, then dividing each side by $y_1 - y_2$, and finally multiplying both sides by $2y_2/y_1^2$, we get

$$\frac{2V_1^2}{gy_1} = \left(\frac{y_2}{y_1}\right)^2 + \frac{y_2}{y_1} \qquad (12\text{–}27)$$

Since the Froude number at section 1 is $\text{Fr}_1 = V_1/\sqrt{gy_1}$, then

$$2\text{Fr}_1^2 = \left(\frac{y_2}{y_1}\right)^2 + \frac{y_2}{y_1}$$

Using the quadratic formula, and solving for the positive root of y_2/y_1, we have

$$\frac{y_2}{y_1} = \frac{1}{2}\left(\sqrt{1 + 8\text{Fr}_1^2} - 1\right) \qquad (12\text{–}28)$$

Notice that if critical flow occurs upstream, then $\text{Fr}_1 = 1$, and this equation shows $y_2 = y_1$. In other words, there is no jump. When rapid flow occurs upstream, $\text{Fr}_1 > 1$, and so $y_2 > y_1$, as expected. Consequently, tranquil flow will occur downstream.

Energy Equation. If we apply the energy equation between points on top of the water surface at the open control surfaces 1 (in) and 2 (out), Fig. 12–30, noting that $p_1 = p_2 = 0$, we have

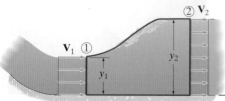

$$\frac{p_1}{\gamma} + \frac{V_1^2}{2g} + z_1 + h_{\text{pump}} = \frac{p_2}{\gamma} + \frac{V_2^2}{2g} + z_2 + h_{\text{turb}} + h_L$$

$$0 + \frac{V_1^2}{2g} + y_1 + 0 = 0 + \frac{V_2^2}{2g} + y_2 + 0 + h_L$$

Datum

Or, written in terms of the specific energy, the head loss is

Fig. 12–30

$$h_L = E_1 - E_2 = \left(\frac{V_1^2}{2g} + y_1\right) - \left(\frac{V_2^2}{2g} + y_2\right)$$

This loss reflects the turbulent mixing of the liquid within the jump, which is dissipated in the form of heat. Using the continuity equation, $V_2 = V_1(y_1/y_2)$, we can also write this expression as

$$h_L = \frac{V_1^2}{2g}\left[1 - \left(\frac{y_1}{y_2}\right)^2\right] + (y_1 - y_2)$$

Finally, if Eq. 12–27 is solved for V_1^2, and the result is substituted into the above equation, it can be shown that after simplification the head loss across the jump is

Hydraulic jump in a small stream

$$h_L = \frac{(y_2 - y_1)^3}{4y_1 y_2} \qquad (12\text{–}29)$$

For any real flow, it is always necessary for h_L to be *positive*. A negative value would violate the second law of thermodynamics, since frictional forces only dissipate energy, they never add energy to the fluid. As shown by Eq. 12–29, h_L is positive only if $y_2 > y_1$, and so a hydraulic jump *occurs only* when the flow changes from rapid flow upstream to tranquil flow downstream.

Important Point

- Nonuniform flow can occur suddenly in a channel in the form of a hydraulic jump. This process removes energy from rapid flow and thereby converts it to tranquil flow over a very short distance. The description of the jump can be determined by satisfying the continuity, momentum, and energy equations.

EXAMPLE | **12.14**

Water flows down a dam spillway and then forms a hydraulic jump, Fig. 12–31. Just before the jump, the velocity of flow is 8 m/s, and the water depth is 0.5 m. Determine the average velocity of the flow downstream in the channel.

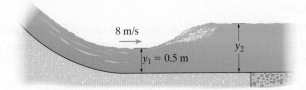

8 m/s

y_2

$y_1 = 0.5$ m

Fig. 12–31

SOLUTION

Fluid Description.   Steady uniform flow occurs before and after the jump. The water is considered incompressible, and average velocity profiles will be used.

Analysis.   The Froude number for the flow before the jump is

$$\mathrm{Fr}_1 = \frac{V_1}{\sqrt{gy_1}} = \frac{8 \text{ m/s}}{\sqrt{(9.81 \text{ m/s}^2)(0.5 \text{ m})}} = 3.6122 > 1$$

Thus the flow is rapid, as expected. After the jump the depth of the water is

$$\frac{y_2}{y_1} = \frac{1}{2}\left(\sqrt{1 + 8\mathrm{Fr}_1^2} - 1 \right)$$

$$\frac{y_2}{0.5 \text{ m}} = \frac{1}{2}\left(\sqrt{1 + 8(3.6122)^2} - 1 \right)$$

$$y_2 = 2.316 \text{ m}$$

We can now obtain the velocity V_2 by applying the continuity equation, Eq. 12–26.

$$V_1 y_1 = V_2 y_2$$

$$(8 \text{ m/s})(0.5 \text{ m}) = V_2(2.316 \text{ m})$$

$$V_2 = 1.727 \text{ m/s} = 1.73 \text{ m/s} \qquad Ans.$$

12

EXAMPLE | 12.15

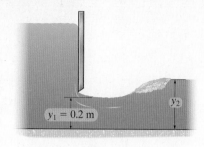

Fig. 12–32

The sluice gate in Fig. 12–32 is partially opened in a 2-m-wide channel, and water passing under the gate forms a hydraulic jump. At the low level, just before the jump, the water depth is 0.2 m, and the measured flow is 1.30 m³/s. Determine the depth of flow in the channel farther downstream and also the head loss across the jump.

SOLUTION

Fluid Description. We will assume the water is incompressible and the level in the reservoir is maintained, so steady flow occurs past the sluice gate.

Analysis. Just before the jump, the Froude number is

$$\mathrm{Fr}_1 = \frac{V_1}{\sqrt{gy_1}} = \frac{Q/A_1}{\sqrt{gy_1}} = \frac{(1.30 \ \mathrm{m^3/s})/[2 \ \mathrm{m}(0.2 \ \mathrm{m})]}{\sqrt{9.81 \ \mathrm{m/s^2}(0.2 \ \mathrm{m})}} = 2.320 > 1$$

Thus the flow is rapid, so indeed a jump can occur. Applying Eq. 12–28 to determine the water height after the jump, we have

$$\frac{y_2}{y_1} = \frac{1}{2}\left(\sqrt{1 + 8\mathrm{Fr}_1^2} - 1\right); \qquad \frac{y_2}{0.2 \ \mathrm{m}} = \frac{1}{2}\left(\sqrt{1 + 8(2.320)^2} - 1\right)$$

$$y_2 = 0.5638 \ \mathrm{m} = 0.564 \ \mathrm{m} \qquad \textit{Ans.}$$

At this depth,

$$\mathrm{Fr}_2 = \frac{Q/A_2}{\sqrt{gy_2}} = \frac{(1.30 \ \mathrm{m^3/s})/(2 \ \mathrm{m})(0.5638 \ \mathrm{m})}{\sqrt{9.81 \ \mathrm{m/s^2}(0.5638 \ \mathrm{m})}} = 0.4902 < 1$$

The flow is tranquil, as expected. The head loss can be determined from Eq. 12–29,

$$h_L = \frac{(y_2 - y_1)^3}{4y_1y_2} = \frac{(0.5638 \ \mathrm{m} - 0.2 \ \mathrm{m})^3}{4(0.2 \ \mathrm{m})(0.5638 \ \mathrm{m})} = 0.1068 \ \mathrm{m} = 0.107 \ \mathrm{m} \quad \textit{Ans.}$$

Note that since the original specific energy of the flow is

$$E_1 = \frac{Q^2}{2gb^2y_1^2} + y_1 = \frac{(1.30 \ \mathrm{m^3/s})^2}{2(9.81 \ \mathrm{m/s^2})(2 \ \mathrm{m})^2(0.2 \ \mathrm{m})^2} + 0.2 \ \mathrm{m} = 0.7384 \ \mathrm{m}$$

then, after the jump, the specific energy of the flow becomes

$$E_2 = E_1 - h_L = 0.7384 \ \mathrm{m} - 0.1068 \ \mathrm{m} = 0.6316 \ \mathrm{m}$$

This produces the following percent of energy lost within the jump,

$$E_L = \frac{h_L}{E_1} \times 100\% = \frac{0.1068 \ \mathrm{m}}{0.7384 \ \mathrm{m}}(100\%) = 14.46\%$$

12.9   Weirs

Most open-channel flow is measured using a *weir*. This device consists of a sharp-edged obstruction that is placed within the channel, causing the water to back up and then flow over it. Generally there are two types of weirs: sharp-crested weirs and broad-crested weirs.

Sharp-Crested Weirs.

A ***sharp-crested weir*** normally takes the form of a rectangular or triangular plate that has a sharp edge at the upstream side to minimize contact with the water, Fig. 12–33. As the water flows over the weir, it forms a vena contracta called a ***nappe***. To maintain this shape it is necessary to provide proper air ventilation underneath the nappe so that the water will fall clear of the weir plate. This is especially true for rectangular plates that extend the entire width of the channel, as in Fig. 12–33.*

Streamlines within the nappe are *curved*, and so the accelerations that occur here will cause *nonuniform* flow. Also, in the channel near the weir plate, the flow involves the effects of turbulence and vortex motion. Upstream from this region, however, the streamlines remain approximately parallel, the pressure varies hydrostatically, and the flow is uniform. And so, if we assume the liquid is an ideal fluid, then we can show that the flow over the weir is a function *only* of the *upstream depth* of the liquid— something that makes the weir a convenient device for measuring the flow.

Flood control of water flowing from a reservoir

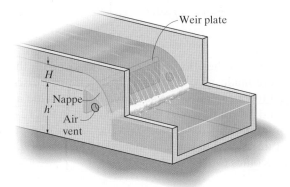

Flow over a rectangular weir

Fig. 12–33

---

*It may be of interest to note that the spillway of a large dam often has the same profile as a free-falling nappe, because then the water is only in slight contact with the surface, and so the pressure distribution over the surface will be approximately atmospheric.

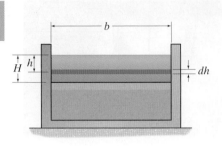

Rectangular weir

(a)

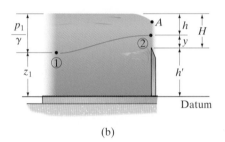

(b)

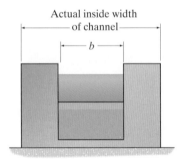

Actual inside width
of channel

Suppressed rectangular weir

(c)

Fig. 12–34

Rectangle. If the weir has a rectangular opening that extends over the entire width of the channel, Fig. 12–34a, then the Bernoulli equation can be applied between points 1 and 2 on the streamline shown in Fig. 12–34b. Assuming the approach velocity V_1 is *small* compared to V_2, so that V_1 can be neglected, we have

$$\frac{p_1}{\gamma} + \frac{V_1^2}{2g} + z_1 = \frac{p_2}{\gamma} + \frac{V_2^2}{2g} + z_2$$

$$\frac{p_1}{\gamma} + 0 + z_1 = 0 + \frac{V_2^2}{2g} + (h' + y)$$

Here $p_2 = 0$ because within the nappe the liquid is in free fall, so the pressure is atmospheric. Also, note from Fig. 12–34b that the hydraulic head $p_1/\gamma + z_1 = h' + y + h$. Substituting and solving for V_2, we obtain

$$V_2 = \sqrt{2gh} \tag{12–30}$$

Since the velocity is a function of h, the *theoretical discharge* through the entire cross section of the nappe must be determined by integration, Fig. 12–34a. We have

$$Q_t = \int_A V_2\, dA = \int_0^H \sqrt{2gh}\,(b\, dh) = \sqrt{2g}\, b \int_0^H h^{1/2} dh$$

$$= \frac{2}{3}\sqrt{2g}\, bH^{3/2} \tag{12–31}$$

To account for the effects of frictional loss, and for the other assumptions that were made, an experimentally determined **discharge coefficient** C_d is used to calculate the *actual discharge*. Its value also accounts for the smaller depth of flow over the weir, point A in Fig. 12–34b, and the amount of contraction of the nappe. Specific values of C_d can be found in the literature related to open-channel flow. See Ref. [9]. Using C_d, Eq.12–31 then becomes

$$Q_{\text{actual}} = C_d \frac{2}{3}\sqrt{2g}\, bH^{3/2} \tag{12–32}$$

To further restrict the upstream flow V_1, a **suppressed rectangular weir** can be used, Fig. 12–34c. Care must be taken, however, in selecting the width, b, since the nappe will also contract horizontally for very narrow widths.

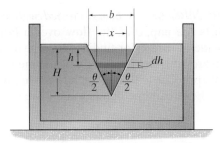

Triangular weir

Fig. 12–35

Triangle. When the discharge is small, it is convenient to use a weir plate with a triangular opening, Fig. 12–35. The Bernoulli equation yields the same result for the velocity as before, which is expressed by Eq. 12–30. Using the differential area $dA = x\, dh$, the theoretical discharge now becomes

$$Q_t = \int_A V_2 \, dA = \int_0^H \sqrt{2gh}\; x\, dh$$

Here the value of x is related to h by similar triangles,

$$\frac{x}{H - h} = \frac{b}{H}$$

$$x = \frac{b}{H}(H - h)$$

so that

$$Q_t = \sqrt{2g}\,\frac{b}{H} \int_0^H h^{1/2}(H - h)\,dh = \frac{4}{15}\sqrt{2g}\; bH^{3/2}$$

Because $\tan(\theta/2) = (b/2)/H$, then

$$Q_t = \frac{8}{15}\sqrt{2g}\; H^{5/2}\, \tan\frac{\theta}{2} \qquad\qquad (12\text{--}33)$$

Using a discharge coefficient, C_d, determined from experiment, the actual flow through a triangular weir is therefore

$$Q_{\text{actual}} = C_d\frac{8}{15}\sqrt{2g}\; H^{5/2}\, \tan\frac{\theta}{2} \qquad\qquad (12\text{--}34)$$

12

Broad-Crested Weirs.

A *broad-crested weir* consists of a weir block that supports the nappe of the flow over a horizontal distance, Fig. 12–36. It provides a means for finding both the critical depth y_c and the flow. As in the previous two cases, we will assume the fluid to be ideal and the velocity of approach to be negligible. Then applying the Bernoulli equation between the streamline points 1 and 2, we have

$$\frac{p_1}{\gamma} + \frac{V_1^2}{2g} + z_1 = \frac{p_2}{\gamma} + \frac{V_2^2}{2g} + z_2$$

$$0 + 0 + H = 0 + \frac{V_2^2}{2g} + y$$

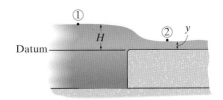

Datum

H

①

② y

Broad-crested weir

Fig. 12–36

Thus,

$$V_2 = \sqrt{2g(H - y)}$$

Provided the flow is originally tranquil, then as it begins to pass over the weir block, it will accelerate until the depth drops to the critical depth $y = y_c$, where the maximum discharge will occur and the specific energy will be a minimum. Here Eq. 12–9 applies.

$$V_2 = V_c = \sqrt{g y_c}$$

Substituting this result into the above equation yields

$$y_c = \frac{2}{3}H \qquad (12\text{–}35)$$

If the channel is rectangular and has a width b, then using these results, the theoretical discharge is

$$Q_t = V_2 A = \sqrt{g\left(\frac{2}{3}H\right)}\left[b\left(\frac{2}{3}H\right)\right]$$

$$= b\sqrt{g}\left(\frac{2}{3}H\right)^{3/2} \qquad (12\text{–}36)$$

This result is rather close to the actual flow obtained by experiment; however, to bring the two into better agreement, so as to account for the assumption of an ideal fluid, an experimentally determined *broad-crested-weir coefficient* C_w is used. See Ref. [9]. Thus,

$$Q_{\text{actual}} = C_w b\sqrt{g}\left(\frac{2}{3}H\right)^{3/2} \qquad (12\text{–}37)$$

Weir used to measure the flow of a river

Important Points

- Sharp-crested and broad-crested weirs are used to measure the flow in an open channel.

- Sharp-crested weirs are plates that have either rectangular or triangular shapes and are placed perpendicular to the flow in the channel.

- Broad-crested weirs support the flow over a horizontal distance. They are used to determine the critical depth and the discharge.

- The flow or discharge over a weir can be determined using the Bernoulli equation, and the result is adjusted to account for frictional loss and other effects by using an experimentally determined discharge coefficient.

EXAMPLE | 12.16

Water in the channel in Fig. 12–37 is 2 m deep, and the depth from the bottom of the triangular weir to the bottom of the channel is 1.75 m. If the discharge coefficient is $C_d = 0.57$, determine the volumetric flow in the channel.

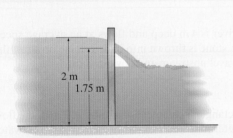
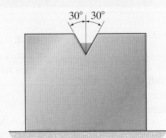

Fig. 12–37

SOLUTION

Fluid Description. The water level in the channel is assumed to be constant, so we have steady flow. Also, water is considered to be incompressible.

Analysis. Here $H = 2$ m -1.75 m $= 0.25$ m. Applying Eq. 12–34, the flow is therefore

$$Q_{actual} = C_d \frac{8}{15} \sqrt{2g}\, H^{5/2} \tan\frac{\theta}{2}$$

$$= (0.57)\frac{8}{15}\sqrt{2(9.81 \text{ m/s}^2)}\,(0.25 \text{ m})^{5/2} \tan 30° = 0.0243 \text{ m}^3/\text{s} \quad Ans.$$

12

References

1. R. W. Carter et al., "Friction factors in open channels," *Journal Hydraulics Division, ASCE,* Vol. 89, No. AY2, 1963, pp. 97–143.

2. V. T. Chow, *Open Channel Hydraulics,* McGraw-Hill, New York, NY, 2009.

3. R. French, *Open Channel Hydraulics,* McGraw-Hill, New York, NY, 1992.

4. C. E. Kindsater and R. W. Carter, "Discharge characteristics of rectangular thin-plate weirs," *Trans ASCE,* 124, 1959, pp. 772–822.

5. R. Manning, "The flow of water in open channels and pipes," *Trans Inst of Civil Engineers of Ireland,* Vol. 20, Dublin, 1891, pp. 161–201.

6. H. Rouse, *Fluid Mechanics for Hydraulic Engines*, Dover Publications, Inc., New York, N. Y.

7. A. L. Prasuhn, *Fundamentals of Fluid Mechanics*, Prentice-Hall, NJ, 1980.

8. E. F. Brater, *Handbook of Hydraulics,* 7th ed., McGraw-Hill, New York, NY.

9. P. Ackers et al., *Weirs and Flumes for Flow Measurement,* John Wiley, New York, NY, 1978.

10. M. H. Chaudhry, *Open Channel Flow,* 2nd ed., Springer-Verlag, New York, NY, 2007.

PROBLEMS

Sec. 12.1–12.4

12–1. A large tank contains water that has a depth of 4 m. If the tank is on an elevator that is descending, determine the speed of a wave created on its surface if the rate of descent is (a) constant at 8 m/s, (b) accelerated at 4 m/s², (c) accelerated at 9.81 m/s².

12–2. A rectangular channel has a width of 2 m. If the flow is 5 m³/s, determine the Froude number when the water depth is 0.5 m. At this depth, is the flow subcritical or supercritical? Also, what is the critical speed of the flow?

12–3. A rectangular channel transports water at 8 m³/s. The channel width is 3 m and the water depth is 2 m. Is the flow subcritical or supercritical?

*__12–4.__ A river is 4 m deep and flows at an average speed of 2 m/s. If a stone is thrown into it, determine how fast the waves will travel upstream and downstream.

12–5. A rectangular channel has a width of 2 m. If the flow is 5 m³/s, determine the Froude number when the water depth is 1.5 m. At this depth, is the flow subcritical or supercritical? Also, what is the critical speed of the flow?

12–6. Water flows in a rectangular channel with a speed of 3 m/s and depth of 1.25 m. What other possible depth of flow provides the same specific energy?

12–7. A rectangular channel having a width of 3 m is required to transport 40 m³/s of water. Determine the critical depth and critical velocity of the flow. What is the specific energy at the critical depth, and also when the depth is 2 m?

*12–8. Water flows in a rectangular channel with a mean speed of 6 m/s and depth of 4 m. What other possible average velocity of flow provides the same specific energy?

12–9. Water flows within the rectangular channel with a flow of 8 m³/s. Determine the two possible flow depths, and identify the flow as supercritical or subcritical, if the specific energy is 2 m. Also, plot the specific energy diagram.

12–11. The channel transports water at 8 m³/s. If the depth of flow is $y = 1.5$ m, determine if the flow is subcritical or supercritical. What is the critical depth of flow? Compare the specific energy of the flow with its minimum specific energy.

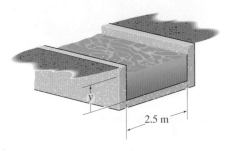

Prob. 12–11

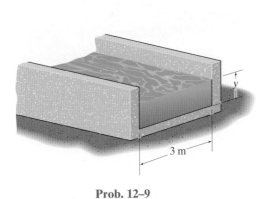

Prob. 12–9

*12–12. The rectangular channel transports water at 4 m³/s. Determine the critical depth y_c and plot the specific energy diagram for the flow. Indicate y for $E = 1.25$ m.

12–10. The rectangular channel transports water at 8 m³/s. Determine the critical depth y_c and plot the specific energy diagram for the flow. Indicate y for $E = 2$ m.

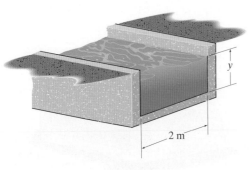

Prob. 12–10

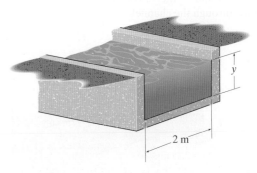

Prob. 12–12

12–13.   The rectangular channel passes through a transition that causes its width to narrow to 1.5 m. If the flow is 5 m³/s and $y_A = 3$ m, determine the depth of flow at B.

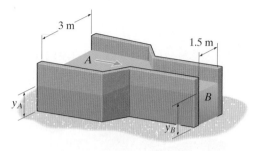

Prob. 12–13

12–14.   The rectangular channel has a transition that causes its width to narrow to 1.5 m. If the flow is 5 m³/s and $y_A = 5$ m, determine the depth of flow at B.

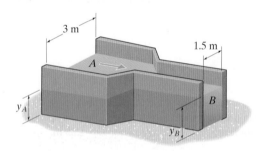

Prob. 12–14

12–15.   The venturi is placed in the channel in order to measure the volumetric flow. If the depth of flow at A is $y_A = 2.50$ m and at the throat B is $y_B = 2.35$ m, determine the flow through the channel.

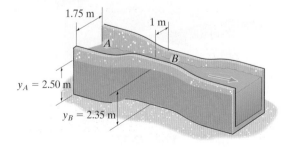

Prob. 12–15

***12–16.**   Water flows within the rectangular channel such that the flow is 4 m³/s. Determine the critical depth of flow and the minimum specific energy. If the specific energy is 8 m, what are the two possible flow depths?

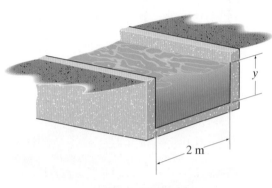

Prob. 12–16

12–17.   The rectangular channel transports water at a flow of 8 m³/s. Plot the specific energy diagram for the flow and indicate y for $E = 3$ m.

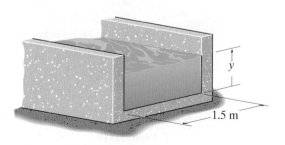

Prob. 12–17

Sec. 12.4–12.5

12–18. The channel is 2 m wide and transports water at a flow of 18 m³/s. If the elevation of the bed is lowered by 0.1 m, determine the new depth y_2 of the water.

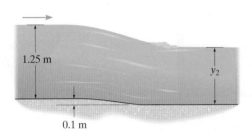

Prob. 12–18

12–19. The channel is 2 m wide and transports water at 18 m³/s. If the elevation of the bed is raised 0.25 m, determine the new depth y_2 of the water and the speed of the flow. Is the new flow subcritical or supercritical?

Prob. 12–19

*12–20. Water flows within the 4-m-wide rectangular channel at 20 m³/s. Determine the depth of flow y_B at the downstream end and the velocity of flow at A and B. Take $y_A = 5$ m.

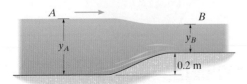

Prob. 12–20

12–21. Water flows within the 4-m-wide rectangular channel at 20 m³/s. Determine the depth of flow y_B at the downstream end and the velocity of flow at A and B. Take $y_A = 0.5$ m.

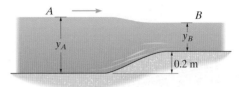

Prob. 12–21

12–22. The rectangular channel is 2 m wide, and the depth of the water is 1.5 m as it flows with an average velocity of 0.5 m/s. Show that the flow is tranquil, and determine the required height h of the bump so that the flow can change to rapid flow after it passes over the bump. What is the new depth y_2 for rapid flow?

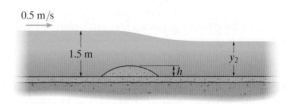

Prob. 12–22

12–23. The rectangular channel is 2 m wide, and the depth of the water is 0.75 m as it flows with an average velocity of 4 m/s. Show that the flow is rapid, and determine the required height h of the bump so that the flow can change to tranquil flow after it passes over the bump. What is the new depth y_2 for tranquil flow?

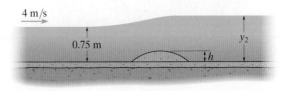

Prob. 12–23

*12–24.   The 2-m-wide sluice gate is used to control the flow of water from a reservoir. If the depths $y_1 = 4$ m and $y_2 = 0.75$ m, determine the volumetric flow through the gate and the depth y_3 just before the gate.

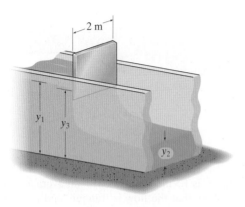

Prob. 12–24

12–25.   The 2-m-wide sluice gate is used to control the flow of water from a reservoir. If the flow is 10 m³/s and $y_1 = 4$ m, determine the depth y_2, and depth y_3 just before the gate.

12–26.   The sluice gate and channel both have a width of 2 m. If the depth of flow at A is $y_1 = 3$ m, determine the volumetric flow through the channel as a function of depth y_2 and specify Q when the depth y_2 is (a) 1 m, (b) 1.5 m.

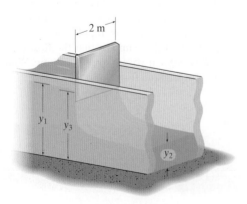

Probs. 12–25/26

Sec. 12.6

12–27.   Determine the hydraulic radius for each channel cross section.

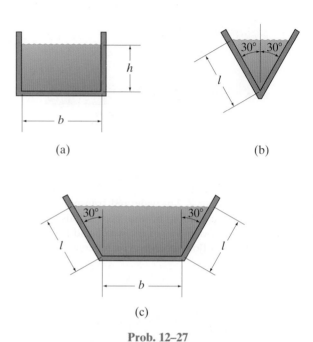

(a)

(b)

(c)

Prob. 12–27

*12–28.**   The channel has a triangular cross section. Determine the critical depth $y = y_c$ in terms of θ and the flow Q.

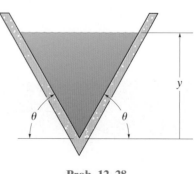

Prob. 12–28

12–29. A rectangular channel has a width of 2 m, is made of unfinished concrete, and is inclined at a slope of 0.0014. Determine the volumetric flow when the depth of flow of the water is 1.5 m.

12–30. Water flows uniformly down the triangular channel having a downward slope of 0.0083. If the walls are made of finished concrete, determine the volumetric flow when $y = 1.5$ m.

12–33. The culvert carries water and is at a downward slope S_0. Determine the depth y that will produce the maximum volumetric flow.

12–34. The culvert carries water and is at a downward slope S_0. Determine the depth y that will produce maximum velocity for the flow.

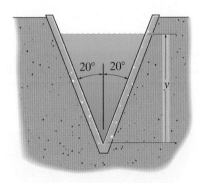

Prob. 12–30

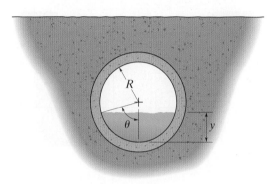

Probs. 12–33/34

12–31. The channel is made of unfinished concrete and has a downward slope of 0.003. Determine the volumetric flow if the depth is $y = 2$ m. Is the flow subcritical or supercritical?

*12–32. The channel is made of unfinished concrete and has a downward slope of 0.003. Determine the volumetric flow if the depth is $y = 3$ m. Is the flow subcritical or supercritical?

12–35. The drainage canal has a downward slope of 0.002. If its bottom and sides have weed growth, determine the volumetric flow of water when the depth of flow is 2.5 m.

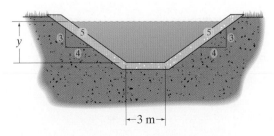

Probs. 12–31/32

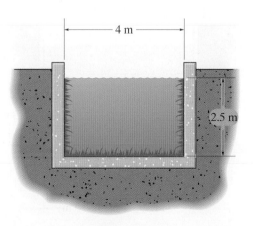

Prob. 12–35

*12–36. A rectangular channel has a downward slope of 0.006 and a width of 3 m. The depth of the water is 4 m. If the volumetric flow through the channel is 30 m³/s, determine the value of n in the Manning formula.

The channel is made of unfinished concrete and has the cross section shown. If the downward slope is 0.0008, determine the flow of water through the channel when $y = 4$ m.

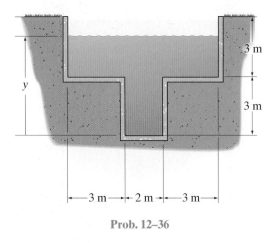

Prob. 12–36

12–37. The channel is made of unfinished concrete and has the cross section shown. If the downward slope is 0.0008, determine the flow of water through the channel when $y = 6$ m.

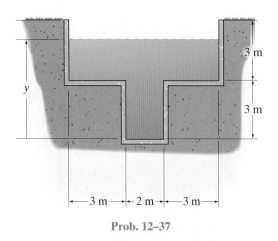

Prob. 12–37

12–38. The channel is made of finished concrete and has a trapezoidal cross section. If the average velocity of the flow is to be 6 m/s when the water depth is 2 m, determine the required slope.

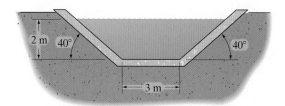

Prob. 12–38

12–39. The unfinished concrete channel is intended to have a downward slope of 0.002 and sloping sides at 60°. If the flow is estimated to be 100 m³/s, determine the base dimension b of the channel bottom.

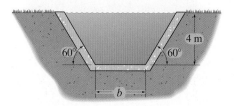

Prob. 12–39

*12–40. A rectangular channel has a width of 2.5 m and is made of unfinished concrete. If it is inclined downward at a slope of 0.0014, what depth of water will produce a discharge of 12 m³/s?

12–41. Determine the length of the sides a of the channel in terms of its base b, so that for the flow at full depth it provides the best hydraulic cross section that uses the minimum amount of material for a given discharge.

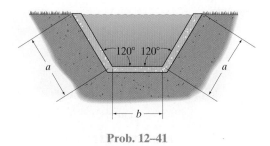

Prob. 12–41

12–42. Determine the volumetric flow of water through the channel if the depth of flow is $y = 1.25$ m and the downward slope of the channel is 0.005. The sides of the channel are finished concrete.

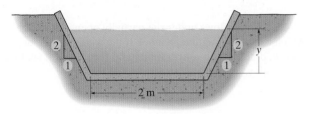

Prob. 12–42

12–43. Determine the normal depth of water in the channel if the flow is $Q = 15$ m³/s. The sides of the channel are finished concrete, and the downward slope is 0.005.

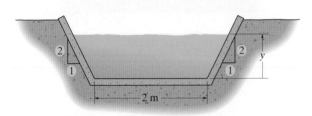

Prob. 12–43

*12–44. Determine the angle θ of the channel so that it has the best hydraulic triangular cross section that uses the minimum amount of material for a given discharge.

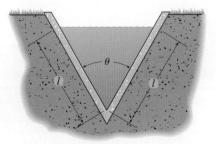

Prob. 12–44

12–45. Show that the width $b = 2h(\csc\theta - \cot\theta)$ in order to minimize the wetted perimeter for a given cross-sectional area and angle θ. At what angle θ will the wetted perimeter be the smallest for a given cross-sectional area and depth h?

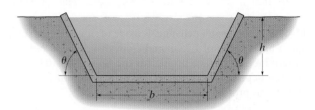

Prob. 12–45

12–46. Show that when the depth of flow $y = R$, the semicircular channel provides the best hydraulic cross section.

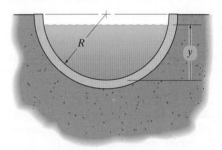

Prob. 12–46

12–47. Determine the angle θ and the length l of its sides so that the channel has the best hydraulic trapezoidal cross section of base b.

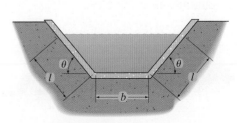

Prob. 12–47

Sec. 12.7–12.9

*12–48. A rectangular channel is made of unfinished concrete, and it has a width of 1.25 m and an upward slope of 0.01. Determine the surface profile for the flow if it is 0.8 m³/s and the depth of the water at a specific location is 0.5 m. Sketch this profile.

12–49. A rectangular channel is made of finished concrete, and it has a width of 1.25 m and a downward slope of 0.01. Determine the surface profile for the flow if it is 0.8 m³/s and the depth of the water at a specific location is 0.6 m. Sketch this profile.

12–50. A rectangular channel is made of finished concrete, and it has a width of 1.25 m and an upward slope of 0.01. Determine the surface profile for the flow if it is 0.8 m³/s and the depth of the water at a specific location is 0.2 m. Sketch this profile.

12–51. Water flows at 4 m³/s along a horizontal channel made of unfinished concrete. If the channel has a width of 2 m, and the water depth at a control section A is 0.9 m, approximate the depth at the section where $x = 2$ m from the control section. Use increments of $\Delta y = 0.004$ m and plot the profile for 0.884 m $\leq y \leq$ 0.9 m.

*12–52. A rectangular channel is made of finished concrete, and it has a width of 1.25 m and a downward slope of 0.01. Determine the surface profile for the flow if it is 0.8 m³/s and the depth of the water at a specific location is 0.3 m. Sketch this profile.

12–53. Water flows at 12 m³/s down a rectangular channel made of unfinished concrete. The channel has a width of 4 m and a downward slope of 0.008, and the water depth is 2 m at the control section A. Determine the distance x from A to where the depth is 2.4 m. Use increments of $\Delta y = 0.1$ m and plot the profile for 2 m $\leq y \leq$ 2.4 m.

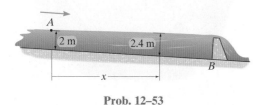

Prob. 12–53

12–54. Water flows at 4 m³/s along a horizontal channel made of unfinished concrete. If the channel has a width of 2 m, and the water depth at a control section A is 0.9 m, determine the approximate distance x from A to where the depth is 0.8 m. Use increments of $\Delta y = 0.025$ m and plot the profile for 0.8 m $\leq y \leq$ 0.9 m.

12–55. Water flows under the partially opened sluice gate, which is in a rectangular channel. If the water has the depth shown, determine if a hydraulic jump forms, and if so, find the depth y_C at the downstream end of the jump.

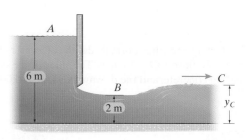

Prob. 12–55

*12–56. Water runs from a sloping channel with a flow of 8 m³/s onto a horizontal channel, forming a hydraulic jump. If the channel is 2 m wide, and the water is 0.25 m deep before the jump, determine the depth of water after the jump. What energy is lost during the jump?

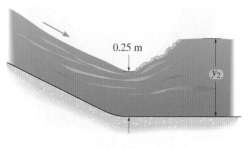

Prob. 12–56

12–57. The hydraulic jump has a depth of 5 m at the downstream end, and the velocity is 1.25 m/s. If the channel is 2 m wide, determine the depth y_1 of the water before the jump and the energy head lost during the jump.

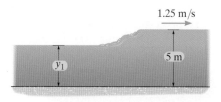

Prob. 12–57

12–58. Water flows at 18 m³/s over the 4-m-wide spillway of the dam. If the depth of the water at the bottom apron is 0.5 m, determine the depth y_2 of the water after the hydraulic jump.

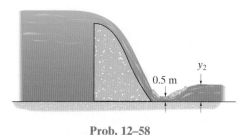

Prob. 12–58

12–59. The sill at A causes a hydraulic jump to form in the channel. If the channel width is 1.5 m, determine the average upstream speed and downstream speed of the water. What amount of energy head is lost in the jump?

Prob. 12–59

***12–60.** The flow of water over the broad-crested weir is 15 m³/s. If the weir and the channel have a width of 3 m, determine the depth of water y within the channel. Take $C_w = 0.80$.

Prob. 12–60

12–61. The rectangular channel has a width of 3 m and the depth of flow is 1.5 m. Determine the volumetric flow of water over the rectangular sharp-crested weir. Take $C_d = 0.83$.

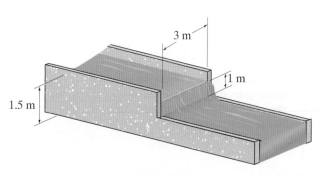

Prob. 12–61

CHAPTER REVIEW

Practically all open-channel flow is turbulent. For analysis, we assume the flow is one-dimensional and use an average velocity over the cross section.	
The speed of a wave traveling over a liquid surface is called the wave celerity. It is a function of the depth of a channel.	$c = \sqrt{gy}$ 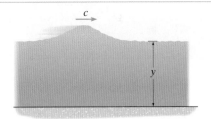
The characteristics of flow along a channel are classified using the Froude number, since gravity is a driving force for open-channel flow. When Fr < 1, the flow is tranquil, and when Fr > 1, the flow is rapid. If Fr = 1, then critical flow occurs, in which case a standing wave can form on the surface of the liquid.	$Fr = \dfrac{V}{\sqrt{gy}}$
Both rapid and tranquil flow can produce the *same* specific energy for a specified flow and channel width. Using a properly designed bump, where critical flow occurs at its peak, the flow can be changed from rapid to tranquil, or from tranquil to rapid.	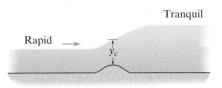
Flow over a channel rise, or under a sluice gate, can be analyzed using the continuity equation and the Bernoulli equation.	

Steady uniform flow in an open channel will occur if the channel has a constant cross section, slope, and surface roughness. The velocity of the flow can be determined using Manning's equation, which uses an empirical coefficient n to specify the surface roughness.

$$V = \frac{kR_h^{2/3}S_0^{1/2}}{n}$$

If nonuniform flow occurs in a channel, then there are twelve possible types of surface profiles. The profile can be determined numerically using a finite difference method derived from the energy equation.

Nonuniform flow can occur suddenly in an open channel, causing a hydraulic jump. The jump removes energy from the flow, causing it to convert from rapid to tranquil flow.

The flow in an open channel can be measured using a sharp-crested or broad-crested weir. In particular, a broad-crested weir can also convert the flow into flow having a critical depth.

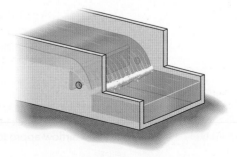

Chapter 13

(© Super Nova Images/Alamy)

Local shock waves form on the surface of this jet plane at places where the airflow approaches the speed of sound. The effect is seen because the air is humid and condensation forms on the shock.

Compressible Flow

CHAPTER OBJECTIVES

■ To present some of the important concepts of thermodynamics that are used to analyze compressible gas flow.

■ To discuss isentropic flow through a variable area, and show how it affects the temperature, pressure, and density of a gas.

■ To show how compressible flow is affected by friction within a duct (Fanno flow), and by the addition or removal of heat from the surface of the duct (Rayleigh flow).

■ To study the formation of shock waves in convergent and divergent nozzles, and the formation of compression and expansion waves on curved or irregular surfaces.

■ To present some of the methods used to measure velocity and pressure for compressible flow.

13.1 Thermodynamic Concepts

Up to this point we have considered applications of fluid mechanics only to problems where the fluid is considered incompressible. In this chapter we will extend our study to include the effects of compressibility, something that is important in the design of jet engines, rockets and aircraft, and high speed flow through long municipal and industrial gas lines, as well as in air ducts used for heating and ventilation.

We will begin the chapter with a presentation of some of the more important concepts in thermodynamics that are involved in compressible gas flow. Thermodynamics plays an important role in understanding compressible flow behavior because any drastic change to the kinetic energy of a gas will be converted into heat, causing large density and pressure variations throughout the gas.

13

Ideal Gas Law.   Throughout this chapter we will consider the gas to be an *ideal gas*, that is, one that is composed of molecules that behave like elastic spheres. The molecules are assumed to have random motion and interact only on occasion, since there is a large distance between them compared to their size. Real gases are quite similar to ideal gases at the pressures and temperatures found on earth, although this approximation becomes more accurate as the density of the gas becomes smaller.

It has been found *through experiment* that the *absolute temperature*, *absolute pressure*, and the density of an ideal gas are related through a single equation, referred to as the **ideal gas law**. It is

$$p = \rho R T \tag{13-1}$$

Here R is the gas constant, which has a unique value for each gas. The ideal gas law is referred to as an **equation of state** since it relates the three **state properties** p, ρ, T of the gas when it is in a specific state or condition.

Internal Energy and the First Law of Thermodynamics. Another important state property is **internal energy**, and for an ideal gas this refers to the collective kinetic and potential energy of its atoms and molecules. A change in internal energy occurs if heat and work are transferred to or from the gas. The *first law of thermodynamics* is a formal statement of this balance. It states that if we consider a *unit mass* of a gas as a contained system, then the change in its internal energy du, as the system moves from one state to another, is equal to the heat (thermal energy) dq transferred *into* the system minus the work dw done *by* the system on its surroundings. The work is defined as *flow work*, discussed in Sec. 5.5, and can be written as $dw = p\,dv$. Thus

$$\underset{\substack{\text{Change in} \\ \text{internal energy}}}{du} = \overset{\substack{\text{Heat} \\ \text{added}}}{dq} - \underset{\substack{\text{Flow work} \\ \text{removed}}}{p\,dv} \tag{13-2}$$

This change in internal energy is *independent* of the process used to move the system from one state to another; however, the values of dq and $p\,dv$ depend upon this process. Typical processes include changes under constant volume, constant pressure, or constant temperature; or, if no heat enters or leaves the system, an adiabatic process occurs.

Specific Heat.   The amount of heat dq is directly related to the change in temperature dT of a gas by the **specific heat** c, which is a *physical property* of the gas. It is defined as the amount of heat needed to raise the temperature of a unit mass of the gas by one degree.

$$c = \frac{dq}{dT} \tag{13-3}$$

This property depends upon the process by which the heat is added. It is measured in $\text{J}/(\text{kg} \cdot \text{K})$. Normally, it is determined by adding the heat at constant volume, c_v, or at constant pressure, c_p.

Constant-Volume Process.   When the volume of a gas remains constant, then $dv = 0$, and no external flow work is done. The first law of thermodynamics then states that $du = dq$. In other words, the internal energy of the gas is raised only because of the amount of heat supplied. Thus, Eq. 13–3 becomes

$$c_v = \frac{du}{dT} \tag{13–4}$$

For the range of most engineering applications, c_v is essentially constant as the temperature changes, and so we can integrate this equation between any two points of state and write it as

$$\Delta u = c_v \Delta T \tag{13–5}$$

Therefore, if c_v is known, this result provides a means of obtaining the change in internal energy for a given temperature change ΔT.

Constant-Pressure Process.   For a constant-pressure process, the gas can expand, and so Eq. 13–2 states that a change in internal energy and flow work can occur. Solving for dq, and using Eq. 13–3, we have

$$c_p = \frac{du + p\,dv}{dT}$$

To *simplify* this expression, we will define another state property of a gas called the **enthalpy** h. Formally, enthalpy is defined as the sum of the internal energy u and the flow work pv of a unit mass of the gas. Since the volume per unit mass is $v = 1/\rho$, then

$$h = u + pv = u + \frac{p}{\rho} \tag{13–6}$$

Using the ideal gas law, the enthalpy can also be written in terms of temperature.

$$h = u + RT \tag{13–7}$$

To find the *change* in the enthalpy, we take the derivative of Eq. 13–6.

$$dh = du + dp\,v + p\,dv \tag{13–8}$$

Since the pressure is constant, $dp = 0$, and so $dh = du + p\,dv$. Therefore, c_p now becomes

$$c_p = \frac{dh}{dT} \tag{13–9}$$

This equation can be integrated between any two points of state, since c_p is only a function of temperature. Within the temperature ranges normally considered in engineering, c_p, like c_v, is essentially constant. Thus,

$$\Delta h = c_p \Delta T \tag{13–10}$$

Therefore if c_p is known, then if a change in temperature ΔT occurs, we can calculate the corresponding change in the enthalpy (internal energy plus flow work).

If we take the derivative of Eq. 13–7 and substitute Eq. 13–4 and Eq. 13–9 into the result, we obtain a relation among c_p, c_v, and the gas constant. It is

$$c_p - c_v = R \tag{13–11}$$

If we now express the ratio of specific heats as

$$k = \frac{c_p}{c_v} \tag{13–12}$$

then, using Eq. 13–11, we can write

$$c_v = \frac{R}{k - 1} \tag{13–13}$$

and also

$$c_p = \frac{kR}{k - 1} \tag{13–14}$$

Appendix A gives values of k and R for common gases, so the values of c_v and c_p can be calculated from these two equations.

Entropy and the Second Law of Thermodynamics.

Entropy s is a state property of a gas, and in thermodynamics, as well as here, we will be interested in how it *changes*. We define the **change in entropy** as the amount of heat that takes place per degree of temperature, as a unit mass of gas moves from one state of pressure, volume, and temperature to another. Thus,

$$ds = \frac{dq}{T} \tag{13–15}$$

For example, when two objects at different temperatures are placed near one another in an insulated container (closed system), eventually they will reach the same temperature due to increments of heat dq moving from the hot to the cold object. One will gain entropy, and the other will lose entropy. This process of heat flowing from the hot to the cold body is *irreversible*. In other words, heat never flows from a cold to a hot body, because there is more thermal agitation of the molecules or internal energy within the hot body than in the cold body.

The **second law of thermodynamics** is based on the change in entropy, and it determines the time order in which a physical phenomenon can take place. It gives a preferred direction to time and it is sometimes called the "arrow of time." For an **isolated system**, meaning one that is closed and insulated, the second law states that the entropy will *always increase*, since the process for change is **irreversible**. In a gas it is the result of viscous friction which increases the agitation of the gas, thereby causing the heat of the gas to increase. If the process is assumed to be **reversible**, that is, without internal friction, no change in entropy will occur. Thus,

$$\begin{aligned} ds &= 0 & \text{Reversible} \\ ds &> 0 & \text{Irreversible} \end{aligned} \tag{13–16}$$

For purposes of calculation, we can obtain a relationship between the change in entropy and the intensive properties T and ρ by first combining Eq. 13–15 with the first law of thermodynamics to eliminate dq. We have

$$T \, ds = du + p \, dv \tag{13–17}$$

Now, if we substitute $v = 1/\rho$ into this equation, and use the ideal gas law $p = \rho RT$ and the definition of c_v, Eq. 13–4, we get

$$ds = c_v \frac{dT}{T} + \frac{R}{1/\rho} d\left(\frac{1}{\rho}\right)$$

Integrating, with c_v remaining constant during the temperature change, yields

$$\boxed{s_2 - s_1 = c_v \ln \frac{T_2}{T_1} - R \ln \frac{p_2}{p_1}} \qquad (13\text{–}18)$$

Also, the change in entropy can be related to T and p. First, the enthalpy can be related to the entropy by substituting Eq. 13–17 into Eq. 13–8. We get

$$T\, ds = dh - v\, dp \qquad (13\text{–}19)$$

Then using the definition of c_p, Eq. 13–9, and the ideal gas law, written as $p = RT/v$, we have

$$ds = c_p \frac{dT}{T} - R \frac{dp}{p}$$

When integrated, this gives

$$\boxed{s_2 - s_1 = c_p \ln \frac{T_2}{T_1} - R \ln \frac{p_2}{p_1}} \qquad (13\text{–}20)$$

T–s Diagram.   When solving compressible gas flow problems, it is sometimes helpful to establish a T–s **state diagram**, which represents a plot of the temperature versus the entropy. For example, a constant-temperature process would plot as a horizontal line, since for every change in entropy the temperature must be constant, T_c, Fig. 13–1. If instead the volume of the gas is held constant at V_c, then $dv = 0$, and eliminating du from Eqs. 13–17 and 13–4, we find

$$T ds = c_v\, dT + p\, dv$$

$$\frac{dT}{ds} = \frac{T}{c_v}$$

which represents the slope of the curve for constant volume, Fig. 13–1. Finally, if the pressure is held constant at p_c, then $dp = 0$, and eliminating dh from Eqs. 13–19 and 13–9, we have

$$T\, ds = c_p\, dT - v\, dp$$

$$\frac{dT}{ds} = \frac{T}{c_p}$$

which is the slope of the curve for constant pressure, Fig. 13–1. Apart from these examples, we will show, in Secs. 13.7 and 13.8, how the T–s diagram provides a graphical aid for the interpretation of analytical results.

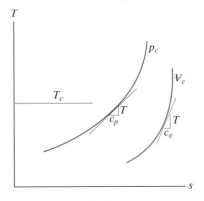

T–s diagrams

Fig. 13–1

Isentropic Process.

Many types of problems involving compressible flow through nozzles and duct transitions occur within a localized region over a *very short period of time*, and when this happens, the changes in the gas can often be modeled as an ***isentropic process***. This process involves *no heat transfer* to the surroundings, while the gas is suddenly brought from one state to another; that is, the proccess is ***adiabatic***, $dq = 0$. Also, it is *reversible*, provided friction within the system is neglected, and as a result, $ds = 0$. Therefore, for an isentropic process, Eqs. 13–17 and 13–19 become

$$0 = du + p\,dv$$

$$0 = dh - v\,dp$$

If we express these equations in terms of the specific heats, we have

$$0 = c_v\,dT + p\,dv$$

$$0 = c_p\,dT - v\,dp$$

Eliminating dT and using Eq. 13–12, $k = c_p/c_v$, yields

$$\frac{dp}{p} + k\frac{dv}{v} = 0$$

Since k is constant, integrating between any two points of state gives

$$\ln\frac{p_2}{p_1} + k\ln\frac{v_2}{v_1} = 0 \qquad \left(\frac{p_2}{p_1}\right)\left(\frac{v_2}{v_1}\right)^k = 1$$

Substituting $v = 1/\rho$ and rearranging terms, the relationship between pressure and density becomes

$$\frac{p_2}{p_1} = \left(\frac{\rho_2}{\rho_1}\right)^k \tag{13–21}$$

Also, we can express the pressure in terms of the absolute temperature by using the ideal gas law, $p = \rho RT$. We obtain

$$\frac{p_2}{p_1} = \left(\frac{T_2}{T_1}\right)^{k/(k-1)} \tag{13–22}$$

Throughout the chapter, we will use several of the equations developed in this section to describe either isentropic or adiabatic compressible flow.

13

Important Points

- For most engineering applications, a gas can be considered as ideal, that is, composed of molecules in random motion, with large distances between them compared to their size. Ideal gases obey the ideal gas law, $p = \rho RT$.

- A system in equilibrium has certain state properties, such as pressure, density, temperature, internal energy, entropy, and enthalpy.

- The change in internal energy per unit mass of a system is *increased* when *heat is added* to the system, and it is *decreased* when the *system does flow work*. This is the first law of thermodynamics, $du = dq - p\,dv$.

- Enthalpy is a state property of a gas that is defined in terms of three other state properties, namely, $h = u + p/\rho$.

- The specific heat at constant volume relates the change in internal energy of a gas to the change in its temperature, $\Delta u = c_v \Delta T$.

- The specific heat at constant pressure relates the change in enthalpy of a gas to the change in its temperature, $\Delta h = c_p \Delta T$.

- Change in entropy ds provides a measure of the amount of heat that takes place per degree of temperature, $ds = dq/T$.

- The second law of thermodynamics states that for an *irreversible process*, entropy will always increase because of friction, $ds > 0$. If the process is *reversible*, or frictionless, then the change in entropy will be zero, $ds = 0$.

- The specific heat c_v can be used to relate the change in entropy Δs to the change in T and ρ, Eq. 13–18, and c_p can be used to relate Δs to the change in T and p, Eq. 13–20.

- An isentropic process is one where no heat is gained or lost and the flow is frictionless. This means that the process is adiabatic and reversible, $ds = 0$.

EXAMPLE 13.1

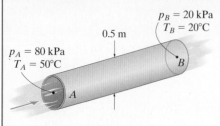

$p_B = 20$ kPa
$T_B = 20°C$

0.5 m

$p_A = 80$ kPa
$T_A = 50°C$

A

B

Fig. 13–2

Air flows at 5 kg/s through the duct in Fig. 13–2, such that the gage pressure and temperature at A are $p_A = 80$ kPa and $T_A = 50°C$, and at B, $p_B = 20$ kPa and $T_B = 20°C$. Determine the changes in the enthalpy, internal energy, and entropy of the air between these points.

SOLUTION

Fluid Description. Because the temperature and pressure change between A and B, the density of the air will also change, and so we have steady compressible flow. For air, $k = 1.4$ and $R = 286.9$ J/kg·K.

Change in Enthalpy. The change in the enthalpy is determined from Eq. 13–10. However, first we must determine the specific heat at constant pressure using Eq. 13–14.

$$c_p = \frac{kR}{k-1} = \frac{1.4(286.9 \text{ J/kg} \cdot \text{K})}{(1.4-1)} = 1004.15 \text{ J/kg} \cdot \text{K}$$

The temperature in thermodynamic calculations in SI units must be expressed in kelvins, and so we have

$$\Delta h = c_p(T_B - T_A) = 1004.15 \text{ J/kg} \cdot \text{K}[(273+20)\text{ K} - (273+50)\text{ K}]$$
$$= -30.1 \text{ kJ/kg} \qquad\qquad\qquad Ans.$$

The negative sign indicates a decrease in enthalpy.

Change in Internal Energy. In this case Eq. 13–13 applies. First we determine the specific heat at constant volume.

$$c_v = \frac{R}{k-1} = \frac{286.9 \text{ J/kg} \cdot \text{K}}{(1.4-1)} = 717.25 \text{ J/kg} \cdot \text{K}$$

$$\Delta u = c_v(T_B - T_A) = (717.25 \text{ J/kg} \cdot \text{K})[(273+20)\text{ K} - (273+50)\text{ K}]$$
$$= -21.5 \text{ kJ/kg} \qquad\qquad\qquad Ans.$$

Here the gas moving from a hot to a cooler temperature will decrease the internal energy of the gas.

Change in Entropy. Since both the temperature and pressure are known at points A and B, we will use Eq. 13–20 to find Δs. *Always remember p and T must be absolute values.*

$$s_B - s_A = c_p \ln\frac{T_B}{T_A} - R\ln\frac{p_B}{p_A}$$

$$\Delta s = (1004.15 \text{ J/kg} \cdot \text{K}) \ln\frac{(273+20)\text{ K}}{(273+50)\text{ K}} - (286.9 \text{ J/kg} \cdot \text{K}) \ln\frac{(101.3+20)\text{ kPa}}{(101.3+80)\text{ kPa}}$$
$$= 17.4 \text{ J/kg} \cdot \text{K} \qquad\qquad\qquad Ans.$$

As expected, $\Delta s > 0$.

NOTE: If needed, the velocity at A can be determined from the mass flow, $\dot{m} = \rho_A V_A A_A$, where the density of the air ρ_A is found from $p_A = \rho_A R T_A$. The results are $\rho_A = 1.956$ kg/m³ and $V_A = 13.0$ m/s.

EXAMPLE | 13.2

The closed container with the movable lid in Fig. 13–3 contains 4 kg of helium at a gage pressure of 100 kPa and a temperature of 20°C. Determine the temperature and density of the gas if the force **F** compresses the gas isentropically to a pressure of 250 kPa.

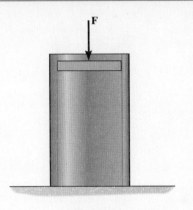

F

Fig. 13–3

SOLUTION

Fluid Description. Since we have an isentropic process, during the change no heat or friction losses occur.

Temperature. The new temperature can be determined using Eq. 13–22 since the initial and final pressures are known. From Appendix A, for helium $k = 1.66$, and so

$$\frac{p_2}{p_1} = \left(\frac{T_2}{T_1}\right)^{k/(k-1)}$$

$$\frac{(101.3 + 250)\ \text{kPa}}{(101.3 + 100)\ \text{kPa}} = \left(\frac{T_2}{(273 + 20)\ \text{K}}\right)^{\frac{1.66}{(1.66-1)}}$$

$$T_2 = 365.6\ \text{K} = 366\ \text{K} \qquad\qquad Ans.$$

Density. The initial density of helium can be determined from the ideal gas law. Since $R = 2077\ \text{J/kg} \cdot \text{K}$,

$$p_1 = \rho_1 R T_1$$

$$(101.3 + 100)(10^3)\ \text{Pa} = \rho_1(2077\ \text{J/kg} \cdot \text{K})(273 + 20)\ \text{K}$$

$$\rho_1 = 0.3308\ \text{kg/m}^3$$

Applying Eq. 13–21, to determine the final density, we have

$$\frac{p_2}{p_1} = \left(\frac{\rho_2}{\rho_1}\right)^k$$

$$\frac{(101.3 + 250)\ \text{kPa}}{(101.3 + 100)\ \text{kPa}} = \left[\frac{\rho_2}{\left(0.3307\ \text{kg/m}^3\right)}\right]^{1.66}$$

$$\rho_2 = 0.463\ \text{kg/m}^3 \qquad\qquad Ans.$$

Or, we can use the ideal gas law.

$$p = \rho R T; \qquad (101.3 + 250)\ \text{kPa} = \rho_2(2077\ \text{J/kg} \cdot \text{K})(365.6\ \text{K})$$

$$\rho_2 = 0.463\ \text{kg/m}^3 \qquad\qquad Ans.$$

This represents about a 40% change in the density.

13.2 Wave Propagation through a Compressible Fluid

If a fluid is assumed to be *incompressible*, then any *pressure disturbance* will be noticed *instantly* at all points within the fluid. All fluids, however, are *compressible*, so they will propagate a pressure disturbance through the fluid at a *finite speed*. This *speed c* is referred to as the **speed of sound**, or the **sonic velocity**.

We can determine the sonic velocity by considering the fluid contained within the long open tube shown in Fig. 13–4a. If the piston moves a slight distance to the right at velocity ΔV, a sudden increase in pressure Δp will be developed in the fluid just next to the piston. The molecular collisions within this region will propagate to adjacent fluid molecules on the right, and the momentum exchange that occurs will in turn be transmitted down the tube in the form of a very thin wave that will separate from the piston and travel along the tube with the sonic velocity c, where $c \gg \Delta V$. As noted on the differential control volume of this wave, Fig. 13–4b, as the wave travels down the tube, *behind* the wave the movement of the piston has caused the fluid density, the pressure, and the velocity of the fluid to *increase* by $\Delta \rho$, Δp, and ΔV, respectively. In front of the wave the fluid is still undisturbed, and so the density is ρ, the pressure is p, and the velocity is zero.

If the wave is seen by a fixed observer, then the flow past the observer will appear as unsteady flow since the fluid, originally at rest, will begin to change with time as the wave passes by. Instead, we will consider the observer to be fixed to the wave and moving with the same speed c, Fig. 13–4c. From this viewpoint we have *steady flow*, so that the fluid appears to enter the control volume on the right with a velocity c, and leave it on the left with a velocity $c - \Delta V$.

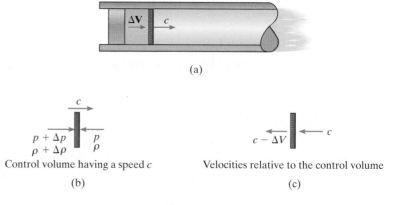

(a)

Control volume having a speed c

(b)

Velocities relative to the control volume

(c)

Fig. 13–4

Continuity Equation.

Since the cross-sectional area A on each side of the control volume remains the same, the continuity equation for the one-dimensional steady flow of the wave is

$$\frac{\partial}{\partial t} \int_{cv} \rho \, d\forall + \int_{cs} \rho \mathbf{V}_{f/cs} \cdot d\mathbf{A} = 0$$

$$0 - \rho c A + (\rho + \Delta\rho)(c - \Delta V)A = 0$$

$$-\rho c A + \rho c A - \rho A \, \Delta V + c \Delta \rho A - \Delta \rho \Delta V A = 0$$

As ΔV and $\Delta \rho$ approach zero, the last term goes to zero since it is of the second order. Therefore, this equation reduces to

$$c \, d\rho = \rho \, dV$$

Linear Momentum Equation.

As shown on the free-body diagram, Fig. 13–4d, the only forces acting on the open control surfaces are those caused by pressure. If we now apply the momentum equation to the control volume, for steady flow, we have

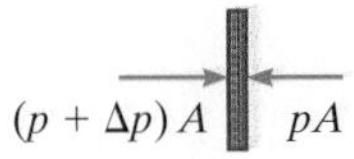

$$\overset{+}{\rightarrow} \Sigma \mathbf{F} = \frac{\partial}{\partial t} \int_{cv} \mathbf{V}_{f/cv} \rho \, d\forall + \int_{cs} \mathbf{V}_{f/cs} \rho \mathbf{V}_{f/cs} \cdot d\mathbf{A}$$

$$(p + \Delta p)A - pA = 0 + \left[-c\rho(-cA) - (c - \Delta V)(\rho + \Delta\rho)(c - \Delta V)A \right]$$

Neglecting the second- and third-order terms, in the limit,

$$dp = 2\rho c \, dV - c^2 \, d\rho$$

Using the continuity equation, and solving for c, we get

$$c = \sqrt{\frac{dp}{d\rho}} \qquad\qquad (13\text{--}23)$$

$(p + \Delta p)A \quad pA$

Free-body diagram

(d)

Fig. 13–4 (cont.)

We can express c in terms of the absolute temperature by noting that since the wave is very thin, *no heat* is transferred into or out of the control volume *during* the very short time the wave passes through the fluid. In other words, the process is *adiabatic*. Also, frictional losses within the "thin" wave can be neglected, and so the pressure and density changes involve a process that is *reversible*. Consequently, sound waves or pressure disturbances form an *isentropic process*. Therefore we can relate the pressure to the density using Eq. 13–21, which can be written in the form

$$p = C\rho^k$$

where C is a constant. Taking the derivative, the change in p and ρ gives

$$\frac{dp}{d\rho} = Ck\rho^{k-1} = Ck\left(\frac{\rho^k}{\rho}\right) = Ck\left(\frac{p/C}{\rho}\right) = k\left(\frac{p}{\rho}\right)$$

Using the ideal gas law, where $p/\rho = RT$, Eq. 13–23 becomes

$$\boxed{c = \sqrt{kRT}} \tag{13–24}$$

Sonic velocity

The velocity of sound in the gas therefore depends upon the absolute temperature of the gas. For example, in air at 15°C (288 K), $c = 340$ m/s, which is very close to its value found from experiment.

We can also express the speed of sound in terms of the bulk modulus and the density of the fluid. Recall that the bulk modulus is defined by Eq. 1–12 as

$$E_V = -\frac{dp}{dV/V}$$

For a mass $m = \rho V$, the change in mass is $dm = d\rho\, V + \rho\, dV$. Since the mass is constant, $dm = 0$, and so $-dV/V = d\rho/\rho$. Thus,

$$E_V = \frac{dp}{d\rho/\rho}$$

Finally, from Eq. 13–23, we have

$$c = \sqrt{\frac{E_V}{\rho}} \tag{13–25}$$

This result shows that the speed of sound, or the speed of the pressure disturbance, depends on the elasticity or compressibility (E_V) of the medium and on its inertial property (ρ). The more incompressible the fluid, the *faster* a pressure wave will be transmitted through it; and the larger the fluid's density, the *slower* this wave will travel. For example, the density of water is about a thousand times the density of air, but water's bulk modulus is so much greater than that of air that sound travels a little over four times faster in water than in air. At 20°C, $c_a = 343$ m/s, and $c_w = 1482$ m/s.

13.3   Types of Compressible Flow

To classify a compressible flow, we will use the *Mach number*, M. Recall from Chapter 8 that this is a dimensionless number that represents the square root of the ratio of the inertial force to the compressible force acting on the fluid. There we showed that it can be expressed as a ratio of the velocity of the fluid, V, to the sonic velocity c produced by a pressure wave within the fluid. Using Eq. 13–24, we can therefore write the Mach number as

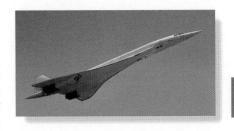

The Concorde was a supersonic commercial aircraft that was capable of flying at speeds up to M = 2.3.

$$M = \frac{V}{c} = \frac{V}{\sqrt{kRT}} \qquad (13\text{–}26)$$

or, if the Mach number is known, then

$$\boxed{V = M\sqrt{kRT}} \qquad (13\text{–}27)$$

Let us now consider a body such as an airfoil moving through a fluid with a velocity V, Fig. 13–5a. During the motion, its forward surface, like the piston in Fig. 13–4, will compress the air in front of it, causing pressure waves to form and move away from the front surface at the sonic velocity c. The effect produced depends upon the magnitude of V.

Subsonic Flow, M < 1.   As long as the body continues to move at a subsonic velocity V, the pressure waves the body creates will always move ahead of the body with a relative speed of $c - V$. In a sense, these pressure disturbances signal the fluid in front of the body that it is advancing and enable the fluid to adjust before the body arrives. Consequently, the fluid molecules begin to move apart, which produces a smooth flow over and around the body's surface, as shown in Fig. 13–5a. As a general rule, the changes in pressure generated by the movement of the body begin to become significant when M > 0.3, or $V > 0.3c$. At the speed $V = 0.3c$, the compressibility of the air creates pressure changes of about 1%; and as we have assumed in the previous chapters, speeds at or below 0.3 or 30% of c can be considered using *incompressible flow analysis*, which is accurate enough for most engineering work. Speeds within the range $0.3c < V < c$ are referred to as *subsonic compressible flow*.

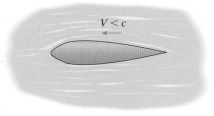

Subsonic flow
(a)

Fig. 13–5

13

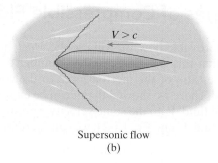

Supersonic flow
(b)

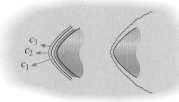

Since $c_1 > c_2 > c_3$ the waves
bunch up and form a shock

(c)

Fig. 13–5 (cont.)

Sonic and Supersonic Flow, $M \geq 1$. When the body is traveling at a speed V that is equal to or faster than the pressure waves it creates, the fluid just ahead of the body cannot sense the presence of the body advancing, and so the fluid does not move out of the way. Instead, the pressure waves bunch up and form a very thin wave in front of the body, Fig. 13–5b.

To understand how this occurs, consider a close-up view of the interaction between the fluid and the surface of the body shown in Fig. 13–5c*. As the fluid impacts the surface, because of the molecular collisions, a temperature gradient is created within the fluid, such that the highest temperature develops at the surface. Recall the sonic velocity is a function of temperature, $c = \sqrt{kRT}$, and so the pressure disturbance or wave formed closest to the surface will have the highest sonic speed as it moves away from the surface at c_1. As a result it will catch up with the wave in front of it since $c_1 > c_2$. The succession of all sonic waves coming off the surface of the body will therefore bunch up, and create an increasing pressure gradient within a localized region. Although each successive pressure disturbance or wave can be considered an isentropic process, as stated in Sec. 13.2, the *collection* of these waves will reach a point where the pressure will get so large that viscous friction and heat conduction effects begin to stabilize their formation. In other words the collective process becomes nonisentropic. In standard atmospheric air, the thickness is on the order of a few times the mean free path of the molecules, about 0.03 μm. It is called a **shock wave**, and its effect will cause an abrupt localized change in the pressure, density, and temperature as it passes through the fluid. If the body and the shock wave near it move at $M = 1$, it is **sonic flow**, and if $M > 1$, it is termed **supersonic**. The motion is further classified as **hypersonic** if a body, such as a missile or low-flying spacecraft, is moving at $M \geq 5$.

* Further details of this process are outlined in Ref. [5].

Mach Cone.

It is important to realize that the formation of a shock wave is a very localized phenomenon that occurs on or near the surface of a moving body. As the body moves from one position to the next, each shock wave formed will move away from the body at the sonic speed, c. To illustrate this, consider a jet plane that is flying horizontally at a supersonic speed V, Fig. 13–6. At each location the plane will produce a spherical shockwave that then travels through the atmosphere at c. As shown, the wave produced when $t = 0$ will travel a distance ct', when the plane travels a distance Vt' in the time $t = t'$. Portions of the waves that are produced at $t = t'/3$ and $t = 2t'/3$ are also shown in the figure. If we summed all the waves produced during the time t' it would have a conical boundary, called a **Mach cone**. The sound produced by the shock wave will be heard by someone within the cone, and outside no sound produced by the plane will be detected. The energy of the waves is mostly concentrated at the cone's surface, where the spherical waves interact, and so when this surface passes by an observer, the large pressure discontinuity created by the wave will produce a loud "crack" or sonic boom.

The half angle α of the Mach cone in Fig. 13–6 depends upon the speed of the plane. It can be determined from the red shaded triangle drawn within the cone. It is

$$\sin \alpha = \frac{c}{V} = \frac{1}{M} \qquad (13\text{–}28)$$

As the jet increases its speed V, then $\sin \alpha$, and therefore α, will become smaller.

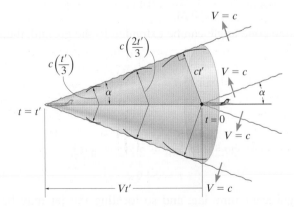

Development of a mach cone

Fig. 13–6

13

Important Points

- The *speed* at which a pressure wave travels through a medium is called the *sonic velocity*, c, because sound is a result of changes in pressure. All pressure waves undergo an isentropic process, and for a particular gas, the sonic velocity is $c = \sqrt{kRT}$.

- Compressible flow is classified by using the Mach number, $M = V/c$. Pressure waves for subsonic flow ($M < 1$) will always move ahead of the body and signal the fluid that the body is advancing, thereby allowing the fluid to adjust. If $M \leq 0.3$, we can generally consider the fluid to be incompressible.

- Pressure waves for sonic ($M = 1$) or supersonic flow ($M > 1$) cannot move ahead of the body, so they bunch up in front of the body and develop a shock wave near or on its surface. As it is formed, it produces a Mach cone having a surface that travels at $M = 1$ away from the body.

EXAMPLE | 13.3

The jet flies at $M = 2.3$ when it is at an altitude of 18 km, Fig. 13–7. Determine the time for someone on the ground to hear the sound of the plane just after the plane passes overhead. Take $c = 295$ m/s.

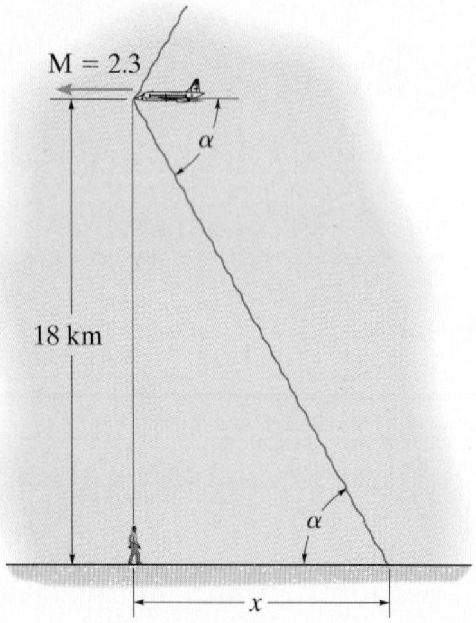

SOLUTION

Fluid Description. The plane compresses the air ahead of it since it is flying faster than $M = 1$. Although the speed of sound (or the Mach number) depends upon the air temperature, which actually varies with elevation, for simplicity, here we have assumed c is constant.

Analysis. The Mach cone, which forms on the jet, has an angle of

$$\sin \alpha = \frac{c}{V} = \frac{1 \text{ M}}{2.3 \text{ M}} = 0.4340 \qquad \alpha = 25.77°$$

Since this same cone angle can be extended to the ground, then from Fig. 13–7,

$$\tan 25.77° = \frac{18 \text{ km}}{x}$$

$$x = 37.28 \text{ km}$$

Thus

$$x = Vt; \qquad 37.28(10^3) \text{ m} = 2.3(295 \text{ m/s})t$$

$$t = 54.9 \text{ s} \qquad\qquad\qquad Ans.$$

This is a significant time lag, and so locating the jet may be a bit difficult.

M = 2.3

18 km

x

Fig. 13–7

13.4   Stagnation Properties

The high-speed gas flow through nozzles, transitions, or venturies can usually be approximated as an *isentropic process*. This is because the flow is over a *short distance*, and as a result, the temperature changes will be small over this distance. Therefore, it is reasonable to assume no heat transfer occurs, and frictional effects can be ignored.

During the flow, the state of the gas can be described by its temperature, pressure, and density. In this section we will show how to obtain these properties at any point in the gas provided we know their values at some other *reference point*. For problems involving compressible flow, we will choose the *stagnation point* in the flow as this reference point, because experimental measurements can conveniently be made from this point, as will be shown in Sec. 13.13.

Stagnation Temperature. The *stagnation temperature* T_0 represents the temperature of the gas when *its velocity is zero*. For example, the temperature of a gas at rest in a reservoir is at the stagnation temperature. For either adiabatic or isentropic flow no heat is lost, and so this temperature T_0 is sometimes referred to as the *total temperature*. It will be the *same at every point* in the flow *after* the flow is brought to rest. However, it is *different* from the *static temperature* T, which is measured by an observer moving *with the flow*.

We can relate the static temperature to the stagnation temperature of a gas by considering the fixed control volume in Fig. 13–8, where point O is in a reservoir where the stagnation temperature is T_0 and the velocity is $V_0 = 0$, and some other point, located in the pipe where the static temperature is T and the velocity is V. If we apply the energy equation, Eq. 5–15, neglecting the changes in the potential energy of the gas, and assuming the flow is adiabatic, so there is no heat transfer, we have

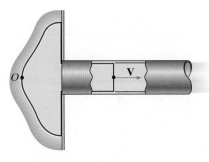

Fig. 13–8

$$\dot{Q}_{in} - \dot{W}_{turb} + \dot{W}_{pump} = \left[\left(h_{out} + \frac{V_{out}^2}{2} + gz_{out}\right) - \left(h_{in} + \frac{V_{in}^2}{2} + gz_{in}\right)\right]\dot{m}$$

$$0 - 0 + 0 = \left[\left(h + \frac{V^2}{2} + 0\right) - (h_0 + 0 + 0)\right]\dot{m}$$

$$h_0 = h + \frac{V^2}{2} \tag{13–29}$$

This result can be written in terms of temperature using Eq. 13–10, $\Delta h = c_p \Delta T$ or $h_0 - h = c_p(T_0 - T)$. Therefore,

$$c_p T_0 = c_p T + \frac{V^2}{2}$$

or

$$T_0 = T\left(1 + \frac{V^2}{2c_p T}\right) \tag{13–30}$$

13

To eliminate c_p and also express this equation in terms of the Mach number, we can use Eq. 13–14, $c_p = kR/(k-1)$, and Eq. 13–24, $c = \sqrt{kRT}$. This gives

$$T_0 = T\left(1 + \frac{k-1}{2}M^2\right) \tag{13–31}$$

To summarize, T is the *static temperature* of the gas at a point because it is measured *relative to the flow*, whereas T_0 is the temperature of the gas after it is brought to rest at this point through an adiabatic process. Notice that where $M = 0$, as expected, Eq. 13–31 indicates that the temperature T of the gas is then equal to its stagnation temperature.

Stagnation Pressure.   The pressure p in a gas at a point is referred to as the *static pressure* because it is *measured relative to the flow*. The stagnation or total pressure p_0 is the pressure of the gas after the flow has been brought *isentropically to rest* at the point. This process must be isentropic since otherwise heat transfer and frictional effects will change the total pressure. For an ideal gas, the temperature and pressure for an isentropic process (adiabatic and reversible) are related using Eq. 13–22.

$$p_0 = p\left(\frac{T_0}{T}\right)^{k/(k-1)}$$

Substituting Eq. 13–31, we get

$$p_0 = p\left(1 + \frac{k-1}{2}M^2\right)^{k/(k-1)} \tag{13–32}$$

Although the static pressure p may change, the stagnation pressure p_0 is the *same* at all points along a streamline, provided the flow is isentropic.

When $k = 1.4$, which is appropriate for air, oxygen, and nitrogen, the ratios T/T_0 and p/p_0 have been calculated from the above two equations for various values of M, and for convenience, they are tabulated in Appendix B, Table B–1.* If you take a moment to scan the data in this appendix, you will notice that the ratios T/T_0 and p/p_0 are always less than one, so that the static values of T and p will always be smaller than the corresponding stagnation values of T_0 and p_0.

Stagnation Density.   If Eq. 13–32 is substituted into Eq. 13–21, $p = C\rho^k$, we obtain a relationship between the stagnation density ρ_0 and the static density ρ of the gas. This is

$$\rho_0 = \rho\left(1 + \frac{k-1}{2}M^2\right)^{1/(k-1)} \tag{13–33}$$

As expected, like T_0 and p_0, this value ρ_0 is the *same* along any streamline, provided the flow is isentropic.

---

*The many equations involving compressible flow can also be solved using programmable pocket calculators, or their calculated values can be found on websites that provide this information.

Important Points

- Stagnation temperature T_0 is the same at each point on a streamline, provided the flow is adiabatic (no heat loss). The stagnation pressure p_0 and density ρ_0 remain the same if the flow is isentropic (adiabatic and no frictional losses). Most often, these properties are measured at a point in a reservoir, where the gas is stagnant or at rest.

- The static temperature T, pressure p, and density ρ are measured within the gas, while moving with the flow.

- The value of T is related to T_0 using the energy equation, assuming an adiabatic process. Since the flow of gas through a nozzle or transition is essentially isentropic, then p and ρ can be related to p_0 and ρ_0. For each case the corresponding values depend upon the Mach number and the ratio k of the specific heats for the gas.

EXAMPLE | 13.4

Air at a temperature of 100°C is under pressure in the large tank shown in Fig. 13–9. If the nozzle is open, the air flows out at M = 0.6. Determine the temperature of the air at the exit.

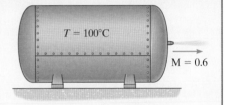

Fig. 13–9

SOLUTION

Fluid Description. The Mach number, M = 0.6 < 1, indicates that this problem involves subsonic compressible flow. We will assume the flow is steady.

Analysis. The stagnation temperature is $T_0 = (273 + 100)\,\text{K} = 373\,\text{K}$ since the air is *at rest* within the tank. Assuming the flow to be adiabatic through the nozzle, this is the same stagnation temperature throughout the flow since no heat is lost in the process. Applying Eq. 13–31 yields

$$T_0 = T\left(1 + \frac{k-1}{2}M^2\right)$$

$$373\,\text{K} = T\left(1 + \frac{1.4-1}{2}(0.6)^2\right)$$

$$T = 348\,\text{K} = 75°\text{C} \qquad\qquad Ans.$$

We can also obtain this result using the ratio T/T_0 listed in Table B–1 for M = 0.6. We have

$$T = 373\,\text{K}(0.9328) = 348\,\text{K} \qquad\qquad Ans.$$

This lower temperature, as measured relative to the flow, is actually the result of a drop in pressure that occurs as the air emerges from the tank.

EXAMPLE | 13.5

Nitrogen flows isentropically through the pipe in Fig. 13–10 such that its gage pressure is $p = 200$ kPa, the temperature is 80°C, and the velocity is 150 m/s. Determine the stagnation temperature and stagnation pressure for this gas. The atmospheric pressure is 101.3 kPa.

150 m/s

$T = 80°C$
$p = 200$ kPa

Fig. 13–10

SOLUTION

Fluid Description. The Mach number for the flow is first determined. The speed of sound for nitrogen at $T = (273 + 80)$ K $= 353$ K is

$$c = \sqrt{kRT} = \sqrt{1.40(296.8 \text{ J/kg} \cdot \text{K})(353 \text{ K})} = 383.0 \text{ m/s}$$

Thus,

$$\text{M} = \frac{V}{c} = \frac{150 \text{ m/s}}{383.0 \text{ m/s}} = 0.3917 < 1$$

Here we have steady subsonic compressible flow.

Stagnation Temperature. Applying Eq. 13–31, we have

$$T_0 = T\left(1 + \frac{k-1}{2}\text{M}^2\right) = 353 \text{ K}\left(1 + \frac{1.4-1}{2}(0.3917)^2\right)$$

$$T_0 = 363.8 \text{ K} = 364 \text{ K} \hspace{2cm} Ans.$$

Stagnation Pressure. The static pressure $p = 200$ kPa is measured relative to the flow. Applying Eq. 13–32 and reporting the result as an absolute stagnation pressure, we have

$$p_0 = p\left(1 + \frac{k-1}{2}M^2\right)^{k/(k-1)}$$

$$p_0 = (101.3 + 200)\text{ kPa}\left(1 + \frac{1.4-1}{2}(0.3917)^2\right)^{\frac{1.4}{1.4-1}}$$

$$p_0 = 334.9\text{ kPa} = 335\text{ kPa} \hspace{3cm} \textit{Ans.}$$

Since $k = 1.4$ for nitrogen, these values for T_0 and p_0 can also be determined by using Table B–1. Throughout this chapter, to improve numerical accuracy, we will use linear interpolation when using any of the tables in Appendix B. For example, the temperature ratio in Appendix B–1 is determined as follows: $M = 0.39$, $T/T_0 = 0.9705$, and $M = 0.40$, $T/T_0 = 0.9690$. Therefore, for $M = 0.3917$,

$$\frac{0.4 - 0.39}{0.4 - 0.3917} = \frac{0.9690 - 0.9705}{0.9690 - T/T_0}$$

$$0.009690 - 0.01T/T_0 = 0.00001251$$

$$T/T_0 = 0.97025$$

Thus,

$$T_0 = \frac{353\text{ K}}{0.97025} = 364\text{ K}$$

Because we have an isentropic process, realize there will be no change in the entropy. This can be shown by applying Eq. 13–20.

$$s - s_0 = c_p \ln\frac{T}{T_0} - R\ln\frac{p}{p_0}$$

$$\Delta s = \left(\frac{1.4(296.8\text{ J/kg} \cdot \text{K})}{1.4 - 1}\right)\ln\left(\frac{353\text{ K}}{363.8\text{ K}}\right) - (296.8\text{ J/kg} \cdot \text{K})\ln\left(\frac{301.3\text{ kPa}}{334.9\text{ kPa}}\right)$$

$$\Delta s = 0$$

13

EXAMPLE | 13.6

The absolute pressure within the entrance to the pipe in Fig. 13–11 is 98 kPa. Determine the mass flow into the pipe once the valve is opened. The outside air is at rest, at a temperature of 20°C and atmospheric pressure of 101.3 kPa. The pipe has a diameter of 50 mm.

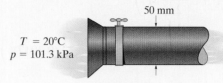

Fig. 13–11

SOLUTION

Fluid Description. We will assume steady isentropic flow through the entrance. The Mach number can be obtained using Eq. 13–32 since the pressures are known. The stagnation pressure is $p_0 = 101.3$ kPa since the outside air is at rest, and the lower (static) pressure $p = 98$ kPa is within the pipe. Since $k = 1.4$ for air, we have

$$p_0 = p\left(1 + \frac{k-1}{2}M^2\right)^{k/(k-1)}$$

$$101.3 \text{ kPa} = 98 \text{ kPa}\left(1 + \frac{1.4-1}{2}M^2\right)^{\frac{1.4}{1.4-1}}$$

$$M = 0.218 < 1 \text{ subsonic flow}$$

Although this value is less than what we classify as compressible flow (M > 0.3), all flow, regardless of its speed, is *actually* compressible flow. Also, this result can be obtained by interpolating in Table B–1, using $p/p_0 = 98 \text{ kPa}/101.3 \text{ kPa} = 0.967$.

Analysis. The mass flow is determined using $\dot{m} = \rho A V$, and so we must first find the density and velocity of the gas.

The stagnation density of the air at $T = 20°C$ is determined from Appendix A. It is $\rho_0 = 1.202$ kg/m³. Therefore, using Eq. 13–33, the density of the air in the pipe is

$$\rho_0 = \rho\left(1 + \frac{k-1}{2}M^2\right)^{1/(k-1)}$$

$$1.202 \text{ kg/m}^3 = \rho\left(1 + \frac{1.4-1}{2}(0.218)^2\right)^{\frac{1}{1.4-1}}$$

$$\rho = 1.1739 \text{ kg/m}^3$$

Note that we can also obtain this value from Eq. 13–21, $p/p_0 = (\rho/\rho_0)^k$.

The velocity of the flow into the entrance is determined using Eq. 13–27, $V = M\sqrt{kRT}$, which depends upon the temperature within the flow. The temperature can be found from Table B–1 or using Eq. 13–31 for $M = 0.218$.

$$T_0 = T\left(1 + \frac{k-1}{2}M^2\right)$$

$$(273 + 20)\ \text{K} = T\left(1 + \frac{1.4-1}{2}(0.218)^2\right)$$

$$T = 290.24\ \text{K}$$

Thus,

$$V = M\sqrt{kRT} = 0.218\sqrt{1.4(286.9\ \text{J/kg}\cdot\text{K})(290.24\ \text{K})} = 74.44\ \text{m/s}$$

The mass flow is therefore

$$\dot{m} = \rho VA = 1.1739\ \text{kg/m}^3\,(74.44\ \text{m/s})\left[\pi(0.025\ \text{m})^2\right]$$

$$= 0.172\ \text{kg/s} \qquad\qquad\qquad\qquad\qquad\qquad Ans.$$

NOTE: If this problem is solved assuming the air is an *ideal fluid* (incompressible and frictionless), then with steady flow, the Bernoulli equation can be used to determine the velocity. In this case,

$$\frac{p_0}{\rho} + \frac{V_0^2}{2} = \frac{p_1}{\rho} + \frac{V_1^2}{2}$$

$$\frac{101.3\left(10^3\right)\ \text{N/m}^2}{1.202\ \text{kg/m}^3} + 0 = \frac{98\left(10^3\right)\ \text{N/m}^2}{1.202\ \text{kg/m}^3} + \frac{V_1^2}{2}$$

$$V_1 = 74.10\ \text{m/s}$$

This value is in error by about 0.46% from the value $V = 74.44$ m/s, found by accounting for the compressibility of the air.

13.5   Isentropic Flow through a Variable Area

Compressible flow analysis is often applied to gases that pass through ducts in jet engines and rocket nozzles. For these applications we will generalize the discussion, and show how the pressure, velocity, and density of the gas are affected by varying the cross-sectional area of the duct through which the gas flows, Fig. 13–12a. For short distances, we will require the flow to be steady and the process to be isentropic. Also, the cross-sectional area of the duct is assumed to change gradually, so that the flow can be considered one-dimensional and average gas properties can be used. The fixed control volume shown in Fig. 13–12a contains a portion of the gas in the duct.

Continuity Equation.   Since the velocity, density, and cross-sectional area all change, the continuity equation gives

$$\frac{\partial}{\partial t}\int_{cv} \rho \, d\forall + \int_{cs} \rho \mathbf{V} \cdot d\mathbf{A} = 0$$

$$0 - \rho VA + (\rho + \Delta\rho)(V + \Delta V)(A + \Delta A) = 0$$

After multiplying and taking the limit as $\Delta x \to 0$, the second- and third-order terms will drop out. Simplifying, we get

$$\rho V \, dA + VA \, d\rho + \rho A \, dV = 0$$

Solving for the velocity change yields

$$dV = -V\left(\frac{d\rho}{\rho} + \frac{dA}{A}\right) \qquad (13\text{–}34)$$

Linear Momentum Equation.   As shown on the free-body diagram of the control volume, Fig. 13–12b, the surrounding gas exerts pressure on the front and back open control surfaces. Since the sides of the duct increase the cross-sectional area by ΔA, the *average pressure*, $p + \Delta p/2$, will act horizontally on this increased area. Applying the linear momentum equation in the direction of flow, we have

$$\xrightarrow{+}\ \Sigma \mathbf{F} = \frac{\partial}{\partial t}\int_{cv} \mathbf{V}\rho \, d\forall + \int_{cs} \mathbf{V}\rho\mathbf{V} \cdot d\mathbf{A}$$

$$pA + \left(p + \frac{\Delta p}{2}\right)\Delta A - (p + \Delta p)(A + \Delta A) =$$

$$0 + V\rho(-VA) + (V + \Delta V)(\rho + \Delta\rho)(V + \Delta V)(A + \Delta A)$$

Expanding, and again realizing that the higher-order terms drop out in the limit, we get

$$dp = -\left(2\rho V \, dV + V^2 \, d\rho + \rho V^2 \frac{dA}{A}\right) \qquad (13\text{–}35)$$

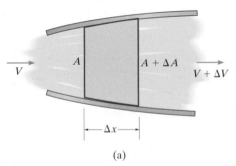

V A $A + \Delta A$ $V + \Delta V$

Δx

(a)

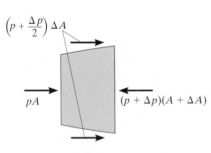

$\left(p + \dfrac{\Delta p}{2}\right)\Delta A$

pA $(p + \Delta p)(A + \Delta A)$

Free-body diagram

(b)

Fig. 13–12

13

Substituting the value for dV from Eq. 13–34 yields

$$dp = \rho V^2 \left(\frac{d\rho}{\rho} + \frac{dA}{A} \right)$$

Eliminating $d\rho$ by substituting Eq. 13–23, $d\rho = dp/c^2$, and then expressing the result in terms of the Mach number $M = V/c$, we find that the change in pressure related to the change in area becomes

$$dp = \frac{\rho V^2}{1 - M^2} \frac{dA}{A} \qquad (13\text{–}36)$$

The change in the velocity related to the change in the area can be determined by equating Eq. 13–36 to Eq. 13–35, and then eliminating the term dp/ρ using Eq. 13–34. This gives

$$dV = -\frac{V}{1 - M^2} \frac{dA}{A} \qquad (13\text{–}37)$$

Finally, the change in the density related to the change in the area is determined by equating Eq. 13–37 and Eq. 13–34. The result is

$$d\rho = \frac{\rho M^2}{1 - M^2} \frac{dA}{A} \qquad (13\text{–}38)$$

Subsonic Flow. When the flow is *subsonic*, $M < 1$, then the term $(1 - M^2)$ in the above three equations is *positive*. As a result, an *increase* in area, dA, or a divergent duct, will cause the pressure and density to *increase* and the velocity to *decrease*, Fig. 13–13a. Likewise, a decrease in area, or a convergent duct, will cause the pressure and density to decrease and the velocity to increase, Fig. 13–13b. These results for pressure and velocity are similar to those for incompressible flow, as noted from the Bernoulli equation. For example, if the pressure increases, then the velocity will decrease, and vice versa.

Supersonic Flow. When the flow is supersonic, $M > 1$, then the term $(1 - M^2)$ will be *negative*. Now the *opposite effect* occurs; that is, an *increase* in duct area, Fig. 13–13c, will cause the pressure and density to *decrease* and the velocity to *increase*, whereas a *decrease* in area will cause the pressure and density to *increase* and the velocity to *decrease*, Fig. 13–13d. This seems contrary to what we might expect, but experimental results have shown that indeed it occurs. In a sense, supersonic flow behaves in a manner similar to traffic flow. When cars come to a widening of a roadway, their speed increases (higher velocity) and so they begin to spread out (lower pressure and density). A narrowing of the roadway causes congestion (higher pressure and density) and a lowering in speed (lower velocity).

M < 1 (subsonic flow)
Divergent duct + dA
Pressure and density *increase*.
Velocity *decreases*.
(a)

M < 1 (subsonic flow)
Convergent duct − dA
Pressure and density *decrease*.
Velocity *increases*.
(b)

M > 1 (supersonic flow)
Divergent duct + dA
Pressure and density *decrease*.
Velocity *increases*.
(c)

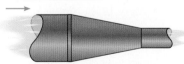

M > 1 (supersonic flow)
Convergent duct − dA
Pressure and density *increase*.
Velocity *decreases*.
(d)

Fig. 13–13

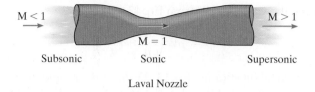

M < 1   M > 1

M = 1

Subsonic   Sonic   Supersonic

Laval Nozzle

Fig. 13–14

Laval Nozzle. From these comparisons, we see that to produce supersonic flow ($M > 1$), a nozzle can be fashioned as in Fig. 13–14 so that it has an initial convergent section to increase the subsonic velocity until it becomes sonic at the throat, $M = 1$, and then add a divergent section to *further increase* the now sonic velocity to supersonic speed, $M > 1$. This type of nozzle is termed a ***Laval nozzle***, named after the Swedish engineer Carl de Laval, who designed it in 1893 for use in a steam turbine. It is important to realize that the flow can *never be greater than sonic speed* ($M = 1$) *through the throat*, because once sonic speed is attained, a pressure wave at this speed *can no longer move upstream* (against the flow) to send a signal of any changes downstream that may cause a further increase of the flow into the nozzle.

Area Ratios. We can determine the cross-sectional area at any point along a nozzle by expressing it in terms of the Mach number using the continuity equation. If sonic conditions occur at the throat, then the cross section at the throat A^* can be used as a *reference*, where $T = T^*, \rho = \rho^*$, and $M = 1$. At any other point, since $V = Mc = M\sqrt{kRT}$, continuity of mass flow requires

$$\dot{m} = \rho V A = \rho^* V^* A^*; \quad \rho\left(M\sqrt{kRT}\right)A = \rho^*\left(1\sqrt{kRT^*}\right)A^*$$

or

$$\frac{A}{A^*} = \frac{1}{M}\left(\frac{\rho^*}{\rho}\right)\sqrt{\frac{T^*}{T}} \tag{13–39}$$

This result can also be expressed in terms of the stagnation density and temperature by introducing the appropriate ratios.

$$\frac{A}{A^*} = \frac{1}{M}\left(\frac{\rho^*}{\rho_0}\right)\left(\frac{\rho_0}{\rho}\right)\sqrt{\frac{T^*}{T_0}}\sqrt{\frac{T_0}{T}} \tag{13–40}$$

Substituting Eqs. 13–31 and 13–33 for each ratio, and realizing that $M = 1$ for the ratios ρ^*/ρ_0 and T^*/T_0, we get, after simplifying,

$$\frac{A}{A^*} = \frac{1}{M}\left[\frac{1 + \frac{1}{2}(k - 1)M^2}{\frac{1}{2}(k + 1)}\right]^{\frac{k + 1}{2(k - 1)}} \tag{13–41}$$

A graph of this equation for a given value of k is shown in Fig. 13–15. Except when $A = A^*$, there is a *double value* for the Mach number for each value of A/A^*. One value, M_1, is for the area A' in the region where subsonic flow exists, and the other, M_2, is for A' in the region of supersonic flow. Rather than solving Eq. 13–41 for M_1 and M_2, for convenience, if $k = 1.4$, we can use Table B–1. It is worth noting how the values within this table follow the shape of the curve in Fig. 13–15. Scan down the table and notice that as M increases, A/A^* decreases, until $M = 1$, and then A/A^* increases again.

With Eq. 13–41, or Table B–1, we can now determine the required cross-sectional area A_2 of a nozzle at a point where the flow must be M_2, provided we know M_1 and the cross-sectional area A_1 at some other point. To show how to use the table in this manner, take the case shown in Fig. 13–16, where $M_1 = 0.5$, $A_1 = \pi(0.03\,\text{m})^2$, and $M_2 = 1.5$. To determine A_2 we must *reference* A_1 and A_2 to the throat area A^*, since the ratio A/A^* is used in the table. Using $M_1 = 0.5$, the ratio $A_1/A^* = 1.3398$ is determined from Table B–1 (or Eq. 13–41). Likewise, using $M_2 = 1.5$, the ratio $A_2/A^* = 1.176$ is also found from Table B–1. With these two area ratios, we can then write

$$\frac{A_1}{A_2} = \frac{A_1/A^*}{A_2/A^*} = \frac{1.3398}{1.176}$$

so

$$A_2 = \frac{1}{4}\pi d_2^2 = \pi(0.03\,\text{m})^2\left(\frac{1.176}{1.3398}\right)$$

$$d_2 = 56.2\,\text{mm}$$

This method will also provide a valid solution *even if flow through the throat is not at* $M = 1$. In this case, we can imagine that the nozzle has a location where the throat narrows to A^*. This is a *reference* where $M = 1$, and in the above area ratios A^* cancels and is never actually calculated. In other words, we have considered A^* only as a *reference area*, where the area ratios in the above fraction actually cause A^* to *cancel out*.

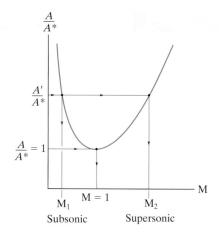

Flow through a Laval nozzle area ratio vs. Mach number

Fig. 13–15

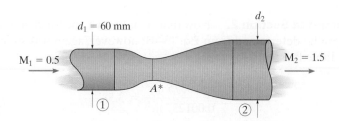

Fig. 13–16

EXAMPLE | 13.7

Air is drawn through the 50-mm-diameter pipe in Fig. 13–17 and passes section 1 with a speed of $V_1 = 150\,\text{m/s}$, while it has an absolute pressure of $p_1 = 400\,\text{kPa}$ and absolute temperature of $T_1 = 350\,\text{K}$. Determine the required area of the throat of the nozzle to produce sonic flow at the throat. Also, if supersonic flow is to occur at section 2, find the velocity, temperature, and required pressure at this location.

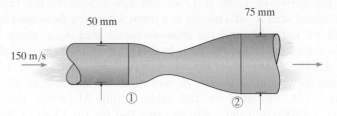

Fig. 13–17

SOLUTION

Fluid Description.   We assume isentropic steady flow through the nozzle.

Throat Area.   The Mach number at section 1 is first calculated. For air $k = 1.4$, $R = 286.9\,\text{J/kg}\cdot\text{K}$. Therefore,

$$M_1 = \frac{V_1}{\sqrt{kRT_1}} = \frac{150\,\text{m/s}}{\sqrt{1.4(286.9\,\text{J/kg}\cdot\text{K})(350\,\text{K})}} = 0.40$$

Although we can use Eq. 13–41 with M_1 and A_1 to determine the throat area A^*, since $k = 1.4$ it is simpler to use Table B–1. Thus, for $M_1 = 0.40$, we find

$$\frac{A_1}{A^*} = 1.5901$$

$$A^* = \frac{\pi(0.025\,\text{m})^2}{1.5901}$$

$$= 0.001235\,\text{m}^2 \qquad\qquad Ans.$$

Properties at Section 2.   Now that A^* is known, the Mach number at A_2 can be determined from Eq. 13–41; however, again it is simpler to use Table B–1, with the ratio

$$\frac{A_2}{A^*} = \frac{\pi(0.0375\,\text{m})^2}{0.001235\,\text{m}^2} = 3.58$$

We get $M_2 = 2.8230$ (approximately), because supersonic flow must occur at the end of the divergent section. (The other root, $M_1 = 0.164$ (approximately) refers to subsonic flow at the exit.)

The temperature and pressure at the exit can be determined using $M = 2.8230$ and Eqs. 13–31 and 13–32; however, first we must know the stagnation values T_0 and p_0. To find them we can again use Eqs. 13–31 and 13–32 with $M_1 = 0.40$ and T_1 and p_1. A simpler method is to use Table B–1 by referencing the ratios T_2/T_1 and p_2/p_1 in terms of the stagnation ratios as follows:

$$\frac{T_2}{T_1} = \frac{T_2/T_0}{T_1/T_0} = \frac{0.38552}{0.9690} = 0.39785$$

$$\frac{p_2}{p_1} = \frac{p_2/p_0}{p_1/p_0} = \frac{0.035578}{0.8956} = 0.039725$$

Therefore, without having to find T_0 and p_0, we have

$$T_2 = 0.39785\, T_1 = 0.39785(350 \text{ K}) = 139.25 \text{ K} \qquad \textit{Ans.}$$

$$p_2 = 0.039725\, p_1 = 0.039725(400 \text{ kPa}) = 15.9 \text{ kPa} \qquad \textit{Ans.}$$

This lower pressure at the exit is what draws the air through the 50-mm-diameter pipe at 150 m/s.

The average velocity of the air at section 2 is

$$V_2 = M_2 \sqrt{kRT_2} = 2.8230\sqrt{1.4(286.9 \text{ J/kg} \cdot \text{K})(139.25 \text{ K})}$$

$$= 668 \text{ m/s} \qquad \textit{Ans.}$$

13.6   Isentropic Flow through Converging and Diverging Nozzles

In this section we will study the compressible flow through a nozzle that is attached to a large vessel or reservoir of stagnant gas, Fig. 13–18a. Here the pipe on the end of the nozzle is connected to a tank and vacuum pump. By operating the pump and opening the valve on the pipe, we can regulate the *backpressure* p_b in the tank and thus the flow through the nozzle. We will be interested in how this backpressure affects the pressure along the nozzle and the mass flow through the nozzle.

13

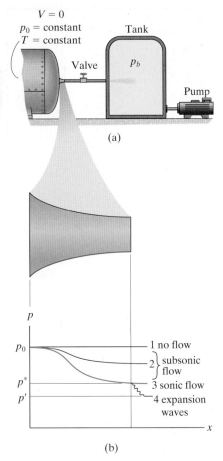

(a)

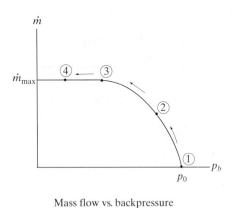

Mass flow vs. backpressure

(c)

Fig. 13–18

Converging Nozzle.   Here we will consider attaching a converging nozzle to the reservoir, Fig. 13–18a.

- When the backpressure is equal to the pressure within the reservoir, $p_b = p_0$, no flow occurs through the nozzle, as indicated by curve 1, Fig. 13–18b.

- If p_b is slightly *lower* than p_0, then flow through the nozzle will remain subsonic. Here the velocity will increase through the nozzle, causing the pressure to decrease, as shown by curve 2.

- Further drops in pressure p_b will cause the flow at the nozzle exit to eventually reach sonic flow, M = 1. The backpressure at this point is called the **critical pressure** p^*, curve 3. The value of this pressure can be determined from Eq. 13–32 with M = 1. Since p_0 is the stagnation pressure, the result is

$$\frac{p^*}{p_0} = \left(\frac{2}{k+1} \right)^{k/(k-1)} \tag{13-42}$$

For example, for air $k = 1.4$, and so $p^*/p_0 = 0.5283$ (also see Table B–1). In other words, the pressure just outside the nozzle must be approximately one-half that within the reservoir to achieve sonic flow. For backpressures from $p_b = p_0$ down to $p_b = p^*$ the flow can be considered isentropic without appreciable error. This is because the absolute pressure cannot be negative within the nozzle, and the flow is rapid, resulting in boundary layers that are thin so as to produce minimal frictional losses as the flow is accelerated through the nozzle.

- If the backpressure is lowered further, say to $p' < p^*$, then neither the pressure distribution through the nozzle nor the mass flow out of it will be affected. The nozzle is said to be **choked** since the pressure at the exit of the nozzle, "the throat," must remain at p^*. Remember that at sonic velocity, a pressure *less* than p^* *cannot* be transmitted back into the *upstream* flow to draw more gas through the nozzle. Outside and just beyond the nozzle exit, the pressure will suddenly decrease to this lower backpressure p'; however, this occurs only through the formation of three-dimensional expansion waves, curve 4. Within this region the isentropic process ceases, since the expansion of the gas causes an increase in entropy due to friction and heat loss.

The mass flow as a function of the backpressure for each of these four cases is shown in Fig. 13–18c.

Converging–Diverging Nozzle.

Let's now consider the same test using a converging–diverging, or Laval, nozzle, Fig. 13–19a.

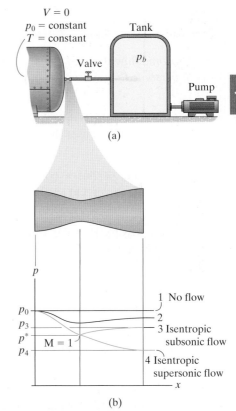

(a)

- As before, if the backpressure is equal to the pressure in the chamber, $p_b = p_0$, then no flow occurs through the nozzle because the pressure through the nozzle is constant, curve 1, Fig. 13–19b.

- If the backpressure drops somewhat, then subsonic flow occurs. Through the convergent section the velocity at the throat increases to its maximum, while the pressure decreases to a minimum. Through the divergent section the velocity decreases, while the pressure increases, curve 2.

- When the backpressure becomes p_3, the pressure at the throat drops to p^* so that the flow finally reaches sonic speed, $M = 1$, at the throat. This is a limiting case where subsonic flow continues to occur in *both* the convergent and divergent sections, curve 3. Any further *slight* decrease in backpressure *will not* cause an increase in *mass flow* through the nozzle since the velocity through the throat is at its maximum ($M = 1$). Hence, the nozzle becomes *choked*, and the mass flow remains *constant*.

- To further accelerate the flow *isentropically* within the divergent section, it is necessary to decrease the backpressure *all the way* until it reaches p_4, as shown by curve 4. Again, this will not affect the mass flow since the nozzle is choked.

- Because of the branching of the pressure curves 3 and 4, *isentropic flow* through the nozzle occurs for $p \geq p_3$ and $p = p_4$. This is because for a given ratio of A/A^* (exit area versus throat area), Eq. 13–41 gives only *two* possible Mach numbers at the exit, as noted in Fig. 13–15 (M_1 subsonic and M_2 supersonic). Thus, if the backpressure is somewhere in between these isentropic exit pressures of p_3 and p_4, or is lower than p_4, then the exit pressure will suddenly convert to this pressure only through a *shock wave*, formed within the nozzle or just outside of it. This is *nonisentropic* since a shock involves frictional losses and will result in an inefficient use of the nozzle. We will discuss this phenomenon further in Sec. 13.10.

Fig. 13–19

Military jet aircraft have nozzles that can be flared or closed so that they alter the efficiency of their thrust.

Important Points

13

- *Subsonic flow* through a *converging duct* will cause the velocity to increase and the pressure to decrease. *Supersonic flow* causes the opposite effect; the velocity will decrease and the pressure will increase.

- *Subsonic flow* through a *diverging duct* will cause the velocity to decrease and the pressure to increase. *Supersonic flow* causes the opposite effect; the velocity will increase and the pressure will decrease.

- A Laval nozzle has a convergent section to accelerate subsonic flow to sonic speed at the throat, M = 1, and a divergent section to further accelerate the flow to supersonic speed.

- It is not possible to make a gas flow faster than sonic speed, M = 1, through the throat of any nozzle, since at this speed the pressure at the throat cannot be transmitted back upstream to signal an increase in flow.

- The Mach number at a cross-sectional area A of a nozzle is a function of the throat area, where M = 1.

- A nozzle becomes choked provided M = 1 at the throat. When this occurs, the pressure at the throat is called the critical pressure p*. This condition provides maximum mass flow through the nozzle.

- For a Laval nozzle, when M = 1 at the throat there are two possible backpressures that produce *isentropic flow* within the nozzle. One produces subsonic speeds within the divergent section, M < 1, and the other produces supersonic speeds within this section, M > 1. No shock wave is produced in either case.

EXAMPLE | 13.8

50 mm

⓪

①

Fig. 13–20

Determine the required pressure at the entrance of the 50-mm-diameter pipe at section 1 in Fig. 13–20, in order to produce the greatest flow of air through the pipe. Outside the pipe, the air is at standard atmospheric pressure and temperature. What is the mass flow?

SOLUTION

Fluid Description. We assume steady isentropic flow through the nozzle.

Analysis. The stagnation pressure, temperature, and density are equal to the "standard atmospheric values" since the outside air is at rest. From Appendix A, we have $p_0 = 101.3\,\text{kPa}$, $T_0 = 15°\text{C}$, and $\rho_0 = 1.23\,\text{kg/m}^3$. The *maximum flow* into the pipe occurs when $M = 1$ at the pipe entrance. In other words, once this occurs, the air flowing through the entrance cannot transmit a reduced pressure to the air behind it any faster than $M = 1$. Using Table B–1 or Eq. 13–32 to obtain this required pressure, we have

$$p_0 = p_1\left(1 + \frac{k-1}{2}M_1^2\right)^{k/(k-1)} \qquad 101.3\,\text{kPa} = p_1\left[1 + \left(\frac{1.4-1}{2}\right)(1)^2\right]^{1.4/(1.4-1)}$$

$$p_1 = 53.5\,\text{kPa} \qquad\qquad Ans.$$

   To obtain the mass flow $\dot{m} = \rho V A$, we must first determine the density of the air and the velocity of the flow to produce $M = 1$. The density of the air within the pipe at location 1 can be determined from Eq. 13–33 or Eq. 13–21, $p_2/p_1 = (\rho_2/\rho_1)^k$. Using Eq. 13–33.

$$\rho_0 = \rho_1\left(1 + \frac{k-1}{2}M_1^2\right)^{1/(k-1)}$$

$$1.23\,\text{kg/m}^3 = \rho_1\left[1 + \left(\frac{1.4-1}{2}\right)(1)^2\right]^{1/(1.4-1)}$$

$$\rho_1 = 0.7797\,\text{kg/m}^3$$

The velocity is a function of the air temperature in the pipe, that is, $V = M\sqrt{kRT}$. We can obtain this temperature using Table B–1, Eq. 13–31 with $M = 1$, or Eq. 13–22, $p_2/p_1 = (T_2/T_1)^{k(k-1)}$. Using Eq. 13–31, we have

$$T_0 = T_1\left(1 + \frac{k-1}{2}M_1^2\right) \qquad (273 + 15°\text{C})\,\text{K} = T_1\left[1 + \left(\frac{1.4-1}{2}\right)(1)^2\right]$$

$$T_1 = 240\,\text{K}$$

Thus,

$$V_1 = M_1\sqrt{kRT_1} = (1)\sqrt{1.4(286.9\,\text{J/kg·K})(240\,\text{K})} = 310.48\,\text{m/s}$$

The mass flow is therefore

$$\dot{m} = \rho_1 V_1 A_1$$
$$= (0.7797\,\text{kg/m}^3)(310.48\,\text{m/s})[\pi(0.025\,\text{m})^2]$$
$$= 0.475\,\text{kg/s} \qquad\qquad Ans.$$

EXAMPLE 13.9

300 mm

Fig. 13–21

The converging nozzle on the tank in Fig. 13–21 has a 300-mm exit diameter. If nitrogen within the tank has an absolute pressure of 500 kPa and an absolute temperature of 1200 K, determine the mass flow from the nozzle if the absolute pressure in the pipe at the nozzle is 300 kPa.

SOLUTION

Fluid Description. We assume steady isentropic flow through the nozzle.

Analysis. Since the nitrogen within the tank is at rest, the stagnation pressure and temperature are $p_0 = 500$ kPa and $T_0 = 1200$ K. For the *greatest* mass flow through the nozzle, it is necessary that M = 1 at the exit, and so Eq. 13–32 or Table B–1 requires

$$\frac{p^*}{p_0} = 0.5283 \quad \text{or} \quad p^* = (500\,\text{kPa})(0.5283) = 264.15\,\text{kPa}$$

For this case, however, $p = 300$ kPa, which is *greater* than 264.15 kPa. Therefore, the nozzle is *not choked* at its exit.

Since we know both p and p_0, then $p/p_0 = 300\,\text{kPa}/500\,\text{kPa} = 0.6$, and so we can determine M from Eq. 13–32 or Table B–1. We get M = 0.8864.

To obtain the mass flow, $\dot{m} = \rho V A$, we must obtain the density and velocity at the exit. First the temperature is determined from Eq. 13–31 or from Table B–1 for M = 0.8864 or $p/p_0 = 0.6$.

$$\frac{T}{T_0} = 0.8642$$

$$T = 0.8642(1200\,\text{K}) = 1037\,\text{K}$$

Therefore, the exit velocity of the nitrogen is

$$V = M\sqrt{kRT} = (0.8864)\sqrt{1.4\,(296.8\,\text{J/kg}\cdot\text{K})(1037\,\text{K})} = 581.9 \text{ m/s}$$

The density can be found using the ideal gas law. The mass flow from the nozzle is therefore

$$\dot{m} = \rho V A = \left(\frac{p}{RT}\right)VA = \left(\frac{300(10^3)\,\text{N/m}^2}{(296.8\,\text{J/kg}\cdot\text{K})(1037\,\text{K})}\right)(581.9 \text{ m/s})\left[\pi(0.15\,\text{m})^2\right]$$

$$\dot{m} = 40.1 \text{ kg/s} \qquad\qquad Ans.$$

EXAMPLE   13.10

The Laval nozzle in Fig. 13–22 is connected to a large chamber containing air at an absolute pressure of 350 kPa. Determine the backpressure in the pipe at B that will cause the nozzle to choke and yet produce isentropic *subsonic* flow through the pipe. Also, what backpressure is needed to cause isentropic *supersonic* flow?

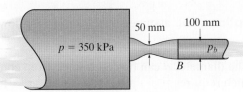

Fig. 13–22

SOLUTION

Fluid Description.   We assume steady isentropic flow through the nozzle.

Analysis.   Here we must find the two backpressures, p_3 and p_4, in Fig. 13–19b, required to produce $M = 1$ at the throat. The area ratio for the nozzle between the exit and the throat is

$$\frac{A_B}{A^*} = \frac{\pi(0.05 \text{ m})^2}{\pi(0.025 \text{ m})^2} = 4$$

If this ratio is used in Eq. 13–41, two roots for M at the exit can be determined, Fig. 13–15. However, we can also solve this problem using Table B–1. With $A_B/A^* = 4$, we get $M_1 = 0.1467$ (subsonic flow) and $p_B/p_0 = 0.9851$. Thus, the *higher* backpressure at B that will cause *subsonic flow* is

$$(p_B)_{\text{max}} = 0.9851(350 \text{ kPa}) = 345 \text{ kPa} \qquad Ans.$$

Further in the table, the alternative solution, where $A_B/A^* = 4$, gives $M_2 = 2.940$ (supersonic flow) and $p_B/p_0 = 0.02980$. Thus, the *lower* backpressure at B is for *supersonic flow*.

$$(p_B)_{\text{min}} = 0.02980(350 \text{ kPa}) = 10.4 \text{ kPa} \qquad Ans.$$

EXAMPLE | 13.11

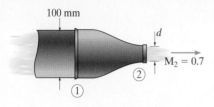

100 mm

d

$M_2 = 0.7$

①

②

Fig. 13–23

Air flows through the 100-mm-diameter pipe in Fig. 13–23 having an absolute pressure of $p_1 = 90\,\text{kPa}$. Determine the diameter d at the end of the nozzle so that isentropic flow occurs out of the nozzle at $M_2 = 0.7$. The air within the pipe is taken from a large reservoir at standard atmospheric pressure and temperature.

SOLUTION I

Fluid Description.   We assume steady isentropic flow through the nozzle.

Analysis.   The diameter d can be determined from continuity of the mass flow, which requires

$$\dot{m} = \rho_1 V_1 A_1 = \rho_2 V_2 A_2 \tag{1}$$

We must first find M_1 and then find the densities and velocities at 1 and 2.
   The stagnation values for atmospheric air are determined from Appendix A as $p_0 = 101.3$ kPa, $T_0 = 15°\text{C}$, and $\rho_0 = 1.23$ kg/m³. Knowing p_1 and p_0, we can now determine M_1 at the entrance 1 of the nozzle using Eq. 13–32.

$$p_0 = p_1\left(1 + \frac{k-1}{2}M_1{}^2\right)^{k/(k-1)}$$

$$101.3\ \text{kPa} = (90\ \text{kPa})\left(1 + \frac{1.4-1}{2}M_1{}^2\right)^{\frac{1.4}{1.4-1}}$$

$$M_1 = 0.4146$$

As expected, $M_1 < M_2 = 0.7$.
   Since $V = M\sqrt{kRT}$, applying Eq. 13–31 to find the temperatures at the entrance and exit, we have

$$T_0 = T_1\left(1 + \frac{k-1}{2}M_1{}^2\right)$$

$$(273 + 15)\ \text{K} = T_1\left(1 + \frac{1.4-1}{2}(0.4146)^2\right)$$

$$T_1 = 278.4\ \text{K}$$

$$T_0 = T_2\left(1 + \frac{k-1}{2}M_2{}^2\right)$$

$$(273 + 15)\ \text{K} = T_2\left(1 + \frac{1.4-1}{2}(0.7)^2\right)$$

$$T_2 = 262.3\ \text{K}$$

Thus, the velocities at the entrance and exit are

$$V_1 = M_1\sqrt{kRT_1} = 0.4146\sqrt{1.4\,(286.9 \text{ J/kg}\cdot\text{K})(278.4 \text{ K})} = 138.6 \text{ m/s}$$

$$V_2 = M_2\sqrt{kRT_2} = 0.7\sqrt{1.4\,(286.9 \text{ J/kg}\cdot\text{K})(262.3 \text{ K})} = 227.2 \text{ m/s}$$

The density of the air at the entrance and exit of the nozzle is determined using Eq. 13–33.

$$\rho_0 = \rho_1\left(1 + \frac{k-1}{2}M_1^2\right)^{1/(k-1)} \qquad 1.23 \text{ kg/m}^3 = \rho_1\left[1 + \frac{1.4-1}{2}(0.4146)^2\right]^{\frac{1}{1.4-1}}$$

$$\rho_1 = 1.130 \text{ kg/m}^3$$

$$\rho_0 = \rho_2\left(1 + \frac{k-1}{2}M_2^2\right)^{1/(k-1)} \qquad 1.23 \text{ kg/m}^3 = \rho_2\left[1 + \frac{1.4-1}{2}(0.7)^2\right]^{\frac{1}{1.4-1}}$$

$$\rho_2 = 0.9736 \text{ kg/m}^3$$

Finally, applying Eq. 1,

$$\rho_1 V_1 A_1 = \rho_2 V_2 A_2$$

$$\left(1.130 \text{ kg/m}^3\right)(138.6 \text{ m/s})\left[\pi(0.05 \text{ m})^2\right] = \left(0.9736 \text{ kg/m}^3\right)(227.2 \text{ m/s})\pi\left(\frac{d}{2}\right)^2$$

$$d = 84.1 \text{ mm} \qquad\qquad\qquad Ans.$$

SOLUTION II

We can also solve this problem in a direct manner using Table B–1, even though subsonic flow occurs at the end of the nozzle. To do so, we will make reference to a phantom *extension* of the nozzle, where $M = 1$ and $A = A^*$, and then relate the area ratios A_1 for $M_1 = 0.4146$ and A_2 for $M_2 = 0.7$ to this reference. Using Table B–1, we therefore have

$$\frac{A_2}{A_1} = \frac{A_2/A^*}{A_1/A^*}$$

Thus,

$$A_2 = A_1\left(\frac{A_2/A^*}{A_1/A^*}\right)$$

$$\left(\pi\frac{d^2}{4}\right) = \pi(0.05 \text{ m})^2\left(\frac{1.0944}{1.5450}\right)$$

$$d = 84.2 \text{ mm} \qquad\qquad\qquad Ans.$$

High volumetric gas flows within industrial pipes can be studied using compressible flow analysis. (© Kodda/Shutterstock)

13.7 The Effect of Friction on Compressible Flow

In most real situations, the conduit or duct through which a gas flows will have a rough surface, and so frictional effects will cause heating of the gas and thereby alter the characteristics of the flow. This typically occurs in exhaust and compressed-air pipes. In this section we will consider how the flow will change if the conduit is a pipe having a constant cross section, and a wall *friction factor f*, as determined from the Moody diagram.* We will assume the gas is ideal and has a constant specific heat, and the flow is steady. Also, the heat that is generated in the gas is assumed not to escape through the walls of the duct, and so the *process is adiabatic.* This type of flow is sometimes called **Fanno flow**, named after Gino Fanno, who was the first to investigate it.

To study how the flow is affected by friction and the Mach number, we will apply the fundamental equations of fluid mechanics to the fixed differential control volume in Fig. 13–24a. The flow properties are listed at each open control surface.

Continuity Equation.   Since the flow is steady, the continuity equation becomes

$$\frac{\partial}{\partial t} \int_{cv} \rho \, d\mathcal{V} + \int_{cs} \rho \mathbf{V} \cdot d\mathbf{A} = 0$$

$$0 + (\rho + \Delta\rho)(V + \Delta V)A + \rho(-VA) = 0$$

In the limit,

$$\frac{d\rho}{\rho} + \frac{dV}{V} = 0 \tag{13–43}$$

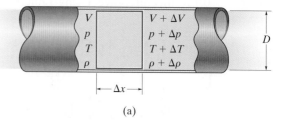

(a)

Fig. 13–24

---

*Although most duct cross sections are circular, if other geometries are to be considered, then we can replace the pipe diameter D by the *hydraulic diameter* of the duct, defined as $D_h = 4A/P$. Here A is the cross-sectional area, and P is the perimeter of the duct. Note that for a circular duct, $D_h = 4(\pi D^2/4)/(\pi D) = D$, as required.

Linear Momentum Equation. As shown on the free-body diagram, Fig. 13–24b, the friction force ΔF_f acts on the closed control surface, and is the result of wall shear stress τ_w, discussed in Chapter 9. It is defined by Eq. 9–16, $\tau_w = \dfrac{r}{2}\dfrac{\partial}{\partial x}(p + \gamma h)$. Since the fluid is a gas, its weight can be neglected, and so we get

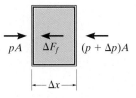

Free-body diagram

(b)

Fig. 13–24 (cont.)

$$\tau_w = \left(\frac{D}{4}\right)\left(\frac{\Delta p}{\Delta x}\right)$$

We can eliminate Δp by noting that the head loss from Eq. 10–1 is $\Delta h_L = \Delta p/\rho g$, or from Eq. 10–3, $\Delta h_L = f(\Delta x/D)\left(V^2/2g\right)$. If we equate the right sides of these two equations and solve for Δp, we get $\Delta p = f(\Delta x/D)\left(\rho V^2/2\right)$. Therefore,

$$\tau_w = \left(\frac{D}{4}\right)\left(\frac{f}{D}\right)\left(\frac{\rho V^2}{2}\right) = \frac{f\rho V^2}{8}$$

Finally, since τ_w acts on the control surface area $\pi D\,\Delta x$, and since the open control surfaces have an area of $A = \pi D^2/4$, the friction force becomes

$$\Delta F_f = \tau_w\left[\pi D\Delta x\right] = \frac{fA}{D}\left(\frac{\rho V^2}{2}\right)\Delta x$$

Using this result, the momentum equation for the control volume is therefore

$$\overset{+}{\to}\Sigma\mathbf{F} = \frac{\partial}{\partial t}\int_{cv}\mathbf{V}\rho\,d\mathcal{V} + \int_{cs}\mathbf{V}\rho\mathbf{V}\cdot d\mathbf{A}$$

$$-\frac{fA}{D}\left(\frac{\rho V^2}{2}\right)\Delta x - (p + \Delta p)A + pA = 0 + (V + \Delta V)(\rho + \Delta\rho)(V + \Delta V)A + V\rho(-VA)$$

In the limit, where $\Delta x \to 0$, neglecting the second- and third-order terms, and using Eq. 13–43, we get

$$-\frac{f}{D}\left(\frac{\rho V^2}{2}\right)dx - dp = \rho V\,dV \qquad\qquad (13\text{–}44)$$

Our goal is to now use this result, along with the ideal gas law and the energy equation, to relate $f\,dx/D$ to the Mach number for the flow.

Ideal Gas Law.   This law is $p = \rho RT$, but its differential form is

$$dp = d\rho RT + \rho R \, dT$$

or

$$dp = \left(\frac{d\rho}{\rho}\right)p + \frac{p \, dT}{T}$$

Here ρ can be eliminated by using the continuity equation, Eq. 13–43, so that

$$\frac{dp}{p} = \frac{dT}{T} - \frac{dV}{V} \tag{13–45}$$

Energy Equation.   Since the flow is adiabatic, the stagnation temperature throughout the pipe will remain *constant*, and so application of the energy equation produces Eq. 13–31, which is

$$T_0 = T\left(1 + \frac{k-1}{2}M^2\right) \tag{13–46}$$

Taking the derivative, we get after simplification

$$\frac{dT}{T} = -\frac{2(k-1)\,M}{2 + (k-1)\,M^2}\,dM \tag{13–47}$$

Also, since $V = M\sqrt{kRT}$, its derivative becomes

$$\frac{dV}{V} = \frac{dM}{M} + \frac{1}{2}\frac{dT}{T} \tag{13–48}$$

Now eliminating ρ in Eq. 13–44 using the ideal gas law, realizing that $V = M\sqrt{kRT}$, we obtain

$$\frac{1}{2}f\frac{dx}{D} + \frac{dp}{kM^2 p} + \frac{dV}{V} = 0$$

Substituting Eqs. 13–45, 13–47, and 13–48 into this equation and simplifying the algebra gives our final result.

$$f\frac{dx}{D} = \frac{\left(1 - M^2\right)d\left(M^2\right)}{kM^4\left(1 + \frac{1}{2}(k-1)M^2\right)} \tag{13–49}$$

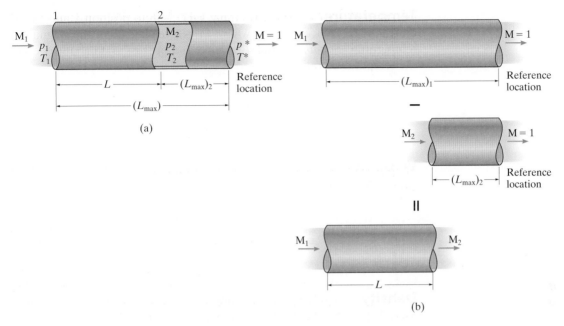

Fig. 13–25

Pipe Length versus Mach Number.

When Eq. 13–49 is integrated along the pipe, from position 1 to position 2, Fig. 13–25a, it will result in a complicated expression, and additional work will be required to apply it numerically. However, if the pipe is actually long enough (or imagined to be long enough), the effect of friction will tend to change the flow to the sonic speed M = 1. This occurs at the **critical location**, and we will use it as a *reference point* to apply the limits of integration from position 1 to this position $x_{cr} = L_{max}$, where M = 1, $p = p^*$, $T = T^*$, and $\rho = \rho^*$, Fig. 13–25a. Along the length L_{max} the friction factor will actually vary, because it is a function of the Reynolds number, but since the Reynolds number is generally high, for our purposes we will use a *mean value*[*] for f. Therefore,

$$\frac{f}{D} \int_0^{L_{max}} dx = \int_M^1 \frac{\left(1 - M^2\right) d\left(M^2\right)}{kM^4\left(1 + \frac{1}{2}(k - 1)M^2\right)}$$

$$\frac{fL_{max}}{D} = \frac{1 - M^2}{kM^2} + \frac{k + 1}{2k} \ln\left[\frac{\left[(k + 1)/2\right] M^2}{1 + \frac{1}{2}(k - 1) M^2}\right] \qquad (13\text{–}50)$$

With this equation, we can now determine the length L of pipe needed to change the Mach number from M_1 to M_2 if the pipe length $L \leq L_{max}$. As shown in Fig. 13–25b, to do this we simply require

$$\frac{fL}{D} = \frac{f(L_{max})_1}{D}\bigg|_{M_1} - \frac{f(L_{max})_2}{D}\bigg|_{M_2} \qquad (13\text{–}51)$$

---

[*]High values of Re produce almost constant values for f, because the curves of the Moody diagram tend to flatten out.

Temperature.   If we now apply Eq. 13–46 to position 1 and to the critical or reference location where M = 1, realizing that the stagnation temperature *remains constant* because the process is adiabatic, we get the temperature ratio expressed in terms of the Mach number.

$$\frac{T}{T^*} = \frac{T/(T_0)_1}{T^*/(T_0)_1} = \frac{\frac{1}{2}(k + 1)}{1 + \frac{1}{2}(k - 1)M^2} \tag{13–52}$$

Velocity.   Relating the velocity to the Mach number, we can use Eq. 13–52 to express the velocity ratio as

$$\frac{V}{V^*} = \frac{M\sqrt{kRT}}{(1)\sqrt{kRT^*}} = M\left[\frac{\frac{1}{2}(k + 1)}{1 + \frac{1}{2}(k - 1)M^2}\right]^{1/2} \tag{13–53}$$

Density.   Applying the continuity equation, $\rho VA = \rho^* V^* A$, and using Eq. 13–53, the density ratio becomes $\rho/\rho^* = V^*/V$, or

$$\frac{\rho}{\rho^*} = \frac{1}{M}\left[\frac{1 + \frac{1}{2}(k - 1)M^2}{\frac{1}{2}(k + 1)}\right]^{1/2} \tag{13–54}$$

Pressure.   From the ideal gas law, $p = \rho RT$, we have $p/p^* = (\rho/\rho^*)(T/T^*)$. Therefore, from Eqs. 13–52 and 13–54, we obtain the pressure ratio

$$\frac{p}{p^*} = \frac{1}{M}\left[\frac{\frac{1}{2}(k + 1)}{1 + \frac{1}{2}(k - 1)M^2}\right]^{1/2} \tag{13–55}$$

Finally, the stagnation pressure ratio will vary along the pipe since the process is nonisentropic. It can be obtained by realizing that $p_0/p_0^* = (p_0/p)(p/p^*)(p^*/p_0^*)$. And so, using Eqs. 13–32 and 13–55, we get

$$\frac{p_0}{p_0^*} = \frac{1}{M}\left[\left(\frac{2}{k + 1}\right)\left(1 + \frac{k - 1}{2}M^2\right)\right]^{(k+1)/2(k-1)} \tag{13–56}$$

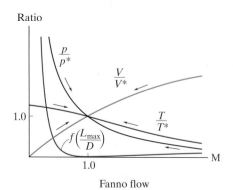

Ratio

1.0

1.0

M

Fanno flow

Fig. 13–26

Graphs of the variation of the ratios T/T^*, V/V^*, p/p^*, and $f(L_{max}/D)$ versus M are shown in Fig. 13–26, and their numerical values can be determined from the equations, or by using calculated values found on the Internet, or if $k = 1.4$, by interpolation using Table B–2 of Appendix B.

The Fanno Line.   Although we can describe the flow completely using the previous equations, it is instructive to show how the fluid behaves by considering how the entropy will vary along the pipe as a function of temperature. To do this, we must first express the change in entropy between the initial location 1 and some arbitrary location along the duct.

Starting with Eq. 13–18,

$$s - s_1 = c_v \ln \frac{T}{T_1} + R \ln \frac{\rho_1}{\rho} \tag{13-57}$$

we will want to express ρ_1/ρ in terms of temperature. Since A is constant, the continuity equation requires $\rho_1/\rho = V/V_1$, and since the stagnation temperature remains <u>constant for</u> an adiabatic process, then from Eq. 13–30, we have $V = \sqrt{2c_p(T_0 - T)}$. Using these expressions, Eq. 13–57 now becomes

$$s - s_1 = c_v \ln T - c_v \ln T_1 + R \ln \sqrt{2c_p(T_0 - T)} - R \ln V_1$$

$$= c_v \ln T + \frac{R}{2} \ln (T_0 - T) + \left[-c_v \ln T_1 + \frac{R}{2} \ln 2c_p - R \ln V_1 \right] \tag{13-58}$$

The last three terms are constant and are evaluated at the initial location of the pipe, where $T = T_1$ and $V = V_1$. If we plot Eq. 13–58, it represents the **Fanno line** for the flow (T–s diagram) and looks like that shown in Fig. 13–27.

The point of maximum entropy is found by taking the derivative of the above expression and setting it equal to zero, $ds/dT = 0$. This occurs when the flow is sonic, that is, $M = 1$. The region above $M = 1$ is for subsonic flow ($M < 1$), and the lower region is for supersonic flow ($M > 1$). For both cases, *friction increases the entropy* as the gas travels down the pipe. As expected, for *supersonic flow* the Mach number *decreases* until it reaches $M = 1$, where the flow becomes choked at the critical length. For *subsonic flow*, however, the Mach number *increases*. Although this may seem counterintuitive, it happens because the pressure drops rapidly, as noted in Fig. 13–26, for $M \leq 1$. This drop *increases* the velocity of the flow *more* than friction can provide resistance to slow the flow.

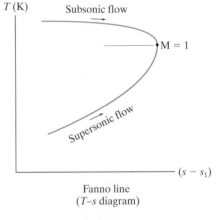

Fanno line
(T–s diagram)

Fig. 13–27

Important Points

- Ideal gas flow through a pipe or duct that includes the effect of friction along the wall of the pipe without heat loss is called Fanno flow. Using an average value for the friction factor f, the gas properties T, V, ρ, and p can be determined at a location along the pipe where the Mach number is known, provided these properties are known at the reference or critical location, where $M = 1$.

- Friction in the pipe will cause the Mach number to *increase* for *subsonic flow* until it reaches $M = 1$, and to *decrease* for *supersonic flow* until it reaches $M = 1$.

13

EXAMPLE   13.12

Air enters the 30-mm-diameter pipe with a velocity of 153 m/s, and a temperature of 300 K, Fig. 13–28. If the average friction factor is $f = 0.040$, determine how long, L_{max}, the pipe should be so that sonic flow occurs at the exit. Also, what is the velocity of the flow in the pipe at L_{max}, and at the location $L = 0.8$ m?

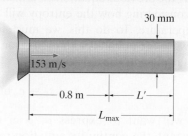

Fig. 13–28

SOLUTION

Fluid Description.   We assume that adiabatic steady compressible (Fanno) flow occurs along the pipe.

Maximum Pipe Length.   The critical length of pipe, L_{max}, is determined using Eq. 13–50 or Table B–2. First we need to determine the initial Mach number.

$$V = M\sqrt{kRT}; \qquad 153 \text{ m/s} = M_1\sqrt{1.4(286.9 \text{ J/kg} \cdot \text{K})(300 \text{ K})}$$

$$M_1 = 0.4408 < 1 \text{ subsonic flow}$$

Using Table B–2, we get $(f/D)(L_{max}) = 1.6817$, so that

$$L_{max} = \left(\frac{0.03 \text{ m}}{0.040}\right)(1.6817) = 1.2613 \text{ m} = 1.26 \text{ m} \qquad Ans.$$

At this exit, $M = 1$. The velocity of the gas is determined from the tabulated ratio for $M_1 = 0.4408$. It is $V_1/V^* = 0.47371$. Thus,

$$V^* = \frac{V_1}{V_1/V^*} = \left(\frac{1}{0.47371}\right)(153 \text{ m/s}) = 322.98 \text{ m/s} = 323 \text{ m/s} \quad Ans.$$

Flow Properties at $L = 0.8$ m.   Since the table and equations are referenced from the critical location, we must calculate $(f/D)L$ from this location, Fig. 13–28. Thus,

$$\frac{f}{D}L' = \frac{0.04}{0.03 \text{ m}}(1.2613 \text{ m} - 0.8 \text{ m}) = 0.6150$$

Using the table, this time with interpolated values of the ratio for V/V^*, we have

$$V = \frac{V}{V^*}V^* = (0.60667)(322.98 \text{ m/s}) = 196 \text{ m/s} \qquad Ans.$$

As the air travels 800 mm down the pipe, notice how the velocity has increased from 153 m/s to 196 m/s. As an exercise, show that the temperature decreases from 300 K to 260 K, and the velocity V^* at the end of the pipe can also be calculated using $V^* = (1)\sqrt{kRT^*}$, where $T^* = 259.71$ K.

EXAMPLE 13.13

Air within a large reservoir flows into the 50-mm-diameter pipe in Fig. 13–29a, at M = 0.5, determine the Mach number of the air when it exits the pipe. Take L = 1 m. Explain what happens if the pipe is extended, so that L = 2 m. The average friction factor for the pipe is f = 0.030.

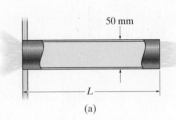

50 mm

L

(a)

SOLUTION

Fluid Description. We assume that adiabatic steady compressible (Fanno) flow occurs within the pipe.

L = 1. First we will calculate L_{max} for the pipe, so that sonic flow (M = 1) chokes the flow at the exit when M_1 = 0.5 at the entrance. From Table B–2 or Eq. 13–50, we get

$$\frac{f\,L_{max}}{D} = 1.0691; \quad L_{max} = \frac{1.0691(0.05\text{ m})}{0.030} = 1.782\text{ m}$$

Since L = 1 m < 1.782 m, then at the exit,

$$\frac{f}{D}L' = \frac{0.030}{(0.05\text{ m})}(1.782\text{ m} - 1\text{ m}) = 0.4691$$

Using Table B–2,

$$M_2 = 0.606 \qquad\qquad Ans.$$

L = 2 m. The length L_{max} = 1.782 m produces sonic flow (M = 1) at the exit when M_1 = 0.5. If the pipe is extended to L = 2 m, then friction will cause a *reduced flow* into the pipe, so that sonic flow chokes the exit of the pipe. In this case,

$$\frac{f\,L_{max}}{D} = \frac{(0.03)(2\text{ m})}{(0.05\text{ m})} = 1.2$$

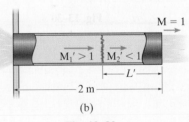

M = 1

$M_1' > 1$ $M_2' < 1$

L'

2 m

(b)

Fig. 13–29

Then using Table B–2, the new Mach number at the entrance becomes

$$M_1 = 0.485 \qquad\qquad Ans.$$

NOTE: Consider what would happen if we required supersonic flow (M_1 > 1) through the entrance of the *extended pipe*. In this case, sonic flow (M = 1) will still occur at the pipe exit; however, a normal shock wave will form within the pipe, Fig. 13–29b. This wave will *convert* the supersonic flow on the left side of the wave to subsonic flow on the right side. In Sec. 13.9 we will show how to relate the Mach numbers M_1' and M_2' on each side of this wave. With this relationship, the specific location L' of the wave can then be determined, since it must be where L_1' gives M = 1 at the exit. As the pipe is extended further, the wave will be located further towards the entrance, and then within the supersonic supply nozzle. If it reaches the throat of this nozzle it will choke it (M = 1) and thereby reduce the mass flow.

EXAMPLE | **13.14**

A room is at atmospheric pressure, 101 kPa, and a temperature of 293 K. If the air from the room is drawn into a 100-mm-diameter pipe isentropically, such that it has an absolute pressure of $p_1 = 80$ kPa as it enters the pipe, determine the mass flow, and the stagnation temperature and stagnation pressure, at the location $L = 0.9$ m. The average friction factor is $f = 0.03$. Also, what is the total friction force acting on this 0.9-m length of pipe?

SOLUTION

Fluid Description. We assume that adiabatic steady compressible (Fanno) flow occurs along the pipe.

Mass Flow. The mass flow can be determined at the entrance to the pipe using $\dot{m} = \rho_1 V_1 A_1$, but we must first determine V_1 and ρ_1. Since the flow into the pipe is isentropic, and the pressure is $p_1 = 80$ kPa, while the stagnation pressure is $p_0 = 101$ kPa, we can determine the Mach number of the air and its temperature at the entrance using Eq. 13–32 and Eq. 13–31.

$$\frac{p_1}{p_0} = \frac{80 \text{ kPa}}{101 \text{ kPa}} = 0.792$$

$$M_1 = 0.5868 \quad \text{and} \quad \frac{T_1}{T_0} = 0.93557$$

Therefore, $T_1 = 0.93557(293 \text{ K}) = 274.12$ K, and so

$$V_1 = M_1 \sqrt{kRT_1} = 0.5868\sqrt{1.4(286.9 \text{ J/kg} \cdot \text{K})(274.12 \text{ K})}$$

$$= 194.71 \text{ m/s}$$

Using the ideal gas law to obtain ρ_1, we have

$$p_1 = \rho_1 RT_1; \qquad 80(10^3) \text{ Pa} = \rho_1(286.9 \text{ J/kg} \cdot \text{K})(274.12 \text{ K})$$

$$\rho_1 = 1.0172 \text{ kg/m}^3$$

The mass flow is then

$$\dot{m} = \rho_1 V_1 A_1 = (1.0172 \text{ kg/m}^3)(194.71 \text{ m/s})\left[\pi(0.05 \text{ m})^2\right]$$

$$\dot{m} = 1.5556 \text{ kg/s} = 1.56 \text{ kg/s} \qquad \qquad Ans.$$

Stagnation Temperature and Pressure. Because the flow is adiabatic through the pipe, the stagnation temperature remains constant at

$$(T_0)_2 = (T_0)_1 = 293 \text{ K} \qquad \qquad Ans.$$

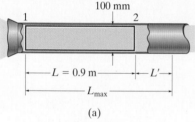

100 mm

$L = 0.9$ m ⟶ L'

L_{max}

(a)

Fig. 13–30

Friction will change the *stagnation pressure* throughout the pipe because the flow is nonisentropic. We can determine $(p_0)_2$ at $L = 0.9$ m by using Eq. 13–56 (or Table B–2).* First we must find the length of duct L_{max} needed to choke the flow. Using $M_1 = 0.5868$, Eq. 13–50 gives $fL_{max}/D = 0.03\,L_{max}/0.1 = 0.5455$, and so $L_{max} = 1.8183$ m. At this location, Eqs. 13–53, 13–56, and 13–55 give

$$\frac{V_1}{V^*} = 0.6218 \qquad\qquad V^* = \frac{194.71 \text{ m/s}}{0.6218} = 313.16 \text{ m/s}$$

$$\frac{(p_0)_1}{p_0{}^*} = 1.2043; \qquad\qquad p_0{}^* = \frac{101 \text{ kPa}}{1.2043} = 83.87 \text{ kPa}$$

$$\frac{p_1}{p^*} = 1.8057; \qquad\qquad p^* = \frac{80 \text{ kPa}}{1.8057} = 44.30 \text{ kPa}$$

Since L_{max} is the reference point, then at section 2, Fig. 13–30a, $fL'/D = 0.03(1.8183 \text{ m} - 0.9 \text{ m})/0.1 \text{ m} = 0.27548$. From Eq. 13–56 the stagnation pressure at this location is

$$\frac{(p_0)_2}{p_0{}^*} = 1.1188; \qquad (p_0)_2 = 1.1188(83.87 \text{ kPa}) = 93.8 \text{ kPa} \qquad\qquad Ans.$$

Friction Force. The resultant friction force is obtained using the momentum equation applied to the free-body diagram of the control volume, shown in Fig. 13–30b. First we must determine the static pressure p_2 and the velocity V_2. At $fL'/D = 0.27548$,

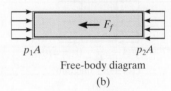

p_1A p_2A

Free-body diagram

(b)

$$\frac{p_2}{p^*} = 1.5689 \qquad p_2 = 1.5689(44.30 \text{ kPa}) = 69.51 \text{ kPa}$$

$$\frac{V_2}{V^*} = 0.7021; \quad V_2 = 0.7021(313.16 \text{ m/s}) = 219.9 \text{ m/s}$$

Fig. 13–30 (cont.)

Therefore,

$$\xrightarrow{+}\ \Sigma\mathbf{F} = \frac{\partial}{\partial t}\int_{cv} \mathbf{V}\rho\, d\mathbf{V} + \int_{cs} \mathbf{V}\rho\mathbf{V}\cdot d\mathbf{A}$$

$$-F_f + p_1 A - p_2 A = 0 + V_2\dot{m} + V_1(-\dot{m})$$

$$-F_f + \left[80\left(10^3\right) \text{N/m}^2\right]\left[\pi(0.05 \text{ m})^2\right] - \left[69.51\left(10^3\right) \text{N/m}^2\right]\left[\pi(0.05 \text{ m})^2\right]$$

$$= 0 + 1.5556 \text{ kg/s}\,(219.9 \text{ m/s} - 194.71 \text{ m/s})$$

$$F_f = 43.4 \text{ N} \qquad\qquad Ans.$$

* More accuracy is obtained from the equation, rather than using linear interpolation from the table.

Pipes in chemical processing plants are sometimes heated along their lengths, resulting in the conditions for Rayleigh flow. (© Eric Gevaert / Alamy)

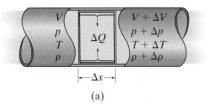

(a)

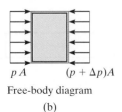

pA $(p + \Delta p)A$

Free-body diagram

(b)

Fig. 13–31

13.8 The Effect of Heat Transfer on Compressible Flow

In this section we will consider how heat transfer through the walls of a straight pipe of constant cross-sectional area will affect the steady compressible flow of an ideal gas having a constant specific heat. This type of flow can typically occur in the pipes and ducts of a combustion chamber of a turbojet engine, where heat transfer is significant and friction can be ignored. The heat can also be added within the gas itself, and not through the walls of the pipe. For example, this can occur by a chemical process or nuclear radiation. However the heat is added, this type of flow is sometimes called **Rayleigh flow**, named after the British physicist Lord Rayleigh. To simplify the numerical work, we will do the same as for Fanno flow, and develop the necessary equations in terms of the Mach number, making reference to the gas properties T^*, p^*, ρ^*, and V^* at the location in the pipe where the *critical* or *choked condition* $M = 1$ occurs. A differential control volume for this situation is shown in Fig. 13–31a. Here ΔQ is *positive* if heat is supplied to the gas, and it is *negative* if cooling occurs.

Continuity Equation. The continuity equation is the same as Eq. 13–43, namely

$$\frac{d\rho}{\rho} + \frac{dV}{V} = 0 \qquad (13\text{–}59)$$

Linear Momentum Equation. Only a pressure force acts on the open control surfaces, as shown on the free-body diagram, Fig. 13–31b. We have

$$\overset{+}{\rightarrow} \Sigma \mathbf{F} = \frac{\partial}{\partial t}\int_{cv} \mathbf{V}\rho \, d\forall + \int_{cs} \mathbf{V}\rho\mathbf{V}\cdot d\mathbf{A}$$

$$-(p + \Delta p)A + p(A) = 0 + (V + \Delta V)(\rho + \Delta\rho)(V + \Delta V)A + V(-\rho VA)$$

Taking the limit, and eliminating $d\rho$ using Eq. 13–59, we obtain

$$dp + \rho V dV = 0$$

If we divide this equation by p and use the ideal gas law, $p = \rho RT$ and $V = M\sqrt{kRT}$ to eliminate ρ and T, we get

$$\frac{dp}{p} + kM^2\frac{dV}{V} = 0 \qquad (13\text{–}60)$$

Ideal Gas Law.   When the ideal gas law, $p = \rho RT$, is expressed in differential form, combined with the continuity equation, we get Eq. 13–45.

$$\frac{dp}{p} = \frac{dT}{T} - \frac{dV}{V} \qquad (13\text{–}61)$$

Energy Equation.   No shaft work is performed on the gas, and there is no change in its potential energy. Therefore, the energy equation becomes

$$\dot{Q}_{in} - \dot{W}_{turb} + \dot{W}_{pump} = \left[\left(h_{out} + \frac{V_{out}^2}{2} + gz_{out}\right) - \left(h_{in} + \frac{V_{in}^2}{2} + gz_{in}\right)\right]\dot{m}$$

$$\dot{Q} - 0 + 0 = \left[\left(h + \Delta h + \frac{(V + \Delta V)^2}{2} + 0\right) - \left(h + \frac{V^2}{2}\right)\right]\dot{m}$$

Dividing both sides by $\dot{m}$ in the limit yields

$$\frac{dQ}{dm} = dh + V\,dV$$

$$= d\left(h + \frac{V^2}{2}\right)$$

At the stagnation point, $h + V^2/2 = h_0$, and so using Eq. 13–10, $dh = c_p\,dT$, we have for a finite application of heat,

$$\frac{dQ}{dm} = d(h_0) = c_p\,dT_0$$

$$\boxed{\frac{\Delta Q}{\Delta m} = c_p\left[(T_0)_2 - (T_0)_1\right]} \qquad (13\text{–}62)$$

As expected, because we do not have an adiabatic process, the result indicates that the *stagnation temperature will not remain constant;* rather, it will *increase* as heat is applied.

We will now combine, and then integrate, the above equations to show how the velocity, pressure, and temperature are related to the Mach number.

Velocity.   Since $V = M\sqrt{kRT}$, its derivative produces Eq. 13–48, that is,

$$\frac{dV}{V} = \frac{dM}{M} + \frac{1}{2}\frac{dT}{T} \qquad (13\text{–}63)$$

13

If we combine this equation with Eqs. 13–60 and 13–61, we obtain

$$\frac{dV}{V} = \frac{2}{M(1 + kM^2)} dM \tag{13–64}$$

Integrating between the limits $V = V^*$, $M = 1$ to $V = V$, $M = M$, we get

$$\frac{V}{V^*} = \frac{M^2(1 + k)}{1 + kM^2} \tag{13–65}$$

Density. From the continuity equation, for a pipe of *finite* length, $\rho^* V^* A = \rho V A$ or $V/V^* = \rho^*/\rho$, and so the densities are related by

$$\frac{\rho}{\rho^*} = \frac{1 + kM^2}{M^2(1 + k)} \tag{13–66}$$

Pressure. For the pressure, combining Eqs. 13–60 and 13–64, we get

$$\frac{dp}{p} = -\frac{2kM}{-1 + kM^2} dM$$

which, when integrated from $p = p^*$, $M = 1$ to $p = p$, $M = M$, yields

$$\frac{p}{p^*} = \frac{1 + k}{1 + kM^2} \tag{13–67}$$

Temperature. Finally, the temperature ratio is determined by substituting Eq. 13–64 into Eq. 13–63. This gives

$$\frac{dT}{T} = \frac{2(1 - kM^2)}{M(1 + kM^2)} dM$$

And upon integrating from $T = T^*$, $M = 1$ to $T = T$, $M = M$, we obtain

$$\frac{T}{T^*} = \frac{M^2(1 + k)^2}{(1 + kM^2)^2} \tag{13–68}$$

The variation of the ratios V/V^*, p/p^*, and T/T^* versus the Mach number is shown in Fig. 13–32, and for $k = 1.4$ their numerical values are given in Appendix B, Table B–3.

Stagnation Temperature and Pressure. The ratios of the stagnation temperatures and pressures at a location in the pipe, and at the critical or reference location, are sometimes needed for calculations. They can be determined using Eqs. 13–68 and 13–31,

$$\frac{T_0}{T_0^*} = \frac{T_0}{T} \frac{T}{T_0^*} \frac{T^*}{T_0^*} = \left(1 + \frac{k - 1}{2} M^2\right) \left[\frac{M^2(1 + k)^2}{(1 + kM^2)^2}\right] \frac{2}{k + 1}$$

$$\frac{T_0}{T_0^*} = \frac{2(k + 1)M^2\left(1 + \dfrac{k - 1}{2} M^2\right)}{(1 + kM^2)^2} \tag{13–69}$$

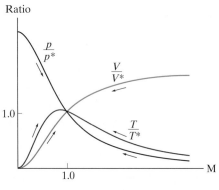

Ratio

$\dfrac{p}{p^*}$

$\dfrac{V}{V^*}$

$\dfrac{T}{T^*}$

1.0

1.0

M

Rayleigh Flow

Fig. 13–32

And in a similar manner, using the pressure ratios of Eqs. 13–67 and 13–32, we get

$$\frac{p_0}{p_0^*} = \left(\frac{1 + k}{1 + k\text{M}^2}\right)\left[\left(\frac{2}{k + 1}\right)\left(1 + \frac{k - 1}{2}\text{M}^2\right)\right]^{k/(k-1)} \quad (13\text{–}70)$$

For convenience these ratios are also given in Table B–3.

Rayleigh Line.   To better understand Rayleigh flow, we will show, just as we did for Fanno flow, how the entropy of the gas changes with temperature. Making reference to the critical state, where M = 1, the change in entropy in terms of the temperature and pressure ratios is expressed by Eq. 13–20; that is,

$$s - s^* = c_p \ln\frac{T}{T^*} - R \ln\frac{p}{p^*}$$

The last term can be expressed in terms of the temperature by squaring Eq. 13–67 and then substituting the result into Eq. 13–68. This gives

$$\left(\frac{p}{p^*}\right)^2 = \frac{T}{\text{M}^2 T^*}$$

Finally, when we solve for M^2 in Eq. 13–68 and substitute this into the above equations, the change in entropy becomes

$$s - s^* = c_p \ln\frac{T}{T^*} - R \ln\left[\frac{k + 1}{2} \pm \sqrt{\left(\frac{k + 1}{2}\right)^2 - k\frac{T}{T^*}}\right]$$

When this equation is graphed, it produces the **Rayleigh line** (*T–s* diagram) shown in Fig. 13–33. If we set $ds/dT = 0$, then like Fanno flow, the maximum entropy occurs when M = 1. Also, like Fanno flow, the upper portion of the graph defines subsonic flow (M < 1) and the lower portion supersonic flow (M > 1). Notice that for *supersonic flow*, the addition of heat will cause the temperature of the gas to increase, but its Mach number will decrease until it reaches M = 1, and the flow becomes choked. Therefore, to increase supersonic flow, it is necessary to cool the pipe rather than heat it. For *subsonic flow* the addition of heat will cause the gas to reach a maximum temperature, T_{max}, while its speed is increasing until its Mach number is M = $1/\sqrt{k}$ (at $dT/ds = 0$); then the gas temperature *will drop* as M approaches the limit M = 1. This is also evident in Fig. 13–32.

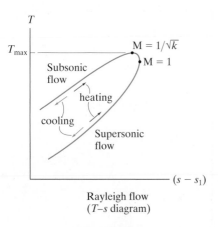

Rayleigh flow
(*T–s* diagram)

Fig. 13–33

13

Important Points

• Rayleigh flow occurs when heat is added or removed as the gas flows through a pipe or duct. Because the process is not adiabatic, the stagnation temperature is not constant.

• The gas properties V, ρ, p, and T, at a specific location in the pipe where M is known, can be determined, provided these properties are also known at the reference or critical location, where M = 1.

EXAMPLE | 13.15

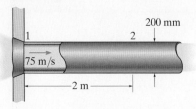

200 mm

1
2

75 m/s

2 m

Fig. 13–34

Outside air is drawn isentropically into the pipe having a diameter of 200 mm, Fig. 13–34. When it arrives at section 1, it has a velocity of 75 m/s, an absolute pressure of 135 kPa, and a temperature of 295 K. If the walls of the pipe supply heat at 100 kJ/kg·m, determine the properties of the air when it reaches section 2.

SOLUTION

Fluid Description.   We assume the air to be inviscid and to have steady compressible flow. Due to the heating, this is Rayleigh flow.

Air Properties at Critical Location.   The air properties at location 2 can be determined using the ratios in Table B–3, provided we first know the properties at the critical location, where M = 1. We can find these using the properties at section 1, but first we need the Mach number at section 1.

$$V_1 = M_1 \sqrt{kRT_1} \qquad 75 \text{ m/s} = M_1 \sqrt{1.4(286.9 \text{ J/kg} \cdot \text{K})(295 \text{ K})}$$

$$M_1 = 0.2179 < 1 \quad \text{Subsonic}$$

Using Table B–3,

$$T^* = \frac{T_1}{T_1/T^*} = \frac{295 \text{ K}}{0.24046} = 1226.84 \text{ K}$$

$$p^* = \frac{p_1}{p_1/p^*} = \frac{135 \text{ kPa}}{2.2504} = 59.99 \text{ kPa}$$

$$V^* = \frac{V_1}{V_1/V^*} = \frac{75 \text{ m/s}}{0.10686} = 701.84 \text{ m/s}$$

Air Properties at Section 2.   We can determine the Mach number at section 2 by using Eq. 13–69. However, before we can do this or use the tables, we must find the stagnation temperatures $(T_0)_2$ and T_0^*. First, $(T_0)_1$ can be determined using Eq. 13–31, or Table B–1 for isentropic flow. For $M_1 = 0.2179$, we get

$$(T_0)_1 = \frac{T_1}{T_1/(T_0)_1} = \frac{295 \text{ K}}{0.9904} = 297.86 \text{ K}$$

Now, using the energy equation, Eq. 13–62, with

$$c_p = \frac{kR}{k-1} = \frac{1.4(286.9 \text{ J/kg} \cdot \text{K})}{1.4 - 1} = 1004.15 \text{ J/kg} \cdot \text{K}$$

$$\frac{\Delta Q}{\Delta m} = c_p[(T_0)_2 - (T_0)_1]$$

$$\frac{100(10^3) \text{ J}}{\text{kg} \cdot \text{m}}(2 \text{ m}) = \left[1.00415(10^3) \text{ J/kg} \cdot \text{K}\right]\left[(T_0)_2 - 297.9 \text{ K}\right]$$

$$(T_0)_2 = 497.03 \text{ K}$$

Also, from Table B–3, for $M_1 = 0.2179$, the stagnation temperature at the critical or reference location is therefore

$$T_0^* = \frac{(T_0)_1}{(T_0)_1/T_0^*} = \frac{297.86 \text{ K}}{0.20229} = 1472.44 \text{ K}$$

Finally, we can find M_2 from the stagnation temperature ratio.

$$\frac{(T_0)_2}{T_0^*} = \frac{497.03 \text{ K}}{1472.44 \text{ K}} = 0.33756$$

Using Table B–3, we get $M_2 = 0.2949$. The other ratios at M_2 give

$$T_2 = T^*\left(\frac{T_2}{T^*}\right) = 1226.84 \text{ K}(0.39813) = 488 \text{ K} \qquad\qquad Ans.$$

$$p_2 = p^*\left(\frac{p_2}{p^*}\right) = 59.99 \text{ kPa}(2.1394) = 128 \text{ kPa} \qquad\qquad Ans.$$

$$V_2 = V^*\left(\frac{V_2}{V^*}\right) = 701.84 \text{ m/s}(0.18612) = 131 \text{ m/s} \qquad\qquad Ans.$$

The results indicate that as the Mach number increases from $M_1 = 0.2179$ to $M_2 = 0.2949$, the pressure decreases from 135 kPa to 128 kPa, and the temperature and velocity increase from 295 K to 488 K and from 75 m/s to 131 m/s. These changes follow the trend shown by the curves for Rayleigh subsonic flow in Fig. 13–32.

13.9   Normal Shock Waves

When designing any nozzle or diffuser used in supersonic wind tunnels or for high-speed aircraft or rockets, it is possible to develop a standing shock wave within the nozzle. As indicated in the previous sections, standing shock waves can also develop in pipes, for both Fanno and Rayleigh flow. In this section we will study how the flow properties change across a shock wave as a function of the Mach number. To do this, we will use the equations of continuity, momentum, and energy, and the ideal gas law.

In Sec. 13.3, we stated that a shock wave is a high-intensity compression wave that is very thin. If the wave is "standing," that is, at rest, then on the *downstream* side the temperature, pressure, and density will be high and the velocity low, whereas the opposite effect occurs on the *upstream* side, Fig. 13–35a. A large amount of heat conduction and viscous friction develops *within* the wave due to the extreme *decelerations* of the gas molecules. As a result, the thermodynamic process within the wave becomes *irreversible*, and so the entropy across the wave will increase. Thus, the process is nonisentropic. If we consider a control volume to surround the wave and extend a slight distance beyond it, then the gas system within this control volume undergoes an *adiabatic process* because no heat passes through the control surfaces. Instead, the changes in temperature are made within the control volume.

The exhaust from this rocket is designed to pass through its nozzles at supersonic *speed*, ideally without forming a shock. However, as the rocket ascends, the ambient pressure will decrease, and so the exhaust will fan or flare off the sides of the nozzles, forming expansion waves. (© Valerijs Kostreckis/Alamy)

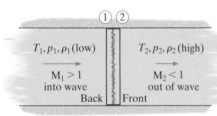

Standing shock wave

(a)

Continuity Equation. When the wave is standing, the flow is steady,* and so the continuity equation becomes

$$\frac{\partial}{\partial t} \int_{cv} \rho \, d\forall + \int_{cs} \rho \mathbf{V} \cdot d\mathbf{A} = 0$$

$$0 - \rho_1 V_1 A + \rho_2 V_2 A = 0$$

$$\rho_1 V_1 = \rho_2 V_2 \tag{13–71}$$

Linear Momentum Equation. As shown on the free-body diagram of the control volume, Fig. 13–35b, only pressure forces act on each side of the wave, and it is the *difference* in these forces that causes the gas to decelerate and thereby lose its momentum. Applying the linear momentum equation, we have

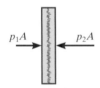

Free-body diagram

(b)

Fig. 13–35

$$\xrightarrow{+} \Sigma \mathbf{F} = \frac{\partial}{\partial t} \int_{cv} \mathbf{V} \rho \, d\forall + \int_{cs} \mathbf{V} \rho \mathbf{V} \cdot d\mathbf{A}$$

$$p_1 A - p_2 A = 0 + V_1 \rho_1(-V_1 A) + V_2 \rho_2(V_2 A)$$

$$p_1 + \rho_1 V_1^2 = p_2 + \rho_2 V_2^2 \tag{13–72}$$

With this equation we can relate the Mach numbers on each side of the wave, and from this, obtain the ratios of the gas properties T, p_1, and ρ across the wave.

*If the wave is moving, then steady flow occurs if we attach our reference to the wave. This will produce the same results.

Ideal Gas Law. Considering the gas to be ideal with constant specific heats, then using the ideal gas law, $p = \rho RT$, we can write Eq. 13–72 as

$$p_1\left(1 + \frac{V_1^2}{RT_1}\right) = p_2\left(1 + \frac{V_2^2}{RT_2}\right)$$

Since $V = M\sqrt{kRT}$, the pressure ratio, written in terms of the Mach numbers, becomes

$$\frac{p_2}{p_1} = \frac{1 + kM_1^2}{1 + kM_2^2} \tag{13–73}$$

Energy Equation. Since an adiabatic process occurs, the stagnation temperature will remain constant across the wave. Therefore, $(T_0)_1 = (T_0)_2$, and from Eq. 13–31, which was derived from the energy equation, we have

$$\frac{T_2}{T_1} = \frac{1 + \dfrac{k-1}{2}M_1^2}{1 + \dfrac{k-1}{2}M_2^2} \tag{13–74}$$

The ratio of the velocities on each side of the shock can be determined from Eq. 13–27, $V = M\sqrt{kRT}$. It is

$$\frac{V_2}{V_1} = \frac{M_2\sqrt{kRT_2}}{M_1\sqrt{kRT_1}} = \frac{M_2}{M_1}\sqrt{\frac{T_2}{T_1}} \tag{13–75}$$

Using Eq. 13–74, our result is

$$\frac{V_2}{V_1} = \frac{M_2}{M_1}\left[\frac{1 + \dfrac{k-1}{2}M_1^2}{1 + \dfrac{k-1}{2}M_2^2}\right]^{1/2} \tag{13–76}$$

From the continuity equation, Eq. 13–71, we obtain the density ratio

$$\frac{\rho_2}{\rho_1} = \frac{V_1}{V_2} = \frac{M_1}{M_2}\left[\frac{1 + \dfrac{k-1}{2}M_2^2}{1 + \dfrac{k-1}{2}M_1^2}\right]^{1/2} \tag{13–77}$$

We can establish a relationship between the Mach numbers M_1 and M_2 by first forming the temperature ratio using the ideal gas law.

$$\frac{T_2}{T_1} = \frac{p_2/\rho_2 R}{p_1/\rho_1 R} = \frac{p_2}{p_1}\left(\frac{\rho_1}{\rho_2}\right)$$

If we now substitute Eqs. 13–73, 13–74, and 13–77 into this expression and equate the result to Eq. 13–74, we will be able to solve for M_2 in terms of M_1. Two solutions are possible. The first one leads to the trivial solution $M_2 = M_1$, which refers to isentropic flow with no shock. The other solution leads to irreversible flow, which gives the desired relationship.

$$M_2^2 = \frac{M_1^2 + \dfrac{2}{k-1}}{\dfrac{2k}{k-1}M_1^2 - 1} \tag{13–78}$$

Thus, if M_1 is known, then M_2 can be found from this equation, and then the ratios p_2/p_1, T_2/T_1, V_2/V_1, and ρ_2/ρ_1 that occur in front of and behind the shock can be determined from the previous equations.

Finally, the increase in entropy occurring across the shock can be found from Eq. 13–20 (or Eq. 13–19),

$$s_2 - s_1 = c_p \ln \frac{T_2}{T_1} - R \ln \frac{p_2}{p_1} \tag{13–79}$$

The stagnation pressure across the shock will *decrease* due to this increase, $s_2 - s_1$. To determine $(p_0)_2/(p_0)_1$, we use Eq. 13–32 and write

$$\frac{(p_0)_2}{(p_0)_1} = \left(\frac{(p_0)_2}{p_2}\right)\left(\frac{p_2}{p_1}\right)\left(\frac{p_1}{(p_0)_1}\right) = \frac{p_2}{p_1}\left[\frac{1 + \dfrac{k-1}{2}M_2^2}{1 + \dfrac{k-1}{2}M_1^2}\right]^{k/(k-1)} \tag{13–80}$$

If Eqs. 13–73 and 13–78 are combined and simplified, we obtain

$$\frac{p_2}{p_1} = \frac{2k}{k+1}M_1^2 - \frac{k-1}{k+1} \tag{13–81}$$

Substituting this and Eq. 13–78 into Eq. 13–80 and simplifying, we obtain our result,

$$\frac{(p_0)_2}{(p_0)_1} = \frac{\left[\dfrac{\dfrac{k+1}{2}M_1^2}{1 + \dfrac{k-1}{2}M_1^2}\right]^{k/(k-1)}}{\left[\dfrac{2k}{k+1}M_1^2 - \dfrac{k-1}{k+1}\right]^{1/(k-1)}} \tag{13–82}$$

For convenience, this ratio along with p_2/p_1, ρ_2/ρ_1, T_2/T_1, and M_2 are tabulated in Appendix B, Table B–4, for $k = 1.4$.

For any specific value of k, it can be shown, using Eq. 13–78, that when *supersonic* flow occurs *behind* the shock, Fig. 13–36a, $M_1 > 1$, *subsonic flow* will *always* occur in *front* of the shock, $M_2 < 1$. This happens because the entropy, determined from Eq. 13–79, increases, which is in accordance with the second law of thermodynamics. Realize that subsonic

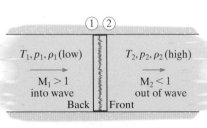

T_1, p_1, ρ_1 (low) T_2, p_2, ρ_2 (high)

$M_1 > 1$ $M_2 < 1$
into wave   out of wave
Back | Front

Standing shock wave

(a)

Fig. 13–35

flow cannot occur behind the shock, $M_1 < 1$, because Eq. 13–78 would predict that supersonic flow must occur in front of the shock, $M_2 > 1$. This is *not possible* since Eq. 13–79 would indicate a decrease in entropy, which is a violation of the second law of thermodynamics.

13.10   Shock Waves in Nozzles

Propulsion nozzles used, for example, on rockets are subjected to *changing* conditions of outside pressure and temperature as the rocket ascends through the atmosphere. Consequently, the thrust, which is a function of this pressure, will also be changing. This can lead to a shock wave forming within the nozzle, which in turn will cause the nozzle to lose its efficiency. To understand this process, let us again review the pressure variations through a converging–diverging (Laval) nozzle in Fig. 13–36a for various changes in the backpressure.

- When the backpressure is at the stagnation pressure, $p_1 = p_0$, there is no flow through the nozzle, curve 1 in Fig. 13–36b.

- Lowering the backpressure to p_2 causes subsonic flow through the nozzle, with the lowest pressure and maximum velocity occurring at the throat. This flow is isentropic, curve 2.

- When the backpressure is lowered to p_3, sonic velocity (M = 1) develops at the throat, and subsonic isentropic flow continues to occur through both the convergent and the divergent sections of the nozzle. At this point, *maximum mass flow* occurs through the nozzle and is independent of a further drop in the backpressure, curve 3.

- As the backpressure is lowered to p_5, a standing normal shock wave will develop within the divergent portion of the nozzle, Fig. 13–36c. This is nonisentropic flow. Across the shock the pressure rises suddenly from A to B, curve 5, Fig. 13–36b, causing subsonic flow to occur from the shock to the exit plane. In other words, the pressure follows the curve from B and reaches the backpressure p_5 at the exit.

- A further lowering of the backpressure to p_6 will bring the shock wave to the exit plane, Fig. 13–36d. Here the divergent section has supersonic flow throughout its length so that it reaches a pressure of p_4 to the *left* of the wave. At the exit the shock wave will suddenly change the pressure from p_4 to p_6 so that the exit flow is *subsonic*, curve 6.

- An even further lowering of the backpressure from p_6 to p_7 will not affect the pressure to the left of the exit plane from the nozzle. It will remain at $p_4 < p_7$. Under these conditions, at p_4 the gas molecules are farther apart than when they are at p_7, and so the gas is said to be **overexpanded**. As a result, the gas will develop a series of *oblique compression* shock waves forming a series of **shock diamonds** *outside* the nozzle, as the flow extends from the nozzle and the pressure of the gas rises to equal the backpressure, p_7, Fig. 13–36e.

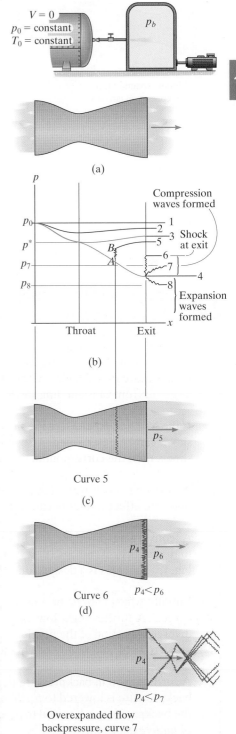

(a)

(b)

Curve 5

(c)

Curve 6

(d)

Overexpanded flow
backpressure, curve 7

(e)

Fig. 13–36

13

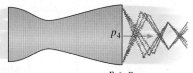

$p_4 > p_8$

Underexpanded flow
backpressure, curve 8
(f)

Fig. 13–36 (cont.)

- When the backpressure is lowered to p_4, it has reached the isentropic design condition for the nozzle, with subsonic flow through the converging section, sonic flow at the throat, and supersonic flow through the divergent section, curve 4. *No shock will be produced, and so no energy is lost.* Efficiency is at a maximum.

- A further reduction in backpressure to p_8 will cause the flow within the divergent section to be ***underexpanded***, since the pressure to the left of the exit plane of the nozzle, p_4, will now be *higher* than the backpressure, $p_4 > p_8$. As a result the gas will undergo a series of *expansion* shock waves, again forming a pattern of shock diamonds outside the nozzle until the pressure equals the backpressure, Fig. 13–36f.

The effect of overexpanded and underexpanded flow is further discussed in Sec. 13.12. Also, these details are treated more thoroughly in books related to gas dynamics. For example, see Ref. [3].

Important Points

- A shock wave is very thin. It is a nonisentropic process that causes the entropy to increase due to frictional effects within the wave. Because the process is adiabatic, no heat is gained or lost, and so the stagnation temperature on each side of the shock is the same. The stagnation pressure and density, however, will be larger in front of a standing shock, due to the change in entropy.

- If the Mach number M_1 of the flow behind a standing shock is known, the Mach number M_2 in front of the shock can be determined. Also, if the temperature, pressure, and density, T_1, p_1, ρ_1, are known behind the shock, then the corresponding values T_2, p_2, ρ_2 in front of the shock can be found.

- A standing shock wave always requires supersonic flow behind it and subsonic flow in front of it. The opposite effect cannot occur, since it would violate the second law of thermodynamics.

- A converging–diverging, or Laval, nozzle will be most efficient when it operates at a backpressure that produces isentropic flow, with M = 1 in the throat and supersonic flow at the exit plane, curve 4, Fig. 13–36b. This is a design condition in which no heat transfer or friction loss will occur.

- Initially, when a nozzle becomes choked, then M = 1 at the throat and the flow is subsonic at the exit plane, curve 3. A further slight lowering of the backpressure will cause a standing shock wave to form within the divergent portion of the nozzle, curve 5. Further lowering of the backpressure to p_6 will cause this wave to move toward, and finally reach, the exit plane, curve 6.

- A further drop in backpressure to p_7 results in shock waves forming off the edges of the nozzle. When the backpressure is lowered to p_4, then supersonic isentropic flow occurs throughout the nozzle. And finally, if the backpressure is lowered to p_8, expansion waves will form on the edge of the nozzle, creating conditions of *underexpansion*.

EXAMPLE | 13.16

The pipe in Fig. 13–37 transports air at a temperature of 20°C having an absolute pressure of 30 kPa and a speed of 550 m/s, measured just behind a standing shock wave. Determine the temperature, pressure, and speed of the air just in front of the wave.

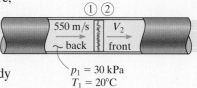

$p_1 = 30$ kPa
$T_1 = 20°C$

Fig. 13–37

SOLUTION

Fluid Description.   The shock wave is an adiabatic process. Steady flow occurs in back and in front of the wave.

Analysis.   The control volume contains the shock, as shown in Fig. 13–37. Since $k = 1.4$ for air, it is easier to solve this problem using Table B–4, rather than using the equations. First, though, we must determine M_1. We have

$$M_1 = \frac{V_1}{\sqrt{kRT_1}} = \frac{550 \text{ m/s}}{\sqrt{1.4(286.9 \text{ J/kg} \cdot \text{K})(273 + 20) \text{ K}}} = 1.6032$$

Since $M_1 > 1$, we expect $M_2 < 1$ and for both the pressure and temperature to *increase* in front of the wave.

Using $M_1 = 1.6032$, the value of M_2 in front of the shock (subsonic) and the ratios of pressure and temperature are taken from Table B–4. They are

$$M_2 = 0.66747$$

$$\frac{p_2}{p_1} = 2.8322$$

$$\frac{T_2}{T_1} = 1.3902$$

Thus,

$$p_2 = 2.8322(30 \text{ kPa}) = 85.0 \text{ kPa} \qquad Ans.$$

$$T_2 = 1.3902(273 + 20) \text{ K} = 407.3 \text{ K} \qquad Ans.$$

The speed of the air in front of the shock wave can be determined from Eq. 13–76, or, since T_2 is known, we can use Eq. 13–27.

$$V_2 = M_2\sqrt{kRT_2} = 0.66747\sqrt{1.4(286.9 \text{ J/kg} \cdot \text{K})(407.3 \text{ K})} = 270 \text{ m/s} \qquad Ans.$$

13

EXAMPLE | 13.17

A jet plane is traveling at M = 1.5, where the absolute air pressure is 50 kPa and the temperature is 8°C. At this speed, a shock forms at the intake of the engine, as shown in Fig. 13–38a. Determine the pressure and velocity of the air just to the right of the shock.

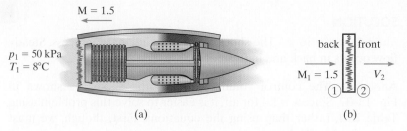

(a)   (b)

Fig. 13–38

SOLUTION

Fluid Description. The control volume that contains the shock moves with the engine so that steady flow occurs through the open control surfaces, Fig. 13–38b. An adiabatic process occurs within the shock.

Analysis. The velocity V_2 will be obtained from $V_2 = M_2\sqrt{kRT_2}$, and so we must first obtain M_2 and T_2. Since $k = 1.4$, using Table B–4, or Eqs. 13–78, 13–81, and 13–74, for $M_1 = 1.5$ (supersonic), we have

$$M_2 = 0.70109 \quad \text{Subsonic}$$

$$\frac{p_2}{p_1} = 2.4583$$

$$\frac{T_2}{T_1} = 1.3202$$

Therefore, just to the right of the shock,

$$p_2 = 2.4583(50 \text{ kPa}) = 123 \text{ kPa} \qquad \qquad Ans.$$

$$T_2 = 1.3202(273 + 8) \text{ K} = 370.98 \text{ K}$$

Thus, relative to the engine, the velocity of the air is

$$V_2 = M_2\sqrt{kRT_2} = 0.70109\sqrt{1.4(286.9 \text{ J/kg} \cdot \text{K})(370.98 \text{ K})}$$

$$V_2 = 271 \text{ m/s} \qquad \qquad Ans.$$

EXAMPLE 13.18

The nozzle in Fig. 13–39a is connected to a large reservoir where the absolute air pressure is 350 kPa. Determine the range of outside backpressures that cause a shock wave to form within the nozzle and just outside of it.

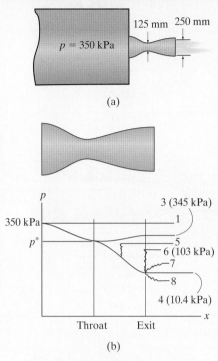

(a)

SOLUTION

Fluid Description.   We assume steady flow through the nozzle.

Analysis.   First we will establish the backpressures that produce subsonic and supersonic isentropic flow through the nozzle, curves 3 and 4 in Fig. 13–39b. The area ratio between the exit and the throat of the nozzle (divergent section) is $A/A^* = \pi(0.125 \text{ m})^2/\pi(0.0625 \text{ m})^2 = 4$. This ratio gives two values for the Mach number at the exit using Table B–1 ($k = 1.4$). Choosing the *isentropic subsonic flow* within the divergent section (curve 3) for $A/A^* = 4$, we get $M \approx 0.1467 < 1$ and $p/p_0 = 0.9851$. Since there is no shock wave, the stagnation pressure throughout the flow is 350 kPa. Then, at the exit, $p_3 = 0.9851(350 \text{ kPa}) = 345 \text{ kPa}$. In other words, this backpressure will cause $M = 1$ at the throat and subsonic isentropic flow of $M = 0.1467$ at the exit.

For *isentropic supersonic flow* within the divergent region (curve 4), from Table B–1, for $A/A^* = 4$, we get $M \approx 2.9402 > 1$ and $p/p_0 = 0.02979$. Thus, at the exit, $p_4 = 0.02979(350 \text{ kPa}) = 10.43 \text{ kPa} = 10.4 \text{ kPa}$.

This pressure is lower than the previous value since it must produce the required supersonic flow at the exit of $M = 2.9402$. The two solutions are for isentropic flow with backpressures that produce *no shock* within the nozzle; however, for both conditions the nozzle is choked since $M = 1$ at the throat.

If a standing shock is formed *at the exit* of the nozzle, Fig. 13–39b, curve 6, then this will occur when the pressure *in the nozzle* at the exit, to the left or *behind the shock*, is 10.4 kPa. To find the backpressure in front of the shock, that is, just outside the exit plane, we must use Table B–4 with $M = 2.9402$, in which case $p_6/p_4 = 9.9176$, so that $p_6 = 9.9176 p_4 = 9.9176(10.43 \text{ kPa}) = 103 \text{ kPa}$.

Thus, a normal shock is created *within* the divergent portion of the nozzle (such as curve 5, which is between curves 3 and 6) when the backpressure is in the following range:

$$345 \text{ kPa} > p_b > 103 \text{ kPa} \qquad \textit{Ans.}$$

For *compression waves* to occur at the exit, curve 7, the backpressure (between curves 6 and 4) should be in the range

$$103 \text{ kPa} > p_b > 10.4 \text{ kPa} \qquad \textit{Ans.}$$

Finally, *expansion waves* will form if the backpressure is anywhere below curve 4 (as in curve 8), that is,

$$10.4 \text{ kPa} > p_b \qquad \textit{Ans.}$$

Fig. 13–39

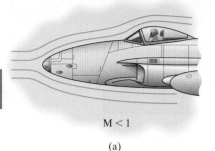

M < 1

(a)

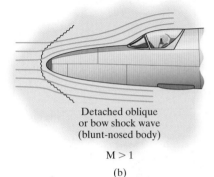

Detached oblique
or bow shock wave
(blunt-nosed body)

M > 1

(b)

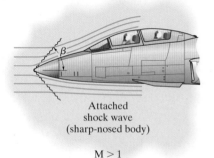

Attached
shock wave
(sharp-nosed body)

M > 1

(c)

Fig. 13–40

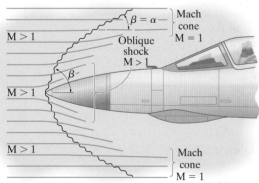

Fig. 13–41

13.11   Oblique Shock Waves

In Sec. 13.2 we showed that when a jet plane or other fast-moving body encounters the surrounding air in front of it, the pressure created by the body pushes the air to flow around it. At subsonic speeds, M < 1, the streamlines adjust and follow the contour of the surface, Fig. 13–40a. However, as the speed of the body increases to supersonic, so that M ≥ 1, the pressure created in front of the surface cannot communicate to the air upstream fast enough to move out of the way. Instead the air molecules bunch up and create an ***oblique shock wave***. This shock begins to bend just in front of the surface, Fig. 13–40b. Here it is detached from its surface. At higher speeds, and if the body has a sharp nose, Fig. 13–40c, the shock can become attached to the surface, causing the wave to turn at a sharp angle β. As the speed increases further, this angle β will continue to decrease. Farther removed from the surface, the effect of this shock weakens and develops into a Mach cone having an angle α that travels at M = 1 through the atmosphere, Fig. 13–41. The result of all this will alter the direction of the streamlines for the flow. Near the origin of the oblique shock, the streamlines are turned the most, as they become almost parallel to the surface of the body.* Farther away they essentially remain unchanged when they pass through the weaker Mach cone.

Oblique shock waves can be studied in the same manner as normal shocks, although here the change in direction of the streamlines becomes important. To analyze the situation, we will use two angles to define the geometry. As shown in Fig. 13–42a, β defines the angle of the shock wave, and θ defines the angle of the deflected streamline or the direction of the velocity $\mathbf{V}_2$ in front of the shock. For convenience we will reference the flow normal and tangential to the wave. Resolving $\mathbf{V}_1$ and $\mathbf{V}_2$ into their n and t components, we have

$$V_{1n} = V_1 \sin \beta \qquad V_{2n} = V_2 \sin (\beta - \theta)$$

$$V_{1t} = V_1 \cos \beta \qquad V_{2t} = V_2 \cos (\beta - \theta)$$

Or, since M = V/c, we can also write

$$M_{1n} = M_1 \sin \beta \qquad M_{2n} = M_2 \sin (\beta - \theta) \qquad (13\text{–}83)$$

$$M_{1t} = M_1 \cos \beta \qquad M_{2t} = M_2 \cos (\beta - \theta) \qquad (13\text{–}84)$$

For analysis, we will consider a standing wave and select a fixed control volume that includes an arbitrary portion of the wave, having a front (on the right) and back (on the left), each with an area A, Fig. 13–42a.

*At supersonic speeds the boundary layer is very thin, and so it has little effect on the direction of the streamlines.

Continuity Equation.   Since the flow is steady, measured relative to the surface, and no flow is assumed to occur through the wave in the t direction, we have

$$\frac{\partial}{\partial t} \int_{cv} \rho \, d\mathcal{V} + \int_{cs} \rho \mathbf{V} \cdot d\mathbf{A} = 0$$

$$0 - \rho_1 V_{1n} A + \rho_2 V_{2n} A = 0$$

$$\rho_1 V_{1n} = \rho_2 V_{2n} \tag{13–85}$$

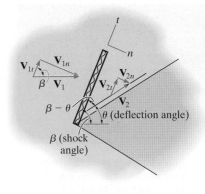

(a)

Momentum Equation.   As shown on the free-body diagram of the control volume, Fig. 13–42b, the pressure only acts in the n direction. Also, the flow $\rho \mathbf{V} \cdot \mathbf{A}$ is caused only by the normal components of $\mathbf{V}_1$ and $\mathbf{V}_2$. Therefore, applying the momentum equation in the t direction, we have

$$+\nearrow \Sigma F_t = \frac{\partial}{\partial t} \int_{cv} V_t \rho \, d\mathcal{V} + \int_{cs} V_t \rho \mathbf{V} \cdot d\mathbf{A}$$

$$0 = 0 + V_{1t}(-\rho_1 V_{1n} A) + V_{2t}(\rho_2 V_{2n} A)$$

Using Eq. 13–85, we obtain

$$V_{1t} = V_{2t} = V_t$$

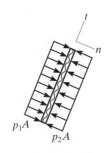

In other words, the tangential velocity components remain *unchanged* on each side of the shock.

Free-body diagram

(b)

Fig. 13–42

In the n direction, we have

$$\searrow^+ \Sigma F_n = \frac{\partial}{\partial t} \int_{cv} V_n \rho \, d\mathcal{V} + \int_{cs} V_n \rho \mathbf{V} \cdot d\mathbf{A}$$

$$p_1 A - p_2 A = 0 + V_{1n}(-\rho_1 V_{1n} A) + V_{2n}(\rho_2 V_{2n} A)$$

or

$$p_1 + \rho_1 V_{1n}^2 = p_2 + \rho_2 V_{2n}^2 \tag{13–86}$$

Energy Equation.   Applying the energy equation, neglecting the effect of gravity, and assuming an adiabatic process occurs,

$$\left(\frac{dQ}{dt}\right)_{in} - \left(\frac{dW_s}{dt}\right)_{out} = \left[\left(h_{out} + \frac{V_{out}^2}{2} + gz_{out}\right) - \left(h_{in} + \frac{V_{in}^2}{2} + gz_{in}\right)\right]\dot{m}$$

$$0 - 0 = \left[\left(h_2 + \frac{V_{2n}^2 + V_{2t}^2}{2} + 0\right) - \left(h_1 + \frac{V_{1n}^2 + V_{1t}^2}{2} + 0\right)\right]\dot{m}$$

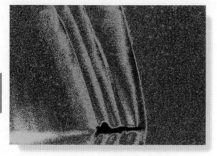

A Schlieren image showing the development of oblique shocks formed on a model that is being tested in a wind tunnel (© L. Weinstein/ Science Source)

Since $V_{1t} = V_{2t}$, we get

$$h_1 + \frac{V_{1n}^2}{2} = h_2 + \frac{V_{2n}^2}{2} \tag{13-87}$$

Equations 13–85, 13–86, and 13–87 are the same as Eqs. 13–71, 13–72, and 13–29. As a result, for oblique shocks we can describe the flow in the normal direction using the normal shock equations (Table B–4) developed previously. These equations become

$$M_{2n}^2 = \frac{M_{1n}^2 + \frac{2}{k-1}}{\frac{2k}{k-1} M_{1n}^2 - 1} \tag{13-88}$$

$$\frac{p_2}{p_1} = \frac{2k}{k+1} M_{1n}^2 - \frac{k-1}{k+1} \tag{13-89}$$

$$\frac{(p_0)_2}{(p_0)_1} = \frac{\left[\frac{\frac{k+1}{2} M_{1n}^2}{1 + \frac{k-1}{2} M_{1n}^2} \right]^{k/(k-1)}}{\left[\frac{2k}{k+1} M_{1n}^2 - \frac{k-1}{k+1} \right]^{1/(k-1)}} \tag{13-90}$$

$$\frac{T_2}{T_1} = \frac{1 + \frac{k-1}{2} M_{1n}^2}{1 + \frac{k-1}{2} M_{2n}^2} \tag{13-91}$$

$$\frac{\rho_2}{\rho_1} = \frac{V_{1n}}{V_{2n}} = \frac{M_{1n}}{M_{2n}} \left[\frac{1 + \frac{k-1}{2} M_{2n}^2}{1 + \frac{k-1}{2} M_{1n}^2} \right]^{1/2} \tag{13-92}$$

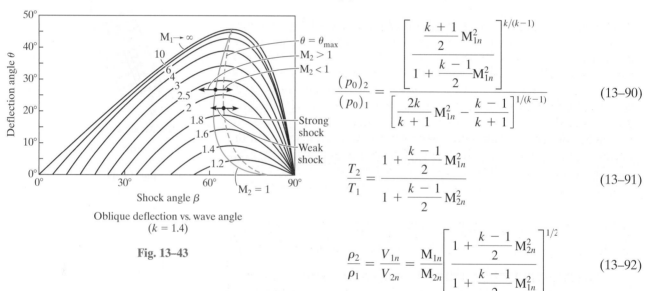

Oblique deflection vs. wave angle
($k = 1.4$)

Fig. 13–43

If M_1 and M_2 on each side of the shock wave are *both supersonic*, and the *component* M_{1n} is also supersonic, then M_{2n} must be *subsonic*, so it does not violate the second law of thermodynamics. For many practical problems, the initial flow properties V_1, p_1, T_1, ρ_1 and the angle θ will be known.

We can relate the angles θ and β and the Mach number M_1 in the following manner. Since $V_{1t} = V_{2t}$, then by Eq. 13–27,

$$M_{1t} \sqrt{kRT_1} = M_{2t} \sqrt{kRT_2}$$

$$M_{2t} = M_{1t} \sqrt{\frac{T_1}{T_2}}$$

Like velocity, in Fig. 13–42a, $M_{2n} = M_{2t} \tan(\beta - \theta)$. Therefore,

$$M_{2n} = M_{1t}\sqrt{\frac{T_1}{T_2}}\tan(\beta - \theta) = M_1 \cos \beta \sqrt{\frac{T_1}{T_2}}\tan(\beta - \theta)$$

Squaring this equation and substituting the temperature ratio, Eq. 13–91, we obtain

$$M_{2n}^2 = M_1^2 \cos^2 \beta \left(\frac{1 + \dfrac{k-1}{2}M_{2n}^2}{1 + \dfrac{k-1}{2}M_{1n}^2} \right) \tan^2(\beta - \theta)$$

Combining this equation with Eq. 13–88 and Eq. 13–83, and noting that in Fig. 13–42a, like velocity, $M_{1n} = M_1 \sin \beta$, we get our final result.

$$\tan\theta = \frac{2\cot\beta\left(M_1^2\sin^2\beta - 1\right)}{M_1^2(k + \cos 2\beta) + 2} \tag{13–93}$$

A plot of this equation, for several different values of M_1, produces curves of θ vs. β, shown in Fig. 13–43. Notice, for example, that for the curve $M_1 = 2$, when the deflection angle $\theta = 20°$, there are two values for the shock angle β. The lower value $\beta = 53°$ corresponds to a weak shock, and the higher value $\beta = 74°$ to a strong shock. Most often a weak shock will form before a strong shock because its pressure ratio will be smaller.* Also, notice that solutions are not possible for all deflection angles θ. For the higher deflection values, the wave will become detached from the surface and instead form in front of it, producing a higher drag, Fig. 13–41b. For the extreme case of *no deflection*, $\theta = 0°$, Eq. 13–93 gives $\beta = \sin^{-1}(1/M_1) = \alpha$, which produces a Mach cone, Fig. 13–41.

Important Points

- An oblique shock wave will form on the front surface of a body traveling at $M \geq 1$. As the speed increases, the wave will begin to attach itself to the surface of the body, and the angle of alignment β will begin to decrease. Farther from the localized region of the wave, a Mach cone will form and will travel at $M = 1$.

- The tangential component of velocity or its Mach number for an oblique shock remains the same on each side of the wave. The normal component can be analyzed using the same equations as for normal shocks. An equation is also available to determine the deflection angle θ of the streamlines that pass through the wave.

---

*An increase in pressure can occur downstream if the flow is blocked by a sudden change in the shape of the surface. If this occurs, the pressure ratio will be higher and a strong shock will be produced.

EXAMPLE | 13.19

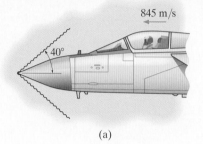

(a)

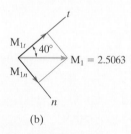

(b)

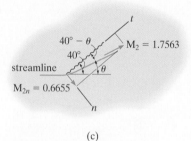

(c)

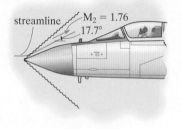

(d)

Fig. 13–44

A jet plane is flying horizontally at 845 m/s, at an altitude where the air temperature is 10°C and the absolute pressure is 80 kPa. If an oblique shock forms on the nose of the plane, at the angle shown in Fig. 13–44a, determine the pressure and temperature, and the direction of the air just behind the shock.

SOLUTION

Fluid Description. The air is considered compressible, and the shock is an adiabatic nonisentropic process. Steady flow occurs as viewed from the plane.

Analysis. The Mach number for the jet must first be determined.

$$M_1 = \frac{V_1}{c} = \frac{V_1}{\sqrt{kRT}} = \frac{845 \text{ m/s}}{\sqrt{1.4(286.9 \text{ J/kg} \cdot \text{K})(273 + 10) \text{ K}}} = 2.5063$$

From the geometry shown in Fig. 13–44b, M_1 is resolved into its normal and tangential components relative to the wave. The normal component is $M_{1n} = 2.5063 \sin 40° = 1.6110$. We can now use Table B–4 or Eqs. 13–88, 13–89, and 13–91 to obtain the speed, temperature, and pressure in front of the shock. Using the table, we get $M_{2n} = 0.6651$. This is subsonic, which is to be expected so as not to violate the second law of thermodynamics. Also from Table B–4,

$$\frac{T_2}{T_1} = 1.3956; \quad T_2 = 1.3956(273 + 10) \text{ K} = 394.96 \text{ K} = 395 \text{ K} \quad Ans.$$

$$\frac{p_2}{p_1} = 2.8619; \quad p_2 = 2.8619(80 \text{ kPa}) = 228.91 \text{ kPa} = 229 \text{ kPa} \quad Ans.$$

The angle θ can be obtained directly from Eq. 13–93 since M_1 and β are known. However, another way of calculating it is to first find M_2. This can be done by first finding the stagnation temperature *in front of the shock*. Using Table B–1, for $M_1 = 2.5063$, it is

$$\frac{T_1}{(T_0)_1} = 0.4432; \quad (T_0)_1 = \frac{(273 + 10) \text{ K}}{0.4432} = 638.54 \text{ K}$$

Since flow across the shock is adiabatic, the stagnation temperature is constant, and so $(T_0)_1 = (T_0)_2$. Therefore, using Table B–1,

$$\frac{T_2}{(T_0)_2} = \frac{394.96 \text{ K}}{638.54 \text{ K}} = 0.6185; \quad M_2 = 1.7563$$

Finally, from Fig. 13–44c, the streamline deflection can now be determined from

$$M_{2n} = M_2 \sin(\beta - \theta); \quad 0.66551 = 1.7563 \sin(40° - \theta)$$

$$\theta = 40° - \sin^{-1}\frac{0.6651}{1.7563} = 17.7° \quad Ans.$$

The resulting flow is shown in Fig. 13–44d.

13.12   Compression and Expansion Waves

When an airfoil or other body is moving at supersonic speed, not only will it form an oblique shock at its front surface, but in addition, if the surface is curved, the flow must follow this surface, and in doing so it may also form compression or expansion waves as the flow is redirected. For example, in Fig. 13–45a the concave surface on the jet plane causes compression waves to form, which, when extended, become concurrent and merge into an oblique shock—something we studied in the previous section. However, if the surface is convex, then the air has room to *expand* as it follows the surface, and so a multitude of divergent expansion waves will form, creating a "fan" of an infinite succession of Mach waves, Fig. 13–45b. This behavior also occurs at a sharp corner, as shown in Fig. 13–45c. Notice that during the expansion process, the Mach number for each succeeding wave will *increase*, while the Mach angle α of each wave will *decrease*. Because the changes in the gas properties that are made through the formation of *each* of these waves are *infinitesimal*, the process of creating each wave can be considered isentropic, and for the entire "fan" of waves it is also isentropic.

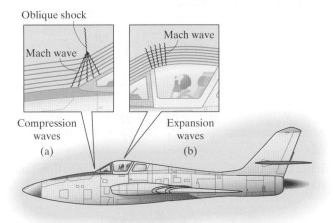

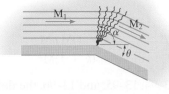

Expansion
waves

(c)

Fig. 13–45

To study this wave fan, we will isolate one wave and assume it alters the streamline that passes through the wave by the differential *turning angle* $d\theta$, Fig. 13–46a. Our goal is to express $d\theta$ in terms of the Mach number M that is just in front of the wave. Here M > 1 and the wave acts at the Mach angle α. Resolving the velocities **V** and **V** + d**V** into their normal and tangential components, and realizing that the velocity components in the tangential direction remain the same, we require $V_{t1} = V_{t2}$ or

$$V \cos \alpha = (V + dV) \cos (\alpha + d\theta) = (V + dV)(\cos \alpha \cos d\theta - \sin \alpha \sin d\theta)$$

Since $d\theta$ is small, $\cos d\theta \approx 1$ and $\sin d\theta \approx d\theta$. Therefore, our equation becomes $dV/V = (\tan \alpha)d\theta$. Using Eq. 13–28, $\sin \alpha = 1/M$, the $\tan \alpha$ is

$$\tan \alpha = \frac{\sin \alpha}{\cos \alpha} = \frac{\sin \alpha}{\sqrt{1 - \sin^2 \alpha}} = \frac{1}{\sqrt{M^2 - 1}}$$

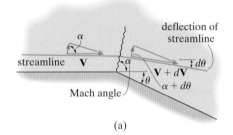

deflection of streamline

streamline   **V**

Mach angle

(a)

Therefore,

$$\frac{dV}{V} = \frac{d\theta}{\sqrt{M^2 - 1}} \tag{13–94}$$

Since $V = M\sqrt{kRT}$, its derivative is

$$dV = dM\sqrt{kRT} + M\left(\frac{1}{2}\right)\frac{kR}{\sqrt{kRT}} dT$$

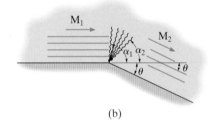

(b)

Fig. 13–46

or

$$\frac{dV}{V} = \frac{dM}{M} + \frac{1}{2}\frac{dT}{T} \tag{13–95}$$

For adiabatic flow, the relationship between the stagnation temperature and the static temperature is determined from Eq. 13–31, that is,

$$T_0 = T\left(1 + \frac{k-1}{2}M^2\right)$$

Taking the derivative of this expression, and rearranging the terms, we obtain

$$\frac{dT}{T} = -\frac{2(k-1)M\,dM}{2 + (k-1)M^2} \tag{13–96}$$

Finally, combining Eqs. 13–94, 13–95, and 13–96, the deflection angle for the wave can now be expressed in terms of the Mach number and its change. It is

$$d\theta = \frac{2\sqrt{M^2 - 1}}{2 + (k-1)M^2}\frac{dM}{M}$$

For a finite turning angle θ, we must integrate this expression between the initial and final waves, each having a different Mach number. Normally, the final Mach number for the flow is unknown, so it is more convenient to integrate this expression from a *reference position*, where at $\theta = 0°$, M = 1. Then from this reference, we can determine the turning angle $\theta = \omega$, through which the flow expands, so that the Mach number changes from M = 1 to M. We have

$$\int_0^\omega d\theta = \int_1^M \frac{2\sqrt{M^2 - 1}}{2 + (k - 1)M^2} \frac{dM}{M}$$

$$\omega = \sqrt{\frac{k + 1}{k - 1}} \tan^{-1}\left(\sqrt{\frac{k - 1}{k + 1}(M^2 - 1)}\right) - \tan^{-1}\left(\sqrt{M^2 - 1}\right) \quad (13\text{--}97)$$

This equation is referred to as the ***Prandtl–Meyer expansion function***, named after Ludwig Prandtl and Theodore Meyer. With it we can find the total turning angle of the flow caused by the fan of isentropic expansion waves. For example, if the flow in Fig. 13–46b has an initial Mach number of M_1 and the surface deflects θ, then we can determine the resulting Mach number M_2 by first applying Eq. 13–97 to obtain ω_1 for M_1. Since $\theta = \omega_2 - \omega_1$, then $\omega_2 = \omega_1 + \theta$. With this angle, ω_2, we can then reapply Eq. 13–97 to determine M_2. Although this would require a numerical procedure, it is convenient to use tabular values for Eq. 13–97. They are listed in Appendix B, Table B–5.

Important Points

- As the flow passes over a curve or edge of a surface, it can cause compression or expansion waves when $M \geq 1$.

- The compression waves merge into an oblique shock.

- The expansion waves form a "fan" of an infinite succession of Mach waves. The streamline deflection angle caused by the expansion waves can be determined using the Prandtl–Meyer expansion function.

EXAMPLE | 13.20

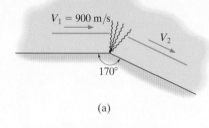

(a)

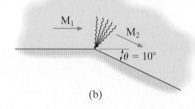

(b)

Fig. 13–47

Air flows over a surface at a speed of 900 m/s, where the absolute pressure is 100 kPa and the temperature is 30°C. Expansion waves occur at the sharp transition shown in Fig. 13–47a. Determine the velocity, temperature, and pressure of the flow just to the right of the wave.

SOLUTION

Fluid Description. We have steady compressible flow that expands isentropically.

The initial or upstream Mach number for the flow is

$$M_1 = \frac{V_1}{c} = \frac{V_1}{\sqrt{kRT}} = \frac{900 \text{ m/s}}{\sqrt{1.4(286.9 \text{ J/kg} \cdot \text{K})(273 + 30) \text{ K}}} = 2.5798$$

Analysis. Since the Prandtl–Meyer expansion function, Eq. 13–97, has been referenced from $M = 1$, the turning angle from this reference is

$$\omega_1 = \sqrt{\frac{k+1}{k-1}} \tan^{-1} \sqrt{\frac{k-1}{k+1}\left(M_1^2 - 1\right)} - \tan^{-1}\sqrt{M_1^2 - 1} \quad (1)$$

$$= \sqrt{\frac{1.4+1}{1.4-1}} \tan^{-1} \sqrt{\frac{1.4-1}{1.4+1}\left((2.5798)^2 - 1\right)} - \tan^{-1}\sqrt{(2.5798)^2 - 1}$$

$$= 40.96°$$

This same value (or one close to it) can also be determined from Table B–5. Since the boundary layer over the inclined surface is very thin, the deflection angle for the streamlines is defined by the same angle as the deflection of the surface, that is, $180° - 170° = 10°$, Fig. 13–47b. The downstream wave must therefore have a Mach number M_2 that produces a turning angle of $40.96° + 10° = 50.96°$ from the $M = 1$ reference.

Rather than using this angle and trying to solve for M_2 by trial and error using the Prandtl–Meyer function, Eq. 13–97, we will use the Table for $\omega_2 = 50.796°$. We get

$$M_2 = 3.0631 \qquad\qquad\qquad Ans.$$

In other words, the expansion waves increase the Mach number from $M_1 = 2.5798$ to $M_2 = 3.0631$. Realize, however, that this increase does not violate the second law of thermodynamics, because only the *normal component* of the wave converts from supersonic to subsonic.

Since we have isentropic expansion, we can determine the temperature on the right side of the waves using Eq. 13–74.

$$\frac{T_2}{T_1} = \frac{1 + \dfrac{k-1}{2}M_1^2}{1 + \dfrac{k-1}{2}M_2^2} = \frac{1 + \dfrac{1.4-1}{2}(2.5798)_1^2}{1 + \dfrac{1.4-1}{2}(3.0631)^2} = 0.8104$$

$$T_2 = 0.8104(273 + 30)\ \text{K} = 245.54\ \text{K} = 246\ \text{K} \qquad Ans.$$

We can also use Table B–1 as follows:

$$\frac{T_2}{T_1} = \frac{T_2}{T_0}\frac{T_0}{T_1} = (0.34764)\left(\frac{1}{0.42894}\right) = 0.8104$$

so that again $T_2 = 0.8104(273 + 30)\ \text{K} = 246\ \text{K}$.

The pressure is obtained in a similar manner using Eq. 13–22, or Table B–1. Here

$$\frac{p_2}{p_1} = \frac{p_2}{p_0}\frac{p_0}{p_1} = (0.02478)\left(\frac{1}{0.052170}\right) = 0.4793$$

so

$$p_2 = 0.4793(100\ \text{kPa}) = 47.9\ \text{kPa} \qquad Ans.$$

The velocity of the flow after the expansion is

$$V_2 = M_2\sqrt{kRT_2} = 3.0631\sqrt{1.4\,(286.9\ \text{J/kg}\cdot\text{K})(245.54\ \text{K})} = 962\ \text{m/s} \quad Ans.$$

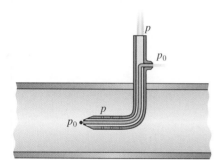

Pitot-static tube
Subsonic flow

(a)

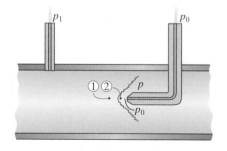

Pitot tube and piezometer
Sonic or supersonic flow

(b)

Fig. 13–48

13.13   Compressible Flow Measurement

Pressure and velocity in compressible gas flows can be measured in a variety of ways. Here we will discuss a few of them.

Pitot Tube and Piezometer.   As in the case of incompressible flow, discussed in Sec. 5.3, a Pitot-static tube, such as the one shown in Fig. 13–48a, can also be used for compressible flow measurement. The static pressure p *within the flow* is measured at the side opening of the tube, whereas the stagnation or total pressure p_0 is measured at the stagnation point, which occurs at the front opening. At the stagnation point the flow comes rapidly to a standstill, with no significant frictional loss, and so the process can be assumed isentropic.

Subsonic Flow.   For subsonic compressible flow, the pressures are related by Eq. 13–32. Since $V = M\sqrt{kRT}$, solving for M, substituting this into Eq. 13–32, and then solving for V, we get

$$V = \sqrt{\frac{2kRT}{k-1}\left[\left(\frac{p_0}{p}\right)^{(k-1)/k} - 1\right]} \qquad (13\text{–}98)$$

Provided we know the temperature T, we can then use this equation to determine the velocity of the flow.

In practice, it is generally easier to measure the stagnation temperature T_0 at the stagnation point, rather than T within the flow, since the flow is easily disturbed. The relationship used for this case can be determined by combining Eqs. 13–30 and 13–98, and after a bit of algebra, we obtain

$$V = \sqrt{2c_pT_0\left[1 - \left(\frac{p_0}{p}\right)^{(k-1)/k} - 1\right]} \qquad (13\text{–}99)$$

Substitution of the measured quantities then gives the undisturbed, free-flow velocity of the gas.

Supersonic Flow.   If the gas is flowing at supersonic speeds, then it will form a shock just before it strikes the nose of the Pitot tube, Fig. 13–48b. The shock changes the flow from supersonic, point 1, to subsonic, point 2. To obtain the velocity of flow in this case, we must therefore use Eq. 13–73 to relate the pressures p_1 and p_2 across the shock. Also, the relationship between p_2 and the measured stagnation pressure p_0 is determined using Eq. 13–32. We can combine these two equations and express the result in terms of the upstream Mach number using Eq. 13–78. Doing this, we get

$$\frac{p_0}{p_1} = \frac{\left(\dfrac{k+1}{2}M_1^2\right)^{\frac{k}{k-1}}}{\left(\dfrac{2k}{k+1}M_1^2 - \dfrac{k-1}{k+1}\right)^{\frac{1}{k-1}}} \qquad (13\text{–}100)$$

The pressure p_1 can be measured independently, using a piezometer on the boundary of the flow, well upstream of the shock, Fig. 13–48b. With p_1 and p_0 known, we can then obtain the Mach number $M_1 > 1$ for the flow from Eq. 13–100. Since the stagnation temperature remains constant on both sides of the shock, combining Eqs. 13–30 and 13–27, we can then determine the velocity of the flow.

$$V_1 = \left[\frac{2c_p T_0}{1 + \left(2c_p / kRM_1^2 \right)} \right]^{1/2} \tag{13–101}$$

As an alternative procedure, rather than using a Pitot tube and piezometer, we can measure the velocity of the gas using a hot-wire anemometer, which was described in Sec. 10.5.

Venturi Meter.   If a venturi meter, as in Fig. 13–49, is used, then the change in density of the gas must be taken into account when making measurements of the flow at the throat. The flow can be assumed isentropic, since the passage through the meter occurs rapidly and with little frictional loss. Between points 1 and 2, the continuity equation is

$$\frac{\partial}{\partial t} \int_{cv} \rho \, d\forall + \int_{cs} \rho \mathbf{V} \cdot d\mathbf{A} = 0; \quad 0 - \rho_1 A_1 V_1 + \rho_2 A_2 V_2$$

And the energy equation, Eq. 13–29, when applied between any two points within the flow, becomes

$$h_1 + \frac{V_1^2}{2} = h_2 + \frac{V_2^2}{2}$$

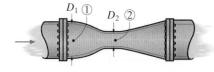

Fig. 13–49

For isentropic flow, the pressure is related to the density by Eq. 13–21, $p_1/\rho_1^k = p_2/\rho_2^k$, and the enthalpy change is related to the temperature change by Eq. 13–10, $h_2 - h_1 = c_p (T_2 - T_1)$. If all four of these equations are combined, the theoretical mass flow at the throat of the nozzle can be determined. Any losses not accounted for are considered by multiplying this result by a **velocity coefficient**, C_v, determined experimentally as a function of the Reynolds number determined at the throat. The actual compressible mass flow rate then becomes

$$\dot{m} = C_v A_2 \sqrt{ \frac{2k}{k-1} \frac{p_1 \rho_1 \left[(p_2/p_1)^{2/k} - (p_2/p_1)^{(k+1)/k} \right]}{\left[1 - (A_2/A_1)^2 (p_2/p_1)^{2/k} \right]} }$$

If the flow has a low velocity ($V < 0.3M$), then the gas can be considered incompressible, and, as shown in Sec. 5.3, the mass flow can be determined from

$$\dot{m} = C_v V_2 \rho A_2 = C_v A_2 \sqrt{ \frac{2\rho(p_1 - p_2)}{1 - (D_2/D_1)^4} } \tag{13–102}$$

13

In practice, Eq. 13–102 is sometimes modified to compare it with the compressible case. This is done by using an experimentally determined **expansion factor**, Y, so that for compressible flow we can then use

$$\dot{m} = C_v Y A_2 \sqrt{\frac{2\rho_1(p_1 - p_2)}{[1 - (D_2/D_1)^4]}} \tag{13–103}$$

Values of the velocity coefficient or expansion factor are available from graphs for various pressure ratios. See Ref. [6]. Equations 13–102 and 13–103 can also be used with a flow nozzle and orifice meter since the basic principles of operation are the same.

References

1. J. E. A. John, *Gas Dynamics*, Prentice Hall, Upper Saddle River, NJ, 2005.

2. F. M. White, *Fluid Mechanics*, McGraw-Hill, New York, NY, 2011.

3. H. Liepmann, *Elements of Gasdynanics*, Dover, New York, NY, 2002.

4. P. H. Oosthuizen and W. E. Carsvallen, *Compressible Fluid Flow*, McGraw-Hill, New York, NY, 2003.

5. S. Schreier, *Compressible Flow*, Wiley-Interscience Publication, New York, N.Y., 1982.

6. W. B. Brower, *Theory, Table, and Data for Compressible Flow*, Taylor and Francis, New York, NY, 1990.

PROBLEMS

Sec. 13.1–13.4

13–1. Oxygen is decompressed from an absolute pressure of 600 kPa to 100 kPa, with no change in temperature. Determine the changes in the entropy and enthalpy.

13–2. Helium is contained in a closed vessel under an absolute pressure of 400 kPa. If the temperature increases from 20°C to 85°C, determine the changes in pressure and entropy.

13–3. If a pipe contains helium at a gage pressure of 100 kPa and a temperature of 20°C, determine the density of the helium. Also, determine the temperature if the helium is compressed isentropically to a gage pressure of 250 kPa. The atmospheric pressure is 101.3 kPa.

*13–4. The closed tank contains helium at 200°C and under an absolute pressure of 530 kPa. If the temperature is increased to 250°C, determine the changes in density and pressure, and the change per unit mass in the internal energy and enthalpy of the helium.

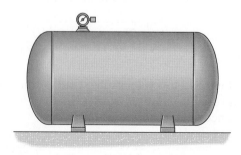

Prob. 13–4

13–5. A gas has a specific heat that varies with the absolute temperature, such that $c_p = \left(1256 + 36\,728/T^2\right)$ J/kg·K. If the temperature rises from 300 K to 400 K, determine the change in enthalpy per unit mass.

13–6. Air flows in a horizontal duct at 20°C with a velocity of 180 m/s. If the velocity increases to 250 m/s, determine the corresponding temperature of the air. *Hint*: Use the energy equation to find Δh.

13–7. The half-angle α on the Mach cone of a rocket is 20°. If the air temperature is 15°C, determine the speed of the rocket.

*13–8. A ship is located where the depth of the ocean is 3 km. Determine the time needed for a sonar signal to bounce off the bottom and return to the ship. Assume the water temperature is 10°C. Take $\rho = 1030$ kg/m³ and $E_V = 2.11\left(10^9\right)$ Pa for sea water.

13–9. Determine how fast a race car must travel in 20°C weather in order for M = 0.3.

13–10. Compare the speed of sound in water and air at a temperature of 20°C. The bulk modulus of water at $T = 20°C$ is $E_V = 2.2$ GPa.

13–11. A jet plane is flying at Mach 2.2. Determine its speed in kilometers per hour. The air is at 10°C.

*13–12. A jet plane has a speed of 600 m/s. If the air has a temperature of 10°C, determine the Mach number and the half-angle α of the Mach cone.

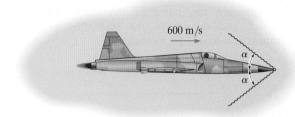

600 m/s

Prob. 13–12

13–13. Determine the half-angle α of the Mach cone at the nose of a jet if it is flying at 1125 m/s in air at 5°C.

13–14. A jet plane passes 5 km directly overhead. If the sound of the plane is heard 6 s later, determine the speed of the plane. The average air temperature is 10°C.

13–15. The Mach number of the airflow in the wind tunnel at B is to be M = 2.0 with an air temperature of 10°C and absolute pressure of 25 kPa. Determine the required absolute pressure and temperature within the large reservoir at A.

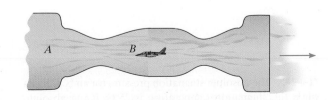

A B

Prob. 13–15

*13–16. Determine the pressure of air if it is flowing at 1600 km/h. When the air is still, the temperature is 20°C and the absolute pressure is 101.3 kPa.

13–17. What are the ratios of the critical pressure, temperature, and density to the stagnation pressure, temperature, and density for methane?

13–18. The temperature and absolute pressure of air within the circular duct are 40°C and 800 kPa, respectively. If the mass flow is 30 kg/s, determine the Mach number.

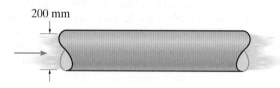

Prob. 13–18

13–19. The flow at a point in a wind tunnel has a speed of M = 2.5 when the absolute pressure of the air is 16 kPa and the temperature is 200 K. Determine the speed of the air at the point, and also find the temperature and pressure of the air in the supply reservoir.

*13–20. The absolute stagnation pressure for air is 875 kPa when the stagnation temperature is 25°C. If the absolute pressure for the flow is 630 kPa, determine the velocity of the flow.

Sec. 13.5–13.6

13–21. Nitrogen in the reservoir is at a temperature of 20°C and an absolute pressure of 300 kPa. Determine the mass flow through the nozzle. The atmospheric pressure is 100 kPa.

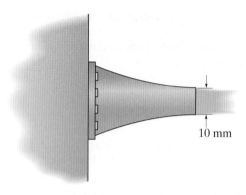

10 mm

Prob. 13–21

13–22. Nitrogen, at an absolute pressure of 600 kPa and temperature of 800 K, is contained in the large tank. Determine the backpressure in the hose to choke the nozzle and yet maintain isentropic supersonic flow through the divergent portion of the nozzle. The nozzle has an outer diameter of 40 mm, and the throat has a diameter of 20 mm.

13–23. Nitrogen, at an absolute pressure of 600 kPa and temperature of 800 K, is contained in the large tank. Determine the backpressure in the hose to choke the nozzle and maintain isentropic subsonic flow through the divergent portion of the nozzle. The nozzle has an outer diameter of 40 mm, and the throat has a diameter of 20 mm.

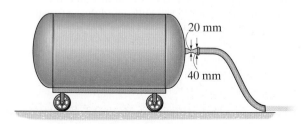

20 mm

40 mm

Probs. 13–22/23

*13–24. The large tank contains air at an absolute pressure of 150 kPa and temperature of 20°C. The 5-mm-diameter nozzle at A is opened to let air out of the tank. Determine the mass flow and the horizontal force that must be applied to the tank to prevent it from moving. The atmospheric pressure is 100 kPa.

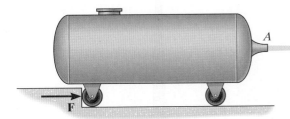

Prob. 13–24

13–25. The large tank contains air at an absolute pressure of 600 kPa and temperature of 70°C. The Laval nozzle has a throat diameter of 20 mm and an exit diameter of 50 mm. Determine the absolute pressure within the connected pipe so that the nozzle chokes but also maintains isentropic subsonic flow within the divergent portion of the nozzle. Also, what is the mass flow from the tank if the absolute pressure within the pipe is 150 kPa?

13–26. The large tank contains air at an absolute pressure of 600 kPa and temperature of 70°C. The nozzle has a throat diameter of 20 mm and an exit diameter of 50 mm. Determine the absolute pressure within the connected pipe, and the corresponding mass flow through the pipe, when the nozzle chokes and maintains isentropic supersonic flow within the divergent portion of the nozzle.

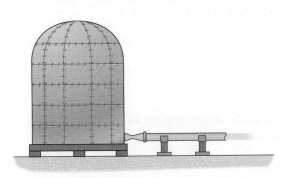

Probs. 13–25/26

13–27. The large tank contains air at an absolute pressure of 700 kPa and temperature of 400 K. Determine the mass flow from the tank into the pipe if the converging nozzle has an exit diameter of 40 mm and the absolute pressure in the pipe is 150 kPa.

*13–28. The large tank contains air at an absolute pressure of 700 kPa and temperature of 400 K. Determine the mass flow from the tank into the pipe if the converging nozzle has an exit diameter of 40 mm and the absolute pressure in the pipe is 400 kPa.

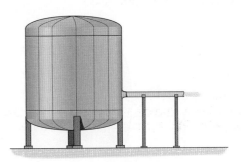

Probs. 13–27/28

13–29. Air exits a large tank through a converging nozzle having an exit diameter of 20 mm. If the temperature of the air in the tank is 35°C and the absolute pressure in the tank is 600 kPa, determine the velocity of the air as it exits the nozzle. The absolute pressure outside the tank is 101.3 kPa.

13–30. Air exits a large tank through a converging nozzle having an exit diameter of 20 mm. If the temperature of the air in the tank is 35°C and the absolute pressure in the tank is 150 kPa, determine the mass flow of the air as it exits the nozzle. The absolute pressure outside the tank is 101.3 kPa.

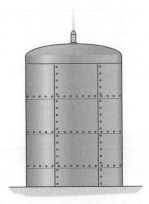

Probs. 13–29/30

13–31. The absolute pressure is 400 kPa and the temperature is 20°C in the large tank. If the pressure at the entrance A of the nozzle is 300 kPa, determine the mass flow out of the tank through the exit of the nozzle.

*13–32. Atmospheric air at an absolute pressure of 103 kPa and temperature of 20°C flows through the converging nozzle into the tank where the absolute pressure at A is 30 kPa. Determine the mass flow into the tank.

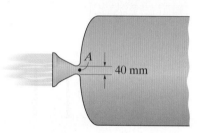

Probs. 13–31/32

13–33. The diameter of the exit of a converging nozzle is 50 mm. If its entrance is connected to a large tank containing air at an absolute pressure of 500 kPa and temperature of 125°C, determine the mass flow through the nozzle. The ambient air is at an absolute pressure of 101.3 kPa.

13–34. Air flows at $V_A = 100$ m/s at 1200 K and has an absolute pressure of $p_A = 6.25$ MPa. Determine the diameter d of the pipe at B so that M = 1 at B.

13–35. Air flows at $V_A = 100$ m/s at 1200 K and has an absolute pressure of $p_A = 6.25$ MPa. Determine the diameter d of the pipe at B so that M = 0.8 at B.

*13–36. If the fuel mixture within the chamber of the rocket is under an absolute pressure of 1.30 MPa, determine the Mach number of the exhaust if the area ratio of the exit to the throat is 2.5. Assume that fully expanded supersonic flow occurs. Take $k = 1.40$ for the fuel mixture. The atmosphere has a pressure of 101.3 kPa.

Prob. 13–36

13–37. The diameter of the exit of a converging nozzle is 50 mm. If its entrance is connected to a large tank containing air at an absolute pressure of 180 kPa and temperature of 125°C, determine the mass flow from the tank. The ambient air is at an absolute pressure of 101.3 kPa.

13–38. The converging–diverging nozzle at the end of a supersonic jet engine is to be designed to operate efficiently when the absolute outside air pressure is 25 kPa. If the absolute stagnation pressure within the engine is 400 kPa and the stagnation temperature is 1200 K, determine the exit plane diameter and the throat diameter for the nozzle if the mass flow is 15 kg/s. Take $k = 1.40$ and $R = 256$ J/kg·K.

13–39. Air flows at 200 m/s through the pipe. Determine the Mach number of the flow and the mass flow if the temperature is 500 K and the absolute stagnation pressure is 200 kPa. Assume isentropic flow.

*13–40. Air flows at 200 m/s through the pipe. Determine the pressure within the flow if the temperature is 400 K and the absolute stagnation pressure is 280 kPa. Assume isentropic flow.

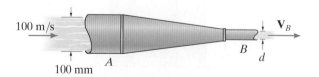

Probs. 13–34/35

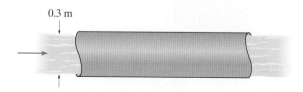

Probs. 13–39/40

13–41. Natural gas (methane) has an absolute pressure of 400 kPa and flows through the pipe at A at $M = 0.1$. Determine the diameter of the throat of the nozzle so that $M = 1$ at the throat. Also, what are the stagnation pressure, the pressure at the throat, and the subsonic and supersonic Mach numbers of the flow through pipe B?

*13–42. Air has an absolute pressure of 400 kPa and flows through the pipe at A at $M = 0.5$. Determine the Mach number at the throat of the nozzle where $d_t = 110$ mm, and the Mach number in the pipe at B. Also, what are the stagnation pressure and the pressure in the pipe at B?

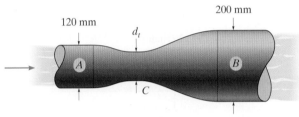

Probs. 13–41/42

13–43. The tank contains oxygen at a temperature of 70°C and absolute pressure of 800 kPa. If the converging nozzle at the exit has a diameter of 6 mm, determine the initial mass flow out of the tank if the outside absolute pressure is 100 kPa.

*13–44. The tank contains helium at a temperature of 80°C and absolute pressure of 175 kPa. If the converging nozzle at the exit has a diameter of 6 mm, determine the initial mass flow out of the tank if the outside absolute pressure is 98 kPa.

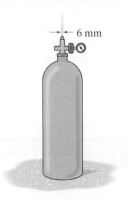

Probs. 13–43/44

13–45. The large tank contains air at 250 K under an absolute pressure of 1.20 MPa. When the valve is opened, the nozzle chokes. The outside atmospheric pressure is 101.3 kPa. Determine the mass flow from the tank. The nozzle has an exit diameter of 40 mm and a throat diameter of 20 mm.

13–46. The large tank contains air at 250 K under an absolute pressure of 150 kPa. When the valve is opened, determine if the nozzle is choked. The outside atmospheric pressure is 90 kPa. Determine the mass flow from the tank. Assume the flow is isentropic. The nozzle has an exit diameter of 40 mm and a throat diameter of 20 mm.

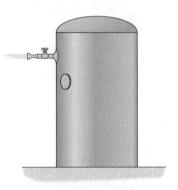

Probs. 13–45/46

13–47. Air at A flows into the nozzle at $M = 0.4$. Determine the Mach number at C and at B.

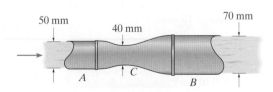

Prob. 13–47

*13–48. Air flows isentropically into the nozzle at $M_A = 0.2$ and out at $M_B = 2$. If the diameter of the nozzle at A is 30 mm, determine the diameter of the throat and the diameter at B. Also, if the absolute pressure at A is 300 kPa, determine the stagnation pressure and the pressure at B.

13–49. Air flows isentropically into the nozzle at $M_A = 0.2$ and out at $M_B = 2$. If the diameter of the nozzle at A is 30 mm, determine the diameter of the throat and the diameter at B. Also, if the temperature at A is 300 K, determine the stagnation temperature and the temperature at B.

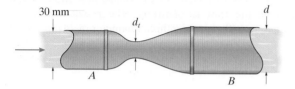

30 mm

d_t

d

A

B

Probs. 13–48/49

13–50. Air flows through a pipe having a diameter of 50 mm. Determine the mass flow if the stagnation temperature of the air is 20°C, the absolute pressure is 300 kPa, and the stagnation pressure is 375 kPa.

13–51. Air at A flows into the nozzle at $M = 0.4$. If $p_A = 125$ kPa and $T_A = 300$ K, determine the pressure at B and the velocity at B.

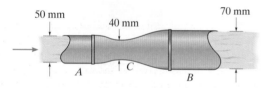

50 mm

40 mm

70 mm

A

C

B

Prob. 13–51

*13–52. Air at a temperature of 25°C and standard atmospheric pressure of 101.3 kPa flows through the nozzle into the pipe where the absolute internal pressure is 80 kPa. Determine the mass flow into the pipe. The nozzle has a throat diameter of 10 mm.

13–53. Air at a temperature of 25°C and standard atmospheric pressure of 101.3 kPa flows through the nozzle into the pipe where the absolute internal pressure is 30 kPa. Determine the mass flow into the pipe. The nozzle has a throat diameter of 10 mm.

10 mm

Probs. 13–52/53

Sec. 13.7

13–54. The duct has a diameter of 200 mm. If the average friction factor is $f = 0.003$, and air is drawn into the duct with an inlet velocity of 200 m/s, temperature of 300 K, and absolute pressure of 180 kPa, determine these properties at the exit.

13–55. The duct has a diameter of 200 mm. If the average friction factor is $f = 0.003$, and air is drawn into the duct with an inlet velocity of 200 m/s, temperature of 300 K, and absolute pressure of 180 kPa, determine the mass flow through the duct and the resultant friction force acting on the 30-m length of duct.

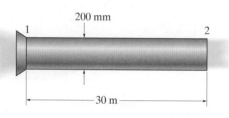

200 mm

1

2

30 m

Probs. 13–54/55

*13–56. Air in a large room has a temperature of 24°C and absolute pressure of 101 kPa. If it is drawn into the 200-mm-diameter duct isentropically such that the absolute pressure at section 1 is 90 kPa, determine the critical length of duct L_{max} where the flow becomes choked, and the Mach number, temperature, and pressure at section 2. Take the average friction factor to be $f = 0.002$.

13–57. Air in a large room has a temperature of 24°C and absolute pressure of 101 kPa. If it is drawn into the 200-mm-diameter duct isentropically such that the absolute pressure at section 1 is 90 kPa, determine the mass flow through the duct and calculate the resultant friction acting on the duct. Also, what is the required length L_{max} to choke the flow? Take the average friction factor to be $f = 0.002$.

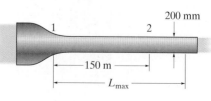

200 mm

1

2

150 m

L_{max}

Probs. 13–56/57

Sec. 13.7

13–58. A large reservoir contains air at a temperature of $T = 20°C$ and absolute pressure of $p = 300\,kPa$. The air flows isentropically through the nozzle and then through the 1.5-m-long, 50-mm-diameter pipe having an average friction factor of 0.03. Determine the mass flow through the pipe and the corresponding velocity, pressure, and temperature at the inlet 1 and outlet 2 if the flow is choked at section 2.

13–59. A large reservoir contains air at a temperature of $T = 20°C$ and absolute pressure of $p = 300\,kPa$. The air flows isentropically through the nozzle and then through the 1.5-m-long, 50-mm-diameter pipe having an average friction factor of 0.03. Determine the stagnation temperature and pressure at outlet 2 and the change in entropy between the inlet 1 and outlet 2 if the pipe is choked at section 2.

13–62. The 100-mm-diameter pipe is connected by a nozzle to a large reservoir of air that is at a temperature of 40°C and absolute pressure of 450 kPa. If the absolute pressure at section 1 is 30 kPa, determine the mass flow through the pipe and the length L of the pipe so that a backpressure of 90 kPa in the tank maintains supersonic flow through the pipe. Assume a constant friction factor of 0.0085 throughout the pipe.

13–63. The 100-mm-diameter pipe is connected by a nozzle to a large reservoir of air that is at a temperature of 40°C and absolute pressure of 450 kPa. If the absolute pressure at section 1 is 30 kPa, determine the mass flow through the pipe and the length L of the pipe so that sonic flow occurs into the tank. What is the required backpressure in the tank for this to occur? Assume a constant friction factor of 0.0085 throughout the pipe.

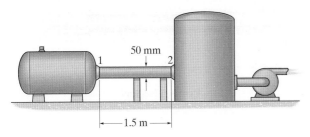

Probs. 13–58/59

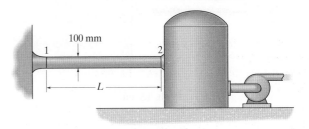

Probs. 13–62/63

13–60. The 40-mm-diameter pipe has a friction factor of $f = 0.015$. A nozzle on the large tank A delivers nitrogen isentropically to the pipe at section 1 with a velocity of 1200 m/s, temperature of 460 K, and absolute pressure of 750 kPa. Determine the mass flow. Show that a normal shock forms within the pipe. Take $L' = 1.35\,m$.

13–61. The 40-mm-diameter pipe has a friction factor of $f = 0.015$. A nozzle on the large tank A delivers nitrogen isentropically to the pipe at section 1 with a velocity of 200 m/s and temperature of 460 K. Determine the velocity and temperature of the nitrogen at $L=2\,m$ if $L' = 3\,m$.

13–64. The 100-mm-diameter pipe is connected by a nozzle to a large reservoir of air that is at a temperature of 40°C and absolute pressure of 450 kPa. If the backpressure causes $M_1 > 1$, and the flow is choked at the exit, section 2, when $L = 5\,m$, determine the mass flow through the pipe. Assume a constant friction factor of 0.0085 throughout the pipe.

13–65. The 100-mm-diameter pipe is connected by a nozzle to a large reservoir of air that is at a temperature of 40°C and absolute pressure of 450 kPa. If the backpressure causes $M_1 < 1$, and the flow is choked at the exit, section 2, when $L = 5\,m$, determine the mass flow through the pipe. Assume a constant friction factor of 0.0085 throughout the pipe.

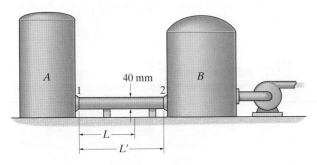

Probs. 13–60/61

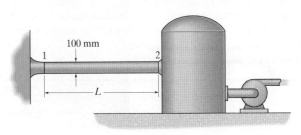

Probs. 13–64/65

13

Sec. 13.8

13–66. Air from a large reservoir is at a temperature of 275 K and absolute pressure of 101 kPa. It isentropically enters the duct at section 1. If 80 kJ/kg of heat is added to the flow, determine the greatest possible velocity it can have at section 1. The backpressure at 2 causes $M_1 < 1$.

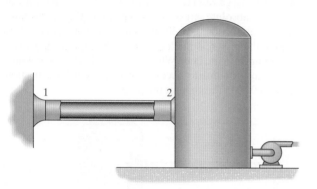

Prob. 13–66

13–67. Air from a large reservoir is at a temperature of 275 K and absolute pressure of 101 kPa. It isentropically enters the duct at section 1. If 80 kJ/kg of heat is added to the flow, determine the temperature and pressure at the entrance of the duct. The backpressure at 2 causes $M_1 = 1$.

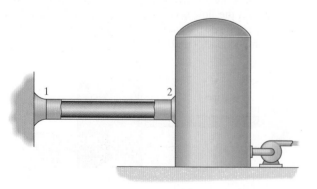

Prob. 13–67

*_**13–68.**_ Nitrogen having a temperature of $T_1 = 270\,\text{K}$ and absolute pressure of $p_1 = 330\,\text{kPa}$ flows into the smooth pipe at $M_1 = 0.3$. If it is heated at 100 kJ/kg·m, determine the velocity and pressure of the nitrogen when it exits the pipe at section 2.

13–69. Nitrogen having a temperature of $T_1 = 270\,\text{K}$ and absolute pressure of $p_1 = 330\,\text{kPa}$ flows into the smooth pipe at $M_1 = 0.3$. If it is heated at 100 kJ/kg·m, determine the stagnation temperatures at sections 1 and 2, and the change in entropy per unit mass between these two sections.

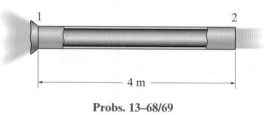

4 m

Probs. 13–68/69

13–70. Nitrogen having a temperature of 300 K and absolute pressure of 450 kPa flows from a large reservoir into a 100-mm-diameter duct. As it flows, 100 kJ/kg of heat is added. Determine the temperature, pressure, and density at section 1 if the backpressure causes $M_1 > 1$ and the flow becomes choked at section 2.

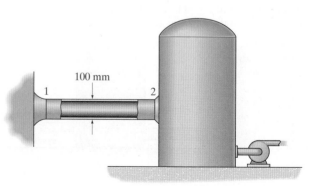

100 mm

Prob. 13–70

13–71. Nitrogen having a temperature of 300 K and absolute pressure of 450 kPa flows from a large reservoir into a 100-mm-diameter duct. As it flows, 100 kJ/kg of heat is added. Determine the mass flow if the backpressure causes $M_1 < 1$ and the flow becomes choked at section 2.

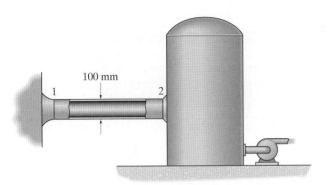

100 mm

1   2

Prob. 13–71

*13–72.** Air is drawn isentropically into the pipe at $V_1 = 640$ m/s, $T_1 = 80°C$, and absolute pressure of $p_1 = 250$ kPa. If it exits the pipe having a speed of 470 m/s, determine the amount of heat per unit mass that the pipe supplies to the air.

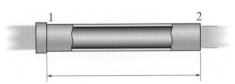

1   2

Prob. 13–72

13–73. Air is drawn into the 100-mm-diameter pipe at $V_1 = 640$ m/s, $T_1 = 80°C$, and absolute pressure of $p_1 = 250$ kPa. If it exits the pipe having a speed of 470 m/s, determine the stagnation temperatures at sections 1 and 2 and the change in entropy per unit mass between these sections.

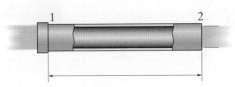

1   2

Prob. 13–73

Sec. 13.9–13.10

13–74. The converging nozzle has an exit diameter of 0.25 m. If the fuel-oxidizer mixture within the large tank has an absolute pressure of 4 MPa and temperature of 1800 K, determine the mass flow from the nozzle when the backpressure is a vacuum. The mixture has $k = 1.38$ and $R = 296$ J/kg·K.

13–75. The converging nozzle has an exit diameter of 0.25 m. If the fuel-oxidizer mixture within the large tank has an absolute pressure of 4 MPa and temperature of 1800 K, determine the mass flow from the nozzle if the atmospheric pressure is 100 kPa. The mixture has $k = 1.38$ and $R = 296$ J/kg·K.

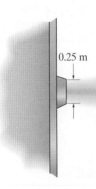

0.25 m

Probs. 13–74/75

*13–76.** The nozzle is attached onto the end of the pipe. The air supplied from the pipe is at a stagnation temperature of 120°C and an absolute stagnation pressure of 800 kPa. Determine the mass flow from the nozzle if the backpressure is 60 kPa.

13–77. The nozzle is attached onto the end of the pipe. The air in the pipe is at a stagnation temperature of 120°C and an absolute stagnation pressure of 800 kPa. Determine the two values of the backpressure that will choke the nozzle yet produce isentropic flow. Also, what is the maximum velocity of the isentropic flow?

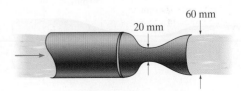

60 mm

20 mm

Probs. 13–76/77

13

13–78. A standing shock occurs in the pipe when the upstream conditions for air have an absolute pressure of $p_1 = 80$ kPa, temperature $T_1 = 75°C$, and velocity $V_1 = 700$ m/s. Determine the downstream pressure, temperature, and velocity of the air. Also, what are the upstream and downstream Mach numbers?

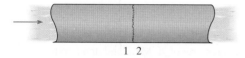

Prob. 13–78

13–79. The jet plane creates a shock that forms in air having a temperature of 20°C and absolute pressure of 80 kPa. If it travels at 1200 m/s, determine the pressure and temperature just behind the shock wave.

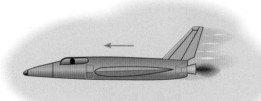

Prob. 13–79

*13–80. Air at a temperature of 20°C and an absolute pressure of 180 kPa flows from a large tank through the nozzle. Determine the backpressure at the exit that causes a shock wave to form at the location where the nozzle diameter is 50 mm.

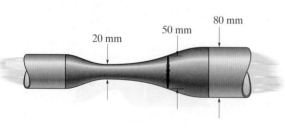

Prob. 13–80

13–81. The large tank supplies air at a temperature of 350 K and an absolute pressure of 600 kPa to the nozzle. If the throat diameter is 0.3 m and the exit diameter is 0.5 m, determine the range of backpressures that will cause expansion shock waves to form at the exit.

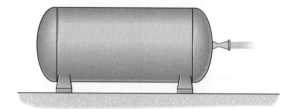

Prob. 13–81

13–82. The jet engine is tested on the ground at standard atmospheric pressure of 101.3 kPa. If the fuel–air mixture enters the inlet of the 300-mm-diameter nozzle at 250 m/s, with an absolute pressure of 300 kPa and temperature of 800 K, and exits with supersonic flow, determine the velocity of the exhaust developed by the engine. Take $k = 1.4$ and $R = 249$ J/kg·K. Assume isentropic flow.

13–83. The jet engine is tested on the ground at standard atmospheric pressure of 101.3 kPa. If the fuel–air mixture enters the inlet of the 300-mm-diameter nozzle at 250 m/s, with an absolute pressure of 300 kPa and temperature of 800 K, determine the required diameter of the throat d_t, and the exit diameter d_e, so that the flow exits with isentropic supersonic flow. Take $k = 1.4$ and $R = 249$ J/kg·K.

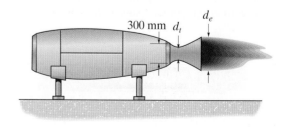

Probs. 13–82/83

*13–84. The large tank supplies air at a temperature of 350 K and an absolute pressure of 600 kPa to the nozzle. If the throat diameter is 0.3 m and the exit diameter is 0.5 m, determine the range of backpressures that will cause oblique shock waves to form at the exit.

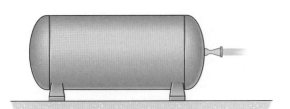

Prob. 13–84

13–85. A 200-mm-diameter pipe contains air at a temperature of 10°C and an absolute pressure of 100 kPa. If a shock is formed in the pipe and the speed of the air in front of the shock is 1000 m/s, determine the speed of the air behind the shock.

13–86. The bottle tank contains 0.13 m³ of oxygen at an absolute pressure of 900 kPa and temperature of 20°C. If the nozzle has an exit diameter of 15 mm, determine the time needed to drop the absolute pressure in the tank to 300 kPa once the valve is opened. Assume the temperature remains constant in the tank during the flow and the ambient air is at an absolute pressure of 101.3 kPa.

Prob. 13–86

13–87. The large tank supplies air at a temperature of 350 K and an absolute pressure of 600 kPa to the nozzle. If the throat diameter is 30 mm and the exit diameter is 60 mm, determine the range of backpressures that will cause a standing shock to form within the nozzle.

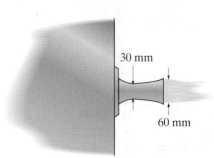

30 mm

60 mm

Prob. 13–87

*13–88. The jet is flying at M = 1.3, where the absolute air pressure is 50 kPa. If a shock is formed at the inlet of the engine, determine the Mach number of the air flow just within the engine where the diameter is 0.6 m. Also, what are the pressure and the stagnation pressure in this region? Assume isentropic flow within the engine.

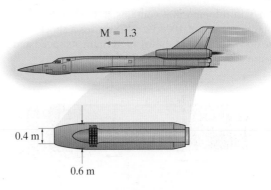

M = 1.3

0.4 m

0.6 m

Prob. 13–88

13–89. The large tank supplies air at a temperature of 350 K and an absolute pressure of 600 kPa to the nozzle. If the throat diameter is 30 mm and the exit diameter is 60 mm, determine the range of backpressures that will cause oblique shock waves to form at the exit.

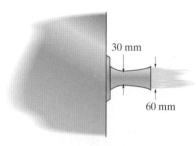

30 mm

60 mm

Prob. 13–89

13–90. A shock is formed in the nozzle at C, where the diameter is 100 mm. If the air flows through the pipe at A at $M_A = 3.0$ and the absolute pressure is $p_A = 15$ kPa, determine the pressure in the pipe at B.

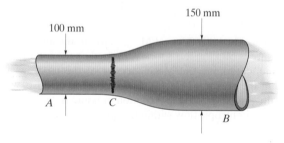

100 mm

150 mm

A   C

B

Prob. 13–90

13–91. The cylindrical plug is fired with a speed of 150 m/s in the pipe that contains still air at 20°C and an absolute pressure of 100 kPa. This causes a shock wave to move down the pipe as shown. Determine its speed and the pressure acting on the plug.

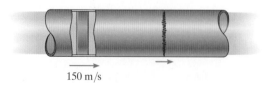

150 m/s

Prob. 13–91

Sec. 13.11–13.12

*13–92. Air flows at 800 m/s through a long duct in a wind tunnel, where the temperature is 20°C and the absolute pressure is 90 kPa. The leading edge of a wing in the tunnel is represented by the 7° wedge. Determine the pressure created on its top surface if the angle of attack is set at $\alpha = 2°$.

13–93. Air flows at 800 m/s through a long duct in a wind tunnel, where the temperature is 20°C and the pressure is 90 kPa. The leading edge of a wing in the tunnel is represented by the 7° wedge. Determine the pressure created on its bottom surface if the angle of attack is set at $\alpha = 2°$.

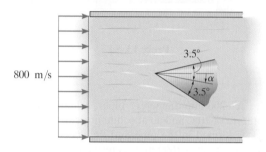

800 m/s

3.5°
α
3.5°

Probs. 13–92/93

13–94. A jet plane is flying in air that has a temperature of 8°C and absolute pressure of 90 kPa. The leading edge of the wing has the wedge shape shown. If the plane has a speed of 800 m/s and the angle of attack is 2°, determine the pressure and temperature of the air at the upper surface A just in front or to the right of the oblique shock wave that forms at the leading edge.

13–95. A jet plane is flying in air that has a temperature of 8°C and absolute pressure of 90 kPa. The leading edge of the wing has the wedge shape shown. If the plane has a speed of 800 m/s and the angle of attack is 2°, determine the pressure and temperature of the air at the lower surface B just in front or to the right of the oblique shock wave that forms at the leading edge.

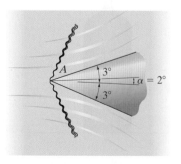

A
3°
$\alpha = 2°$
3°

Probs. 13–94/95

*13–96. Air flows at 800 m/s through a long duct in a wind tunnel, where the temperature is 20°C and the absolute pressure is 90 kPa. The leading edge of a wing in the tunnel can be represented by the 7° wedge. Determine the pressure created on its top surface if the angle of attack is set at $\alpha = 5°$.

13–97. Air flows at 800 m/s through a long duct in a wind tunnel, where the temperature is 20°C and the absolute pressure is 90 kPa. The leading edge of a wing in the tunnel can be represented by the 7° wedge. Determine the pressure created on its bottom surface if the angle of attack is set at $\alpha = 5°$.

13–99. Nitrogen gas at a temperature of 30°C and an absolute pressure of 150 kPa flows through the large rectangular duct at 1200 m/s. When it comes to the transition, it is redirected as shown. Determine the angle β of the oblique shock that forms at A, and the temperature and pressure of the nitrogen just in front or to the right of the wave.

*13–100. Nitrogen gas at a temperature of 30°C and an absolute pressure of 150 kPa flows in the large rectangular duct at 1200 m/s. When it comes to the transition, it is redirected as shown. Determine the temperature and pressure just in front or to the right of the expansion waves that form in the duct at B.

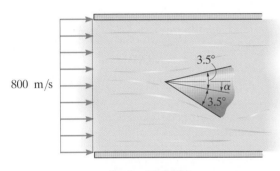

Probs. 13–96/97

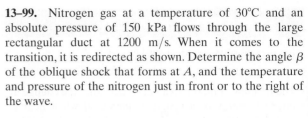
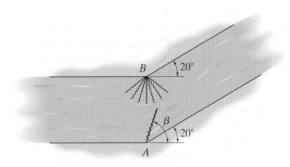

Probs. 13–99/100

13–98. The jet plane is flying upward such that its wings make an angle of attack of 15° with the horizontal. The plane is traveling at 700 m/s, in air having a temperature of 8°C and absolute pressure of 90 kPa. If the leading edge of the wing has an angle of 8°, determine the pressure and temperature of the air just in front or to the right of the expansion waves.

13–101. The wing of a jet plane is assumed to have the profile shown. It is traveling horizontally at 900 m/s, in air having a temperature of 8°C and absolute pressure of 85 kPa. Determine the pressure that acts on the top surface in front or to the right of the oblique shock at A and in front or to the right of the expansion waves at B.

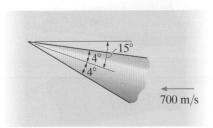

Prob. 13–98

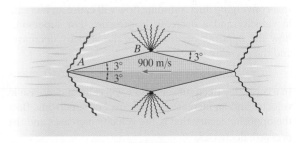

Prob. 13–101

CHAPTER REVIEW

Ideal gases obey the ideal gas law.	$p = \rho RT$
The first law of thermodynamics states that the change in internal energy of a system is increased when heat is added to the system, and decreased when the system does flow work.	$du = dq - \rho dv$
The change in entropy ds indicates the heat energy transfer per unit mass that occurs at a specific temperature during the change in state of a gas. The second law of thermodynamics states that this change will always increase due to friction.	$ds = \dfrac{dq}{T} > 0$
A pressure wave travels through a medium at the maximum velocity c, called the sonic velocity. This is an isentropic process; that is, no heat is lost (adiabatic) and the process is frictionless, so $ds = 0$.	$c = \sqrt{kRT}$
Compressible flow is classified according to the Mach number, $M = V/c$. If $M < 1$ the flow is subsonic, if $M = 1$ it is sonic, and if $M > 1$ it is supersonic. If $V \leq 0.3c$, we can assume the flow is incompressible.	
The stagnation temperature remains the same if the flow is adiabatic, and the stagnation pressure and stagnation density are the same if the flow is isentropic. These properties can be measured at a point where the gas is at rest.	
Specific values of the static temperature T, pressure p, and density ρ of the gas, measured while moving with the flow, are obtained from formulas that relate them to the stagnation values T_0, p_0, ρ_0. They depend on the Mach number and the ratio k of the specific heats for the gas.	
In a converging duct, subsonic flow will cause an increase in velocity and a decrease in pressure. For supersonic flow the opposite occurs; that is, the velocity decreases and the pressure increases.	
In a diverging duct, subsonic flow will cause a decrease in velocity and an increase in pressure. For supersonic flow the opposite occurs; that is, the velocity increases and pressure decreases.	

A Laval nozzle can be used to convert subsonic flow in the convergent section to sonic speed at the throat, and then to further accelerate the flow through the divergent section to supersonic flow at the exit.

A Laval nozzle becomes choked when M = 1 at the throat. If this occurs, there are two backpressures that will produce isentropic flow throughout the divergent section. One produces subsonic flow downstream of the throat, and the other produces supersonic flow.

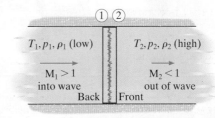

M < 1

M = 1

M > 1

Subsonic   Sonic   Supersonic

13

Fanno flow considers the effect of pipe friction as the gas undergoes an adiabatic process. Equations are used to determine the properties of the gas at any location, provided these properties are known at the reference or critical location, where M = 1.

Rayleigh flow considers the effect of heating or cooling the gas as it flows through a pipe. The properties of the gas can be determined at any location, provided they are known at the reference or critical location, where M = 1.

A shock wave is a nonisentropic process that occurs over a very small thickness. Because the process is adiabatic, the stagnation temperature will remain the same on each side of the shock.

The Mach number, temperature, pressure, and density of a gas on each side of a shock can be related, and the results from these equations are presented in tabular form.

① ②

T_1, p_1, ρ_1 (low) T_2, p_2, ρ_2 (high)

$M_1 > 1$ $M_2 < 1$
into wave   out of wave

Back | Front

If the backpressure does not create isentropic flow through a nozzle, then a shock can form within the nozzle or at its exit. This is an inefficient use of the nozzle.

An oblique shock will form at the point where the speed over a surface is the greatest. The properties of temperature, pressure, and density on each side of the shock are related to the normal component of the Mach number, in the same manner as they are for normal shocks.

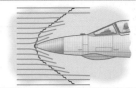

On curved surfaces or at sharp corners, compression waves will merge into an oblique shock, or they can produce expansion waves. The deflection angle of the flow caused by expansion waves can be calculated using the Prandtl–Meyer expansion function.

Compressible flow can be measurement using a Pitot tube and piezometer, or a venturi meter. Hot-wire anemometers and other meters can also be used for this purpose.

Chapter 14

(© Liunian/Shutterstock)

Pumps play an important role in transporting fluids in chemical processing plants, and distributing water as shown in this water treatment plant.

Turbomachines

CHAPTER OBJECTIVES

■ To discuss how axial-flow and radial-flow pumps and turbines operate by adding energy to or removing it from the fluid.

■ To study the flow kinematics, torque, power, and performance characteristics of a turbomachine.

■ To discuss the effects of cavitation and show how it can be reduced or eliminated.

■ To show how to select a pump so that it meets the requirements of the flow system.

■ To present some important pump-scaling laws that are related to turbomachine similitude.

14.1 Types of Turbomachines

Turbomachines consist of various forms of pumps and turbines, which transfer energy between the fluid and the rotating blades of the machine. *Pumps*, which include fans, compressors, and blowers add energy to fluids, whereas *turbines* remove energy. Each of these turbomachines can be categorized by the way the fluid flows through it. If the fluid flows along its axis, it is called an ***axial-flow machine***. Examples include the compressor and turbine for a jet engine and the axial-flow pump in Fig. 14–1a. A ***radial-flow machine*** directs the flow mainly in the radial direction to the rotating blades. This occurs in a centrifugal pump, Fig. 14–1b. Finally, a ***mixed-flow machine*** changes the direction of axial flow a moderate amount in the radial direction, as in the case of the mixed-flow pump in Fig. 14–1c.

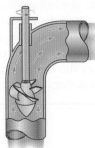

Axial-flow pump
(a)

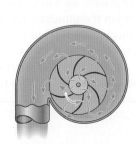

Radial-flow pump
(b)

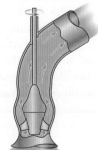

Mixed-flow pump
(c)

Fig. 14–1

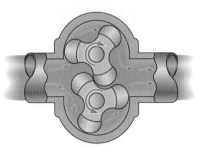

Positive-displacement pump

(d)

Fig. 14–1 (cont.)

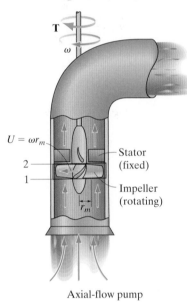

Axial-flow pump

(a)

These three types of turbomachines are generally referred to as *dynamic fluid devices* because the flow is changed by its dynamic interaction with a series of rotating blades as it freely passes through the machine. A device called a *positive-displacement pump* will not be discussed here. It works like your heart, transferring a *specific volume* of fluid by moving the *boundary* of the volume in contact with the fluid. Examples include a gear pump for an internal combustion engine, Fig. 14–1*d*, and a pump composed of a piston and cylinder or screw. Although these devices can produce pressures that are much higher than those produced by a dynamic fluid device, the flow through them is generally much smaller.

14.2   Axial-Flow Pumps

Axial-flow pumps can produce a high flow, but they have a disadvantage of delivering the fluid at a relatively low pressure. As a result, this type of pump works well for removing water from a low-lying area. They are usually designed so that in the process of pumping the fluid, there is no change in the direction of the flow. The fluid enters and then exits the pump in the axial direction, Fig. 14–2*a*. Energy is added to the fluid by using an *impeller* consisting of a series of vanes or blades that are fixed to a rotating shaft. Fixed diffuser vanes, called *stator vanes*, are often located at the downstream side to remove the swirl component of the velocity of the fluid as it exits the impeller. These vanes are sometimes also located at the upstream side if the fluid has an initial whirl, although in most cases straight upward flow occurs, and so initial vanes are not used.

We can show how an axial-flow pump works by considering the impeller blade shown in Fig. 14–2*b*. As the fluid is scooped upwards by the impeller, this removal of fluid *lowers the pressure* in the downstream region, and so more fluid is drawn upwards with a velocity V_a. Once on the blade, the velocity of the fluid is V_1. When the fluid moves off the impeller, it will have a greater velocity V_2.

To analyze the flow through the pump, we will assume the liquid to be an ideal fluid that is guided smoothly onto and over the blades of the impeller. With these assumptions, we will consider the control volume to be the contained liquid within the impeller, Fig. 14–2*a*, and then write the basic equations of fluid mechanics related to the flow. Although within the impeller the flow is unsteady, it is cyclic, and on average it can be considered *quasi steady* as it enters and then leaves the open control surfaces, which are a short distance away from the impeller.

(b)

Fig. 14–2

Continuity.

As the pump draws incoming axial flow through the open control surface 1 in Fig. 14–2b, it must equal the outgoing axial flow through the open control surface 2. Since these surfaces have the same cross-sectional area, applying the continuity equation for steady flow, we have

$$\frac{\partial}{\partial t} \int_{cv} \rho \, d\Psi + \int_{cs} \rho \mathbf{V} \cdot d\mathbf{A} = 0$$

$$0 - \rho V_{a1} A + \rho V_{a2} A = 0$$

$$V_{a1} = V_{a2} = V_a$$

This result is to be expected; that is, the mean velocity of the flow in the axial direction remains constant.

Angular Momentum.

The torque $\mathbf{T}$ that is applied to the liquid by the impeller changes the liquid's angular momentum as it passes over the blades. If we assume that the blades are relatively short, then as a *first approximation*, the angular momentum of the liquid can be determined using the impeller's mean radius r_m, Fig. 14–2c. Applying the angular momentum equation, Eq. 6–4, about the center axis of the control volume yields

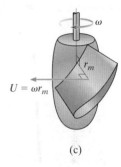

(c)

Fig. 14–2 (cont.)

$$\Sigma \mathbf{M} = \frac{\partial}{\partial t} \int_{cv} (\mathbf{r} \times \mathbf{V}) \rho \, d\Psi + \int_{cs} (\mathbf{r} \times \mathbf{V}) \rho \, \mathbf{V} \cdot d\mathbf{A}$$

$$= 0 + \int_{cs} (\mathbf{r} \times \mathbf{V}) \rho \mathbf{V} \cdot d\mathbf{A}$$

The component of $\mathbf{V}$ that produces the moment in the term $\mathbf{r} \times \mathbf{V}$ is $\mathbf{V}_t$, and the component that contributes to the flow $\rho \mathbf{V} \cdot d\mathbf{A}$ is $\mathbf{V}_a$. Therefore,

$$T = \int_{cs} (r_m V_t) \rho V_a \, dA \qquad (14\text{–}1)$$

Integrating over the control surfaces at sections 1 and 2, using $Q = V_a A$, yields $T = r_m V_{t2} \rho Q - r_m V_{t1} \rho Q$ or

$$\boxed{T = \rho Q r_m (V_{t2} - V_{t1})} \qquad (14\text{–}2)$$

This equation is often referred to as the ***Euler turbomachine equation***. The terms on the right are the product of the mass flow produced by the pump, ρQ, and the moment, $r_m V_t$. If the blades are long, then the torque on the liquid can be determined to a closer approximation by dividing the blades into small segments, each having its own small width Δr and mean radius r_m. In this way, numerical integration of Eq. 14–1 can be carried out.

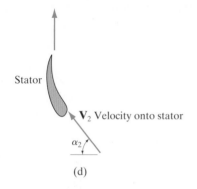

Stator

$\mathbf{V}_2$ Velocity onto stator

α_2

(d)

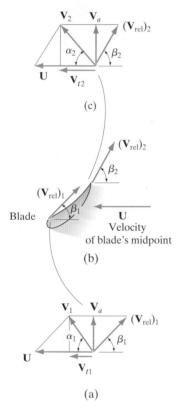

$\mathbf{V}_2$ $\mathbf{V}_a$ $(\mathbf{V}_{rel})_2$

α_2 β_2

$\mathbf{U}$ $\mathbf{V}_{t2}$

(c)

$(\mathbf{V}_{rel})_2$

β_2

$(\mathbf{V}_{rel})_1$

β_1

Blade $\mathbf{U}$
Velocity
of blade's midpoint

(b)

$\mathbf{V}_1$ $\mathbf{V}_a$ $(\mathbf{V}_{rel})_1$

α_1 β_1

$\mathbf{U}$ $\mathbf{V}_{t1}$

(a)

Fig. 14–3

Power.

Power.   The *shaft power* developed by the pump on the fluid is defined as the product of the applied torque and the angular velocity ω (omega) of the impeller. Using Eq. 14–2, we have

$$\dot{W}_{pump} = T\omega = \rho Q r_m (V_{t2} - V_{t1})\omega \qquad (14\text{–}3)$$

We can also write this equation in terms of the velocity of the *midpoint* of the impeller, rather than the impeller's angular velocity. At this point, the blade has a speed of $U = \omega r_m$, Fig. 14–2c, and so

$$\dot{W}_{pump} = \rho Q U (V_{t2} - V_{t1}) \qquad (14\text{–}4)$$

The above equations imply that the torque developed, or the rate of energy transferred to the liquid, is independent of the geometry of the pump or the number of blades that are on the impeller.* Instead, it depends only upon the motion U of the midpoint of the impeller and the tangential component V_t of the velocity of the liquid as it enters and leaves the impeller.

Flow Kinematics.   To properly define the velocity components in the above equations, it is convenient to establish *velocity kinematic diagrams* of the flow as it enters and then exits each impeller blade, Fig. 14–3. The center of the blade has a velocity $U = r_m \omega$, so the velocity of the liquid relative to the blade, $(\mathbf{V}_{rel})_1$, will be at the blade tangent angle β_1, Fig. 14–3a. This interaction changes the velocity of the liquid to $\mathbf{V}_1$, acting at the angle α_1. Note carefully the convention used to establish α_1 and β_1. The angle α_1 is directed between $\mathbf{V}_1$ and $\mathbf{U}$, whereas β_1 is between $\mathbf{V}_{rel}$ and $-\mathbf{U}$. From the parallelogram law of vector addition, we can therefore express the velocity $\mathbf{V}_1$ in terms of *two* different sets of components,

$$\mathbf{V}_1 = \mathbf{U} + (\mathbf{V}_{rel})_1 \qquad (14\text{–}5)$$

and

$$\mathbf{V}_1 = \mathbf{V}_{t1} + \mathbf{V}_a \qquad (14\text{–}6)$$

where $V_{t1} = V_a \cot \alpha_1$.

For the most efficient performance of the pump, the blade angle β_1 should be designed so that $\mathbf{V}_1$ is directed *upwards*, where $\alpha_1 = 90°$. When this is the case, the initial upward direction of the liquid velocity will be maintained so that $\mathbf{V}_1 = \mathbf{V}_a$. Also, $V_{t1} = 0$, so that the power, calculated from Eq. 14–4, will be a maximum.

---

*There are disadvantages to having too many blades because the flow becomes restricted and frictional losses increase.

A similar situation occurs just as the liquid is directed off the impeller. Here, the relative velocity of the liquid at the tail of the impeller blade is $(\mathbf{V}_{rel})_2$, which is directed at the blade angle β_2, Fig. 14–3b. The components of $\mathbf{V}_2$ are shown in Fig. 14–3c, where

$$\mathbf{V}_2 = \mathbf{U} + (\mathbf{V}_{rel})_2 \qquad (14\text{--}7)$$

and

$$\mathbf{V}_2 = \mathbf{V}_{t2} + \mathbf{V}_a \qquad (14\text{--}8)$$

To reduce turbulence and frictional losses, proper design requires that $\mathbf{V}_2$ be directed tangentially onto the stator vanes at the angle α_2, as shown in Fig. 14–3d.

Procedure for Analysis

The following procedure provides a method for analyzing the flow through the blades of an axial-flow pump.

- For flow *onto* a blade, first establish the direction of the midpoint velocity of the blade, $\mathbf{U}$. Its magnitude is determined from $U = \omega r_m$. Then show the velocity of the flow $\mathbf{V}_1$ at an angle α_1 from $\mathbf{U}$, Fig. 14–3b.

- The axial velocity through the pump $\mathbf{V}_a$ is always perpendicular to $\mathbf{U}$. Its magnitude is determined from the flow $Q = V_a A$, where A is the open cross-sectional area through the blades. The relative velocity of the flow onto a blade, $(\mathbf{V}_{rel})_1$, is tangent to the blade at the design angle β_1, measured from the direction of $-\mathbf{U}$, Fig. 14–3b.

- Analyzing flow *off* the blades follows the same procedure, as noted by the kinematic diagram in Fig. 14–3d.

- When $\mathbf{U}$, $\mathbf{V}$, $\mathbf{V}_a$, and $\mathbf{V}_{rel}$ are established, then the tangential component of velocity $\mathbf{V}_t$ can be constructed by the vector resolution of $\mathbf{V}$ into its rectangular components. Here $\mathbf{V} = \mathbf{V}_a + \mathbf{V}_t$, but also $\mathbf{V} = \mathbf{U} + \mathbf{V}_{rel}$. Depending upon the problem, the various magnitudes and/or angles can be determined from these vector resolutions using trigonometry.

- Once the tangential components of velocity, $\mathbf{V}_{t1}$ and $\mathbf{V}_{t2}$, are known, then the torque and power requirements for the pump can be found from Eqs. 14–2 through 14–4.

14

EXAMPLE | **14.1**

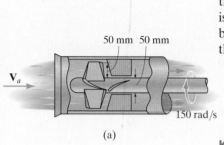

50 mm 50 mm

V_a

150 rad/s

(a)

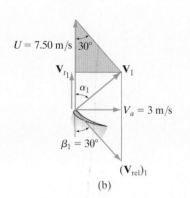

$U = 7.50$ m/s   30°

V_{t_1} V_1

α_1

$V_a = 3$ m/s

$\beta_1 = 30°$

$(V_{rel})_1$

(b)

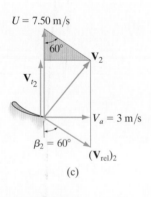

$U = 7.50$ m/s

60° V_2

V_{t_2}

$V_a = 3$ m/s

$\beta_2 = 60°$

$(V_{rel})_2$

(c)

Fig. 14–4

The axial-flow pump in Fig. 14–4a has an impeller that is rotating at 150 rad/s. The blades are 50 mm long and are fixed to the 50-mm-diameter shaft. If the pump produces a flow of 0.06 m³/s, and the lead blade angle of each blade is $\beta_1 = 30°$ and the tail blade angle is $\beta_2 = 60°$, determine the velocity of the water when it is just on a blade and when it just leaves it. The average cross-sectional area of the open region within the impeller is 0.02 m².

SOLUTION

Fluid Description.   We assume steady ideal flow, using average velocities.

Kinematics.   Using the mean radius to determine the velocity of the midpoint of the blades, we have

$$U = \omega r_m = (150 \text{ rad/s})\left(0.025 \text{ m} + \frac{0.05 \text{ m}}{2} \right) = 7.50 \text{ m/s}$$

Also, since the flow is known, the axial velocity of the liquid through the blades is

$$Q = VA; \qquad 0.06 \text{ m}^3/\text{s} = V_a(0.02 \text{ m}^2)$$
$$V_a = 3 \text{ m/s}$$

The kinematic diagram for the water as it just encounters a blade is shown in Fig. 14–4b. As usual, two sets of components for V_1 are established. They are $\mathbf{V}_1 = \mathbf{V}_{t1} + \mathbf{V}_a$ and $\mathbf{V}_1 = \mathbf{U} + (\mathbf{V}_{rel})_1$. Using trigonometry, one way we can determine V_1 is as follows:

$$V_{t1} = 7.50 \text{ m/s} - (3 \text{ m/s}) \cot 30° = 2.304 \text{ m/s}$$
$$\tan \alpha_1 = \frac{3 \text{ m/s}}{2.304 \text{ m/s}}, \qquad \alpha_1 = 52.47°$$
$$3 \text{ m/s} = V_1 \sin 52.47°, \quad V_1 = 3.78 \text{ m/s} \qquad\qquad Ans.$$

The kinematic diagram for the water just leaving the blade is shown in Fig. 14–4c. We can determine V_2 the same way we determined V_1, but here is another way to do it.

$$\tan 60° = \frac{3 \text{ m/s}}{7.50 \text{ m/s} - V_{t2}}, \quad V_{t2} = 5.768 \text{ m/s}$$

Therefore,

$$V_2 = \sqrt{(V_a)^2 + (V_{t2})^2} = \sqrt{(3 \text{ m/s})^2 + (5.768 \text{ m/s})^2} = 6.50 \text{ m/s} \; Ans.$$

EXAMPLE 14.2

The blades on an impeller of the axial-flow pump in Fig. 14–5a have a mean radius of $r_m = 125$ mm and rotate at 1000 rev/min. If the pump is required to produce a flow of 0.2 m³/s, determine the initial blade angle β_1 so that the pump runs efficiently. Also, find the average torque that must be applied to the shaft of the impeller, and the average power output of the pump if the fluid is water. The average open cross-sectional area through the impellers is 0.03 m².

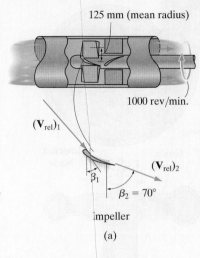

125 mm (mean radius)

1000 rev/min.

Impeller

(a)

SOLUTION

Fluid Description. We assume steady ideal flow through the pump and will use average velocities. $\rho = 1000$ kg/m³.

Kinematics. The midpoint velocity of the blades has a magnitude of

$$U = \omega r_m = \left(\frac{1000 \text{ rev}}{\text{min}}\right)\left(\frac{1 \text{ min}}{60 \text{ s}}\right)\left(\frac{2\pi \text{ rad}}{1 \text{ rev}}\right)(0.125 \text{ m}) = 13.09 \text{ m/s}$$

And the axial velocity of the flow through the impeller is

$$Q = V_a A; \qquad 0.2 \text{ m}^3/\text{s} = V_a(0.03 \text{ m}^2); \qquad V_a = 6.667 \text{ m/s}$$

The pump will run efficiently when the velocity of the water onto the blades is $V_1 = V_a = 6.667$ m/s as shown in Fig. 14–5b. Here $\alpha_1 = 90°$ and so

$$\tan \beta_1 = \frac{6.667 \text{ m/s}}{13.09 \text{ m/s}} \qquad \beta_1 = 27.0° \qquad Ans.$$

Also, because $\alpha_1 = 90°$,

$$V_{t1} = 0$$

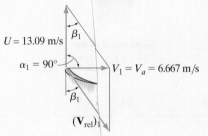

$U = 13.09$ m/s

$\alpha_1 = 90°$

$V_1 = V_a = 6.667$ m/s

$(\mathbf{V}_{rel})_1$

Entrance

(b)

At the exit, Fig. 14–5c,

$$V_{t2} = 13.09 \text{ m/s} - (6.667 \text{ m/s}) \cot 70° = 10.664 \text{ m/s}$$

Torque and Power. Since the tangential components of velocity are known, Eq. 14–2 can be applied to determine the torque.

$$T = \rho Q r_m(V_{t2} - V_{t1})$$
$$= (1000 \text{ kg/m}^3)(0.2 \text{ m}^3/\text{s})(0.125 \text{ m})(10.664 \text{ m/s} - 0)$$
$$= 267 \text{ N} \cdot \text{m} \qquad Ans.$$

From Eq. 14–4, the power supplied to the water by the pump is

$$\dot{W}_{pump} = \rho Q U(V_{t2} - V_{t1})$$
$$= (1000 \text{ kg/m}^3)(0.2 \text{ m}^3/\text{s})(13.09 \text{ m/s})(10.664 \text{ m/s} - 0)$$
$$= 27.9 \text{ kW} \qquad Ans.$$

Note: If $\alpha_1 < 90°$, V_1 would have a component V_{t1} as in the previous example and this would decrease the power.

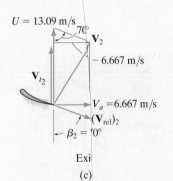

$U = 13.09$ m/s

$70°$

V_2

-6.667 m/s

V_{t2}

$V_a = 6.667$ m/s

$(\mathbf{V}_{rel})_2$

$\beta_2 = 70°$

Exit

(c)

Fig. 14–5

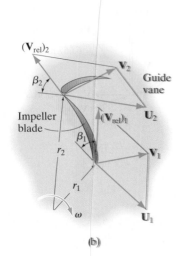

Casing

Guide vane

Impeller blade

T

ω 1   2

b

Centrifugal pump

(a)

$(\mathbf{V}_{rel})_2$

β_2

$\mathbf{V}_2$ Guide vane

Impeller blade

$(\mathbf{V}_{rel})_1$ $\mathbf{U}_2$

β_1

r_2

$\mathbf{V}_1$

r_1

ω

$\mathbf{U}_1$

(b)

Fig. 14–6

14.3   Radial-Flow Pumps

A **radial-flow pump** is probably the most common type of pump used in industry. It operates at slower speeds than an axial-flow pump, so it produces a lower flow, but at a higher pressure. Radial-flow pumps are designed so that the liquid enters the pump in the axial direction at the center of the pump, and then it is directed onto the impeller in the radial direction, Fig. 14–6a. To show how this pump works, consider the impeller blade in Fig. 14–6b. Due to its rotation, the liquid is scooped up by the blade. Just as in the axial-flow pump, this will lower the pressure and draw more liquid onto the blade. As the liquid flows off each blade, it then moves onto, then later off, the guide vanes and accumulates around the entire circumference of the pump's casing, Fig. 14–6a. Because of the way the liquid flows, this type of pump is also called a **centrifugal** or a **volute pump**.

Kinematics. The kinematics of the flow over the impeller can be analyzed in a manner similar to that used for an axial-flow pump, as described in the Procedure for Analysis given in Sec. 14.2.

A typical blade is shown in Fig. 14–6c, where the front end of the blade has a velocity of $U_1 = \omega r_1$, and the tail has a higher velocity of $U_2 = \omega r_2$. Note carefully how the front and tail blade angles β are established between $-\mathbf{U}$ and the relative velocity $\mathbf{V}_{rel}$, and the angles α are established between the velocity of the fluid $\mathbf{V}$ and $\mathbf{U}$. Similar to Eqs. 14–5 and 14–6, we can express $\mathbf{V}$ in terms of two different sets of its components, namely, $\mathbf{V} = \mathbf{U} + \mathbf{V}_{rel}$, or its tangential and radial components, $\mathbf{V} = \mathbf{V}_t + \mathbf{V}_r$.

Continuity. To analyze the flow, we will again assume that friction can be neglected, the fluid is incompressible, and the flow is guided smoothly over the blades of the impeller, each of which has a constant width b. The control volume contains the fluid within the impeller as shown in Fig. 14–6a. Since steady flow through the open control surfaces 1 and 2 occurs in the radial direction, then the continuity equation applied to these surfaces becomes

$$\frac{\partial}{\partial t}\int_{cv} \rho\, d\forall + \int_{cs} \rho\mathbf{V}\cdot d\mathbf{A} = 0$$

$$0 - \rho V_{r1}(2\pi r_1 b) + \rho V_{r2}(2\pi r_2 b) = 0$$

or

$$V_{r1}r_1 = V_{r2}r_2 \qquad (14\text{–}9)$$

Note that since $r_2 > r_1$, the velocity component $V_{r2} < V_{r1}$.

Angular Momentum.

The torque on the shaft of the impeller can be related to the angular momentum of the fluid using the angular momentum equation.

$$\Sigma \mathbf{M} = \frac{\partial}{\partial t} \int_{cv} (\mathbf{r} \times \mathbf{V}) \rho \, d\mathbb{V} + \int_{cv} (\mathbf{r} \times \mathbf{V}) \rho \mathbf{V} \cdot d\mathbf{A}$$

The component of $\mathbf{V}$ that produces the *moment* in the factor $\mathbf{r} \times \mathbf{V}$ is $\mathbf{V}_t$, and the component that contributes to the flow $\rho \mathbf{V} \cdot d\mathbf{A}$ is $\mathbf{V}_r$. Therefore, we have

$$T = 0 + r_1 V_{t1} \rho \left[-(V_{r1})(2\pi r_1 b) \right] + r_2 V_{t2} \rho \left[(V_{r2})(2\pi r_2 b) \right]$$

$$= \rho(V_{r2})(2\pi r_2 b) r_2 V_{t2} - \rho(V_{r1})(2\pi r_1 b) r_1 V_{t1} \qquad (14\text{--}10)$$

Since $Q = V_{r1}(2\pi r_1 b) = V_{r2}(2\pi r_2 b)$, we can also express this equation as

$$T = \rho Q(r_2 V_{t2} - r_1 V_{t1}) \qquad (14\text{--}11)$$

From the impeller kinematics in Fig. 14–6c, for any r, $V_t = V_r \cot \alpha$. Therefore, the torque can also be expressed in terms of V_r and α. It is

$$\boxed{T = \rho Q(r_2 V_{r2} \cot \alpha_2 - r_1 V_{r1} \cot \alpha_1)} \qquad (14\text{--}12)$$

Power.

The power developed by a pump is often called the **shaft or brake power**, since it refers to the actual power going to the pump shaft and *not* the power going to the motor. Power is measured in watts, where $1\,\text{W} = 1\,\text{N} \cdot \text{m/s} = 1\,\text{J/s}$.

With Eq. 14–12, the power delivered to the fluid can be expressed in terms of the speed of the head and tail of the impeller blades. Since $U_1 = \omega r_1$ and $U_2 = \omega r_2$, we have

$$\dot{W}_{pump} = T\omega = \rho Q(U_2 V_{r2} \cot \alpha_2 - U_1 V_{r1} \cot \alpha_1) \qquad (14\text{--}13)$$

Radial-flow pumps are often designed so that the blade angle β_1 provides an initial flow that is directed onto the impeller blades in the *radial direction*, that is, the tangential component of velocity $V_{t1} = 0$, and so $\alpha_1 = 90°$, Fig. 14–6d. If this is the case, then Eq. 14–13 will provide a greater power because the last term in the above equation will be zero, i.e., $\cot 90° = 0$.

We can also express Eq. 14–13 in terms of the tangential component of velocity. Since in general $V_t = V_r \cot \alpha$, then

$$\dot{W}_{pump} = \rho Q(U_2 V_{t2} - U_1 V_{t1}) \qquad (14\text{--}14)$$

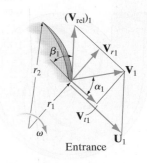

Entrance

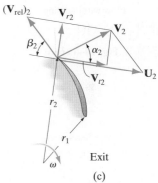

Exit

(c)

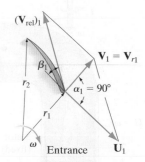

Entrance

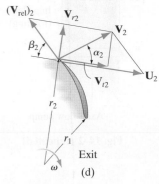

Exit

(d)

Fig. 14–6 (cont.)

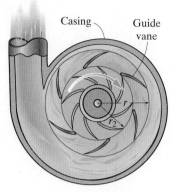

Casing   Guide
            vane

Flow within casing

(e)

Fig. 14–6 (cont.)

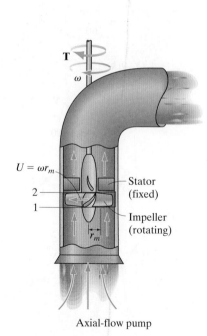

T

ω

$U = \omega r_m$

Stator
(fixed)

2

1

Impeller
(rotating)

r_m

Axial-flow pump

Fig. 14–2 (repeated)

Flow within the Casing.   The flow that occurs within the pump casing can be obtained by applying the angular momentum equation to the fluid within a control volume having open control surfaces just after the head of the guide vanes, 2, and at a general point within the casing, where the liquid is free flowing, Fig. 14–6e. Since here there is no torque on the liquid, $T = 0$, Eq. 14–11 becomes

$$0 = rV_t - r_2V_{t2}$$

or

$$V_t = \frac{r_2V_{t2}}{r} = \frac{\text{const.}}{r}$$

This represents a case of *free-vortex flow*, discussed in Sec. 7.9. As a result, the flow pattern within the casing looks like that shown in Fig. 14–6e. It is this *accumulation* of the flow, as it passes off the guide vanes, that requires the casing to have the growing spiral or volute shape; hence the name "volute pump."

14.4   Ideal Performance for Pumps

The performance of any pump depends upon a balance of energy as the fluid flows through the impeller. For example, in the case of an axial-flow pump, Fig. 14–2a, assuming we have incompressible steady flow through the open control surfaces, and neglecting frictional losses, the energy equation, Eq. 5–13, applied between points 1 (in) and 2 (out) on the open control surfaces becomes

$$\frac{p_{in}}{\gamma} + \frac{V_{in}^2}{2g} + z_{in} + h_{pump} = \frac{p_{out}}{\gamma} + \frac{V_{out}^2}{2g} + z_{out} + h_{turb} + h_L$$

$$h_{pump} = \left(\frac{p_{out}}{\gamma} + \frac{V_{out}^2}{2g} + z_{out}\right) - \left(\frac{p_{in}}{\gamma} + \frac{V_{in}^2}{2g} + z_{in}\right)$$

This is an *ideal pump head* because it neglects any frictional losses that may occur. It represents the change in the total head of the fluid and it applies to both axial- and radial-flow pumps. Using Eq. 5–16, the ideal power produced by the pump is then

$$\dot{W}_{pump} = Q\gamma h_{pump} \tag{14–15}$$

Finding h_{pump} for a pump is important, because it refers to the additional height to which the liquid can be raised when the pump is operating.

Head Loss and Efficiency.

If we express the ideal pump head in terms of the impeller's tangential velocities using Eq. 14–4 for an axial-flow pump or Eq. 14–14 for a radial-flow pump, we have

$$h_{\text{pump}} = \frac{U(V_{t2} - V_{t1})}{g} \qquad (14\text{--}16)$$

Axial-flow pump

$$h_{\text{pump}} = \frac{U_2 V_{t2} - U_1 V_{t1}}{g} \qquad (14\text{--}17)$$

Radial-flow pump

The actual head, $(h_{\text{pump}})_{\text{act}}$, delivered by the pump will be less than this ideal value, because it must account for the mechanical head losses h_L within the pump. These losses are a result of friction developed at the shaft bearings, fluid friction within the pump casing and the impeller, and additional fluid flow losses that occur because of inefficient circulation into and out of the impeller. The head loss is therefore

$$h_L = h_{\text{pump}} - (h_{\text{pump}})_{\text{act}}$$

The **hydraulic** or **pump efficiency** η_{pump} (eta) is the ratio of the actual head delivered by the pump to its ideal head, that is,

$$\eta_{\text{pump}} = \frac{(h_{\text{pump}})_{\text{act}}}{h_{\text{pump}}}(100\%) \qquad (14\text{--}18)$$

Head-Discharge Curve — Radial-Flow Pump.

As stated earlier, radial-flow pumps generally have impeller blades designed so that the fluid has no inlet swirl, so that $V_{t1} = 0$ since $\alpha_1 = 90°$, Fig. 14–6d. Also, for this case, the component V_{t2} can be related to V_{r2} at the tail of the blade by noting that $V_{t2} = U_2 - V_{r2} \cot \beta_2$. The ideal pump head then becomes

$$h_{\text{pump}} = \frac{U_2 V_{t2} - U_1 V_{t1}}{g} = \frac{U_2(U_2 - V_{r2} \cot \beta_2) - U_1(0)}{g}$$

Since $Q = V_r A = V_{r2}(2\pi r_2 b)$, where b is the width of the blades, Fig. 14–6b, we get

$$h_{\text{pump}} = \frac{U_2^2}{g} - \frac{U_2 Q \cot \beta_2}{2\pi r_2 b g} \qquad (14\text{--}19)$$
$$(\alpha_1 = 90°)$$

This equation is plotted in Fig. 14–7a for two different blade angles, $\beta_2 < 90°$, Fig. 14–7b, and $\beta_2 = 90°$, Fig. 14–7c. Notice from the graph that when the blades of the impeller are *curved backward*, $\beta_2 < 90°$, which is the usual case, then the ideal pump head will decrease as the flow Q increases. If $\beta_2 = 90°$, then the tails of the blades are in the radial direction, and so $\cot 90° = 0$, and h_{pump} does not depend on the flow Q. It is simply $h_{\text{pump}} = U_2^2/g$. Engineers generally do not design radial-flow pumps with forward curved blades, for which $\beta_2 > 90°$, because the flow within the pump has a tendency to become unstable and cause the pump to **surge**. This can cause rapid pressure changes that make the impeller oscillate back and forth in an attempt to find its operating point.

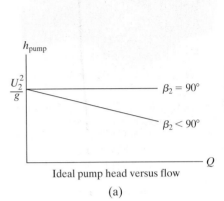

Ideal pump head versus flow

(a)

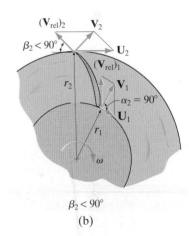

$\beta_2 < 90°$

(b)

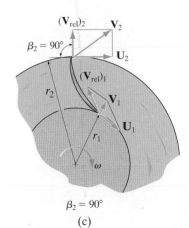

$\beta_2 = 90°$

(c)

Fig. 14–7

14

This leaf blower acts as a radial-flow pump. Notice the volute shape of the casing as it accumulates air and expels it out of the pipe.

Important Points

- The continuity equation requires that the velocity of the flow of liquid through an axial-flow pump remains constant along its axis, whereas for radial-flow pumps, the radial component of velocity of the liquid must decrease as the liquid flows outward.

- Blade angles for axial- and radial-flow pumps are defined in the same way; that is, α is measured between the velocity $\mathbf{U}$ of the blade and the velocity $\mathbf{V}$ of the flow, and β is measured between $-\mathbf{U}$ and the velocity $\mathbf{V}_{rel}$ of the flow measured relative to the blade, Fig. 14–3 and Fig. 14–6b.

- The torque and power developed by axial- and radial-flow pumps depend on the motion of the blades, and on the *tangential components* of the flow as it just enters and just leaves the blades.

- Axial- and radial-flow pumps are usually designed so that the entrance flow onto the blades is only in the axial or radial direction. Therefore, $V_{t1} = 0$ since $\alpha_1 = 90°$.

- The performance of axial- and radial-flow pumps is measured by the increase in energy, or the increase in head h_{pump} produced by the pump once the fluid passes over the impeller. This head depends on the impeller's speed U and the difference in the tangential components of velocity, V_t, of the fluid as it just enters and just leaves the impeller.

- Hydraulic efficiency of axial- and radial-flow pumps is the ratio of the actual pump head divided by the ideal pump head.

EXAMPLE   14.3

Determine the hydraulic efficiency for the axial-flow pump in Example 14.2 if frictional head losses produced by the pump are 0.8 m.

SOLUTION

Fluid Description.   We assume steady, incompressible flow through the pump.

Pump Head.   We can use Eq. 14–16 and the results of Example 14.2 to determine the ideal pump head.

$$h_{pump} = \frac{U(V_{t2} - V_{t1})}{g} = \frac{7.50 \text{ m/s}(5.768 \text{ m/s} - 2.304 \text{ m/s})}{9.81 \text{ m/s}^2} = 2.648 \text{ m}$$

The actual pump head is determined from Eq. 14–11.

$$(h_{pump})_{act} = (h_{pump}) - h_L = 2.648 \text{ m} - 0.8 \text{ m} = 1.848 \text{ m}$$

Hydraulic Efficiency.   Applying Eq. 14–18 yields

$$\eta_{pump} = \frac{(h_{pump})_{act}}{h_{pump}}(100\%) = \frac{1.848 \text{ m}}{2.648 \text{ m}}(100\%) = 69.8\% \qquad Ans.$$

EXAMPLE 14.4

The impeller on the radial-flow pump in Fig. 14–8a has an average inlet radius of 50 mm and outlet radius of 150 mm, and an average width of 30 mm. If the blade angles are $\beta_1 = 20°$ and $\beta_2 = 10°$, determine the flow through the pump, and the ideal pump head when the impeller is rotating at 400 rev/min. The flow onto the impeller is in the radial direction.

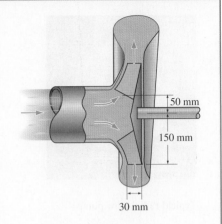

SOLUTION

Fluid Description.   We will assume steady, incompressible flow and use average velocities.

Kinematics.   To find the flow, we must first determine the speed of the fluid as it moves *onto* the blades. We will also need the speed of the blades at its entrance and its exit.

$$U_1 = \omega r_1 = \left(\frac{400 \text{ rev}}{\text{min}}\right)\left(\frac{1 \text{ min}}{60 \text{ s}}\right)\left(\frac{2\pi \text{ rad}}{1 \text{ rev}}\right)(0.05 \text{ m}) = 2.094 \text{ m/s}$$

$$U_2 = \omega r_2 = \left(\frac{400 \text{ rev}}{\text{min}}\right)\left(\frac{1 \text{ min}}{60 \text{ s}}\right)\left(\frac{2\pi \text{ rad}}{1 \text{ rev}}\right)(0.150 \text{ m}) = 6.283 \text{ m/s}$$

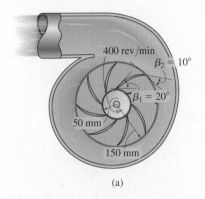

(a)

The kinematic diagram for the flow onto the impeller is shown in Fig. 14–8b. Since $\mathbf{V}_1$ is in the radial direction ($\alpha_1 = 90°$),

$$V_1 = V_r = U_1 \tan \beta_1 = (2.094 \text{ m/s}) \tan 20° = 0.7623 \text{ m/s}$$

Flow.   The flow into as well as out of the pump is

$$\begin{aligned}
Q = V_1 A_1 &= V_1(2\pi r_1 b_1) \\
&= 0.7623 \text{ m/s}\left[2\pi(0.05 \text{ m})(0.03 \text{ m})\right] \\
&= 0.007184 \text{ m}^3/\text{s} = 0.00718 \text{ m}^3/\text{s} \qquad Ans.
\end{aligned}$$

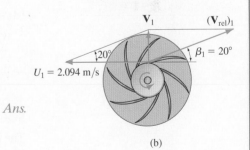

(b)

Fig. 14–8

Ideal Pump Head.   The ideal pump head is

$$\begin{aligned}
h_{\text{pump}} &= \frac{U_2^2}{g} - \frac{U_2 Q \cot \beta_2}{2\pi r_2 bg} \\
&= \frac{(6.283 \text{ m/s})^2}{9.81 \text{ m/s}^2} - \frac{(6.283 \text{ m/s})(0.00718 \text{ m}^3/\text{s}) \cot 10°}{2\pi(0.150 \text{ m})(0.03 \text{ m})(9.81 \text{ m/s}^2)} \\
&= 3.10 \text{ m} \qquad Ans.
\end{aligned}$$

EXAMPLE 14.5

Typical radial-flow pump

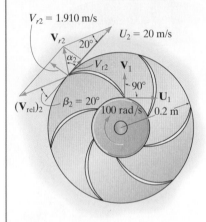

Fig. 14–9

The impeller of a radial-flow water pump has an outer radius of 200 mm, average width of 50 mm, and a tail angle $\beta_2 = 20°$, Fig. 14–9. If the flow onto the blades is in the radial direction, and the blades are rotating at 100 rad/s, determine the ideal power. The discharge is 0.12 m³/s.

SOLUTION I

Fluid Description.   We assume steady, incompressible flow and will use average velocities. Here $\rho_w = 1000$ kg/m³.

Kinematics.   The power can be determined using Eq. 14–13, with $\alpha_1 = 90°$. First we must find U_2, V_{r2}, and α_2. The speed of the impeller at the exit is

$$U_2 = \omega r_2 = (100 \text{ rad/s})(0.2 \text{ m}) = 20 \text{ m/s}$$

Also, the radial component of velocity V_{r2} can be found from $Q = V_{r2}A_2$.

$$0.12 \text{ m}^3/\text{s} = V_{r2}[2\pi(0.2 \text{ m})(0.05 \text{ m})] \qquad V_{r2} = 1.910 \text{ m/s}$$

As shown in Fig. 14–9,

$$V_{t2} = U_2 - V_{r2} \cot 20° = 20 \text{ m/s} - (1.910 \text{ m/s}) \cot 20° = 14.75 \text{ m/s}$$

Therefore,

$$\tan \alpha_2 = \frac{V_{r2}}{V_{t2}} = \left(\frac{1.910 \text{ m/s}}{14.75 \text{ m/s}}\right); \quad \alpha_2 = 7.376°$$

Ideal Power

$$\dot{W}_{\text{pump}} = \rho Q(U_2 V_{r2} \cot \alpha_2 - U_1 V_{r1} \cot \alpha_1)$$

$$= (1000 \text{ kg/m}^3)(0.12 \text{ m}^3/\text{s})[(20 \text{ m/s})(1.910 \text{ m/s}) \cot 7.376° - 0]$$

$$= 35.41 (10^3) \text{ W} = 35.4 \text{ kW} \qquad\qquad Ans.$$

SOLUTION II

The ideal power can also be related to the ideal pump head by Eq. 14–15, $\dot{W}_{\text{pump}} = Q\gamma h_{\text{pump}}$. First we must find h_{pump} by using Eq. 14–19. This gives

$$h_{\text{pump}} = \frac{U_2^2}{g} - \frac{U_2 Q \cot \beta_2}{2\pi r_2 b g}$$

$$= \frac{(20 \text{ m/s})^2}{9.81 \text{ m/s}^2} - \frac{(20 \text{ m/s})(0.12 \text{ m}^3/\text{s}) \cot 20°}{2\pi(0.2 \text{ m})(0.05 \text{ m})(9.81 \text{ m/s}^2)} = 30.08 \text{ m}$$

Therefore,

$$\dot{W}_{\text{pump}} = Q\gamma h_{\text{pump}} = \left(0.12 \text{ m}^3/\text{s}\right)\left(1000 \text{ kg/m}^3\right)\left(9.81 \text{ m/s}^2\right)\left(30.08 \text{ m}\right)$$
$$= 35.41(10^3) \text{ W} = 35.4 \text{ kW} \qquad\qquad Ans.$$

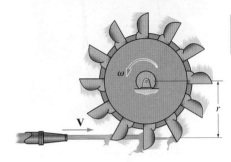

Pelton wheel

(a)

14.5   Turbines

Unlike a pump that delivers energy to a fluid, a *turbine* is a turbomachine that withdraw energy from a fluid. Turbines can be classified into two different types: impulse turbines and reaction turbines. Each type removes the fluid's energy in a specific way.

Impulse Turbines.   An *impulse turbine* consists of a series of "buckets" attached to a wheel as shown in Fig. 14–10a. A high-velocity water jet strikes the buckets, and the momentum of the water is then converted into an angular impulse acting on the wheel. If the flow is split uniformly into two directions by using double-cupped buckets, as shown in Fig. 14–10b, then the device is generally referred to as a **Pelton wheel**, named after Lester Pelton, who designed this turbine in the late 1870's. An impulse turbine such as this is often used in mountainous regions, where water is delivered with a high velocity and with small volume.

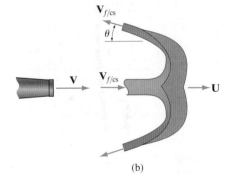

(b)

The force created by the fluid striking the buckets of a Pelton wheel can be determined by applying the equation of linear momentum to the control volume in Fig. 14–10b, which is attached to the bucket and moving with a constant speed **U**. If **V** is the velocity of flow from the nozzle jet, then $\mathbf{V}_{f/cs} = \mathbf{V} - \mathbf{U}$ is the *relative velocity* onto, and then off, each bucket, Fig. 14–10c. Therefore, from the free-body diagram, Fig. 14–10d, for steady flow, with $Q = VA$, we have

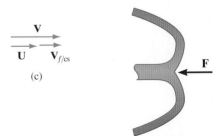

(c)

$$\Sigma \mathbf{F} = \frac{\partial}{\partial t}\int_{cv} \mathbf{V}_{f/cv}\, \rho\, d\forall + \int_{cs} \mathbf{V}_{f/cs}\, \rho\, \mathbf{V}_{f/cs} \cdot d\mathbf{A}$$

$$\left(\overset{+}{\rightarrow}\right) \qquad -F = 0 + V_{f/cs}\,\rho\left(-V_{f/cs}A\right) + \left(-V_{f/cs}\cos\theta\right)\rho\left(V_{f/cs}A\right)$$

Since $Q = V_{f/cs}A$,

$$F = \rho Q V_{f/cs}(1 + \cos\theta) \qquad\qquad (14\text{–}20)$$

Free-body diagram

(d)

Notice that when the exit angle $0° \le \theta < 90°$, the $\cos\theta$ remains positive, thereby producing a larger force than when $90° \le \theta \le 180°$.

Fig. 14–10

Pelton wheels are most efficient at high hydraulic heads and low flows. This one has a diameter of 2.5 m, and was used for a head of 700 m, flow of 4 m³/s, and a rotation of 500 rpm.

Torque. The torque developed on the wheel is the moment of this impulsive force about the axis of the wheel. Here we have a *succession of buckets* that receives the flow, and so for continuous turning, we have

$$T = Fr = \rho Q V_{f/cs}(1 + \cos \theta)r \tag{14–21}$$

Power. Since each bucket has an average speed of $U = \omega r$, Fig. 14–10a, the shaft power developed by the wheel is

$$\dot{W}_{turb} = T\omega = \rho Q V_{f/cs} U(1 + \cos \theta) \tag{14–22}$$

Maximum power from the fluid requires the bucket angle $\theta = 0°$, since $\cos 0° = 1$. Also, since $V_{f/cs} = V - U$, then $\dot{W}_{turb} = \rho Q(V - U)U(2)$. The product $(V - U)U$ must also be a maximum, which requires

$$\frac{d}{dU}(V - U)U = (0 - 1)U + (V - U)(1) = 0$$

$$U = \frac{V}{2}$$

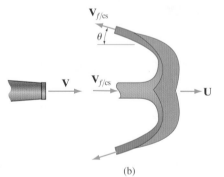

$\mathbf{V}_{f/cs}$

θ

$\mathbf{V}$ $\mathbf{V}_{f/cs}$ $\mathbf{U}$

(b)

Fig. 14–10 (repeated)

Therefore, to generate maximum power, the relative velocity of the fluid *onto* the buckets is $V_{f/cs} = V - V/2 = V/2$, so as the fluid comes *off* the buckets, it will have a velocity of $V = U + V_{f/cs} = V/2 - V/2 = 0$. In other words, the fluid has no kinetic energy, and instead, the power of the water jet will be converted completely into the power of the wheel.

Substituting these results into Eq. 14–22, we get

$$(\dot{W}_{turb})_{max} = \rho Q\left(\frac{V^2}{2}\right) \tag{14–23}$$

This, of course, is a theoretical value, which unfortunately is not practical since the fluid coming off one bucket would actually splash off the back of the next bucket that strikes the water, thus causing a reverse impulse. To avoid this back splashing, engineers normally design the exit angle θ to be about 20°, Fig. 14–10b. Considering other losses due to splashing, and viscous and mechanical friction, a Pelton wheel will have an efficiency of about 85% in converting the energy of the fluid into rotational energy of the wheel.

EXAMPLE | 14.6

The Pelton wheel in Fig. 14–11 has a diameter of 3 m and bucket deflection angles of 160°. If the diameter of the water jet striking the wheel is 150 mm, and the velocity of the jet is 8 m/s, determine the power developed by the wheel when it is rotating at 3 rad/s.

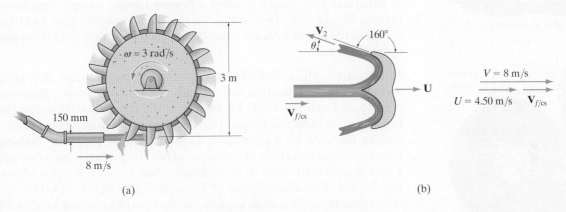

(a)   (b)

Fig. 14–11

SOLUTION

Fluid Description. We have steady flow onto the blades, and we will assume water to be an ideal fluid for which $\rho_w = 1000 \text{ kg/m}^3$.

Kinematics. The average speed of the buckets is

$$U = \omega r = 3 \text{ rad/s}(1.5 \text{ m}) = 4.50 \text{ m/s}$$

The kinematic diagram in Fig. 14–11b shows the flow of the water onto and then off each bucket. Here the relative speed of the flow onto the bucket is $V_{f/cs} = 8 \text{ m/s} - 4.50 \text{ m/s} = 3.50 \text{ m/s}$.

Power. We will use Eq. 14–22, with $\theta = 20°$ and $Q = VA$. Thus,

$$\dot{W}_{turb} = \rho_w Q V_{f/cs} U(1 + \cos \theta)$$

$$= (1000 \text{ kg/m}^3)[(8 \text{ m/s})\pi(0.075 \text{ m})^2](3.50 \text{ m/s})(4.50 \text{ m/s})(1 + \cos 20°)$$

$$= 4.32 \text{ kW} \qquad\qquad\qquad Ans.$$

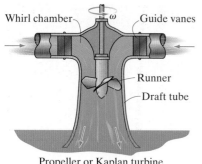

Whirl chamber — ω — Guide vanes

Runner

Draft tube

Propeller or Kaplan turbine

(a)

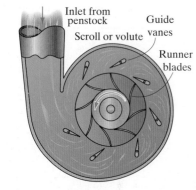

Inlet from penstock — Guide vanes

Scroll or volute

Runner blades

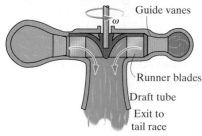

Guide vanes

ω

Runner blades

Draft tube

Exit to tail race

Francis turbine

(b)

Fig. 14–12

Reaction Turbines.   Unlike an impulse turbine, the flow delivered to blades or **runners** of a **reaction turbine** is relatively slow, but it does handle a large volume of fluid. As a result, the casing for the turbine will be completely filled. If the turbine is designed to receive axial flow, it is known as a **propeller turbine**. A **Kaplan turbine**, named after Viktor Kaplan, Fig. 14–12a, is a type of propeller turbine, where the blades can be adjusted to accommodate various flows. Propeller turbines are used for low flows and low hydraulic heads. If the flow through the turbine is radial or a mix of radial and axial flow, it is called a **Francis turbine**, named after James Francis, Fig. 14–12b. This turbine is the most common type used for hydroelectric power because it can be designed for a variety of flows and hydraulic heads.

Kinematics.   The analysis of reaction turbines follows the same methods used to analyze axial- and radial-flow pumps, discussed previously. Even the blade kinematics is the same, as outlined in the Procedure for Analysis of Sec. 14.2. As a review, consider the case of a turbine fan used on a jet engine. This is a type of **propeller turbine**, where constant axial flow of a mixture of hot air and fuel passes through a sequence of stationary or *stator blades*, then rotating or *rotor blades*, Fig. 14–13a. Considering one of these interactions, Fig. 14–13b, the stator directs the flow at an angle α_1, so that it strikes a rotor blade with a velocity $\mathbf{V}_1$. The rotor blade then delivers the flow with a velocity $\mathbf{V}_2$ at the angle α_2 tangent to the blades on the next stator, etc. Since the turbine *removes energy* from the flow, the fluid velocity (kinetic energy) will *decrease* ($V_2 < V_1$) as the flow passes over each set of rotors. Following the same convention as for pumps, note carefully how the velocity components are established. By convention, α is measured between $\mathbf{U}$ and $\mathbf{V}$, and β is measured between $-\mathbf{U}$ and $\mathbf{V}_{\text{rel}}$.

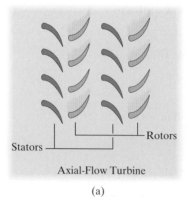

Stators

Rotors

Axial-Flow Turbine

(a)

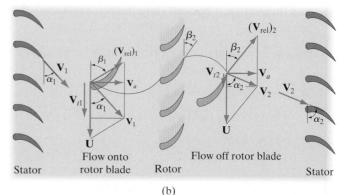

$\mathbf{V}_1$ α_1 β_1 $(\mathbf{V}_{\text{rel}})_1$ $\mathbf{V}_a$

$\mathbf{V}_{t1}$ α_1 $\mathbf{V}_1$

$\mathbf{U}$

Stator

Flow onto rotor blade

Rotor

β_2 $(\mathbf{V}_{\text{rel}})_2$

$\mathbf{V}_{t2}$ β_2 $\mathbf{V}_a$

α_2 $\mathbf{V}_2$ $\mathbf{V}_2$

$\mathbf{U}$ α_2

Flow off rotor blade

Stator

(b)

Fig. 14–13

Torque.   If we apply the angular momentum equation to the rotor, then we can determine the torque applied to the rotor from Eq. 14–2 or Eq. 14–11.

$$T = \rho Q r_m (V_{t2} - V_{t1}) \qquad (14\text{–}24)$$
$$\text{Propeller turbine}$$

$$T = \rho Q (r_2 V_{t2} - r_1 V_{t1}) \qquad (14\text{–}25)$$
$$\text{Francis turbine}$$

Here the torque will be negative since $V_{t1} > V_{t2}$, Fig. 14–13b.

Power.   Knowing the torque, the power produced by the turbine is then

$$\dot{W}_{\text{turb}} = T\omega \qquad (14\text{–}26)$$

Head and Efficiency.   The *ideal turbine head* removed from the fluid can be expressed as a function of the power by using Eq. 14–15.

$$h_{\text{turb}} = \frac{\dot{W}_{\text{turb}}}{Q\gamma} \qquad (14\text{–}27)$$

Finally, since the turbine *extracts energy*, the *actual turbine head* removed from the fluid will be *greater* than the ideal head, because the actual head *also* accounts for frictional losses. Thus,

$$(h_{\text{turb}})_{\text{act}} = h_{\text{turb}} + h_L$$

Therefore, the *turbine efficiency* based on the frictional losses is

$$\eta_{\text{turb}} = \frac{h_{\text{turb}}}{(h_{\text{turb}})_{\text{act}}} (100\%) \qquad (14\text{–}28)$$

The example which follows provides a typical application of these equations.

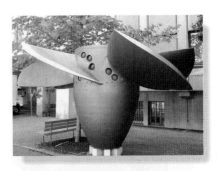

This propeller turbine has a diameter of 4.6 m, and was used for a head of 7.65 m, flow of 87.5 m³/s, and had a rotation of 75 rpm.

This Francis turbine has a diameter of 4.6 m, and was used for a head of 69 m, flow of 447 mm³/s, and had a rotation of 125 rpm.

Important Points

* A Pelton wheel acts as an impulse turbine, where a high-velocity jet of fluid strikes the buckets on the wheel. By changing the momentum of the jet, this creates a torque, causing the wheel to turn, thereby producing power.

* Maximum power produced by a Pelton wheel occurs when the flow is completely reversed by the buckets and the wheel turns so that the velocity of the fluid leaving the buckets is zero.

* Reaction turbines, such as Kaplan and Francis turbines, are similar to axial-flow and radial-flow pumps, respectively. All turbines *remove* energy from the fluid.

EXAMPLE 14.7

Linkage used to operate guide vanes for a Francis turbine

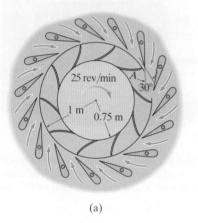

(a)

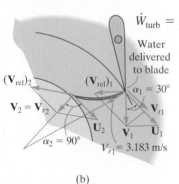

(b)

Fig. 14–14

The guide vanes of a Francis turbine in Fig. 14–14a direct water onto the 200-mm-wide runner blades at an angle of $\alpha_1 = 30°$. The blades are rotating at 25 rev/min, and they discharge water at 4 m³/s in the radial direction, that is, toward the center of the turbine, $\alpha_2 = 90°$, Fig. 14–14b. Determine the power developed by the turbine and the ideal head loss.

SOLUTION

Fluid Description. We have steady flow onto the blades and assume water to be an ideal fluid for which $\rho_w = 1000 \text{ kg/m}^3$.

Kinematics. In order to determine the power, we must first find the *tangential components* of velocity onto and off the blades, Fig. 14–14b. Using the flow, the radial component of velocity of the water *onto* each blade, at $r_1 = 1$m, is

$$Q = V_{r1}A_1$$

$$4 \text{ m}^3/\text{s} = V_{r1}\left[2\pi(1 \text{ m})(0.20 \text{ m})\right]$$

$$V_{r1} = 3.183 \text{ m/s}$$

Thus, from Fig. 14–14b, the tangential component is

$$V_{t1} = (3.183 \text{ m/s}) \cot 30° = 5.513 \text{ m/s}$$

At the tail of the runner, $V_{t2} = 0$, since only radial flow occurs.

Power. Applying Eqs. 14–25 and 14–26, we get $\dot{W}_{\text{turb}} = \rho_w Q(r_2 V_{t2} - r_1 V_{t1})\omega$. Since we have a Francis turbine, the power output is therefore

$$\dot{W}_{\text{turb}} = (1000 \text{ kg/m}^3)(4 \text{ m}^3/\text{s})\left[0 - (1 \text{ m})(5.513 \text{ m/s})\right]\left(\frac{25 \text{ rev}}{\text{min}}\right)\left(\frac{1 \text{ min}}{60 \text{ s}}\right)\left(\frac{2\pi \text{ rad}}{1 \text{ rev}}\right)$$

$$\dot{W}_{\text{turb}} = -57.74(10^3) \text{ W} = -57.7 \text{ kW} \qquad \textit{Ans.}$$

The negative sign indicates the removal of energy from the water, which is expected.

Head Loss. The ideal head loss is determined from Eq. 14–27.

$$h_{\text{turb}} = \frac{\dot{W}_{\text{turb}}}{Q\gamma} = \frac{-57.74(10^3) \text{ W}}{(4 \text{ m}^3/\text{s})(1000 \text{ kg/m}^3)(9.81 \text{ m/s}^2)}$$

$$= -1.47 \text{ m} \qquad \textit{Ans.}$$

14.6   Pump Performance

When selecting a specific pump to use for an intended application, an engineer must have some idea about its **performance characteristics**. For any given flow, these include the shaft power requirements $\dot{W}_{pump}$, the actual pump head $(h_{pump})_{act}$ that can be developed, and the pump's efficiency η. In past sections, we have studied how these characteristics are calculated based on an analytical treatment; however, for any real application, where fluid and mechanical losses occur, the data must be determined *experimentally*. In order to give some idea as to how this is accomplished, we will consider an experimental test as it applies to a radial-flow pump.

Pumps and pipe systems play an important role in refineries and chemical processing plants.

The testing of this pump follows a standardized procedure, which is outlined in Ref. [12]. The setup is shown in Fig. 14–15a, where the pump is made to circulate water (or the intended fluid) from tank A to another tank B through a constant-diameter pipe. Pressure gages are located on each side of the pump, and the valve is used to control the flow, while a meter measures the flow before the water is returned from B to A. The pump's impeller is turned by an electric motor and so the power input $\dot{W}_{pump}$ can also be measured.

The test begins with the valve closed. The pump is then turned on and its impeller begins to always run at a fixed nominal speed ω_0. The valve is then opened slightly, and the rate of flow Q, the pressure difference across the pump $(p_2 - p_1)$, and the power supply $\dot{W}_{pump}$ are all measured. The actual pump head is then calculated using the energy equation, applied between points 1 and 2. Here $V_2 = V_1 = V$, and any elevation difference $z_2 - z_1$ and head losses within the pump are factored in as part of the pressure head, i.e., $(h_{pump})_{act} = h_{pump} - h_L$. We have

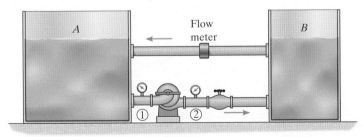

Pump test

(a)

Fig. 14–15

$$\frac{p_{in}}{\gamma} + \frac{V_{in}^2}{2g} + z_{in} + h_{pump} = \frac{p_{out}}{\gamma} + \frac{V_{out}^2}{2g} + z_{out} + h_{turb} + h_L$$

$$\frac{p_1}{\gamma} + \frac{V^2}{2g} + 0 + h_{pump} = \frac{p_2}{\gamma} + \frac{V^2}{2g} + 0 + 0 + h_2$$

$$(h_{pump})_{act} = h_{pump} - h_2 = \frac{p_2 - p_1}{\gamma}$$

Once $(h_{pump})_{act}$ is calculated, and Q and $\dot{W}_{pump}$ are measured, we can then determine the hydraulic efficiency using Eqs. 14–15 and 14–18; that is,

$$\eta_{pump} = \frac{(h_{pump})_{act}}{h_{pump}} = \frac{Q\gamma(h_{pump})_{act}}{\dot{W}_{pump}}$$

If we continue to change Q in increments, until the pump is running at its maximum capacity, and plot the successive values of $(h_{pump})_{act}$, $\dot{W}_{pump}$, and η_{pump} versus Q, we will produce three **performance curves** that look similar to those shown in Fig. 14–15b. Included on this graph is the straight-line ideal pump head, as established previously in Fig. 14–7. Both this and the actual pump head blue curve assume the impeller blade has a backward curve, $\beta < 90°$, which, as we stated earlier, is most often the case.

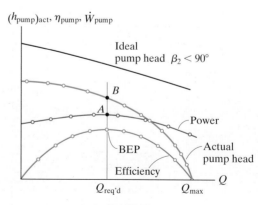

Pump Performance

(b)

Fig. 14–15 cont.

Note that the actual pump head in blue is always *below* its ideal head. Several factors are responsible for this. The most important is the loss of head that occurs because the impeller has a finite, rather than an infinite, set of blades. As a result, the flow will actually leave each blade at an angle that will be slightly different from the blade's design angle β_2. Hence, the pump head will be proportionally reduced. Along with this effect, there are additional losses due to fluid friction and mechanical friction at the shaft bearings and seals, and turbulent losses due to improper flow along the impeller.

The green curve for efficiency shows that it increases as Q increases, until it reaches a maximum called the ***best efficiency point*** (BEP), and then it begins to decrease to zero at Q_{max}. If the required design flow happens to be $Q_{req'd}$, then based on these curves, this pump *should be chosen*, because it will operate at maximum efficiency (BEP) for this flow. When used, it will require a power defined at point A on the red curve, and a head defined at point B on the blue curve.

Manufacturer's Pump Performance Curves.

From similar types of experimental tests, manufacturers provide pump performance curves for many of their pumps. Normally, each pump is designed to run at a constant nominal speed ω_0 and is designed to accommodate several different diameter impellers within its casing. An example is shown in Fig. 14–16, which is for an impeller that runs at $\omega_0 = 1750\,\text{RPM}$ and can be fitted with 125 mm, 150 mm, and 175 mm diameter impellers as indicated by the blue curves. Following these curves one can then obtain the efficiency (green) and brake horsepower (red) for a required flow. For example, when a pump with a 175-mm-diameter impeller is required to produce a flow of 1400 liters/min, it then has an efficiency of 86% (point A) and produces a total head of about 46 m with a power of 16 kW.

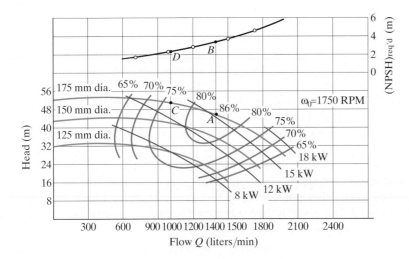

Fig. 14–16

14

14.7 Cavitation and the Net Positive Suction Head

One important phenomenon that can actually limit the performance of a pump occurs when the pressure within its casing drops below a certain limit, so that it causes cavitation. Recall from Sec. 1.8 that *cavitation* is the result of the pressure within a liquid falling to or below the *vapor pressure* p_v for the liquid. For a radial-flow pump this usually occurs on the suction side, within the center or eye of the impeller, since there the pressure is the lowest. When cavitation does occur, the liquid will boil, and bubbles or cavities will form within the liquid. As these bubbles are transported along the blades of the impeller, they reach regions of higher pressure, where they suddenly collapse. This violence causes outward pressure waves that produce a repeated pounding on any adjacent hard surface, which eventually causes material fatigue and wearing away of the surface. The process of wear is further aggravated by corrosion, or other electromechanical effects. Cavitation is associated with vibration and noise, which usually resembles the sound of rocks or pebbles striking the sides of the casing, and once it occurs, the pump's efficiency falls dramatically.

For any specific pump there will be a *critical suction head*, developed at the *suction side* of the pump, where cavitation within the pump will begin to occur. Its value can be determined *experimentally* by maintaining a specified flow and *varying the elevation* of the pump from a reservoir. As the vertical length of pipe needed to draw the fluid up into the pump increases, a critical elevation will be reached where the pump's efficiency will suddenly drop. Using the energy equation, with the datum at the pump inlet, we can obtain this critical suction head. It is represented by the sum of the pressure and velocity heads at the inlet, $\left(p/\gamma + V^2/2g\right)$. If the *vapor pressure head* p_v/γ for the liquid is *subtracted* from this head, the result is referred to as the *net positive suction head* (NPSH). Since cavitation or p_v actually occurs *within* the pump, and not at the inlet, the NPSH is the additional head needed to move the fluid from the inlet to the point of cavitation within the pump.

Manufacturers normally perform experiments of this sort, and for various values of the flow, the results are plotted on the same graph as the performance curves. The values are termed the *required* NPSH, or $(NPSH)_{req'd}$. Notice that for the pump in Fig. 14–16, the $(NPSH)_{req'd}$ is about 3.5 m (point *B*) when the flow is 1400 liters/min and the pump with the 175-mm-diameter impeller is at its maximum efficiency (86%). Once $(NPSH)_{req'd}$ is determined from the graph, it is then compared to the *available* net positive suction head, $(NPSH)_{avail}$. To obtain this value, we must apply the energy equation to the inlet or suction side of the pump, and to some other point in the actual flow system for which the pump is being used. The result is then subtracted from the vapor pressure head, p_v. To *prevent* cavitation, it is necessary that

$$(NPSH)_{avail} \geq (NPSH)_{req'd} \qquad\qquad (14\text{–}29)$$

For further clarification, the following example provides an application of this concept.

EXAMPLE 14.8

The pump shown in Fig. 14–17 is used to transfer 20°C sewage water from the wet well to the sewage treatment plant. If the flow through the 75-mm-diameter pipe is to be $\frac{1}{60}$ m³/s, determine if cavitation occurs when the pump in Fig. 14–16 is selected. The pump turns off just after the water reaches its lowest level of $h = 4$ m. Take the friction factor for the pipe to be $f = 0.02$, and neglect minor losses.

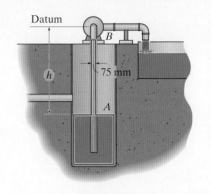

Fig. 14–17

SOLUTION

Fluid Description. We will assume steady flow of an incompressible fluid. From table in Appendix A, $\rho_w = 998.3$ kg/m³ for water at $T = 20°C$.

Inlet Pressure. We can determine the *available suction head* at the pump's impeller inlet by applying the energy equation. The control volume consists of the water in the vertical pipe and the well, Fig. 14–17. The largest suction at B occurs when $h = 4$ m. Here we will work with *absolute pressures* since the vapor pressure is generally reported as an absolute pressure. At A the atmospheric pressure is $p_A = 101.3$ kPa. Since

$$V_B = V = \frac{Q}{A} = \frac{\frac{1}{60} \text{ m}^3/\text{s}}{\pi(0.0375 \text{ m})^2} = 3.7726 \text{ m/s}$$

then

$$\frac{p_A}{\gamma} + \frac{V_A^2}{2g} + z_A + h_{\text{pump}} = \frac{p_B}{\gamma} + \frac{V_B^2}{2g} + z_B + h_{\text{turb}} + f\frac{L}{D}\frac{V^2}{2g} + \Sigma K_L \frac{V^2}{2g}$$

$$\frac{101.3(10^3) \text{ N/m}^3}{(998.3 \text{ kg/m}^3)(9.81 \text{ m/s}^2)} + 0 - 4 \text{ m} + 0 =$$

$$\frac{p_B}{(998.3 \text{ kg/m}^3)(9.81 \text{ m/s}^2)} + \frac{(3.7726 \text{ m/s})^2}{2(9.81 \text{ m/s}^2)} + 0 + 0 + 0.02\left(\frac{4 \text{ m}}{0.075 \text{ m}}\right)\left[\frac{(3.7726 \text{ m/s})^2}{2(9.81 \text{ m/s}^2)}\right] + 0$$

$$p_B = 47.445(10^3) \text{ Pa}$$

The available suction head at the pump inlet is therefore

$$\frac{p_B}{\gamma} + \frac{V_B^2}{2g} = \frac{47.445(10^3) \text{ N/m}^2}{(998.3 \text{ kg/m}^3)(9.81 \text{ m/s}^2)} + \frac{(3.7726 \text{ m/s})}{2(9.81 \text{ m/s}^2)} = 5.570 \text{ m}$$

From Appendix A, at 20°C, the (absolute) vapor pressure for water is 2.34 kPa. Therefore, the available NPSH is

$$(\text{NPSH})_{\text{avail}} = 5.570 \text{ m} - \frac{2.34(10^3) \text{ N/m}^2}{(998.3 \text{ kg/m}^3)(9.81 \text{ m/s}^2)} = 5.331 \text{ m}$$

The flow in liters per minute is

$$Q = \left(\frac{\frac{1}{60} \text{ m}^3}{\text{s}}\right)\left(\frac{1000 \text{ liters}}{1 \text{ m}^3}\right)\left(\frac{60 \text{ s}}{1 \text{ min}}\right) = 1000 \text{ liters/min}$$

From Fig. 14–16, at $Q = 1000$ liters/min, $(\text{NPSH})_{\text{req'd}} = 2.33$ m (point D), and since $(\text{NPSH})_{\text{avail}} > (\text{NPSH})_{\text{req'd}}$, cavitation will not occur in the pump.

Also note from Fig. 14–16 that if a 175-mm-diameter impeller is used in this pump (point C), then the pump will have an efficiency of about 77% and will have a shaft or brake power of about 13.5 kW.

14.8 Pump Selection Related to the Flow System

A flow system may consist of reservoirs, pipes, fittings, and a pump that is used to convey the fluid. If it is required to provide a specific flow for the system, then this should be accomplished in the most economical and efficient way. For example, consider the system shown in Fig. 14–18a. If we apply the energy equation between points 1 and 2 to the system, we have

$$\frac{p_{in}}{\gamma} + \frac{V_{in}^2}{2g} + z_{in} + h_{pump} = \frac{p_{out}}{\gamma} + \frac{V_{out}^2}{2g} + z_{out} + h_{turb} + h_L$$

$$0 + 0 + z_1 + \left(h_{pump}\right)_{act} = 0 + 0 + z_2 + 0 + h_L$$

$$\left(h_{pump}\right)_{act} = (z_2 - z_1) + h_L$$

Here $\left(h_{pump}\right)_{act}$ is the *actual head* supplied by the pump to the system. It is a function of Q^2, since the head loss in the above equation is of the form $h_L = C(V^2/2g)$, where C is a constant. Since $V = Q/A$, then we have $h_L = C\left(Q^2/2gA^2\right) = C'Q^2$. If we plot this equation, $\left(h_{pump}\right)_{act} = (z_2 - z_1) + C'Q^2$, it will be a parabola and will look something like the solid curve A shown in Fig. 14–18b.

If the pump in Fig. 14–18a produces a pump head $\left(h_{pump}\right)_{act}$, like the blue head performance curve in Fig. 14–15b, then the flow required for the system will have to be $Q_{req'd}$. This is represented by point O on the system curve. In other words, if this pump is selected, then the actual pump head $\left(h_{pump}\right)_{act}$ it produces will satisfy the flow requirement $Q_{req'd}$ for *the system*. This is the **operating point** for the system, and if it is close to the best efficiency point (BEP) for the pump (Fig. 14–15b), it justifies choosing this pump for this application. In doing so, realize that over time the pump's characteristics will change. For example, the pipes in the system may corrode, causing an increased frictional head loss, and this would raise the system curve, Fig. 14–18b. Also, the pump can deteriorate, causing its performance curve to lower. Both effects would shift the operating point to O', and lower the pump's efficiency. For best engineering design, the consequences of these changes should be considered when choosing a pump for any particular application.

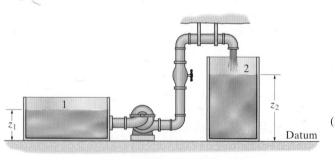

(a)

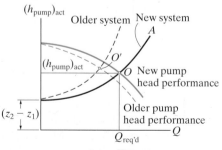

Pump System Curves

(b)

Fig. 14–18

EXAMPLE | 14.9

The radial-flow pump in Fig. 14–19a is used to transfer water from the lake at A into a large storage tank, B. This is done through a 100-mm-diameter pipe that is 100 m long and has a friction factor of 0.015. The manufacturer's data for the performance of the pump is given in Fig. 14–19b. Determine the flow if this pump is selected with a 175-mm impeller to do the job. Neglect minor losses.

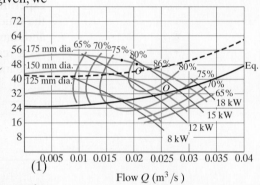

(a)

SOLUTION

Fluid Description. We assume steady, incompressible flow while the pump is operating.

System Equation. We can relate the pump head to the flow Q by applying the energy equation between the water levels at A and B. The control volume contains the water within the pipe and a portion of the water in the lake and tank. Since the friction factor has been given, we do not need to obtain its value from the Moody diagram.

$$\frac{p_A}{\gamma} + \frac{V_A^2}{2g} + z_A + h_{pump} = \frac{p_B}{\gamma} + \frac{V_B^2}{2g} + z_B + h_{turb} + h_L$$

$$\left(h_{pump}\right)_{act} = 25 \text{ m} + (0.015)\left(\frac{100 \text{ m}}{0.1 \text{ m}}\right)\left[\frac{V^2}{2(9.81 \text{ m/s}^2)}\right]$$

Also,

$$Q = V\left[\pi(0.05 \text{ m})^2\right]$$

When the above two equations are combined, we get

$$\left(h_{pump}\right)_{act} = \left[25 + 12.394(10^3)\, Q^3\right] \text{ m} \qquad (1)$$

A plot of this equation is shown in Fig. 14–19b as Eq. 1. It shows the actual pump head $\left(h_{pump}\right)_{act}$ that must be supplied by the pump to provide a flow $Q_{req'd}$ for the system. Along with this curve are the manufacturer's curves for the pump. (For convenience Q is reported in m³/s; however, in practice it is commonly in liters/min.) We see that for the pump with a 175-mm-diameter impeller, the operating point O is where the performance and system head curves intersect, graphically at about $Q = 0.0295$ m³/s (*Ans.*) and $\left(h_{pump}\right)_{act} = 36.0$ m. Therefore, if this pump is selected, then the efficiency for this flow is determined from the efficiency curves to be about $\eta = 74\%$.* By comparison, this is far from the best efficiency point for the pump (86%), and so the choice of this pump with a 175-mm impeller would not be appropriate. Instead, performance curves for other pumps would have to be considered to match a more efficient pump to the system.

It is interesting to note that if the elevation difference between the water levels in the lake and storage tank were 40 m, rather than 25 m, then Eq. 1 would plot as the dashed curve in Fig. 14–29b. In this case, the operating point O' gives a flow of about 0.0225 m³/s; however, the operating efficiency is about $\eta = 86\%$, making the pump the best choice for this condition.

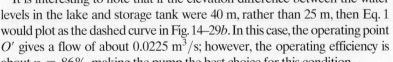

Flow Q (m³/s)
$\omega_0 = 1750$ RPM

(b)

Fig. 14–19

*As noted by the curves, use of a smaller diameter impeller would give a higher efficiency.

14.9   Turbomachine Similitude

In the previous two sections we showed how to select a radial-flow pump to produce a required flow. However, if we are to select a *type* of pump, such as an axial-, radial-, or mixed-flow pump, that works best for a specific job, it then becomes convenient to use dimensional analysis, and express each machine's performance parameters in terms of dimensionless groups of variables that involve its geometry and the fluid properties. We can also use these dimensionless groups to compare the performance of one type of pump with that of a similar type of pump, or if we build a model of a pump, to test its performance characteristics and thereby predict the characteristics for a prototype.

In the previous section we found that when producing performance curves, it was convenient to consider the *dependent variables* for the pump to be the pump head h, the power $\dot{W}$, and the efficiency η. From experiments, it has been found that these three variables all depend upon the fluid properties ρ and μ, the flow Q, the rotation ω of the impeller, and some "characteristic length"—usually the diameter D of the impeller. As a result, the three dependent variables are related in some way to these independent variables by the functions

$$gh = f_1(\rho, \mu, Q, \omega, D)$$

$$\dot{W} = f_2(\rho, \mu, Q, \omega, D)$$

$$\eta = f_3(\rho, \mu, Q, \omega, D)$$

Here we have considered gh, the energy per unit mass, for the convenience of a dimensional analysis. If we apply the Buckingham Pi theorem, as described in Chapter 8, we can create dimensional groups of the variables in these three functions. They are*

$$\frac{gh}{\omega^2 D^2} = f_4\left(\frac{Q}{\omega D^3}, \frac{\rho \omega D^2}{\mu}\right)$$

$$\frac{\dot{W}}{\rho \omega^3 D^5} = f_5\left(\frac{Q}{\omega D^3}, \frac{\rho \omega D^2}{\mu}\right)$$

$$\eta = f_6\left(\frac{Q}{\omega D^3}, \frac{\rho \omega D^2}{\mu}\right)$$

*See, for example, Probs. 8–37 and 8–43.

The dimensionless parameters on the *left* of these three equations are called the **head coefficient**, the **power coefficient**, and, as noted previously, the **efficiency**. On the right, $Q/\omega D^3$ is the **flow coefficient**, and $\rho \omega D^2/\mu$ is a form of the Reynolds number that considers the viscous effects within the pump. Experiments have shown that for either pumps or turbines, this number $\left(\rho \omega D^2/\mu\right)$ does not affect the magnitudes of the three dependent variables as much as the flow coefficient. As a result, we will neglect its effect and write

$$\frac{gh}{\omega^2 D^2} = f_7\!\left(\frac{Q}{\omega D^3}\right) \qquad \frac{\dot{W}}{\rho \omega^3 D^5} = f_8\!\left(\frac{Q}{\omega D^3}\right) \qquad \eta = f_9\!\left(\frac{Q}{\omega D^3}\right)$$

Pump Scaling Laws.   For a particular *type* of pump, these three functional relationships can be determined by experiment, which requires varying the flow coefficient, and then plotting the resultant head, power coefficients, and the efficiency. The resulting curves will all have shapes something like those shown in Fig. 14–20. And, because these shapes are all similar for any family of pumps (axial-, radial-, or mixed-flow), the coefficients become **pump scaling laws**, sometimes called **pump affinity laws**. In other words, they can be used to design or *compare any two pumps of the same type.* For example, the flow coefficients for two pumps of the same type must be the same, so that

$$\frac{Q_1}{\omega_1 D_1^3} = \frac{Q_2}{\omega_2 D_2^3} \tag{14–30}$$

Flow coefficient

Likewise, the other two coefficients can be equated between, say, two radial-flow pumps, or a model and its prototype, to determine their scaled characteristics. If the same fluid is used, then

$$\frac{h_1}{\omega_1^2 D_1^2} = \frac{h_2}{\omega_2^2 D_2^2} \tag{14–31}$$

Head coefficient

$$\frac{\dot{W}_1}{\omega_1^3 D_1^5} = \frac{\dot{W}_2}{\omega_2^3 D_2^5} \tag{14–32}$$

Power coefficient

$$\eta_1 = \eta_2$$

Efficiency

As stated previously, a pump can often house different-diameter impellers within its casing, or it can be made to run at different angular velocities. As a result, these scaling laws can also be used to determine Q, h, and $\dot{W}$ for the pump when D or ω changes. For example, if the pump can produce a head of h_1 when the impeller diameter is D_1, then from the head coefficient, the head produced by the same pump with an impeller diameter D_2 would be $h_2 = h_1\!\left(D_2^2/D_1^2\right)$, provided it has the same ω.

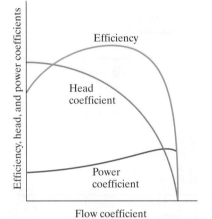

Fig. 14–20

14

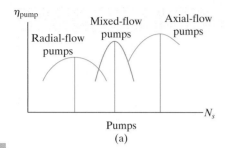

Pumps
(a)

Fig. 14–21

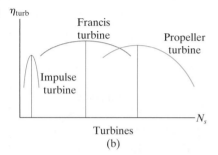

Turbines
(b)

Fig. 14–21

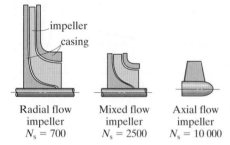

Radial flow impeller $N_s = 700$	Mixed flow impeller $N_s = 2500$	Axial flow impeller $N_s = 10\,000$

Impeller selection for various values of N_s

Fig. 14–22

Specific Speed.

To select the *type* of turbomachine to use for a specific job, it is sometimes helpful to use another dimensionless parameter that does not involve the dimensions of the machine. This parameter is called the ***specific speed***, N_s, and it can be determined either through a dimensional analysis or by simply eliminating the impeller diameter D from the ratio of the flow and head coefficients. This gives

$$N_s = \frac{\left(Q/\omega D^3\right)^{1/2}}{\left(gh/\omega^2 D^2\right)^{3/4}} = \frac{\omega Q^{1/2}}{(gh)^{3/4}} \qquad (14\text{–}33)$$

For *each type* of turbomachine we can plot N_s versus efficiency as shown in Fig. 14–21. See Ref. [15]. Notice that maximum efficiency for a particular type of turbomachine occurs at the peak of each curve, and this peak is located at a particular value of specific speed N_s. For example, it is inherent in their design that radial-flow pumps operate at low specific speeds, so they produce low flows and large heads (high pressure), Fig. 14–21a. On the other hand, axial-flow pumps produce high flows and develop low heads (low pressure). They operate well at high specific speeds, although this does make them susceptible to cavitation. Pumps designed for mixed flow operate in the intermediate range of specific speeds. Typical profiles of the impellers used for these three types of pumps are shown in Fig. 14–22. The same trend occurs for turbines that work on the same principles as pumps, Fig. 14–21b.

Important Points

- Due to both mechanical and fluid frictional losses in turbomachines, the actual performance of the machine must be determined by experiment.

- Cavitation can occur within turbomachines wherever the liquid pressure falls below the vapor pressure for the liquid. To avoid this, it is important to select a turbomachine that has a $(\text{NPSH})_{\text{avail}}$ that is greater than its required $(\text{NPSH})_{\text{req'd}}$.

- Performance curves of efficiency, total head, and power versus flow provide a means for selecting a proper-size pump for a particular application. The pump selected must match the required flow and total head demands for the fluid system, and it must operate with high efficiency.

- The performance characteristics of a pump or turbine can be compared to a geometrically similar pump or model by using the dimensionless parameters of the head, power, and flow coefficients.

- The selection of a *type of turbomachine* to be used for a specific task is based upon the machine's specific speed. For example, radial-flow pumps are efficient for low flows and delivering high heads, whereas axial-flow pumps are efficient at high flows and delivering low heads.

EXAMPLE | 14.10

A turbine for the dam operates under a hydraulic head of 90 m, producing a discharge of 50 m³/s. If the reservoir level drops so that the hydraulic head becomes 60 m, determine the discharge from the turbine.

SOLUTION

Here $h_1 = 90$ m and $Q_1 = 50$ m³/s. Because we also know $h_2 = 60$ m, then we will eliminate the unknowns ω_1 and ω_2 from the flow and head coefficients to obtain a relationship between h and Q. Using Eqs. 14–30 and 14–31, we have

$$\frac{\omega_1}{\omega_2} = \frac{Q_1 D_2^3}{Q_2 D_1^3} \tag{1}$$

$$\frac{\omega_1^2}{\omega_2^2} = \frac{h_1 D_2^2}{h_2 D_1^2}$$

so that

$$\frac{Q_1^2 D_2^4}{Q_2^2 D_1^4} = \frac{h_1}{h_2} \tag{2}$$

Since $D_1 = D_2$,

$$Q_2 = Q_1 \sqrt{\frac{h_2}{h_1}}$$

$$= (50 \text{ m}^3/\text{s})\sqrt{\frac{60 \text{ m}}{90 \text{ m}}} = 40.8 \text{ m}^3/\text{s} \qquad \textit{Ans.}$$

EXAMPLE | 14.11

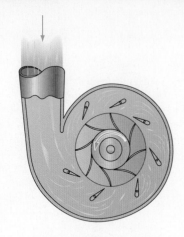

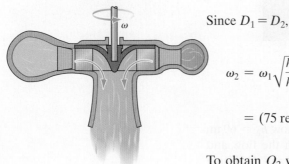

Fig. 14–23

The Francis turbine in Fig. 14–23 is rotating at 75 rev/min under a hydraulic head of 10 m and develops 85 kW with a discharge of 0.10 m³/s. If the guide vanes remain in their fixed position, what is the rotation of this turbine when the hydraulic head is 3 m? Also, what is the corresponding discharge and the power of the turbine?

SOLUTION

Here $\omega_1 = 75$ rev/min, $h_1 = 10$ m, $\dot{W}_1 = 85$ kW, and $Q_1 = 0.10$ m³/s. For $h_2 = 3$ m, the rotation ω_2 can be determined from the head coefficient similitude, Eq. 14–31,

$$\frac{h_1}{\omega_1^2 D_1^2} = \frac{h_2}{\omega_2^2 D_2^2}$$

Since $D_1 = D_2$,

$$\omega_2 = \omega_1 \sqrt{\frac{h_2}{h_1}}$$

$$= (75 \text{ rev/min})\sqrt{\frac{3 \text{ m}}{10 \text{ m}}} = 41.08 \text{ rev/min} = 41.1 \text{ rev/min} \quad Ans.$$

To obtain Q_2 we use the flow coefficient similitude, Eq. 14–30, with $D_1 = D_2$; that is,

$$Q_2 = Q_1 \frac{\omega_2}{\omega_1}$$

$$= (0.10 \text{ m}^3/\text{s})\left(\frac{41.08 \text{ rev/min}}{75 \text{ rev/min}}\right) = 0.0548 \text{ m}^3/\text{s} \quad Ans.$$

Finally, $\dot{W}_2$ is determined from the power coefficient similitude, Eq. 14–32, with $D_1 = D_2$. We have

$$\dot{W}_2 = \dot{W}_1 \left(\frac{\omega_2}{\omega_1}\right)^3$$

$$= (85 \text{ kW})\left(\frac{41.08 \text{ rev/min}}{75 \text{ rev/min}}\right)^3 = 14.0 \text{ kW} \quad Ans.$$

EXAMPLE | 14.12

The pump has an impeller diameter of 250 mm and when operating discharges 0.15 m³/s of water while producing a hydraulic head of 6 m. The power requirement is 9 kW. Determine the required impeller diameter of a similar type of pump that must deliver a discharge of 0.25 m³/s and produce a 10-m head. What is the power requirement for this pump?

SOLUTION

Here we have $D_1 = 250$ mm, $Q_1 = 0.15$ m³/s, $h_1 = 6$ m, and $\dot{W}_1 = 9$ kW. Since $Q_2 = 0.25$ m³/s and $h_2 = 10$ m, we can eliminate the angular velocity ratio ω_1/ω_2 by using Eq. 2 in Example 14.10 to determine D_2.

$$\frac{Q_1^2 D_2^4}{Q_2^2 D_1^4} = \frac{h_1}{h_2}$$

$$D_2 = D_1\left(\frac{Q_2}{Q_1}\right)^{1/2}\left(\frac{h_1}{h_2}\right)^{1/4}$$

$$= (250 \text{ mm})\left(\frac{0.25 \text{ m}^3/\text{s}}{0.15 \text{ m}^3/\text{s}}\right)^{1/2}\left(\frac{6 \text{ m}}{10 \text{ m}}\right)^{1/4} = 284.05 \text{ mm} = 284 \text{ mm } \textit{Ans.}$$

From the flow coefficient, or Eq. 1 of Example 14.10, the angular velocity ratio is

$$\frac{\omega_1}{\omega_2} = \frac{Q_1 D_2^3}{Q_2 D_1^3}$$

Thus, the power coefficient similitude, for constant ρ, becomes

$$\frac{\dot{W}_1}{\omega_1^3 D_1^5} = \frac{\dot{W}_2}{\omega_2^3 D_2^5}$$

$$\frac{\dot{W}_1}{\dot{W}_2} = \left(\frac{\omega_1}{\omega_2}\right)^3\left(\frac{D_1}{D_2}\right)^5 = \left(\frac{Q_1}{Q_2}\right)^3\left(\frac{D_2}{D_1}\right)^9\left(\frac{D_1}{D_2}\right)^5 = \left(\frac{Q_1}{Q_2}\right)^3\left(\frac{D_2}{D_1}\right)^4$$

Therefore,

$$\dot{W}_2 = \dot{W}_1\left(\frac{Q_2}{Q_1}\right)^3\left(\frac{D_1}{D_2}\right)^4 = (9 \text{ kW})\left(\frac{0.25 \text{ m}^3/\text{s}}{0.15 \text{ m}^3/\text{s}}\right)^3\left(\frac{250 \text{ mm}}{284.05 \text{ mm.}}\right)^4 = 25.0 \text{ kW } \textit{Ans.}$$

References

1. W. Janna, *Introduction to Fluid Mechanics*, Brooks/Cole, 1983.

2. F. Yeaple, *Fluid Power Design Handbook*, Marcel Dekker, New York, NY, 1984.

3. R. Warring, *Pumping Manual*, 7th ed., Gulf Publishing, Houston, TX, 1984.

4. O. Balje, *Turbomachines: A Guide to Design, Selection and Theory*, John Wiley, New York, NY.

5. I. J. Karassick, *Pump Handbook*, 2th ed., McGraw-Hill, New York, NY, 1995.

6. R. Evans et al., *Pumping Plant Performance Evaluation*, North Carolina Cooperative Extension Service, Publ. No. AG 452-6.

7. R. Wallis, *Axial Flow Fans and Ducts*, John Wiley, New York, NY.

8. I. J. Karassick, *Pump Handbook*, McGraw-Hill, New York, NY.

9. *Hydraulic Institute Standards*, 14th ed., Hydraulic Institute, Cleveland, OH.

10. P. N. Garay, *Pump Application Desk Book*, Fairmont Press, Lilburn, GA. 1990.

11. G. F. Wislicenus, *Fluid Mechanics of Turbomachinery* 2nd ed., Dover Publications, New York, NY, 1965.

12. *Performance Test Codes: Centrifugal Pumps*, ASME PTC 8.2-1990, New York, NY, 1990.

13. *Equipment Testing Procedure: Centrifugal Pumps*, American Institute of Chemical Engineers, New York, NY.

14. E. S. Logan and R. Roy, *Handbook of Turbomachinery*, 2nd ed., Marcel Dekker, New York, NY, 2003.

15. J. A. Schetz and A. E. Fuhs, *Handbook of Fluid Dynamics and Fluid Machinery*, John Wiley, New York, NY, 1996.

16. L. Nelik, *Centrifugal and Rotary Pumps*, CRC Press, Boca Raton, FL, 1999.

PROBLEMS

Sec. 14.1–14.2

14–1. Water flows at 5 m/s towards the impeller of the axial-flow pump. If the impeller is rotating at 60 rad/s, and it has a mean radius of 200 mm, determine the initial blade angle β_1 so that $\alpha_1 = 90°$. Also, find the relative velocity of the water as it flows onto the blades of the impeller.

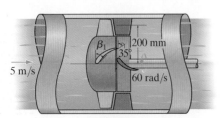

Prob. 14–1

14–2. Water flows at 5 m/s towards the impeller of the axial-flow pump. If the impeller is rotating at 60 rad/s, and it has a mean radius of 200 mm, determine the velocity of the water as it exits the blades, and the relative velocity of the water as it flows off the blades of the impeller.

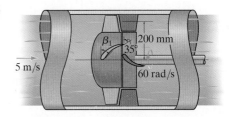

Prob. 14–2

14–3. Water flows through the axial-flow pump at $4(10^{-3})$ m³/s, while the impeller has an angular velocity of 30 rad/s. If the blade tail angle is 35°, determine the velocity and tangential velocity component of the water when it leaves the blade. $\rho_w = 1000$ kg/m³.

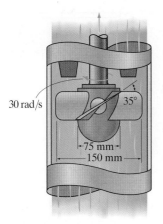

Prob. 14–3

*14–4. If the water velocity onto the impeller at $\alpha_1 = 90°$ is 3 m/s, determine the required initial blade angle β_1. Also, what is the power supplied to the water by the pump? The impeller blades have a mean radius of 50 mm and $\omega = 180$ rad/s. The volumetric flow is 0.9 m³/s.

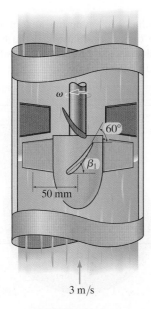

Prob. 14–4

14–5. An axial-flow pump has an impeller with a mean radius of 100 mm that rotates at 1200 rev/min. At the exit the stator blade angle $\alpha_2 = 70°$. If the velocity of the water leaving the impeller is 8 m/s, determine the tangential component of the velocity and the relative velocity of the water at this instant.

14–6. Water flows at 6 m/s through an axial-flow pump. If the impeller blade has an angular velocity of $\omega = 100$ rad/s, determine the velocity of the water when it is delivered to the stator blades. The impeller blades have a mean radius of 100 mm and the angles shown.

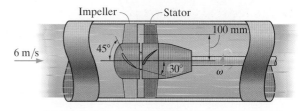

Prob. 14–6

14–7. Water flows at 6 m/s through an axial-flow pump. If the impeller blade angles are 45° and 30° as shown, determine the power supplied to the water by the pump when $\omega = 100$ rad/s. The impeller blades have a mean radius of 100 mm. The volumetric flow is 0.4 m³/s.

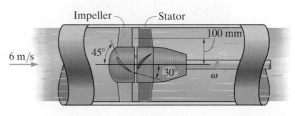

Prob. 14–7

Sec. 14.3–14.4

*14–8.** The radial ventilation fan is used to force air into the ducts of a building. If the air is at a temperature of 20°C, and the shaft is rotating at 60 rad/s, determine the power output of the motor. The blades have a width of 30 mm. Air enters the blades in the radial direction and is discharged with a velocity of 50 m/s at the angle shown.

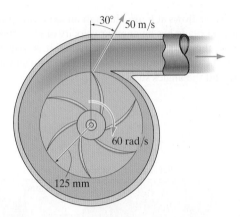

Prob. 14–8

14–9. Water flows through the centrifugal pump impeller such that the entrance velocity is $V_1 = 6$ m/s and the exit velocity is $V_2 = 10$ m/s. If the discharge is 0.04 m³/s and the width of each blade is 20 mm, determine the torque that must be applied to the pump shaft.

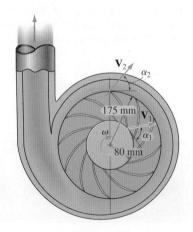

Prob. 14–9

14–10. The blades of the centrifugal pump are 30 mm wide and are rotating at 60 rad/s. Water enters the blades in the radial direction and flows off the blades with a velocity of 20 m/s as shown. If the discharge is 0.4 m³/s, determine the torque that must be applied to the shaft of the pump.

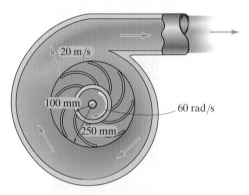

Prob. 14–10

14–11. The radial-flow pump has a 60-mm-wide impeller with the dimensions shown. If the blades rotate at 160 rad/s, determine the discharge if the water enters each blade in the radial direction.

*14–12.** The radial-flow pump has a 60-mm-wide impeller with the radial dimensions shown. If the blades rotate at 160 rad/s and the discharge is 0.3 m³/s, determine the power supplied to the water.

14–13. The radial-flow pump has a 60-mm-wide impeller with the radial dimensions shown. If the blades rotate at 160 rad/s and the discharge is 0.3 m³/s, determine the ideal head developed by the pump.

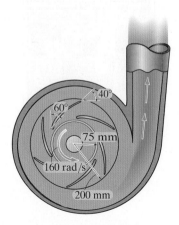

Probs. 14–11/12/13

14–14. Water flows radially onto the 40-mm-wide blades of the centrifugal pump impeller and exits with a velocity V_2 at an angle of 20°, as shown. If the impeller is turning at 10 rev/s and the flow is 0.04 m³/s, determine the ideal head supplied to the water, the torque required to turn the impeller, and the power supplied to the pump if it has an efficiency of $\eta = 0.65$.

14–18. Water flows through the centrifugal pump impeller at the rate of 0.04 m³/s. If the blades are 20 mm wide and the velocities at the entrance and exit are directed at the angles $\alpha_1 = 45°$ and $\alpha_2 = 10°$, respectively, determine the torque that must be applied to the pump shaft.

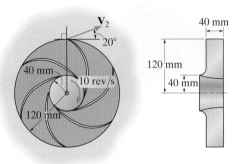

Prob. 14–14

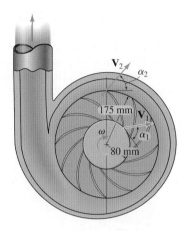

Prob. 14–18

14–15. Show that the ideal head for a radial-flow pump can be determined from $\Delta H = (U_2 V_2 \cos \alpha_2)/g$, where V_2 is the velocity of the water leaving the impeller blades. Water enters the impeller blades in the radial direction.

*14–16. The velocity of water flowing onto the 40-mm-wide impeller blades of the radial-flow pump is directed at 20° as shown. If the flow leaves the blades at the blade angle of 40°, determine the torque the pump must exert on the impeller.

14–17. The velocity of water flowing onto the 40-mm-wide impeller blades of the radial-flow pump is directed at 20° as shown. If the flow leaves the blades at the blade angle of 40°, determine the total head developed by the pump.

14–19. The impeller of the centrifugal pump is rotating at 1200 rev/min and produces a flow of 0.03 m³/s. Determine the speed of the water as it exits the blade, and the ideal power and the ideal head produced by the pump.

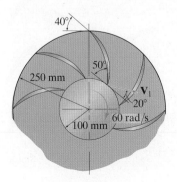

Probs. 14–16/17

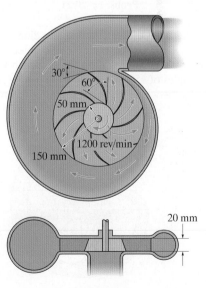

Prob. 14–19

Sec. 14.5

*14–20. The buckets of the Pelton wheel deflect the 100-mm-diameter water jet 140° as shown. If the velocity of the water from the nozzle is 30 m/s, determine the torque needed to hold the wheel in a fixed position and the torque that maintains an angular velocity of 10 rad/s.

14–21. The buckets of the Pelton wheel deflect the 100-mm-diameter water jet 140° as shown. If the velocity of the water from the nozzle is 30 m/s, determine the power that is delivered to the shaft when the wheel is rotating at a constant angular velocity of 2 rad/s. How fast must the wheel be turning to maximize the power developed by the wheel?

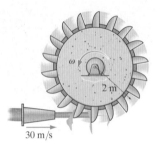

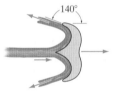

Probs. 14–20/21

14–22. Water flows through the 400-mm-diameter delivery pipe at 2 m/s. Each of the four 50-mm-diameter nozzles is aimed tangentially at the Pelton wheel, which has bucket deflection angles of 150°. Determine the torque and power developed by the wheel when it is rotating at 10 rad/s.

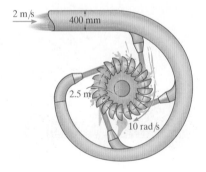

Prob. 14–22

14–23. Water flows from a lake through a 300-m-long pipe having a diameter of 300 mm and a friction factor of $f = 0.015$. The flow from the pipe passes through a 60-mm-diameter nozzle and is used to drive the Pelton wheel, where the bucket deflection angles are 160°. Determine the power and torque produced when the wheel is turning under optimum conditions. Neglect minor losses.

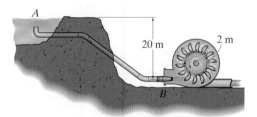

Prob. 14–23

*14–24. Water flowing at 4 m/s is directed from the stator onto the blades of an axial-flow turbine at an angle of $\alpha_1 = 28°$ and exits at an angle of $\alpha_2 = 43°$. If the blades are rotating at 80 rad/s, determine the required angles β_1 and β_2 of the turbine blades so that they properly accept, and then deliver the flow to the adjacent stator. The turbine has a mean radius of 600 mm.

14–25. Water flowing at 4 m/s is directed from the stator onto the blades of the axial-flow turbine, where the mean radius of the blades is 0.75 m. If the blades are rotating at 80 rad/s and the flow is 7 m³/s, determine the torque produced by the water.

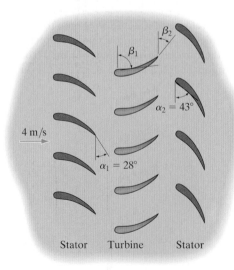

Probs. 14–24/25

14–26. A stator directs 8 kg/s of gas onto the blades of a gas turbine that is rotating at 20 rad/s. If the mean radius of the turbine blades is 0.8 m, and the velocity of the flow entering the blades is 12 m/s as shown, determine the exit velocity of the gas from the blades. Also, what is the required angle β_1 at the entrance of the blades?

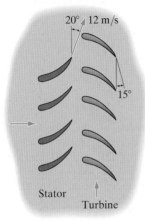

Prob. 14–26

14–27. The dimensions of the blades on the axial-flow water turbine are shown. Water passes through the guide vanes at an angle of 60°. If the flow is 0.85 m³/s, determine the velocity of the water as it strikes the mean radius of the blades. *Hint:* Within the free passage from the guide vanes to the turbine, free-vortex flow occurs; that is, $V_t r = $ constant.

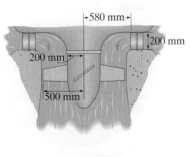

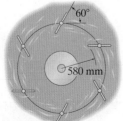

Prob. 14–27

***14–28.** The velocities on and off the 90-mm-wide blades of the turbine are directed as shown. If $V_1 = 18$ m/s and the blades are rotating at 80 rad/s, determine the relative velocity of the flow off the blades. Also, determine the blade angles, β_1 and β_2.

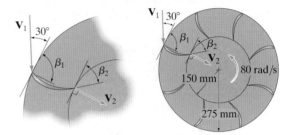

Prob. 14–28

14–29. The velocities on and off the 90-mm-wide blades of a turbine are directed as shown. If the blades are rotating at 80 rad/s and the discharge is 1.40 m³/s, determine the power that the turbine withdraws from the water.

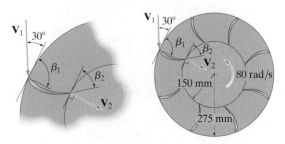

Prob. 14–29

14–30. Water is directed at $\alpha_1 = 50°$ onto the blades of the Kaplan turbine and leaves the blades in the axial direction. Each blade has an inner radius of 200 mm and outer radius of 600 mm. If the blades are rotating at $\omega = 28$ rad/s, and the flow is 8 m³/s, determine the power the water supplies to the turbine.

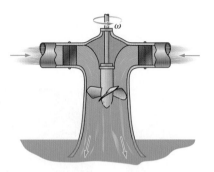

Prob. 14–30

14–31. Water enters the 50-mm-wide blades of the turbine with a velocity of 20 m/s as shown. If the blades are rotating at 75 rev/min and the flow off the blades is radial, determine the power the water supplies to the turbine.

*14–32. Water enters the 50-mm-wide blades of the turbine with a velocity of 20 m/s as shown. If the blades are rotating at 75 rev/min and the flow off the blades is radial, determine the ideal head the turbine draws from the water.

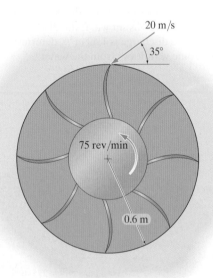

Probs. 14–31/32

14–33. The blades of the Francis turbine rotate at 40 rad/s as they discharge water at 0.5 m³/s. Water enters the blades at an angle of $\alpha_1 = 30°$ and leaves in the radial direction. If the blades have a width of 0.3 m, determine the torque and power the water supplies to the turbine shaft.

14–34. The blades of the Francis turbine rotate at 40 rad/s as they discharge water at 0.5 m³/s. Water enters the blades at an angle of $\alpha_1 = 30°$ and leaves in the radial direction. If the blades have a width of 0.3 m and the turbine operates under a total head of 3 m, determine the hydraulic efficiency.

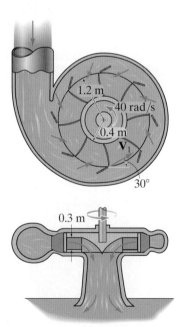

Probs. 14–33/34

Sec. 14.6–14.9

14–35. Water at 20°C is pumped from a lake into the tank on the truck using a 50-mm-diameter galvanized iron pipe. If the pump performance curve is as shown, determine the maximum flow the pump will generate. The total length of the pipe is 50 m. Include the minor losses of the five elbows.

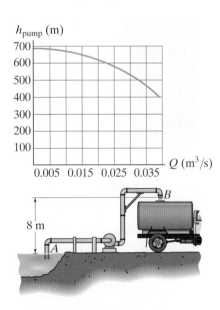

Prob. 14–35

*14–36.** A 200-mm-diameter impeller of a radial-flow water pump rotates at 150 rad/s and has a discharge of 0.3 m³/s. Determine the discharge for a similar pump that has an impeller diameter of 100 mm and operates at 80 rad/s.

14–37. A radial-flow pump has a 175-mm-diameter impeller, and the performance curves for it are shown in Fig. 14–16. Determine the approximate flow it provides to pump water from the reservoir tank to the fill tank, where $h = 35$ m. Neglect minor losses, and use a friction factor of $f = 0.02$ for the 30-m-long, 75-mm-diameter hose.

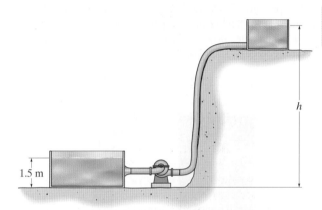

Prob. 14–37

14–38. The radial-flow pump having an impeller diameter of 125 mm and the performance curves shown in Fig. 14–16 is to be used to pump water from the reservoir into the tank. Determine the efficiency of the pump if the flow is 1200 liters/min. Also, what is the maximum height h to which the tank can be filled? Neglect any losses.

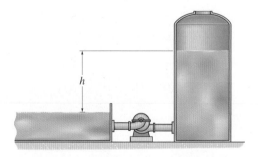

Prob. 14–38

14–39. A 200-mm-diameter impeller of a radial-flow water pump rotates at 150 rad/s and produces a change in ideal head of 0.3 m. Determine the change in head for a geometrically similar pump that has an impeller diameter of 100 mm and operates at 80 rad/s.

*14–40. The temperature of benzene in a processing tank is maintained by recycling this liquid through a heat exchanger, using a pump that has an impeller speed of 1750 rpm and produces a flow of 3600 liters/min. If it is found that the heat exchanger can maintain the temperature only when the flow is 2600 liters/min, determine the required angular speed of the impeller.

14–41. The temperature of benzene in a processing tank is maintained by recycling this liquid through a heat exchanger, using a pump that has an impeller diameter of 150 mm and produces a flow of 3600 liters/min. If it is found that the heat exchanger can maintain the temperature only when the flow is 2600 liters/min, determine the required diameter of the impeller if it maintains the same angular speed.

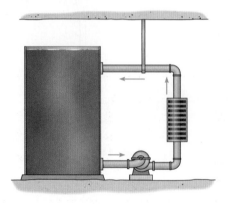

Prob. 14–40

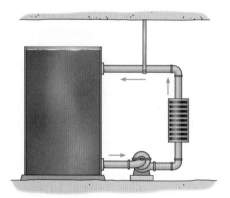

Prob. 14–41

CHAPTER REVIEW

Axial-flow pumps maintain the direction of the flow as it passes through the impeller of the pump. Radial-flow pumps direct the flow from the center of the impeller radially outward into a volute. The kinematic analysis of both types of pumps is similar. It depends upon the speed of the impeller and its blade angles.	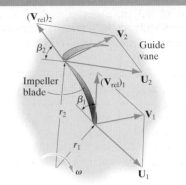
The torque, the power, and the head produced by an axial- or radial-flow pump depend on the motion of the impeller, and on the *tangential components* of the velocity of the flow as it enters and leaves the impeller.	
A Pelton wheel acts as an impulse turbine. It produces power by changing the momentum of the flow as it strikes the buckets of the turbine wheel. Kaplan and Francis turbines are called reaction turbines. The analysis of these devices is similar to that of axial- and radial-flow pumps, respectively.	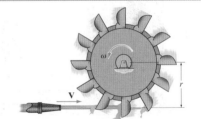
The actual performance characteristics of any turbomachine are determined from experiment, to account for both mechanical and fluid frictional losses in the machine. Cavitation within the flow can occur in a turbomachine. It can be avoided if the $(NPSH)_{avail}$ is greater than the required $(NPSH)_{req'd}$, which is determined by experiment. A pump that is selected to work within a fluid system must produce a required flow and total head, and must operate with a high efficiency.	
To compare the performance characteristics of two turbomachines of the same type, the flow, head, and power coefficients must be similar.	$$\frac{Q_1}{\omega_1 D_1^3} = \frac{Q_2}{\omega_2 D_2^3}$$ $$\frac{h_1}{\omega_1^2 D_1^2} = \frac{h_2}{\omega_2^2 D_2^2}$$ $$\frac{\dot{W}_1}{\omega_1^3 D_1^5} = \frac{\dot{W}_2}{\omega_2^3 D_2^5}$$
Turbomachines can be selected for a specific job based on their specific speed, which is a function of the angular velocity ω the flow Q, and the pump head h. Radial-flow pumps are efficient for low specific speeds, and axial-flow pumps are efficient for high specific speeds.	$$N_s = \frac{\omega Q^{1/2}}{(gh)^{3/4}}$$

Physical Properties of Fluids

Physical Properties of Liquids at Standard Atmospheric Pressure 101.3 kPa, and 20°C

Liquid	Density ρ (kg/m^3)	Dynamic Viscosity μ (N·s/m^2)	Kinematic Viscosity ν (m^2/s)	Surface Tension σ (N/m)
Ethyl alcohol	789	$1.19(10^{-3})$	$1.51(10^{-6})$	0.0229
Gasoline	726	$0.317(10^{-3})$	$0.437(10^{-6})$	0.0221
Carbon tetrachloride	1590	$0.958(10^{-3})$	$0.603(10^{-6})$	0.0269
Kerosene	814	$1.92(10^{-3})$	$2.36(10^{-6})$	0.0293
Glycerin	1260	1.50	$1.19(10^{-3})$	0.0633
Mercury	13 550	$1.58(10^{-3})$	$0.177(10^{-6})$	0.466
Crude oil	880	$30.2(10^{-3})$	$0.0344(10^{-3})$	

Physical Properties of Gases at Standard Atmospheric Pressure 101.3 kPa

Gas	Density ρ (kg/m^3)	Dynamic Viscosity μ (N·s/m^2)	Kinematic Viscosity ν (m^2/s)	Gas Constant R (J/[kg·K])	Specific Heat Ratio $k = c_p/c_v$
Air (15°C)	1.23	$17.9(10^{-6})$	$14.6(10^{-6})$	286.9	1.40
Oxygen (20°C)	1.33	$20.4(10^{-6})$	$15.2(10^{-6})$	259.8	1.40
Nitrogen (20°C)	1.16	$17.5(10^{-6})$	$15.1(10^{-6})$	296.8	1.40
Hydrogen (20°C)	0.0835	$8.74(10^{-6})$	$106(10^{-6})$	4124	1.41
Helium (20°C)	0.169	$19.2(10^{-6})$	$114(10^{-6})$	2077	1.66
Carbon dioxide (20°C)	1.84	$14.9(10^{-6})$	$8.09(10^{-6})$	188.9	1.30
Methane (20°C) (natural gas)	0.665	$11.2(10^{-6})$	$16.8(10^{-6})$	518.3	1.31

Physical Properties of Water vs. Temperature

Temperature T (°C)	Density ρ (kg/m³)	Dynamic Viscosity μ (N·s/m²)	Kinematic Viscosity ν (m²/s)	Vapor Pressure p_v (kPa)
0	999.8	$1.80(10^{-3})$	$1.80(10^{-6})$	0.681
5	1000.0	$1.52(10^{-3})$	$1.52(10^{-6})$	0.872
10	999.7	$1.31(10^{-3})$	$1.31(10^{-6})$	1.23
15	999.2	$1.15(10^{-3})$	$1.15(10^{-6})$	1.71
20	998.3	$1.00(10^{-3})$	$1.00(10^{-6})$	2.34
25	997.1	$0.897(10^{-3})$	$0.898(10^{-6})$	3.17
30	995.7	$0.801(10^{-3})$	$0.804(10^{-6})$	4.25
35	994.0	$0.723(10^{-3})$	$0.727(10^{-6})$	5.63
40	992.3	$0.659(10^{-3})$	$0.664(10^{-6})$	7.38
45	990.2	$0.599(10^{-3})$	$0.604(10^{-6})$	9.59
50	988.0	$0.554(10^{-3})$	$0.561(10^{-6})$	12.4
55	985.7	$0.508(10^{-3})$	$0.515(10^{-6})$	15.8
60	983.2	$0.470(10^{-3})$	$0.478(10^{-6})$	19.9
65	980.5	$0.437(10^{-3})$	$0.446(10^{-6})$	25.0
70	977.7	$0.405(10^{-3})$	$0.414(10^{-6})$	31.2
75	974.8	$0.381(10^{-3})$	$0.390(10^{-6})$	38.6
80	971.6	$0.356(10^{-3})$	$0.367(10^{-6})$	47.4
85	968.4	$0.336(10^{-3})$	$0.347(10^{-6})$	57.8
90	965.1	$0.318(10^{-3})$	$0.329(10^{-6})$	70.1
95	961.6	$0.300(10^{-3})$	$0.312(10^{-6})$	84.6
100	958.1	$0.284(10^{-3})$	$0.296(10^{-6})$	101

Properties of Air vs. Altitude

Altitude (km)	Temperature T (°C)	Pressure p (kPa)	Density ρ (kg/m³)	Dynamic Viscosity μ (Pa·s)	Kinematic Viscosity ν (m²/s)
0	15.00	101.3	1.225	$17.89(10^{-6})$	$14.61(10^{-6})$
1	8.501	89.88	1.112	$17.58(10^{-6})$	$15.81(10^{-6})$
2	2.004	79.50	1.007	$17.26(10^{-6})$	$17.15(10^{-6})$
3	−4.491	70.12	0.9092	$16.94(10^{-6})$	$18.63(10^{-6})$
4	−10.98	61.66	0.8194	$16.61(10^{-6})$	$20.28(10^{-6})$
5	−17.47	54.05	0.7364	$16.28(10^{-6})$	$22.11(10^{-6})$
6	−23.96	47.22	0.6601	$15.95(10^{-6})$	$24.16(10^{-6})$
7	−30.45	41.10	0.5900	$15.61(10^{-6})$	$26.46(10^{-6})$
8	−36.94	35.65	0.5258	$15.27(10^{-6})$	$29.04(10^{-6})$
9	−43.42	30.80	0.4671	$14.93(10^{-6})$	$31.96(10^{-6})$
10	−49.90	26.45	0.4135	$14.58(10^{-6})$	$35.25(10^{-6})$
11	−56.38	22.67	0.3648	$14.22(10^{-6})$	$39.00(10^{-6})$
12	−56.50	19.40	0.3119	$14.22(10^{-6})$	$45.57(10^{-6})$
13	−56.50	16.58	0.2666	$14.22(10^{-6})$	$53.32(10^{-6})$
14	−56.50	14.17	0.2279	$14.22(10^{-6})$	$62.39(10^{-6})$
15	−56.50	12.11	0.1948	$14.22(10^{-6})$	$73.00(10^{-6})$
16	−56.50	10.35	0.1665	$14.22(10^{-6})$	$85.40(10^{-6})$
17	−56.50	8.850	0.1423	$14.22(10^{-6})$	$99.90(10^{-6})$
18	−56.50	7.565	0.1217	$14.22(10^{-6})$	$0.1169(10^{-3})$
19	−56.50	6.468	0.1040	$14.22(10^{-6})$	$0.1367(10^{-3})$
20	−56.50	5.529	0.08891	$14.22(10^{-6})$	$0.1599(10^{-3})$
21	−55.57	4.729	0.07572	$14.27(10^{-6})$	$0.1884(10^{-3})$
22	−54.58	4.048	0.06451	$14.32(10^{-6})$	$0.2220(10^{-3})$
23	−53.58	3.467	0.05501	$14.38(10^{-6})$	$0.2614(10^{-3})$
24	−52.59	2.972	0.04694	$14.43(10^{-6})$	$0.3074(10^{-3})$
25	−51.60	2.549	0.04008	$14.48(10^{-6})$	$0.3614(10^{-3})$

Properties of Air vs. Temperature

Temperature T (°C)	Density ρ (kg/m³)	Dynamic Viscosity μ (N·s/m²)	Kinematic Viscosity ν (m²/s)
−50	1.582	$14.6(10^{-6})$	$9.21(10^{-6})$
−40	1.514	$15.1(10^{-6})$	$9.98(10^{-6})$
−30	1.452	$15.6(10^{-6})$	$10.8(10^{-6})$
−20	1.394	$16.1(10^{-6})$	$11.6(10^{-6})$
−10	1.342	$16.7(10^{-6})$	$12.4(10^{-6})$
0	1.292	$17.2(10^{-6})$	$13.3(10^{-6})$
10	1.247	$17.6(10^{-6})$	$14.2(10^{-6})$
20	1.202	$18.1(10^{-6})$	$15.1(10^{-6})$
30	1.164	$18.6(10^{-6})$	$16.0(10^{-6})$
40	1.127	$19.1(10^{-6})$	$16.9(10^{-6})$
50	1.092	$19.5(10^{-6})$	$17.9(10^{-6})$
60	1.060	$20.0(10^{-6})$	$18.9(10^{-6})$
70	1.030	$20.5(10^{-6})$	$19.9(10^{-6})$
80	1.000	$20.9(10^{-6})$	$20.9(10^{-6})$
90	0.973	$21.3(10^{-6})$	$21.9(10^{-6})$
100	0.946	$21.7(10^{-6})$	$23.0(10^{-6})$
150	0.834	$23.8(10^{-6})$	$28.5(10^{-6})$
200	0.746	$25.7(10^{-6})$	$34.5(10^{-6})$
250	0.675	$27.5(10^{-6})$	$40.8(10^{-6})$

Table B–1 Isentropic Relations (*k* = 1.4)

M	$\dfrac{T}{T_0}$	$\left(\dfrac{p}{p_0}\right)$	$\dfrac{A}{A^*}$
0	1.0000	1.0000	∞
0.10	0.9980	0.9930	5.8218
0.11	0.9976	0.9916	5.2992
0.12	0.9971	0.9900	4.8643
0.13	0.9966	0.9883	4.4969
0.14	0.9961	0.9864	4.1824
0.15	0.9955	0.9844	3.9103
0.16	0.9949	0.9823	3.6727
0.17	0.9943	0.9800	3.4635
0.18	0.9936	0.9776	3.2779
0.19	0.9928	0.9751	3.1123
0.20	0.9921	0.9725	2.9635
0.21	0.9913	0.9697	2.8293
0.22	0.9901	0.9668	2.7076
0.23	0.9895	0.9638	2.5968
0.24	0.9886	0.9607	2.4956
0.25	0.9877	0.9575	2.4027
0.26	0.9867	0.9541	2.3173
0.27	0.9856	0.9506	2.2385
0.28	0.9846	0.9470	2.1656
0.29	0.9835	0.9433	2.0979
0.30	0.9823	0.9395	2.0351
0.31	0.9811	0.9355	1.9765
0.32	0.9799	0.9315	1.9219
0.33	0.9787	0.9274	1.8707
0.34	0.9774	0.9231	1.8229
0.35	0.9761	0.9188	1.7780
0.36	0.9747	0.9143	1.7358
0.37	0.9733	0.9098	1.6961
0.38	0.9719	0.9052	1.6587
0.39	0.9705	0.9004	1.6234
0.40	0.9690	0.8956	1.5901
0.41	0.9675	0.8907	1.5587
0.42	0.9659	0.8857	1.5289
0.43	0.9643	0.8807	1.5007
0.44	0.9627	0.8755	1.4740
0.45	0.9611	0.8703	1.4487
0.46	0.9594	0.8650	1.4246
0.47	0.9577	0.8596	1.4018
0.48	0.9560	0.8541	1.3801
0.49	0.9542	0.8486	1.3595
0.50	0.9524	0.8430	1.3398
0.51	0.9506	0.8374	1.3212
0.52	0.9487	0.8317	1.3034
0.53	0.9468	0.8259	1.2865
0.54	0.9449	0.8201	1.2703
0.55	0.9430	0.8142	1.2550
0.56	0.9410	0.8082	1.2403
0.57	0.9390	0.8022	1.2263
0.58	0.9370	0.7962	1.2130
0.59	0.9349	0.7901	1.2003
0.60	0.9328	0.7840	1.1882
0.61	0.9307	0.7778	1.1767
0.62	0.9286	0.7716	1.1657
0.63	0.9265	0.7654	1.1552
0.64	0.9243	0.7591	1.1452
0.65	0.9221	0.7528	1.1356
0.66	0.9199	0.7465	1.1265
0.67	0.9176	0.7401	1.1179
0.68	0.9153	0.7338	1.1097
0.69	0.9131	0.7274	1.1018
0.70	0.9107	0.7209	1.0944
0.71	0.9084	0.7145	1.0873
0.72	0.9061	0.7080	1.0806
0.73	0.9037	0.7016	1.0742
0.74	0.9013	0.6951	1.0681
0.75	0.8989	0.6886	1.0624
0.76	0.8964	0.6821	1.0570
0.77	0.8940	0.6756	1.0519
0.78	0.8915	0.6691	1.0471
0.79	0.8890	0.6625	1.0425
0.80	0.8865	0.6560	1.0382
0.81	0.8840	0.6495	1.0342
0.82	0.8815	0.6430	1.0305
0.83	0.8789	0.6365	1.0270
0.84	0.8763	0.6300	1.0237
0.85	0.8737	0.6235	1.0207
0.86	0.8711	0.6170	1.0179
0.87	0.8685	0.6106	1.0153
0.88	0.8659	0.6041	1.0129
0.89	0.8632	0.5977	1.0108
0.90	0.8606	0.5913	1.0089
0.91	0.8579	0.5849	1.0071
0.92	0.8552	0.5785	1.0056
0.93	0.8525	0.5721	1.0043
0.94	0.8498	0.5658	1.0031
0.95	0.8471	0.5595	1.0022
0.96	0.8444	0.5532	1.0014
0.97	0.8416	0.5469	1.0008
0.98	0.8389	0.5407	1.0003
0.99	0.8361	0.5345	1.0001
1.00	0.8333	0.5283	1.000
1.01	0.8306	0.5221	1.000
1.02	0.8278	0.5160	1.000
1.03	0.8250	0.5099	1.001
1.04	0.8222	0.5039	1.001
1.05	0.8193	0.4979	1.002
1.06	0.8165	0.4919	1.003
1.07	0.8137	0.4860	1.004
1.08	0.8108	0.4800	1.005
1.09	0.8080	0.4742	1.006
1.10	0.8052	0.4684	1.008
1.11	0.8023	0.4626	1.010
1.12	0.7994	0.4568	1.011
1.13	0.7966	0.4511	1.013
1.14	0.7937	0.4455	1.015
1.15	0.7908	0.4398	1.017
1.16	0.7879	0.4343	1.020
1.17	0.7851	0.4287	1.022
1.18	0.7822	0.4232	1.025
1.19	0.7793	0.4178	1.026
1.20	0.7764	0.4124	1.030
1.21	0.7735	0.4070	1.033
1.22	0.7706	0.4017	1.037
1.23	0.7677	0.3964	1.040
1.24	0.7648	0.3912	1.043
1.25	0.7619	0.3861	1.047
1.26	0.7590	0.3809	1.050
1.27	0.7561	0.3759	1.054
1.28	0.7532	0.3708	1.058
1.29	0.7503	0.3658	1.062
1.30	0.7474	0.3609	1.066

Table B–1	Isentropic Relations ($k = 1.4$)		
M	$\dfrac{T}{T_0}$	$\left(\dfrac{p}{p_0}\right)$	$\dfrac{A}{A^*}$
1.31	0.7445	0.3560	1.071
1.32	0.7416	0.3512	1.075
1.33	0.7387	0.3464	1.080
1.34	0.7358	0.3417	1.084
1.35	0.7329	0.3370	1.089
1.36	0.7300	0.3323	1.094
1.37	0.7271	0.3277	1.099
1.38	0.7242	0.3232	1.104
1.39	0.7213	0.3187	1.109
1.40	0.7184	0.3142	1.115
1.41	0.7155	0.3098	1.120
1.42	0.7126	0.3055	1.126
1.43	0.7097	0.3012	1.132
1.44	0.7069	0.2969	1.138
1.45	0.7040	0.2927	1.144
1.46	0.7011	0.2886	1.150
1.47	0.6982	0.2845	1.156
1.48	0.6954	0.2804	1.163
1.49	0.6925	0.2764	1.169
1.50	0.6897	0.2724	1.176
1.51	0.6868	0.2685	1.183
1.52	0.6840	0.2646	1.190
1.53	0.6811	0.2608	1.197
1.54	0.6783	0.2570	1.204
1.55	0.6754	0.2533	1.212
1.56	0.6726	0.2496	1.219
1.57	0.6698	0.2459	1.227
1.58	0.6670	0.2423	1.234
1.59	0.6642	0.2388	1.242
1.60	0.6614	0.2353	1.250
1.61	0.6586	0.2318	1.258
1.62	0.6558	0.2284	1.267
1.63	0.6530	0.2250	1.275
1.64	0.6502	0.2217	1.284
1.65	0.6475	0.2184	1.292
1.66	0.6447	0.2151	1.301
1.67	0.6419	0.2119	1.310
1.68	0.6392	0.2088	1.319
1.69	0.6364	0.2057	1.328
1.70	0.6337	0.2026	1.338
1.71	0.6310	0.1996	1.347
1.72	0.6283	0.1966	1.357
1.73	0.6256	0.1936	1.367
1.74	0.6229	0.1907	1.376
1.75	0.6202	0.1878	1.386
1.76	0.6175	0.1850	1.397
1.77	0.6148	0.1822	1.407
1.78	0.6121	0.1794	1.418
1.79	0.6095	0.1767	1.428
1.80	0.6068	0.1740	1.439
1.81	0.6041	0.1714	1.450
1.82	0.6015	0.1688	1.461
1.83	0.5989	0.1662	1.472
1.84	0.5963	0.1637	1.484
1.85	0.5936	0.1612	1.495
1.86	0.5910	0.1587	1.507
1.87	0.5884	0.1563	1.519
1.88	0.5859	0.1539	1.531
1.89	0.5833	0.1516	1.543
1.90	0.5807	0.1492	1.555
1.91	0.5782	0.1470	1.568
1.92	0.5756	0.1447	1.580
1.93	0.5731	0.1425	1.593
1.94	0.5705	0.1403	1.606
1.95	0.5680	0.1381	1.619
1.96	0.5655	0.1360	1.633
1.97	0.5630	0.1339	1.646
1.98	0.5605	0.1318	1.660
1.99	0.5580	0.1298	1.674
2.00	0.5556	0.1278	1.688
2.01	0.5531	0.1258	1.702
2.02	0.5506	0.1239	1.716
2.03	0.5482	0.1220	1.730
2.04	0.5458	0.1201	1.745
2.05	0.5433	0.1182	1.760
2.06	0.5409	0.1164	1.775
2.07	0.5385	0.1146	1.790
2.08	0.5361	0.1128	1.806
2.09	0.5337	0.1111	1.821
2.10	0.5313	0.1094	1.837
2.11	0.5290	0.1077	1.853
2.12	0.5266	0.1060	1.869
2.13	0.5243	0.1043	1.885
2.14	0.5219	0.1027	1.902
2.15	0.5196	0.1011	1.919
2.16	0.5173	0.09956	1.935
2.17	0.5150	0.09802	1.953
2.18	0.5127	0.09649	1.970
2.19	0.5104	0.09500	1.987
2.20	0.5081	0.09352	2.005
2.21	0.5059	0.09207	2.023
2.22	0.5036	0.09064	2.041
2.23	0.5014	0.08923	2.059
2.24	0.4991	0.08785	2.078
2.25	0.4969	0.08648	2.096
2.26	0.4947	0.08514	2.115
2.27	0.4925	0.08382	2.134
2.28	0.4903	0.08251	2.154
2.29	0.4881	0.08123	2.173
2.30	0.4859	0.07997	2.193
2.31	0.4837	0.07873	2.213
2.32	0.4816	0.07751	2.233
2.33	0.4794	0.07631	2.254
2.34	0.4773	0.07512	2.273
2.35	0.4752	0.07396	2.295
2.36	0.4731	0.07281	2.316
2.37	0.4709	0.07168	2.338
2.38	0.4688	0.07057	2.359
2.39	0.4668	0.06948	2.381
2.40	0.4647	0.06840	2.403
2.41	0.4626	0.06734	2.425
2.42	0.4606	0.06630	2.448
2.43	0.4585	0.06527	2.471
2.44	0.4565	0.06426	2.494
2.45	0.4544	0.06327	2.517
2.46	0.4524	0.06229	2.540
2.47	0.4504	0.06133	2.564
2.48	0.4484	0.06038	2.588
2.49	0.4464	0.05945	2.612
2.50	0.4444	0.05853	2.637
2.51	0.4425	0.05762	2.661
2.52	0.4405	0.05674	2.686
2.53	0.4386	0.05586	2.712
2.54	0.4366	0.05500	2.737
2.55	0.4347	0.05415	2.763
2.56	0.4328	0.05332	2.789
2.57	0.4309	0.05250	2.815

M	$\dfrac{T}{T_0}$	$\left(\dfrac{p}{p_0}\right)$	$\dfrac{A}{A^*}$
Table B–1 Isentropic Relations ($k = 1.4$)			
2.58	0.4289	0.05169	2.842
2.59	0.4271	0.05090	2.869
2.60	0.4252	0.05012	2.896
2.61	0.4233	0.04935	2.923
2.62	0.4214	0.04859	2.951
2.63	0.4196	0.04784	2.979
2.64	0.4177	0.04711	3.007
2.65	0.4159	0.04639	3.036
2.66	0.4141	0.04568	3.065
2.67	0.4122	0.04498	3.094
2.68	0.4104	0.04429	3.123
2.69	0.4086	0.04362	3.153
2.70	0.4068	0.04295	3.183
2.71	0.4051	0.04229	3.213
2.72	0.4033	0.04165	3.244
2.73	0.4015	0.04102	3.275
2.74	0.3998	0.04039	3.306
2.75	0.3980	0.03978	3.338
2.76	0.3963	0.03917	3.370
2.77	0.3945	0.03858	3.402
2.78	0.3928	0.03799	3.434
2.79	0.3911	0.03742	3.467
2.80	0.3894	0.03685	3.500
2.81	0.3877	0.03629	3.534
2.82	0.3860	0.03574	3.567
2.83	0.3844	0.03520	3.601
2.84	0.3827	0.03467	3.636
2.85	0.3810	0.03415	3.671
2.86	0.3794	0.03363	3.706
2.87	0.3777	0.03312	3.741
2.88	0.3761	0.03263	3.777
2.89	0.3745	0.03213	3.813
2.90	0.3729	0.03165	3.850
2.91	0.3712	0.03118	3.887
2.92	0.3696	0.03071	3.924
2.93	0.3681	0.03025	3.961
2.94	0.3665	0.02980	3.999
2.95	0.3649	0.02935	4.038
2.96	0.3633	0.02891	4.076
2.97	0.3618	0.02848	4.115
2.98	0.3602	0.02805	4.155
2.99	0.3587	0.02764	4.194
3.00	0.3571	0.02722	4.235
3.01	0.3556	0.02682	4.275
3.02	0.3541	0.02642	4.316
3.03	0.3526	0.02603	4.357
3.04	0.3511	0.02564	4.399
3.05	0.3496	0.02526	4.441
3.06	0.3481	0.02489	4.483
3.07	0.3466	0.02452	4.526
3.08	0.3452	0.02416	4.570
3.09	0.3437	0.02380	4.613
3.10	0.3422	0.02345	4.657
3.11	0.3408	0.02310	4.702
3.12	0.3393	0.02276	4.747
3.13	0.3379	0.02243	4.792
3.14	0.3365	0.02210	4.838
3.15	0.3351	0.02177	4.884
3.16	0.3337	0.02146	4.930
3.17	0.3323	0.02114	4.977
3.18	0.3309	0.02083	5.025
3.19	0.3295	0.02053	5.073
3.20	0.3281	0.02023	5.121
3.21	0.3267	0.01993	5.170
3.22	0.3253	0.01964	5.219
3.23	0.3240	0.01936	5.268
3.24	0.3226	0.01908	5.319
3.25	0.3213	0.01880	5.369
3.26	0.3199	0.01853	5.420
3.27	0.3186	0.01826	5.472
3.28	0.3173	0.01799	5.523
3.29	0.3160	0.01773	5.576
3.30	0.3147	0.01748	5.629
3.31	0.3134	0.01722	5.682
3.32	0.3121	0.01698	5.736
3.33	0.3108	0.01673	5.790
3.34	0.3095	0.01649	5.845
3.35	0.3082	0.01625	5.900
3.36	0.3069	0.01602	5.956
3.37	0.3057	0.01579	6.012
3.38	0.3044	0.01557	6.069
3.39	0.3032	0.01534	6.126
3.40	0.3019	0.01512	6.184
3.41	0.3007	0.01491	6.242
3.42	0.2995	0.01470	6.301
3.43	0.2982	0.01449	6.360
3.44	0.2970	0.01428	6.420
3.45	0.2958	0.01408	6.480
3.46	0.2946	0.01388	6.541
3.47	0.2934	0.01368	6.602
3.48	0.2922	0.01349	6.664
3.49	0.2910	0.01330	6.727
3.50	0.2899	0.01311	6.790
3.51	0.2887	0.01293	6.853
3.52	0.2875	0.01274	6.917
3.53	0.2864	0.01256	6.982
3.54	0.2852	0.01239	7.047
3.55	0.2841	0.01221	7.113
3.56	0.2829	0.01204	7.179
3.57	0.2818	0.01188	7.246
3.58	0.2806	0.01171	7.313
3.59	0.2795	0.01155	7.382
3.60	0.2784	0.01138	7.450
3.61	0.2773	0.01123	7.519
3.62	0.2762	0.01107	7.589
3.63	0.2751	0.01092	7.659
3.64	0.2740	0.01076	7.730
3.65	0.2729	0.01062	7.802
3.66	0.2718	0.01047	7.874
3.67	0.2707	0.01032	7.947
3.68	0.2697	0.01018	8.020
3.69	0.2686	0.01004	8.094
3.70	0.2675	0.009903	8.169
3.71	0.2665	0.009767	8.244
3.72	0.2654	0.009633	8.320
3.73	0.2644	0.009500	8.397
3.74	0.2633	0.009370	8.474
3.75	0.2623	0.009242	8.552
3.76	0.2613	0.009116	8.630
3.77	0.2602	0.008991	8.709
3.78	0.2592	0.008869	8.789
3.79	0.2582	0.008748	8.870
3.80	0.2572	0.008629	8.951
3.81	0.2562	0.008512	9.032
3.82	0.2552	0.008396	9.115
3.83	0.2542	0.008283	9.198
3.84	0.2532	0.008171	9.282

B

Table B–1	Isentropic Relations ($k = 1.4$)		
M	$\dfrac{T}{T_0}$	$\left(\dfrac{p}{p_0}\right)$	$\dfrac{A}{A^*}$
3.85	0.2522	0.008060	9.366
3.86	0.2513	0.007951	9.451
3.87	0.2503	0.007844	9.537
3.88	0.2493	0.007739	9.624
3.89	0.2484	0.007635	9.711
3.90	0.2474	0.007532	9.799
3.91	0.2464	0.007431	9.888
3.92	0.2455	0.007332	9.977
3.93	0.2446	0.007233	10.07
3.94	0.2436	0.007137	10.16
3.95	0.2427	0.007042	10.25
3.96	0.2418	0.006948	10.34
3.97	0.2408	0.006855	10.44
3.98	0.2399	0.006764	10.53
3.99	0.2390	0.006675	10.62
4.00	0.2381	0.006586	10.72
4.01	0.2372	0.006499	10.81
4.02	0.2363	0.006413	10.91
4.03	0.2354	0.006328	11.01
4.04	0.2345	0.006245	11.11
4.05	0.2336	0.006163	11.21
4.06	0.2327	0.006082	11.31
4.07	0.2319	0.006002	11.41
4.08	0.2310	0.005923	11.51
4.09	0.2301	0.005845	11.61
4.10	0.2293	0.005769	11.71
4.11	0.2284	0.005694	11.82
4.12	0.2275	0.005619	11.92
4.13	0.2267	0.005546	12.03
4.14	0.2258	0.005474	12.14
4.15	0.2250	0.005403	12.24
4.16	0.2242	0.005333	12.35
4.17	0.2233	0.005264	12.46
4.18	0.2225	0.005195	12.57
4.19	0.2217	0.005128	12.68
4.20	0.2208	0.005062	12.79
4.21	0.2200	0.004997	12.90
4.22	0.2192	0.004932	13.02
4.23	0.2184	0.004869	13.13
4.24	0.2176	0.004806	13.25
4.25	0.2168	0.004745	13.36
4.26	0.2160	0.004684	13.48
4.27	0.2152	0.004624	13.60
4.28	0.2144	0.004565	13.72
4.29	0.2136	0.004507	13.83
4.30	0.2129	0.004449	13.95
4.31	0.2121	0.004393	14.08
4.32	0.2113	0.004337	14.20
4.33	0.2105	0.004282	14.32
4.34	0.2098	0.004228	14.45
4.35	0.2090	0.004174	14.57
4.36	0.2083	0.004121	14.70
4.37	0.2075	0.004069	14.82
4.38	0.2067	0.004018	14.95
4.39	0.2060	0.003968	15.08
4.40	0.2053	0.003918	15.21
4.41	0.2045	0.003868	15.34
4.42	0.2038	0.003820	15.47
4.43	0.2030	0.003772	15.61
4.44	0.2023	0.003725	15.74
4.45	0.2016	0.003678	15.87
4.46	0.2009	0.003633	16.01
4.47	0.2002	0.003587	16.15
4.48	0.1994	0.003543	16.28
4.49	0.1987	0.003499	16.42
4.50	0.1980	0.003455	16.56
4.51	0.1973	0.003412	16.70
4.52	0.1966	0.003370	16.84
4.53	0.1959	0.003329	16.99
4.54	0.1952	0.003288	17.13
4.55	0.1945	0.003247	17.28
4.56	0.1938	0.003207	17.42
4.57	0.1932	0.003168	17.57
4.58	0.1925	0.003129	17.72
4.59	0.1918	0.003090	17.87
4.60	0.1911	0.003053	18.02
4.61	0.1905	0.003015	18.17
4.62	0.1898	0.002978	18.32
4.63	0.1891	0.002942	18.48
4.64	0.1885	0.002906	18.63
4.65	0.1878	0.002871	18.79
4.66	0.1872	0.002836	18.94
4.67	0.1865	0.002802	19.10
4.68	0.1859	0.002768	19.26
4.69	0.1852	0.002734	19.42
4.70	0.1846	0.002701	19.58
4.71	0.1839	0.002669	19.75
4.72	0.1833	0.002637	19.91
4.73	0.1827	0.002605	20.07
4.74	0.1820	0.002573	20.24
4.75	0.1814	0.002543	20.41
4.76	0.1808	0.002512	20.58
4.77	0.1802	0.002482	20.75
4.78	0.1795	0.002452	20.92
4.79	0.1789	0.002423	21.09
4.80	0.1783	0.002394	21.26
4.81	0.1777	0.002366	21.44
4.82	0.1771	0.002338	21.61
4.83	0.1765	0.002310	21.79
4.84	0.1759	0.002283	21.97
4.85	0.1753	0.002255	22.15
4.86	0.1747	0.002229	22.33
4.87	0.1741	0.002202	22.51
4.88	0.1735	0.002177	22.70
4.89	0.1729	0.002151	22.88
4.90	0.1724	0.002126	23.07
4.91	0.1718	0.002101	23.25
4.92	0.1712	0.002076	23.44
4.93	0.1706	0.002052	23.63
4.94	0.1700	0.002028	23.82
4.95	0.1695	0.002004	24.02
4.96	0.1689	0.001981	24.21
4.97	0.1683	0.001957	24.41
4.98	0.1678	0.001935	24.60
4.99	0.1672	0.001912	24.80
5.00	0.1667	0.001890	25.00
6.00	0.1220	0.0006334	53.18
7.00	0.09259	0.0002416	104.1
8.00	0.07246	0.0001024	190.1
9.00	0.05814	0.00004739	327.2
10.00	0.04762	0.00002356	535.9

B

Table B–2   Fanno Flow ($k = 1.4$)

M	$\dfrac{fL_{max}}{D}$	$\dfrac{T}{T^*}$	$\dfrac{V}{V^*}$	$\dfrac{p}{p^*}$	$\dfrac{p_0}{p_0^*}$
0.0	∞	1.2000	0.0	∞	∞
0.1	66.9216	1.1976	0.1094	10.9435	5.8218
0.2	14.5333	1.1905	0.2182	5.4554	2.9635
0.3	5.2993	1.1788	0.3257	3.6191	2.0351
0.4	2.3085	1.1628	0.4313	2.6958	1.5901
0.5	1.0691	1.1429	0.5345	2.1381	1.3398
0.6	0.4908	1.1194	0.6348	1.7634	1.1882
0.7	0.2081	1.0929	0.7318	1.4935	1.0944
0.8	0.0723	1.0638	0.8251	1.2893	1.0382
0.9	0.0145	1.0327	0.9146	1.1291	1.0089
1.0	0.0000	1.0000	1.0000	1.0000	1.0000
1.1	0.0099	0.9662	1.0812	0.8936	1.0079
1.2	0.0336	0.9317	1.1583	0.8044	1.0304
1.3	0.0648	0.8969	1.2311	0.7285	1.0663
1.4	0.0997	0.8621	1.2999	0.6632	1.1149
1.5	0.1360	0.8276	1.3646	0.6065	1.1762
1.6	0.1724	0.7937	1.4254	0.5568	1.2502
1.7	0.2078	0.7605	1.4825	0.5130	1.3376
1.8	0.2419	0.7282	1.5360	0.4741	1.4390
1.9	0.2743	0.6969	1.5861	0.4394	1.5553
2.0	0.3050	0.6667	1.6330	0.4082	1.6875
2.1	0.3339	0.6376	1.6769	0.3802	1.8369
2.2	0.3609	0.6098	1.7179	0.3549	2.0050
2.3	0.3862	0.5831	1.7563	0.3320	2.1931
2.4	0.4099	0.5576	1.7922	0.3111	2.4031
2.5	0.4320	0.5333	1.8257	0.2921	2.6367
2.6	0.4526	0.5102	1.8571	0.2747	2.8960
2.7	0.4718	0.4882	1.8865	0.2588	3.1830
2.8	0.4898	0.4673	1.9140	0.2441	3.5001
2.9	0.5065	0.4474	1.9398	0.2307	3.8498
3.0	0.5222	0.4286	1.9640	0.2182	4.2346

B

Table B–3   Rayleigh Flow ($k = 1.4$)

M	$\dfrac{T}{T^*}$	$\dfrac{V}{V^*}$	$\dfrac{p}{p^*}$	$\dfrac{T_0}{T_0^*}$	$\dfrac{p_0}{p_0^*}$
0.0	0.0	0.0	2.4000	0.0	1.2679
0.1	0.0560	0.0237	2.3669	0.0468	1.2591
0.2	0.2066	0.0909	2.2727	0.1736	1.2346
0.3	0.4089	0.1918	2.1314	0.3469	1.1985
0.4	0.6151	0.3137	1.9608	0.5290	1.1566
0.5	0.7901	0.4444	1.7778	0.6914	1.1140
0.6	0.9167	0.5745	1.5957	0.8189	1.0753
0.7	0.9929	0.6975	1.4235	0.9085	1.0431
0.8	1.0255	0.8101	1.2658	0.9639	1.0193
0.9	1.0245	0.9110	1.1246	0.9921	1.0049
1.0	1.0000	1.0000	1.0000	1.0000	1.0000
1.1	0.9603	1.0780	0.8909	0.9939	1.0049
1.2	0.9118	1.1459	0.7958	0.9787	1.0194
1.3	0.8592	1.2050	0.7130	0.9580	1.0437
1.4	0.8054	1.2564	0.6410	0.9343	1.0776
1.5	0.7525	1.3012	0.5783	0.9093	1.1215
1.6	0.7017	1.3403	0.5236	0.8842	1.1756
1.7	0.6538	1.3746	0.4756	0.8597	1.2402
1.8	0.6089	1.4046	0.4335	0.8363	1.3159
1.9	0.5673	1.4311	0.3964	0.8141	1.4033
2.0	0.5289	1.4545	0.3636	0.7934	1.5031
2.1	0.4936	1.4753	0.3345	0.7741	1.6162
2.2	0.4611	1.4938	0.3086	0.7561	1.7434
2.3	0.4312	1.5103	0.2855	0.7395	1.8860
2.4	0.4038	1.5252	0.2648	0.7242	2.0450
2.5	0.3787	1.5385	0.2462	0.7101	2.2218
2.6	0.3556	1.5505	0.2294	0.6970	2.4177
2.7	0.3344	1.5613	0.2142	0.6849	2.6343
2.8	0.3149	1.5711	0.2004	0.6738	2.8731
2.9	0.2969	1.5801	0.1879	0.6635	3.1359
3.0	0.2803	1.5882	0.1765	0.6540	3.4244

Table B–4   Normal Shock Relations ($k = 1.4$)

M_1	M_2	$\dfrac{p_2}{p_1}$	$\dfrac{\rho_2}{\rho_1}$	$\dfrac{T_2}{T_1}$	$\dfrac{(p_0)_2}{(p_0)_1}$
1.00	1.000	1.000	1.000	1.000	1.000
1.01	0.9901	1.023	1.017	1.007	1.000
1.02	0.9805	1.047	1.033	1.013	1.000
1.03	0.9712	1.071	1.050	1.020	1.000
1.04	0.9620	1.095	1.067	1.026	0.9999
1.05	0.9531	1.120	1.084	1.033	0.9999
1.06	0.9444	1.144	1.101	1.059	0.9997
1.07	0.9360	1.169	1.118	1.016	0.9996
1.08	0.9277	1.194	1.135	1.052	0.9994
1.09	0.9196	1.219	1.152	1.059	0.9992
1.10	0.9118	1.245	1.169	1.065	0.9989
1.11	0.9041	1.271	1.186	1.071	0.9986
1.12	0.8966	1.297	1.203	1.078	0.9982
1.13	0.8892	1.323	1.221	1.084	0.9978
1.14	0.8820	1.350	1.238	1.090	0.9973
1.15	0.8750	1.376	1.255	1.097	0.9967
1.16	0.8682	1.403	1.272	1.103	0.9961
1.17	0.8615	1.430	1.290	1.109	0.9953
1.18	0.8549	1.458	1.307	1.115	0.9916
1.19	0.8485	1.485	1.324	1.122	0.9937
1.20	0.8422	1.513	1.342	1.128	0.9928
1.21	0.8360	1.541	1.359	1.134	0.9918
1.22	0.8300	1.570	1.376	1.141	0.9907
1.23	0.8241	1.598	1.394	1.147	0.9896
1.24	0.8183	1.627	1.411	1.153	0.9884
1.25	0.8126	1.656	1.429	1.159	0.9871
1.26	0.8071	1.686	1.446	1.166	0.9857
1.27	0.8016	1.715	1.463	1.172	0.9842
1.28	0.7963	1.745	1.481	1.178	0.9827
1.29	0.7911	1.775	1.498	1.185	0.9811
1.30	0.7860	1.805	1.516	1.191	0.9794
1.31	0.7809	1.835	1.533	1.197	0.9776
1.32	0.7760	1.866	1.551	1.204	0.9758
1.33	0.7712	1.897	1.568	1.210	0.9738
1.34	0.7664	1.928	1.585	1.216	0.9718
1.35	0.7618	1.960	1.603	1.223	0.9697
1.36	0.7572	1.991	1.620	1.229	0.9676
1.37	0.7527	2.023	1.638	1.235	0.9653
1.38	0.7483	2.055	1.655	1.242	0.9630
1.39	0.7440	2.087	1.672	1.248	0.9607
1.40	0.7397	2.120	1.690	1.255	0.9582
1.41	0.7355	2.153	1.707	1.261	0.9557
1.42	0.7314	2.186	1.724	1.268	0.9531
1.43	0.7274	2.219	1.742	1.274	0.9504
1.44	0.7235	2.253	1.759	1.281	0.9476
1.45	0.7196	2.286	1.776	1.287	0.9448
1.46	0.7157	2.320	1.793	1.294	0.9420
1.47	0.7120	2.354	1.811	1.300	0.9390
1.48	0.7083	2.389	1.828	1.307	0.9360
1.49	0.7047	2.423	1.845	1.314	0.9329
1.50	0.7011	2.458	1.862	1.320	0.9298
1.51	0.6976	2.493	1.879	1.327	0.9266
1.52	0.6941	2.529	1.896	1.334	0.9233
1.53	0.6907	2.564	1.913	1.340	0.9200
1.54	0.6874	2.600	1.930	1.347	0.9166
1.55	0.6841	2.636	1.947	1.354	0.9132
1.56	0.6809	2.673	1.964	1.361	0.9097
1.57	0.6777	2.709	1.981	1.367	0.9061
1.58	0.6746	2.746	1.998	1.374	0.9026
1.59	0.6715	2.783	2.015	1.381	0.8989
1.60	0.6684	2.820	2.032	1.388	0.8952
1.61	0.6655	2.857	2.049	1.395	0.8915
1.62	0.6625	2.895	2.065	1.402	0.8877
1.63	0.6596	2.933	2.082	1.409	0.8538
1.64	0.6568	2.971	2.099	1.416	0.8799
1.65	0.6540	3.010	2.115	1.423	0.8760
1.66	0.6512	3.048	2.132	1.430	0.8720
1.67	0.6485	3.087	2.148	1.437	0.8680
1.68	0.6458	3.126	2.165	1.444	0.8640

Table B–4   Normal Shock Relations ($k = 1.4$)

M_1	M_2	$\dfrac{p_2}{p_1}$	$\dfrac{\rho_2}{\rho_1}$	$\dfrac{T_2}{T_1}$	$\dfrac{(p_0)_2}{(p_0)_1}$
1.69	0.6431	3.165	2.181	1.451	0.8598
1.70	0.6405	3.205	2.198	1.458	0.8557
1.71	0.6380	3.245	2.214	1.466	0.8516
1.72	0.6355	3.285	2.230	1.473	0.8474
1.73	0.6330	3.325	2.247	1.480	0.8431
1.74	0.6305	3.366	2.263	1.487	0.8389
1.75	0.6281	3.406	2.279	1.495	0.8346
1.76	0.6257	3.447	2.295	1.502	0.8302
1.77	0.6234	3.488	2.311	1.509	0.8259
1.78	0.6210	3.530	2.327	1.517	0.8215
1.79	0.6188	3.571	2.343	1.524	0.8171
1.80	0.6165	3.613	2.359	1.532	0.8127
1.81	0.6143	3.655	2.375	1.539	0.8082
1.82	0.6121	3.698	2.391	1.547	0.8038
1.83	0.6099	3.740	2.407	1.554	0.7993
1.84	0.6078	3.783	2.422	1.562	0.7948
1.85	0.6057	3.826	2.438	1.569	0.7902
1.86	0.6036	3.870	2.454	1.577	0.7857
1.87	0.6016	3.913	2.469	1.585	0.7811
1.88	0.5996	3.957	2.485	1.592	0.7765
1.89	0.5976	4.001	2.500	1.600	0.7720
1.90	0.5956	4.045	2.516	1.608	0.7674
1.91	0.5937	4.089	2.531	1.616	0.7627
1.92	0.5918	4.134	2.546	1.624	0.7581
1.93	0.5899	4.179	2.562	1.631	0.7535
1.94	0.5880	4.224	2.577	1.639	0.7488
1.95	0.5862	4.270	2.592	1.647	0.7442
1.96	0.5844	4.315	2.607	1.655	0.7395
1.97	0.5826	4.361	2.622	1.663	0.7349
1.98	0.5808	4.407	2.637	1.671	0.7302
1.99	0.5791	4.453	2.652	1.679	0.7255
2.00	0.5774	4.500	2.667	1.688	0.7209
2.01	0.5757	4.547	2.681	1.696	0.7162
2.02	0.5740	4.594	2.696	1.704	0.7115
2.03	0.5723	4.641	2.711	1.712	0.7069
2.04	0.5707	4.689	2.725	1.720	0.7022
2.05	0.5691	4.736	2.740	1.729	0.6975
2.06	0.5675	4.784	2.755	1.737	0.6928
2.07	0.5659	4.832	2.769	1.745	0.6882
2.08	0.5643	4.881	2.783	1.754	0.6835
2.09	0.5628	4.929	2.798	1.762	0.6789
2.10	0.5613	4.978	2.812	1.770	0.6742
2.11	0.5598	5.027	2.826	1.779	0.6696
2.12	0.5583	5.077	2.840	1.787	0.6649
2.13	0.5568	5.126	2.854	1.796	0.6603
2.14	0.5554	5.176	2.868	1.805	0.6557
2.15	0.5540	5.226	2.882	1.813	0.6511
2.16	0.5525	5.277	2.896	1.822	0.6464
2.17	0.5511	5.327	2.910	1.821	0.6419
2.18	0.5498	5.378	2.924	1.839	0.6373
2.19	0.5484	5.429	2.938	1.848	0.6327
2.20	0.5471	5.480	2.951	1.857	0.6281
2.21	0.5457	5.531	2.965	1.866	0.6236
2.22	0.5444	5.583	2.978	1.875	0.6191
2.23	0.5431	5.636	2.992	1.883	0.6145
2.24	0.5418	5.687	3.005	1.892	0.6100
2.25	0.5406	5.740	3.019	1.901	0.6055
2.26	0.5393	5.792	3.032	1.910	0.6011
2.27	0.5381	5.845	3.045	1.919	0.5966
2.28	0.5368	5.898	3.058	1.929	0.5921
2.29	0.5356	5.951	3.071	1.938	0.5877
2.30	0.5344	6.005	3.085	1.947	0.5833
2.31	0.5332	6.059	3.098	1.956	0.5789
2.32	0.5321	6.113	3.110	1.965	0.5745
2.33	0.5309	6.167	3.123	1.974	0.5702
2.34	0.5297	6.222	3.136	1.984	0.5658
2.35	0.5286	6.276	3.149	1.993	0.5615
2.36	0.5275	6.331	3.162	2.002	0.5572

Table B-4   Normal Shock Relations ($k = 1.4$)

M_1	M_2	$\dfrac{p_2}{p_1}$	$\dfrac{\rho_2}{\rho_1}$	$\dfrac{T_2}{T_1}$	$\dfrac{(p_0)_2}{(p_0)_1}$
2.37	0.5264	6.386	3.174	2.012	0.5529
2.38	0.5253	6.442	3.187	2.021	0.5486
2.39	0.5242	6.497	3.199	2.031	0.5444
2.40	0.5231	6.553	3.212	2.040	0.5401
2.41	0.5221	6.609	3.224	2.050	0.5359
2.42	0.5210	6.666	3.237	2.059	0.5317
2.43	0.5200	6.722	3.249	2.069	0.5276
2.44	0.5189	6.779	3.261	2.079	0.5234
2.45	0.5179	6.836	3.273	2.088	0.5193
2.46	0.5169	6.894	3.285	2.098	0.5152
2.47	0.5159	6.951	3.298	2.108	0.5111
2.48	0.5149	7.009	3.310	2.118	0.5071
2.49	0.5140	7.067	3.321	2.128	0.5030
2.50	0.5130	7.125	3.333	2.138	0.4990
2.51	0.5120	7.183	3.345	2.147	0.4950
2.52	0.5111	7.242	3.357	2.157	0.4911
2.53	0.5102	7.301	3.369	2.167	0.4871
2.54	0.5092	7.360	3.380	2.177	0.4832
2.55	0.5083	7.420	3.392	2.187	0.4793
2.56	0.5074	7.479	3.403	2.198	0.4754
2.57	0.5065	7.539	3.415	2.208	0.4715
2.58	0.5056	7.599	3.426	2.218	0.4677
2.59	0.5047	7.659	3.438	2.228	0.4639
2.60	0.5039	7.720	3.449	2.238	0.4601
2.61	0.5030	7.781	3.460	2.249	0.4564
2.62	0.5022	7.842	3.471	2.259	0.4526
2.63	0.5013	7.903	3.483	2.269	0.4489
2.64	0.5005	7.965	3.494	2.280	0.4452
2.65	0.4996	8.026	3.505	2.290	0.4416
2.66	0.4988	8.088	3.516	2.301	0.4379
2.67	0.4980	8.150	3.527	2.311	0.4343
2.68	0.4972	8.213	3.537	2.322	0.4307
2.69	0.4964	8.275	3.548	2.332	0.4271
2.70	0.4956	8.338	3.559	2.343	0.4236
2.71	0.4949	8.401	3.570	2.354	0.4201
2.72	0.4941	8.465	3.580	2.364	0.4166
2.73	0.4933	8.528	3.591	2.375	0.4131
2.74	0.4926	8.592	3.601	2.386	0.4097
2.75	0.4918	8.656	3.612	2.397	0.4062
2.76	0.4911	8.721	3.622	2.407	0.4028
2.77	0.4903	8.785	3.633	2.418	0.3994
2.78	0.4896	8.850	3.643	2.429	0.3961
2.79	0.4889	8.915	3.653	2.440	0.3928
2.80	0.4882	8.980	3.664	2.451	0.3895
2.81	0.4875	9.045	3.674	2.462	0.3862
2.82	0.4868	9.111	3.684	2.473	0.3829
2.83	0.4861	9.177	3.694	2.484	0.3797
2.84	0.4854	9.243	3.704	2.496	0.3765
2.85	0.4847	9.310	3.714	2.507	0.3733
2.86	0.4840	9.376	3.724	2.518	0.3701
2.87	0.4833	9.443	3.734	2.529	0.3670
2.88	0.4827	9.510	3.743	2.540	0.3639
2.89	0.4820	9.577	3.753	2.552	0.3608
2.90	0.4814	9.645	3.763	2.563	0.3577
2.91	0.4807	9.713	3.773	2.575	0.3547
2.92	0.4801	9.781	3.782	2.586	0.3517
2.93	0.4795	9.849	3.792	2.598	0.3487
2.94	0.4788	9.918	3.801	2.609	0.3457
2.95	0.4782	9.986	3.811	2.621	0.3428
2.96	0.4776	10.06	3.820	2.632	0.3398
2.97	0.4770	10.12	3.829	2.644	0.3369
2.98	0.4764	10.19	3.839	2.656	0.3340
2.99	0.4758	10.26	3.848	2.667	0.3312
3.00	0.4752	10.33	3.857	2.679	0.3283

Table B–5 Prandtl – Meyer Expansion ($k = 1.4$)	
M	ω (degrees)
1.00	0.00
1.02	0.1257
1.04	0.3510
1.06	0.6367
1.08	0.9680
1.10	1.336
1.12	1.735
1.14	2.160
1.16	2.607
1.18	3.074
1.20	3.558
1.22	4.057
1.24	4.569
1.26	5.093
1.28	5.627
1.30	6.170
1.32	6.721
1.34	7.279
1.36	7.844
1.38	8.413
1.40	8.987
1.42	9.565
1.44	10.146
1.46	10.730
1.48	11.317
1.50	11.905
1.52	12.495
1.54	13.086
1.56	13.677
1.58	14.269
1.60	14.860
1.62	15.452
1.64	16.043
1.66	16.633
1.68	17.222
1.70	17.810
1.72	18.396
1.74	18.981
1.76	19.565
1.78	20.146
1.80	20.725
1.82	21.302
1.84	21.877
1.86	22.449
1.88	23.019
1.90	23.586
1.92	24.151
1.94	24.712
1.96	25.271
1.98	25.827
2.00	26.380
2.02	26.930
2.04	27.476
2.06	28.020
2.08	28.560
2.10	29.097
2.12	29.631
2.14	30.161
2.16	30.688
2.18	31.212
2.20	31.732
2.22	32.249
2.24	32.763
2.26	33.273
2.28	33.780
2.30	34.283
2.32	34.782
2.34	35.279
2.36	35.772
2.38	36.261
2.40	36.746
2.42	37.229
2.44	37.708
2.46	38.183
2.48	38.655
2.50	39.124
2.52	39.589
2.54	40.050
2.56	40.508
2.58	40.963
2.60	41.415
2.62	41.863
2.64	42.307
2.66	42.749
2.68	43.187
2.70	43.622
2.72	44.053
2.74	44.481
2.76	44.906
2.78	45.328
2.80	45.746
2.82	46.161
2.84	46.573
2.86	46.982
2.88	47.388
2.90	47.790
2.92	48.190
2.94	48.586
2.96	48.980
2.98	49.370
3.00	49.757
3.02	50.14
3.04	50.52
3.06	50.90
3.08	51.28
3.10	51.65
3.12	52.01
3.14	52.39
3.16	52.75
3.18	53.11
3.20	53.47
3.22	53.83
3.24	54.18
3.26	54.53
3.28	54.88
3.30	55.22
3.32	55.56
3.34	55.90
3.36	56.24
3.38	56.58
3.40	56.90
3.42	57.24
3.44	57.56
3.46	57.89
3.48	58.21
3.50	58.53

Fundamental Solutions

Chapter 2

F2–1. $p_B + \rho_w g h_w = 400(10^3)$ Pa

$p_B + (1000 \text{ kg/m}^3)(9.81 \text{ m/s}^2)(0.3 \text{ m}) = 400(10^3)$ Pa

$p_B = 397.06(10^3)$ Pa

$+\uparrow F_R = \Sigma F_y;\ F_R = [397.06(10^3) \text{ N/m}^2][\pi(0.025 \text{ m})^2]$
$\qquad - [101(10^3) \text{ N/m}^2][\pi(0.025 \text{ m})^2]$

$\qquad = 581.31 \text{ N} = 581 \text{ N}$ *Ans.*

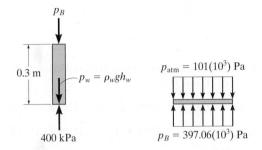

F2–2. The pressures at A, B and C can be obtained by writing the manometer equation. For point A, referring to Fig. *a*,

$p_{atm} + \rho_{wg} h_w + \rho_{og} h_o = p_A$

$0 + (1000 \text{ kg/m}^3)(9.81 \text{ m/s}^2)(1.25 \text{ m}) +$
$(830 \text{ kg/m}^3)(9.81 \text{ m/s}^2)(1.25 \text{ m}) = p_A$

$p_A = 22.44(10^3) \text{ Pa} = 22.4 \text{ kPa}$ *Ans.*

For point B, referring to Fig. *b*,

$p_{atm} + \rho_{og} h_o - \rho_{wg} h_w = p_B$

$0 + (830 \text{ kg/m}^3)(9.81 \text{ m/s}^2)(1.25 \text{ m}) -$
$(1000 \text{ kg/m}^3)(9.81 \text{ m/s}^2)(0.25 \text{ m}) = p_B$

$p_B = 7.725(10^3) \text{ Pa} = 7.73 \text{ kPa}$ *Ans.*

For point C, referring to Fig. *c*,

$p_{atm} + \rho_{og} h_o + \rho_{wg} h_w = p_C$

$0 + (830 \text{ kg/m}^3)(9.81 \text{ m/s}^2)(1.25 \text{ m}) +$
$(1000 \text{ kg/m}^3)(9.81 \text{ m/s}^2)(1 \text{ m}) = p_C$

$p_C = 19.987(10^3) \text{ Pa} = 20.0 \text{ kPa}$ *Ans.*

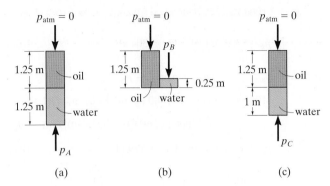

F2–3. Referring to the figure,

$p_{atm} + \rho_{wg} h_w - \rho_{Hg} g h_{Hg} = p_{atm}$

$\rho_{wg} h_w - \rho_{Hg} g h_{Hg} = 0$

$(1000 \text{ kg/m}^3)(9.81 \text{ m/s}^2)(2 + h)$
$\qquad - (13\,550 \text{ kg/m}^3)(9.81 \text{ m/s}^2)h = 0$

$\qquad 2000 + 1000h - 13\,550h = 0$

$h = 0.1594 \text{ m} = 159 \text{ mm}$ *Ans.*

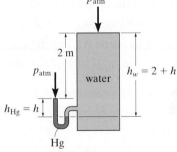

F2–4. $p_{atm} + \rho_{wg} h_w - \rho_{Hg} g h_{Hg} = p_{atm}$

$\rho_w h_w = \rho_{Hg} h_{Hg}$

$h_w = \left(\dfrac{\rho_{Hg}}{\rho_w}\right) h_{Hg}$

$(h - 0.3 \text{ m}) = \left(\dfrac{13\,550 \text{ kg/m}^3}{1000 \text{ kg/m}^3}\right)(0.1 \text{ m} + 0.5 \sin 30° \text{ m})$

$h = 5.0425 \text{ m} = 5.04 \text{ m}$

F2–5. $p_B - \rho_w g h_w = p_A$

$p_B - (1000 \text{ kg/m}^3)(9.81 \text{ m/s}^2)(0.4 \text{ m}) = 300(10^3) \text{ N/m}^2$

$p_B = 303.92(10^3) \text{ Pa} = 304 \text{ kPa}$ *Ans.*

F2–6. $p_{atm} + \rho_{\;co} g h_{co} + \rho_w g h_w = p_B$

$[101(10^3) \text{ N/m}^2] + (880 \text{ kg/m}^3)(9.81 \text{ m/s}^2)(1.1 \text{ m})$

$+ (1000 \text{ kg/m}^3)(9.81 \text{ m/s}^2)(0.9 \text{ m}) = p_B$

$p_B = 119.33(10^3) \text{ Pa} = 119 \text{ kPa}$ *Ans.*

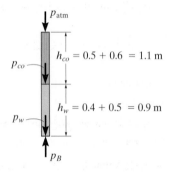

$h_{co} = 0.5 + 0.6 = 1.1 \text{ m}$

$h_w = 0.4 + 0.5 = 0.9 \text{ m}$

F2–7.   The intensity of the distributed load at A is

$w_A = \rho_w g h_A b = 1000(9.81)(2.5)(1.5) = 36.7875(10^3) \text{ N/m}$

The resultant forces on AB and BC are

$(F_R)_{AB} = \frac{1}{2}[36.7875(10^3)](2.5) = 45.98(10^3) \text{ N}$

$= 46.0 \text{ kN} \quad Ans.$

$(F_R)_{BC} = [36.7875(10^3)](2) = 73.575(10^3) \text{ N} = 73.6 \text{ kN } Ans.$

Alternatively,

$(F_R)_{AB} = \rho_w g \bar{h}_{AB} A_{AB} = 1000(9.81)(2.5/2)[2.5(1.5)]$

$= 45.98(10^3) \text{ N} = 46.0 \text{ kN}$ *Ans.*

$(F_R)_{BC} = \rho_w g \bar{h}_{BC} A_{BC} = 1000(9.81)(2.5)[2(1.5)]$

$= 73.575(10^3) \text{ N} = 73.6 \text{ kN}$ *Ans.*

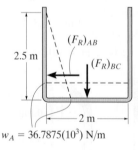

2.5 m

$(F_R)_{AB}$

$(F_R)_{BC}$

2 m

$w_A = 36.7875(10^3) \text{ N/m}$

F2–8.   The intensity of the distributed load is

$w_A = \rho_o g h_A b = 900(9.81)(3)(2) = 52.974(10^3) \text{ N/m}$

Here, $L_{AB} = 3/\sin 60° = 3.464 \text{ m}$

$F_R = \frac{1}{2}[52.974(10^3)](3.464) = 91.8 \text{ kN}$ *Ans.*

Alternatively,

$F_R = \rho_o g \bar{h} A = 900(9.81)(1.5)[(3/\sin 60°)(2)] = 91.8 \text{ kN} \quad Ans.$

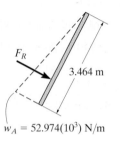

F_R

3.464 m

$w_A = 52.974(10^3) \text{ N/m}$

F2–9.   The intensities of the distributed loads at the bottom of A and B are

$w_1 = \rho_w g h_1 b = 1000(9.81)(0.9)(2) = 17.658(10^3) \text{ N/m}$

$w_2 = \rho_w g h_2 b = 1000(9.81)(1.5)(2) = 29.43(10^3) \text{ N/m}$

Then, the resultant forces are

$(F_R)_A = \frac{1}{2}[17.658(10^3)](0.9) = 7.9461(10^3) \text{ N} = 7.94 \text{ kN} \quad Ans.$

$(F_R)_{B_1} = [17.658(10^3)](0.6) = 10.5948(10^3) \text{ N}$

$(F_R)_{B_2} = \frac{1}{2}[29.43(10^3) - 17.658(10^3)](0.6) = 3.5316(10^3)$

$(F_R)_B = (F_R)_{B_1} + (F_R)_{B_2} = 10.5948(10^3) + 3.5816(10^3)$

$= 14.13(10^3) \text{ N} = 14.1 \text{ kN}$ *Ans.*

And they act at

$(y_p)_A = \frac{2}{3}(0.9) = 0.6 \text{ m}$ *Ans.*

$(y_p)_B = \dfrac{[10.5948(10^3)](0.9 + 0.6/2) + 3.5316(10^3)[0.9 + \frac{2}{3}(0.6)]}{14.1264(10^3)}$

$= 1.225 \text{ m}$ *Ans.*

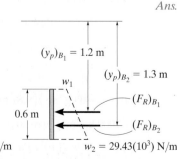

0.9 m

$(y_p)_A$

$(F_R)_A$

$w_1 = 17.658(10^3) \text{ N/m}$

$(y_p)_{B_1} = 1.2 \text{ m}$

$(y_p)_{B_2} = 1.3 \text{ m}$

w_1

0.6 m

$(F_R)_{B_1}$

$(F_R)_{B_2}$

$w_2 = 29.43(10^3) \text{ N/m}$

Alternatively, the resultant force on A is

$$(F_R)_A = \rho_w g \bar{h}_A A_A = 1000(9.81)(0.45)(0.9)(2)$$

$$= 7.9461(10^3) = 7.94 \text{ kN} \qquad Ans.$$

And acts at

$$(y_p)_A = \frac{(\bar{I}_x)_A}{\bar{y}_A A_A} + \bar{y}_A = \frac{\frac{1}{12}(2)(0.9^3)}{0.45[0.9(2)]} + 0.45 = 0.6 \text{ m} \qquad Ans.$$

The resultant force on B is

$$(F_p)_B = \rho_w g \bar{h}_B A_B = 1000(9.81)(0.9 + 0.6/2)(0.6)(2)$$

$$= 14.1264(10^3) = 14.1 \text{ kN} \qquad Ans.$$

And acts at

$$(y_p)_B = \frac{(\bar{I}_x)_B}{\bar{y}_B A_B} + \bar{y}_B = \frac{\frac{1}{12}(2)(0.6^3)}{(0.9 + 0.6/2)(0.6)(2)} + (0.9 + 0.6/2)$$

$$= 1.225 \text{ m} \qquad Ans.$$

F2–10. Here, $\bar{y}_A = \bar{h}_A = \frac{2}{3}(1.2) = 0.8 \text{ m}$
$A_A = \frac{1}{2}(0.6)(1.2) = 0.36 \text{ m}^2$. Then

$$F_R = \rho_w g \bar{h}_A A_A = 1000(9.81)(0.8)(0.36) = 2825.28 \text{ N}$$

$$= 2.83 \text{ kN} \qquad Ans.$$

Also, $(\bar{I}_x)_A = \frac{1}{36}(0.6)(1.2^3) = 0.0288 \text{ m}^4$. Then

$$y_p = \frac{(\bar{I}_x)_A}{\bar{y}_A A_A} + \bar{y}_A = \frac{0.0288}{0.8(0.36)} + 0.8 = 0.9 \text{ m} \qquad Ans.$$

F2–11. $\bar{y} = 2 \text{ m}, \bar{h} = 2 \sin 60° = \sqrt{3} \text{ m}$,
$A = \pi(0.5^2) = 0.25\pi \text{ m}^2$

$$(\bar{I}_x) = \frac{\pi}{4}(0.5^4) = 0.015625\pi \text{ m}^4. \text{ Then}$$

$$F_R = \rho_w g \bar{h} A = 1000(9.81)(\sqrt{3})(0.25\pi) = 13.345(10^3) \text{ N}$$

$$= 13.3 \text{ kN} \qquad Ans.$$

$$y_p = \frac{\bar{I}_x}{\bar{y} A} + \bar{y} = \frac{0.015625\pi}{2(0.25\pi)} + 2 = 2.03125 \text{ m} = 2.03 \text{ m} \; Ans.$$

F2–12. The intensities of the distributed loads are

$$w_1 = \rho_k g h_k b = 814(9.81)(1 \sin 60°)(2) = 13.831(10^3) \text{ N/m}$$

$$w_2 = w_1 + \rho_w g h_w b = 13.831(10^3) + 1000(9.81)(3 \sin 60°)(2)$$

$$= 64.805(10^3) \text{ N/m}$$

Thus, the resultant force is

$$F_R = \frac{1}{2}[13.831(10^3)](1) + 13.831(10^3)(3)$$

$$+ \frac{1}{2}[64.805(10^3) - 13.831(10^3)](3)$$

$$= 124.87(10^3) \text{ N} = 125 \text{ kN} \qquad Ans.$$

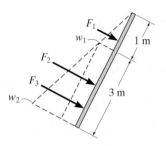

F2–13.   Horizontal Component:

$$w_A = \rho_w g h_A b$$

$$= (1000 \text{ kg/m}^3)(9.81 \text{ m/s}^2)(3 \sin 30° \text{ m})(0.5 \text{ m}) = 7357.5 \text{ N/m}$$

$$F_R = \frac{1}{2} w_A h_A$$

$$= \frac{1}{2}(7357.58 \text{ N/m})(3 \sin 30° \text{ m}) = 5518.125 \text{ N}$$

$$= 5.518 \text{ kN}$$

Vertical Component:

$$F_v = \rho_w g V$$

$$= 1000 \text{ kg/m}^3(9.81 \text{ m/s}^2)[\frac{1}{2}(3 \cos 30° \text{ m})(3 \sin 30° \text{ m})(0.5 \text{ m})]$$

$$= 9557.67 \text{ N} = 9.558 \text{ kN}$$

$$\xrightarrow{+} \Sigma F_x = 0; \quad A_x - 5.518 \text{ kN} = 0 \qquad A_x = 5.52 \text{ kN} \quad Ans.$$

$$+\uparrow \Sigma F_y = 0; \quad 9.558 \text{ kN} - A_y = 0 \qquad A_y = 9.56 \text{ kN} \quad Ans.$$

$$\zeta + \Sigma M_A = 0; \quad (9.558 \text{ kN})[\frac{1}{3}(3 \cos 30° \text{ m})]$$

$$+ (5.518 \text{ kN})\frac{1}{3}(3 \sin 30° \text{ m}) - M_A = 0$$

$$M_A = 11.0 \text{ kN} \cdot \text{m} \qquad Ans.$$

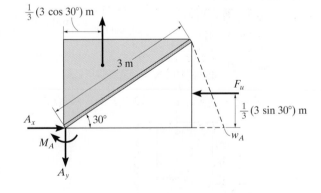

F2–14. The resultant force is equal to the weight of the oil block above surface AB.

$$F_v = \rho_o g(A_{ACB} + A_{ABDE})b$$

$$= \left(900 \text{ kg/m}^3\right)\left(9.81 \text{ m/s}^2\right)\left[\frac{\pi}{2}(0.5 \text{ m})^2 + (1)(1.5 \text{ m})\right](3 \text{ m})$$

$$= 50.13\left(10^3\right) \text{ N} = 50.1 \text{ kN} \qquad\qquad Ans.$$

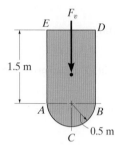

F2–15.   **Side AB**

Horizontal Component

$$w_B = \rho_w g h_B b = \left(1000 \text{ kg/m}^3\right)\left(9.81 \text{ m/s}^2\right)(2 \text{ m})(0.75 \text{ m})$$

$$= 14.715\left(10^3\right) \text{ N/m}$$

$$F_h = \frac{1}{2}w_B h_B = \frac{1}{2}\left[14.715\left(10^3\right) \text{ N/m}\right](2 \text{ m})$$

$$= 14.715\left(10^3\right) \text{ N} = 14.7 \text{ kN} \rightarrow \quad Ans.$$

Vertical Component

$$F_v = \rho_w g V\!\!\!/$$

$$= \left(1000 \text{ kg/m}^3\right)\left(9.81 \text{ m/s}^2\right)\left[\frac{1}{2}\left(\frac{2 \text{ m}}{\tan 60°}\right)(2 \text{ m})(0.75 \text{ m})\right]$$

$$= 8495.71 \text{ N} = 8.50 \text{ kN} \uparrow \qquad\qquad Ans.$$

Side CD

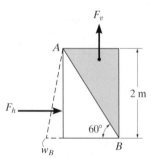

Horizontal Component

$$w_D = \rho_w g h_D b = \left(1000 \text{ kg/m}^3\right)\left(9.81 \text{ m/s}^2\right)(2 \text{ m})(0.75 \text{ m})$$

$$= 14.715\left(10^3\right) \text{ N/m}$$

$$F_h = \frac{1}{2}w_D h_D = \frac{1}{2}\left[14.715\left(10^3\right) \text{ N/m}\right](2 \text{ m})$$

$$= 14.715\left(10^3\right) \text{ N} = 14.7 \text{ kN} \leftarrow \qquad Ans.$$

Vertical Component

$$F_v = \rho_w g V\!\!\!/$$

$$= \left(1000 \text{ kg/m}^3\right)\left(9.81 \text{ m/s}^2\right)\left[\frac{1}{2}\left(\frac{2 \text{ m}}{\tan 45°}\right)(2 \text{ m})(0.75 \text{ m})\right]$$

$$= 14.715\left(10^3\right) \text{ N} = 14.7 \text{ kN} \downarrow \qquad\qquad Ans.$$

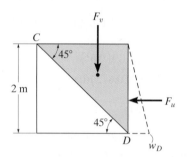

F2–16.   **Plate AB**

Horizontal Component

$$w_B = \rho_w g h_B b$$

$$= \left(1000 \text{ kg/m}^3\right)\left(9.81 \text{ m/s}^2\right)(0.5 \text{ m})(2\text{m})$$

$$= 9.81\left(10^3\right) \text{ N/m}$$

$$w_A = \rho_w g h_A b$$

$$= \left(1000 \text{ kg/m}^3\right)\left(9.81 \text{ m/s}^2\right)(2 \text{ m})(2 \text{ m})$$

$$= 39.24\left(10^3\right) \text{ N/m}$$

$$(F_h)_1 = \left[9.81\left(10^3\right) \text{ N/m}\right](1.5 \text{ m}) = 14.715\left(10^3\right) \text{ N} = 14.715 \text{ kN}$$

$$(F_h)_2 = \tfrac{1}{2}\left[39.24\left(10^3\right) \text{ N/m} - 9.81\left(10^3\right) \text{ N/m}\right](1.5 \text{ m})$$

$$= 22.0725\left(10^3\right) \text{ N} = 22.0725 \text{ kN}$$

Thus,

$$F_h = (F_h)_1 + (F_h)_2 = 14.715 \text{ kN} + 22.0725 \text{ kN}$$

$$= 36.8 \text{ kN} \leftarrow \qquad\qquad Ans.$$

Vertical Component

$$(F_v)_1 = \rho_w g V\!\!\!/_1$$

$$= \left(1000 \text{ kg/m}^3\right)\left(9.81 \text{ m/s}^2\right)\left[\left(\frac{1.5 \text{ m}}{\tan 60°}\right)(0.5 \text{ m})(2 \text{ m})\right]$$

$$= 8.4957\left(10^3\right) \text{ N} = 8.4957 \text{ kN}$$

$$(F_v)_2 = \rho_w g V_2$$

$$= \left(1000 \text{ kg/m}^3\right)\left(9.81 \text{ m/s}^2\right)\left[\frac{1}{2}\left(\frac{1.5 \text{ m}}{\tan 60°}\right)(1.5 \text{ m})(2 \text{ m})\right]$$

$$= 12.7436\left(10^3\right) \text{ N} = 12.7436 \text{ kN}$$

Thus,

$$F_v = (F_v)_1 + (F_v)_2$$

$$= 8.4957 \text{ kN} + 12.7436 \text{ kN}$$

$$= 21.2 \text{ kN} \uparrow \qquad\qquad Ans.$$

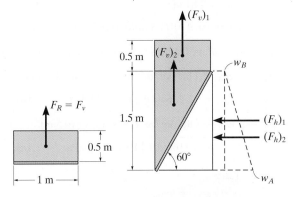

F2–17.   Plate *AB*

Horizontal Component

$$w_B = \rho_w g h_B b = \left(1000 \text{ kg/m}^3\right)\left(9.81 \text{ m/s}^2\right)(2 \text{ m})(1.5 \text{ m})$$

$$= 29.43\left(10^3\right) \text{ N/m}$$

$$F_h = \frac{1}{2}w_B h_B = \frac{1}{2}\left[29.43\left(10^3\right) \text{ N/m}\right](2 \text{ m})$$

$$= 29.43\left(10^3\right) \text{ N} = 29.4 \text{ kN} \rightarrow \qquad Ans.$$

Vertical Component

$$F_v = \rho_w g V = \left(1000 \text{ kg/m}^3\right)\left(9.81 \text{ m/s}^2\right)\left[\frac{1}{2}\left(\frac{2 \text{ m}}{\tan 45°}\right)(2 \text{ m})(1.50 \text{ m})\right]$$

$$= 29.43\left(10^3\right) \text{ N} = 29.4 \text{ kN} \uparrow \qquad Ans.$$

Plate *BC*

Horizontal Component

$$w_B = \rho_w g h_B b = \left(1000 \text{ kg/m}^3\right)\left(9.81 \text{ m/s}^2\right)(2 \text{ m})(1.5 \text{ m})$$

$$= 29.43\left(10^3\right) \text{ N/m}$$

$$F_h = \frac{1}{2}w_B h_B = \frac{1}{2}\left[29.43\left(10^3\right) \text{ N/m}\right](2 \text{ m})$$

$$= 29.43\left(10^3\right) \text{ N} = 29.4 \text{ kN} \leftarrow \qquad Ans.$$

Vertical Component

$$F_v = \rho_w g V$$

$$= \left(1000 \text{ kg/m}^3\right)\left(9.81 \text{ m/s}^2\right)\left[(2 \text{ m})(2 \text{ m})(1.5 \text{ m}) - \frac{\pi}{4}(2 \text{ m})^2(1.5 \text{ m})\right]$$

$$= 12.631\left(10^3\right) \text{ N} = 12.6 \text{ kN} \uparrow \qquad Ans.$$

F2–18.   Horizontal Component:

$$w_C = \rho_w g h_C b$$

$$= \left(1000 \text{ kg/m}^3\right)\left(9.81 \text{ m/s}^2\right)(5 \text{ m})(2 \text{ m})$$

$$= 98.1\left(10^3\right) \text{ N/m}$$

$$F_h = \tfrac{1}{2}w_C h_C$$

$$= \tfrac{1}{2}\left[98.1\left(10^3\right) \text{ N/m}\right](5 \text{ m}) = 245.25\left(10^3\right) \text{ N}$$

$$= 245.25 \text{ kN} \leftarrow$$

Vertical Component:

$$F_v = \rho_w g V = \left(1000 \text{ kg/m}^3\right)\left(9.81 \text{ m/s}^2\right)\left[\frac{1}{2}\left(\frac{5 \text{ m}}{\tan \theta}\right)(5 \text{ m})(2 \text{ m})\right]$$

$$= \frac{245.25\left(10^3\right)}{\tan \theta} \text{ N} = \frac{245.25}{\tan \theta} \text{ kN} \uparrow$$

$$\zeta + \Sigma M_A = 0;$$

$$\left(\frac{245.25}{\tan \theta} \text{ kN}\right)\left[\frac{1}{3}\left(\frac{5 \text{ m}}{\tan \theta}\right)\right] - (245.25 \text{ kN})\left[\frac{2}{3}(5 \text{ m})\right] = 0$$

$$\frac{1}{\tan^2 \theta} = 2, \quad \theta = 35.3° \qquad Ans.$$

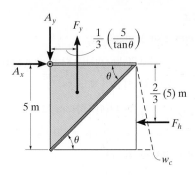

F2–19. $F_b = \rho_w g \forall b$

$\qquad = \left(1000 \text{ kg/m}^3\right)\left(9.81 \text{ m/s}^2\right)\left[\pi(0.1 \text{ m})^2 d\right] = 308.2d$

$+\uparrow \Sigma F_y = 0; \quad 308.2d - \left[2(9.81) \text{ N}\right] = 0; \quad d = 0.06366 \text{ m}$

$\qquad \forall_w = \forall' - \forall_{wb}$

$\pi(0.2 \text{ m})^2(0.5 \text{ m}) = \pi\left(0.2 \text{ m}^2\right)h - \pi(0.1 \text{ m})^2(0.0636 \text{ m})$

$\qquad\qquad h = 0.516 \text{ m}$ \hfill *Ans.*

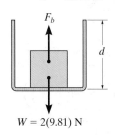

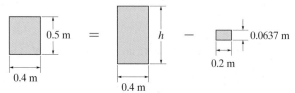

F2–20. $\tan \theta = \dfrac{a_c}{g} = \dfrac{4 \text{ m/s}^2}{9.81 \text{ m/s}^2} = 0.4077$

$\qquad\qquad \theta = 22.18° = 22.2°$ \hfill *Ans.*

$\qquad h_B = 1.5 \text{ m} + (1 \text{ m}) \tan 22.18°$

$\qquad\quad = 1.9077 \text{ m}$

$w_B = \rho_w g h_B b = \left(1000 \text{ kg/m}^3\right)\left(9.81 \text{ m/s}^2\right)(1.9077 \text{ m})(3 \text{ m})$

$\qquad\quad = 56.145\left(10^3\right) \text{ N/m} = 56.143 \text{ kN/m}$

$F_R = \dfrac{1}{2}(56.145 \text{ kN/m})(1.9077 \text{ m}) = 53.56 \text{ kN} = 53.6 \text{ kN} \, Ans.$

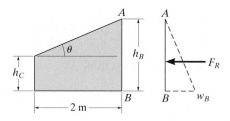

F2–21. $\tan \theta = \dfrac{a_c}{g} = \dfrac{6 \text{ m/s}^2}{9.81 \text{ m/s}^2} = 0.6116$

$\quad h' = (1.5 \text{ m}) \tan \theta = (1.5 \text{ m})(0.6116) = 0.9174 \text{ m}$

$\quad h_A = h_B + h' = 0.5 \text{ m} + 0.9174 \text{ m} = 1.4174 \text{ m}$

$\quad p_A = \rho_o g h_A = \left(880 \text{ kg/m}^3\right)\left(9.81 \text{ m/s}^2\right)(1.4174 \text{ m})$

$\qquad\quad = 12.2364\left(10^3\right) \text{ Pa} = 12.2 \text{ kPa}$ \hfill *Ans.*

$\quad p_B = \rho_o g h_B = \left(880 \text{ kg/m}^3\right)\left(9.81 \text{ m/s}^2\right)(0.5 \text{ m})$

$\qquad\quad = 4.3164\left(10^3\right) \text{ Pa} = 4.32 \text{ kPa}$ \hfill *Ans.*

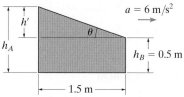

F2–22. $(\forall_{\text{air}})_i = (\forall_{\text{air}})_f$

$\qquad \pi(1 \text{ m})^2(3 \text{ m} - 2 \text{ m}) = \dfrac{1}{2}\left[\pi(1 \text{ m})^2\right]$

$\qquad h = 2 \text{ m}$

$\qquad h = \dfrac{\omega^2}{2g} r^2; \qquad 2 \text{ m} = \left[\dfrac{\omega^2}{2\left(9.81 \text{ m/s}^2\right)}\right](1 \text{ m})^2$

$\qquad \omega = 6.26 \text{ rad/s}$ \hfill *Ans.*

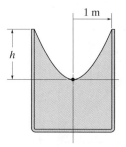

F2–23. $h = \dfrac{\omega^2}{2g}r^2;$ $h = \left[\dfrac{(8\ \text{rad/s})^2}{2(9.81\ \text{m/s}^2)}\right](1\ \text{m})^2$

$$= 3.2620\ \text{m}$$

$$V_w = V' - V_{\text{par}}$$

$$\pi(1\ \text{m})^2(2\ \text{m}) = \pi(1\ \text{m})^2(h_0 + 3.2620\ \text{m}) - \tfrac{1}{2}\big[\pi(1\ \text{m})^2(3.2620\ \text{m})\big]$$

$$h_0 = 0.3690\ \text{m}$$

Thus,

$$h_{\text{max}} = h + h_0 = 3.2620\ \text{m} + 0.3690\ \text{m} = 3.6310\,\text{m}$$

$$h_{\text{min}} = h_0 = 0.3690\ \text{m}$$

$$p_{\text{max}} = \rho_w g h_{\text{max}} = (1000\ \text{kg/m}^3)(9.81\ \text{m/s}^2)(3.6310\ \text{m})$$

$$= 35.62(10^3)\ \text{Pa} = 35.6\ \text{kPa} \qquad \textit{Ans.}$$

$$p_{\text{min}} = \rho_w g h_{\text{min}} = (1000\ \text{kg/m}^3)(9.81\ \text{m/s}^2)(0.3690\ \text{m})$$

$$= 3.62(10^3)\ \text{Pa} = 3.62\ \text{kPa} \qquad \textit{Ans.}$$

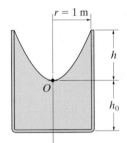

F2–24. $h = \dfrac{\omega^2}{2g}r^2;$ $h_{\text{max}} = \left[\dfrac{(4\ \text{rad/s})^2}{2(9.81\ \text{m/s}^2)}\right](1.5\ \text{m})^2$

$$= 1.8349\ \text{m}$$

$$p_{\text{max}} = \rho_o g h_{\text{max}} = (880\ \text{kg/m}^3)(9.81\ \text{m/s}^2)(1.8349\ \text{m})$$

$$= 15.84(10^3)\ \text{Pa}$$

$$= 15.8\ \text{kPa} \qquad \textit{Ans.}$$

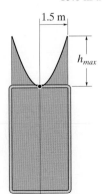

Chapter 3

F3–1. Since $x = 2$ m, $y = 6$ m when $t = 0$,

$$\frac{dx}{dt} = u = \frac{1}{4x}; \qquad \int_2^x \frac{dx}{x} = \int_0^t \frac{1}{4}dt$$

$$\ln x \Big|_2^x = \frac{1}{4}t\Big|_0^t; \qquad \ln\frac{x}{2} = \frac{1}{4}(t)$$

$$x = 2e^{\frac{1}{4}(t)}$$

$$\frac{dy}{dt} = v = 2t; \qquad \int_6^y dy = \int_0^t 2t\,dt$$

$$y\Big|_6^y = t^2\Big|_0^t; \quad y - 6 = t^2$$

$$y = t^2 + 6$$

When $t = 2$ s, the position of the particles is

$$x = 2\,e^{\frac{1}{4}(2)} = 3.30\ \text{m}, \quad y = (2)^2 + 6 = 10\ \text{m} \quad \textit{Ans.}$$

F3–2. $\dfrac{dy}{dx} = \dfrac{v}{u}; \qquad \dfrac{dy}{dx} = \dfrac{8y}{2x^2} = \dfrac{4y}{x^2}$

$$\int_{3\,\text{m}}^y \frac{dy}{y} = \int_{2\,\text{m}}^y \frac{4\,dx}{x^2}; \quad \ln y\Big|_{3\,\text{m}}^y = -4\left(\frac{1}{x}\right)\Big|_{2\,\text{m}}^x$$

$$\ln\frac{y}{3} = -4\left(\frac{1}{x} - \frac{1}{2}\right)$$

$$y = 3e^{\frac{2(x-2)}{x}} \qquad \textit{Ans.}$$

F3–3. $a = \dfrac{\partial V}{\partial t} + v\dfrac{\partial V}{\partial x}$

$$\frac{\partial v}{\partial t} = 20t; \qquad \frac{\partial v}{\partial x} = 600x^2$$

$$a = \big[20t + (200x^3 + 10t^2)(600x^2)\big]\ \text{m/s}^2$$

when $t = 0.2$ s, $x = 0.1$ m

$$a = 20(0.2) + \big[200(0.1^3) + 10(0.2^2)\big]\big[600(0.1^2)\big]$$

$$= 7.60\ \text{m/s}^2 \qquad \textit{Ans.}$$

F3–4. $a = \dfrac{\partial u}{\partial t} + u\dfrac{\partial u}{\partial x}$

$$= 0 + 3(x + 4)(3)$$

$$= 9(x + 4)\ \text{m/s}^2$$

At $x = 0.1$ m,

$$a = 9(0.1 + 4) = 36.9\ \text{m/s}^2 \qquad \textit{Ans.}$$

$$\frac{dx}{dt} = u = 3(x + 4); \qquad \int_0^x \frac{dx}{3(x + 4)} = \int_0^t dt$$

$$\frac{1}{3} \ln(x + 4)\Big|_0^x = t; \qquad \frac{1}{3} \ln\left(\frac{x + 4}{4}\right) = t$$

$$x = 4e^{3t} - 4; \qquad x = \left[4\left(e^{3t} - 1\right)\right] \text{m}$$

When $t = 0.025$ s

$$x = 4\left[e^{3(0.025)} - 1\right] = 0.2473 \text{ m} = 247 \text{ mm} \qquad Ans.$$

F3–5. $(a_x)_{\text{local}} = \dfrac{\partial u}{\partial t} = (4t) \text{ m/s}^2$

When $t = 2$ s,

$$(a_x)_{\text{local}} = \left[4(2)\right] \text{ m/s}^2 = 8 \text{ m/s}^2$$

$$(a_x)_{\text{conv}} = u\frac{\partial u}{\partial x} + v\frac{\partial u}{\partial y}$$

$$= \left(3x + 2t^2\right)(3) + \left(2y^3 + 10t\right)(0)$$

$$= \left[3\left(3x + 2t^2\right)\right] \text{m/s}^2$$

When $t = 2$ s, $x = 3$ m

$$(a_y)_{\text{conv}} = \left[3\left(3(3) + 2\left(2^2\right)\right)\right] \text{m/s}^2 = 51 \text{ m/s}^2$$

$$(a_y)_{\text{local}} = \frac{\partial v}{\partial t} = 10 \text{ m/s}^2$$

$$(a_y)_{\text{conv}} = u\frac{\partial v}{\partial x} + v\frac{\partial v}{\partial y}$$

$$= \left(3x + 2t^2\right)(0) + \left(2y^3 + 10t\right)\left(6y^2\right)$$

$$= \left[6y^2\left(2y^3 + 10t\right)\right] \text{m/s}^2$$

When $t = 2$ s, $y = 1$ m.

$$(a_y)_{\text{conv}} = \left[6\left(1^2\right)\right]\left[2\left(1^3\right) + 10(2)\right] \text{m/s}^2$$

$$= 132 \text{ m/s}^2$$

Thus

$$a_{\text{local}} = \sqrt{(a_x)^2_{\text{local}} + (a_y)^2_{\text{local}}} = \sqrt{\left(8 \text{ m/s}^2\right)^2 + \left(10 \text{ m/s}^2\right)^2}$$

$$= 12.8 \text{ m/s}^2 \qquad Ans.$$

$$a_{\text{conv}} = \sqrt{(a_x)^2_{\text{conv}} + (a_y)^2_{\text{conv}}} = \sqrt{\left(51 \text{ m/s}^2\right)^2 + \left(132 \text{ m/s}^2\right)^2}$$

$$= 142 \text{ m/s}^2 \qquad Ans.$$

F3–6. $a_s = \left(\dfrac{\partial V}{\partial t}\right)_s + V\dfrac{\partial V}{\partial s}$

$$\left(\frac{\partial V}{\partial t}\right)_s = 0 \text{ (steady flow)} \qquad \frac{\partial v}{\partial s} = (40s) \text{ s}^{-1}$$

$$a_s = 0 + (20s^2 + 4)(40s) = \left[40s(20s^2 + 4)\right] \text{m/s}^2$$

At A, $s = r\theta = (0.5 \text{ m})\left(\dfrac{\pi}{4} \text{ rad}\right) = 0.125\pi$ m

$$a_s = 40(0.125\pi)\left[20(0.125\pi)^2 + 4\right] = 111.28 \text{ m/s}^2$$

$$a_n = \left(\frac{\partial V}{\partial t}\right)_n + \frac{V^2}{\rho}$$

Here, $\left(\dfrac{\partial v}{\partial t}\right)_n = 0$, $\rho = 0.5$ m, and at A,

$$V = \left[20(0.125\,\pi)^2 + 4\right] \text{m/s} = 7.084 \text{ m/s}$$

$$a_n = 0 + \frac{(7.0842)^2}{0.5} = 100.37 \text{ m/s}^2$$

Then,

$$a = \sqrt{a_s^2 + a_n^2} = \sqrt{\left(111.28 \text{ m/s}^2\right)^2 + \left(100.37 \text{ m/s}^2\right)^2}$$

$$= 150 \text{ m/s}^2 \qquad Ans.$$

F3–7. Since the speed of the particle is constant,

$$a_s = 0$$

Since the streamline does not rotate, $(\partial V / \partial t)_n = 0$.

Therefore,

$$a_n = \left(\frac{\partial V}{\partial t}\right)_n + \frac{V^2}{R} = 0 + \frac{\left(3 \text{ m/s}\right)^2}{0.5\,\text{m}} = 18 \text{ m/s}^2$$

Thus,

$$a = \sqrt{a_s^2 + a_n^2} = \sqrt{0^2 + \left(18 \text{ m/s}^2\right)^2}$$

$$= 18 \text{m/s}^2 \qquad Ans.$$

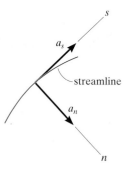

streamline

F3–8. $a_s = \left(\dfrac{\partial V}{\partial t}\right)_s + V\dfrac{\partial V}{\partial s}$

$$\left(\frac{\partial V}{\partial t}\right)_s = \left[\frac{3}{2}(1000)\, t^{1/2}\right] \text{m/s}^2 = \left(1500\, t^{1/2}\right) \text{m/s}^2$$

$$\frac{\partial V}{\partial s} = (40s) \text{ s}^{-1}$$

$$a_s = \left[1500\,t^{1/2} + \left(20s^2 + 1000t^{3/2} + 4\right)(40s)\right] \text{m/s}^2$$

At A, $s = 0.3$ m and $t = 0.02$ s.

$$a_s = \left\{1500\left(0.02^{1/2}\right) + \left[20\left(0.3^2\right) + 1000\left(0.02^{3/2}\right) + 4\right]\right.$$
$$\left.\left[40(0.3)\right]\right\} \text{m/s}^2$$

$$= 315.67 \text{ m/s}^2$$

$$a_n = \left(\frac{\partial V}{\partial t}\right)_n + \frac{V^2}{\rho}$$

At A, $\left(\dfrac{\partial V}{\partial t}\right)_n = 0$, $\rho = 0.5$ m and

$$V = \left[20\left(0.3^2\right) + 1000\left(0.02^{3/2}\right) + 4\right] \text{m/s} = 8.628 \text{ m/s}$$

Then,

$$a_n = 0 + \frac{(8.628 \text{ m/s})^2}{0.5 \text{ m}} = 148.90 \text{ m/s}^2$$

Thus,

$$a = \sqrt{a_s^2 + a_n^2} = \sqrt{\left(315.67 \text{ m/s}^2\right)^2 + \left(148.90 \text{ m/s}^2\right)^2}$$

$$= 349 \text{ m/s}^2 \qquad\qquad Ans.$$

Chapter 4

F4–1.

$$\dot{m} = \rho_w \mathbf{V} \cdot \mathbf{A} = \left(1000 \text{ kg/m}^3\right)(16 \text{ m/s})(0.06 \text{ m})\left[(0.05 \text{ m}) \sin 60°\right]$$

$$= 41.6 \text{ kg/s} \qquad\qquad Ans.$$

F4–2. $\quad p = \rho RT;$

$$(70 + 101)\left(10^3\right) \text{N/m}^2 = \rho(286.9 \text{ J/kg} \cdot \text{K})(15 + 273) \text{ K}$$

$$\rho = 2.0695 \text{ kg/m}^3$$

$$\dot{m} = \rho VA; \quad 0.7 \text{ kg/s} = \left(2.0695 \text{ kg/m}^3\right)(V)\left[\tfrac{1}{2}(0.3 \text{ m})(0.3 \text{ m})\right]$$

$$V = 7.52 \text{ m/s} \qquad\qquad Ans.$$

F4–3. $\quad Q = VA = (8 \text{ m/s})\left[\pi(0.15 \text{ m})^2\right]$

$$= 0.565 \text{ m}^3/\text{s} \qquad\qquad Ans.$$

$$\dot{m} = \rho_w Q = \left(1000 \text{ kg/m}^3\right)\left(0.565 \text{ m}^3/\text{s}\right) = 565 \text{ kg/s} \qquad Ans.$$

F4–4. $\quad Q = \displaystyle\int_A V dA;$

$$0.02 \text{ m}^3/\text{s} = \int_0^{0.2 \text{ m}} V_0\left(1 - 25r^2\right)(2\pi r\, dr)$$

$$\frac{0.01}{\pi} = V_0\left(\frac{r^2}{2} - \frac{25r^4}{4}\right)\Big|_0^{0.2 \text{ m}}$$

$$\frac{0.01}{\pi} = V_0\left(\frac{0.2^2}{2} - \frac{25\left(0.2^4\right)}{4}\right)$$

$$V_0 = 0.318 \text{ m/s} \qquad\qquad Ans.$$

Also, $\displaystyle\int_A V dA$ is equal to the volume of the paraboloid under the velocity profile.

$$0.02 \text{ m}^3/\text{s} = \frac{1}{2}\pi(0.2 \text{ m})^2 V_0$$

$$V_0 = 0.318 \text{ m/s} \qquad\qquad Ans.$$

The average velocity can be determine from

$$V_{avg} = \frac{Q}{A} = \frac{0.02 \text{ m}^3/\text{s}}{\pi(0.2 \text{ m})^2} = 0.159 \text{ m/s} \qquad Ans.$$

F4–5. $\quad p = \rho RT;$

$$(80 + 101)\left(10^3\right) \text{N/m}^2 = \rho(286.9 \text{ J/kg} \cdot \text{K})(20 + 273) \text{ K}$$

$$\rho = 2.1532 \text{ kg/m}^3$$

$$\dot{m} = \rho VA = \left(2.1532 \text{ kg/m}^3\right)\left(3 \text{ m/s}\right)\left[\pi(0.2 \text{ m})^2\right]$$

$$= 0.812 \text{ kg/s} \qquad\qquad Ans.$$

F4–6. $\quad Q = \displaystyle\int_A V dA = \int_0^{0.5 \text{ m}} 6y^2(0.5 dy) = \int_0^{0.5 \text{ m}} 3y^2 dy$

$$= 0.125 \text{ m}^3/\text{s} \qquad\qquad Ans.$$

Also the volume of the parabolic block under the velocity profile is

$$Q = \tfrac{1}{3}(0.5 \text{ m})\left[6\left(0.5^2\right) \text{ m/s}\right](0.5 \text{ m})$$

$$= 0.125 \text{ m}^3/\text{s} \qquad\qquad Ans.$$

F4–7. $\quad \dfrac{\partial}{\partial t}\displaystyle\int_{cv} \rho\, d\mathcal{V} + \int_{cs} \rho \mathbf{V} \cdot d\mathbf{A} = 0$

$$0 - V_A A_A + V_B A_B + V_C A_C = 0$$

$$0 - (6 \text{ m/s})\left(0.1 \text{ m}^2\right) + (2 \text{ m/s})\left(0.2 \text{ m}^2\right) + V_C\left(0.1 \text{ m}^2\right) = 0$$

$$V_C = 2 \text{ m/s} \qquad\qquad Ans.$$

F4–8. Choose a changing control volume.

$$\mathcal{V} = (3 \text{ m})(2 \text{ m})(y) = (6y) \text{ m}^3; \quad \frac{\partial \mathcal{V}}{\partial t} = 6\frac{\partial y}{\partial t}$$

Thus,

$$\frac{\partial}{\partial t}\int_{cv} \rho_l\, d\mathcal{V} + \int_{cs} \rho_l \mathbf{V} \cdot d\mathbf{A} = 0$$

$$\rho_l \frac{\partial \mathcal{V}}{\partial t} - \rho_l V_A A_A = 0$$

$$\frac{\partial \mathcal{V}}{\partial t} = V_A A_A; \quad 6\frac{\partial y}{\partial t} = \left(4 \text{ m/s}\right)\left(0.1 \text{ m}^2\right)$$

$$\frac{\partial y}{\partial t} = 0.0667 \text{ m/s} \qquad\qquad Ans.$$

F4–9. $\dfrac{\partial}{\partial t}\displaystyle\int_{cv} \rho \, d\forall + \int_{cs} \rho \mathbf{V} \cdot d\mathbf{A} = 0$

$$0 - \dot{m}_a - \dot{m}_w + \rho_m V_A = 0$$

$$0 - 0.05 \text{ kg/s} - 0.002 \text{ kg/s} + (1.45 \text{ kg/m}^3)(V)[\pi(0.01 \text{ m}^2)] = 0$$

$$V = 114 \text{ m/s} \qquad\qquad Ans.$$

F4–10. $p = \rho_A R T_A;$

$$(200 + 101)(10^3) \text{ N/m}^2 = \rho_A (286.9 \text{ J/kg} \cdot \text{K})(16 + 273) \text{ K}$$

$$\rho_A = 3.6303 \text{ kg/m}^3$$

$$p = \rho_B R T_B;$$

$$(200 + 101)(10^3) \text{ N/m}^2 = \rho_B (286.9 \text{ J/kg} \cdot \text{K})(70 + 273) \text{ K}$$

$$\rho_B = 3.0587 \text{ kg/m}^3$$

$$\frac{\partial}{\partial t}\int_{cv} \rho \cdot d\forall + \int_{cs} \rho \mathbf{V} \cdot d\mathbf{A} = 0$$

$$0 - \rho_A V_A A_A + \rho_B V_B A_B = 0$$

$$A_A = A_B$$

$$V_B = \left(\frac{\rho_A}{\rho_B}\right) V_A = \left(\frac{3.6303 \text{ kg/m}^3}{3.0587 \text{ kg/m}^3}\right)(12 \text{ m/s})$$

$$= 14.2 \ m/s \qquad\qquad Ans.$$

F4–11. Choose a changing control volume.

$$\forall = \frac{1}{2}(2h \tan 30°)(h)(1 \text{ m}) = h^2 \tan 30°$$

$$\frac{\partial \forall}{\partial t} = 2 \tan 30° h \frac{\partial h}{\partial t}$$

Thus,

$$\frac{\partial}{\partial t}\int_{cv} \rho_w d\forall + \int_{cs} \rho_w \mathbf{V} \cdot d\mathbf{A} = 0$$

$$\rho_w \frac{\partial \forall}{\partial t} - \rho_w VA = 0$$

$$\frac{\partial \forall}{\partial t} = VA$$

$$2 \tan 30° h \frac{\partial h}{\partial t} = (6 \text{ m/s})[\pi(0.025 \text{ m})^2]$$

$$\frac{\partial h}{\partial t} = \frac{0.01020}{h} \qquad\qquad (1)$$

$$\int_{0.1m}^{h} h \, dh = 0.01020 \int_{0}^{t} dt$$

$$h^2 - (0.1 \text{ m})^2 = 0.020405$$

$$h = \sqrt{0.020405t + 0.01}$$

When $t = 10$ s, $h = 0.46266$

Substituting this result into Eq. 1 yields

$$\frac{\partial h}{\partial t} = \frac{0.01020}{0.46266} = 0.0220 \text{ m/s} \qquad Ans.$$

F4–12. $\dfrac{\partial}{\partial t}\displaystyle\int_{cv} \rho \, d\forall + \int_{cs} \rho \mathbf{V} \cdot d\mathbf{A} = 0$

$$0 - V_A A_A - V_B A_B + V_C A_C = 0$$

$$-(1.5 \text{ m/s})[\pi(0.02 \text{ m})^2] - (2 \text{ m/s})[\pi(0.015 \text{ m})^2]$$

$$+ (7 \text{ m/s})\left(\frac{\pi}{4} d^2\right) = 0$$

$$d = 0.02449 \text{ m} = 24.5 \text{ mm} \qquad\qquad Ans.$$

F4–13. $\dfrac{\partial}{\partial t}\displaystyle\int_{cv} \rho_o \, d\forall + \int_{cs} \rho_o \mathbf{V} \cdot d\mathbf{A} = 0$

$$\rho_o \frac{\partial}{\partial t}\forall - \rho_o V_A A_A + \rho_o V_B A_B = 0$$

$$\rho_o \frac{\partial}{\partial t}[y(2 \text{ m})(3 \text{ m})] - \rho_o(4 \text{ m/s})[\pi(0.025 \text{ m})^2]$$

$$+ \rho_o(2 \text{ m/s})[\pi(0.01 \text{ m})^2] = 0$$

$$\left(6 \text{ m}^2\right)\frac{\partial y}{\partial t} = 7.226(10^{-3}) \text{ m}^3/\text{s}$$

$$v_y = \frac{\partial y}{\partial t} = 1.20 \text{ mm/s} \qquad\qquad Ans.$$

Since the result is positive, the oil surface is rising in the tank.

Chapter 5

F5–1. Since the pipe has a constant diameter,

$$V_B = V_A = 6 \text{ m/s} \qquad\qquad Ans.$$

$$\frac{p_A}{\rho_w} + \frac{V_A^2}{2} + gz_A = \frac{p_B}{\rho_w} + \frac{V_B^2}{2} + gz_B$$

$$\frac{p_A}{1000 \text{ kg/m}^3} + \frac{V^2}{2} + (9.81 \text{ m/s}^2)(3 \text{ m}) = 0 + \frac{V^2}{2} + 0$$

$$p_A = -29.43(10^3) \text{ Pa} = -29.4 \text{ kPa} \qquad Ans.$$

The negative sign indicates that the pressure at A is a partial vacuum.

F5–2. $\dfrac{\partial}{\partial t}\displaystyle\int_{cv} \rho \, d\forall + \int_{cs} \rho \mathbf{V} \cdot d\mathbf{A} = 0$

$$0 - V_A A_A + V_B A_B = 0$$

$$-(7 \text{ m/s})[\pi(0.06 \text{ m})^2] + V_B[\pi(0.04 \text{ m})^2] = 0$$

$$V_B = 15.75 \text{ m/s} \qquad\qquad Ans.$$

$$\frac{p_A}{\rho_0} + \frac{V_A^2}{2} + gz_A = \frac{p_B}{\rho_0} + \frac{V_B^2}{2} + gz_B$$

Select A and B on the same horizontal streamline, $z_A = z_B = z$.

$$\frac{300(10^3) \text{ N/m}^2}{940 \text{ kg/m}^3} + \frac{(7 \text{ m/s})^2}{2} + gz$$

$$= \frac{p_s}{940 \text{ kg/m}^3} + \frac{(15.75 \text{ m/s})^2}{2} + gz$$

$$p_B = 206.44(10^3) \text{ Pa} = 206 \text{ kPa} \qquad \text{Ans.}$$

F5–3. Here $p_C = p_B = 0$ and $V_C = 0$.

$$\frac{p_B}{\rho_w} + \frac{V_B^2}{2} + gz_B = \frac{p_C}{\rho_w} + \frac{V_C^2}{2} + gz_C$$

$$0 + \frac{V_B^2}{2} + 0 = 0 + 0 + (9.81 \text{ m/s}^2)(2 \text{ m})$$

$$V_B = 6.264 \text{ m/s}$$

$$\frac{\partial}{\partial t} \int_{cv} \rho \, d\forall + \int_{cs} \rho \mathbf{V} \cdot d\mathbf{A} = 0$$

$$0 - V_A A_A + V_B A_B = 0$$

$$0 - V_A[\pi(0.025 \text{ m})^2] + (6.264 \text{ m/s})[\pi(0.005 \text{ m})^2] = 0$$

$$V_A = 0.2506 \text{ m/s}$$

Since AB is a short distance,

$$\frac{p_A}{\rho_w} + \frac{V_A^2}{2} + gz_A = \frac{p_B}{\rho_w} + \frac{V_B^2}{2} + gz_B$$

$$\frac{p_A}{1000 \text{ kg/m}^3} + \frac{(0.2506 \text{ m/s})^2}{2} + 0 = 0 + \frac{(6.264 \text{ m/s})^2}{2} + 0$$

$$p_A = 19.59(10^3) \text{ Pa} = 19.6 \text{ kPa} \qquad \text{Ans.}$$

F5–4. Here $V_A = 8 \text{ m/s}$, $V_B = 0$ (B is a stagnation point), and $z_A = z_B = 0$ (AB is a horizontal streamline).

$$\frac{p_A}{\rho_w} + \frac{V_A^2}{2} + gz_A = \frac{p_B}{\rho_w} + \frac{V_B^2}{2} + gz_B$$

$$\frac{80(10^3) \text{ N/m}^2}{1000 \text{ kg/m}^3} + \frac{(8 \text{ m/s})^2}{2} + 0 = \frac{p_B}{1000 \text{ kg/m}^3} + 0 + 0$$

$$p_B = 112(10^3) \text{ Pa}$$

The manometer equation gives

$$p_B + \rho_w g h_w = p_C$$

$$112(10^3) \text{ Pa} + (1000 \text{ kg/m}^3)(9.81 \text{ m/s}^2)(0.3 \text{ m}) = p_C$$

$$p_C = 114.943(10^3) \text{ Pa} = 115 \text{ kPa} \qquad \text{Ans.}$$

F5–5.

$$\frac{\partial}{\partial t} \int_{cv} \rho_w \, d\forall + \int_{cs} \rho_w \mathbf{V} \cdot d\mathbf{A} = 0$$

$$\rho_w \frac{\partial \forall}{\partial t} + \rho_w V_B A_B = 0$$

$$\frac{\partial \forall}{\partial t} = V_B A_B$$

However, $\forall = (2 \text{ m})(2 \text{ m})y = 4y$

$$\frac{\partial \forall}{\partial t} = 4 \frac{\partial y}{\partial t}$$

Thus,

$$4 \frac{\partial y}{\partial t} = V_B[\pi(0.01 \text{ m})^2]$$

$$V_A = \frac{\partial y}{\partial t} = 25\pi(10^{-6}) V_B \qquad (1)$$

$p_A = p_B = 0$, V_A can be neglected since $V_B \gg V_A$ (Eq. 1).

$$\frac{p_A}{\rho_w} + \frac{V_A^2}{2} + gz_A = \frac{p_B}{\rho_w} + \frac{V_B^2}{2} + gz_B$$

$$0 + 0 + (9.81 \text{ m/s}^2)y = 0 + \frac{V_B^2}{2} + 0$$

$$V_B = \sqrt{19.62 \, y}$$

At $y = 0.4$ m,

$$V_B = \sqrt{19.62(0.4)} = 2.801 \text{ m/s}$$

$$Q = V_B A_B = (2.801 \text{ m/s})[\pi(0.01 \text{ m})^2]$$

$$= 0.88(10^{-3}) \text{ m}^3/\text{s} \qquad \text{Ans.}$$

At $y = 0.2$ m,

$$V_B = \sqrt{19.62(0.2)} = 1.981 \text{ m/s}$$

$$Q = V_B A_B = (1.981 \text{ m/s})[\pi(0.01 \text{ m})^2]$$

$$= 0.622(10^{-3}) \text{ m}^3/\text{s} \qquad \text{Ans.}$$

F5–6.

$$\frac{\partial}{\partial t} \int_{cv} \rho \, d\forall + \int_{cs} \rho \mathbf{V} \cdot d\mathbf{A} = 0$$

$$0 - V_A A_A + V_B A_B = 0$$

$$0 - (4 \text{ m/s})[\pi(0.1 \text{ m})^2] + V_B[\pi(0.025 \text{ m})^2] = 0$$

$$V_B = 64 \text{ m/s}$$

Between A and B, $z_A = z_B = 0$, $\rho_a = 1.000 \text{ kg/m}^3$ at $T = 80°C$ (Appendix A).

$$\frac{p_A}{\rho_a} + \frac{V_A^2}{2} + gz_A = \frac{p_B}{\rho_a} + \frac{V_B^2}{2} + gz_B$$

$$\frac{20(10^3) \text{ N/m}^2}{1.000 \text{ kg/m}^3} + \frac{(4 \text{ m/s})^2}{2} + 0$$

$$= \frac{p_B}{1.000 \text{ kg/m}^3} + \frac{(64 \text{ m/s})^2}{2} + 0$$

$$p_B = 17.96(10^3) \text{ Pa} = 18.0 \text{ kPa} \qquad Ans.$$

F5–7. $p_A = p_B = 0$, $V_A \cong 0$ (large reservoir), $z_A = 6 \text{ m}$ and $z_B = 0$.

$$\frac{p_A}{\gamma_w} + \frac{V_A^2}{2g} + z_A = \frac{p_B}{\gamma_w} + \frac{V_B^2}{2g} + z_B$$

$$0 + 0 + 6 \text{ m} = 0 + \frac{V_B^2}{2(9.81 \text{ m/s}^2)} + 0$$

$$V_B = 10.85 \text{ m/s}$$

$$Q = V_B A_B = (10.85 \text{ m/s})[\pi(0.05 \text{ m})^2]$$

$$= 0.0852 \text{ m}^3/\text{s} \qquad Ans.$$

The velocity head is

$$\frac{V_B^2}{2g} = \frac{(10.85 \text{ m/s})^2}{2(9.81 \text{ m/s}^2)} = 6 \text{ m}$$

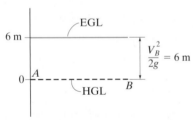

F5–8. $V_A = V_B = V$ (constant pipe diameter), $z_A = 2 \text{ m}$, $z_B = 1.5 \text{ m}$, and $\rho_{co} = 880 \text{ kg/m}^3$ (Appendix A).

$$\frac{p_A}{\gamma_{co}} + \frac{V_A^2}{2g} + z_A = \frac{p_B}{\gamma_{co}} + \frac{V_B^2}{2g} + z_B$$

$$\frac{300(10^3) \text{ N/m}^2}{(880 \text{ kg/m}^3)(9.81 \text{ m/s}^2)} + \frac{V^2}{2g} + 2 \text{ m}$$

$$= \frac{p_B}{(880 \text{ kg/m}^3)(9.81 \text{ m/s}^2)} + \frac{V^2}{2g} + 1.5 \text{ m}$$

$$p_B = 304.32(10^3) \text{ Pa} = 304 \text{ kPa} \qquad Ans.$$

$$H = \frac{p_A}{\gamma_{co}} + \frac{V_A^2}{2g} + z_A$$

$$= \frac{300(10^3) \text{ N/m}^2}{(880 \text{ kg/m}^3)(9.81 \text{ m/s}^2)} + \frac{(4 \text{ m/s})^2}{2(9.81 \text{ m/s}^2)} + 2 \text{ m}$$

$$= 37.567 \text{ m}$$

$$\frac{V^2}{2g} = \frac{(4 \text{ m/s})^2}{2(9.81 \text{ m/s}^2)} = 0.815 \text{ m}$$

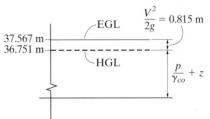

F5–9. $$\frac{\partial}{\partial t}\int_{cv} \rho \, dV + \int_{cs} \rho \mathbf{V} \cdot d\mathbf{A} = 0$$

$$0 - V_A A_A + V_B A_B = 0$$

$$0 - (3 \text{ m/s})[\pi(0.075 \text{ m})^2] + V_B[\pi(0.05 \text{ m})^2] = 0$$

$$V_B = 6.75 \text{ m/s} \qquad Ans.$$

Also,

$$0 - V_A A_A + V_C A_C = 0$$

$$-(3 \text{ m/s})[\pi(0.075 \text{ m})^2] + V_C[\pi(0.025 \text{ m})^2] = 0$$

$$V_C = 27 \text{ m/s} \qquad Ans.$$

Between A and B,

$$\frac{p_A}{\gamma_w} + \frac{V_A^2}{2g} + z_A = \frac{p_B}{\gamma_w} + \frac{V_B^2}{2g} + z_B$$

$$\frac{400(10^3) \text{ N/m}^2}{9810 \text{ N/m}^3} + \frac{(3 \text{ m/s})^2}{2(9.81 \text{ m/s}^2)} + 0$$

$$= \frac{p_B}{9810 \text{ N/m}^3} + \frac{(6.75 \text{ m/s})^2}{2(9.81 \text{ m/s}^2)} + 0$$

$$p_B = 381.72(10^3) \text{ Pa} = 382 \text{ kPa} \qquad Ans.$$

Between A and C,

$$\frac{p_A}{\gamma_m} + \frac{V_A^2}{2g} + z_A = \frac{p_C}{\gamma_w} + \frac{V_C^2}{2g} + z_C$$

$$\frac{400(10^3) \text{ N/m}^2}{9810 \text{ N/m}^3} + \frac{(3 \text{ m/s})^2}{2(9.81 \text{ m/s}^2)} + 0$$

$$= \frac{p_C}{9810 \text{ N/m}^3} + \frac{(27 \text{ m/s})^2}{2(9.81 \text{ m/s}^2)} + 0$$

$$p_C = 40.0(10^3) \text{ Pa} = 40 \text{ kPa} \qquad Ans.$$

$$H = \frac{p_A}{\gamma_w} + \frac{V_A^2}{2g} + z_A$$

$$= \frac{400\left(10^3\right) \text{ N/m}^2}{9810 \text{ N/m}^3} + \frac{(3 \text{ m/s})^2}{2\left(9.81 \text{ m/s}^2\right)} + 0$$

$$= 41.233 \text{ m}$$

Velocity heads at A, B, and C are

$$\frac{V_A^2}{2g} = \frac{(3 \text{ m/s})^2}{2\left(9.81 \text{ m/s}^2\right)} = 0.459 \text{ m}$$

$$\frac{V_B^2}{2g} = \frac{(6.75 \text{ m/s})^2}{2\left(9.81 \text{ m/s}^2\right)} = 2.322 \text{ m}$$

$$\frac{V_C^2}{2g} = \frac{(27 \text{ m/s})^2}{2\left(9.81 \text{ m/s}^2\right)} = 37.156 \text{ m}$$

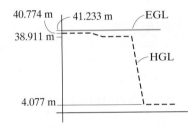

F5–10. $V_A \cong 0$, $p_A = p_C = 0$, $z_A = 40$ m, and $z_C = 0$;

$$\frac{p_A}{\gamma_w} + \frac{V_A^2}{2g} + z_A + h_{\text{pump}} = \frac{p_C}{\gamma_w} + \frac{V_C^2}{2g} + z_C + h_{\text{turb}} + h_L$$

$$0 + 0 + 40 \text{ m} + 0 = 0 + \frac{(8 \text{ m/s})^2}{2\left(9.81 \text{ m/s}^2\right)} + 0 + h_{\text{turb}} + \left(\frac{150}{100}\right)(1.5 \text{ m})$$

$$h_{\text{turb}} = 34.488 \text{ m}$$

$$Q = V_C A_C = (8 \text{ m/s})\left[\pi(0.025 \text{ m})^2\right] = 5\pi\left(10^{-3}\right) \text{ m}^3/\text{s}$$

$$\dot{W}_i = Q\gamma_w h_s = \left[5\pi\left(10^{-3}\right) \text{ m}^3/\text{s}\right]\left(9810 \text{ N/m}^3\right)(34.488 \text{ m})$$

$$= 5314.43 \text{ W} = 5.314 \text{ kW}$$

$$\varepsilon = \frac{\dot{W}_o}{\dot{W}_i}; \qquad 0.6 = \frac{\dot{W}_o}{5.314 \text{ kW}} \qquad \dot{W}_o = 3.19 \text{ kW} \qquad Ans.$$

F5–11. $Q = V_B A_B$; $0.02 \text{ m}^3/\text{s} = V_B\left[\pi(0.025 \text{ m})^2\right]$

$$V_B = 10.19 \text{ m/s}$$

$$p_B = 0, \ z_A = 0, \ z_B = 8 \text{ m}$$

$$\frac{p_A}{\gamma_w} + \frac{V_A^2}{2g} + z_A + h_{\text{pump}} = \frac{p_B}{\gamma_w} + \frac{V_B^2}{2g} + z_B + h_{\text{turb}} + h_L$$

$$\frac{80\left(10^3\right) \text{ N/m}^2}{9810 \text{ N/m}^3} + \frac{(2 \text{ m/s})^2}{2\left(9.81 \text{ m/s}^2\right)} + 0 + h_{\text{pump}}$$

$$= 0 + \frac{(10.19 \text{ m/s})^2}{2\left(9.81 \text{ m/s}^2\right)} + 8 \text{ m} + 0 + 0.75 \text{ m}$$

$$h_{\text{pump}} = 5.6793 \text{ m}$$

There the power supplied by the pump to the water is

$$\dot{W} = Q\gamma_w h_{\text{pump}} = \left(0.02 \text{ m}^3/\text{s}\right)\left(9810 \text{ N/m}^3\right)(5.6793 \text{ m})$$

$$= 1.114\left(10^3\right) \text{ W} = 1.11 \text{ kW} \qquad Ans.$$

F5–12. Here, $\dot{Q}_m = -1.5$ kJ/s. Then

$$\dot{Q}_{\text{in}} - \dot{W}_{\text{out}} = \left[\left(h_B + \frac{V_B^2}{2} + gz_B\right) - \left(h_A + \frac{V_A^2}{2} + gz_A\right)\right]\dot{m}$$

$$-1.5\left(10^3\right) \text{ J/s} - \dot{W}_{\text{out}} = \left\{\left[450\left(10^3\right) \text{ J/kg} + \frac{(48 \text{ m/s})^2}{2} + 0\right]\right.$$

$$\left. - \left[600\left(10^3\right) \text{ J/kg} + \frac{(12 \text{ m/s})^2}{2} + 0\right]\right\}(2 \text{ kg/s})$$

$$\dot{W}_{\text{out}} = 298 \text{ kW} \qquad Ans.$$

The negative sign indicates that the power is being added to the flow by the engine.

Chapter 6

F6–1. $Q = VA$; $\qquad 0.012 \text{ m}^3/\text{s} = V\left[\pi(0.02 \text{ m})^2\right]$;
$V = 9.549 \text{ m/s}$

$$V_A = V_B = V = 9.549 \text{ m/s and } p_B = 0$$

$$\Sigma F = \frac{\partial}{\partial t} \int_{\text{cv}} V\rho \, dV + \int_{\text{cs}} V\rho \mathbf{V} \cdot d\mathbf{A}$$

$$\xrightarrow{+} \Sigma F_x = 0 + V_B \rho_w (V_B A_B)$$

$$F_x = (9.549 \text{ m/s})\left(1000 \text{ kg/m}^3\right)\left(0.012 \text{ m}^3/\text{s}\right) = 114.59 \text{ N} \rightarrow$$

$$+ \uparrow \Sigma F_y = 0 + V_A(-\rho_w V_A A_A)$$

$$\left(160\left(10^3\right) \text{ N/m}^2\right)\left(\pi(0.02 \text{ m})^2\right) - F_y$$

$$= (9.549 \text{ m/s})\left[-\left(1000 \text{ kg/m}^3\right)\left(0.012 \text{ m}^2/\text{s}\right)\right]$$

$$F_y = 315.65 \text{ N} \downarrow$$

$$F = \sqrt{F_x^2 + F_y^2} = \sqrt{(114.59 \text{ N})^2 + (315.65)^2} = 336 \text{ N} \qquad Ans.$$

$$\theta = \tan^{-1}\left(\frac{F_y}{F_x}\right) = \tan^{-1}\left(\frac{315.65 \text{ N}}{114.59 \text{ N}}\right) = 70.0° \qquad Ans.$$

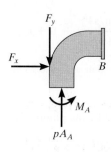

F6–2. $Q_A = V_A A_A$; $0.02 \text{ m}^3/\text{s} = V_A[\pi(0.02 \text{ m})^2]$;

$V_A = 15.915 \text{ m/s}$

$Q_B = 0.3 Q_A = 0.3(0.02 \text{ m}^3/\text{s}) = 0.006 \text{ m}^3/\text{s}$

$Q_C = 0.7 Q_A = 0.7(0.02 \text{ m}^3/\text{s}) = 0.014 \text{ m}^3/\text{s}$

Also, $V_B = V_C = V_A = 15.915 \text{ m/s}$ and $p_A = p_B = p_C = 0$.

$$\Sigma F = \frac{\partial}{\partial t} \int_{cv} V\rho \, d\Psi + \int_{cs} V\rho V \cdot d\mathbf{A}$$

$\xrightarrow{+} \Sigma F_x = 0 + V_A[-(\rho_w V_A A_A)] + V_B \cos 60°(\rho_w V_B A_B)$
$\qquad\qquad\qquad + (-V_C \cos 60°)(\rho_w V_C A_C)$

$- F_x = (15.915 \text{ m/s})\left[-\left(1000 \text{ kg/m}^3\right)\left(0.02 \text{ m}^3/\text{s}\right)\right]$
$\qquad + (15.915 \text{ m/s})(\cos 60°)\left(1000 \text{ kg/m}^3\right)\left(0.006 \text{ m}^3/\text{s}\right)$
$\qquad + (-15.915 \text{ m/s})(\cos 60°)\left(1000 \text{ kg/m}^3\right)\left(0.014 \text{ m}^3/\text{s}\right)$

$F_x = 381.97 \text{ N}$

$+\uparrow \Sigma F_y = 0 + V_B \sin 60°(\rho_w V_B A_B) + (V_C \sin 60°)[-(\rho_w V_C A_C)]$

$-F_y = (15.915 \text{ m/s}) \sin 60°\left(1000 \text{ kg/m}^3\right)\left(0.006 \text{ m}^3/\text{s}\right)$
$\qquad + (15.915 \text{ m/s}) \sin 60°\left[-\left(1000 \text{ kg/m}^3\right)\left(0.014 \text{ m}^3/\text{s}\right)\right]$

$F_y = 110.27 \text{ N}$

$F = \sqrt{F_x^2 + F_y^2} = \sqrt{(381.97 \text{ N})^2 + (110.27 \text{ N})^2} = 397.57 \text{ N}$

$\qquad\qquad\qquad\qquad\qquad = 398 \text{ N}$ *Ans.*

$\theta = \tan^{-1}\left(\frac{F_y}{F_x}\right) = \tan^{-1}\left(\frac{110.27 \text{ N}}{381.97 \text{ N}}\right) = 16.1°$ *Ans.*

F6–3. $\Sigma \mathbf{F} = \dfrac{\partial}{\partial t} \int_{cv} \mathbf{V}\rho \, d\Psi + \int_{cs} \mathbf{V} \rho \mathbf{V} \cdot d\mathbf{A}$

$\xrightarrow{+} \Sigma F_x = \dfrac{dV}{dt}\rho_w V_0 + V_A[-(\rho_w V_A A_A)] + V_B (\rho_w V_B A_B)$

$F_A = p_A A_A = p_A[\pi(0.025 \text{ m})^2] = 0.625\pi(10^{-3})p_A$

$F_B = 0$

$\Psi_0 = [\pi(0.025 \text{ m})^2](0.2 \text{ m}) = 0.125\pi(10^{-3}) \text{ m}^3$

$A_A = A_B$ then $V_A = V_B$

$\xrightarrow{+} \Sigma F_x = \dfrac{dV}{dt}\rho_w \Psi_0$; where $\dfrac{dV}{dt} = 3 \text{ m/s}^2$

$0.625\pi(10^{-3})p_A = (3 \text{ m/s}^2)(1000 \text{ kg/m}^3)[0.125\pi(10^{-2})\text{m}^3]$

$p_A = 600 \text{ Pa}$ *Ans.*

F6–4.

$Q_A = V_A A_A = (6 \text{ m/s})[\pi(0.015 \text{ m})^2] = 1.35\pi(10^{-3}) \text{ m}^3/\text{s}$

$Q_B = Q_C = \dfrac{1}{2}Q_A = \dfrac{1}{2}[1.35\pi(10^{-3}) \text{ m}^3/\text{s}]$

$\qquad = 0.675\pi(10^{-3}) \text{ m}^3/\text{s}$

$Q_B = V_B A_B$; $0.675\pi(10^{-3}) \text{ m}^3/\text{s} = V_B[\pi(0.01 \text{ m})^2]$

$\qquad\qquad\qquad V_B = 6.75 \text{ m/s}$

$\qquad\qquad V_C = V_B = 6.75 \text{ m/s}$

$$\Sigma F = \frac{\partial}{\partial t} \int_{cv} V\rho \, d\Psi + \int_{cs} V\rho V \cdot d\mathbf{A}$$

$\xrightarrow{+} \Sigma F_x = 0 + (V_C \cos 45°)(\rho_{co} V_C A_C) + (-V_B \cos 45°)(\rho_{co} V_B A_B)$

$\qquad F_x = (V_C \cos 45°)\rho_{co} Q_C - (V_B \cos 45°)\rho_{co} Q_B$

$\qquad F_x = 0$

$+\uparrow \Sigma F_y = 0 + V_A(-\rho_{co} V_A A_A) + V_B \sin 45°(\rho_{co} V_B A_B)$
$\qquad\qquad\qquad\qquad\qquad + V_C \sin 45°(\rho_{co} V_C A_C)$

$[80(10^3) \text{ N/m}^2](\pi(0.015 \text{ m})^2) - F_y$

$\qquad = (6 \text{ m/s})\left[-\left(880 \text{ kg/m}^3\right)\left(1.35\pi(10^{-3}) \text{ m}^3/\text{s}\right)\right]$
$\qquad + 2(6.75 \text{ m/s}) \sin 45°\left[\left(880 \text{ kg/m}^3\right)0.675\pi(10^{-3}) \text{ m}^3/\text{s}\right]$

$\qquad F_y = 61.1 \text{ N}$

Since $F_x = 0$,

$F = F_y = 61.1 \text{ N}\downarrow$ *Ans.*

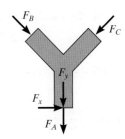

F6–5. $\Sigma F = \dfrac{\partial}{\partial t}\displaystyle\int_{cv} V\rho d\mathcal{V} + \int_{cs} V\rho \mathbf{V}\cdot d\mathbf{A}$

$\overset{+}{\to}\Sigma F_x = 0 + (-V_{out})[\rho_a V_{out} A_{out}]$

$-F_f = (-20 \text{ m/s})\left[\left(1.22 \text{ kg/m}^3\right)(20 \text{ m/s})\,\pi(0.125 \text{ m})^2\right]$

$F_f = 24.0 \text{ N}$ *Ans.*

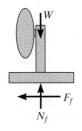

F6–6. $\left(\overset{+}{\to}\right)\mathbf{V}_f = \mathbf{V}_{cv} + \mathbf{V}_{f/cs}$

$20 \text{ m/s} = -1.5 \text{ m/s} + V_{f/cs}$

$V_{f/cs} = 21.5 \text{ m/s}$

$\Sigma F = \dfrac{\partial}{\partial t}\displaystyle\int_{cv} V\rho d\mathcal{V} + \int_{cs} V\rho \mathbf{V}\cdot d\mathbf{A}$

$\overset{+}{\to}\Sigma F_x = 0 + (V_{f/cs})_{in}\left[-\rho(V_{f/cs})_{in} A_{in}\right]$

$-F_x = (21.5 \text{ m/s})\left[-(1000 \text{ kg/m}^3)(21.5 \text{ m/s})\,\pi(0.01 \text{ m})^2\right]$

$F_x = 145.2 \text{ N}$

$+\uparrow\Sigma F_y = 0 + (V_{f/cs})_{out}\left[\rho\,(V_{f/cs})_{out}\,(A_{out})\right]$

$F_y = (21.5 \text{ m/s})\left[\left(1000 \text{ kg/m}^3\right)(21.5 \text{ m/s})\,\pi(0.01 \text{ m})^2\right]$

$= 145.2 \text{ N}$

$F = \sqrt{F_x^2 + F_y^2} = \sqrt{(145.2)^2 + (145.2)^2} = 205 \text{ N}$
           Ans.

$\theta = \tan^{-1}\left(\dfrac{F_y}{F_x}\right) = \tan^{-1}\left(\dfrac{145.2}{145.2}\right) = 45° \angle$ *Ans.*

Answers to Selected Problems

Chapter 1

1–1.
a. 0.181 N^2
b. $4.53(10^3) \text{ s}^2$
c. 26.9 m

1–2.
a. 11.9 mm/s
b. $9.86 \text{ Mm} \cdot \text{s/kg}$
c. $1.26 \text{ Mg} \cdot \text{m}$

1–3.
a. $\text{kN} \cdot \text{m}$
b. Gg/m
c. $\mu\text{N/s}^2$
d. GN/s

1–5.
$\rho_{Hg} = 13.6 \text{ Mg/m}^3$
$S_{Hg} = 13.6$

1–6. 91.5 N

1–7. 18.3 kg

1–9. 55.7 N/m^3

1–10. $T_c = 26.9°\text{C}$

1–11. 0.629 kg

1–13. $p = (0.619 \, T_c + 169) \text{ kPa, where } T_c \text{ is in C}°$

1–14. 220 kPa

1–15. 541 kPa

1–17. 0.362 kg/m^3

1–18. $1.06(10^3) \text{ kg/m}^3$

1–19. 47.5 kN

1–21. $548(10^9) \text{ kg}$

1–22. 19.5 kN

1–23. $\rho_2 = 0.986 \text{ kg/m}^3, \, \forall_2 = 0.667 \text{ m}^3$

1–25. 6.60 MPa

1–26. 91.5 MPa

1–27. non-Newtonian

1–29. $0.408(10^{-3}) \text{ m}^2/\text{s}$

1–30.
$C = 1863.10 = 1863 \text{ K}$
$B = 1.73(10^{-6}) \text{ N} \cdot \text{s/m}^2$

1–31.
At $T = 283 \text{ K}, \mu = 1.25(10^{-3}) \text{ N} \cdot \text{s/m}^2$
At $T = 353 \text{ K}, \mu = 0.339(10^{-3}) \text{ N} \cdot \text{s/m}^2$

1–33. $B = 1.36 \left(10^{-6} \right) \text{ N} \cdot \text{s}/(\text{m}^2 \cdot \text{K}^{\frac{1}{2}}), C = 78.8 \text{ K}$

1–34. $8.93(10^{-3}) \text{ N} \cdot \text{s/m}^2$

1–35. $\mu = 0.025 \text{ N} \cdot \text{s/m}^2$

1–37. 0.1875 mPa

1–38. $\tau_p = 4.26 \text{ Pa}, \tau_{fs} = 5.32 \text{ Pa}$

1–39.
$\mu = 0.849 \text{ N} \cdot \text{s/m}^2$
$v = 2.00 \text{ m/s}$

1–41.
$\tau_{y=0} = 31.7 \text{ N/m}^2$
$\tau = 0 \text{ when } y = 5 \text{ μm}$

1–42.
$y = 1.25 \, (10^{-6}) \text{ m}, u = 0.109 \text{ m/s},$
$\tau = 23.8 \text{ N/m}^2$

1–43. 21.3 mN

1–45. 1.35 mN

1–46. $7.00 \text{ mN} \cdot \text{m}$

1–47. $0.942 \text{ N} \cdot \text{m}$

1–49. $T = \dfrac{4\pi\mu\omega r^3 L}{t}$

1–50.
At $r = 50 \text{ mm}, \tau = 3.28 \text{ Pa}$
At $r = 100 \text{ mm}, \tau = 6.57 \text{ Pa}$

1–51. $10.3 \text{ mN} \cdot \text{m}$

1–53. $T = 0.218\pi \, \mu\text{N} \cdot \text{m}$

1–54. $94.6°\text{C}$

1–55. 7.38 kPa

1–57. 4.25 kPa

1–58. 1.71 kPa

1–59. 3.17 kPa

1–61. $D = 1.0 \text{ mm}, h = 28.6 \text{ mm}$

1–62. 588 Pa

1–63. 4.67 mm

1–65. $L = (0.0154/\sin \theta) \text{ m}$

1–66. 24.3 mm

1–67. 0.158 N/m

1–69. 0.0716 N/m

Chapter 2

2–2. $\rho_m = 1390 \text{ kg/m}^3$

2–3.
$F_B = 827 \text{ N}$
$F_C = 883 \text{ N}$

2–5. $h_{Hg} = 301 \text{ mm}$

2–6.
$p_s = 16.0 \text{ kPa}$
$p_d = 10.6 \text{ kPa}$

2–7. 5.33 m

2–9. $p = -E_\forall \ln \left(1 - \dfrac{\rho_o g h}{E_\forall} \right)$

2–10.
Incompressible: $p = 2.943 \text{ MPa}$
Compressible: $p = 2.945 \text{ MPa}$

2–11. $p = p_0 \left(\dfrac{T_0 - Cz}{T_0} \right)^{g/RC}$

2–13. 167 kPa

2–14.
$p_A = 0$
$p_B = p_C = 7.16 \text{ kPa}$
$p_D = p_E = 21.5 \text{ kPa}$

2–15.
a. $p_A = 101.3(10^3) \text{ N/m}^2 + 0 = 101.3 \text{ kPa}$
b. 108 kPa
c. 123 kPa

2–17. $p = p_0\left(\dfrac{T_0 - Cz}{T_0}\right)^{g/RC}$

2–18. $p = p_0 e^{-(z-z_0)g/RT_0}$

2–19. 5.43 kPa

2–21. $p = 95.2$ kPa
$\rho = 1.18$ kg/m^3

2–22. **a.** 2.945 MPa
b. 2.943 MPa

2–23. 329 kPa

2–25. 736 Pa

2–26. 1.32 kPa

2–27. 246 mm

2–29. $p_A = 29.4$ kPa, $p_B = 14.7$ kPa

2–30. 9.81 kPa

2–31. 39.9 kPa

2–33. 12.7 kPa

2–34. 17.8 kPa

2–35. 135 mm

2–37. 18.2 kPa

2–38. 893 mm

2–39. 15.1 kPa

2–41. $p_A - p_B = e\left[\gamma_t - \left(1 - \dfrac{A_t}{A_R}\right)\gamma_R - \left(\dfrac{A_t}{A_R}\right)\gamma_L\right]$

2–42. **a.** 0.319 mm
b. 18.0 mm

2–43. 297 mm < 450 mm

2–45. $F_R = 29.3$ kN, $\bar{y}_P = 1.51$ m

2–46. 77.2°

2–47. 937 N

2–49. $F_C = 25.3$ kN
$h = 3.5$ m

2–50. $F_C = (9.81\ h^2 - 0.545\ h^3 - 96.8)$ kN where h is in meters

2–51. 4.11 m

2–53. 2.16 m

2–54. $F_R = 72.8$ kN
$y_P = 3.20$ m

2–55. $F_R = 72.8$ kN
$y_P = 3.20$ m

2–57. $F_R = 40.4$ kN
$d = 2.44$ m

2–58. 40.4 kN
2.44 m

2–59. $F = 17.3$ kN

2–61. 1.65 m

2–62. $F_R = 4.83$ kN
$Y_P = 1.77$ m

2–63. 4.83 kN
1.77 m

2–65. $F_R = 6.89$ kN
$y_P = 1.80$ m

2–66. $F_R = 9.18$ kN
$y_P = 1.75$ m

2–67. 9.18 kN
1.75 m

2–69. $F_R = 3.68$ kN
$y_P = 442$ mm

2–70. 442 mm

2–71. 442 kN

2–73. $N_B = 194$ kN
$A_x = 177$ kN
$A_y = 31.9$ kN

2–74. $N_B = 397$ kN
$A_x = 0, A_y = 473$ kN

2–75. 326 kN

2–77. $F_R = \left[\sqrt{601\left(10^6\right)h^4 + 16.7\left(10^6\right)h^6}\right]$ N
where h is in m

2–78. $F_R = 368$ kN
$\theta = 53.1°$ ⟋

2–79. $F_R = 163.5$ kN
$\theta = 53.1°$ ⟍

2–81. 179 kN

2–82. $F_R = 73.1$ kN
$\theta = 57.5°$ ⟋

2–83. $F_{AB} = 58.9$ kN
$F_{BC} = 147$ kN
$F_{CD} = 70.7$ kN

2–85. 196 kN · m

2–86. $N_B = 50.3$ kN
$A_y = 55.3$ kN
$A_x = 298$ kN

2–87. $B_x = 583$ kN
$B_y = 53.2$ kN

2–89. 37.9 N

2–90. $F_R = 53.8$ kN
$T = 42.3$ kN · m

2–91. 53.8 kN

2–93. $m_b = 225$ kg
$h = 0.367$ m

2–94. 136 mm

2–95. $T_{AB} = 2.70$ kN
Remains the same

2–97. $h = \dfrac{2\sigma \cos \theta}{gr(\rho_A - \rho_B)}$

2–98. 265 mm, 196 mm

2–99. 89.7 N

2–101. The barge will restore itself.

2–102. The barge will restore itself.

2–103. $r = 1.01$ m
$m = 15.6$ kg

2–105. 4.12 kPa

2–106. $h'_A = 0.171$ m

$h'_B = 0.629$ m

2–107. $a_c = 0.2(9.81 \text{ m/s}^2) = 1.96 \text{ m/s}^2$

2–109. $a_c = 3.92 \text{ m/s}^2$

2–110. $a_c = 4.45 \text{ m/s}^2$

The safer location is at the bottom of the tank.

2–111. $\tan \theta = \dfrac{a \cos \phi}{a \sin \phi + g}$

2–113. 23.6 kPa

13.6 kPa

2–114. $p_A = 21.6$ kPa

$p_B = 11.6$ kPa

2–117. 28.1 kPa

2–118. 26.4 rad/s

2–119. 132 mm

2–121. 4.05 kPa

2–122. 37.7 rad/s

2–123. 8.09 rad/s

2–125. $p_B = -3.06$ kPa

$p_C = 24.6$ kPa

Chapter 3

3–2. $\ln y^2 + y = 2x - 2.61$

$V = 5.66$ m/s

$\theta = 45°\ \measuredangle$

3–3. For $t = 1$ s, $y = 4e^{(x^2 + x - 2)/15}$

For $t = 2$ s, $y = 4e^{2(x^2 + x - 2)/15}$

For $t = 3$ s, $y = 4e^{(x^2 + x - 2)/5}$

3–5. $V = 2.19$ m/s

$\theta = 43.2°$

3–6. $y = 1.6x$

3–7. $y = \left(\dfrac{1}{16} \ln^2 \dfrac{x}{2} + \dfrac{1}{2} \ln \dfrac{x}{2} + 6 \right)$

3–9. $V = 10.3$ m/s

$\theta = 299°$

3–10. $V = 19.8$ m/s

$\theta = 40.9°$

3–11. $2y^3 - 1.5x^2 - y - 2x + 52 = 0$

$V = 30.5$ m/s

3–13. $x = 0.4t^2$

$y = 0.4t$

3–14. $y = x$

3–15. $4y^2 = 3x^2 + 13$

3–17. $y = \dfrac{4}{3}(e^{1.2x} - 1)$

3–18. $y = 6e^{\frac{1}{900}(x^2 + x + 54)^2 - 4}$

3–19. For $0 \le t < 5$ s, $y = \dfrac{8}{8 - 2 \ln x}$

For 5 s $< t \le 10$ s, $y = 2.67e^{(1/x - 0.0821)}$

3–21. $y^3 = 8x - 15, y = 9$ m

$x = 93$ m

3–22. $y = 3.65 - (2x + 1)^{\frac{1}{2}}$

$V = 7.07$ m/s

3–23. $y = \dfrac{2}{3} \ln \left(\dfrac{2}{2 - x} \right)$

3–25. $y = \dfrac{27(10^3)}{(x^2 + 5)^3}$

$V = \sqrt{u^2 + v^2} = \sqrt{(30 \text{ m/s})^2 + (30 \text{ m/s})^2}$

$= 42.4$ m/s

$\theta = \tan^{-1}\left(\dfrac{v}{u}\right) = \tan^{-1}\left(\dfrac{30 \text{ m/s}}{30 \text{ m/s}}\right) = 45°\ \measuredangle$

3–26. $u = 3.43$ m/s

$v = 3.63$ m/s

3–27. $y = \left(\dfrac{1}{2} \ln \dfrac{x}{2} + 6 \right)$ m

3–29. $a = 36.1 \text{ m/s}^2$

$\theta = 33.7°\ \measuredangle$

3–30. $V = V_A + \left(\dfrac{V_B - V_A}{L_{AB}} \right) x = 8 + \left(\dfrac{2 - 8}{3} \right) x$

$= (8 - 2x) \text{ m/s}$

$= 4(x - 4) \text{ m/s}^2$

$x = 4(1 - e^{-2t}) \text{ m}$

3–31. 339 m/s^2

3–32. $V = 23.3$ m/s

$a = 343 \text{ m/s}^2$

3–33. $x = 2.72$ m

$y = 1.5$ m

$z = 0.634$ m

$\mathbf{a} = \{10.9\mathbf{i} + 32.6\mathbf{j} + 24.5\mathbf{k}\} \text{ m/s}^2$

3–34. $y = 2x$

$a = 286 \text{ m/s}^2$

$\theta = 63.4°\ \measuredangle$

3–35. $y = 1.25$ mm

$x = 15.6$ mm

$a = 0.751 \text{ m/s}^2$

$\theta = 2.29°\ \measuredangle$

3–37. -1.23 m/s^2

3–38. For point (2 m, 0),

$a = 0.5 \text{ m/s}^2$

$y = \pm\sqrt{4x^2 - 16}$

For point (4 m, 0),

$a = 0.0625 \text{ m/s}^2$

$y = \pm\sqrt{4x^2 - 64}$

3–39. 100 m/s^2

3–41. $V = 33.5$ m/s

$\theta_V = 17.4°$

$a = 169$ m/s^2

$\theta_a = 79.1°$ ⟋

3–42. $V = 0.601$ m/s

$a = 0.100$ m/s^2

$x^2/(4.24)^2 + y^2/(2.83)^2 = 1$

3–43. $V = 4.47$ m/s, $a = 16$ m/s^2

$y = \dfrac{1}{2}\ln x + 2$

3–45. $V = 5.66$ m/s

17.9 m/s^2

3–46. $V = 4.12$ m/s

$a = 17.0$ m/s^2

$x^2 = \dfrac{y^3 + 5}{6y}$

3–47. At point (0, 0),

$a = 64$ m/s^2 ↓

At point (1 m, 0),

$a = 89.4$ m/s$^2, \theta = 63.4°$ ⬊

For the streamline passing through point (0, 0),

$y = \left[2\ln\left(\dfrac{8}{2x^2 + 8}\right)\right]$ m

For the streamline passing through point (1 m, 0),

$y = \left[2\ln\left(\dfrac{10}{2x^2 + 8}\right)\right]$ m

3–49. 72 m/s^2

3–50. 3.38 m/s^2

3–51. $a = a_n = 4$ m/s^2

3–53. $a_s = 3.20$ m/s^2

$a_n = 7.60$ m/s^2

3–54. $a = 3.75$ m/s^2

$\theta = 36.9°$ ⬊

3–55. $a_n = \dfrac{V^2}{R} = 0$

$a_s = 286$ m/s^2

3–57. $a_s = 78.7$ m/s^2

$a_n = 73.0$ m/s^2

Chapter 4

4–13. 8.51 m/s

8.69(10^{-3}) slug/s

4–14. $Q = 0.225$ m^3/s

$\dot{m} = 225$ kg/s

4–15. 0.0493 m^3/s

4–17. $Q_A = 0.036$ m^3/s

$Q_B = 0.0072$ m^3/s

4–18. 0.955 m/s

4–19. $t = \left(\dfrac{0.382}{D^2}\right)$ min, where d is in m

4–22. 12.3 mm/s

4–23. 119 m^3/s

4–25. $\dfrac{49}{60}U$

4–26. $\dot{m} = \dfrac{49\pi}{60}\rho UR^2$

4–27. 0.0294 m^3/s

4–29. $V_A = 3.36$ m/s, $V_B = 1.49$ m/s

$a_A = 0.407$ m/s$^2, a_B = 0.181$ m/s^2

4–30. $7.07(10^9)$

4–31. 0.413 s

4–33. $V_A = 12.2$ m/s, $V_B = 9.98$ m/s

4–34. $u = 3.56$ m/s, $a = 112$ m/s^2

4–35. 3.76 m/s

4–37. $V_A = 3.64$ m/s

4–38. 279 m/s^2

4–39. 55.4 m/s^2

4–41. $v_{out} = \left[\dfrac{2(10^{-6})d^2}{0.04 - (10^{-6})d^2}\right]$ m/s where d is in mm

4–42. $V = \dfrac{0.0637}{(0.02 - 0.141x)^2}$

4–43. $\left[\dfrac{1.14(10^{-3})}{(0.02 - 0.141x)^5}\right]$ m/s^2

4–45. 0.6 m/s

4–46. 3.01 m/s

4–47. 10.0 m/s

4–49. 0.647 kg/s

4–50. 31.9 m/s

4–51. 9.42 m/s

4–53. $V_B = 28.8$ m/s

$a_A = 3.20$ m/s^2

4–54. 48.5 s

4–55. $t = \left(\dfrac{6814}{d^{\frac{4}{3}}}\right)$ s where d is in mm

4–57. 2.48 hr

4–58. $V_s = \left(\dfrac{993}{t}\right)$ m/s, where t is in hrs

4–59. 3.18 min

4–61. $V = (6.25V_p)$ m/s

4–62. $V_P = 0.0312$ m/s

$V_C = -0.0104$ m/s

4–63. 0.0509 m/s

4–65. -0.00356 kg/(m$^3 \cdot$ s)

4–66. 0.00530 kg/$(m^3 \cdot s)$

4–67. -0.101 kg/$(m^3 \cdot s)$, unsteady

4–69. $\dfrac{dh}{dt} = 0.890 \left(10^{-3}\right)$ m/s

$\rho_{avg} = 923$ kg/m^3

4–70. 6.77 m^3

4–71. 10.2 s

4–73. $\dfrac{dy_c}{dt} = \left(\dfrac{0.157\left(10^{-3}\right)}{0.173y_c + 0.0146}\right)$ m/s

4–74. -0.0183 m/s

4–75. $w = (0.02e^{-t/450})$ m

4–77. 0.0144 m/s

4–78. 6.43 m/s

4–79. 3.11 m/s

Chapter 5

5–1. -2 kPa

5–2. $\Delta p = \rho V^2 \ln\left(r/r_i\right)$

5–3. $p_C = -15.4$ Pa

$\dot{W} = 151$ mW

5–5. 10.7 m/s

5–6. $V_n = \left(3.283 \sqrt{F}\right)$ m/s, where F is in N

5–7. 73.0 kPa

5–9. -36.7 kPa

5–10. $P_A = \left[19.6 - \dfrac{0.730(10^9)}{d^4}\right]$ kPa where d is in mm

5–11. 0.0329 m^3/s

5–13. $V = 12.7$ m/s, $p = -60.8$ kPa

5–14. $p = 0.5\left[40.5 - \dfrac{6.48\left(10^6\right)}{r^2}\right]$ kPa,

where r is in mm.

5–15. $Q = 2.78\left(10^{-3}\right)$ m^3/s

$p_E = 38.8$ kPa

5–17. $dy/dt = 0.496$ m/s, $p = -2.39$ kPa

5–18. 3.32 m

5–19. 159 Pa

5–21. 3.10 kPa

5–22. $3.11\left(10^{-3}\right)$ m^3/s

5–23. 0.0374 m^3/s

5–25. 1.23 kPa

5–26. 7.63 kPa

5–27. 3.51 m/s

5–29. 0.130 m^3/s

5–30. 12.8 kPa

5–31. $V_B = 5.33$ m/s

$p_B = 121$ kPa

5–33. 1.65 m^3/s

5–34. 90 mm

5–35. 28.2 kg/s

5–37. 260 m

5–38. $p(x) - p_A = \left(30x - 4.5x^2\right)$ kPa

5–39. 34.9 kPa

5–41. 101 kPa

5–42. 275.025 kPa

5–43. 567 Pa

5–45. 41.0 mm

5–46. $V_C = 1.401$ m/s $= 1.40$ m/s

$d_D = 0.02378$ m $= 23.8$ mm

5–47. 0.0163 m^3/s

5–49. $p_A = -96.4$ kPa

$Q = 0.0277$ m^3/s

5–50. $V_A = 14.6$ m/s

$V_B = 7.58$ m/s

5–51. -3.11 mm/s

5–53. 2

5–54. 1.06

5–55. $V_B = 8$ m/s

$p_B = 52.9$ kPa

5–57. $V_B = 2$ m/s

$p_B = 94.1$ kPa

5–59. 83.1 kPa

5–61. 0.00116 m^3/s

5–62. 22.6 kW

5–63. 21.0 kW

5–65. 314 kW

5–66. 13.9 kW

5–67. 20.7 kW

5–69. 14.0 kW

5–71. $V = 3.56$ m/s, $p = 278$ kPa

5–73. 2.66 kJ/kg

5–74. 146 J/kg

5–75. 1.67 kW

5–77. 2.91 kW

5–78. 4.45 MW

5–79. 116 kW

5–81. 32.5 kW

5–82. 3.03 kW

Chapter 6

6–1. $L = 10.1$ kg · m/s by either method

6–3. 526 N

6–5. 0.103 mN

6–6. $F_x = 1.21$ kN

$F_y = 697$ N

6–7. 11.3 kPa

6–9. $Q_A = 0.00460$ m^3/s

$Q_B = 0.0268$ m^3/s

6–10. 858 N

6–11. $F = \left[160\pi\left(1 + \sin\theta\right)\right]$ N

6–13. $h = \left[\dfrac{8.26(10^6)Q^4 - 0.307(10^{-6})}{Q^2}\right]$ m

6–14. 2.26 kN

6–15. $F_x = 125$ N
$F_y = 232$ N

6–17. $V_A = 15.7$ m/s
$T = 2.31$ kN

6–18. 0.659 N

6–19. $F_x = 349$ N $\leftarrow$

6–21. 251 N

6–22. 117 mm

6–23. 1.43 N

6–25. $F = \left[\dfrac{2.55(10^{-3})}{x^2} + 25.5 \right]$ N

6–26. $F_x = 542$ N
$F_y = 183$ N$\uparrow$

6–27. 418 Pa

6–29. $F_x = 0$
$F_y = 53.1$ N

6–30. 63.8 N

6–31. $F = \left[\dfrac{97.7x^2 - 11.1x + 0.157}{(3.69x - 32.6x^2)^2} + 14.3 \right]$ N

6–33. 1.57 kN

6–34. 72.3 N

6–35. 104 N

6–37. 1.92 kN

6–38. $F_1 = F_2 = F_3 = \rho_w A V^2$

6–39. $V = 10.2$ m/s

6–41. 141 N

6–42. 62.8 N
$V_b = 10$ m/s $\rightarrow$

6–43. $A_x = 222$ N $\leftarrow$
$A_y = 141$ N
$M_A = 65.6$ N·m

6–45. $A_x = 0$
$A_y = 520$ N
$M_A = 109$ N·m

6–46. $B_x = 4$ kN
$A_x = 1.33$ kN
$A_y = 5.33$ kN

6–47. 482 N·m

6–49. $C_x = 45.0$ N
$C_y = 168$ N
$M_C = 33.6$ N·m

6–50. $V_w = 5.48$ m/s
$M = 0.211$ mN·m

6–51. $N_D = 60.0$ N
$C_x = 77.6$ N
$C_y = 125$ N

6–53. $P = \rho_w bh\omega RV(V - \omega R)$

6–54. 300 kN

6–55. $V_2 = 25.53$ m/s
$\dot{W}_{in} = 23.9$ MW

6–57. 68.2 kN
662 kW

6–58. $V = 9.71$ m/s
$\Delta p = 54.3$ Pa

6–59. $F = 79.4$ kN
$\dot{W} = 5.51$ MW
$e = 0.5$

6–61. 8.80 kN

6–62. 24.9 m/s

6–63. 3.50 MW

6–65. 18.2 kN

6–66. 22 kN

6–67. 100 kN

6–69. $T_1 = 600$ N
$T_2 = 900$ N
$e_1 = 0.5$
$e_2 = 0.6$

6–70. 13.7°

6–71. 0.00584 kg/s

6–73. $V_e = 450$ m/s
$F_D = 11.5$ kN

6–74. $V_{max} = \dfrac{\rho V_e^2 A - F}{\rho V_e A} \ln\left(\dfrac{M + m_0}{M} \right)$

6–75. $T = 74.2$ kN
$e = 0.502$

Chapter 7

7–1. $\omega_z = \dfrac{U}{2h}$

$\dot{\gamma}_{xy} = \dfrac{U}{h}$

7–2. $\Gamma_r = 1440\pi$ m²/s
$\Gamma_r = 2560\pi$ m²/s

7–3. $\Gamma = 0$

7–6. $\psi = V_0 \left[(\cos\theta_0)y - (\sin\theta_0)x \right]$
$\phi = V_0 \left[(\cos\theta_0)x + (\sin\theta_0)y \right]$

7–7. $V = 3.61$ m/s
$\phi = 2x - 3y$

7–9. ϕ cannot be established.

7–10. irrotational

7–11. $\psi = 50y^2 + 0.2y$, ϕ cannot be established.

7–13. rotational

$\psi = \dfrac{1}{2}(3y^2 - 9x^2)$

$y = \sqrt{3x^2 - 39}$

7–14. $\psi = 4y - 3x$
$\phi = 4x + 3y$

7–15. $q_{AB} = 14 \text{ m}^2/\text{s}$
$q_{BC} = 32 \text{ m}^2/\text{s}$
$q_{AC} = 18 \text{ m}^2/\text{s}$

7–17. $\phi = -4x - 8y$

7–18. 5.57 m/s

7–19. $p_A = 34 \text{ kPa}$
$\psi_2 - \psi_1 = 0.5 \text{ m}^2/\text{s}$

7–21. 2 m/s

7–22. $\phi = V_0\left[(\sin\theta_0)x - (\cos\theta_0)y\right]$

7–23. $v_r = 0$
$v_\theta = 8r$
$v_x = -8y$
$v_y = 8x$

7–25. $V = 6.72 \text{ m/s}$
irrotational

7–26. -1.45 kPa

7–27. $p_B = 832 \text{ Pa}$

7–29. $u = 16 \text{ m/s}\downarrow$
$a_x = 128 \text{ m/s}^2 \uparrow$

7–30. -101 kPa

7–31. $v_r = 4 \text{ m/s}$
$v_\theta = -6.93 \text{ m/s}$
$\phi = 8(x^2 - y^2)$

7–33. $\psi = \dfrac{1}{2}(y^2 - x^2)$

7–34. $\phi = xy^2 - \dfrac{x^3}{3}$
$p_B = 628 \text{ kPa}$

7–35. $\psi = \dfrac{5}{3}x(3y^2 - x^2)$

7–37. $V_A = 114 \text{ m/s}$
$p_O - p_A = 6.02 \text{ MPa}$

7–41. $\phi = \dfrac{5 \text{ m}^2/\text{s}}{2\pi}\ln\dfrac{r_2}{r_1}$

The equipotential line for $\phi = 0$ is along the y axis.

7–42. $\theta = \pi$
$r = \dfrac{3}{16\pi} \text{ m}$

7–43. $u = \dfrac{q}{2\pi}\left[\dfrac{x-4}{(x-4^2)+y^2} + \dfrac{x+4}{(x+4^2)+y^2}\right]$

$v = \dfrac{q}{2\pi}\left[\dfrac{y}{(x-4^2)+y^2} + \dfrac{y}{(x+4^2)+y^2}\right]$

The flow is irrotational.
$x = 0$

7–45. $V = 60.0 \text{ m/s}$
$p = -2.16 \text{ kPa}$

7–46. $\psi = \dfrac{q}{2\pi}(\theta_1 - \theta_2)$

7–47. $\psi = \dfrac{q}{2\pi}(\theta_1 + \theta_2)$

7–49. $\dfrac{y}{x^2 + y^2 - 0.25} = \tan 40\pi y$

7–50. $p = p_0 - \dfrac{\rho U^2}{2(\pi - \theta)^2}\left[\sin^2\theta + (\pi - \theta)\sin 2\theta\right]$

7–51. $r_0 = 0.398 \text{ m}$
$h = 1.25 \text{ m}$
$V = 0.447 \text{ m/s}$
$p = 283 \text{ Pa}$

7–53. $z = \dfrac{\Gamma^2}{8\pi^2 g r^2}$

7–54. $h = 0.609 \text{ m}$
$V = 11.2 \text{ m/s}$
$p = 29.2 \text{ kPa}$

7–57. 81.0 m/s; yes

7–58. $p_{max} = 535 \text{ Pa}$

7–59. $F = 568 \text{ kN/m}$
$p = -99.3 \text{ Pa}$

7–61. $(F_R)_y = \dfrac{5}{3}\rho RLU^2$

7–62. $V_A = 2U$
$(p_g)_A = -\dfrac{3}{2}\rho U^2$

7–63.

r (m)	0.1	0.2	0.3	0.4	0.5
V (m/s)	12	7.5	6.67	6.375	6.24
p (kPa)	69	135	144	147	148

7–65. $y = x\tan\left[\pi(1 - 32y)\right]$

7–66. $\theta = 113°$
or
$\theta = 247°$

7–67. $F_R = 6.64 \text{ N/m}$ outward (suction)

7–69. $p(x) = \left[-\dfrac{25}{32x^6}(16x^3 + 1) + 4.2032\right] \text{ kPa}$

7–71. $v_\theta = \dfrac{\omega r_i^2}{r_0^2 - r_i^2}\left(\dfrac{r_0^2 - r^2}{r}\right)$

Chapter 8

8–1. a. $\dfrac{Vt}{L}$

b. $\dfrac{E_V L}{\sigma}$

c. $\dfrac{V^2}{gL}$

8–2.   **a.** yes
     b. no
     c. no
     d. no

8–3.   **a.** $\dfrac{L^6}{FT^2}$

     b. $\dfrac{1}{L}$

     c. L

     d. $\dfrac{F}{L^2}$

8–5. $\left(\dfrac{M}{LT^2}\right)^{\frac{1}{2}}$

8–6.   6.79

8–7. $F = k\gamma\,\cancel{V}$

8–9. $\dfrac{\rho VL}{\mu}$ or $\dfrac{\mu}{\rho VL}$

8–10. $W_O = 1$

8–13. $\tau = k\sqrt{\dfrac{m}{\gamma A}}$

8–14. $Q = k\left[\dfrac{D^4}{\mu}\left(\dfrac{\Delta p}{\Delta x}\right)\right]$

8–15. $Q = k\sqrt{gD^5}$

8–17. $T = FDf\left(\dfrac{\mu D^2\omega}{F}\right)$

8–18. $h = df\left(\dfrac{\sigma}{\rho d^2 g}\right)$

8–19. $c = k\sqrt{\dfrac{\sigma}{\rho\lambda}}$

     18.4%

8–21. $h_L = Df(\text{Re})$

8–22. $p = k\dfrac{\sigma}{r}$

8–23. $f(\text{Re, M, Eu, Fr, We}) = 0$

8–25. $t = \sqrt{\dfrac{d}{g}}\left(\dfrac{\mu}{\rho d^{\frac{3}{2}} g^{\frac{1}{2}}}\right)$

8–26. $\Delta p = \rho\omega^2 D^2 f\left(\dfrac{Q}{\omega D^3}\right)$

8–27. $\tau = \sqrt{\dfrac{\lambda}{g}}\,f\left(\dfrac{h}{\lambda}\right)$

8–29. $c = \sqrt{g\lambda}\,f\left(\dfrac{\lambda}{h}\right)$

8–30. $Q = k\dfrac{T}{\omega\rho D^2}$

8–31. $T = \rho\omega^2 D^4 f\left(\dfrac{\mu}{\rho\omega D^2}, \dfrac{V}{\omega D}\right)$

8–33. $T = \rho Q^{\frac{5}{3}}\omega^{\frac{1}{3}} f\left[\left(\dfrac{\omega}{Q}\right)^{\frac{1}{3}} h\right]$

8–34. $P = D^3\omega^2\mu f\left(\dfrac{Q}{D^3\omega}, \dfrac{\rho D^2\omega}{\mu}\right)$

8–35. $Q = D^3\omega f\left(\dfrac{\rho D^5\omega^3}{\dot{W}}, \dfrac{D^3\omega^2\mu}{\dot{W}}\right)$

8–37. $\delta = xf(\text{Re})$

8–38. $\Delta p = \rho V^2 f(\text{Re})$

8–39. $F_D = \rho V^2 L^2[f(\text{Re})]$

8–41. $\dot{W} = CQ\Delta p$

8–42. $p = p_0 f\left(\dfrac{\rho r^3}{m}, \dfrac{E_V}{p_0}\right)$

8–43. $g\left(\text{Re, We, }\dfrac{D}{h}, \dfrac{d}{h}\right) = 0$

8–45. $Q = \sqrt{gH^5}\,f\left(\dfrac{b}{H}, \dfrac{h}{H}, \dfrac{\mu}{\rho\sqrt{gH^3}}, \dfrac{\sigma}{\rho gH^2}\right)$

8–46. $V = 1.00$ m/s

8–47.   2.83 Pa

8–49.   0.775 m/s

8–50.   18 Mm/h

8–51. $V_m = 14.2$ m/s

8–53.   12 m/s

8–54.   12 km/h

8–55.   0.894 m/s

8–57. $V_p = 20.8$ m/s

8–58.   4.33 s

8–59. $Q_m = 0.133$ m^3/s
     $(\Delta P)_m = 11.5$ MPa
     $\dot{W} = 150$ kW

8–61.   11.0 kg/m^3

8–62.   7 kN

8–63.   684 m/s

8–65.   200 kN

8–66.   20.0 kN

Chapter 9

9–1. $0.141(10^{-3})$ m^3/s

9–2. $F = 0.0258$ N

9–3. $u_{max} = 0.132$ m/s
$\tau_t = -8$ Pa
$\tau_b = 8$ Pa

9–5. $Q = 0.0224$ m^3/s
9–6. 4.88 m/s
9–7. $0.490(10^{-6})$ m^3/s
9–9. 1.79 mm
9–10. 0.326 N·m

9–11. $\tau = \dfrac{dp}{dx} y$

$u = \dfrac{a^2}{8\mu} \dfrac{dp}{dx}\left[4\left(\dfrac{y}{a}\right)^2 - 1\right]$

9–13. $\tau = -27.5$ N/m^2
$u = (0.3 - 125y)$ m/s

9–14. $\tau_w = \tau_o = \dfrac{\mu_w \mu_o U}{(\mu_o + \mu_w)a}$

$u_w = \dfrac{\mu_o U}{(\mu_o + \mu_w)a} y$

$u_o = \dfrac{U}{a(\mu_o + \mu_w)}[\mu_w y + a(\mu_o - \mu_w)]$

9–17. $u = \left(\dfrac{U_t - U_b}{a}\right)y + U_b$

$\tau_{xy} = \dfrac{\mu(U_t - U_b)}{a}$

9–18. laminar
9–19. 112 Pa
9–21. $(V)_{max} = 0.304$ m/s
Turbulence will not occur.
9–22. 11.9 kPa
9–23. 953 Pa
9–25. $u_{max} = 6.35$ m/s
$\tau_{max} = 250$ N/m^2
9–26. $\Delta p = 6.14$ Pa
$\tau_{max} = 7.68\ (10^{-3})$ Pa
9–27. $0.849(10^{-3})$ m^3/s
9–29. $\tau = 183$ N/m^2
$u_{max} = 1.40$ m/s
$Q = 0.494(10^{-3})$ m^3/s
9–30. 0.0264 m^3/s
9–31. $\tau_{max} = 6.15$ Pa
$u_{max} = 2.55$ m/s
9–33. 0.970 N·s/m^2
9–34. $R = \dfrac{128\mu_a}{\pi D^4}$
9–35. $\tau_w = 350V$
9–37. Re $= 79.14\ (10^3)$
Since Re > 2300, then **the flow is not laminar.**

9–39. 41.8 mm
9–41. $t = \dfrac{32\mu l D^2}{\rho g d^4} \ln\left(\dfrac{h_1}{h_2}\right)$
9–45. $Q = 0.0583$ m^3/s
9–47. $Q = 0.215$ m^3/s
$L = 13.8$ m
9–49. $u = 18.5$ m/s
$y = 0.203$ mm
9–50. $\tau_{visc} = 1.50$ Pa $t_{turb} = 11.6$ Pa
9–51. $\tau_0 = 21.9$ Pa
$u_{max} = 2.658$ m/s
$y = 1.09$ mm
9–53. At $r = 0.05$ m,
$\tau_{lam} = 2.50$ Pa
$\tau_{turb} = 0$
At $r = 0.025$ m,
$\tau_{lam} = 0.0239$ Pa
$\tau_{turb} = 1.23$ Pa
9–54. $\tau = 13.1$ Pa
$u = 2.30$ m/s
$\tau_{turb} = 11.6$ Pa
9–55. 278 mm^3/s

Chapter 10

10–1. $\Delta p = 0.288$ Pa
10–2. $Q = 17.7$ liters/s
10–3. 0.056
10–5. 9.50 m
10–6. 123 mm
10–7. 0.0368
10–9. $Q = 0.00608$ m^3/s
$\Delta p = 0.801$ Pa
10–10. $D = 61.6$ mm
10–11. 17.7 kPa
10–13. $\dot{W}_{co} = 13.5$ kW
$\dot{W}_w = 10.9$ kW
10–14. 76.7 mm
10–15. 0.00529 m^3/s
10–17. 0.642 kg/s
10–18. 300%
10–19. 17.4 kPa
10–21. 133 kPa
10–22. $\Delta p = 6.38$ kPa
10–23. 5.87 kPa
10–25. $\Delta p_2 = 10$ kPa
10–26. 22.9 kPa
10–27. 19.5 liter/s
10–29. 0.0145 m^3/s
10–30. 0.00804 m^3/s

10–31. 427 W

10–33. $Q_1 = 0.00357 \text{ m}^3/\text{s}$

10–34. 51.3 W

10–35. 2.83 kW

10–37. 106 mm

10–38. 412 W

10–39. 0.797 W

10–41. $0.00431 \text{ m}^3/\text{s}$

10–42. $0.0148 \text{ m}^3/\text{s}$

10–43. $p_A = 742 \text{ kPa}$

10–45. 34.4 kPa

10–46. $0.0124 \text{ m}^3/\text{s}$

10–47. $0.0111 \text{ m}^3/\text{s}$

10–49. $0.00265 \text{ m}^3/\text{s}$

10–50. 0.821

10–51. 201 kPa

10–53. $0.00290 \text{ m}^3/\text{s}$

10–54. $Q_D = 0.00333 \text{ m}^3/\text{s}$

10–55. $Q_C = 0.00588 \text{ m}^3/\text{s}$
$Q_D = 0.00412 \text{ m}^3/\text{s}$

10–57. $0.0335 \text{ m}^3/\text{s}$

10–58. $0.0430 \text{ m}^3/\text{s}$

Chapter 11

11–1. 24.3 m

11–2. 2.25 Pa

11–3. 242 mm

11–5. 2.57 N

11–6. $u|_{y=0.1 \text{ m}} = 13.6 \text{ m/s}$
$u|_{y=0.3 \text{ m}} = 14.5 \text{ m/s}$

11–7. 0.585 m/s

11–9. $\delta_w = 3.54 \text{ mm}$
$\delta_a = 13.7 \text{ mm}$

11–10. 1.51 N

11–11. $F_D = 4.97 \text{ N}$

11–13. $x = \dfrac{\rho U a^2}{100\mu}$

11–14. $\delta|_{x=0.2 \text{ m}} = 3.33 \text{ mm}$
$\delta|_{x=0.4 \text{ m}} = 4.70 \text{ mm}$

11–15. $\tau_0 = 0.289\mu\left(\dfrac{U}{x}\right)\sqrt{\text{Re}_x}$

11–17. $a = 1.02 \text{ m}$

11–18. 7.98 kN

11–19. 92.8 N

11–21. $\delta = \dfrac{0.343x}{(\text{Re}_x)^{\frac{1}{5}}}$

11–22. $\delta = 4.36 \text{ mm}$
$F = 1.39 \text{ kN}$

11–23. 1.22 kN

11–25. $C_1 = 0 \quad C_2 = \dfrac{3}{2} \quad C_3 = -\dfrac{1}{2}$

11–26. $\delta = 0.0118 \text{ mm}$
$\Theta = 1.02 \text{ mm}$

11–27. 6.22 m/s

11–29. $\delta = \dfrac{4.80x}{\sqrt{\text{Re}_x}}$

11–30. $\delta^* = \dfrac{1.74x}{\sqrt{\text{Re}_x}}$

11–31. $C_2 = 0$
$C_1 = \dfrac{3}{2}$
$C_3 = \dfrac{1}{2}$
$\delta = \dfrac{4.64x}{\sqrt{\text{Re}_x}}$

11–33. $\tau_0 = 0.323\mu\left(\dfrac{U}{x}\right)\sqrt{\text{Re}_x}$

11–34. 206 mm

11–35. 0.961 Pa

11–37. $F = 448 \text{ N}$

11–38. 6.43 N

11–39. 4.82 N

11–41. $F = 495 \text{ N}$

11–42. $F = 2.14 \text{ kN}$

11–43. 240 N

11–45. 24.9 kN

11–46. 43.7 kN

11–47. 124 N

11–49. $F_D = 17.3 \text{ N}$

11–50. 2.71 kN·m

11–51. 4.25 kN·m

11–53. $(F_D)_{AB} = 20.9 \text{ N}$
$(F_D)_{BC} = 40.5 \text{ N}$

11–54. 2.20 kN

11–55. $\dot{W} = 80.5 \text{ kW}$

11–57. 1.75 kN·m

11–58. $U = 9.27 \text{ m/s}$

11–59. 3.71 m

11–61. $F_D = 17.6 \text{ N}$

11–62. 6.70 m/s

11–63. 15.7 m/s

11–65. 0.374 m/s

11–66. $T = \dfrac{1}{4}C_D\rho w \omega^2 L^4$

11–67. $U = 0.00128 \text{ m/s}$

11–69. 81.8 N

11–70. 114 days

11–71. 5.03 m/s

11–73. 5.45 m/s

11–74. $U = 3.87$ m/s

11–75. 5.50 m

11–77. 40.6 min

11–78. $t = 0.805$ s

$V_t = 12.0$ m/s

11–79. 13.7 m/s^2

11–81. $\dot{W} = 156$ kW

11–82. $F_D = 1.05$ kN

11–83. 227 km/h

11–85. 5° (approx.)

11–86. 0.274

11–87. 31.8 mm

11–89. 0.00816 N

11–90. 3.27°

Chapter 12

12–1. **a.** 6.26 m/s

b. 4.82 m/s

c. 0

12–2. Fr = 2.26

supercritical

$V_c = 2.91$ m/s

12–3. subcritical

12–5. $F_r = 0.434$

$V_c = 2.91$ m/s

12–6. 1.02 m

12–7. $y_c = 2.63$ m

$V_c = 5.08$ m/s

$E_{min} = 3.94$ m

At $y = 2$ m, $E = 4.27$ m.

12–9. $y = 1.90$ m (subcritical)

$y = 0.490$ m (supercritical)

12–10. $y_c = 1.18$ m

$y = 1.73$ m (subcritical)

$y = 0.838$ m (supercritical)

12–11. $y_c = 1.01$ m

$E_{min} = 1.52$ m

At $y = 1.5$ m, $E = 1.73$ m.

12–13. $y_B = 2.95$ m

12–14. 4.98 m

12–15. 4.78 m^3/s

12–17. $y = 2.82$ m (subcritical)

$y = 0.814$ m (supercritical)

12–18. 1.22 m

12–19. supercritical

$y_2 = 1.78$ m

$V_2 = 5.07$ m/s

12–21. $y_B = 0.511$ m

$V_A = 10$ m/s

$V_B = 9.79$ m/s

12–22. $y_2 = 0.145$ m

$h = 0.935$ m

12–23. $y_2 = 0.75$ m < 0.97168 m (rapid)

$y_2 = 1.29$ m > 0.97168 m (tranquil)

$y_2 = -0.4742$ m (unrealistic)

$h = 108$ mm

12–25. $y_2 = 0.613$ m

$y_3 = 3.92$ m

12–26. $Q = \left[\sqrt{78.48 y_2^2 (3 - y_2)} \right]$ m^3/s

a. 12.5 m^3/s

b. 16.3 m^3/s

12–27. **a.** $R_h = \dfrac{bh}{2h + b}$

b. $R_h = \dfrac{\sqrt{3}}{8} l$

c. $R_h = \dfrac{\sqrt{3} l (l + 2b)}{4 (2l + b)}$

12–29. 5.70 m^3/s

12–30. 2.51 m^3/s

12–31. $Q = 49.3$ m^3/s

12–33. 1.88R

12–34. 1.63R

12–35. $Q = 16.0$ m^3/s

12–37. 79.4 m^3/s

12–38. 0.00422

12–39. 3.08 m

12–41. $a = b$

12–42. 15.0 m^3/s

12–43. $y = 1.25$ m

12–45. $\theta = 60°$

12–47. $\theta = 60°$

$l = b$

12–49. $S_c = 0.00362$

12–50. $A3$

12–51. 0.888 m

12–53. $x = 47.7$ m

12–54. 11.0 m

12–55. 4.74 m

12–57. $y_1 = 0.300$ m

$h_L = 17.3$ m

12–58. $y_2 = 2.63$ m

12–59. $V_1 = 7.14$ m/s

$V_2 = 4.46$ m/s

$h_L = 0.0844$ m

12–61. 7.35 m^3/s

Chapter 13

13–1. $\Delta h = 0$
$\Delta s = 465 \text{ J}/(\text{kg} \cdot \text{K})$

13–2. $\Delta p = 88.7 \text{ kPa}$
$\Delta s = 631 \text{ J}/(\text{kg} \cdot \text{K})$

13–3. $\rho_1 = 0.331 \text{ kg}/\text{m}^3$
$T_2 = 366 \text{ K}$

13–5. $c_p = 126 \text{ kJ}/\text{kg}$

13–6. $5.01° \text{ C}$

13–7. $3580 \text{ m}/\text{s}$

13–9. $103 \text{ m}/\text{s}$

13–10. $c_{air} = 343 \text{ m}/\text{s}$
$c_w = 1485 \text{ m}/\text{s}$

13–11. $2.67(10^3) \text{ km}/\text{h}$

13–13. $17.3°$

13–14. $V = 369 \text{ m}/\text{s}$

13–15. $T_0 = 509 \text{ K}$
$p_0 = 196 \text{ kPa}$

13–17. $T^*/T_0 = 0.866$
$p^*/p_0 = 0.544$
$\rho^*/\rho_0 = 0.628$

13–18. 0.302

13–19. $V = 709 \text{ m}/\text{s}$
$T_0 = 450 \text{ K}$
$p_0 = 273 \text{ kPa}$

13–21. $0.0547 \text{ kg}/\text{m}^3$

13–22. 17.9 kPa

13–23. 591 kPa

13–25. For isentropic flow, $p = 596 \text{ kPa}$
When $p = 150 \text{ kPa}, \dot{m} = 0.411 \text{ kg}/\text{s}$

13–26. $p = 8.93 \text{ kPa}$
$\dot{m} = 0.411 \text{ kg}/\text{s}$

13–27. $1.78 \text{ kg}/\text{s}$

13–29. $321 \text{ m}/\text{s}$

13–30. $0.103 \text{ kg}/\text{s}$

13–31. $1.05 \text{ kg}/\text{s}$

13–33. $\dot{m} = 1.99 \text{ kg}/\text{s}$

13–34. 49.6 mm

13–35. 50.5 mm

13–37. $0.714 \text{ kg}/\text{s}$

13–38. $d_t = 197 \text{ mm}$
$d_e = 313 \text{ mm}$

13–39. $M = 0.446$
$\dot{m} = 17.2 \text{ kg}/\text{s}$

13–41. $p_0 = 403 \text{ kPa}$
$d_t = 49.5 \text{ mm}$
$p_t = 219 \text{ kPa}$
$M_B = 0.0358 < 1 \text{ (subsonic)}$
$M_B = 4.07 > 1 \text{ (supersonic)}$

13–42. $p_c = 474 \text{ kPa}$
$M_C = 0.661$
$M_B = 0.158$
$p_B = 466 \text{ kPa}$

13–43. $0.0519 \text{ kg}/\text{s}$

13–45. $0.964 \text{ kg}/\text{s}$

13–46. $p = 90 < p_B,$ **the nozzle will choke.**
$\dot{m} = 0.120 \text{ kg}/\text{s}$

13–47. $M_C = 0.861$
$M_B = 0.190$

13–49. $d_t = 17.4 \text{ mm}$
$d_B = 22.6 \text{ mm}$
$T_O = 302 \text{ K}$
$T_B = 168 \text{ K}$

13–50. $1.42 \text{ kg}/\text{s}$

13–51. $p_B = 136 \text{ kPa}$
$V_B = 66.7 \text{ m}/\text{s}$

13–53. $0.0186 \text{ kg}/\text{s}$

13–54. $T_e = 288 \text{ K}$
$V_e = 251 \text{ m}/\text{s}$
$p_e = 138 \text{ kPa}$

13–55. $\dot{m} = 13.1 \text{ kg}/\text{s}$
$F_f = 654 \text{ N}$

13–57. $\dot{m} = 4.77 \text{ kg}/\text{s}$
$L_{max} = 215 \text{ m}$
$F_f = 913 \text{ N}$

13–58. $T_1 = 278 \text{ K}$
$p_1 = 249 \text{ kPa}$
$\rho_1 = 3.12 \text{ kg}/\text{m}^3$
$\dot{m} = 1.07 \text{ kg}/\text{s}$
$T^* = 244 \text{ K}$
$V^* = 313 \text{ m}/\text{s}$
$p^* = 122 \text{ kPa}$

13–59. $p_0^* = 231 \text{ kPa}$
$T_0^* = 293 \text{ K}$
$75.1 \text{ J}/(\text{kg} \cdot \text{K})$

13–61. $T = 452 \text{ K}$
$V = 238 \text{ m}/\text{s}$

13–62. $\dot{m} = 3.31 \text{ kg}/\text{s}$
$L = 4.81 \text{ m}$

13–63. $\dot{m} = 3.31 \text{ kg}/\text{s}$
$L = 4.87 \text{ m}$
$p_2 = 97.4 \text{ kPa}$

13–65. $\dot{m} = 6.91 \text{ kg}/\text{m}^3$

13–66. $T_1 = 259 \text{ K}$
$V_1 = 181 \text{ m}/\text{s}$

13–67. $T_1 = 147 \text{ K}$
$p_1 = 11.2 \text{ kPa}$

13–69. $\Delta s = 943 \text{ J}/\text{kg} \cdot \text{K}$
$(T_0)_1 = 275 \text{ K}$
$(T_0)_2 = 660 \text{ K}$

13–70. $T_1 = 153$ K
$p_1 = 42.4$ kPa
$\rho_1 = 0.936$ kg/m^3

13–71. 6.45 kg/s

13–73. $(T_0)_1 = 557$ K
$(T_0)_2 = 648$ K
$\Delta s = 213$ J/(kg $\cdot$ K)

13–74. 183 kg/s

13–75. $\dot{m} = 183$ kg/s

13–77. For subsonic flow,
$p_e = 798$ kPa
For supersonic flow,
$p_e = 6.85$ kPa
$V = 766$ m/s

13–78. $M_1 = 1.87$
$M_2 = 0.601$
$T_2 = 552$ K
$p_2 = 314$ MPa
$V_2 = 283$ m/s

13–79. $p_2 = 1.13$ MPa
$T_2 = 971$ K

13–81. $p_b < 32.2$ kPa

13–82. $V_e = 659$ m/s

13–83. $d_t = 254$ mm
$d_e = 272$ mm

13–85. 261 m/s

13–86. 4.28 s

13–87. 177 kPa $< p_b <$ 591 kPa

13–89. 17.9 kPa $< p_b <$ 177 kPa

13–90. 176 kPa

13–91. $V_s = 445$ m/s
$p_2 = 179$ kPa

13–93. 126 kPa

13–94. $p_A = 95.9$ kPa
$T_A = 286$ K

13–95. $p_B = 123$ kPa
$T_B = 307$ K

13–97. 149 kPa

13–98. $T_2 = 231$ K
$p_2 = 45.3$ kPa

13–99. $\beta = 35.2°$
$p_2 = 641$ kPa
$T_2 = 499$ K

13–101. $p_A = 105$ kPa
$p_B = 68.5$ kPa

Chapter 14

14–1. $\beta_1 = 22.6°$
$(V_{rel})_1 = 13.0$ m/s

14–2. $(V_{rel})_2 = 8.72$ m/s
$V_2 = 6.97$ m/s

14–3. $(V_t)_2 = 1.26$ m/s
$V_2 = 1.29$ m/s

14–5. 12.4 m/s

14–6. $V_2 = 8.87$ m/s

14–7. 10.1 kW

14–9. 54.5 N $\cdot$ m

14–10. 1.81 kN $\cdot$ m

14–11. 0.588 m^3/s

14–13. 81.7 m

14–14. $T = 17.5$ N $\cdot$ m
$(\dot{W}_s)_{pump} = 1.69$ kW
$h_{pump} = 2.80$ m

14–17. 18.9 m

14–18. 59.5 N $\cdot$ m

14–19. $V_2 = 16.2$ m/s
$\dot{W}_s = 8.44$ kW
$h_{pump} = 28.7$ m

14–21. $W = 43.3$ kW
$\omega = 7.5$ rad/s

14–22. $T = 8.21$ kN $\cdot$ m
$\dot{W}_{turb} = 82.1$ kW

14–23. $T = 2.10$ kN $\cdot$ m
$\dot{W}_{turb} = 10.3$ kW

14–25. $T = -17.0$ kN $\cdot$ m

14–26. $\beta_1 = 41.0°$
$V_2 = 4.16$ m/s

14–27. 1.70 m/s

14–29. $\dot{W}_s = -480$ kW

14–30. $\dot{W} = 598$ kW

14–31. 167 kW

14–33. $T = 230$ N $\cdot$ m
$\dot{W}_s = 9.19$ kW

14–34. 0.624

14–35. 0.03375 m^3/s

14–37. $Q = 1560$ liters/min

14–38. $h = 26.8$ m

14–39. $\Delta H_2 = 0.0213$ m

14–41. 135 mm

Index

Absolute and gage pressure, 48–49
   standard atmospheric pressure, 48
   zero absolute pressure, 48
Absolute temperature, 7
   international system of units, 7
Acceleration, streamline coordinates, 149–150
   convective change, 150
   local change, 149
   resultant acceleration, 150
Adiabatic process, 662
Adverse pressure gradient, 555
Air at U.S. standard atmospheric pressure
   vs. altitude, physical properties of, 792
   vs. temperature, physical properties of, 792
Airfoil lift, 572–573
Airfoils, 569
Analysis and design for pipe flow, 479–523
   flow measurement, 508–512
   losses occurring from pipe fittings and transitions, 490–495
   pipe systems, 502–507
   resistance to flow in rough pipes, 479–489
   single-pipeline flow, 496–501
Anemometer, 510
Angular distortion, 327
Angular momentum, 755
Angular momentum equation, 286–293
   definition, 286
   steady flow, 287
Angular motion, 749
Apparent shear stress, 458
Applications of Bernoulli Equation, 214–225
   flow around a curved boundary, 215
   flow from a large reservoir, 214
   flow in closed conduit, 216
   flow in an open channel, 215
   Toricelli's law, 214
   venture meter, 217

Average angular velocity, 326
Average velocity, 175–176
Axial-flow machine, 747
Axial-flow pump, 748–753
   angular motion, 749
   continuity, 749
   Euler turbomachine equation, 749
   flow kinematics, 750
   impeller, 748
   power, 750
   stator vanes, 748
   velocity kinematic diagrams, 750

Barometer, 56
Basic two-dimensional flows, 349–359
   doublet, 354–355
   forced-vortex flow, 357–359
   free-vortex flow, 356
   line sink flow, 354
   line source flow, 352–353
   uniform flow, 350–351
Bends, 492
Bernoulli equation, 211–213, 337–339
   definition, 211
   and differential fluid flow, 337–339
   interpretation of terms, 212
   limitations, 213
   propellers, 271
Best efficiency point (BEP), 769
Best hydraulic cross section, 622
Betz's law, 273
Bluff body, 554
Bodies at rest, 271–281
Bodies having constant velocity, 281–286
Boundary layer, 525–530
   classification, 530
   description, 526
   thickness, 527–528
Bourdon gage, 60
Brake horsepower, 767
Broad-crested weir, 642
Broad-crested weir coefficient, 642
Brookfield viscometer, 22
Buckingham pi theorem, 399–407

Bulk modulus, 14
   gas, 15
   liquid, 14
Bump, channel flow, 612–616
Buoyancy, 85–87
   buoyant force, 85
   center of buoyancy, 85
   hydrometer, 86
   principle of buoyancy, 85
Buoyant force, 85
Butterfly valve, 494

Calculations, fluid mechanics, 8–9
   accuracy, 9
   dimensional homogeneity, 8–9
   procedure, 9
Canals, 601
Cavitation and net positive suction head, 770–771
   critical suction head, 770
   net positive suction head, 770
   vapor pressure, 770
   vapor pressure head, 770
Center of buoyancy, 85
Centrifugal pump, 754
Characteristics of matter, 5–6
   continuum, 6
   fluid, 5
   gas, 5
   liquid, 5
   solid, 5
Chézy equation, 621
Chézy formula, 621
Circulation, 328, 574
Coanda effect, 556
Compressible flow, 657–745
   compression and expansion waves, 723–727
   friction, effect on, 694–703
   heat transfer, effect on, 704–709
   isentropic flow through converging and diverging nozzles, 685–693
   isentropic flow through variable area, 680–685

Compressible flow (continued)
  measurement, 728–730
  normal shock waves, 710–713
  oblique shock waves, 74–772
  shock waves in nozzles, 713–717
  stagnation properties, 673–679
  thermodynamics concepts,
    657–665
  types, 669–672
  wave propagation through
    compressible fluid, 666–668
Compressible fluid, 238
Compression and expansion waves,
    723–727
  Prandtl-Meyer expansion
    function, 725
Computational fluid dynamics, 137,
    374–375
Conservation of mass, 165–205,
    332–333
  continuity equation, 178
  cylindrical coordinates, 333
  finite control volumes, 165–167
  requirements, 178–185
  Reynolds transport theorem,
    168–173
  special cases, 178
  two-dimensional steady flow of
    ideal fluid, 333
  volumetric flow, mass
    flow, and average
    velocity, 174–177
Constant horizontal acceleration, 91
Constant-temperature
    anemometer, 511
Constant temperature, compressible
    fluids, 53–55
Constant translational acceleration of
    a liquid, 91–95
  constant horizontal
    acceleration, 91
  constant vertical acceleration,
    92–93
Constant vertical acceleration, 92–93
Constant-volume process, 659
Continuity equation, 634, 667,
    694, 710, 719
Continuum, characteristics, 6

Control volumes having accelerated
    motions, 275
Convective change, 150
Converging-diverging nozzle,
    isentropic flow, 687–688
Converging nozzle, isentropic
    flow, 686
Corrective acceleration, 143
Couette flow, 438
Critical flow, 604
Critical slope, 623
Critical suction head, 770
Culverts, 601
Curl, 327
Cylinder, drag coefficient, 560

Darcy-Weisbach equation, 481
Density, 12
  gas, 12
  liquid, 12
Differential fluid flow, 323–391
  basic two-dimensional flows,
    349–359
  Bernoulli equation, 337–339
  circulation, 328
  computational fluid dynamics,
    374–375
  conservation of mass, 332–333
  differential analysis, 323–324
  equations of motion for a fluid
    particle, 334–335
  Euler equations of motion, 336
  irrotational flow, 329
  kinematics of differential fluid
    elements, 324–327
  Navier–Stokes equations, 370–373
  other applications, 367–369
  potential function, 345–348
  rotational flow, 329
  stream function, 340–344
  superposition of flows, 360–369
  two-dimensional steady
    flow, 336
  vorticity, 329
Differential manometer, 59
Dimensional analysis and similitude,
    393–431
  Buckingham pi theorem, 399–407

dimensional analysis,
    definition, 393
  dimensional homogeneity, 394
  dimensionless groups, 393
  dimensionless numbers,
    396–398
  general considerations, 408
  similitude, 409–420
Dimensional flow, 132
Dimensional homogeneity, 8–9, 394
Dimensionless groups, 393
Dimensionless numbers, 396–398
  Euler number, 396
  Froude number, 397
  Mach number, 398
  Reynolds number, 397
  Weber number, 398
Displacement thickness, 532
Disturbance thickness, 532
Divergence, 325
Doublet, 354–355
Drag and lift, 552–554
Drag coefficient, 558–562
  cylinder, 560
  Froude number, 561
  Mach number, 561
  Reynolds number, 559–560
  sphere, 560
  wave drag, 561
Drag coefficients for bodies
    having various
    shapes, 562–568
Dynamic fluid devices, 748
Dynamic force, 270
Dynamic pressure, 215
Dynamic similitude, 410–411
Dynamic viscosity, 19

Eddy currents, 462
Energy and hydraulic grade lines,
    226–234
  energy grade line, 226
  hydraulic grade line, 226
  hydraulic head, 226
  kinetic head, 226
  pressure head, 226
  total head, 226
  velocity head, 226

Energy equation, 234–247, 636
  compressible flow, 711–713, 719
  compressible fluid, 238
  gravitational energy, 234
  heat energy, 235
  incompressible flow, 237
  internal energy, 234
  kinetic energy, 234
  nonuniform velocity, 238–239
  power and efficiency, 238
  work, 235
Energy grade line, 226
Enthalpy, 238, 659
Entropy, 660
Equation of state, 658
Equivalent length ratio, 495
Euler number, 396
Euler turbomachine equation, 749
Eulerian description control volume
    approach, 130
Euler's equations of motion,
    207–210, 336
  differential fluid flow, 336
  n direction, 208
  s direction, 208
  steady horizontal flow of ideal
      liquid, 209
Expansion and contraction, 492
Extensive property, 168

Fanno flow, 694, 797
Fanno line, 699
Favorable pressure gradient, 555
Finite control volumes, 165–167
  open control surfaces, 166
  steady flow, 167
  velocity, 167
First law of thermodynamics, 658
Flow
  in a closed conduit, 216
  around a curved boundary, 215
  around a cylinder, 364–365
  past a half body, 360–361
  from a large reservoir, 214
  around a Rankine oval, 362–363
  in an open channel, 215
Flow coefficient, 490, 775
Flow kinematics, 750

Flow measurement, 508–512
  anemometer, 510
  laser Doppler flow meter, 512
  magnetic flow meter, 512
  nozzle meter, 509
  nutating disk meter, 511
  orifice discharge coefficient, 509
  orifice meter, 509
  particle image velocimetry, 512
  piezometer rings, 508
  positive displacement flow
      meter, 511
  rotometer, 510
  shedder bar, 510
  Strouhal number, 510
  thermal mass flow meter, 511
  turbine flow meter, 510
  ultrasonic flow meter, 512
  Venturi discharge coefficient, 508
  Venturi meter, 508
  Von Kármán vortex street, 510
  vortex flow meter, 510
  wafer-style magnetic flow
      meter, 512
Flow net, 346
Flow within casing, 756
Flow work, 235
Fluid acceleration, 142–148
  corrective acceleration, 143
  local acceleration, 143
  material derivative, 142, 144
  three-dimensional flow, 143
Fluid, characteristics, 5
Fluid dynamics, 4
Fluid flow classifications, 131–133
  dimensional flow, 132
  frictional effects, 131
  space and time, 133
Fluid flow descriptions, 129–130
  Eulerian description control
      volume approach, 130
  Lagrangian description systems
      approach, 130
Fluid kinematics, 4
Fluid mechanics, 3–43
  basic fluid properties, 12–17
  branches, 4
  calculations, 8–9

characteristics of matter, 5–6
historical development, 4
international system of units, 6–9
introduction, 3–4
problem solving, 10–11
surface tension and capillarity,
    27–30
vapor pressure, 26
viscosity, 17–21
viscosity measurement, 22–25
Fluid momentum, 269–321
  angular momentum equation,
      286–293
  applications for control volumes
      having accelerated
      motions, 275
  applications to bodies at rest,
      271–281
  applications to bodies having
      constant velocity, 281–286
  linear momentum equation,
      269–270
  propellers, 270–272
  rockets, 277–279
  turbojets and turbofans, 276–277
  wind turbine, 273–274
Fluid properties, 12–17
  bulk modulus, 14
  density, 12
  ideal gas law, 13
  specific gravity, 13
  specific weight, 12–13
Fluid statics, 45–127
  absolute and gage pressure, 48–49
  buoyancy, 85–87
  constant translational accelera-
      tion of a liquid, 91–95
  hydrostatic force on a plane
      surface formula method, 64–69
  hydrostatic force on a plane sur-
      face geometric method, 70–74
  hydrostatic force on a plane sur-
      face integration method, 75–77
  hydrostatic forces on an inclined
      plane or curved surface, 78–84
  measurement of static pressure,
      56–63
  pressure, 45–47

Fluid statics (continued)
  pressure variation for compress-
    ible fluids, 53–55
  pressure variation for incom-
    pressible fluids, 51–52
  stability, 88–90
  static pressure variation, 50
  steady rotation of a liquid, 96–99
Flume, 601
Forced vortex, 96
Forced-vortex flow, 357–359
Formula method, hydrostatic force on
    a plane surface, 64–69
  center of pressure, 65
  resultant force, 64
  resultant force, location, 65
  x_P coordinate, 66
Free-body diagram, 270
Free-surface energy, 27
Friction drag, 534
Friction drag coefficient, 534
Friction, effect on compressible
    flow, 694–703
  continuity equation, 694
  density, 699
  energy equation, 696
  Fanno flow, 694
  Fanno line, 699
  ideal gas law, 696
  linear momentum equation, 695
  pipe length versus Mach
    number, 697
  pressure, 698
  temperature, 698
Friction factor, 480
Frictional effects, flow, 131
Froude number, 397, 561, 604
Froude's theorem, 271
Fully developed flow from an
    entrance, 455–456
  definition, 455
  entrance length, 455
  turbulent flow, 455
Fused quartz force-balance Bourdon
    tube, 60

Gas, 5, 12, 13
  bulk modulus, 13

characteristics, 5
density, 12
Gases at standard atmospheric
    pressures, physical properties
    of, 790
Gate valve, 494
Geometric method, hydrostatic force
    on a plane surface, 70–74
  plate having constant width, 71
  resultant force, 70
Geometric similitude, 409
Globe valve, 494
Gradual flow with varying depth,
    627–633
  rectangular cross section, 628
  surface profile calculation, 630
  surface profiles, 628
Graphical descriptions, kinematic fluid
    motion, 134–141
  computational fluid dynamics, 137
  optical methods, 136
  pathlines, 135
  streaklines, 135
  streamlines, 134
  streamtubes, 135
Gravitational energy, 234

Half body, 361
Head coefficient, 775
Head-discharge curve, 757
Head loss, 237
Head loss, efficiency, 757
Heat energy, 235
Heat transfer, effect on compressible
    flow, 704–709
  energy equation, 705
  ideal gas law, 705
  linear-momentum equation, 704
  Rayleigh flow, 695
  Rayleigh line, 707
  stagnation temperature and
    pressure, 706
Horizontal flow caused by constant
    pressure gradient/both plates
    moving, 436
Horizontal flow caused by constant
    pressure gradient/top plate
    moving, 437

Horizontal flow caused only by
    motion of top plate, 438
Hot-film anemometer, 511
Hydraulic diameter, 483
Hydraulic efficiency, 757
Hydraulic grade line, 226
Hydraulic head, 226
Hydraulic jump, 602, 634–638
  continuity equation, 634
  energy equation, 636
  momentum equation, 635
Hydraulics, 4
Hydrodynamics, 4, 324
Hydrometer, 86
Hydrostatic force on a plane surface
    formula method, 64–69
  center of pressure, 65
  resultant force, 64
  resultant force, location, 65
  x_P coordinate, 66
Hydrostatic force on a plane surface
    geometric method, 70–74
  plate having constant
    width, 71
  resultant force, 70
Hydrostatic force on a plane surface
    integration method, 75–77
  location, 75
  resultant force, 75
Hydrostatic forces on an inclined
    plane or curved
    surface, 78–84
  horizontal component, 78
  vertical component, 79
Hydrostatics, 4
Hypersonic motion, 670

Ideal flow around a cylinder, 555
  adverse pressure gradient, 555
  Coanda effect, 556
  favorable pressure gradient, 555
  pressure gradient, 555
  real flow around a cylinder,
    556–557
  vortex shedding, 557
Ideal gas law, 13
  and compressible flow,
    696, 705, 711

Ideal performance for pumps, 756–761
    head-discharge curve, 757
    head loss, efficiency, 757
    hydraulic efficiency, 757
    ideal pump head, 756
    pump efficiency, 757
Ideal pump head, 756
Impeller, 748
Impulse turbine, 761
Incompressible flow, 237
    head loss, 237
    pump head, 237
    turbine head, 237
Induced drag, 576
Induced drag coefficient, 577
Inlet and exit transitions, 491
Integration method, hydrostatic force
    on a plane surface, 75–77
    location, 75
    resultant force, 75
Intensive properties, 168
Internal energy, 234, 658
International system of units, 6–9
    absolute temperature, 7
    temperature, 7
    weight, 6
    weight, 7
Inviscid and ideal fluids, 20
Irrotational flow, 329
Isentropic flow through
    converging and diverging
    nozzles, 685–693
    converging, 686
    converging-diverging, 687–688
Isentropic flow through variable
    area, 680–685
    area ratios, 682–683
    laval nozzle, 682
    linear momentum equation,
    680–681
    subsonic flow, 681
    supersonic flow, 681
Isentropic process, 662
Isentropic relations, 793–796

Kaplan turbine, 764
Kinematic similitude, 410
Kinematic viscosity, 21

Kinematics of differential fluid
    elements, 324–327
    angular distortion, 327
    rotation, 326–327
    translation and linear
    distortion, 325
Kinematics of fluid motion, 129–163
    fluid acceleration, 142–148
    fluid flow descriptions, 129–130
    graphical descriptions, 134–141
    streamline coordinates, 149–152
    types of fluid flow, 131–133
Kinetic energy, 234
Kinetic energy coefficient, 238
Kinetic head, 226
Kutta-Joukowski theorem, 574

Lagrangian description systems
    approach, 130
Laminar and turbulent boundary
    layers, 546–551
Laminar and turbulent shear stress
    within smooth pipe, 457–460
    laminar flow, 457
    turbulent flow, 457–458
    turbulent shear stress, 458–460
Laminar boundary layers, 531–539
    displacement thickness, 532
    disturbance thickness, 532
    friction drag, 534
    friction drag coefficient, 534
    momentum thickness, 533
    shear stress, 533
    skin friction coefficient, 534
Laminar flow, 457, 480, 482, 602
Laminar viscous sublayer, 458
Laser Doppler flow meter, 512
Laval nozzle, 682
Lift and drag on an airfoil, 572–581
    airfoil lift, 572–573
    circulation, 574
    experimental data, 575
    Kutta-Joukowski theorem, 574
    lift coefficient, 575
    Magnus effect, 578
    race cars, 575
    spinning ball, 578
    stall, 575

trailing vortices and induced
    drag, 576
Lift coefficient, 575
Line sink flow, 354
Line source flow, 352–353
Linear momentum, 270–271
Linear momentum equation, 269–270,
    667, 680–681
    and compressible flow, 695,
    704, 710
    dynamic force, 270
    free-body diagram, 270
    steady flow, 270
Liquid, 5, 12, 13
    bulk modulus, 14
    characteristics, 5
    density, 12
Liquid drops, 28
Liquids at standard atmospheric
    pressure, physical properties
    of, 790
Local acceleration, 143
Local change, 149
Loss coefficient, 490
Losses occurring from pipe fittings
    and transitions, 490–495
    bends, 492
    equivalent length ratio, 495
    expansion and contraction, 492
    flow coefficient, 490
    inlet and exit transitions, 491
    loss coefficient, 490
    pipe connections, 495
    resistance, 490
    secondary flow, 492
    threaded fittings, 493
    valves, 494
    vena contracta, 491

Mach cone, 671
Mach number, 398, 561
Magnetic flow meter, 512
Magnus effect, 578
Major head loss, 479
Major loss, 479
Manning equation, 622
Manometer, 57
Manometer rule, 58

Manufacturer's pump performance curves, 769
Mass flow, 176
Material derivative, 142, 144
Mean steady flow, 458
Measurement of static pressure, 56–63
  barometer, 56
  Bourdon gage, 60
  differential manometer, 59
  fused quartz force-balance Bourdon tube, 60
  manometer, 57
  manometer rule, 58
Meniscus, 28
Metacenter, 89
Methods for reducing drag, 569–572
  airfoils, 569
  road vehicles, 571
  section drag, 570
  section drag coefficient, 570
Mixed-flow machine, 747
Momentum equation, 540–541, 636
Momentum integral equation, 540–543
  continuity equation, 540
  momentum equation, 540–541
  momentum integral equation, 541
Momentum thickness, 533
Moody diagram, 482
  laminar flow, 482
  surface roughness, 482
  transitional flow, 482
  turbulent flow, 482

N direction, 208
Nappe, 639
Navier–Stokes equations, 370–373
  cylindrical coordinates, 372
  definition, 371
Navier–Stokes solution for steady laminar flow between parallel plates, 439–444
Navier–Stokes solution for steady laminar flow within a smooth pipe, 448–449
Net positive suction head, 770
Neutral equilibrium, 88
Newtonian fluids, 20

Newton's law, viscosity, 18–19
  dynamic viscosity, 19
  shear strain, 19
  shear stress, 19
  velocity gradient, 19
Non-Newtonian fluids, 20
Noncircular conduits, 483
Nonuniform velocity, 238–239
  kinetic energy coefficient, 238
  quasi-steady flow, 238
Nonwetting liquid, 28
Normal shock relations, 799–801
Normal shock waves, 710–713
  continuity equation, 710
  energy equation, 711–713
  ideal gas law, 711
  linear momentum equation, 710
Normal stress, 334
Nozzle meter, 509
Nutating disk meter, 511

Oblique shock waves, 74–772
  continuity equation, 719
  energy equation, 719
Open-channel flow, 413, 601–655
  bump, 612–616
  classifications, 603–604
  gradual flow with varying depth, 627–633
  hydraulic jump, 634–638
  under sluice gate, 616–619
  specific energy, 604–612
  steady uniform, 620–626
  types, 601–602
  weirs, 639–643
Open control surfaces, 166
Operating point, 772
Optical methods, description of kinematic fluid motion, 136
  schlieren photography, 136
  shadowgraph, 136
Orifice discharge coefficient, 509
Orifice meter, 509
Ostwald viscometer, 23

Paraboloid of revolution, 97
Particle image velocimetry, 512
Pascal's law, 46–47

Pathlines, 135
Pelton wheel, 761–763
Performance characteristics, 767
Performance curves, 768
Physical properties
  air at U.S. standard atmospheric pressure vs. altitude, 792
  air at U.S. standard atmospheric pressure vs. temperature, 792
  gases at standard atmospheric pressures, 790
  liquids at standard atmospheric pressure, 790
  water versus temperature (SI units), 791
Piezometer, 216, 728
Piezometer rings, 508
Pipe connections, 495
Pipe length versus Mach number, 697
Pipe systems, 502–507
  parallel, 503
  series, 502
Pitot-static tube, 216
Pitot tube, 215, 728
Positive displacement flow meter, 511
Positive-displacement pump, 748
Potential function, 345–348
  flow net, 346
Power and efficiency, 238
Power coefficient, 775
Power LAW approximation, 462–463
Prandtl-Meyer expansion, 802
Prandtl-Meyer expansion function, 725
Prefixes, measurement units, 7–8
Pressure, 45–47
  definition, 45
  Pascal's law, 46–47
Pressure and temperature effects, viscosity, 20–21
Pressure drag, 554
Pressure gradient effects, 554–558
  bluff body, 554
  ideal flow around a cylinder, 555
  pressure drag, 554
  shear drag, 554
Pressure head, 52, 226

Pressure variation for compressible fluids, 53–55
constant temperature, 53–55
definition, 53
Pressure variation for incompressible fluids, 51–52
definition, 51
pressure head, 52
Principle of buoyancy, 85
Prismatic channel, 601
Problem solving, fluid mechanics, 10–11
Propeller turbine, 764
Propellers, 270–272
Bernouli equation, 271
Froude's theorem, 271
linear momentum, 270–271
power and efficiency, 272
Pump affinity laws, 775
Pump efficiency, 757
Pump head, 237
Pump performance, 767–769
best efficiency point (BEP), 769
brake horsepower, 767
manufacturer's pump performance curves, 769
performance characteristics, 767
performance curves, 768
shaft horsepower, 767
Pump scaling laws, 775
Pump selection related to flow system, 772–773
operating point, 772

Quasi-steady flow, 238

Race cars, 575
Radial-flow machine, 747
Radial-flow pumps, 754–756
angular momentum, 755
centrifugal pump, 754
continuity, 754
flow within casing, 756
power, 755
volute pump, 754
Rankine oval, 363
Rapid flow, 604
Rayleigh flow, 695, 798

Rayleigh line, 707
Reaction turbines, 764–766
Real flow around a cylinder, 556–557
Rectangular opening, 640
Resistance, 490
Resistance to flow in rough pipes, 479–489
Darcy-Weisbach equation, 481
empirical solutions, 483
friction factor, 480
hydraulic diameter, 483
laminar flow, 480
major head loss, 479
major loss, 479
Moody diagram, 482
noncircular conduits, 483
turbulent flow, 480
Resultant acceleration, 150
Reynolds number, 397, 450–454, 620
critical Reynolds number, 451
critical velocity, 451
drag coefficient, 559–560
Reynolds stress, 459
Reynolds transport theorem, 168–173
applications, 171
definition, 171
extensive property, 168
intensive properties, 168
Rigid-body translation, 325
Rise, channel flow, 612–616
Road vehicles, 571
Rockets, 277–279
Rotation, 326–327
average angular velocity, 326
curl, 327
Rotational flow, 329
Rotational viscometer, 22
Rotometer, 510
Runners, 764

S direction, 208
Schlieren photography, 136
Second law of thermodynamics, 660
Secondary flow, 492
Section drag, 570
Section drag coefficient, 570
Shadowgraph, 136
Shaft horsepower, 767

Shaft work, 235
Sharp-crested weir, 639
Shear drag, 554
Shear strain, 19
Shear stress, 19, 334, 533
Shear velocity, 461
Shear work, 235
Shedder bar, 510
Ships, 414–415
Shock wave, 670
Shock waves in nozzles, 713–717
Similitude, 409–420
definition, 409
dynamic similitude, 410–411
geometric similitude, 409
kinematic similitude, 410
open-channel flow, 413
ships, 414–415
steady flow through a pipe, 412
Single-pipeline flow, 496–501
Skin friction coefficient, 534
Sluice gate, 616–619
Solid, characteristics, 5
Sonic and supersonic flow, 670
hypersonic motion, 670
shock wave, 670
Sonic velocity, 666
Space and time, flow, 133
Specific energy, channel flow, 604–612
definition, 605
nonrectangular cross section, 607
specific-energy diagram, 605
specific head, 605
standing waves, 607
undulations, 607
Specific-energy diagram, 605
Specific gravity, 13
Specific head, 605
Specific heat, 658
Specific speed, 776
Speed of sound, 666
Sphere, drag coefficient, 560
Spinning ball, 578
Stability, 88–90
metacenter, 89
neutral equilibrium, 88
stable equilibrium, 88
unstable equilibrium, 88

Stable equilibrium, 88
Stagnation pressure, 215
Stagnation properties, 673–679
 density, 674
 pressure, 674
 temperature, 673
Stagnation tube, 215
Stall, 575
Standard atmospheric pressure, 48
Standing waves, 607
State diagram, 661
State properties, 658
Static pressure variation, 50
Stator vanes, 748
Steady flow, 167, 602
Steady flow through a pipe, 412
Steady horizontal flow of ideal liquid, 209
Steady laminar flow between parallel
  plates, 433–438
 Couette flow, 438
 horizontal flow caused by con-
  stant pressure gradient/both
  plates moving, 436
 horizontal flow caused by con-
  stant pressure gradient/top
  plate moving, 437
 horizontal flow caused only by
  motion of top plate, 438
 limitations, 438
Steady rotation of a liquid, 96–99
 forced vortex, 96
 paraboloid of revolution, 97
Steady turbulent flow, 458
Steady uniform channel flow, 620–626
 best hydraulic cross section, 622
 Chézy equation, 621
 Chézy formula, 621
 critical slope, 623
 Manning equation, 622
 Reynolds number, 620
 surface roughness coefficient, 622
Streaklines, 135
Stream function, 340–344
 velocity components, 340–341
 volumetric flow, 341–344
Streamline coordinates, 149–152
 acceleration, 149
 velocity, 149

Streamlines, 134
Streamtubes, 135
Stress field, 334
Strouhal number, 510, 558
Subsonic flow, 669, 681
Superposition of flows, 360–369
 flow around a cylinder, 364–365
 flow around a Rankine oval,
  362–363
 flow past a half body, 360–361
 half body, 361
 Rankine oval, 363
 uniform and free-vortex flow
  around a cylinder, 366–367
Supersonic flow, 681, 728
Suppressed rectangular weir, 640
Surface roughness, 482
Surface roughness coefficient, 622
Surface tension and capillarity, 27–30
 definition, 27
 free-surface energy, 27
 liquid drops, 28
 meniscus, 28
 nonwetting liquid, 28
 wetting liquid, 28, 29
Swing check valve, 494
Systems of units
 prefixes, 7–9

T-s diagram, 661
Temperature, 7
 international system of units, 7
Thermal mass flow meter, 511
 constant-temperature
  anemometer, 511
 hot-film anemometer, 511
Thermodynamics concepts, 657–665
 adiabatic process, 662
 constant-volume process, 659
 enthalpy, 659
 entropy, 660
 equation of state, 658
 first law of thermodynamics, 658
 internal energy, 658
 isentropic process, 662
 second law of
  thermodynamics, 660
 specific heat, 658

 state diagram, 661
 state properties, 658
Threaded fittings, 493
Three-dimensional flow, 143
Toricelli's law, 214
Total head, 226
Total pressure, 215
Trailing vortices and induced drag, 576
 induced drag, 576
 wing vortex trail, 576
 winglets, 576
Tranquil flow, 604
Transitional and turbulent flow
  region, 462
Transitional flow, 482
Translation and linear distortion, 325
 divergence, 325
 rigid-body translation, 325
 volumetric dilatation rate, 325
Turbine flow meter, 510
Turbine head, 237
Turbines, 761–766
 impulse turbine, 761
 Kaplan turbine, 764
 Pelton wheel, 761–763
 propeller turbine, 764
 reaction turbines, 764–766
 runners, 764
Turbojets and turbofans, 276–277
Turbomachine similitude, 774–779
 efficiency, 775
 flow coefficient, 775
 head coefficient, 775
 power coefficient, 775
 pump affinity laws, 775
 pump scaling laws, 775
 specific speed, 776
Turbomachines, 746–789
 axial-flow pump, 748–753
 cavitation and net positive
  suction head, 770–771
 ideal performance for pumps,
  756–761
 pump performance, 767–769
 pump selection related to flow
  system, 772–773
 radial-flow pumps, 754–756
 turbines, 761–766

Turbomachines (continued)
     turbomachine similitude, 774–779
     types, 747–748
Turbulent boundary layers, 544–545
     drag on plate, 545
     shear stress along plate, 545
Turbulent flow, 455, 457–458, 480, 482
     laminar viscous sublayer, 458
     mean steady flow, 458
     steady turbulent flow, 458
Turbulent flow within smooth
          pipe, 460–465
     eddy currents, 462
     power LAW approximation,
          462–463
     shear velocity, 461
     transitional and turbulent flow
          region, 462
     viscous sublayer, 461
Turbulent shear stress, 458–460
     apparent shear stress, 458
     Reynolds stress, 459
Two-dimensional steady flow, 336
Two-dimensional steady flow of ideal
          fluid, 333

Ultrasonic flow meter, 512
Undulations, 607
Uniform and free-vortex flow around
          a cylinder, 366–367
Uniform flow, 602
Unstable equilibrium, 88

Valves, 494
     butterfly valve, 494
     gate valve, 494
     globe valve, 494
     swing check valve, 494
Vapor pressure, 26, 770
Vapor pressure head, 770
Velocity, 167
Velocity coefficient, 729
Velocity components, 340–341
Velocity gradient, 19
Velocity head, 226
Velocity kinematic diagrams, 750
Velocity, streamline coordinates, 149
Vena contracta, 491

Venture meter, 217
Venturi discharge coefficient, 508
Venturi meter, 508, 729
Viscosity, 17–21
     inviscid and ideal fluids, 20
     kinematic viscosity, 21
     Newtonian fluids, 20
     Newton's law, 18–19
     non-Newtonian fluids, 20
     physical cause, 18
     pressure and temperature effects,
          20–21
Viscosity measurement, 22–25
     Brookfield viscometer, 22
     Ostwald viscometer, 23
     rotational viscometer, 22
Viscous flow over external surfaces,
          525–599
     boundary layer, 525–530
     drag and lift, 552–554
     drag coefficient, 558–562
     drag coefficients for bodies hav-
          ing various shapes, 562–568
     induced drag coefficient, 577
     laminar and turbulent boundary
          layers, 546–551
     laminar boundary layers, 531–539
     lift and drag on an airfoil,
          572–581
     methods for reducing drag,
          569–572
     momentum integral equation,
          540–543
     pressure gradient effects, 554–558
     turbulent boundary layers, 56–589
Viscous flow within enclosed surfaces,
          433–477
     fully developed flow from an
          entrance, 455–456
     laminar and turbulent shear
          stress within smooth pipe,
          457–460
     Navier–Stokes solution for steady
          laminar flow between parallel
          plates, 439–444
     Navier–Stokes solution for steady
          laminar flow within a smooth
          pipe, 448–449

Reynolds number, 450–454
     steady laminar flow between
          parallel plates, 433–438
     steady laminar flow within a
          smooth pipe, 444–447
     turbulent flow within a smooth
          pipe, 460–465
Viscous sublayer, 461
Volumetric dilatation rate, 325
Volumetric flow, 174–175, 341–344
Volute pump, 754
Von Kármán vortex street, 510, 558
Von Kármán vortex trail, 558
Vortex flow meter, 510
Vortex shedding, 557
     Strouhal number, 558
     Von Kármán vortex street, 558
     Von Kármán vortex trail, 558
Vorticity, 329

Wafer-style magnetic flow meter, 512
Water vs. temperature, physical prop-
          erties of, 791
Wave celerity, 603
Wave drag, 561
Wave propagation through compress-
          ible fluid, 666–668
     continuity equation, 667
     linear momentum equation, 667
     sonic velocity, 666
     speed of sound, 666
Weber number, 398
Weight, 7
     international system of units, 7
Weirs, 639–643
     broad-crested weir, 642
     broad-crested weir coefficient,
          642
     nappe, 639
     rectangular opening, 640
     sharp-crested weir, 639
     suppressed rectangular weir, 640
Wetting liquid, 28, 29
Wind turbine, 273–274
     Bette's law, 273
     capacity factor, 273
     power and efficiency, 273
Wing vortex trail, 576

Work, 235
    flow work, 235
    shaft work, 235
    shear work, 235
Work and energy of moving fluids,
        207–267
    applications of Bernoulli
        Equation, 214–225

Bernoulli equation, 211–213
energy and hydraulic grade lines,
        226–234
energy equation, 234–247
Euler's equations of motion,
        207–210

X_P coordinate, 66

Zero absolute pressure, 48

Geometric Properties of an Area

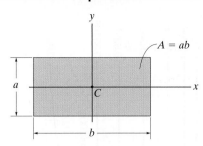

$A = ab$

$$I_x = \frac{1}{12}ba^3$$

$$I_y = \frac{1}{12}ab^3$$

Rectangle

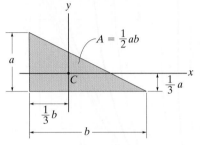

$A = \frac{1}{2}ab$

$$I_x = \frac{1}{36}ba^3$$

$$I_y = \frac{1}{36}ab^3$$

Triangle

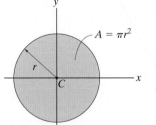

$A = \pi r^2$

$$I_x = \frac{1}{4}\pi r^4$$

$$I_y = \frac{1}{4}\pi r^4$$

Circle

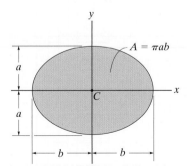

$A = \pi ab$

$$I_x = \frac{1}{4}\pi ba^3$$

$$I_y = \frac{1}{4}\pi ab^3$$

Ellipse

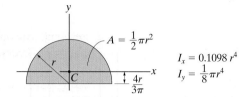

$A = \frac{1}{2}\pi r^2$

$$I_x = 0.1098\, r^4$$

$$I_y = \frac{1}{8}\pi r^4$$

Semicircle

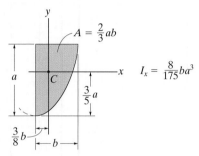

$A = \frac{2}{3}ab$

$$I_x = \frac{8}{175}ba^3$$

Parabola

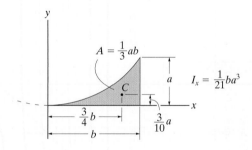

$A = \frac{1}{3}ab$

$$I_x = \frac{1}{21}ba^3$$

Exparabola

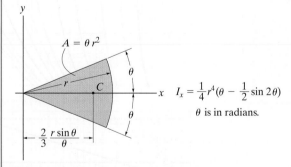

$A = \theta r^2$

$$I_x = \frac{1}{4}r^4\left(\theta - \frac{1}{2}\sin 2\theta\right)$$

θ is in radians.

Circular sector

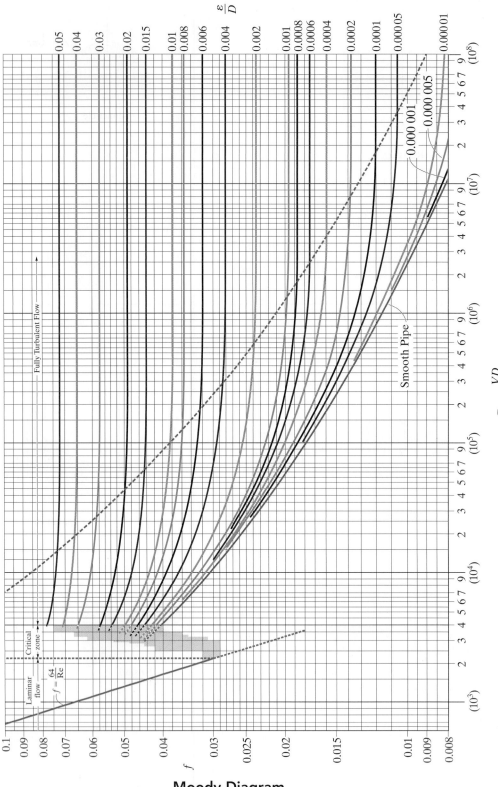

Roughness Factors for New Pipe

Concrete $\varepsilon = 0.3$ mm $- 3$ mm (0.001 ft $-$ 0.01 ft) Commercial steel $\varepsilon = 0.045$ mm (0.000 15 ft)

Cast iron $\varepsilon = 0.26$ mm (0.000 85 ft) Drawn tubing $\varepsilon = 0.0015$ mm (0.000 005 ft)

Galvanized iron $\varepsilon = 0.15$ mm (0.000 5 ft)

$$\frac{\varepsilon}{D}$$

0.05
0.04
0.03
0.02
0.015
0.01
0.008
0.006
0.004
0.002
0.001
0.0008
0.0006
0.0004
0.0002
0.0001
0.00005
0.00001

0.000 001
0.000 005

Smooth Pipe

Fully Turbulent Flow

Laminar flow

$f = \dfrac{64}{\text{Re}}$

Critical zone

$$\text{Re} = \frac{VD}{v}$$

f

Moody Diagram
(Ref. [1], Ch. 10)